Numerical Methods in Fracture Mechanics

Numerical Methods in Fracture Mechanics

Edited by:

D. R. J. Owen
Department of Civil Engineering, University College, Swansea.

A. R. Luxmoore
Department of Civil Engineering, University College, Swansea.

Proceedings of the Second International Conference held at University College, Swansea, 7th - 11th July, 1980.

PINERIDGE PRESS

Swansea, U.K.

First published, 1980 by
Pineridge Press Limited
91, West Cross Lane, West Cross, Swansea, U.K.

ISBN 0–906674–08–5

British Library Cataloguing in Publication Data
Numerical methods in fracture mechanics
1. Fracture mechanics – Mathematics – Congresses
2. Numerical calculations – Congresses
I. Owen, D. R. J. II. Luxmoore, A. R.
620.1'126 TA 409
ISBN 0–906674–08–5

Printed and bound in Great Britain by
Redwood Burn Limited
Trowbridge & Esher

PREFACE

The success of the first conference on this topic indicated a widespread interest in the subject but at the same time illustrated the diversity of opinion that still exists in some areas of study. The numerical analysis of fracture problems is a rapidly developing field, particularly with regard to nonlinear material behaviour and dynamic situations and this conference was conceived as a means of bringing together experts on both fracture mechanics and numerical analysis in an attempt to establish criteria for the application of numerical methods to fracture prediction.

The papers, which have been contributed from authors throughout the world, have been separated into five sections covering the major areas of interest, namely: linear elastic fracture mechanics, nonlinear fracture problems, dynamic fracture analysis, fatigue life prediction and the fracture of brittle materials and composites.

The proceedings are printed from direct lithographs of authors manuscripts and the editors cannot accept responsibility for any inaccuracies, comments or opinions expressed in the papers. It is hoped that a summary of the numerical solutions of the specified set of benchmark problems presented at the conference will be published subsequently.

Finally we would like to thank the authors for their efforts and co-operation in making the conference possible.

D. R. J. OWEN A. R. LUXMOORE

Swansea, April 1980.

CONTENTS

SECTION 2 NONLINEAR FRACTURE ANALYSIS

SECTION 4 FATIGUE LIFE PREDICTION

SECTION 5 FRACTURE OF BRITTLE MATERIALS AND COMPOSITES

SECTION 1

LINEAR ELASTIC FRACTURE MECHANICS

Numerical Methods and Linear Elastic Fracture Mechanics as related to Engineering Structures and Metallic Alloys

Dr. J.F. Knott
Department of Metallurgy and Materials Science,
University of Cambridge.

SUMMARY

The paper summarises linear elastic, non-linear elastic and elastic/plastic behaviour and briefly describes some standard solutions for stresses and displacements around crack tips. The relationships between "linear elastic" K_{IC} toughness values and micromechanisms of fracture are discussed for both cracking and rupture processes. Reference is made to some concepts in elastic/plastic fracture mechanics. Attention is finally focussed on sub-critical crack growth, particularly fatigue, and three specific areas are treated: thumbnail cracks; the growth of cracks from notches; and effects of residual-stress distributions.

INTRODUCTION

The techniques of fracture mechanics have evolved as a result of the need to develop quantitative methods of engineering design to prevent fracture in structures or components. Conventional design codes permit operating stresses which are only a fraction (typically two-thirds) of the yield stress, so that fracture in service can occur only if the piece contains a stress-concentrator, such as a notch or a sharp crack. In this context, it should be noted that the "yield stress" of a standard uniaxial testpiece represents general yield of the testpiece and that localised plastic flow can occur at lower applied stresses, due to the presence of microscopic stress-concentrators, such as non-metallic inclusions. This may have consequences, for example, with respect to the initiation of fatigue-cracks, below the "yield stress".

In structural metals, all fracture mechanisms are preceded by localised plastic flow, so that, even in the so-called "linear elastic" regime, the appropriate notch-tip or crack-tip stress analysis should include non-linear, non-recoverable components. It is precisely for such situations that simple analytical solutions reveal their limitations and recourse must be made to numerical methods of solution. Conversely, numerical solutions must reduce to the analytical solutions for idealised forms of materials behaviour.

ELASTIC BEHAVIOUR

In linear elastic material, each component of the stress tensor, σ_{ij}, is linearly proportional to each component of the strain tensor, ε_{ij}:

$$\sigma_{ij} = c_{ijkl} \varepsilon_{ij} \quad \ldots 1)$$

where c_{ijkl} is the fourth-order tensor of crystal constants. Symmetry of $\sigma_{ij} = \sigma_{ji}$ and $\varepsilon_{ij} = \varepsilon_{ji}$ reduces c_{ijkl} to 36 independent terms, and in matrix form:

$$\sigma_i = c_{ij} \varepsilon_j \quad \ldots 2)$$

where i,j take values 1 to 6. Uniqueness of the strain-energy function:

$$W = \int_0^{\varepsilon} c_{ij} \varepsilon_j d\varepsilon_i \quad \ldots 3)$$

reduces the number of coefficients, c_{ij}, to 21 and operation of crystal symmetry further reduces the number to three: c_{11}, c_{12}, c_{44}; for the cubic system, to which iron, aluminium, nickel and copper belong. Assumption of isotropy gives $(c_{11} - c_{12}) = 2c_{44}$ and it is then possible to equate the coefficients to Lamé's constants, λ and μ:

$$c_{11} = (\lambda+2\mu) \qquad c_{12} = \lambda \qquad c_{44} = \mu \quad \ldots 4)$$

or to Young's modulus, E, and Poisson's ratio, ν, using the forms for λ and μ:

$$\lambda = \nu E/(1+\nu).(1-2\nu) \; ; \qquad \mu = E/2(1+\nu) \quad \ldots 5)$$

In terms of Lamé's constants, the stress-strain relationships are:

$$\sigma_{ii} = \lambda\Delta + 2\mu\varepsilon_{ii} \; ; \qquad \sigma_{ij} = 2\mu\varepsilon_{ij} \quad \ldots 6)$$

where $\Delta = (\varepsilon_{11} + \varepsilon_{22} + \varepsilon_{33})$. Numerical analysis may express stress and strain as column vectors, $\underline{\sigma}$ and $\underline{\varepsilon}$ related by a matrix of elastic coefficients, $\underline{\underline{D}}$: $\underline{\sigma} = \underline{\underline{D}}\,\underline{\varepsilon}$. For example, for a two-dimensional state:

$$\begin{vmatrix} \sigma_{11} \\ \sigma_{22} \\ \sigma_{12} \end{vmatrix} = \begin{vmatrix} (\lambda+2\mu) & \lambda & 0 \\ \lambda & (\lambda+2\mu) & 0 \\ 0 & 0 & \mu \end{vmatrix} \begin{vmatrix} \varepsilon_{11} \\ \varepsilon_{22} \\ 2\varepsilon_{12} \end{vmatrix} \quad \ldots 7)$$

In combination with the equations for stress-equilibrium:

$$\partial \sigma_{ij} / \partial x_j = 0 \qquad \ldots 8)$$

and those for strain-compatibility, a two-dimensional linear elastic solution may be sought in terms of Airy's stress function, $\Phi\ (x_1,x_2)$, from which stresses are defined (in Cartesian coordinates) as:

$$\sigma_{11} = \partial^2\Phi / \partial x_2^2 , \qquad \sigma_{22} = \partial^2\Phi / \partial x_1^2$$
$$\sigma_{12} = \sigma_{21} = - \partial^2\Phi / \partial x_1 \partial x_2 \qquad \ldots 9)$$

The stress function Φ must be chosen to satisfy boundary conditions and the requirement:

$$\nabla^2(\nabla^2\Phi) = \nabla^4\Phi = 0 \qquad \ldots 10)$$

In three dimensions, a combination of three stress functions is required to give a proper solution(1,2).

For non-linear elastic material, equations for stress-equilibrium and strain-compatibility remain unchanged, but linear relationships between stress and strain (equations 1,2, 6,7) no longer hold. Instead, a "power-law hardening" relationship may be invoked and this form is also used to model elastic/plastic material: the dependence of strain, ε_{ij}, on stress deviator, s_{ij}, being given by:-

$$\varepsilon_{ij}/\varepsilon_o = (3\alpha/2)\ (\sigma_e/\sigma_o)^{n-1}\ s_{ij}/\sigma_o \qquad \ldots 11)$$

where σ_e is the effective stress = $(3\ s_{ij}s_{ij}/2)^{\frac{1}{2}}$, σ_o and ε_o are reference (flow) stress and (yield) strain, and α is a material constant (3). It is important to note that the strain-energy function, W, (equation 3), remains unique for non-linear elastic material. If the work is put in isothermally, W is the change in free energy of the system, and is a function of state. If the applied forces are removed, the material unloads elastically and all the stored strain energy is released. This should be contrasted with elastic/plastic material, in which non-recoverable plastic offsets are obtained, once the yield stress has been exceeded.

STANDARD ANALYTICAL ELASTIC SOLUTIONS FOR NOTCHES AND CRACKS IN 2-D

It is useful to summarise some standard solutions for stress distributions around notches and cracks. For an elliptical notch, of semi-major axis a, and radius, ρ, normal to a uniform tensile stress, σ_{app}, in an infinite body, Inglis(4) determined the local stresses for a two-dimensional stress-state simply by choosing an appropriate form for the stress function Φ (equations 9,10, but expressed in hyperbolic coordinates). The main result is that the local tensile stress, σ_{22}, at the tips of the ellipse ($x_1 = \pm a$), is given by:

$$\sigma_{22} = \sigma_{app} (1 + 2 \{a/\rho\}^{\frac{1}{2}}) \quad \ldots 12)$$

where the term in parentheses is the elastic stress-concentration factor (ESCF), taking the value 3 for a circle ($\rho = a$). The distribution of σ_{22} as a function of distance, $r = (x-a)$, ahead of the tip, is not given by any simple expression for the general ellipse, although it is easily described for a circular hole(5).

As the ellipse becomes crack-like, $\rho \to 0$, $\sigma_{22} \to \infty$, and it becomes convenient to deal instead with the distribution of stress ahead of the crack tip (which is now a mathematical singularity). Using an appropriate stress-function, due to Westergaard(6) (for a physical justification, see(7)), and writing $r \ll a$;

$$\sigma_{22} = K (2\pi r)^{-\frac{1}{2}}. (\cos \theta/2)(1+\sin \theta/2 \sin 3\theta/2)+\ldots \quad \ldots 13)$$

$$\sigma_{12} = K (2\pi r)^{-\frac{1}{2}} (\sin \theta/2)(\cos \theta/2) (\cos 3\theta/2)+\ldots \quad \ldots 14)$$

where K is the stress-intensity factor and assumes the value:

$$K = \sigma_{app} (\pi a)^{\frac{1}{2}} \quad \ldots 15)$$

for the geometry chosen. The maximum value of the principal tensile stress, σ_2, is obtained for $\theta = 60^o$, and the maximum value of shear stress is obtained for $\theta = 90^o$ (the maximum octahedral shear stress is at $\theta = 70^o$). Simple predictions made from this might be that brittle cracking should occur on 60° planes and yield on planes at some 70-90° to the line of crack extension. Observations on yielding support this prediction, but brittle cracks tend to run directly ahead of a pre-crack, at least until bifurcation occurs.

Considering the line $\theta = 0$, equation 13) gives $\sigma_{22} = K(2\pi r)^{-\frac{1}{2}}$, and it is possible to calculate the change in strain-energy stored in (unit thickness of) the body ($G \equiv \{dW/da\}$), when the crack is advanced by an infinitesimal amount, da, using a virtual-work argument(8)

$$G = \alpha K^2/E \quad \ldots 16)$$

where $\alpha = 1$ in plane stress or $(1-\nu^2)$ in plane strain. The Griffith fracture criterion(9) may then be expressed as a critical value of energy release-rate, $G = G_{crit}$, so that the fracture stress, $\sigma_{app} = \sigma_F$ is given by:

$$\sigma_F = (E\, G_{crit}/\pi\alpha a)^{\frac{1}{2}} \quad \ldots 17)$$

In different materials, G_{crit} may receive different physical interpretations, such as: elastic work of fracture (often taken as 2γ where γ = surface energy) plastic work, or visco-elastic flow. Strictly, only elastic failure should be considered, but, in engineering terms, local non-elastic behaviour is deemed not to affect the calculation of energy release-rate, provided that the extent of such behaviour is small compared with all other dimensions (e.g. "small-scale yielding").

In material which possesses the "power-law hardening" stress-strain relationship (equation 11), the stress distribution is given by (3,10):

$$\sigma_{ij} = \sigma_o \,(EJ/\alpha\sigma^2_o\, I_n r)^{1/n+1} \;\tilde{\sigma}_{ij}(\theta,n) \qquad \ldots\ 18)$$

where I_n is an integration constant of order 4 to 6 and $\tilde{\sigma}_{ij}$ is a dimensionless function of angle and the hardening exponent; $E = \sigma_o/\varepsilon_o$. The term J which determines the strength of the singularity, is obtained from a contour integral around the crack tip(11). For any circuit, Γ, taken from the lower surface to the upper surface of a stressed crack, J is defined as:

$$J = \int_\Gamma \{W\, dx_1 - \underline{T}\,(\partial \underline{u}/\partial x_2)\, ds\} \qquad \ldots\ 19)$$

where W is the elastic strain-energy density within the contour and $\underline{T}$ is the traction on any length, ds, of the contour, having an outward-facing normal, $\underline{n}$, with direction cosines n_j: $\underline{T} = \sigma_{ij} n_j$. For power-hardening material, J characterises both the crack-tip stress-field (equation 18)) and displacement-field, e.g. for θ = ± π

$$u_2 = (\alpha\sigma_o/E)(EJ/\alpha\sigma^2_o I_n)^{n/n+1}\, r^{1/n+1}\, \tilde{u}_2(n) \ \ldots\ 20)$$

where $\tilde{u}_2(n)$ is a dimensionless function of the hardening exponent. J therefore has properties similar to K and in fact reduces to G for linear elastic material. For non-linear elastic material, J is path-independent and may be equated to the potential energy release rate, J = (∂U/∂a).

NUMERICAL METHODS

Early numerical solutions of elastic stress distributions around cracks and notches tended to employ relaxation techniques, based on finite differences. Essentially, values of the stress function, Φ, (equations 9 and 10) were "guessed" for a series of points throughout the body and changes were made to individual values until $\nabla^4\Phi = 0$ (equation 10) was satisfied at every point and the boundary stresses were simultaneously balanced. Unfortunately, the relaxation grid for ∇^4 is large and cumbersome: if the spacing of points is d, and the value of Φ at any point i,i is $\Phi_{i,i}$, the formula for $\nabla^4\Phi$ is:

$$\nabla^4\Phi_{o,o} = (1/d^4)(\Sigma_4\Phi_{2,0} - 8\Sigma_4\Phi_{1,o} + 2\Sigma_4\Phi_{1,1} + 20\Phi_{o,o}) \quad \ldots 21)$$

where a term such as $\Sigma_4\Phi_{2,0}$ denotes the sum of unit value changes in Φ at the four equivalent points: 2,0; -2,0; 0,2; 0, -2. This tends to reduce the applicability of the technique in regions, such as those ahead of a crack, in which the stress-gradient is steep. Nevertheless, acceptable elastic solutions were developed by Southwell(12) and elastic/plastic solutions in plane-stress were obtained by Stimpson and Eaton(13), and in plane-strain by Jacobs(14).

The more useful numerical technique is that of finite elements (FE). Here, the cracked body is treated as a framework. Forces and displacements are determined at each node and are then summed to give the stiffness of the total framework, which is equated to that of the body. As for the relaxation technique, the F.E. method yields a number of discrete values for stresses ahead of a crack tip or displacements within a crack. The stress-intensity-factor as such is determined by a procedure such that the $r^{-\frac{1}{2}}$ stress distribution inherent in analytical solutions, e.g. equation 13), or the equivalent displacement distribution

$$u_2 = \alpha(K/E)\ (2/\pi)^{\frac{1}{2}}\ (a^2 - x_1{}^2)^{\frac{1}{2}} \quad \ldots 22)$$

where $x_1 \leqslant a$, is assumed to hold near the crack tip, so that graphs of stress vs. $(\text{distance})^{-\frac{1}{2}}$ or displacement vs. $(\text{distance})^{\frac{1}{2}}$ may be extrapolated back to the point $r = o$ $(x_1 = a)$. Agreement between the K-values deduced from stress and those deduced from displacement is not always precise, but should hold, to within a few percent. Alternatively, energy calculations may be used to obtain K (cf equation 16)). The power of numerical methods in general, and F.E. techniques in particular, is that K-values may be tabulated for a whole range of different testpiece geometries and structural geometries, which may involve complicated boundary conditions.

RIGID/PERFECTLY PLASTIC BEHAVIOUR: SLIP-LINE FIELDS

The second extreme of material behaviour is modelled by the assumption of rigid/perfectly plastic deformation, in which elastic strains are ignored and material is supposed to deform by slip at a constant shear stress, k. In plane strain ($\varepsilon_3 = 0$; $\sigma_3 = \frac{1}{2}(\sigma_1 + \sigma_2)$), both Von Mises' and Tresca's yield criteria reduce to the form:

$$\sigma_1 - \sigma_2 = 2k \quad \ldots 23)$$

with $k = \sigma_Y/\sqrt{3}$ (Von Mises') or $k = \sigma_Y/2$ (Tresca), where σ_Y is the uniaxial (general) yield stress. Material yields on slip-lines, which are simultaneously lines of constant maximum shear stress, $\tau_{max} = k$, and lines of tangential velocity discontinuity(15). Two orthogonal sets of lines (labelled α and β) at 45° to the directions of principal stress constitute a slip-line field and the pressure, ρ at any point on a slip-line which curves through an angle ϕ is given by Hencky's relationships:

$$\rho + 2k\phi = \text{const. on an } \alpha\text{-line}$$
$$\rho - 2k\phi = \text{const. on a } \beta\text{-line} \quad \ldots 24)$$

Three types of slip-line field have been of particular value in fracture analysis:

a) when yield spreads ahead of a circular hole, of radius a, in plane strain, the slip-line field takes the form of a logarithmic spiral (always at 45° to radial and tangential principal stresses) and the tensile stress, σ, at a distance x (measured from the centre of the hole) is given by:

$$\sigma = 2k \{1 + \ln (1 + x/a)\} \quad \ldots 25)$$

b) the general yield stress of a body containing deep double-edge cracks is given by the Prandtl field and the yield stress is found to be:

$$\sigma = 2k(1 + \pi/2) \quad \ldots 26)$$

$\sigma/2k$ is the constraint factor, L.

c) a bar, containing a V-notch of included angle, θ, in pure (four-point) bending, deforms by rotation around two plastic "hinges" to give a "general yield" bending moment, M_{GY}(16):

$$M_{GY} = 0.63k (W-a)^2B \quad \ldots 27)$$

where a is the notch depth, W is the specimen's depth and B is the specimen's thickness (L = 1.26). The maximum local tensile stress below the notch at the point of general yield is:

$$\sigma_{max} = 2k (1 + \pi/2 - \theta/2) \quad \ldots 28)$$

The ratio $\sigma_{max}/2k$ is the stress intensification.

ANALYTICAL ELASTIC/PLASTIC SOLUTIONS

Any real structural alloy exhibits a mixture of elastic behaviour and plastic behaviour. In an engineering structure, deformation in regions remote from stress-concentrators is elastic: around stress-concentrators it is mixed elastic/plastic, with the plastic component varying from the yield strain at the elastic/plastic interface to much higher values at the tip of the notch or crack. In this mixed region, stresses must satisfy both the elastic relationships (equations 6), 8), 9), 10)) and the yield criterion (e.g. equation 23)).

An analytical solution has been developed for the spread of (non-hardening) plasticity in plane stress ahead of a central, through-thickness, crack of length 2a, situated in an infinite body, normal to an applied stress, σ_{app}, (17,18,19). The crack tip is no longer a singularity, but, instead, develops a finite crack-opening-displacement, δ, given by:

$$\delta = (8/\pi)(\sigma_Y/E)a \ln \sec (\pi\sigma_{app}/2\sigma_Y) \qquad \ldots 29)$$

associated with a spread of plasticity, d_Y, given by:

$$d_Y/a = \sec (\pi\sigma_{app}/2\sigma_Y)-1 \qquad \ldots 30)$$

For quasi-linear-elastic (small-scale yielding) conditions (σ_{app}/σ_Y very small), appropriate first-order terms in the series expansions of the "ln sec" terms may be taken to give:

$$\delta = \sigma_{app}^2 \pi a/\sigma_Y E = K^2/\sigma_Y E \qquad \ldots 31)$$

$$d_Y = (\pi/8)(K/\sigma_Y)^2 = (\pi/8)(E/\sigma_Y)\delta \qquad \ldots 32)$$

If a "J-integral" is defined as the "rate-of-working"(20) (work-input), we may write:

$$"J" \equiv "G" = \sigma_Y\delta \qquad \ldots 33)$$

On this definition, "J" is path-independent, even if the path crosses the plastic zone. Note that the value of "J" represents work expended in introducing plastic deformation and is non-recoverable: it cannot therefore provide any driving-force for crack extension. Examination of the energetics of crack-spreading in this configuration reveals the system to be in neutral equilibrium(21): any increase in crack length under constant stress is simply associated with a larger value of δ, which is precisely accommodated by a larger plastic zone, entailing more work done, but no more recoverable energy.

Similar modelling has led to analytical solutions for "small-scale yielding" (s.s.y.) around a crack in plane strain, for which (following an earlier F.E. analysis) the yielding was supposed to occur on planes inclined at $\pm$ 70.5^{o} to the crack (22). The relationships equivalent to those in 31) to 33) become (in the s.s.y. limit)

$$\delta = 0.44\, K^2/\sigma_Y E \qquad \ldots 34)$$

$$d_Y = 0.17\, (K/\sigma_Y)^2 \qquad \ldots 35)$$

$$"J" = 2.1\, \sigma_Y\delta \qquad \ldots 36)$$

One particular use of this model was to estimate the effect on the extent of yield of a non-singular "T"-stress term, which is required to satisfy boundary conditions in real testpiece geometries. This term was first encountered in the F.E. studies of Larsson and Carlsson (23).

A number of techniques has been employed to incorporate work-hardening into the solutions and as described earlier one often-used method is to simulate the materials' elastic/plastic stress-strain curve by an equivalent curve for non-linear elastic material (equation 11)). This is satisfactory, provided that "we do not call the material's bluff by unloading" (24) i.e. that we do not attach any energy-release-rate meaning to J, as would exist for non-linear-elastic material. Relationships between C.O.D. and J have been obtained, as follows, close to crack tips where the appropriate HRR singularity (equation 18)) dominates:

$$\delta = d_n (J/\sigma_o) \qquad \ldots 37)$$

where $$d_n = (\alpha\sigma_o/E)^{1/n} (\tilde{u}_1+\tilde{u}_2)^{1/n}(2\tilde{u}_2/I_n) \qquad \ldots 38)$$

see also equations 11), 18) and 20). Values of d_n are discussed by Shih(25,41).For perfectly plastic material in plane stress, d_n is unity and equation 33) is confirmed ($\sigma_o = \sigma_Y$); in general d_n varies between 0.2 and 0.78 in plane strain and is strongly dependent on hardening exponent. These values are of course subject to dominance of the HRR(3,10)singularity.

CLEAVAGE FRACTURE

The use of numerical, elastic/plastic, finite-element analysis has been of help in both the interpretation and the application of "linear elastic" toughness values. Two general microscopic modes of failure have been recognised and treated: these may be generally classed as cracking processes and rupture processes. The former is typified by the unstable propagation of a crack nucleus formed in a brittle particle, perhaps less than 1μm in size.

It is intriguing to note that the principles of LEFM have been applied to describe this micro-scale process(26) and that it is possible to characterise the onset of cleavage propagation in a steel of given carbide size and distribution in terms of a critical local tensile stress, σ_F. The practical measurement of σ_F is most easily made, using blunt-notched bars. Initially only slip-line field solutions of the forms in equations 25) and 28) were available(27), but, in 1971, Griffiths and Owen(28) produced an elastic/plastic F.E. solution for a notched bar of particular geometry. This solution was confirmed by a number of other workers(29) and, to date, remains the most reliable "calibrated" geometry, although some results for other geometries are becoming available(30).

A point on this general type of stress analysis which is

worth noting concerns the variation of the through-thickness stress, σ_{33}, in the "plastic enclave" and in the elastic region. Freely-deforming plastic material undergoes simple slip, so that there is no change in volume and the effective value of Poisson's ratio, ν, is 0.5. In elastic material, a typical value of ν is 0.3. Plane-strain deformation in the X_1X_2 plane then engenders through-thickness stresses of different magnitudes, depending on whether the deformation is elastic or plastic:

$$\underline{\text{elastic}} \quad : \quad \sigma_{33} = 0.3\,(\sigma_{11} + \sigma_{22}) \qquad \ldots\ 39)$$

$$\underline{\text{plastic}} \quad : \quad \sigma_{33} = 0.5\,(\sigma_{11} + \sigma_{22}) \qquad \ldots\ 40)$$

At the plastic/elastic interface σ_{11} and σ_{22} must be continuous and the immediate conclusion is that σ_{33} changes discontinuously from the plastic to the elastic region. This is clearly unreasonable. Many F.E. analyses by-pass the problem by assigning the same value of Poisson's ratio to both the plastic and elastic regions, but this would appear to be equally unrealistic. A line of reasoning adapted by Kfouri(31) was to assume plane-strain conditions for the total (elastic + plastic) strain in the "plastic enclave". The consequence of this is an effective Poisson's ratio which varies smoothly from approx. 0.5 near the notch tip (where the deformation is predominantly plastic) to 0.3 in the elastic region. This is perhaps not as physically unreasonable as it might appear if plastic flow in the enclave (at least near the plastic/elastic interface) can continue only to a limit controlled by the restraint of the surrounding elastic material. The point has general applicability to a number of elastic/plastic configurations.

It has proved possible to utilise critical σ_F values in combination with F.E. elastic/plastic crack-tip solutions, obtained primarily by Rice and co-workers, to predict "linear elastic" K_{IC} toughness values for the cleavage fracture mode (32). The principle is essentially one of "stress at a distance". From the microstructure it is possible to deduce the site of a crack nucleus ahead of a fatigue pre-crack. The plastic zone (and hence K) must be increased in size until sufficient stress is available at that site to propagate the micro-crack formed in the nucleus. The increase of K_{IC} with testing temperature then reflects the decrease of σ_Y with increase in temperature. The general model has recently been refined and put in terms of the probability of finding a carbide of given size together with sufficient stress for propagation(33). The approach has also been shown to be consistent with power-law hardening "elastic-plastic" solutions for crack-tip stress fields(34). The model has powerful predictive capabilities, for example, in showing how increases in σ_Y due to neutron irradiation might be expected to alter K_{IC}(35,36).

FIBROUS RUPTURE

Elastic/plastic crack-tip F.E. analyses have also proved useful with respect to rupture processes, in which cracks progress by the coalescence of voids in the region of combined high-strain and high triaxility ahead of the crack tip(26). The crack-tip-opening-displacement at the point of "initiation" (coalescence of the crack-tip with the first void), δ_i, may be expressed as a function of non-metallic inclusion spacing, X_o, and radius, R_o, following Rice and Johnson(37), and experimental support for the relationship may be found in results obtained on a selected number of steels, which possess high work-hardening capacities(26). In s.s.y., various F.E. elastic/plastic analyses(38,39) give the relationship between C.O.D., δ, and stress intensity, K, as:

$$\delta = 0.49\text{-}0.67\ K^2/\sigma_Y E \qquad \ldots 41)$$

c.f. the dislocation model (equation 34)) and results from the power-law-hardening model (equations 37) and 38)). At initiation of fracture, it is therefore possible to characterise the deformation state by δ_i, or equivalent values of "K_i" or "J_i", given by

$$\text{"}J_i\text{"} \simeq 1.4\text{-}1.9\ \sigma_Y \delta_i \qquad \ldots 42)$$

There is no connotation of energy-release or elastic instability in any of these elastic-plastic characterising parameters, and by analogy with analytical results (equation 33) et seq.), it might, indeed, be expected that the system remains in neutral equilibrium if the crack is extended by an infinitesimal amount da.

Similar situations arise as the amount of plastic deformation preceding fibrous fracture initiation increases. Beyond the s.s.y. limit, the use of K is meaningless, but it is still possible to characterise crack-tip deformation fields using C.O.D. or "J" (work input) values. Such values can be obtained only by use of numerical techniques and important results have been obtained by Sumpter(40), Shih et al.(41) and McMeeking and Parks(42). For example, for a generally-yielded deeply-precracked bar in pure bending, McMeeking and Parks give the relationship between δ and J (taking $1/N \equiv n-1 = 10$) as

$$\delta = 0.33\ \text{"}J\text{"}/\sigma_o \qquad \ldots 43)$$

whereas, Shih(25,41) obtains the following values,

$$\left.\begin{array}{l} n = 3 : \delta = 0.22\text{-}0.19\ \text{"}J\text{"}/\sigma_o \\ n = 10 : \delta = 0.48\text{-}0.40\ \text{"}J\text{"}/\sigma_o \\ n = \infty : \delta = 0.65\text{-}0.58\ \text{"}J\text{"}/\sigma_o \end{array}\right\} \qquad \ldots 44)$$

Experimental relationships between "J_i" and δ_i measured on testpieces of appropriate geometry give coefficients in the range 0.33-1 (43,44,45). Again, these are characterising parameters and for application to engineering design, a structure must also be calibrated to give "J" or δ as a function of load(46), so that the load corresponding to crack initiation may be determined: there is, of course, no guarantee that this will, in general, be coincident with instability.

The general problem is that, for a crack to grow spontaneously, more elastic energy must be released from the system than is required to be dissipated in the plastic work involved in the micro-mechanisms of crack advance. Once the amount of plasticity preceding initiation is substantial, it becomes difficult to compute the elastic energy release-rate (there is, for example, no net driving force for the elastic-plastic crack in an infinite body), and the boundary conditions (falling between limits of "fixed-grips" and "constant-load") have a major effect. One postulate might be that the total input "J" is partitioned between a recoverable part ("elastic") and a non-recoverable part ("plastic"). Similar reasoning appears to underlie Sumpter's use of η_{el} and η_{pl} factors (40). Estimation of the recoverable component, using a physical model, does not seem to have been attempted, but the following argument may be worthy of discussion. It is possible to recover some "plastic work", by the running-back of dislocations, in the yielded-zone ahead of an elastic/plastic crack, when the applied stress is reduced to zero. This "running-back" is commonly invoked in mechanisms of fatigue-crack advance. The reversed C.O.D., δ_{rev}, is obtained by substituting $2\sigma_Y$ for σ_Y in equations such as 29) or 31), and it is possible that a term such as $\sigma_Y\delta_{rev}$ equates to a recoverable component of "J" in plane stress. Thus, under s.s.y. conditions (equation 31)) the recoverable portion would be 0.5, but it would vary, following equation 29), as yielding became more extensive.

An alternative method of calculating energy release, developed by Kfouri,Miller(47), Rice (48) and Sorensen(49) is to determine the strain energy stored in a particular elastic/plastic configuration and then to compute the energy change when the crack is advanced by a discrete amount, effected by cutting one of the links of the F.E. network: this energy change in then termed G^{Δ}. Comparison of the value of G^{Δ} with the incremental work of crack advance ($(d\delta/da)\Delta a$ or $(dJ/da)\Delta a$) enables conditions for instability to be deduced. Support for the technique is given by experimental results on high-strength structural steels(50), but for the model to become more realistic, the link size must be equated to the microstructural unit of crack advance.

SUB-CRITICAL CRACK GROWTH

Perhaps the most strictly "linear-elastic" conditions are maintained when characterising the growth of cracks under conditions of fatigue or stress-corrosion. To illustrate the general principles, attention will be focussed on fatigue. Typically, the simplest form for fatigue-crack growth is the Paris law(51), in which the increase in crack length per cycle, da/dN, is expressed as a power of the alternating stress intensity, $\Delta K = K_{max} - K_{min}$, where K_{max} and K_{min} are respectively the maximum and minimum stress intensity values in the cycle:

$$da/dN = A \ \Delta K^m \qquad \ldots 45)$$

where A and m are constants: A being of order 10^{-11} to 10^{-13} (with a in m and ΔK in $MN^{-3/2}$) and m taking values 2-4 for ductile materials(52). More complicated forms exist, including effects of mean stress and a fatigue-crack threshold, ΔK_{th}(53), but even the simple form serves to illustrate the actual and potential use of numerical methods.

The raw data in a fatigue-crack growth experiment comprise the applied-load range (Δσ), which is usually held constant, and the crack length, or ratio of crack length to testpiece width, a/W, plotted vs. the number of cycles, N. At any particular point, given a pair of values (a/W) and N, it is necessary to calculate both ΔK and da/dN. For standard testpieces, compliances are tabulated(54) as functions of (a/W), so that it is merely necessary to take a value for the compliance, corresponding to the instantaneous value of (a/W) to convert Δσ to ΔK. Feed-back control of a servo-hydraulic machine to give constant ΔK may be effected, using an analog computer on which the coefficients of the polynomial compliance function may be set independently, or, possibly, using a micro-processor, with the function expressed as a look-up table.

Crack-growth monitoring often employs a technique such as the potential-drop, in which changes in crack length produce changes in resistance of the testpiece and hence changes in output voltage. Recently, F.E. analyses have been used to calibrate standard CTS and SEN bend testpieces(55) showing agreement with earlier analogue and conformal mapping solutions (56,57). Calculation of (da/dN) demands numerical fitting to the curve of a vs N, followed by differentiation to give the tangent. Recommended numerical methods include second-order orthogonal polynomials fitted to overlapping sets of eight points(58,59), deriving the tangent from the central two points; or spline functions(60). Towards the end of the fatigue life, the curve of a vs N is steep and the tangent may then often be obtained more conveniently by simple numerical differences.

An important, but somewhat neglected, field of application is the integration of crack growth-rates to deduce the number of cycles taken for a crack to grow from an "initial" length, a_i, (typically, that likely to be found in any fabricated structure and often corresponding to the non-destructive-testing limit of detection) to a "critical" length, a_{crit}, given by the material's fracture toughness, K_{IC}, and the maximum stress in the fatigue cycle, σ_{max}. In an infinite body, $\Delta K = \Delta\sigma(\pi a)^{\frac{1}{2}}$, and if the exponent, m, in equation 45) is equal to 4, the equation may be easily integrated to give:

$$N = (A\ \Delta\sigma^4\pi^2)^{-1}\ (1/a_i - 1/a_{crit}) \qquad \ldots\ 46)$$

For a general value of m,

$$N = (A\ \Delta\sigma^m\pi^{m/2})^{-1}\ \Big[(2/2-m)(a)^{(2-m)/2}\Big]_{a_i}^{a_{crit}} \qquad .\ 47)$$

In general, the functional relationship between ΔK and (a/W) in a finite body is a complicated polynomial, and if this is raised to a power, particularly a non-integral power, in a fatigue-crack-growth-rate law, the only realistic integration must be numerical throughout. In principle, the use of Simpson's rule(59), or a similar integration procedure, is straightforward, but is costly in terms of computer time, if every calculation has to be made <u>ab initio</u>. Standard results in tabular, graphical or nomographic form would make the procedures more acceptable to the engineer.

In many practical applications, the cracks are short compared with respect to the dimensions of the bodies in which they are located, so that infinite-body or semi-infinite-body solutions may be appropriate, at least for the early stages of growth. If the cracks were "through-thickness" cracks, the expressions for stress-intensity should have simple analytical forms, but, in practice, there are three considerations which may demand numerical methods, even for very large bodies:

a) if the cracks have irregular fronts: the simplest forms being semi-elliptical "thumbnails",
b) if the cracks are situated at the roots of major stress-concentrators, such as keyways or holes,
c) if the cracks are associated with welds, around which there may be a complicated distribution of residual stress.

In practice, all three situations may co-exist and their effects may be experienced in any form of sub-critical crack growth and in the conditions pertaining to final fast fracture.

THUMBNAIL CRACKS

The stress-intensity factor for a crack which is an embedded ellipse or a semi-elliptical edge crack of depth a and surface width 2c in a plate of finite thickness, t, may be expressed simply for the case of uniform tension as:

$$K = Q\ \sigma_{app}(\pi a)^{\frac{1}{2}} \qquad \ldots\ 48)$$

where Q is derived from the double elliptical integral(61). Essentially, the thumbnail provides less of a perturbation to the flow of lines of force, through the cross-section, than would a through-thickness crack of length a, and so K is expected to be smaller for a given value of σ_{app}. It has not been generally established that the relationships between δ and K or J even for s.s.y. around thumbnail cracks are similar to those for through-thickness cracks.

In recent experiments, comparisons have been made between the effects of short through-thickness cracks and those of small thumbnail fatigue cracks on the toughness values of high-strength marageing steels and NiCrMoSi forging steel, tested in the standard three-point bend SEN bend configuration(62,63). For these rather brittle materials, K_{IC} remained constant down to very short (through-thickness) crack lengths: in the highest strength marageing steel, down to 0.05mm length. Suitable compliance functions were readily available and the K values calculated for vanishingly small crack lengths agreed with the case of an edge crack subjected to a uniform tension equal to the maximum fibre stress in bending.

Evaluation of K-values for the semi-elliptical edge cracks in three-point bending proved difficult. Three analyses(64,65, 66) were available for the finite cross-sectional geometry in four-point bending, however, and these were multiplied by the ratio of the equivalent compliances for short, through-thickness cracks in three- and four-point bending. The values thus obtained agreed, to within 5%, with that for the "vanishingly small" crack subjected to the fibre stress.

The resultant toughness values were, however, some 20-25% higher than those for through-thickness cracks, of both greater and smaller depths. At this stage, it is not clear whether this represents a genuine increase in toughness or whether the estimation procedure for K has led to inaccuracies of the order observed. Attempts were made to measure stretch-zone-widths on the fracture surfaces, to deduce values of C.O.D. The scatter was high, although a trend towards higher C.O.D. was observed. In the absence of K/δ relationships for elliptical cracks, however, the significance of even a well-established, higher value of C.O.D. is not clear.

A number of questions pertaining to crack growth also remain. There is obvious interest in discovering whether the growth of thumbnails under fatigue, for example, follows the same $da/dN : \Delta K$ relationship as that established for through-thickness cracks (equation 45). To establish this in uniform, stress-relieved material necessitates the following:

a) calibration of the potential drop output from specimens containing thumbnail cracks; an analogue method has been suggested(67), but a F.E. analysis would be of great value for confirmation

b) stress-intensity calibrations for thumbnail cracks in real testpiece geometries; accuracy is essential, since ΔK is

raised to the power 3 or 4.
It must be appreciated that the crack may not continue to grow in a manner such that the c/a ratio is preserved, so that much of the K-calibration may have to be a posteriori, using beach-marks or staining to deduce the successive positions of the crack front. Anisotropy in a wrought product, "banding" of the microstructure, or residual stress in a component such as an autofrettaged gun-barrel(68), may, of course, all contribute to changes in crack profile.

In most cases of defect assessment, it is usually the case that a very ragged crack front is deemed to be as potent as the equivalent, circumscribing (semi-)ellipse. This would appear to be a soundly conservative assumption, although it may be over-conservative if the reason for the "raggedness" is pinning of the crack front by metallurgical features.

CRACKS AT STRESS-CONCENTRATORS

The most common sites at which fatigue cracks or incipient fast fractures are likely to be found in an engineering structure are the roots of stress-concentrators, where local stresses exceed the yield stress. It is then, perhaps, surprising that relatively few solutions are available for the "notch + crack" geometry, even for K-values, compared with the plethora of deep-crack solutions. The levels of plastic strain developed in such regions and the effect of these on local fracture criteria have again been left rather unexplored.

Solutions are available for symmetrical cracks emerging from circular or elliptical holes(69,70), subjected to uniform tension and, for mode III (anti-plane strain) stressing, completely analytical solutions may be obtained, using conformal mapping techniques(71). In mode III, it is also possible to determine the stress intensity for a crack emanating from a V-notch of arbitrary included angle in a semi-infinite body(72), because the Schwarz-Cristoffel mapping function, which is in differential form, has a solution in terms of the β-function: the same technique may be used to provide electrical potential calibrations(57). Solutions in mode-I tension or bending for a configuration such as a V-notch with a root radius (say, a Charpy notch), containing cracks of different lengths ahead of it, are, however, generally lacking, although Mubeen and Bandy-opadhyay(73) and Kfouri(31) have examined some specific cases.

The importance of accurate K-values rests in the growth of fatigue cracks from notches. The total fatigue-life of a notched bar, N_f, may be divided into three parts(74):

$$N_f = N_K + N_{K'} + N_i \qquad \ldots 49)$$

where N_K represents the number of cycles spent in growth as a "long" crack (with K given by standard compliance functions),

N_K, is the number of cycles spent in growth in the K-field of notch + crack, and N_i is the number of cycles spent in "initiation". The "initiation" stage is ill-defined and we will return to it shortly, but, assuming "initiation" to cover growth up to approx. 20μm, the second stage N_K, covers growth from 20μm to approx. two root radii.

The problem of determining N_K, has been discussed in (74), following principles used earlier by (75). It is assumed that a short crack at the notch root is loaded by a set of point-forces corresponding to the stress-distribution ahead of an uncracked notch and da/dN is then integrated, as in equation 47) over short segments to determine the numbers of cycles required to grow from 20μm - 100μm, 100μm - 240μm, and so on. The point-loading technique may be shown to give perfect agreement with the analytical solution for a crack ahead of an elliptical notch in mode III (76), but is modified in the present instance by the incorporation of a factor of 1.12 for edge cracks. Agreement with experiment appears to be satisfactory, but it would be useful to have F.E. solutions for comparison. (The results of (77) are expressed in different form). Accuracy is important, because ΔK is raised to a power, in the range 3-4, in growth-rate laws.

It is, however, for the "initiation" stage that F.E. analysis could prove to be of most value, because "initiation", for blunt notches, dominates the fatigue life. Even at an applied stress of 2/3 yield, plastic strains of a few percent may be expected at the roots of notches and, for simple "on-off" loading, information from other sources (Coffin-Manson plots(78)) may enable N_i to be related to plastic-strain amplitude. (Ideally, the material should have been "shaken-down" to uniform cyclic properties before cutting the notch, to obviate effects of cyclic hardening during the test). At present, two (monotonic) strain analyses are available: one for a "Griffith-and-Owen" notch(28) one for a modified CTS geometry, with a blunted crack(79). Experiments are currently in hand to try to determine whether N_i is the same in the two specimen geometries at the same plastic strain amplitude(80), but the CTS geometry is very "crack-like" and information for other notch profiles would be extremely helpful.

The final points on notches + cracks concern elastic/plastic fracture mechanics (EPFM). One is simply the definition of a "shallow" crack. If a crack is short, in relatively soft material, the triaxial stresses around its tip are relieved by the onset of plastic flow from the free surface behind the crack tip, before the fracture mechanisms operate(81). If the shallow crack is located at the root of a deep notch, however, it is by no means clear what the criterion for relaxation of stresses is(50). In BS 5762 (C.O.D. measurement) the notch + crack requirement for the testpiece (presumably to maintain triaxiality) is that the angle subtended at the crack tip shall be not greater than 30°: in a sense this fixes the boundary between "short" and "long" cracks, but the criterion requires numerical support.

The second point concerns the W.I. design curve, which properly concentrates on local strain levels around stress concentrators(82). Estimations of strain made in the design procedure are clearly not ideal and it ought to be possible now to set up a combination of experiments and numerical analysis to obtain better values. In general, calculations, as a function of applied load, of the J-integral, or preferably the C.O.D. (which may be measured independently, e.g. by rubber infiltration(43)) for a crack situated ahead of a stress concentrator, which is a real structural design feature, such as a nozzle in a pressure vessel, represents one of the most useful fields of application for elastic/plastic numerical analyses. A first approach is made in (83).

RESIDUAL STRESSES

In many practical situations, for example, in surface-hardened components or around welds, a defect will be subjected to a stress field arising not only from the applied stress and any geometrical stress-concentrators, but also from a local residual stress distribution. Severe complications would arise if this had a high hydrostatic tensile component, but, in several cases, the main effect may be treated as uniaxial and additive to the applied stress.

The simplest analytical treatment of a residual stress in elastic material would be to treat it simply as a set of point forces, deduced from the residual stress distribution in defect-free material, and to load the crack with these forces, in addition to the applied stress, to obtain the new K-values (84). More properly, the residual stresses must be allowed to redistribute as a crack is introduced and extended. This mechanically-interactive system is more likely to apply if residual stresses have been produced by volume-changes associated with phase-changes in the material and is less likely to apply if the stresses are the result of cold-work (e.g. shot-peening) which may suffer metallurgical disruption by the plastic deformation associated with the growth of a crack, such as a fatigue-crack.

It is interesting to note that if a residual stress distribution is uniaxial in character, it does not apparently affect the alternating stress intensity ΔK, but only the mean stress (or K_{max} or K_{min}). However, a crack in mode I appears not to be able to grow if ΔK is entirely compressive, or even if K_{max} in the cycle does not exceed K_{max} at the threshold, ΔK_{th}. If the residual stress is compressive in sign, such that K_{max} exceeds the threshold, but K_{min} does not, the apparent stress intensity, ΔK_{app} is reduced in its effect on growth-rate to something of order $(\Delta K_{app}{}^m - \Delta K_{th}{}^m)$ or $(\Delta K_{app} - \Delta K_{th})^m$.

This is nicely illustrated by results (75) on fatigue-crack propagation through induction-hardened layers at notch roots, where an increase in applied tensile mean stress (to compensate for the residual compression in the induction-hardened layer) allowed ΔK_{app} to be reduced markedly, but maintain the same growth-rate.

The consequence of residual compression is that it must be balanced elsewhere in the body by residual tension and, in some circumstances, a fatigue-crack can accelerate markedly as it crosses the interface between compression and tension zones. In the limit, K_{max} can exceed K_{IC} and give immediate fast fracture, but high growth-rates may be obtained in a number of alloys, in which, either micro-brittle fracture of individual grains occurs (52), or particles such as intermetallics fracture or decohere more easily (85).

The residual stress distributions around welds are particularly complicated, but well-established crack growth data at low stress intensities in welded configurations are needed to try to establish general principles of behaviour, because it is both expensive and time-consuming to fatigue large welded joints at low stresses for long periods of time. It is precisely at these low stresses, where residual stresses may be important, that the macroscopic design-curves for welded joints (86) show variability and possess rather few points. Experiments are being carried out by Braid(59), in which da/dN data are first established in bulk specimens heat-treated to simulate the various microstructures of the weld/HAZ/parent-plate configuration. These then provide a basis for comparison with growth rates through the parent-plate/HAZ around a deposited weld bead. Initial results are encouraging, in that increases in rate can be associated with increasing residual tensile components. Further work of this type would benefit from more detailed numerical analysis of stress distributions.

CONCLUSIONS

In this paper, it has been possible only to describe relatively few numerical solutions, which are, or have been, of particular interest to specific experimental programmes in fracture mechanics. There are clearly many other problems in which other numerical solutions would assume much more importance. Although, to a metallurgist, the most satisfying use of F.E. analysis has been the link between "linear elastic" K_{IC} values and micro-mechanisms of fracture (32-36) and although there seems to be equal satisfaction promised in the linking of elastic/plastic parameters to processes such as the growth of micro-voids (46,50), it is the fatigue problem which should really be of most concern to structural engineers. Here, three areas have been high-lighted, in each of which, detailed experimental work is being done and would benefit from detailed stress analysis: thumbnail cracks, notches (particularly initiation conditions) and residual stresses. When one considers that a very common situation in a steel pressure-vessel would be to have a thumbnail crack situated in a residual stress field around a welded nozzle, which acts as a stress concentrator, it is perhaps fair to suggest that these three problems, separately and in combination with each other, do warrant the major attention of both experimentalists and numerical analysts.

ACKNOWLEDGEMENTS

Thanks are due to Professor R.W.K. Honeycombe for provision of research facilities and to members of my research group, particularly Mr. J.E.M. Braid, for useful discussions.

REFERENCES

1) NEUBER, H. "Kerbspannunglehre" Springer-Verlag Berlin 1938 translated as "Theory of Notch Stresses" David Taylor Model Basin Translation 74 Nov. 1945.
2) BENTHEM, J.P. "Three-dimensional State of Stress at the Vertex of a Quarter-Infinite Crack in a Half-Space" Report WTHD no.74 Dept. Mechl. Eng. Delft. 1975.
3) HUTCHINSON, J.W. "Singular Behaviour at the End of a Tensile Crack in a Hardening Material" Jnl. Mech. Phys. Solids 16 13-31, 1968 and 337-347, 1968.
4) INGLIS, C.E. "Stresses in a Plate due to the Presence of Cracks and Sharp Corners" Trans. Instn. Nav. Archit. 1913 LV, 1, p.219-240.
5) TIMOSHENKO, S.P. and GOODIER, J.N. Theory of Elasticity McGraw-Hill 3rd Ed 1970 pp.90-95.
6) WESTERGAARD, H.M. "Bearing Pressures and Cracks" Jnl. Appl. Mech. A 1939, 49-53.
7) KNOTT, J.F. "Fundamentals of Fracture Mechanics" revised impression, Butterworths (London) 1979.
8) IRWIN, G. "Analysis of Stresses and Strains Near the End of a Crack Traversing a Plate" Jnl. Appl. Mech. 1957, 24 p.361-364.
9) GRIFFITH, A.A. "The Phenomenon of Rupture and Flow in Solids" Phil. Trans. Roy. Soc.1920, A221 p.163-198.
10) RICE, J.R. and ROSENGREN, G.F. "Plane Strain Deformation Near a Crack Tip in Power-Law Hardening Material" Jnl.Mech. Phys. Solids 16 pp.1-12 1968.
11) RICE, J.R. "Mathematical Analysis in the Mechanics of Fracture" Fracture - An Advanced Treatise (ed. H.Lieberwitz) Acad. Press 1968, p.191.
12) D.N.de G.ALLEN and SOUTHWELL, R.V. "Relaxation Methods Applied to Engineering Problems" Phil.Trans.Roy.Soc.1949, A242 p.379-414.
13) STIMPSON, L.D. and EATON, D.M. "The Extent of Elasto-Plastic Yielding at the Crack Point of an Externally Notched Plane Stress Tensile Specimen" Wright-Patterson Air-Force Base Aero Res. Lab. Report ARL 24 July 1961.
14) JACOBS, J.A. "Relaxation Methods Applied to Problems of Plastic Flow,I: Notched Bar under Tension" Phil.Mag.

Vol.41 pp.349-361 1950.

15) HILL, R. The Mathematical Theory of Plasticity Clarendon Press, Oxford, 1960.

16) GREEN, A.P. and HUNDY, B.B. "Initial Plastic Yielding in Notch Bend Tests" Jnl. Mech. Phys. Solids 1956 4 p.128-144.

17) DUGDALE, D.S. "Yielding of Steel Sheets Containing Slits" Jnl. Mech. Phys. Solids 8 100-104 1960.

18) BILBY, B.A., COTTRELL, A.H. and SWINDEN, K.S. "Plastic Yielding from Sharp Notches" Proc. Roy. Soc. A272 304-314 1963.

19) BURDEKIN, F.M. and STONE, D.W. "The Crack-Opening Displacement Approach to Fracture Mechanics in Yielding Materials" Jnl. Strain Anal. 1 145-153, 1966.

20) RICE, J.R. Discussion on "The Path Dependence of the J-Contour Integral" Intl. Jnl. of Fracture 11 352-353 1975.

21) RICE, J.R. "Plastic Yielding at a Crack Tip" Proc. Ist Intl. Conf. on Fracture, Sendai, 1, pp.283-308 1966.

22) RICE, J.R. "Limitations to the Small-Scale Yielding Approximation for Crack Tip Plasticity" Jnl. Mech. Phys. Solids 22, 17-26 1974.

23) LARSSON, S.G. and CARLSSON, A.J. "Influence of Non-Singular Stress Terms and Specimen Geometry on Small-Scale Yielding at Crack Tips in Elastic-Plastic Materials" Jnl. Mech. Phys. Solids 21 1973 263-277.

24) ESHELBY, J.D. "The Continuum Theory of Lattice Defects" in Progress in Solid-State Physics Vol.3(ed.F.Seitz and D.Turnbull) Academic Press, 1956, p.79.

25) SHIH, C.F. "Relationships between the J-Integral and Crack-Opening Displacement for Stationary and Extending Cracks" General Electric Report 79CRD075 April 1979.

26) KNOTT, J.F. "Micromechanisms of Fracture and the Fracture Toughness of Engineering Alloys" Proc. 4th Int.Cong. Fracture Waterloo 1977 (Pergamon 1978) 61-92.

27) KNOTT, J.F. "Some Effects of Hydrostatic Tension on the Fracture Behaviour of Mild Steel" Jnl. Iron and Steel Inst. 204 pp.104-111.

28) GRIFFITHS, J.R. and OWEN, D.R.J. "An Elastic-Plastic Analysis for a Notched Bar in Plane Strain Bending" Jnl. Mech. Phys. Solids Vol.19 419-431 1971.

29) OWEN, D.R.J., NAYAK, G.C., KFOURI, A.P. and GRIFFITHS, J.R. "Stresses in a Partly-Yielded Notched Bar" Intl. Jnl. Numerical Methods in Engineering 6 63 1973.

30) PHILLIPS, J. "The Failure of Steel Bars Containing Blunt Notches" CEGB Report RD/B/N4713 January 1980.

31) KFOURI, A.P. "The Elastic-Plastic Analysis of Continua" Ph.D thesis Univ. of Cambridge 1970.

32) RITCHIE, R.O., KNOTT, J.F. and RICE, J.R. "On the Relationship between Critical Tensile Stress and Fracture Toughness in Mild Steel" Jnl. Mech. Phys. Solids 21 395-410 1973.

33) CURRY, D.A. and KNOTT, J.F. "Effect of Microstructure on Cleavage Fracture Toughness of Quenched and Tempered Steels" Metal Science 13 341-345 1979.

34) CURRY, D.A. "The Grain Size Dependence of Cleavage Fracture Toughness in Mild Steel" Nature 276 50-51 1978.

35) PARKS, D.M. "Interpretation of Irradiation Effects on the Fracture Toughness of a Pressure-Vessel Steel in Terms of Crack Tip Stress Analysis. Jnl. Eng. Mat. and Tech. ASME H 98 30-36, 1976.

36) RITCHIE, R.O., SERVER, W.L. and WULLAERT, R.A. "Fracture Toughness Predictions for Nuclear Pressure Vessel Steels" Proc. ICM3 Cambridge 1979 3 489-500.

37) RICE, J.R. and JOHNSON, M.A. "The Role of Large Crack Tip Geometry Changes in Plane Strain Fracture" Inelastic Behaviour of Solids (ed. M.F. Kaminen et al.) McGraw-Hill 1970 641-671.

38) TRACEY, D.M. "Finite Element Solutions for Crack Tip Behaviour in Small-Scale Yielding" Trans.ASME H Jnl. Eng. Mat. and Tech. 98 146-151 1976.

39) McMEEKING, R.M. "Finite Deformation Analysis of Crack Tip Opening in Elastic-Plastic Materials and Implications for Fracture Initiation" Report ERDA (E-11) 3084/44 Brown University May 1976.

40) SUMPTER, J.D.G. "Elastic-Plastic Fracture Analysis and Design, using the Finite-Element Method" Ph.D Thesis Imperial College Univ. of London 1973.

41) SHIH, C.F., ANDREWS, W.R. and WILKINSON, J.P.D. "Characterisation of Crack Initiation and Growth with Applications to Pressure-Vessel Steels" Proc. ICM3 Cambridge 1979, 3, 589-601.

42) McMEEKING, R.M. and PARKS, D.M. "On Criteria for J-dominance of Crack-Tip Fields in Large-Scale Yielding" Elastic-Plastic Fracture (ed. J.D. Lander et al.) ASTM STP 668 pp. 175-194 1979.

43) ROBINSON, J.N. "An Experimental Investigation of the Effect of Specimen Type on the Crack-Tip Opening Displacement and J-integral Fracture Criteria" Intl. Jnl. Fracture 12(5) 723-739, 1976.

44) WILLOUGHBY, A.A. "The Influence of Microstructure on Resistance to Slow Crack Growth in Structural Steel" Ph.D Thesis Imperial College Univ. of London 1979.

45) SLATCHER, S. "The Fracture Toughness of Forging Steels" CPGS Dissertation Univ. Cambridge 1979.

46) KNOTT, J.F. "Macroscopic Aspects of Crack Extension" Advances in Elasto-Plastic Fracture Mechanics (ed. L.H. Larsson) Applied Science 1980, 1-21.

47) KFOURI, A.P. and MILLER, K.J. "Crack Energy Separation Rates in Elastic/Plastic Fracture Mechanics" Proc.I.Mech.E. 1977.

48) KFOURI, A.P. and RICE, J.R. "Elastic/Plastic Separation Energy Rate for Crack Advance in Finite Growth Steps" Proc. ICF 5 Waterloo (Pergamon) 1978 43-59.

49) RICE, J.R. and SORENSEN, E.P. "Continuing Crack-Tip Deformation and Fracture for Plane Strain Crack Growth in Elastic-Plastic Solids" Jnl.Mech.Phys.Solids 26 1978 163-186.

50) CLARK, G., EL SOUDANI, S.H., FERGUSON, W.G., SMITH, R.F. and KNOTT, J.F. "Ductile Crack Initiation in Pressure-Vessel Steels" I.Mech.E. Tolerance of Flaws in Pressurized Components 1978 pp.121-132.

51) PARIS, P.C. and ERDOGAN, F. "A Critical Analysis of Crack Propagation Laws" Jnl. of Basic Engineering Trans. ASME D 85 528-539 1963.

52) RITCHIE, R.O. and KNOTT, J.F. "Mechanisms of Fatigue Crack Growth in Low Alloy Steel" Acta Met. 21 639-650 1973.

53) McEVILY, A.J. "Current Aspects of Fatigue" Metal Science 11 274-284, 1977.

54) British Standard 5447 "Methods for Plane Strain Fracture Toughness (K_{IC}) Testing" 1977.

55) RITCHIE, R.O. and BATHE, K.J. "On the Calibration of the Electrical Potential Technique for Monitoring Crack Growth using Finite Element Methods" Intl. Jnl. Fract. 15 47-55 1979.

56) RITCHIE, R.O., GARRETT, G.G. and KNOTT, J.F. "Crack Growth Monitoring: Optimisation of the Electrical Potential Technique using an Analogue Method" Int. Jnl. Fract. Mech. 7 462-467 1971.

57) CLARK, G. and KNOTT, J.F. "Measurement of Fatigue Cracks in Notched Specimens by means of Theoretical Electrical Potential Calibrations" Jnl. Mech. Phys. Solids 23 265-276 1975.

58) RITCHIE, R.O. "Cyclic Crack Growth in Steels" Ph.D thesis Univ. of Cambridge 1973.

59) BRAID, J.E.M. Current Research, Dept. Metallurgy Univ. of Cambridge.

60) McCARTNEY, L.N. and COOPER, Mrs.P. "Computerized Processing of Fatigue Crack Propagation Data" N.P.L. Report Mat Appl. 23, 1972.

61) ROOKE, D.P. and CARTWRIGHT, D.J. Compendium of Stress Intensity Factors HMSO 1976.

62) WILTSHIRE, B. and KNOTT, J.F. "The Toughness of High-Strength Marageing Steels, Containing Short Cracks", Intl. Jnl. Fracture 16 R19-R26 1980.

63) KING, J.E. and KNOTT, J.F. "The Effects of Crack Length and Shape on the Fracture Toughness of a High-Strength Steel" submitted to Jnl.Mech.Phys.Solids.

64) SHAH, R.C. and KOBAYASHI, A.S. "Stress Intensity Factors for an Elliptical Crack Approaching the Surface of a Semi-Infinite Solid" Intl.Jnl.Fracture 9 ASTM STP 513 pp.133-146 1973.

65) RANDALL, P.N. in discussion Plane Strain Crack Toughness Testing ASTM STP 410, p.88-126 1967.

66) PICKARD, A.C. This conference.

67) KNOTT, J.F. "The Use of Analogue and Mapping Techniques, with particular reference to the Detection of Short Cracks" The Measurement of Crack Length and Shape during Fracture and Fracture Ed. C.J. Beevers EMAS pub. 1980.

68) THROOP, J.F. "Fatigue Crack Growth in Thick-Walled Cylinders" National Conference on Fluid Power 115-131 1972.

69) BOWIE, O.L. "Analysis of an Infinite Plate containing Radial Cracks originating at the Boundary of an Internal Circular Hole" Jnl. Math. Phys. 35 60-71 (1956).

70) NEWMAN, J.C. Jnr. "An Improved Method of Collocation for the Stress Analysis of Cracked Plates with Various Shaped Boundaries, N.A.S.A. Tech Note D-7-376 (1971).

71) SMITH, E. "Fracture Criteria in Bodies containing Blunt Notches" Proc. Roy. Soc. A299 455-474 1967.

72) BILBY, B.A. and HEALD, P.T. "Crack Growth in Notch Fatigue" Proc. Roy. Soc. A305 429-439 (1968).

73) MUBEEN, A. and BANDYOPADHYAY, S.N. "Fracture Strength of a Notched Beam when a Small Fatigue Crack Emanates from the Notch Root" Proc. ICF5 (Pergamon 1978) 741-749.

74) LIVESEY, V.B. and KNOTT, J.F. "The Initiation of Fatigue Cracks at Notches" Proc. ICSMA5 Aachen 27-31 Aug.1979 Vol.2 pp.1137-1144.

75) CLARK, G. and KNOTT, J.F. "Effects of Notches and Surface Hardening on the Early Growth of Fatigue Cracks" Metal Science 11 345-350 1977.

76) CLARK, G. "Fatigue-Crack Growth from Notches" Ph.D thesis Univ. Cambridge 1975.

77) SMITH, R.A., JERRAM, K. and MILLER, K.J. "Experimental and Theoretical Fatigue Crack Propagation Lives of Various Notched Plates" Jnl. Strain Anal. 9 61-70 1974.

78) COFFIN, L.F. "Cyclic Straining and Fatigue" Internal Stresses and Fatigue in Metals Ed. Rassweiler and Grube. Elsevier 1959 pp.364-381.

79) WILSON, W.K. "Elastic-Plastic Analysis of a Blunt-Notched CT Specimen and Applications" Trans. ASME Jnl. Press.Vessel Tech 96 293-298 1974.

80) LIVESEY, V.B. research in progress (Univ. of Cambridge, Dept. of Metallurgy).

81) KNOTT, J.F. "Effects of Notch Depth on the Toughness of Mild Steel" Proc. 2nd Intl. Conf. on Fracture Brighton 1969 Chapman and Hall pp.205-240.

82) DAWES, M.G. "The C.O.D. Design Curve" Advances in Elasto-Plastic Fracture Mechanics ed. L.H.Larsson Applied Science 1980 pp.279-300.

83) SUMPTER, J.D.G. and TURNER, C.E. "Design Using Elastic Plastic Fracture Mechanics" Int. Jnl.Fracture 12 861-871, 1976.

84) HALL, W.J., KIHARA, H., SOETE, W. and WELLS, A.A. Brittle Fracture of Welded Plate Prentice-Hall 1967 pp.112-117.

85) GARRETT, G.G. and KNOTT, J.F. "The Influence of Compositiona and Microstructural Variations on the Mechanisms of Static Fracture in Aluminium Alloys" Metall. Trans. 9A 1187-1201 1978.

86) BOOTH, G.S. and WYLDE, J.G. "Fatigue Problems in North Sea Structures" Metal Sci. 11 308-314 1977.

NEW FINITE ELEMENTS FOR FRACTURE ANALYSIS

B.B. Sabir [(1)] and B. Barr [(2)]

SUMMARY

A new finite element which adequately represents the character of the strains near the crack tip is presented. The element is based on displacement functions, derived from assumed simple strain functions, by integrating the strain displacement equations. The resulting displacement functions contain terms associated with the rigid body modes. The element is 'trapezoidal' in shape with two straight and two circular boundaries. A 'triangular' crack tip element based on assumed simple displacement functions is also presented. This element has two of its boundaries straight, meeting at the tip of the crack, and the other circular.

When employed in the singular region of the crack tip using as few as 312 degrees of freedom, these elements give results for displacements and stresses whose accuracies are within 1.5% and 3%, respectively, of 'exact' solutions.

(1) Computer Applications Engineer, Mott, Hay and Anderson Croydon, Surrey (Formerly Lecturer, South Glamorgan Institute of Higher Education, Western Avenue, Cardiff)

(2) Principal Lecturer, Department of Civil Engineering and Building, The Polytechnic of Wales.

INTRODUCTION

The finite element method has been used in many investigations for the solution of linear fracture mechanics problems. Early papers [1],[2],[3],[4] and [5] in this field used conventional elements to estimate stress intensity factors near the crack tip. In these works the cracked structure is represented by constant strain triangular elements [6]. Although acceptable solutions have been obtained for simple geometries, it was found necessary to use a large number of elements around the crack tip.

Byskov [7] appears to have been the first to formulate a finite element based on classical displacement functions using the singular stress fields established by Muskhelishvili [8]. The element is triangular in shape with the crack tip at the interior of the triangle. A high order circular crack tip element, similar in many respects to that of Byskov's, was presented by Wilson [9]. The authors' general formulation is in terms of a stress function, due to Williams [10], which is equivalent to that of Muskhelishvili.

Polynomial type shape functions, rather than classical solution functions, have also been used to produce element stiffness matrices which contain singularities [9], [11]. These normally lead to simpler forms of elements and easier integration of the stiffness equations.

It has been shown that singular transformations from physical to isoparametric coordinates can be made if node points along the sides of the element are positioned in a certain way [12], [13], [14].

Another approach in producing finite elements for fracture analysis is that based on mixed 'hybrid' formulation. This method involves prescribing two or more stress or displacement fields within the element, employing available singular stress fields from classical fracture analysis together with simple polynomial expansions. An advantage of this approach [15] is that the assumed functions contain the stress intensity factors which are regarded as the unknowns and hence solved for directly.

Although there is still new work in fracture analysis employing conventional elements [16], [17], most of the recent efforts have been aimed at incorporating a singular stress field directly into the finite element formulation. A number of such formulations were presented at the first international symposium held at Swansea in 1978.

In this paper, the formulation in terms of polar coordinates of two element stiffness matrices which take account of the singular condition in the vicinity of the crack tip is presented. The first element (designated I) is 'trapezoidal' in shape having two straight and two circular boundaries. The element is formed from two 'triangular' subregions suitable for combination with constant strain triangular elements [6] . The second element (designated II) is 'triangular' and is based on simple displacement functions similar to those used by Wilson [9]. When analysing problems, a cluster of this type of elements, all having a common nodal point at the tip of the crack, is surrounded by elements I. The circular boundary of element II made it possible to preserve the true geometry of the interface between the two types of elements.

TRAPEZOIDAL ELEMENT I

The stiffness matrix formulated here is for the 'trapezoidal' element shown in Fig I. The element is initially considered to be made up of two 'triangular' subregions as shown in Fig 3. The procedure adopted in forming the 'trapezoidal' element from the two triangles will be described later. The element is derived in terms of the polar coordinates system shown.

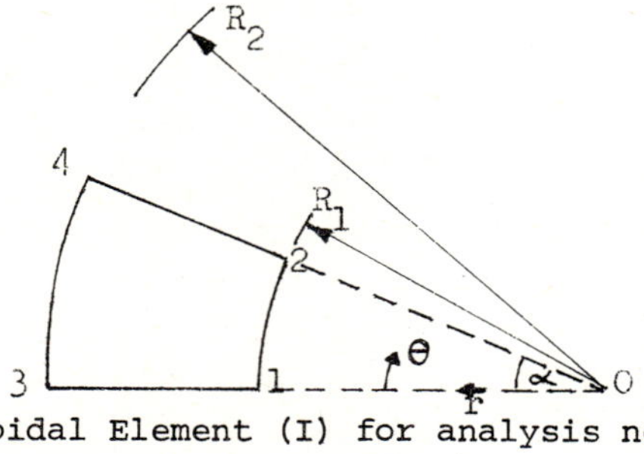

Fig 1. Trapezoidal Element (I) for analysis near crack tip.

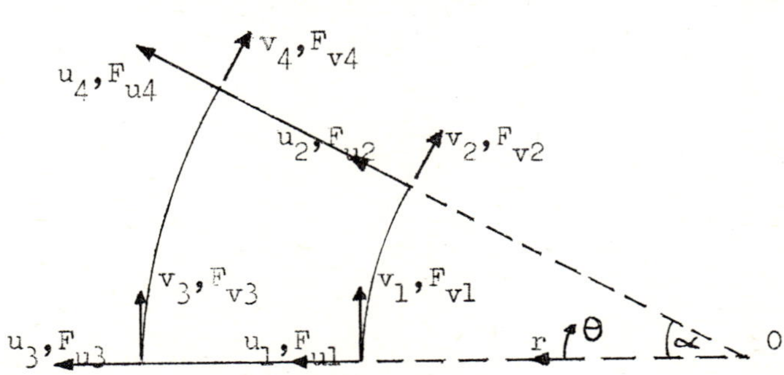

Fig 2. Positive directions of displacements and forces.

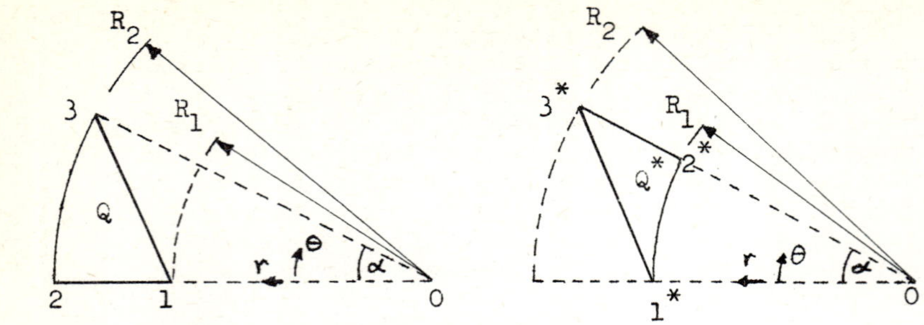

Fig 3. Triangular sub-elements Q and Q*

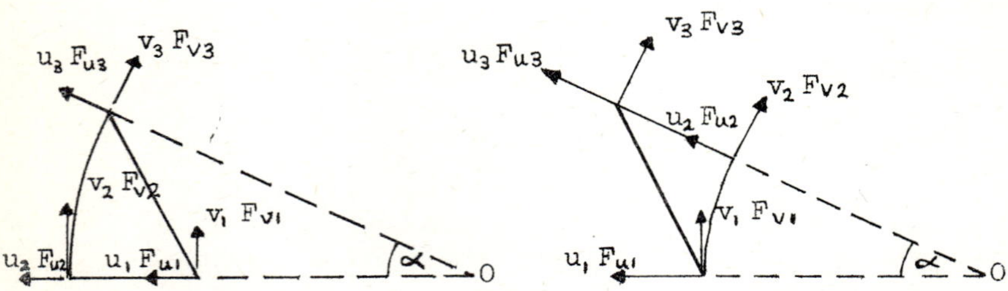

Fig 4. Positive directions of displacements and forces.

Ashwell, Sabir and Roberts [18] demonstrated how a satisfactory displacement shape function can be derived from assumed simple strain functions by integrating the strain-displacement equations. Following their method and using the notation adopted in [19], the strain components for two dimensional analysis are:-

$$\varepsilon_r = \frac{\partial u}{\partial r} = f_r \quad (1)$$

$$\varepsilon_\theta = \frac{u}{r} + \frac{\partial v}{r\partial\theta} = f_\theta \quad (2)$$

$$\gamma_{r\theta} = \frac{\partial u}{r\partial\theta} + \frac{\partial v}{\partial r} - \frac{v}{r} = f_{r\theta} \quad (3)$$

where f_r, f_θ and $f_{r\theta}$ are functions of r and θ.

If $fr = f\theta = fr\theta = 0$, equations (1), (2) and (3) can be integrated to give the following complimentary functions:-

$$\begin{aligned} u &= a_1 . \sin\theta + a_2 . \cos\theta \\ v &= a_1 . \cos\theta - a_2 . \sin\theta + a_3 v \end{aligned} \qquad (4)$$

where a_1, a_2, and a_3 are independent constants of integration. Equations (4) represent displacements with zero strains i.e. they represent rigid body displacements.

If $fr \neq o$, $f\theta \neq o$ and $fr\theta \neq o$, particular integrals can be obtained from equations (1), (2) and (3) by making suitable choices for fr, fθ and frθ. Each 'triangular' subregion will have six degrees of freedom, namely u and v at each node, and thus the shape function is expected to have six independent constants - three are required for the rigid body displacements, equations 4, leaving three constants available for the strain functions. Consideration of the well known crack tip equations suggest strain functions of the form $f(\theta)/r^{\frac{1}{2}}$. The following strain functions are thus assumed:-

$$\begin{aligned} fr &= \frac{1}{r^{\frac{1}{2}}} \cdot \left| b_1 + b_2\theta \right| \\ f\theta &= \frac{1}{r^{\frac{1}{2}}} \cdot \left| b_3 + b_4\theta \right| \\ fr\theta &= \frac{1}{r^{\frac{1}{2}}} \cdot \left| b_5 + b_6\theta \right| \end{aligned} \qquad (5)$$

Where $b_1 \ldots . b_6$ are constants, three of which are independent. The three dependent constants are eliminated by satisfying the compatibility equations (1), (2), (3). By substituting for the strain functions fr and fθ in equations (1) and (2) the following particular integrals may be obtained.

$$u = 2\,(b_1 + b_2\theta)\; r^{\frac{1}{2}} \qquad (6)$$

$$v = \left[(b_3 - b_1)\theta + \tfrac{1}{2}(b_4 - 2b_2)\theta^2 \right] r^{\frac{1}{2}} \qquad (7)$$

Substituting for u, v and $f_{r\theta}$ in equation (3) we have

$$b_4 = b_5 = 2b_2 \tag{8}$$

$$b_6 = b_1 - \frac{1}{2}b_3 \tag{9}$$

Eliminating b_4 from equation (7), we have:-

$$u = 2(b_1 + b_2\theta) \cdot r^{\frac{1}{2}} \tag{10}$$

$$v = (b_3 - 2b_1) \cdot \theta r^{\frac{1}{2}} \tag{11}$$

Equations (10) and (11) can be simplified slightly by observing that:-

(i) u requires an independent term containing $r^{\frac{1}{2}}$ and a term containing $r^{\frac{1}{2}}\theta$

(ii) v requires one independent term containing $r^{\frac{1}{2}}\theta$.

Thus,

$$u = (a_4 + a_5\theta)\, r^{\frac{1}{2}} \tag{12}$$

$$v = a_6 \cdot \theta \cdot r^{\frac{1}{2}} \tag{13}$$

The complete solution for u and v thus become:-

$$u = a_1 \sin\theta + a_2 \cos\theta + a_4 r^{\frac{1}{2}} + a_5 r^{\frac{1}{2}}\theta \tag{14}$$

$$v = a_1 \cos\theta - a_2 \sin\theta + a_3 r + a_6 r^{\frac{1}{2}}\theta \tag{15}$$

The corresponding expressions for the strains are:-

$$\varepsilon_r = \frac{1}{2r^{\frac{1}{2}}} \cdot \left[a_4 + a_5\theta\right] \tag{16}$$

$$\varepsilon_\theta = \frac{1}{r^{\frac{1}{2}}} \cdot \left[a_4 + a_6 + a_5\theta\right] \tag{17}$$

$$\gamma_{r\theta} = \frac{1}{2r^{\frac{1}{2}}} \cdot \left[2\, a_5 - a_6\theta\right] \tag{18}$$

Consideration of these equations indicate that ε_r and ε_θ each contain one independent constant, while $\gamma_{r\theta}$ is completely dependent on ε_r and ε_θ. It is considered that this is the only possibility if terms of $f(\theta)/r^{\frac{1}{2}}$ are to be used while satisfying the compatibility equations. It is also observed from equations (14) and (15) that coupling in the displacement is provided only by the rigid body terms.

The shape functions (equations (14) and (15)) are now used to develop the stiffness matrices for the two sub-elements Q and Q* shown in Fig 3. These are respectively given by:-

$$\underset{\sim}{K}_e = \underset{\sim}{C}^{-1,T} \int_{f(\theta)}^{R_2} \int_{0}^{\alpha} \underset{\sim}{H} . r . d\theta . dr . \underset{\sim}{C}^{-1} \tag{19}$$

$$\underset{\sim}{K}_e = \underset{\sim}{C}^{-1,T*} \int_{R_1}^{f(\theta)} \int_{0}^{\alpha} \underset{\sim}{H} . r . d\theta . dr . \underset{\sim}{C}^{-1*} \tag{20}$$

where $f(\theta) = qR_1/(q \cos \theta - \sin \theta) = r$ which represents the equation of the straight line boundary common to the two sub-elements, and $q = R_2 \sin \alpha/(R_2 \cos \alpha - R_1)$. R_1, R_2 and α are defined in Fig 3. Details of the matrices in equations (19) and (2) are given in [20].

The two sub-elements are now combined to obtain the stiffness matrix for the trapezoidal element I shown in Fig 1. Following the node numbering system shown in Figs 1 and 3, it is observed that the two sub-elements Q and Q* are connected at nodes 1 and 1*, and at nodes 3 and 3* to give nodes 1 and 4, respectively on element I. Nodes 2 and 3 of I are nodes 2 of Q* and 2 of Q respectively. Such observation will determine the positions of the Q_{ij} and Q^*_{ij} $(i,j = 1,2...6)$ coefficients in the over-all (8 x 8) stiffness matrix for element I. Where a contribution is made by both Q and Q*, the two coefficients are superimposed. The resulting matrix in terms of the coefficients of Q and Q* stiffness matrices is given in the Appendix.

TRIANGULAR ELEMENT II

In this section a stiffness matrix for the 'triangular' element shown in Fig 5 is presented. The element stiffness is based on displacement shape functions similar to those suggested by Wilson [9]. These are:-

$$u = (a_1 + a_2\theta)\, r^{\frac{1}{2}}$$

$$v = (a_3 + a_4\theta)\, r^{\frac{1}{2}}$$

The corresponding strain components are obtained from equations (1), (2) and (3). These are:-

$$\varepsilon_r = \frac{1}{2r^{\frac{1}{2}}} \cdot (a_1 + a_2\theta)$$

$$\varepsilon_\theta = \frac{1}{r^{\frac{1}{2}}} \cdot (a_1 + a_4 + a_2\theta)$$

$$\gamma_{r\theta} = \frac{1}{2r^{\frac{1}{2}}} \cdot (2a_1 - a_3 - a_4\theta)$$

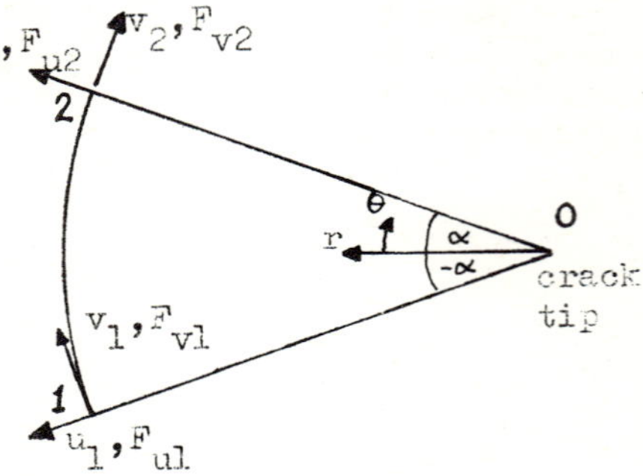

Fig 5. Crack tip element II.

It is observed that ε_θ and $\gamma_{r\theta}$ each contain one independent constant while ε_r is dependent on ε_θ and $\gamma_{r\theta}$.

While Wilson's element has three straight boundaries, the element presented here has one of its boundaries circular. This is done so that when surrounded by the trapezoidal element I, the present element II will preserve the true geometry at the common interface. The element has four degrees of freedom ,namely u and v at each node. The (4 x 4) element stiffness matrix (for plane stress conditions) relating the element nodal forces to the corresponding element nodal displacements is given below:-

$$\begin{bmatrix} F_{u1} \\ F_{v1} \\ F_{u2} \\ F_{v2} \end{bmatrix} = \frac{Et}{24(1-\nu^2)\alpha} \begin{bmatrix} 4J_1+6J_2 & -9J_3 & 2J_1-6J_2 & 3J_4 \\ & 2J_5+12 & -3J_4 & J_5-12 \\ & & 4J_1+6J_2 & 9J_3 \\ \text{symmetrical} & & & 2J_5+12 \end{bmatrix} \begin{bmatrix} u_1 \\ v_1 \\ u_2 \\ v_2 \end{bmatrix}$$

Stiffness Matrix for Element II

where $J_1 = (5 + 4\nu)\alpha^2$, $J_2 = (1-\nu)$, $J_3 = (1 + \nu)\alpha$

$J_4 = (5 + \nu)\alpha$, $J_5 = (1-\nu),\alpha$, ν = Poisson's ratio,

E = elasticity modules, and t = thickness.

NUMERICAL RESULTS

An assessment of the effectiveness of the elements presented here was made by analysing two problems for which 'exact' solutions are available. In the two examples considered, a semi-infinite crack in an infinite plane was analysed by considering a finite circular region of radiu 30.25 (see Fig 6) centred at the crack tip. In the first problem, the outer boundaries of the body were subjected to forces corresponding to the opening Mode 1 conditions and in the second these forces corresponded to the sliding Mode 2 situation. For brevity, the results for the Mode 1 case only are given here. Detailed results appear in [20].

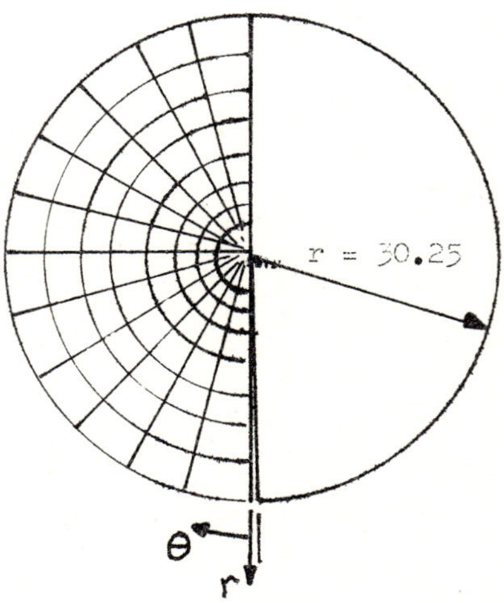

Fig 6. Circular body containing a radial crack along a radius.

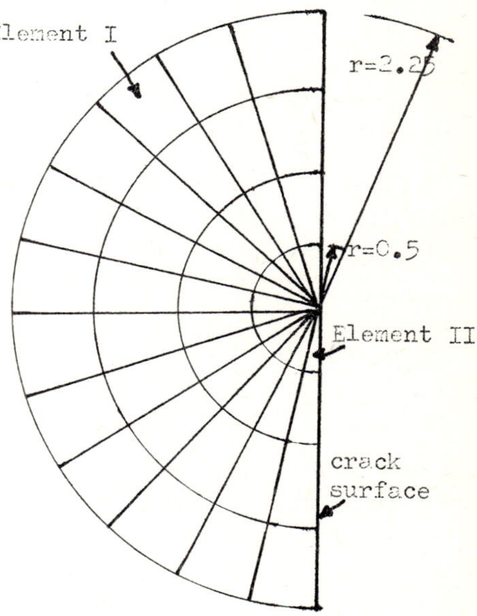

Fig.7 Expanded view at the crack tip region.

Only one half of the body, i.e. $\theta = 0$ to $\theta = \pi$, was considered and this was divided into eleven annular rings as shown in Figs 6 and 7. These rings described radii:-

$$0.5,\ 1.0,\ 1.625, 1.5^2,\ 2.0^2,\ 2.5^2,\ ....5.5^2$$

This was done so that direct comparison with results obtained in [11] could be made. The displacements as calculated from the well known crack tip equations were imposed at the boundaries $\theta = 0$, $\theta = \pi$ and $v = 30.25$.

The body was considered under plane stress conditions with $E = 200$, $v = 0.3$ and $t = 1.0$. In Fig 8 convergence curves for the displacements at $r = 0,5$ and $\theta = 90^\circ$, on the boundary of the triangular element II are shown. The 'exact' results using the crack tip equations are also shown.

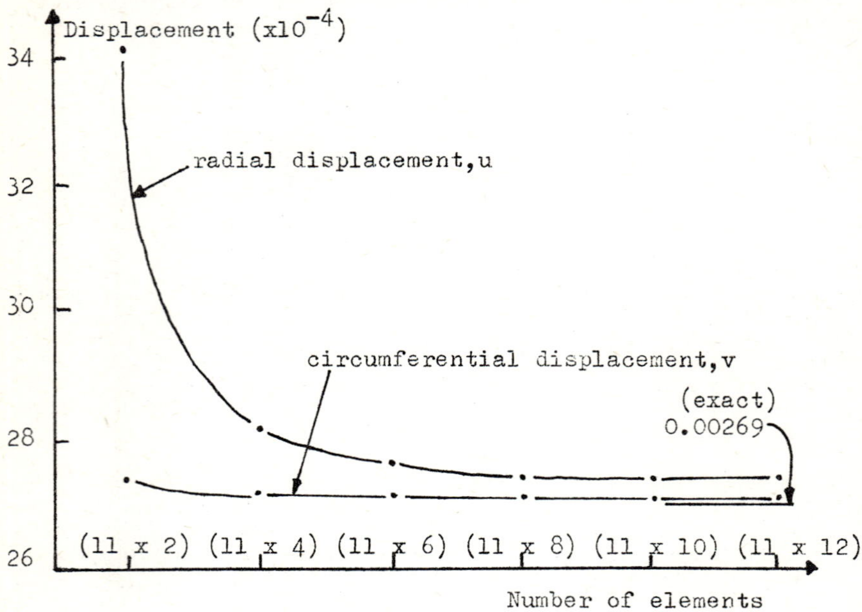

Fig 8. Convergence curves for displacements at $r = 0.5, 0 = 90^\circ$.

It is seen that the v - displacement exhibits faster convergence than the u-displacement. When (11 x 6) elements are used, the value for v is within 1% of the 'exact' solution, while that for u is 2.6% in error.

However, if a criterion for convergence is based on the difference between two successive element refinements being within 1%, then the u-displacement converges to within 1.8% of the 'exact' solution when (11 x 8) elements are used, and the v-displacement converges to within 1.8%, using as few as (11x2) elements.

A study on convergence of stresses was also carried out. The percentage error in u,v, σ_r, σ_θ and $\tau_{r\theta}$ obtained using (11x 12) elements are summerised in Table 1.

	u	v	σ_r	σ_θ	$\tau_{r\theta}$
'exact'	0,00269	0.00269	.0.598	0.200	0.200
(11 x 12) elements	0.00273	0.00270	0.615	0.213	0.195
error (%)	1.5	0.4	2.8	6.5	- 2.5

Table 1. Displacements and stresses at r = 0.5 and $\theta = 90^{\circ}$.

It is observed that the solutions obtained increase in accuracy with increasing distance from the crack tip. In Table 2, the results for the five quantities at r = 2.25 and $\theta = 90^{\circ}$, using (11 x 12) elements are shown.

	u	V	σ_r	σ_θ	$\tau_{r\theta}$
'exact'	0.00571	0.00571	0.282	0.0940	0.0940
(11 x 12) elements	0.00576	0.00574	0.284	0.0994	0.0946
error (%)	0.9	0.4	0.7	5.7	0.6

Table 2. Displacements and stresses at r = 2.25 and $\theta = 90^{\circ}$.

The same problem was analysed by Tracey [11] using an element which embodies the singularity conditions at the crack tip. Generally, his results for the displacements at r = 0.5 were within 2% of the 'exact' values in the range $22^{\circ} \leq \theta \leq 180^{\circ}$. For $\theta = 15^{\circ}$, however, the results were 5% in error. In the present work, although a coarser mesh was used, the results for the displacements were nowhere more than 1.5% in error. While the present results for the stresses had a maximum error of 6.5%, Tracey reports a maximum error in the stresses of 4%.

In Fig 9 both the nodal displacements u and v at r = 0.5 are shown. It is seen that irrespective of the position of the node, the accuracy of the results is maintained on the whole boundary.

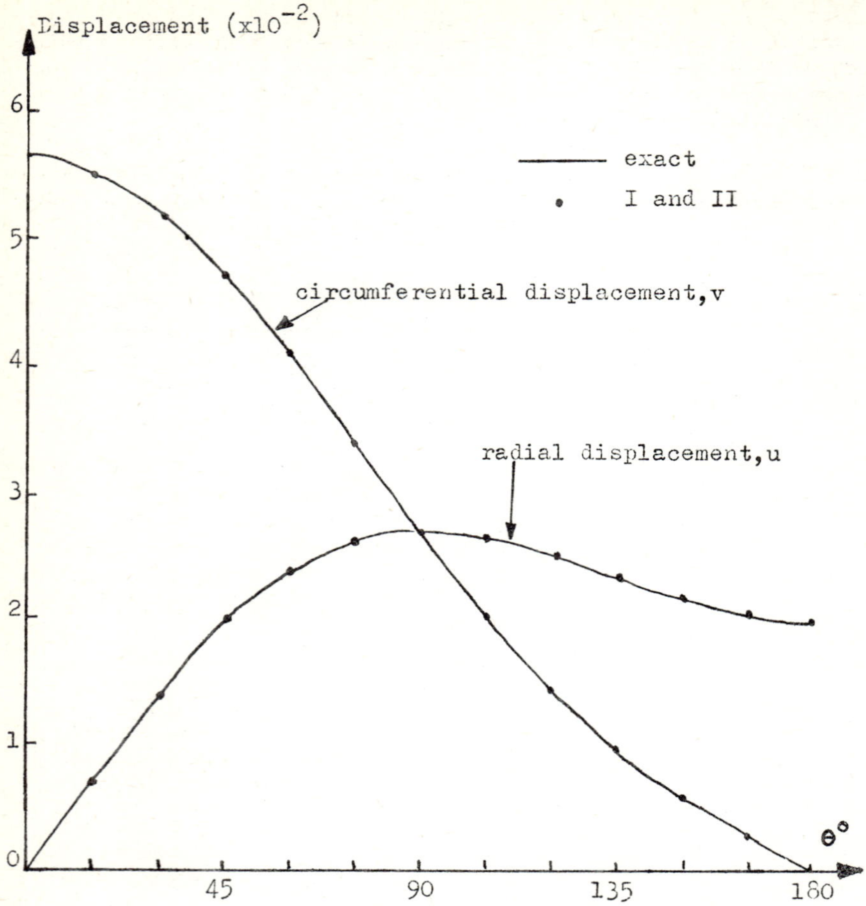

Fig 9. Distribution of displacements u and v at r = 0.5.

The above discussion demonstrates that using as few as 312 degrees of freedom accuracy within 3%, with the exception of σ_θ of 'exact' solutions are obtained for Mode 1 conditions. It was found [20] that similar accuracy was obtained for Mode 2 conditions. In particular the errors in σ_θ associated with the Mode 1 conditions are obviated. In another paper presented at this Conference [21], elements I and II are used together with constant strain triangular elements to analyse a geometry used to determine the fracture toughness of concrete.

CONCLUSIONS

Two elements (I and II) based on displacement functions which represent closely those associated with the singular character of the strains near the crack tip are presented. Using only 312 degrees of freedom (11 x 12 elements) results for displacements and stresses, for the immediate vicinity of the crack tip, within 3% of 'exact' solutions were obtained. The elements work equally well for Mode I and Mode 2 problems.

It is of interest to note that the strains for element I (equations (16) - (18) and those for element II (equations (23 - (25)) vary in an identical manner, with respect to r and θ, at the common interface. Although the displacements for the two elements (equations (14),(15) and (21), (22)) are continuous only at the nodal points, it is argued that (excepting the rigid body terms as the radius of the tip elements decrease indefinitely) continuity at the interface will be approached in the limit. The inclusion of the rigid body terms in the displacement functions of element I may be partly responsible for the generally high degree of accuracy obtained.

REFERENCES

1. KOBAYASHI, A.S. MAIDEN, D.E., SIMON, B.J. and IIDA,S. Applications of Finite Element Analysis Method to Two-dimensional Problems in Fracture Mechanics, A.S.M.E. Winter Annual Meeting 69 - WA/PVP - 12. 1969.

2. CHAN.S.K. TUBA, I.S. and WILSON, W.K. On the Finite Element Method in Linear Fracture Mechanics. 2nd National Fracture Mechanics Symposium, Lehigh University 17 - 19 June 1968.

3. WATWOOD, V.L. The Finite Element Method for Prediction of Crack Behaviour. Nuclear Eng. and Design 11, pp.323 - 332. 1969.

4. DIXON, J.R. and STRANNIGAN, J.S. Determination of energy release rates and stress intensity factors by the finite element method. Journal of strain analysis Vo. 7 No.2, 1972

5. KENDALL, D.P. and HUSSAIN, M.A.
A new facture-toughness test method for thick walled cylinder material, presented at 1971 S.E.S.A. Spring Meeting held in Salt Lake City, Utah on May 18-21, Experimental Mechanics, pp. 184 - 189, 1972.

6. ZIENKIEWICZ, O.C.
The finite element method in engineering science, McGraw-Hill, London 1971.

7. BYSKOV, E. The Calculation of stress intensity factors using the finite element method with cracked elements, Int. J. Fracture Mechanics, 6, pp. 159 - 167, 1970.

8. MUSKHELISHVILI, N. Some basic problems of the mathematical theory of elasticity, Noordhoff, 1963.

9. WILSON, W.K. Some crack-tip finite elements for plane elasticity, proceedings of the 1971 National Symposium on fracture Mechanics, Part 1, A.S.T.M. S.T.P. 513, American Soc. for testing and Materials, pp.90-105,1972.

10. WILLIAMS M.L. On the stress distribution at the base of a stationary crack, Jn. of Applied Mechanics, vol.24, No.1 pp. 109 - 114. 1957.

11. TRACY, D.M. Finite elements for determination of crack-tip elastic stress intensity factors, Engrg. Fracture Mechanics, vol. 3, pp. 255-266, 1971.

12. HENSHELL, R.D. and SHAW, K.G. Crack-tip elements are unnecessary, Int.Jn. for Num.Meth in Engineering, Vol.9 pp. 495-509, 1975

13. HIBBITT, H.D. Some properties of singular isoparametric elements, Int.Jn. for Num.Meth in Engrg. vol.11 No. 1 pp 180-184, 1977.

14. BARSAUM, R.S. On the use of isoparametric finite elements in linear fracture mechanics, Int.Jn. for Num. Meth. in Engrg. Vol. 10. No. 1 pp. 25-37, 1976.

15. PIAN, T.H.H., TONG. P. and LUK.C. Elastic crack analysis by a finite element hybrid method, Proc. 3rd. Air Force Conf. on Matrix Methods in Struct. Mech., Dayton, Ohio, 1971

16. MURAKAMI, Y. A simple procedure for the accurate determination of stress intensity factors by finite element method, Engrg. Fracture Mechanics, Vol. 8, No. 4, pp. 643-656, 1976.

17. BARR. B.I.G. EVANS, W.T. WATKINS, J and RONHI, M.R. Fracture path prediction using numerical and experimental techniques, 1st. Int. Symposium on Num. Methods in Fracture Mechanics, Univ. Coll.Swansea, edited by Owen, D.R.J. and Luxmoore, A.R. 1978.

18. ASHWELL, D.G. SABIR, A.B. and ROBERTS, T.M. Further studies in the application of curved finite elements to circular arches, Int. J. Mech. Sci. Pergammon, Vol. 13, pp. 507-517 .1971.

19. TIMOSHENKO, S.P. and GOODIER, J.N. Theory of elasticity, third edition, McGraw-Hill, Kogakusha Ltd. 1970.

20. SABIR, B.B. The application of the finite element method to fracture mechanics problems, PhD. thesis submitted to the C.N.A.A. London, 1980.

21. BARR.B. EVANS, W.T. DOWERS. R.C. SABIR B.B. The Fracture Toughness of Concrete. 2nd International Conference on Num. Methods in Fracture Mechanics, Swansea 1980.

APPENDIX - Stiffness Matrix for Trapezoidal Element.

$$
\begin{bmatrix} F_{u1} \\ F_{v1} \\ F_{u2} \\ F_{v2} \\ F_{u3} \\ F_{v3} \\ F_{u4} \\ F_{u4} \end{bmatrix}
=
\begin{bmatrix}
Q_{1,1}+Q^*_{1,1} & Q_{1,2}+Q^*_{1,2} & Q^*_{1,3} & Q^*_{1,4} & Q_{1,3} & Q_{1,4} & Q_{1,5}+Q^*_{1,5} & Q_{1,6}+Q^*_{1,6} \\
 & Q_{2,2}+Q^*_{2,2} & Q^*_{2,3} & Q^*_{2,4} & Q_{2,3} & Q_{2,4} & Q_{2,5}+Q^*_{2,5} & Q_{2,6}+Q^*_{2,6} \\
 & & Q^*_{3,3} & Q^*_{3,4} & 0 & 0 & Q^*_{3,5} & Q^*_{3,6} \\
 & & & Q^*_{4,4} & 0 & 0 & Q^*_{4,5} & Q^*_{4,6} \\
 & & & & Q_{3,3} & Q_{3,4} & Q_{3,5} & Q_{3,6} \\
 & \text{symmetrical} & & & & Q_{4,4} & Q_{4,5} & Q_{4,6} \\
 & & & & & & Q_{5,5}+Q^*_{5,5} & Q_{5,6}+Q^*_{5,6} \\
 & & & & & & & Q_{6,6}+Q^*_{6,6}
\end{bmatrix}
\begin{bmatrix} u_1 \\ v_1 \\ u_2 \\ v_2 \\ u_3 \\ v_3 \\ u_4 \\ v_4 \end{bmatrix}
$$

ON A FINITE ELEMENT SCHEME FOR PLANE CRACK PROBLEMS

M. AMARA*, P. DESTUYNDER*, M. DJAOUA*

SUMMARY

We are interested in the calculation of the stress intensity factors for a cracked plate submitted to traction efforts. The stress singularities are taken into account in the finite element scheme here suggested which leads to theoretical error estimates for the factors. For a cheap computational cost, we obtain very accurate numerical results.

NOTATIONS

Ω	plane cracked domain
Γ_0	free boundary
Γ_f	boundary with prescribed tractions
h	linear dimension of finite element
K_1, K_2	stress intensity factors
$\vec{g}$	tractions
$\vec{u}$	displacement fields
σ	stress fields
$\vec{n}$	outward normal to the boundary
ε	strain fields
$\mathcal{R}$	rigid displacements
V	space $(H^1(\Omega))^2/_{\mathcal{R}}$
a	bilinear form defined by $a(\vec{u},\vec{v}) = \int_\Omega \sigma_{\alpha\beta}(\vec{u})\ \varepsilon_{\alpha\beta}(\vec{v})\ ;\ \forall\ \vec{u},\ \vec{v} \in V$

*Ecole Polytechnique, Centre de Mathématiques Appliquées
91128 PALAISEAU Cédex (France)

a_T	bilinear form "a" restricted to a subdomain T of Ω
$\mathcal{T}_h$	regular triangulation of domain Ω
E	Young modulus, $k = \frac{3 - \nu}{1 + \nu}$
ν	Poisson coefficient, $\mu = \frac{1 + \nu}{E}$
γ	characteristic coefficient for crack propagation
W_h	finite displacement space defined by $W_h = \{\vec{v}_h \in (\mathcal{C}^0(\Omega))^2 \ ; \ \vec{v}_{h\mid K} \in (P_1(K))^2 \ \forall K \in \mathcal{T}_h\}_{/\mathcal{R}}$
$\vec{S}_1, \vec{S}_2$	singular displacements
V_h	finite element displacement space defined by $V_h = \{\vec{S}_1, \vec{S}_2\} \oplus W_h$
$\Vert\ \Vert_V$	norm in V
$\vert\vert\vert\ \vert\vert\vert$	norm defined by $(a(\vec{v}, \vec{v}))^{\frac{1}{2}}$
$\Vert\ \Vert_{m,\Omega}$	norm in $H^m(\Omega)$ for any integer m
C	any constant independent of h
Π_h	interpolation operator by piecewise linear functions
A	regular part of the stiffness matrix
F	semi-regular part of the stiffness matrix
P	singular part of the stiffness matrix

Supercripts or subscripts

h	finite element solutions
r	regular parts of solutions
*	reference solutions
α,β	component of a field.

1 - INTRODUCTION

We are dealing with a cracked plate, submitted to traction efforts on a part of its boundary non including the crack. Equations are :

$$(1) \quad \begin{cases} \sigma_{\alpha\beta,\beta} = 0 & \text{in } \Omega \\ \sigma_{\alpha\beta}\, n_\beta = 0 & \text{on } \Gamma_0 \\ \sigma_{\alpha\beta}\, n_\beta = g_\beta & \text{on } \Gamma_f \end{cases}$$

The stress tensor is related to the strain tensor by

$$(2) \quad \sigma_{\alpha\beta}(\vec{u}) = \frac{E}{1+\nu} \ \{\varepsilon_{\alpha\beta}(\vec{u}) + \frac{\nu}{1-\nu}\, \varepsilon_{\mu\mu}(\vec{u})\ \delta_{\alpha\beta}\}$$

A variational formulation of problem (1) is

$$(3) \quad \begin{cases} \text{Find } \vec{u} \in V \text{ such that} \\ a(\vec{u}, \vec{v}) = \int_\Gamma \vec{g}.\vec{v} \quad \forall \vec{v} \in V \end{cases}$$

Problem (3) has a unique solution, which can be expanded on a singular part and a regular part as follows [3]

$$(4) \quad \vec{u} = K_1 \vec{S}_1 + K_2 \vec{S}_2 + \vec{u}^r$$

where singular parts $\vec{S}_1$ and $\vec{S}_2$ are :

$$(5) \quad \vec{S}_1(r,\Theta) = \frac{r^{\frac{1}{2}}}{2\mu\sqrt{2\pi}} \begin{cases} \cos\frac{\Theta}{2}(k - \cos\Theta) \\ \sin\frac{\Theta}{2}(k - \cos\Theta) \end{cases}$$

$$\vec{S}_2(r,\Theta) = \frac{r^{\frac{1}{2}}}{2\mu\sqrt{2\pi}} \begin{cases} \sin\frac{\Theta}{2}(k + \cos\Theta + 2) \\ -\cos\frac{\Theta}{2}(k + \cos\Theta - 2) \end{cases}$$

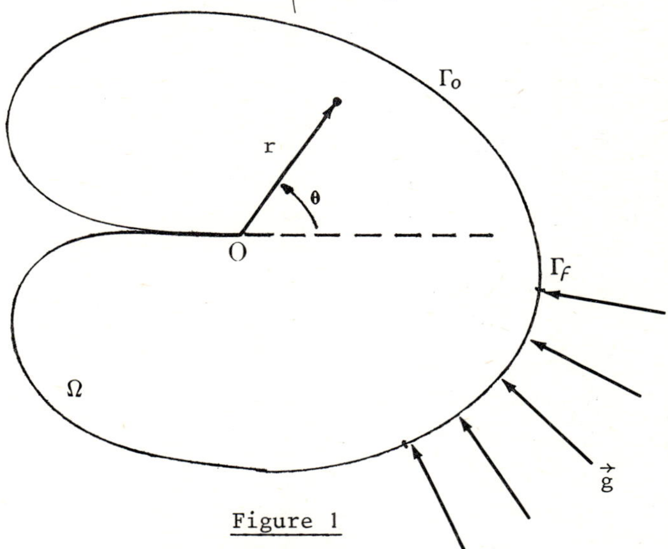

Figure 1

The cracked plate

Stress intensity factors K_1 and K_2 are very important since the energy gradient with respect to the domain can be expressed in term of them [5]

$$(6) \quad \frac{\partial W}{\partial \Omega}(\Omega) = \frac{1}{E}(K_1^2 + K_2^2).$$

The Griffith criterium [6] then sets that the crack will propage if

$$(7) \quad K_1^2 + K_2^2 > \gamma .$$

2 - A FINITE ELEMENT SCHEME

The discrete problem is set as follows

$$(8)\quad \begin{cases} \text{Find } \vec{u}_h \in V_h \text{ such that} \\ a(\vec{u}_h,\vec{v}_h) = \int_{\Gamma_f} \vec{g}.\vec{v}_h \quad \forall\, \vec{v}_h \in \vec{V}_h . \end{cases}$$

One can write :

$$(9)\quad \vec{u}_h = K_1^h\, S_1 + K_2^h\, \vec{S}_2 + \vec{u}_h^r$$

with $\vec{u}_h^r \in W_h$.

PROPOSITION 1

$$(10)\quad \|\vec{u} - \vec{u}_h\|_V \leqslant C\, h\, \|\vec{u}^r\|_{2,\Omega} .$$

Proof

In a very classical way

$$\|\vec{u} - \vec{u}_h\|_V \leqslant C \min_{\vec{v}_h \in V_h} \|\vec{u} - \vec{v}_h\|_V .$$

Choosing $\vec{v}_h = K_1.\vec{S}_1 + K_2\,\vec{S}_2 + \Pi_h\,\vec{u}^r$, and using Ciarlet-Raviart classical estimates [4], one gets (10). ∎

PROPOSITION 2

$$(11)\quad \textit{For } i = 1,2, \textit{ we have} \quad |K_i - K_i^h| \leqslant C\, h^{\frac{1}{2}}\, \|\vec{u}^r\|_{2,\Omega} .$$

Proof

Let us denote by $\alpha\vec{S}_2 + \vec{y}_h$ the projection, with respect to norm $|||\,.\,|||$, of $\vec{S}_1$ onto the subspace $\{\vec{S}_2\} \oplus W_h$, and by $\vec{v}_h$ the element of V_h defined by

$$\vec{v}_h = \vec{S}_1 - \alpha\vec{S}_2 - \vec{y}_h .$$

Thus, using (3) and (8), we obtain

$$(K_1 - K_1^h)\, a(\vec{v}_h,\vec{v}_h) = -\, a(\vec{u}^r - \vec{u}_h^r, \vec{v}_h)$$

$$= -\, a(\vec{u}^r - \vec{w}_h, \vec{v}_h) \quad \forall\, \vec{w}_h \in W_h .$$

Thus, by choosing $\vec{w}_h = \Pi_h\, \vec{u}^r$, we get :

$$(12)\quad |K_1 - K_1^h| \leqslant \frac{C\, h\, \|u^r\|_{2,\Omega}}{|||\, v_h\, |||} .$$

One now has to estimate $|||\,\vec{v}_h\,|||$, which is the "distance" from $\vec{S}_1$ to the space $\{\vec{S}_2\} \oplus W_h$.

Since $\mathcal{T}_h$ is a regular triangulation, there exists some $\Theta_0 > 0$ such that all angles Θ of triangles $K \in \mathcal{T}_h$ verify

$\Theta > \Theta_0$.

We suppose furthermore that dimensions h_K of triangles K remain greater than $\frac{h}{2}$

Let us now define the following domain, included in a triangle K of $\mathcal{T}_h$:

$$T = \{(r,\Theta) \ ; \ r < \frac{h}{3} \ ; \quad -\Pi < \Theta < -\Pi + \Theta_0\}$$

and

$$T^* = \{(r,\Theta) \ ; \ r < \frac{1}{3} \ ; \quad -\Pi < \Theta < -\Pi + \Theta_o\}$$

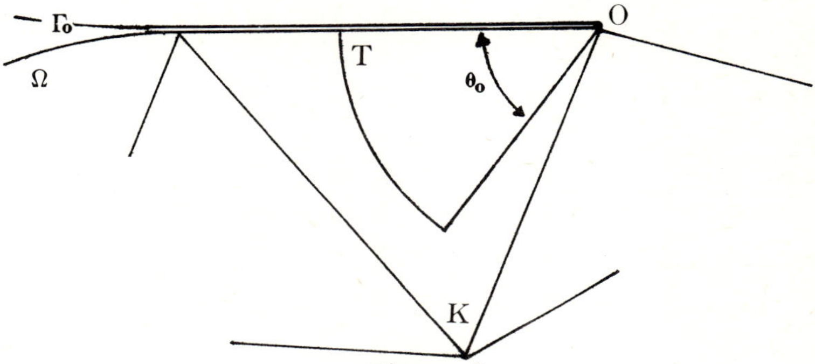

Figure 2

The domain T

Let $\beta\vec{S}_2 + \vec{z}_h$ be the local projection of $\vec{S}_1$ onto $\{\vec{S}_2\} + (P_1(T))^2/\mathbb{R}$, with respect to the norm defined by $a_T(.,.)$. Thus :

(13) $|||\ \vec{v}_h\ ||| \geqslant a_T(\vec{S}_1 - \beta\vec{S}_2 - \vec{z}_h, \vec{S}_1 - \beta\vec{S}_2 - \vec{z}_h)$.

$\beta\vec{S}_2 + \vec{z}_h$ is solution of the 4-dimensional linear problem :

$$\begin{pmatrix} A & F \\ F^t & P \end{pmatrix} \begin{pmatrix} z_h \\ \beta \end{pmatrix} = \begin{pmatrix} B \\ C \end{pmatrix}$$

A sample calculation shows that

$A = h^2\ A^*$

$F = h^{3/2}\ F^*$

$P = h\ P^*$

$C = h\ C^*$

$B = h^{3/2}\ B^*$

where the projection onto $\{\vec{S}_2\} + (P_1(T^*))^2/\mathbb{R}$ is solution of the linear problem :

$$\begin{bmatrix} A^* & F^* \\ F^{*t} & P^* \end{bmatrix} \begin{bmatrix} z^* \\ \beta^* \end{bmatrix} = \begin{bmatrix} B^* \\ c^* \end{bmatrix}$$

Thus

$$\beta = \beta^*$$

and

$$\vec{z}_h = h^{-\frac{1}{2}} \vec{z}^*$$

and

$$a_T(\vec{S}_1 - \beta\vec{S}_2 - \vec{z}_h, \vec{S}_1 - \beta\vec{S}_2 - \vec{z}_h) = C\,h.$$

By using (13), we get

$$|||\vec{v}_h||| \geqslant C\,h^{\frac{1}{2}}$$

and then, by using (12), we obtain (11). ∎

3 - NUMERICAL TRIALS

All these trials were made for the opening mode

3.1. Calculation of the stiffness matrix

Let $(\vec{v}_i)_{i=1,..,N}$ denote the P_1 basis functions ; Thus

$$A_{ij} = a(\vec{v}_i, \vec{v}_j)$$
$$F_i = a(\vec{S}_1, \vec{v}_i)$$
$$P = a(\vec{S}_1, \vec{S}_1).$$

Coefficients A_{ij} are easy to compute

$$F_i = \int_\Omega \sigma_{\alpha\beta}(\vec{S}_1)\ \varepsilon_{\alpha\beta}(\vec{v}_i)$$
$$= -\int_\Omega \sigma_{\alpha\beta,\beta}(\vec{S}_1)\ \ v_i^\alpha + \int_\Gamma \sigma_{\alpha\beta}(\vec{S}_1)\ n_\beta\ v_i^\alpha$$

Since $\vec{S}_1$ is in equilibrium, one has

$$(14)\quad F_i = \int_\Gamma \sigma_{\alpha\beta}(\vec{S}_1)\ n_\beta\ v_i^\alpha\ .$$

And in the same way,

$$(15)\quad P = \int_\Gamma \sigma_{\alpha\beta}(\vec{S}_1)\ n_\beta\ S_1^\alpha\ .$$

Thus, since $\sigma_{\alpha\beta}(\vec{S}_1)\ n_\beta = 0$ at the vicinity of the crack tip, (14) and (15) are easy to compute, for example by a Simpson quadrature formula as we did. It is anyway possible to compute them exactly.

3.2. Resolution of the linear system

The linear system is the following

$$\begin{pmatrix} A & F \\ F^t & P \end{pmatrix} \begin{pmatrix} x \\ K_1 \end{pmatrix} = \begin{pmatrix} b \\ c \end{pmatrix}$$

A is the P_1-linear elasticity matrix, which is band structured. We are interested in the knowledge of K_1

$$(16) \qquad \begin{cases} Ax + F\,K_1 = b \\ {}^tFx + P\,K_1 = c \end{cases}$$

Thus

$$x = A^{-1}\,b - A^{-1}\,F\,K_1$$

and

$$(17) \quad K_1 = \frac{c - F^t\,A^{-1}\,b}{P - F^t\,A^{-1}\,F} \quad .$$

One has only to factorize A, which is band structured, while the band width of the global matrix is much larger.

For a triangulation using 212 triangles (and 264 Degrees of freedom), we used 1.7 second of computer time on IBM 370/168.

3.3. Numerical results

We studied a rectangular cracked plate. The plate is symmetric with respect to the crack, of length d. w is the width of the plate, and L its length. Traction forces of intensity P are applied on two portions of width ℓ of the boundary.

We compare the "normalized" calculated factor

$$(18) \quad K = \frac{wK_1}{P\,\ell\,d^{\frac{1}{2}}}$$

with Bowie's results, obtained by mapping techniques [2].

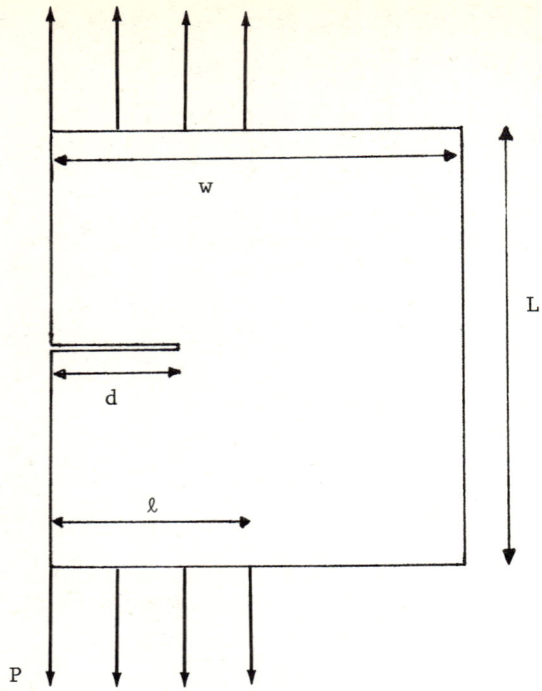

Figure 3

The rectangular cracked plate

For $\frac{d}{w} = 0.5$ and $L = w$, we have the following results :

$\frac{\ell}{w}$	K Bowie	K calculated	Error
1	3.01	3.035	0.8 %
0.9	3.45	3.509	1.7 %
0.8	3.91	4.003	1.7 %
0.7	4.40	4.525	2.4 %
0.6	4.93	5.069	2.8 %
0.5	5.48	5.616	2.5 %
0.4	6.03	6.163	2.2 %
0.3	6.57	6.807	3.6 %
0.2	7.12	7.233	1.6 %
0.1	7.67	7.735	0.8 %

Now, fixing $\frac{\ell}{w} = 1.0$, we obtain, for different values of $\frac{d}{w}$:

$\frac{d}{w}$	K Bowie	K calculated	Error
0.7	6.4	5.95	7 %
0.6	4.15	4.07	2 %
0.5	3.01	3.035	0.8 %
0.4	2.32	2.37	2.3 %

It is important to notice that, when $\frac{d}{w}$ is "large" or "small" (i.e. near from 1. or 0.), then coefficients F_i and P are calculated with greater imprecision, since the crack tip is near from the boundary. It would be worth then to use exact integration to obtain better accuracy.

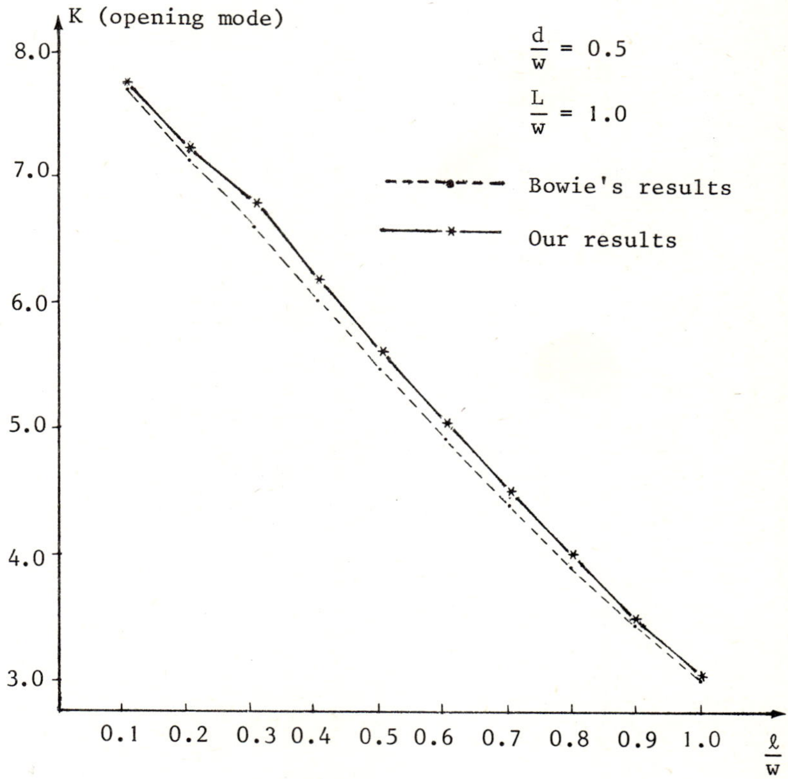

Figure 4

Some numerical results

In conclusion, we can say that this method is very performant, for a cheap computing cost. We have not to worry about the support of the singular functions ([1], [7]) since we take the whole domain. The main advantages of this choice are to provide stable and accurate results, while the calculations are much simplified since the singular integrals are reduced to boundary integrals, far from the crack tip.

REFERENCES

1. BARTOLOMEW, P. - Solutions of elastic crack problems by superposition of finite elements and singular fields, Comp. Methods in Applied Mech. and Eng., 13 (1978) 59-78.

2. BOWIE, O.L. - Solution of crack problems by mapping technique in Methods of Analysis and Solution of Crack Problems, edited by G.C. SIH, Nordhoff Int. (1973)

3. BUI, H.D. - Mécanique de la rupture fragile, Masson (1979)

4. CIARLET, P.G., RAVIART, P.A. - General Lagrange and Hermite interpolation in $\mathbb{R}^n$ with applications to finite element methods, Arch. Rat. Mech. Anal., 46 (1972) 177-199

5. DESTUYNDER, P., DJAOUA M. - Equivalence de l'intégrale de Rice et du critère de Griffith en mécanique de la rupture fragile, CRAS, Paris, Série A 347, t 290, (18 février 1980)

6. GRIFFITH, A.A. - Phil. Trans. Roy. Soc. London, Sev. A 221 (1970)

7. HILTON, P.D., SIH, G.C. - Applications of the finite element method to the calculation of stress intensity factors in Methods of Analysis and Solution Crack Problems, edited by G.C. SIH, Nordhoff Int. (1973).

WEIGHT FUNCTIONS FOR PLANE AND AXISYMMETRIC CRACK PROBLEMS

M.L. Vanderglas

Ontario Hydro Research, Toronto, Canada

ABSTRACT

A finite element technique for the calculation of weight functions for 2D crack problems is presented. The technique is applied to two dimensional mode I and mode II, and axisymmetric problems. It is found that the bulk of the calculation can be performed by a standard finite element program, and that the near-tip leading terms of Williams' series expansion for 2D crack problems may be effectively employed for the calculation of the weight functions in some cases.

INTRODUCTION

The weight function technique described by Rice [1] provides a very convenient means for the calculation of stress-intensity factors for two dimensional elastic crack problems. Rice has demonstrated that, given the geometry and displacement boundary conditions of a body, the weight functions, h_i, are unique and can be extracted from a solution corresponding to any loading. Once the weight functions are known, the stress intensity for any other set of tractions, T_i, can be obtained by integrating over the boundary, S:

$$K = \int h_i T_i dS; \; i=1,2 \text{ (sum over i)} \tag{1}$$

Several methods have been proposed for the calculation of weight functions for specific problems [2-8]. The difficulty in calculating weight functions by numerical techniques appears to stem from their $r^{-1/2}$ singularity approaching the crack tip. A method which avoids this problem by employing Parks' stiffness derivative technique [9] will be described. The proposed method is subsequently extended to axisymmetric and mixed mode problems.

THE NODAL WEIGHT FUNCTIONS

In this section, the finite-element nodal weight functions $\{h\}$, which afford an approximation to the field quantity, h_i, will be identified.

The potential energy of an elastic system, P, may be expressed in terms of the strain energy U and applied work, W as [10,11]

$$P = U - W \qquad (2)$$

Once an equilibrium solution has been found, the strain energy and work are related by [10]

$$2U = W \qquad (3)$$

The equilibrium potential is therefore

$$P = -\frac{1}{2}W \qquad (4)$$

or, in terms of finite element quantities, namely the displacement and force vectors $\{u\}$ and $\{f\}$ [11]

$$P = -\frac{1}{2}\{u\}^T\{f\} \qquad (5)$$

If two equilibrium systems, say $\{u\}$, $\{f\}$ and $\{u^*\}$, $\{f^*\}$ are superimposed, the combined potential is found by applying the previous equation to the sums of the displacement and the force vectors:

$$P_{TOT} = -\frac{1}{2}\{u\}^T\{f\} - \frac{1}{2}\{u^*\}^T\{f^*\} - \{u^*\}^T\{f\} \qquad (6)$$

where the reciprocal theorem [10] has been employed. The stress intensity factor of the combined system will be [12]

$$K_{TOT} = K + K^* \qquad (7)$$

The stress intensity factor may be related to the change in potential energy with crack extension through [12]

$$K^2/H = -\partial P/\partial a \qquad (8)$$

where H is Young's modulus, E, in plane stress or $E/(1-v^2)$ for plane strain, and v is Poisson's ratio.

Combining equations 6 through 8 then gives:

$$K^2/H + K^{*2}/H + 2KK^*/H = -\partial P/\partial a - \partial P^*/\partial a$$

$$+\partial[\{u^*\}^T\{f\}]/\partial a \qquad (9)$$

Identifying like terms and assuming that forces are independent of crack extension, the following expression remains:

$$2KK^*/H = \partial\{u^*\}^T/\partial a \,.\{f\} \qquad (10)$$

and, solving for K:

$$K = \left[\frac{H}{2K^*}\,\frac{\partial\{u^*\}^T}{\partial a}\right]\{f\} \qquad (11)$$

The quantity in square brackets, formally quite similar to Rice's result in [1] will be called the "nodal weight functions".

EVALUATION OF THE NODAL WEIGHT FUNCTIONS

In this section, a technique for the calculation of the nodal weight functions is presented. The finite element stiffness equations are [11]:

$$[K]\{u\} = \{f\} \qquad (12)$$

where [K] is the stiffness matrix. A formal differentiation with respect to crack extension, again assuming constant force, gives:

$$[K]\partial\{u\}/\partial a = -\partial[K]/\partial a.\{u\} \qquad (13)$$

By writing a post-processing computer program to evaluate the quantity on the right-hand side of equation 13 which resembles a pseudo-force vector, the nodal weight functions can be obtained by two submissions to a standard finite element program: once to obtain the displacement vector, followed by post-processing, followed by an execution which gives the displacement derivatives.

Post-multiplying the right-hand side of equation 13 by $-1/2\{u\}^T$ gives the change in potential with crack extension [9] from which the stress intensity may be obtained (equation 8) for use in equation 11 (note that the asterisk has been dropped from $\{u\}$ and K in this section).

In assigning a meaning to the stiffness derivative term in equation 13, Parks' [9] interpretation has been employed. However, this interpretation contains a somewhat hidden property which makes the nodal influence functions slightly different from those discussed by Rice [1]. To illustrate that there is a difference, consider that zero displacements can often be prescribed for the crack-tip node in a symmetrical problem. This node's displacement derivatives (with respect to crack extension) are then also zero, in contrast to the singular behaviour indicated by Rice. The source of this discrepancy is discussed in the following section.

CALCULATION OF THE WEIGHT FUNCTIONS FROM NODAL VALUES

Rice shows that the influence functions may be obtained from the expression [1]

$$h_i = H(\partial u_i/\partial a)/2K \quad (14)$$

which requires the derivatives

$$\partial u_i/\partial a = \partial u_i(x,y,a)/\partial a \quad (15)$$

that is, the change in displacement with crack extension but evaluated at some fixed point (x,y).

In the finite element method, displacements and coordinates at a point may be obtained by interpolating from the nodal values, using the element shape functions. In Parks' method, the change in nodal quantities with crack extension is detected by giving some of the nodes surrounding the crack tip a virtual translation, Δa, and observing the change in the solution. The change of nodal displacement at one of these translated nodes results not only from crack extension, but also very basically from the change in the node's spacial location. Therefore, the displacements are considered to depend on x, y and a:

$$u_i = u_i(x,y,a) \quad (16)$$

The change in displacement resulting from the stiffness derivative procedure is

$$\Delta u_i = (\partial u_i/\partial a).\Delta a + (\partial u_i/\partial x).\Delta x + (\partial u_i/\partial y).\Delta y \quad (17)$$

which is the total change produced both by crack extension, Δa and the change in nodal coordinates, Δx and Δy. Upon dividing by Δa, taking the limit and solving for the partial derivative, there results:

$$\partial u_i/\partial a = du_i/da-(\partial u_i/\partial x)(dx/da)-(\partial u_i/\partial y)(dy/da) \quad (18)$$

If, for example, the crack is aligned with the x-axis, then $dx/da = 1$ and $dy/da = 0$. In this case, the influence functions are

$$h_i = \frac{H}{2K} \cdot \begin{cases} \dfrac{du_i}{da} - \dfrac{\partial u_i}{\partial x} & \text{for elements with translated nodes} \\ \dfrac{du_i}{da} & \text{otherwise} \end{cases} \qquad (19)$$

where the total derivatives (du_i/da) are the values obtained by interpolation from the nodal influence functions of equation 13.

AN EXAMPLE

The procedure above was applied to the single edge notched plate, Figure 1(a). The weight functions for the crack face and end of the plate are shown in Figures 2 and 3. These were obtained by repeated analysis for a series of crack lengths. The results for a/b = 1/2 compare favourably with those of Paris et al [2].

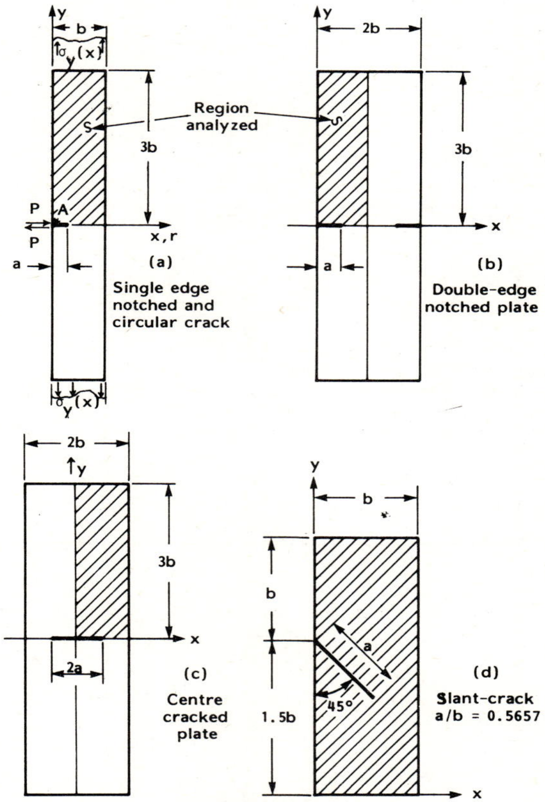

FIGURE 1

SCHEMATICS OF GEOMETRIES USED FOR TEST CASES

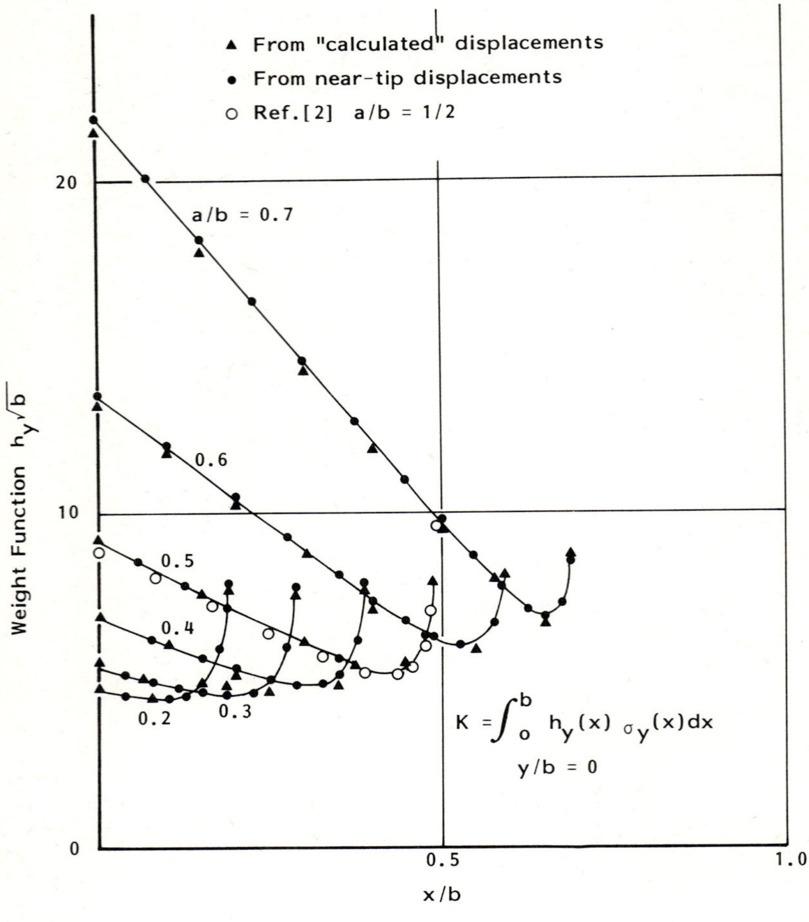

FIGURE 2

MODE I WEIGHT FUNCTIONS FOR VERTICAL TRACTIONS ON CRACK FACE OF S.E.N.

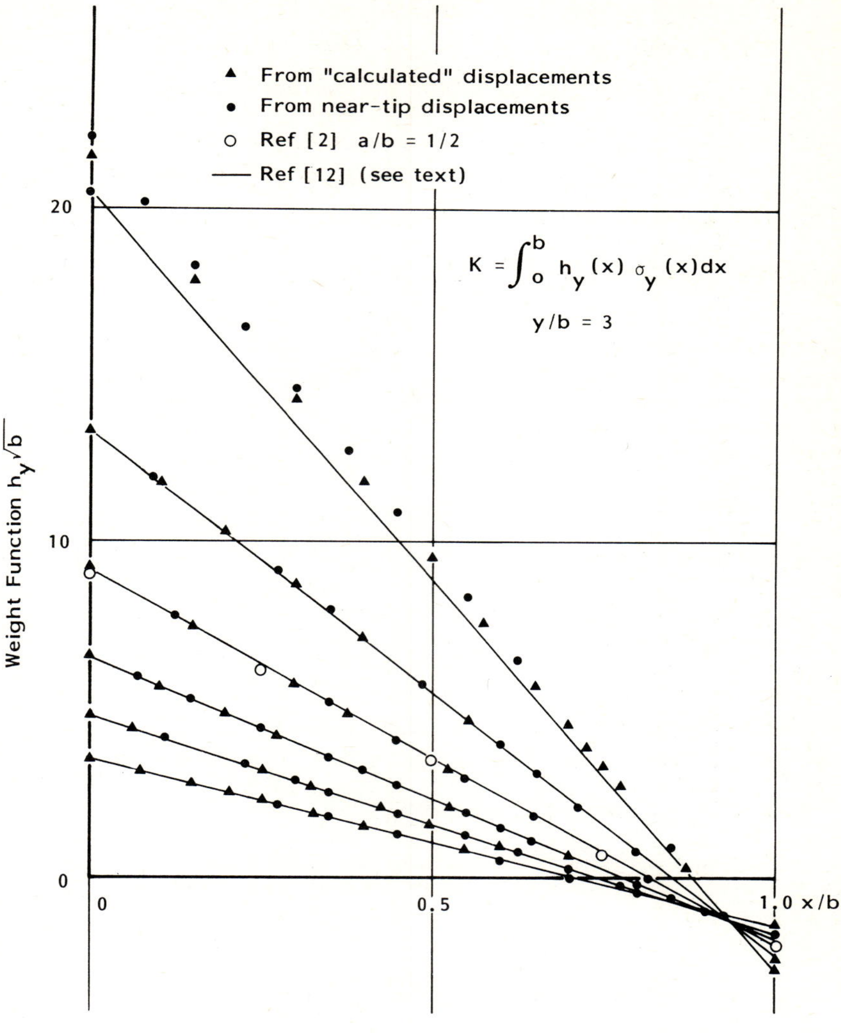

FIGURE 3

MODE I WEIGHT FUNCTIONS FOR VERTICAL TRACTIONS ALONG TOP EDGE OF S.E.N.

It was verified that nearly identical values could be obtained for different loadings by assigning nodal forces more or less at random and repeating the analysis. However, it is noted that the assumption that forces be independent of crack extension means that loads may not be applied to translated nodes.

The calculated distribution of weight functions along the plate's end was found to be nearly linear for a given crack length. Stress intensity factors for tension and bending are available and are usually given in handbooks as

$$K_T = \sigma\sqrt{\pi a}\, F_T; \quad \sigma = \sigma_T = \text{constant for tension} \qquad (20a)$$

$$K_B = \sigma\sqrt{\pi a}\, F_B; \quad \sigma = \sigma_B[1-2x/b] \text{ for bending} \qquad (20b)$$

where F_T and F_B are given explicitly in [12], for example, as polynomials of a/b. Equating these expressions to equation 1 (using $T_1 = T_2 = \sigma_T$ and then $\sigma_B[1-2x/b]$) and solving for the constants A and B, assuming that the weight functions are linear and of form:

$$\sqrt{b}\, h_2 = \sqrt{b}[A + B\,(x/b)] \text{ at } y/b = 3 \qquad (21)$$

gives the result for any crack length:

$$A(a) = \sqrt{\pi a}\,[F_T - 3F_B] \qquad (22a)$$

$$B(a) = -6\sqrt{\pi a}\, F_B \qquad (22b)$$

The weight functions obtained in this manner are also shown in Figure 3; numerical results are in good agreement.

EXTENSION TO AXISYMMETRIC PROBLEMS

The techniques outlined above can be extended to axisymmetric problems. Consider that, for plane problems, unit thickness (t=1) has been tacitly assumed. Strictly, the relation

$$K^2/H = -\partial P/\partial A \; ; \; dA = t.da \qquad (23)$$

should have been written in place of equation 8 where A is the crack area.

In axisymmetric formulations [11] the change in area associated with each point on the circular crack front is the same. The change in area and crack radius, a, are now related by

$$\Delta A = \pm 2\pi a\, \Delta a \qquad (24)$$

the sign depending on whether Δa is directed radially outward or inward. The axisymmetric formulation is then identical to equations 2 to 19 upon replacing $\partial/\partial a$ by $(\pm 1/2\pi a)\partial/\partial a$.

It should be noted, however, that radial rigid body translations are not permissable in the axisymmetric case. Thus, the stiffness derivative term in equation 13 must be evaluated for all elements whose nodes are given a virtual translation. This is in contrast to the plane problems in which only those elements on the periphery of the translated region contribute to this term.

Weight functions for a range of crack diameters were calculated for a solid cylinder containing an internal circular crack. The distribution of weight functions corresponding to vertical tractions on the cylinder's end was found to be nearly independent of radius, implying that the stress intensity for this example depends only on the total load, not its radial distribution. By obtaining a single known solution, an approach similar to equations 21 and 22 can be applied to estimate the weight functions which, following the numerical results, are assumed to be constant for a given crack diameter:

$$K = \sigma\sqrt{\pi a}\,F(a/b) = \sigma h_y \quad ; \; y/b = 3 \qquad (25)$$

where F is given in [12] for uniform tension and σ is the total load applied to one end. The result ($h_y = \sqrt{\pi a}F$) is shown in Figure 4. The numerical results again agree favourably.

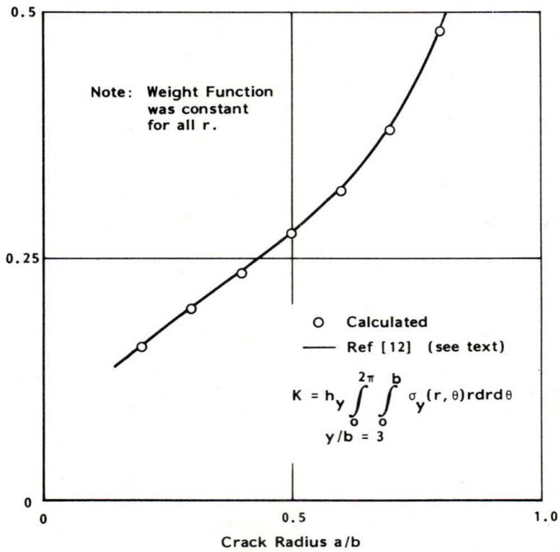

FIGURE 4

MODE I WEIGHT FUNCTIONS ON TOP EDGE OF CYLINDER WITH INTERNAL CIRCULAR CRACK

It may be noted that the technique presented above hinges on the relation between the stress intensity and energy rate, equation 8. This relation applies equally to the cases $K_I = 0$, $K_{II} \neq 0$ and $K_I \neq 0$, $K_{II} = 0$ [12,13,14] assuming "straight-ahead" crack propagation. For a given geometry, two sets of weight functions can therefore be obtained: one for mode I, another for mode II, say h_{iI}, h_{iII}. Having obtained both sets, both stress intensities K_I and K_{II} can subsequently be evaluated for a given mixed-mode loading as:

$$K_I = \int_s h_{iI} T_i ds \qquad (26a)$$

$$K_{II} = \int_s h_{iII} T_i ds \qquad (26b)$$

By first calculating influence functions for the "pure-mode" cases, the inability of energy methods to separate mode I and II stress intensities uniquely for mixed-mode problems [13,14] is circumvented.

The single-edge notched plate of Figure 1(a) was reanalyzed for the mode II case. This choice of test case was unfortunate, however, since there are few published mode II results for this case. The calculated results compare favourably with [12] for one point, as shown in Figure 5 where the weight function for a point load in the x-direction applied to point A in Figure 1(a) has been plotted.

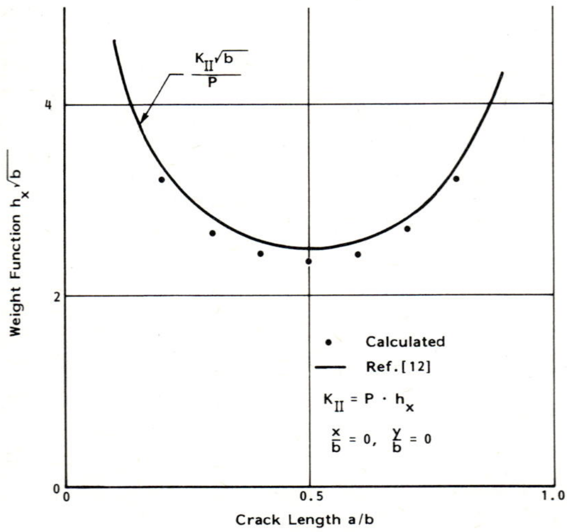

FIGURE 5

MODE II WEIGHT FUNCTION FOR HORIZONTAL POINT LOAD AT CRACK MOUTH OF S.E.N.

A FURTHER IMPROVEMENT

Some of the error in the numerically calculated weight functions can naturally be attributed to the initial solution for the displacements, equation 12. Williams has identified the nature of the stress and displacement fields for plane crack problems [15]. The stress intensities K_I and K_{II}, are associated with the familiar "near-tip" leading terms of his series expansion [16]. The other terms in the series may be viewed as necessary only to satisfy the prescribed boundary conditions. It is surmised that the uniqueness of the weight functions is somehow related to this idea. In any case, there appears to be no reason why the near tip displacements could not be used in equation 13 instead of solving for the displacements corresponding to an arbitrary force vector, equation 12.

As expected, this method works well when the displacement boundary conditions of a particular problem are consistent with the near-tip field. There is practically no difference between the influence functions obtained from a previous solution for displacements and from the near-tip field (Figures 2 and 3). However, it must be mentioned that the same approach was applied to a double-edge-notched and centre-cracked plate, analyzed using two planes of symmetry so that the near-tip solution cannot satisfy all the boundary conditions (Figure 1 b and c). In these two cases, similar but unequal distributions of influence functions on the plate's edges were obtained using a calculated and near-tip displacement field. Typical results are shown in Figure 6.

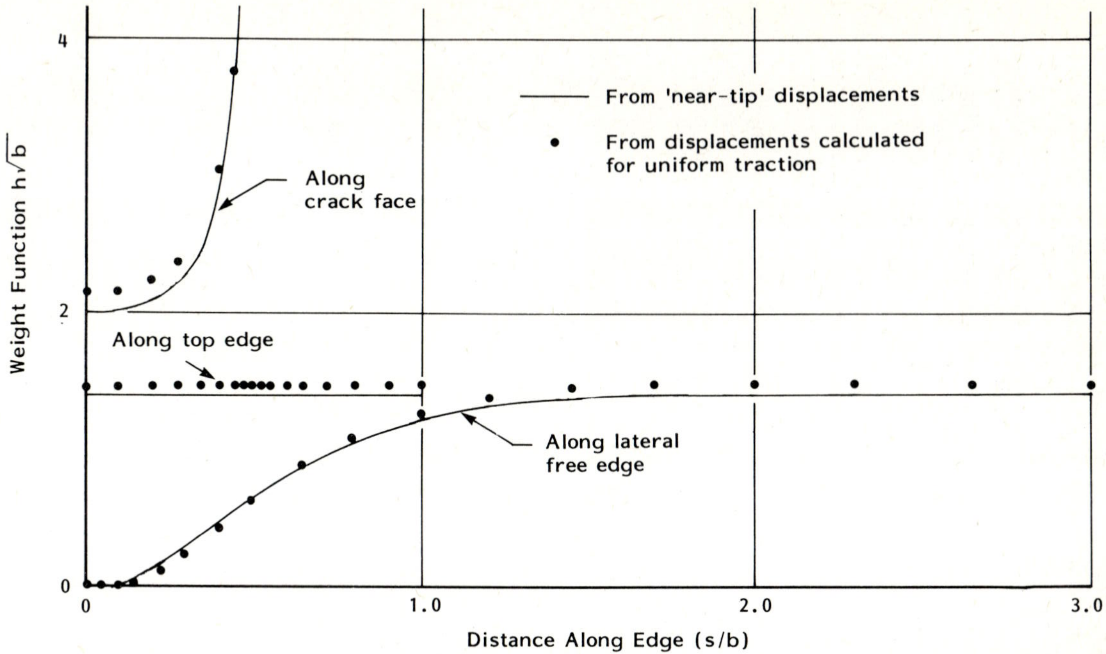

FIGURE 6

COMPARISON OF WEIGHT FUNCTIONS FOR CENTRE-CRACKED PANEL

Results obtained using the calculated field are apparently more accurate (comparing K_I for uniform tension calculated from the weight functions with [17]), but the source of the difference in the solutions has not been clearly identified.

SLANT CRACK EXAMPLE

As a final example, the ideas above were applied to the slant crack geometry, Figure 1(d). First, the near-tip mode I displacement field (suitably aligned with the crack) was used to evaluate the pseudo-force vector in equation 13. For simplicity, the near-tip field for K = 1, H = 1 was used. A solution of equation 13 provided the mode I weight functions. This process was repeated using the mode II field.

Both sets of weight functions corresponding to normal tractions are plotted in Figure 7 for the two ends of the plate.

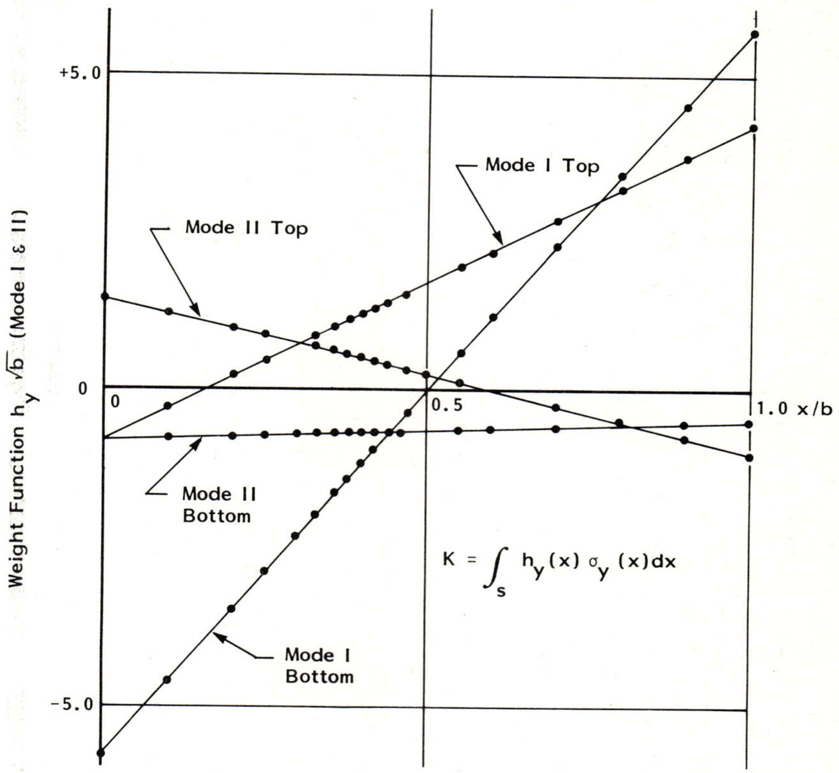

FIGURE 7

45° SLANT CRACK
MODE I & II INFLUENCE FUNCTIONS

The stress intensity factors for uniform tension and bending calculated from these influence functions are compared with [17] below.

Loading	Mode	m* from Weight Functions	m Ref. [26]	Error (%)
Tension	I	1.315	1.34	-2
	II	0.649	0.63	4
Bending	I	0.807	0.85	-5
	II	0.349	0.33	5

*m = $K/\sigma\sqrt{\pi a}$ for tension
$Kb^2/6M\sqrt{\pi a}$ for bending, M = applied moment.

CONCLUSION

A finite element technique for the calculation of two-dimensional mode I, II and axisymmetric weight functions has been presented. The bulk of the calculation can be performed by a standard finite element program. By applying the "near-tip" displacement field to the entire body, the weight functions may be obtained at least in the case when the displacement boundary conditions of the problem at hand conform to the near-tip field.

REFERENCES

1. RICE, J.R. - Some Remarks on Elastic Crack Tip Stress Fields. Int. J. Solids and Structures, Vol 8, pp. 751-758, 1972.

2. PARIS, P.C., McMEEKING, R.M., TADA, H. - The Weight Function Method for Determining Stress Intensity Factors, Cracks and Fracture, ASTM STP601, American Society for Testing and Materials, 1976.

3. PETROSKI, H.J., ACHENBACK, J.D. - Computation of the Weight Function from a Stress Intensity Factor, Eng. Fracture Mechanics, Vol. 10, pp. 249-256, 1978.

4. LABBENS, R.C., HELIOT, J., PELLISSIER-TANON, A. - Weight Functions for Three Dimensional Symmetric Crack Problems, Cracks and Fracture, ASTM STP601, 1976.

5. CHELL, G.G. - The Stress Intensity Factors and Crack Profiles for Center and Edge Cracks in Plates Subject to Arbitrary Stresses, Int. J. Fracture, Vol. 12, pp. 33-46, 1976.

6. CRUSE, T.A., BESUNER, P.M. - Residual Life Prediction for Surface Cracks in Complex Structural Details, AIAA J. Aircraft, Vol. 12, pp. 369-375, 1975.

7. BUCHALET, C.B., BAMFORD, W.H. - Stress Intensity Solutions for Continuous Surface Flows in Reactor Pressure Vessels, Mechanics of Crack Growth, ASTM STP590, American Society for Testing and Materials, 1976.

8. VANDERGLAS, M.L. - A Stiffness Derivative Finite Element Technique for Determination of Influence Functions, Int. J. Fracture, Vol. 14, pp. R291-R295, 1978.

9. PARKS, D.M. - A Stiffness Derivative Finite Element Technique for Determination of Crack Tip Stress Intensity Factors, Int. J. Fracture, Vol. 10, pp. 487-502, 1974.

10. SOKOLNIKOFF, I.S. - Mathematical Theory of Elasticity, McGraw-Hill, New York, 1956.

11. COOKE, R.D. - Concepts and Applications of Finite Element Analysis, John Wiley and Sons, New York, 1974.

12. TADA, H., PARIS, P.C., IRWIN, G.R. - The Stress Analysis of Cracks Handbook, Del Research Corp., Hellertown Pa., 1973.

13. HELLEN, T.K. - Finite Element Energy Methods in Fracture Mechanics, Ph.D. Thesis, University of London, 1976.

14. HELLEN, T.K., BLACKBURN, W.S. - The Calculation of Stress Intensity Factors for Combined Tensile and Shear Loading, Int. J. Fracture, Vol. 11, pp. 605-617, 1975.

15. WILLIAMS, M.L. - On the Stress Distribution at the Base of a Stationary Crack, Trans. ASME, J. Applied Mechanics, Vol. 24, pp. 109-114, 1957.

16. EFTIS, J., SUBRAMONIAN, N., LIEBOWITZ, H. - Crack Border Stress and Displacement Equations Revisited, Eng. Fracture Mechanics, Vol. 9, pp. 189-210, 1977.

17. ROOKE, D.P., CARTWRIGHT, D.J. - Compendium of Stress Intensity Factors, Her Majesty's Stationery Office, London, 1976.

WEIGHT FUNCTIONS FOR CRACKED, CURVED BEAMS

C.P. ANDRASIC* and A.P. PARKER**

SUMMARY

There is a requirement for high accuracy, limited computing stress intensity factor solutions for edge cracks emanating from curved boundaries, under a variety of loading conditions. This paper presents weight functions for such geometries, derived via the accurate Modified-Mapping Collocation process. The weight function data is fitted using cubic splines and may be employed in the derivation of additional stress intensity factor solutions for a variety of loadings. A check on the solution accuracy, and a design example involving complex residual stresses are included.

1. INTRODUCTION

One of the most common problem geometries facing the designer who uses Fracture Mechanics techniques is the case of a crack emanating from a boundary of known radius of curvature in a plate of finite-width. The designer must strike a balance between time, cost and accuracy in selecting a method for the determination of the crack-tip stress intensity factor, K. If crack growth rates, and hence lifetimes, are to be predicted with accuracy, K values with errors of approximately 1% are desirable. The designer who seeks results of this accuracy may be obliged to embark on numerical analyses involving considerable time and sophistication.

---

* Research Scientist ** Principal Lecturer
Civil Engineering Department, Royal Military College of Science, Shrivenham, Swindon, Wilts, UK.

Thus, there does appear to be a requirement for high-accuracy, limited computing solutions which may be obtained by a designer in a matter of hours. Weight functions {1} {2} derived via the Modified-Mapping Collocation (MMC) procedure {3} are well suited to this purpose. The method of 'packaging' the weight-function data is clearly important, since it must retain the accuracy of the basic solution technique, whilst being sufficiently compact to permit economical presentation of data and subsequent processing. Cubic Splines are sufficiently accurate, and convenient in use.

2. MODIFIED MAPPING-COLLOCATION

The MMC method {4} can be used to compute stress intensity factors and displacements for different geometries and loadings. Complex variable methods, due to Muskhelishvili {5} are utilized. Stresses and displacements within a body are given in terms of the complex stress functions $\phi(z)$ and $\psi(z)$ by:

$$\sigma_x + \sigma_y = 4 \text{ Re } \{\phi'(z)\} \tag{1a}$$

$$\sigma_y - \sigma_x + 2i\, \tau_{xy} = 2 \left[\bar{z}\phi''(z) + \psi'(z)\right] \tag{1b}$$

$$2G(u + iv) = \kappa\, \phi(z) - \overline{z\phi'(z)} - \overline{\psi(z)} \tag{1c}$$

where the complex variable $z = x + iy$,* x and y* are the physical coordinates, primes denote differentiation and bars represent the complex conjugate. Also:

$$G = \frac{E}{2(1 + \nu)},$$

$$\kappa = 3 - 4\nu \text{ (plane strain)}, \quad \kappa = \frac{3-\nu}{1+\nu} \text{ (plane stress)}$$

whilst the resultant force over an arc s is:

$$f_1 + if_2 = i \int_s (X_n + iY_n)\, ds = \phi(z) + z\, \overline{\phi'(z)} + \overline{\psi(z)} \tag{2}$$

where $X_n.ds$ and $Y_n.ds$ are the horizontal and vertical components of force acting on ds.

The solution of the cracked ring segment was carried out on similar lines to that of Tracy {3}. The complex mapping:

$$z = \omega_1(\gamma) = iR_1 \, e^{-i\gamma} \tag{3}$$

Maps straight lines parallel to the real axis in the γ-plane to curved lines in the physical (z) plane, in particular the real axis in the γ plane is mapped to an arc of radius R_1, centred at the origin in the z-plane, Figure 1.

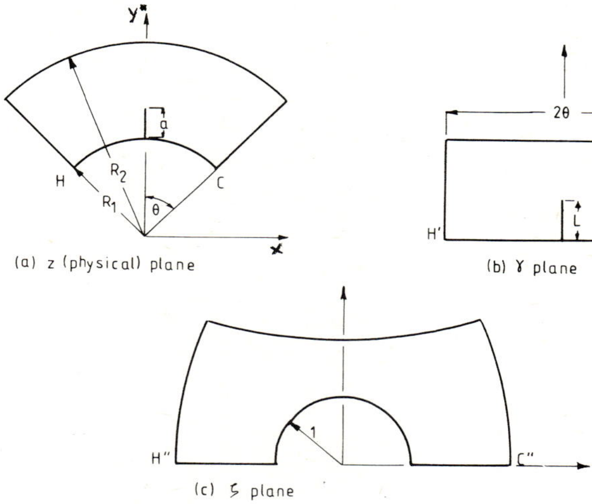

Figure 1: Physical and Mapped Planes

Substitute the mapping (3) into (1c) to obtain:

$$f_1 + if_2 = \phi(\gamma) + \frac{\omega_1(\gamma)}{\overline{\omega_1'(\gamma)}} \overline{\phi'(\gamma)} + \overline{\psi(\gamma)} \tag{4}$$

$$2G(u + iv) = \kappa\phi(\gamma) - \frac{\omega_1(\gamma)}{\overline{\omega_1'(\gamma)}} \overline{\phi'(\gamma)} - \overline{\psi(\gamma)} \tag{5}$$

where, to avoid new notation we define:

$$\phi(\gamma) = \phi(\omega_1(\gamma)) \text{ and } \psi(\gamma) = \psi(\omega_1(\gamma))$$

The analytic continuation arguments of Muskhelishvili are used to ensure traction-free conditions along H'C' and hence HC in the physical plane, Figure 1. Let S^+ denote the region in the γ-plane above the real axis, and define

$$\phi(\gamma) = -\overline{\psi}(\gamma) - \frac{\omega_1(\gamma)}{\overline{\omega}_1'(\gamma)}\,\overline{\phi}'(\gamma), \quad \gamma \epsilon S^- \tag{6}$$

where the bar notation is defined by

$$\overline{f}\,(\gamma) = \overline{f\,(\overline{\gamma})} \tag{7}$$

Then $\psi(\gamma)$ can be written as

$$\psi(\gamma) = -\,\overline{\phi}(\gamma) - \frac{\overline{\omega}_1(\gamma)}{\omega_1'(\gamma)}\,\phi'(\gamma), \quad \gamma \epsilon S^+ \tag{8}$$

and equations (4) and (5) become

$$f_1 + if_2 = \phi(\gamma) - \phi(\overline{\gamma}) + \frac{\left[\omega_1(\gamma) - \omega_1(\overline{\gamma})\right]}{\overline{\omega_1'(\gamma)}}\,\overline{\phi'(\gamma)} \tag{9}$$

$$2G(u + iv) = \kappa\phi(\gamma) + \phi(\overline{\gamma}) - \frac{\left[\omega_1(\gamma) - \omega_1(\overline{\gamma})\right]}{\overline{\omega_1'(\gamma)}}\,\overline{\phi'(\gamma)} \tag{10}$$

When γ is real it is seen that equation (9) gives zero resultant force along H'C' and HC as desired. Note that the analytic continuation argument reduces the problem to finding only one unknown function, $\phi(\gamma)$.

A further mapping is introduced which maps the unit semi-circle plus its exterior in the ζ plane to the crack plus its exterior in the z-plane, see Figure 1 (b,c).

$$\gamma = \omega_2\,(\zeta) = \frac{L}{2}\,(\zeta - \zeta^{-1}) \tag{11}$$

where $L = \ln\,(1 + a/R_1)$ is the length of the crack in the γ-plane. Substituting (11) into (9) and (10):

$$f_1 + if_2 = \phi(\zeta) - \phi(\overline{\zeta}) + \frac{\left[\omega_1(\gamma) - \omega_1(\overline{\gamma})\right]}{\overline{\omega_1'(\gamma)}\;\;\overline{\omega_2'(\zeta)}}\,\overline{\phi'(\zeta)} \tag{12}$$

$$2G(u + iv) = \eta\phi(\zeta) + \phi(\overline{\zeta}) - \frac{\left[\omega_1(\gamma) - \omega_1(\overline{\gamma})\right]}{\overline{\omega_1'(\gamma)}\;\;\overline{\omega_2'(\zeta)}}\,\overline{\phi'(\zeta)} \tag{13}$$

where $\phi(\zeta) = \phi(\omega(\zeta))$.

For certain geometries it is necessary to use partitioning {3} to obtain the desired accuracy. The partitioning of the curved beam is shown in Figure 2.

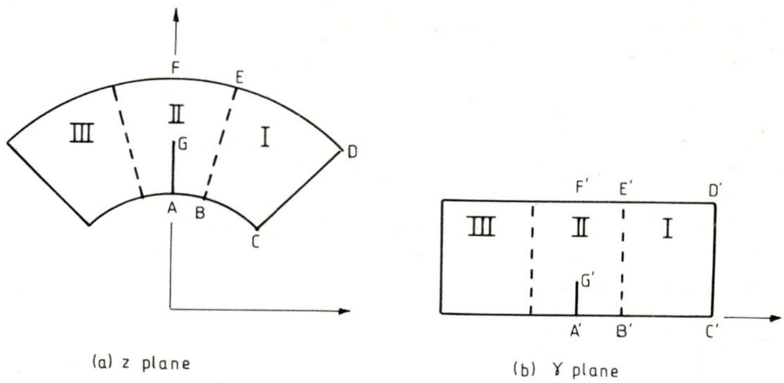

Figure 2: Partitioned Regions

In general each region has its own complex stress and mapping function. For region II the stress function is $\phi(\zeta)$ whilst for regions I and III the stress function is $\phi(\gamma)$.

Since there is symmetry about the imaginary axis, only region I and that part of region II to the right of the imaginary axis need be considered. When partitioning is used it is necessary to "stitch" along common boundaries, by imposing equilibrium and compatibility of displacements, i.e.:

$$(f_1 + if_2)_{II} = (f_1 + if_2)_I \qquad (14)$$

$$(u + iv)_{II} = (u + iv)_I$$

For symmetry about the imaginary axis the stress function for the region containing the crack takes the following form {3}:

$$\phi(\zeta) = \sum_{n=-\infty}^{\infty} \frac{(a_{2n}\,\zeta^{2n} + i\,a_{2n+1}\,\zeta^{2n+1})}{\zeta + i} \qquad (15)$$

where a_{2n} and a_{2n+1} are real coefficients. Also for region I it is assumed that

$$\phi(\gamma) = \sum_{n=0}^{\infty} c_n (\gamma - \gamma_0)^n \qquad (16)$$

where c_n are unknown complex coefficients, and γ_0 is the mid point of the line B'C', Figure 2.

In the MMC method the infinite series representations of $\phi(\zeta)$ and (if appropriate) $\phi(\gamma)$ are truncated to a finite number of terms. Force conditions are imposed at selected boundary points along CD, DF, and AG, via equations (9) and (12) which gives conditions on the unknown coefficients in the stress functions, a_i and c_i. Thus each boundary point produces two rows in the main matrix A, and two corresponding elements in the boundary conditions vector $\underset{\sim}{b}$, where:

$$A \underset{\sim}{x} = \underset{\sim}{b}$$

and $\underset{\sim}{x}$ is the vector of unknown coefficients. The common boundary points along EB are used to obtain conditions relating a_i and c_i, via equations (14), in conjunction with (9), (10), (12) and (13). Each common boundary point gives four rows in A and four corresponding zeros in $\underset{\sim}{b}$. In general A is a matrix of ℓ rows and m columns, where ℓ and m depend upon the number of boundary points and unknown coefficients respectively. It was found that convergence is generally better when $2m < \ell < 2.5m$, this conforms with other workers {7}. A least-squares error minimization procedure was used to solve the overdetermined set of linear equations.

Knowing the coefficients for the stress function in the cracked region the crack shape may be determined from (13). Furthermore the stress intensity factor, K, is given by {6}:

$$K = K_I - iK_{II} = 2(2\pi)^{\frac{1}{2}} \lim_{z \to z_c} (z - z_c)^{\frac{1}{2}} . \phi'(z) \qquad (17)$$

where K_I and K_{II} are opening and sliding mode stress intensity factors respectively, and z_c is the location of the crack tip. Thus, substituting from (15) into (17):

$$K_I - iK_{II} = 2(2\pi)^{\frac{1}{2}} \lim_{\substack{\zeta \to i \\ \gamma \to iL}} (-i(\omega_1(\gamma) - \omega_1(iL))^{\frac{1}{2}} \frac{\phi'(\zeta)}{\omega_1'(\gamma) . \omega_2'(\zeta)} \qquad (18)$$

giving:

$$K_I - i K_{II} = 2 \left(\frac{\pi}{L R_1 e^L} \right)^{\frac{1}{2}} \cdot \phi'(i) \tag{19}$$

where points $\zeta = i$ and $\gamma = iL$ correspond to the location of the crack tip in ζ and γ planes respectively.

3. STRESS INTENSITY FACTOR SOLUTIONS FROM MMC

In all cases considered there is symmetry of loading and geometry about the crack line, the only non-zero stress intensity factor being K_I.

3.1 PURE BENDING OF CURVED BEAM

For this case the appropriate boundary tractions along CD, Figure 3, are {8}:

$$\sigma_\theta = \frac{4M}{N} \left[\frac{R_1^2 R_2^2}{r^2} \ln\left(\frac{R_2}{R_1}\right) - R_2^2 \ln\left(\frac{r}{R_2}\right) - R_1^2 \ln\left(\frac{R_1}{r}\right) - R_2^2 + R_1^2 \right] \tag{20}$$

$$\tau_{r\theta} = 0$$

where M is the applied moment per unit thickness and:

$$N = (R_2^2 - R_1^2)^2 - 4R_1^2 R_2^2 \left[\ln\left(\frac{R_2}{R_1}\right) \right]^2$$

all other tractions are zero.

Alternatively, in order to determine K_I an equal and opposite loading may be applied along the crack surfaces. These two loading options were tested for a limited number of cases, the differences in K_I and crack-opening being negligible. All results reported were obtained with the appropriate loading applied to the crack surfaces.

K_I values were obtained via equation (19) for R_2/R_1 ratios of 1.25, 1.5, 2.0, 2.5 and crack lengths in the range $0.01 \leqslant a/(R_2-R_1) \leqslant 0.7$. The results are presented graphically in Figure 3. These results are compared with those due to Tracy {3},{9}. The two sets of results are generally in agreement to within 1%.

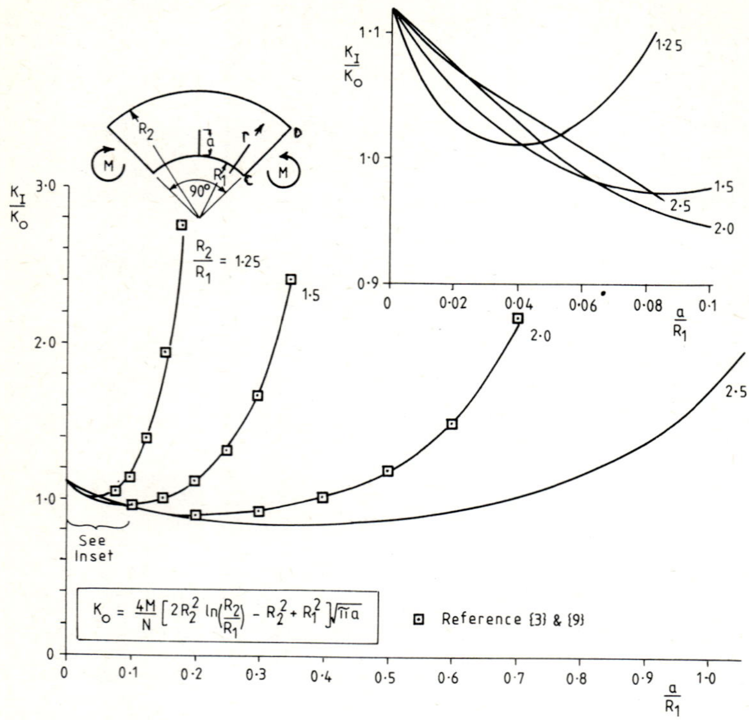

Figure 3: K_I Values for Cracked, Curved Beams, Under Pure Bending

3.2 CONSTANT PRESSURE ON CRACK SURFACES

The loading in this case consists of a constant pressure, p applied to each of the crack surfaces. K_I values for the range of geometrical variables listed in section 3.1 are presented in Figure 4.

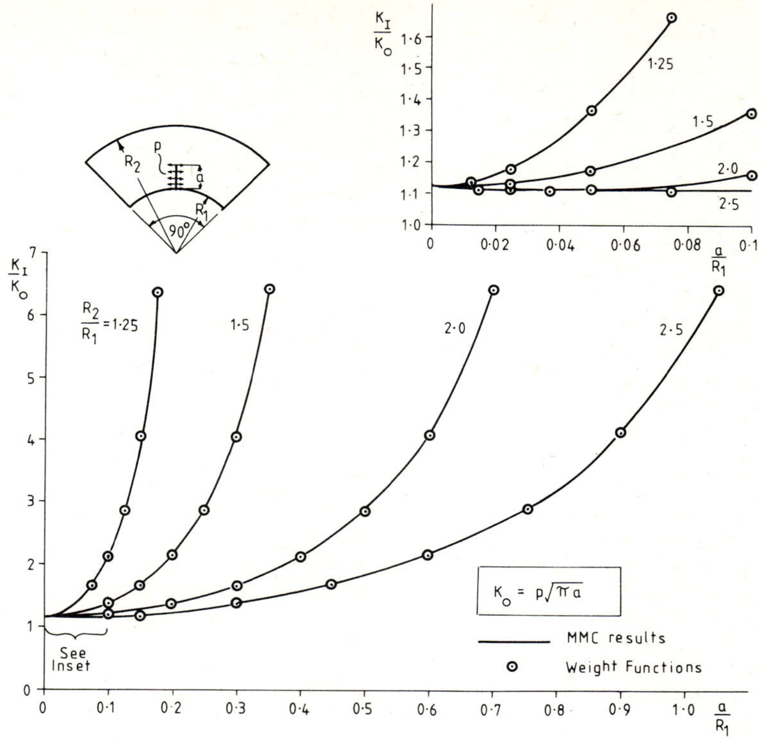

Figure 4: K_I Values for Cracked, Curved Beams, Constant Pressure on Crack Surfaces.

4. WEIGHT FUNCTION (THEORY)

Bueckner {1} and Rice {2} have demonstrated that a particular function, normally termed the weight function, is a property of a cracked geometry, and is independent of the loading. The weight function may be employed in the derivation of additional stress intensity factor solutions, provided details of boundary loading are available.

The weight function m (x,y,a) is defined as:

$$m(x,y,a) = \frac{H}{2K^{(i)}(a)} \cdot \frac{\partial \tilde{u}}{\partial a}(x,y,a) \qquad (21)$$

where H = E (plane stress), H = E/(1-ν^2) (plane strain), $K^{(i)}$ is the stress intensity factor for loading system (i) expressed as a function of crack length, a, and the displacement field for loading system (i) is $\underset{\sim}{u}(x,y,a)$. In the current work attention is restricted to the crack line, so that equation (21) may be written as:

$$m(y,a) = \frac{H}{2K^{(i)}(a)} \cdot \frac{\partial v(y,a)}{\partial a} \tag{22}$$

where v is the x direction crack opening for a crack orientated along the y^* axis, Figure 1(a), and $y = y^* - R_1$.

Additional stress intensity factor solutions for any crack-line loading system may be obtained from {2}:

$$K^{(ii)} = \int_a p(y) \cdot m(y,a) \cdot dy \tag{23}$$

where $K^{(ii)}$ is the stress intensity factor for a crack line loading which consists of a normal pressure p(y) applied to each of the crack faces. (Note that loading system (i) must not possess more symmetries than loading system (ii)). Provided that the loading along the crack line in the unflawed structure is known, superposition techniques will permit the derivation of K solutions via equation (23). Crack line loading may normally be calculated in a straightforward manner for statically determinate structures.

5. CUBIC SPLINES

The cubic spline curve-fitting process consists of defining a set of cubic curves, joined together smoothly with continuity of function and both first and second derivatives at a number of knot points {10}. When the fit is specified in the least squares sense (i.e. by minimising the the sum of the squares of the errors at a number of points) the cubic B spline representation is often used. In this work the curve-fitting is defined by:

$$S(t) = \sum_{j=1}^{N+3} d_j \, N_j(t) \tag{24}$$

where $N_j(t)$ are normalised B-splines given by:

$$N_j(t) = (t_j - t_{j-4}) \sum_{i=j-4}^{j} \frac{(t_i - t)_+^3}{w_j{}'(t_i)}$$

N = number of intervals (i.e. number of internal knots + 1).

d_j are B spline coefficients

$t_{-3}, t_{-2}, \ldots\ldots t_{N+3}$ are knot positions

$t_{-3} = t_{-2} = t_{-1} = t_0$ are at beginning of curve-fitting range

$t_N = t_{N+1} = t_{N+2} = t_{N+3}$ are at end of curve-fitting range

$$(t_j - t)_+^3 = \begin{cases} (t_i - t)^3 & , t < t_i \\ 0 & , t > t_i \end{cases}$$

$$\sum_{i=j-4}^{j} \frac{(t_i - t)_+^3}{w_j'(t_i)} \begin{cases} > 0, & t_{j-4} < t < t_j \\ = 0, & t < t_{j-4}, \; t > t_j \end{cases}$$

$$w_j(t) = (t - t_{j-4})\,(t - t_{j-3})\,(t - t_{j-2})\,(t - t_{j-1})\,(t - t_j)$$

6. DERIVATION AND FITTING OF WEIGHT FUNCTIONS

Weight function data may be derived from the numerical MMC programme by the method of "virtual crack extensions", {11}. The curved beam geometry, subjected to pure bending was used to determine weight functions by this method. With nominal crack length a, two stress intensity factors, K_I and associated crack-opening v(y) were determined, namely:

$$K_I\Big|_{(a - \lambda a)} \quad , \quad v(y)\Big|_{(a - \lambda a)}$$

and $$K_I\Big|_{(a + \lambda a)} \quad , \quad v(y)\Big|_{(a + \lambda a)}$$

and hence mean values of K_I and $\left(\partial v/\partial a\right)$ were calculated as follows:

$$K_I\Big|_a = \frac{1}{2}\left(K_I\Big|_{(a - \lambda a)} + K_I\Big|_{(a + \lambda a)}\right) \tag{25}$$

$$\frac{\partial v(y)}{\partial a} = \frac{v(y)\Big|_{(a + \lambda a)} - v(y)\Big|_{(a - \lambda a)}}{2\lambda a} \tag{26}$$

in most cases $\lambda = 0.01$.

A typical curve of $\frac{\partial v(y)}{\partial a}$ versus (y/a) is shown in Figure 5(a).

Clearly, the function tends asymptotically to infinity as $(y/a) \to 1.0$. In the immediate vicinity of the crack tip, $\frac{\partial v(y)}{\partial a}$ may be written in terms of K as {12}:

$$\frac{\partial v(y)}{\partial a} = \frac{K_I}{H}\left(\frac{2}{\pi}\right)^{\frac{1}{2}} \frac{(2a - y)}{a(a - y)^{\frac{1}{2}}} \tag{27}$$

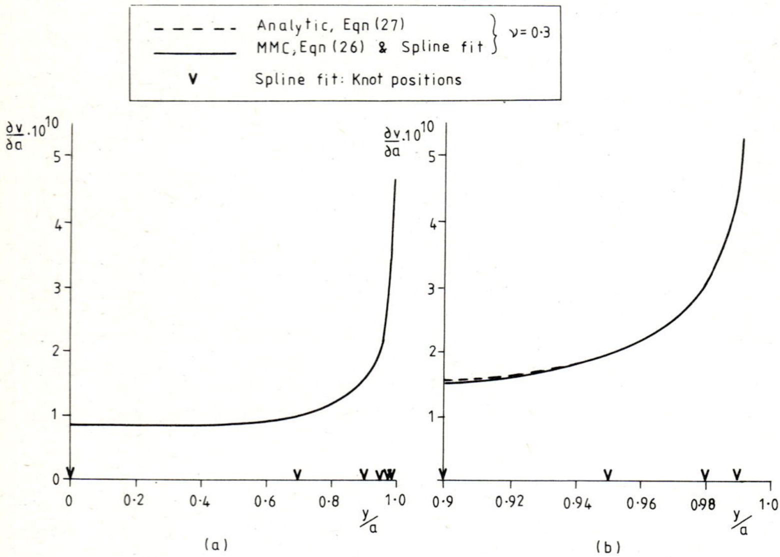

Figure 5: Typical Curve Showing $(\partial v/\partial a)$ versus (y/a).

Figure 5(b) shows $(\partial v/\partial a)$ predicted by equation (27), together with the numerical results in the range $0.9 \leqslant (y/a) \leqslant 1.0$. The $(\partial v/\partial a)$ data was fitted using an available cubic B spline routine {13} in the range $0 \leqslant (y/a) < 0.99$, and the analytic expression (27) was assumed in the range $0.99 \leqslant (y/a) \leqslant 1.0$.

A typical plot showing spline fit appears in Figure 5(a). Numerical spline data, including knot positions and coefficients, is given in Figure 6 for R_2/R_1 values of 1.25, 1.5, 2.0, 2.5, and $a/(R_2 - R_1)$ in the range 0.01 to 0.7. In all cases, the spline fit data is within 0.5% of the original (MMC) results.

Table of B-Spline Coefficients $d_j \times 10^{10}$ (ie multiply tabulated values by 10^{-10} to obtain d_j)

	j \ $\frac{a}{R_2-R_1}$	.01	.025	.05	0.1	0.2	0.3	0.4	0.5	0.6	0.7
$R_2/R_1 = 1.25$	1	12.823	12.596	12.501	13.151	17.107	25.361	41.225	73.711	149.21	365.01
	2	12.990	12.772	12.633	13.125	16.461	23.461	36.774	63.647	125.15	298.36
	3	11.915	11.679	11.553	11.946	14.634	20.119	30.201	49.868	93.391	211.91
	4	17.686	17.374	17.016	16.898	18.288	21.728	28.079	39.916	64.289	124.86
	5	28.487	27.873	27.194	26.582	27.301	30.253	35.952	46.342	66.531	112.52
	6	40.531	39.681	38.808	37.639	38.002	41.065	47.236	58.430	79.358	124.38
	7	59.356	58.193	56.727	54.842	54.459	58.226	65.830	79.997	105.31	157.59
	8	73.226	71.724	69.886	66.690	65.667	70.137	78.714	95.935	125.10	184.74
	9-12	0.0000	0.0000	0.0000	0.0000	0.0000	0.0000	0.0000	0.0000	0.0000	0.0000
	Q^b	1.012	1.086	1.049	1.018	1.014	1.070	1.191	1.406	1.789	2.536
$R_2/R_1 = 1.5$	1	3.4039	3.3163	3.2584	3.3824	4.3722	6.4710	10.498	18.709	37.763	92.582
	2	3.4499	3.3632	3.2945	3.3810	4.2104	5.9857	9.3612	16.148	31.664	75.650
	3	3.1662	3.0840	3.0258	3.0888	3.7528	5.1397	7.6878	12.643	23.607	53.684
	4	4.7071	4.5958	4.4670	4.3894	4.6989	5.5456	7.1327	10.099	16.223	31.562
	5	7.5861	7.3892	7.1646	6.9218	7.0195	7.7194	9.1242	11.711	16.765	28.378
	6	10.795	10.536	10.184	9.8139	9.7705	10.475	11.981	14.745	19.992	31.329
	7	15.811	15.419	14.902	14.302	14.005	14.877	16.712	20.105	26.464	39.536
	8	19.505	19.020	18.383	17.625	16.887	17.980	20.016	23.876	31.299	45.963
	9-12	0.0000	0.0000	0.0000	0.0000	0.0000	0.0000	0.0000	0.0000	0.0000	0.0000
	Q^b	1.098	1.072	1.041	0.995	0.976	1.020	1.129	1.328	1.686	2.386
$R_2/R_1 = 2.0$	1	0.9381	0.8986	0.8651	0.8810	1.1283	1.6651	2.6912	4.7744	9.5986	23.408
	2	0.9531	0.9127	0.8785	0.8836	1.0877	1.5402	2.3985	4.1180	8.0437	19.120
	3	0.8725	0.8400	0.8083	0.8117	0.9739	1.3246	1.9691	3.2208	5.9897	13.558
	4	1.3062	1.2586	1.2074	1.1641	1.2226	1.4273	1.8218	2.5650	4.1061	7.9582
	5	2.1004	2.0302	1.9390	1.8417	1.8290	1.9853	2.3266	2.9702	4.2360	7.1467
	6	2.9921	2.8910	2.7652	2.6095	2.5466	2.6949	3.0535	3.7390	5.0453	7.8872
	7	4.3898	4.2336	4.0558	3.8190	3.6868	3.8491	4.2599	5.1181	6.6593	9.9465
	8	5.4111	5.2241	4.9956	4.7028	4.5280	4.7102	5.1021	6.1384	7.8272	11.562
	9-12	0.0000	0.0000	0.0000	0.0000	0.0000	0.0000	0.0000	0.0000	0.0000	0.0000
	Q^b	1.096	1.059	1.004	0.945	0.907	0.937	1.026	1.201	1.519	2.148
$R_2/R_1 = 2.5$	1	0.4502	0.4244	0.4019	0.4042	0.5167	0.7599	1.2203	2.1526	4.3072	10.488
	2	0.4574	0.4320	0.4091	0.4064	0.4985	0.7028	1.0869	1.8557	3.6080	8.5647
	3	0.4204	0.3984	0.3782	0.3752	0.4474	0.6042	0.8914	1.4497	2.6853	6.0714
	4	0.6294	0.6008	0.5690	0.5412	0.5612	0.6492	0.8222	1.1522	1.8379	3.5589
	5	1.0159	0.9711	0.9167	0.8579	0.8388	0.9013	1.0494	1.3326	1.8952	3.1938
	6	1.4462	1.3836	1.3086	1.2167	1.1673	1.2216	1.3769	1.6776	2.2572	3.5233
	7	2.1071	2.0269	1.9132	1.7781	1.6800	1.7295	1.9343	2.3024	3.0017	4.4624
	8	2.5653	2.5013	2.3601	2.1915	2.0377	2.0732	2.3557	2.7835	3.5929	5.2484
	9-12	0.0000	0.0000	0.0000	0.0000	0.0000	0.0000	0.000	0.0000	0.0000	0.0000
	Q^b	1.082	1.041	0.981	0.907	0.855	0.872	0.950	1.107	1.395	1.968

Knot Details	
Knot No	Position
1-4	y/a = 0.00
5	y/a = 0.70
6	y/a = 0.90
7	y/a = 0.95
8	y/a = 0.98
9-12	y/a = 0.99

NOTES:

(1) $Q^b = K_I{}^b/\sigma^*(\pi a)^{\frac{1}{2}}$

where $\sigma^* = \frac{4M}{N}\left[2R_2^2 \ln\left(\frac{R_2}{R_1}\right) - R_2^2 + R_1^2\right]$

and $N = (R_2^2 - R_1^2)^2 - 4R_1^2 \; R_2^2 \; \left[\ln\left(\frac{R_2}{R_1}\right)\right]^2$

(2) $\nu = 0.3$, $ER_1{}^2 M^{-1} = 210 \times 10^9$, plane strain

(3) To obtain $(\partial v/\partial a)$ substitute:

$\frac{\partial v}{\partial a}(y) = S(t)$, $\left(\frac{y}{a}\right) = t$ into equation (24)

(4) See section 7 for details of integration procedure to obtain stress intensity for new loading.

Figure 6: $(\partial v/\partial a)$ Spline-Fit Data and Associated K_I Values

7. DERIVATION OF STRESS INTENSITY FROM WEIGHT FUNCTION

In order to determine stress intensity solutions from the data presented in Figure 6 it is necessary to integrate this data in accordance with equation (23). This process is described, and results presented for two loading cases.

EXAMPLE 1: Constant Pressure on Crack Faces

This is the loading previously considered in section 3.2. For this particular case, substituting from (22) into (23) produces:

$$K_I^{cp} = \frac{E}{(1-\nu^2)} \frac{p}{K_I^b} \left[\int_o^{.99a} \frac{\partial v^b(y)}{\partial a} . \, dy + \int_{.99a}^{a} \frac{\partial v^b(y)}{\partial a} . \, dy \right] \tag{28}$$

where K_I^{cp} is the stress intensity factor for constant pressure p acting on each of the crack surfaces, K_I^b is the stress intensity factor for the case of pure bending and v^b is the crack-opening for pure bending. The first integral was evaluated from spline data in Figure 6, via an available programme {13} which complements the curve-fitting routine. The second integral was evaluated exactly from the analytic expression (27). The results are presented as discrete points in Figure 4. The continuous line represents the MMC results for this configuration reported in section 3. The maximum difference between the two sets of results is 0.5%.

EXAMPLE 2: Residual Stress Problem, $R_2/R_1 = 2.5$

Consider a configuration in which thermal or residual (e.g. welding) stresses produce a parabolically varying stress field along the axis of symmetry in the uncracked structure, Figure 7. Assuming that the appropriate equal and opposite tractions along the crack line are given by:

$$p(y) = \frac{1}{R_1^2} \left[y^2 - (R_2 - R_1)y + \frac{(R_2 - R_1)^2}{6} \right] \tag{29}$$

it is possible to calculate the stress intensity factor due to the residual stresses from equation (23). On this occasion it was necessary to evaluate $p(y). \frac{\partial v^b(y)}{\partial a}$ for a number of points. This data was then spline-fitted once more, and integrated as in EXAMPLE 1.
(Note: the programme which accepts spline and loading data, and returns stress intensity is given in reference {14})

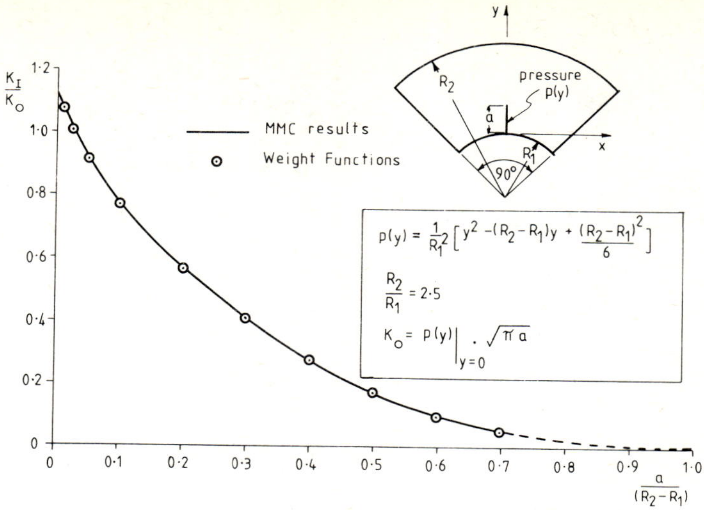

Figure 7: Stress Intensity Factors for a Cracked, Curved Beam with Residual Stresses.

Results for the residual stress Example are shown as discrete points in Figure 7. An additional MMC run with the appropriate loading produced the results shown as a continuous line. The maximum difference between the two sets of results is 1%.

8. SUMMARY

The Modified Mapping Collocation (MMC) method has been used in the derivation of K_I values for a range of cracked, curved beam geometries. The variation in crack-opening resulting from a small crack extension was also calculated from the MMC programme, this data being subsequently curve-fitted using a standard Cubic B-Spline routine. The resulting curve was employed in the derivation of K_I solutions, using Bueckner's weight function approach, for the case of a crack subjected to constant pressure. By comparison with MMC results, the loss of accuracy arising from the curve-fitting process is less than 0.5%.

K_I values were also obtained for a more complex residual stress field. The total loss of accuracy in this case was less than 1%.

Stress intensity and crack opening data in the form described in this paper may be used to obtain accurate K values for any crack-line loading system using packaged spline-fitting and complementary integration routines.

9. REFERENCES

{1} BUECKNER, H.E. - A Novel Principle for the Computation of Stress Intensity Factors. Z. Angew. Math. Mech., Vol. 50, pp. 529-546, 1970.

{2} RICE, J.R. - Some Remarks on Elastic Crack-Tip Stress Fields. Int. J. Solids Structs., Vol. 8, pp. 751-758, 1972.

{3} TRACY, P.G. - Analysis of a Radial Crack in a Circular Ring Segment. Engng. Frac. Mech., Vol. 7, pp. 253-260, 1975.

{4} BOWIE, O.L., NEAL, D.M. - A Modified Mapping Collocation Technique for Accurate Calculation of Stress Intensity Factors. Int. J. Frac. Mech., Vol. 6, pp. 199-206, 1970.

{5} MUSKHELISHVILI, N.I. - Some Basic Problems of the Mathematical Theory of Elasticity, Noordhoff, 1973.

{6} SIH, G.C., PARIS, P.C., ERDOGAN, F. - Crack Tip Stress Intensity Factors for Plane Extension & Plate Bending Problems, J. Appl. Mech., Vol. 29, pp. 306-312, 1962.

{7} EASON, E.D. - A Review of Least Squares Methods for Solving Partial Differential Equations. Int. J. Num. Meth. Engng., Vol. 10, pp. 1021-1046, 1976.

{8} TIMOSHENKO, S.P., GOODIER, J.N. - Theory of Elasticity, McGraw Hill, 1970.

{9} TRACY, P.G. - Private Communication.

{10} COX, M.G. - An Algorithm for Spline Interpolation. J. Inst. Math. Appl., Vol. 15, No. 1, pp. 95-108, 1975.

{11} PARKS, D.M., KAMENETZKY, E.M. - Weight Function from Virtual Crack Extension, Int. J. Num. Meth. Engng., Vol. 14, pp. 1693-1706, 1979.

{12} PARIS, P.C., SIH, G.C. - Stress Analysis of Cracks. Fracture Toughness Testing and its Applications, ASTM STP 381, 1965.

{13} NUMERICAL ALGORITHMS GROUP (NAG) - Fortran Library Manual Mark 7, Vol. 2, Routines E02BAF/E02BBF/E02BDF, 1978.

{14} ANDRASIC, C.P., PARKER, A.P. - A Simple Spline Fit/Weight Function Programme for the Determination of Stress Intensity Factors, Royal Military College of Science, Technical Note MAT/25, 1980.

10. ACKNOWLEDGEMENT

This work was supported by a Ministry of Defence Research Contract.

A FINITE ELEMENT ANALYSIS OF CRACK INSTABILITY

C.W. Woo(1) and C.L. Chow(2)

ABSTRACT

A finite element analysis of the stability conditions of crack propagation under general loading is presented. The method of analysis is based on the quasi-static energy approach and the virtual crack extension formulation. The stability criteria take into account of the specimen geometry and the conditions of the loading system involved.

The computational procedures are outlined and numerical results are given for the case of a centre-cracked plate subject to different boundary conditions. The method is found to be efficient and can readily be incorporated into a standard finite element program.

INTRODUCTION

In a loaded specimen with a pre-existing flaw like defect, crack initiation occurs when the energy release rate becomes equal to the fracture resistance at the crack tip. Crack propagation then takes place either in a stable or unstable manner depending on a number of parameters such as the material properties, crack-tip loading, geometry of specimen and the testing conditions etc. From the consideration of energy balance as advocated by Griffith [1] in linear fracture mechanics, the conditions of crack instability can be characterized by means of the geometric stability factor which defines the relative change in energy release rate to crack propagation [2, 3].

---

(1) Department of Mechanical Engineering, University of Hong Kong, Hong Kong.

(2) The BF Goodrich Research and Development Centre, Ohio, U.S.A.

For an arbitrary cracked specimen, the analysis of the energy release rate and hence the stability condition will have to rely on numerical techniques such as the finite element method. Recently Parks [4, 5] and Hellen [6] proposed a method for the direct computation of the energy release rate using a virtual crack extension formulation. The method was found to yield accurate results and requires less computational effort than the usual approach involving the computation of the strain energy at two slightly different crack lengths.

This paper is intended to present a computational method for the analysis of the stability condition of crack propagation. The quasi-static energy approach is used to formulate the stability criteria in terms of the energy release rate. Using the virtual crack extension method, a finite element scheme is devised to compute directly the geometric stability factor which is the characterizing parameter defining the stability condition of crack propagation in a specimen with arbitrary geometry and loading conditions.

STABILITY OF CRACK PROPAGATION

The basic concepts of crack instability in linear elastic fracture mechanics have been discussed by Srawley and Brown [7] for the special case of instability occurring at maximum load. Recently Gurney [2], Clausing [8] and Glucklich [9] have proposed general stability criteria using energy balance consideration. The quasi-static approach as advocated by Gurney has been found most versatile for general application and for numerical analysis [10, 11]. It is on this basis that a finite element scheme is devised here for the analysis of the stability condition and for the prediction of cracking mode.

Referring to Figure 1 of a simple loaded cracked specimen with a crack length a. The stability criteria based on the quasi-static energy approach can be expressed as (1, 2):

(i) for load-control system (i.e. $dP/P \geq 0$)

$$\frac{1}{G_R}\frac{dG_R}{dA} \geq \frac{1}{G}\left(\frac{\partial G}{\partial A}\right)_P \tag{1}$$

(ii) for displacement-control system (i.e. $dv/v \geq 0$)

$$\frac{1}{G_R}\frac{dG_R}{dA} \geq \frac{1}{G}\left(\frac{\partial G}{\partial A}\right)_v \tag{2}$$

where G_R is the crack resistance force, G is the energy release rate, A is the crack surface area, P and v are respectively the load and displacement at the point of load application. Alternatively the stability criteria can be expressed conveniently in terms of the specimen compliance (v/P) as:

$$\frac{1}{G_R}\frac{dG_R}{dA} \geq \frac{\frac{d^2}{dA^2}(v/P)}{\frac{d}{dA}(v/P)} \qquad (3)$$

and $$\frac{1}{G_R}\frac{dG_R}{dA} \geq \frac{\frac{d}{dA^2}(v/P)}{\frac{d}{dA}(v/P)} - 2\,\frac{\frac{d}{dA}(v/P)}{(v/P)} \qquad (4)$$

for load- and displacement-control system respectively.

The implications of the above equations have been dealt with in details by Gurney [1, 2] and Glucklitch [9], and therefore no attempt is made here to repeat the exercise. However, if at the point of crack instability, the fracture resistance (G_R) becomes equal to the critical energy release rate (G_c) which is assumed to be a material constant, the right hand side of the equations will then be a governing factor for stable and unstable crack propagation depending if the above relationship can be satisfied or not. This governing factor is defined here as the geometric stability factor.

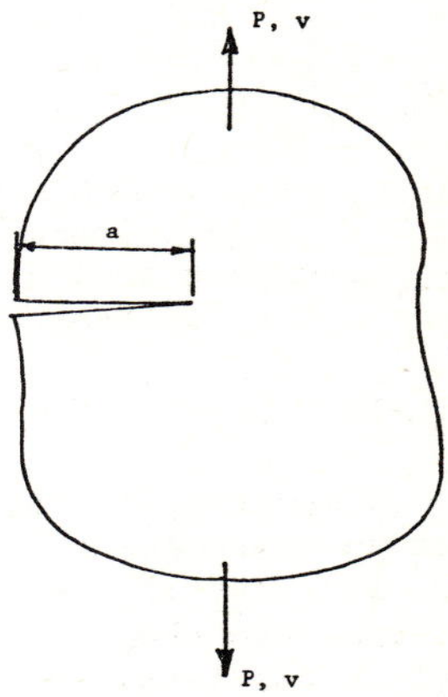

Figure 1. Cracked Specimen

From the practical point of view, crack propagation in a structure under load control is not often possible and accordingly this paper is confined to the case of displacement

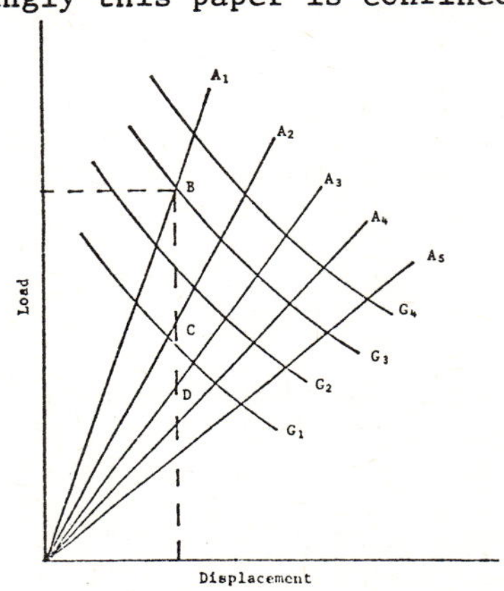

Figure 2. Stable Crack Propagation

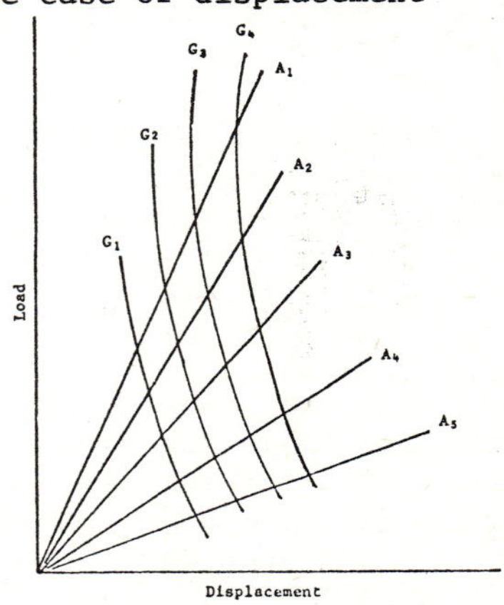

Figure 3. Unstable Crack Propagation

control on the basis of Equations (2) and (4). Consider now a family of curve with constant energy release rate (G) in the load-displacement diagram as shown in Figure 2. It can be deduced from the G curves that the partial derivative $\partial G/\partial A$ being negative would satisfy the stability criterion of Equation (2), and therefore indicating stable crack propagation. On the other hand, for a family of G curves with positive $\partial G/\partial A$ values as shown in Figure 3, unstable crack propagation would be expected.

Similar conclusions may be drawn by considering the changes of strain energy in a cracked specimen with respect to a small crack increment ΔA. Writing Equation (2), in the form of a finite difference, we have

$$\left(\frac{\partial G}{\partial A}\right)_v = \frac{\Lambda_B - 2\Lambda_C + \Lambda_D}{(\Delta A)^2} \tag{5}$$

where Λ_B, Λ_C and Λ_D are the strain energy of the crack areas A_1, A_2 and A_3 respectively at constant displacement (v). Therefore for stable crack propagation the area of triangle OBC must be greater than OCD in Figure 2.

FINITE ELEMENT FORMULATION

The general application of the finite element method to fracture mechanics study is well documented in the literature [11-19]. For linear elastic problems, various computational schemes are used for the determination of the stress intensity factor or the energy release rate. As the stability criteria are formulated on the energy approach, the virtual crack extension method as proposed by Parks and Hellen can readily be extended for the direct computation of the geometric stability factor defined as

$$\frac{1}{G}\left(\frac{\partial G}{\partial A}\right)_u = \frac{1}{G_a}\left(\frac{G_{a+\Delta a} - G_a}{\Delta A}\right) \tag{6}$$

From the consideration of a virtual crack extension, the energy release rate (G) can be expressed in terms of the displacement {u} and the stiffness matrix [K] as

$$G = \frac{1}{2B}\{u\}^T \frac{\partial [K]}{\partial a}\{u\} \tag{7}$$

where a is the crack length and B is the specimen thickness. Since the [K] matrix is the sum of all element stiffness matrices for the entire structure and if a virtual displacement Δa is assumed for the elements enclosed by the inner contour line such as Γ_0 (Figure 4), the derivative $\partial[K]/\partial a$ in Equation (7) only involves the element stiffness matrices between the contour lines Γ_0 and Γ_1 (Figure 4) and may be expressed as

$$G = \frac{1}{2B} \{u\}^T \sum_{i=1}^{N} \frac{\partial [k_i]}{\partial a} \{u\} \tag{8}$$

where $[k_i]$ is the element stiffness matrix of an element between Γ_0 and Γ_1, and N is the total number of such elements. Now the derivative $\partial[k_i]/\partial a$ may be approximated as a forward finite difference as

$$\frac{\partial [k_i]}{\partial a} = \frac{1}{\Delta a} ([k_i]_{a+\Delta a} - [k_i]_a) \tag{9}$$

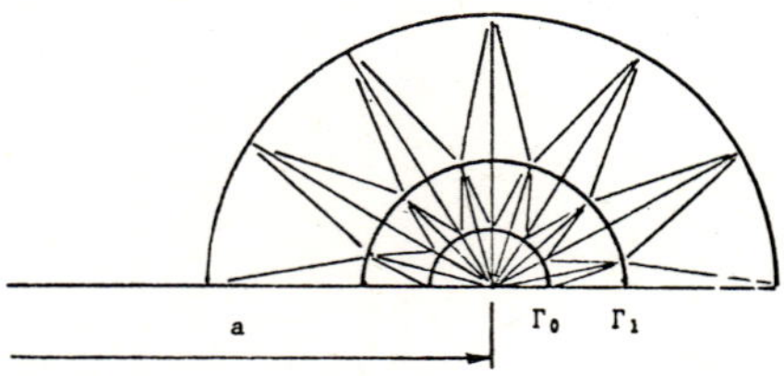

Figure 4. Mesh Geometry at Crack Tip

From Equations (6), (8) and (9), we then obtain the finite element formulation for the geometric stability factor as

$$\frac{1}{G} \left(\frac{\partial G}{\partial A}\right)_u = \frac{1}{B\Delta a} \left(\frac{\{u\}^T_{a+\Delta a} \sum_{i=1}^{N} \{[k_i]_{a+2\Delta a} - [k_i]_{a+\Delta a}\}\{u\}_{a+\Delta a}}{\{u\}^T_a \sum_{i=1}^{N} \{[k_i]_{a+\Delta a} - [k_i]_a\}\{u\}_a} - 1 \right) \tag{10}$$

To compute the geometric stability factor, it is therefore necessary to calculate the variation of the element stiffness matrices $[k_i]$ at three slightly different crack lengths (i.e. a, a+Δa and a+2Δa). And by using a forward elimination and a backward substitution solution procedure, the displacement {u} corresponding to crack lengths a and a+Δa can be readily computed with very little additional computing time.

NUMERICAL EXAMPLES AND RESULTS

The stability conditions of the centre-cracked plate under various loading configurations have been studied using the solution procedures as outlined in the previous section. The finite element mesh for a quarter of the centre-cracked plate is shown in Figure 5, with a total number of 179 triangular elements and 109 nodes. The inner rings of elements at the crack tip are encircled by contour lines at radius of 0.1, 0.2 and 0.5 of crack length a. The inner elements are then generated by sub-dividing the rings into equal sectors of 15°. For the purpose of computation, a value of Δa was chosen to be of the order of 5×10^{-6} of the crack length a as found satisfactory in reference [4].

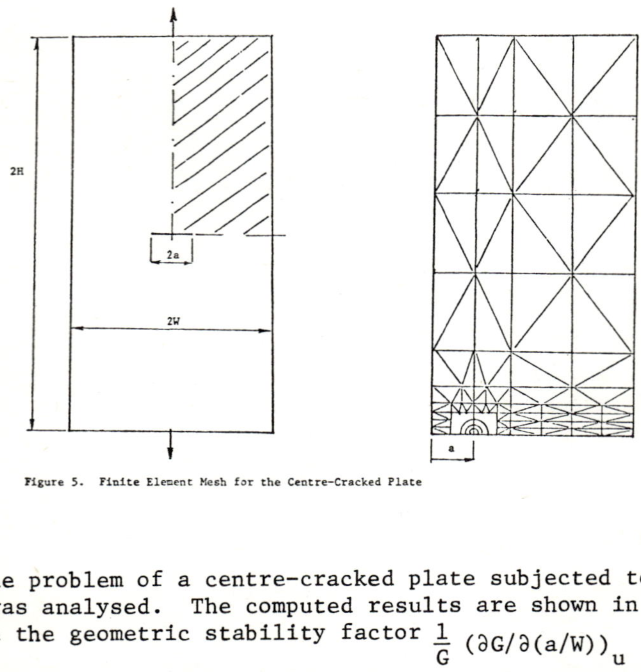

Figure 5. Finite Element Mesh for the Centre-Cracked Plate

The problem of a centre-cracked plate subjected to point loads was analysed. The computed results are shown in Figure 6 where the geometric stability factor $\frac{1}{G}(\partial G/\partial(a/W))_u$ is plotted against the crack length (a/W) ratio. Results compare well with those obtained by Clausing [8] and by experimental measurement on some test specimens made of PMMA material. Experimental observation also confirms that the cracking mode of the point-load centre-cracked plate is one of unstable fracture as predicted for the range of the crack length to specimen width as shown in Figure 6.

For the case of a centre-cracked plate subjected to biaxial loading, the effect of the biaxial stress ratio (α) on the stability conditions is studied. The computed results are summarised as shown in Figure 7. The stability condition is seen to be influenced by the biaxial load. With increasing biaxial stress ratio (α), the centre-cracked plate becomes more stable. However, with biaxial compressible stress introduced parallel to the crack (i.e. negative α values), the plate is made less stable.

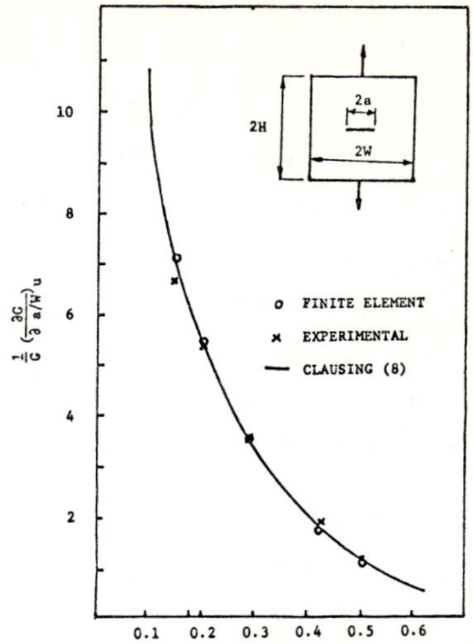

Figure 6. Centre-Cracked Plate under Point Load

As the stability condition of a specimen depends on the relative change in the energy release rate with crack propagation, it is conceivable that the cracking mode can be changed from unstable propagation to that of stable propagation or vise versa if the compliance or the stiffness of the specimen is changed. This can be demonstrated in the problem of a tension plate with a crack between two fastener holes. The stability conditions of this problem are studied with the position of the fastener holes varied thus changing the overall compliance of the cracked plate. The computed results are summarised as shown in Figure 8 where the geometric stability factor is plotted against the crack length (a/W) for various positions of the fastener holes (h/W)

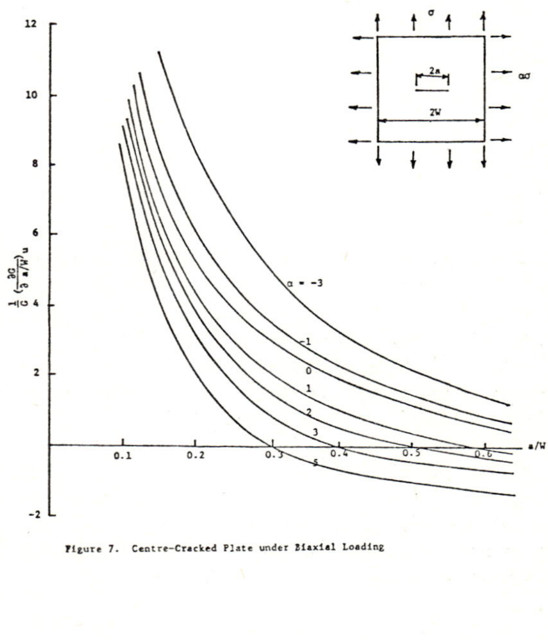

Figure 7. Centre-Cracked Plate under Biaxial Loading

from the crack line. It is of interest to note that as the fastener holes are closed to the crack line the plate becomes less stable. However, the plate is stablised with a ratio (h/W) greater than 0.4 for the given fastener hole size. Experimental testing also confirm the prediction of the cracking mode within the range of crack length ratios (a/W) given in Figure 8.

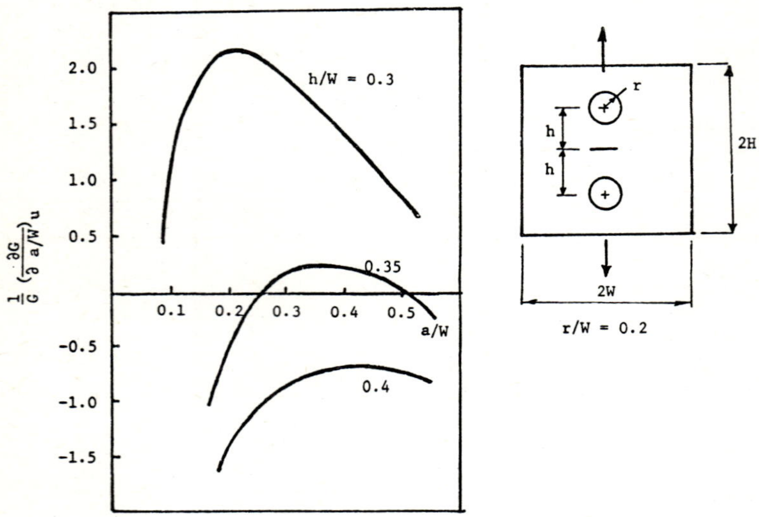

Figure 8. Tension Plate with Fastener Holes

CONCLUSION

The conditions of stable and unstable fracture have been analysed successfully using a finite element technique based on the virtual crack extension formulation. The method is found efficient and can be readily incorporated into any standard finite element program. It does not require special elements to simulate the singularity stress field in the neighbourhood of the crack tip.

The problem of a centre-cracked plate has been chosen to illustrate the effect of the specimen geometry and loading on the stability conditions of a cracked specimen. Experimental observation seems to confirm the computed results for the prediction of the cracking mode.

REFERENCES

1. GRIFFITH, A. A. - The Phenomena of Rupture and Flow in Solids, Phil. Trans. Roy. Soc. (London) A221, pp. 163-198, 1921.

2. GURNEY, C. and HUNT, J. - Quasi-static Crack Propagation, Proc. Royal Society, A299, pp. 508-524, 1967.

3. GURNEY, C. and NGAN, K. M. - Quasi-static Crack Propagation in Nonlinear Structures, Proc. Royal Society, A325, pp. 207-222, 1971.

4. PARKS, D. M. - A Stiffness Derivative Finite Element Technique for Determination of Crack Tip Stress Intensity Factors, Int. J. Fracture, Vol. 10, No. 4, pp. 487-502, 1974.

5. PARKS, D. M. - The Virtual Crack Extension Method for Nonlinear Material Behavior, Comp. Meths. App. Mech. Eng., Vol. 12, pp. 353-364, 1977.

6. HELLEN, T. K. - On the Method of Virtual Crack Extensions, Int. J. Nu. Meths. Eng., Vol. 9, pp. 187-207, 1975.

7. SRAWLEY, J. E. and BROWN, W. F. - Fracture Toughness Testing and Its Applications, ASTM STP 381, Am. Soc. for Testing and Materials, pp. 133-193, 1965.

8. CLAUSING, D. P. - Crack Stability in Linear Elastic Fracture Mechanics, Int. J. Fract. Mech., Vol. 15, pp. 211-227, 1969.

9. GLUCKLICH, J. - On Crack Stability in Some Fracture Tests, Eng. Fract. Mech., Vol. 3, pp. 333-344, 1971.

10. GURNEY, C. and MAI, Y. W. - Stability of Cracking, Eng. Fract. Mech., Vol. 4, pp. 853-863, 1973.

11. CHOW, C. L. and FUNG, W. C. - Prediction of Fracture Instability by a Finite Element Method, Proc. Int. Conf. on Fracture and Technology, Hong Kong, 1977.

12. ZIENKIEWICZ, O. C. - The Finite Element Method in Engineering Science, McGraw-Hill, 1971.

13. CHAN, S. K., TUBA, I. S. and WILSON, W. K. - On the Finite Element Method in Linear Fracture Mechanics, Eng. Fract. Mech., Vol. 2, pp. 1-17, 1970.

14. SWEDLOW, J. L., WILLIAMS, M. L. and YANG, W. H. - Elasto-plastic Stress and Strains in a Cracked Plate, Proc. First Int. Conf. in Fracture, Vol. 1, pp. 259-282, 1966.

15. WATWOOD, JR., V. B. - The Finite Element Method for Prediction of Crack Behavior, Nuclear Eng. and Design, Vol. 11, pp. 323-332, 1969.

16. ANDERSON, A. - A Finite-Element Representation of Stable Crack Growth, J. Mech. Phys. Solids, Vol. 21, pp. 337-356, 1973.

17. HAYES, D. J. - A Practical Formulation for Determining Stress Intensity Factors for Cracked Bodies, Int. J. Fract. Mech., Vol. 8, pp. 157-165, 1972.

18. TRACEY, D. M. - Finite Elements for Determination of Crack Tip Elastic Stress Intensity Factors, Eng. Fract. Mech., Vol. 3, pp. 255-265, 1972.

19. ANDERSON, G. P., RUGGLES, V. L. and STIBOR, G. S. - Use of Finite Element Computer Programs in Fracture Mechanics, Int. J. Fract. Mech., Vol. 7, pp. 63-76, 1971.

ON THE APPLICATION OF THE STIFFNESS DERIVATIVE METHOD TO TWO AND THREE DIMENSIONAL FRACTURE PROBLEMS.

M.A. Astiz, M. Elices

E.T.S. Ingenieros de Caminos, Canales y Puertos. Madrid. Spain

SUMMARY

The stress intensity factor for a thumbnail crack in a cylinder is calculated using the stiffness derivative method. Available results for straight cracks are in good agreement with present results. For a circumferentially cracked cylinder the stress intensity factor is calculated using contour - integrals as well as energy release methods, again the present method is accurate enough in spite of using a coarse mesh. The accuracy of the stiffness derivative method is checked on some plane strain cases. Its usefulness for three dimensional problems is discussed.

INTRODUCTION

The finite element method has been widely used to compute stress intensity factors. Nevertheless, its application to fracture problems is not easy because of the stress singularity which arises at the tip of a crack. Basically one can find three approaches to solve this problem.

The first one consists in performing conventional computations and trying to adapt the stress distribution thus obtained to the theoretical crack tip distribution. This is - usually done by representing the product of the stress or displacement normal to the crack with $r^{1/2}$ or $r^{-1/2}$ respectively as a function of the distance to the crack tip, r. The value of the stress intensity factor is extrapolated for $r = 0$. This approach has been used by some authors [1,2] . Its main advantage is that any conventional code can be used; nevertheless, very fine meshes have to be used in order to obtain acceptable results.

Another group of methods which are essentially the same as the previous one are based on defining a special crack tip element whose shape functions are specially suited to repre-- sent the stress singularity. By varying the interpolation - functions the shape of this element and its boundary conditions different approaches have been derived [2 - 7] . This kind of computations give accurate results but the special - element has to be implemented into an existing finite element code and this is not always easy to do.

Finally the third way to determine stress intensity factors consists in computing the energy release rate, G. This can be done by using the well known compliance method [8 - 12] or by computing the J-integral along any contour line enclo-- sing the crack tip [1, 9]. These methods require very little implementation work and give good results for not too fine -- meshes. The best of these methods is probably the stiffness derivative method which was derived by Parks [13] .

This method consists in simulating a small crack length increment, Δl, by translating uniformly all the nodes of the finite element mesh which are interior to a certain line which encloses the crack tip (Fig. 1). Then the energy release rate is computed by applying the equation

$$G \simeq -\frac{1}{2} \{u\}^T \frac{[K]_{a+\Delta l} - [K]_a}{\Delta A} \{u\} \qquad (1)$$

where $\{u\}$ is the vector of nodal displacements ad $[K]$ is the stiffness matrix which is computed for crack lengths a and $a+\Delta l$, and ΔA is the increment of the crack area corresponding to an increment Δl of the crack length. This equation is valid only if no body forces are applied to the nodes which are translated.

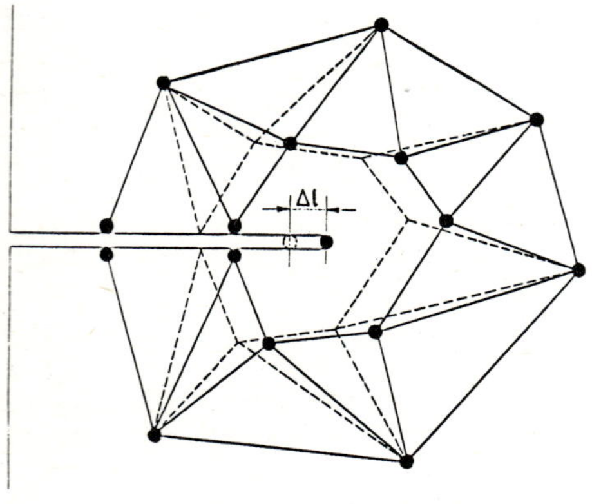

Figure 1
Simulation of a crack advance

This method has two main advantages. In three dimensional problems the energy release rate can be computed at every single point of the crack line instead of getting an average value for the whole crack as it is the case with the compliance method. Moreover it is not necessary to use very fine meshes, as it was the case for the stress limit methods.

According to these considerations the stiffness derivative method seems to be ideally suited for three-dimensional crack problems. We will compare it to other methods in plane problems and then we will apply it to some crack problems in tensioned wires.

AXISYMMETRICAL CRACK PROBLEMS

The circumferentially cracked tensile bar was studied with the finite element mesh shown in figure 2. Only linear quadrilateral and constant strain elements were used. The mesh has 83 nodes and 166 degrees of freedom. It is not a coarse mesh and reasonable results should be obtained either through stress limit methods or through energy methods.

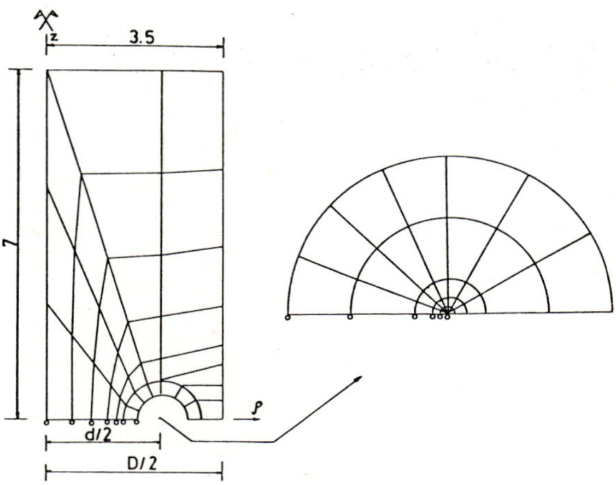

Figure 2
Finite element mesh for axisymmetrical cases

Four crack lengths were studied : d/D = 0.5; 0.55; 0.60; 0.65; where d is the diameter of the crack line and D is the diameter of the bar.

The stress intensity factor was computed through five different ways: the stress limit method, the compliance method, the J and M integrals and the stiffness derivative method.

The stress limit method was used in its two versions: K_I

was computed as the limit of the expressions

$$K_I = \lim_{r \to 0} \sigma_z (2\pi r)^{1/2} \quad (z = 0;\ \rho < d/2) \tag{2}$$

$$K_I = \lim_{r \to 0} \frac{Eu_z}{4(1-\nu^2)} \sqrt{\frac{2\pi}{r}} \quad (z=0;\ \rho > d/2) \tag{3}$$

where the ρ and z directions are defined on figure 2, r is the distance to the crack tip, E is Young's modulus and ν is Poisson's ratio. Equation (3) is only valid for plane strain cases.

The compliance was computed for the four crack lengths previously defined. Then its derivative was obtained at the three intermediate points (d/D = 0.525, 0.575, 0.625) by finite differentiation. Then the energy release rate was obtained by applying the well known equations of the compliance method.

The energy release rate was also computed by using the integral quantities J and M which are defined in axisymmetrical cases by the expressions [14, 15].

$$G = -\frac{J_A}{\pi d} = -\frac{2}{d}\int_C \left(wn_\rho - T_\rho \frac{\delta u_\rho}{\delta \rho} - T_z \frac{\delta u_z}{\delta \rho}\right) \rho \, dl + \frac{2}{d}\int_S \left(-w + T_\theta \frac{u_\rho}{\rho}\right) dS \tag{4}$$

$$G = -\frac{4}{d^2} \int_C \rho (w x_i n_i - T_j u_{j,i} X_i - \frac{1}{2} T_i u_i) \, dl \tag{5}$$

$$i,j = \rho, z$$

where C is any line lying in a meridian plane and enclosing the crack tip, S is the surface enclosed by C, n and T are the normal to C and the tractionvector, w is the strain energy density and θ is the direction perpendicular to the meridian plane.

The contour line, C, was chosen to be the boundary line of the specimen as it can be seen on figure 3. The reason for choosing this line was that the displacements and stresses are better known along this line which is far from the crack tip singularity.

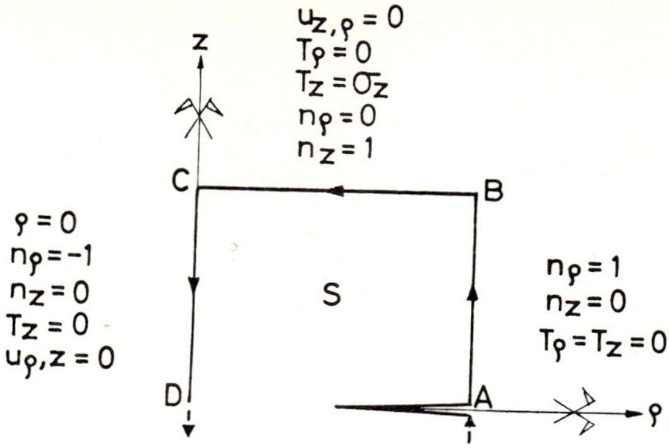

Figure 3
Integration path

Finally the stiffness derivative method was applied by using equation (1) where this matrix product was only performed on the elements whose stiffness matrix was changed by the change in geometry explained on figure 1. The value of Δl was made equal to 10^{-4} times the crack depth as it will be justified later. This translation was applied to the crack tip node and to the first ring of nodes as recommended by Parks [13].

The results are summarized in Table 1 and they are compared in figure 4 to the theoretical solution due to Bueckner [16].

d/D	0.50	0.55	0.60	0.65
stress limit (2)	0.226	0.236	0.252	0.257
displacement limit (3)	0.264	0.280	0.291	0.300
J-integral (4)	0.240	0.257	0.267	0.273
M-integral (5)	0.243	0.257	0.270	0.278
stiffness derivative (1)	0.231	0.242	0.260	0.267
compliance method	0 219	0.232	0.231	

Table 1
Non-dimensional stress intensity factors $\left(\frac{K_I}{\sigma_{NET} (\pi D)^{1/2}} \right)$

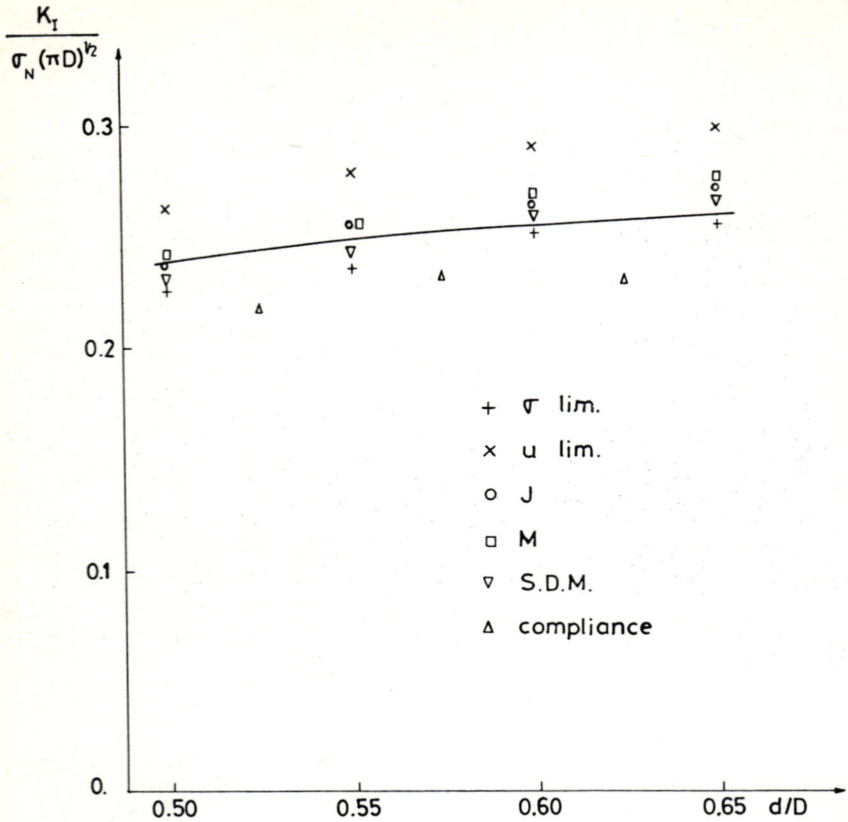

Figure 4
Comparison of stress intensity factors for axisymmetrical cases.

By looking at these results one can appreciate that the most accurate methods seem to be both integral-based methods and the stiffness derivative method for a medium size finite element mesh. The stress intensity results were obtained with less than a 3.5% error. The bad behavior of the displacement limit method is due to the fact that equation (3) is only valid for plane strain cases.

As J and M integral methods can not be applied to three dimensional problems, the stiffness derivative method stays - the best one if we want to compute stress intensity factors at singular points along the crack front in such problems.

THREE DIMENSIONAL CRACK PROBLEMS

The three dimensional problem which will be analyzed is the tensile bar with a straight edge crack or with a semi-elliptical surface crack. Because of the symmetries of the problem, only one fourth of the bar was considered. The nodal points were arranged in several planes parallel to the axis of the bar (see fig. 5). A total of 235 nodes were used. The nodal distribution in each plane was similar to that of figure 2 except that only 47 nodes per plane were used. As this -- mesh is rather coarse to analyze a crack problem, the accuracy of the method was checked on a finite strip with an edge crack and in plane strain conditions. Both problems are similar as it has been pointed out by other authors [11, 17]; the mesh used in the plane strain problem was the same as in any of the planes defined on figure 5.

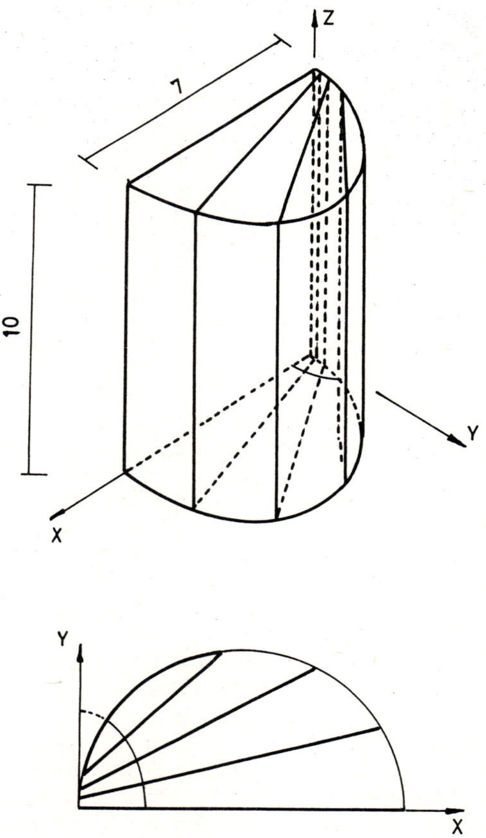

Figure 5
Discretization of the cracked cylinder

In the finite strip problem, the influence of the crack step size, Δl, was studied because this is the most important parameter in the stiffness derivative method. Different values of the crack length, a, and the crack step, Δl, were used to obtain the results which have been represented on figure 6. One can see there that the computed stress intensity factor is slightly variable for small values of Δl but it stays constant (K_{LIM} on fig. 6) over a wide range of values of the crack step size. This is a valuable property of the stiffness derivative method as the results are not really dependent on Δl.

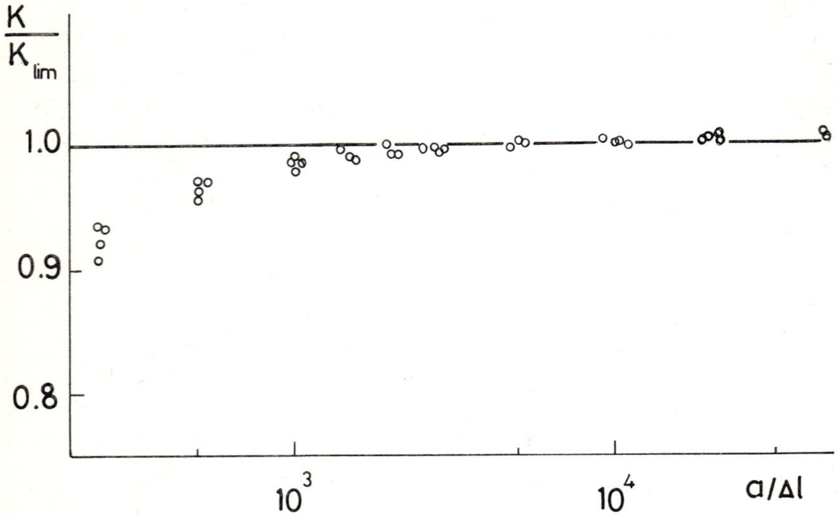

Figure 6
Influence of crack length increment

According to these results, Δl should be less than -- a/2000 to avoid any variation larger than 1% on the computed stress intensity factor. The upper limit for Δl was not studied as it will be fixed mainly by the number of digits used in the computations. In any case, these limitations should also be expressed in terms of the size of the elements whose stiffness is changed by the crack length increment as this is the variable which can directly influenciate the stress intensity factor results according to equation (1). In our case, acceptable values of Δl were found to be less than 1/200 of this element size. The results are summarized in table 2 and compared to the values obtained by Benthem and Koiter [18] . As it is shown in this table the accuracy is better than 8% which is a good result if we take into account that the mesh is very coarse.

a/w	0.057	0.100	0.143	0.200
$K/\sigma\sqrt{a}$ [18]	2.02	2.10	2.22	2.43
$K/\sigma\sqrt{a}$ SDM	1.95	2.00	2.06	2.30
error (%)	3.5	4.8	7.2	5.3

Table - 2
Stress intensity factors for the finite strip

The tensile bar with a straight edge crack has been studied next. This geometry is very interesting as it is used to study fatigue and stress-corrosion problems on high-strength steel wires. The criteria defined on the plane finite strip were applied to this case; the results are summarized on table 3.

a/D	0.057	0.100	0.143	0.200
$K/\sigma\sqrt{a}$	1.90	1.84	1.82	2.00

Table - 3
Stress intensity factors for the
tensile bar with a straight edge crack.

These results are plotted in figure 7 and compared to those obtained by Daoud et al. [11] which give lower values of the stress intensity factor because these authors assume that plane stress conditions are present and that the stress intensity factor is constant along the crack front.

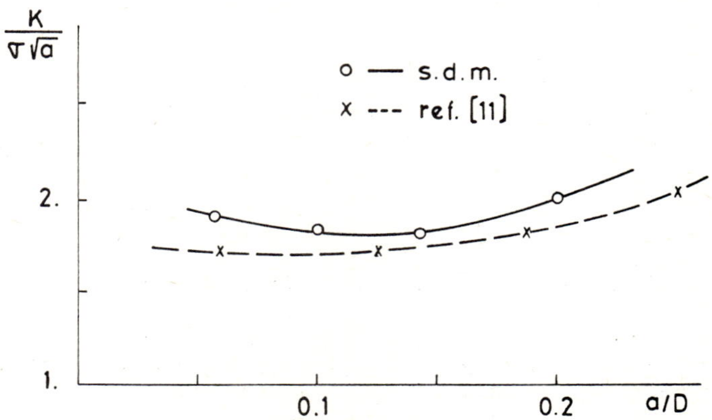

Figure 7
Stress intensity factor for straight edge cracks

The stress intensity factor can be computed by using an interpolation function which has been obtained by the least squares method. This function is

$$\frac{K}{\sigma\sqrt{a}} = 2.156 - 5.836\left(\frac{a}{D}\right) + 25.181\left(\frac{a}{D}\right)^2 \qquad (6)$$

This function gives the value of the stress intensity factor at the center of the crack front and, as it has been already explained it is believed to be accurate within an 8% and it should only be applied for values of a/D comprised between 0.05 and 0.20. This value is the maximum stress intensity factor along the crack front as it has been checked by --- translating not only the nodes which are in the plane of symmetry buy also the nodes situated in the other planes defined on figure 5 according to the scheme of figure 8 where ΔA represents the increment of the crack area which is to be used --- when equation (1) is to be applied.

This idea of studing the variation of the stress intensity factor along the crack front is much more useful when it is applied to semi-elliptical edge cracks because for this -- kind of cracks the maximum stress intensity factor may not necessarily be found at the plane of symmetry. This kind of - research is very useful to study fatigue crack propagation -- phenomena in wires.

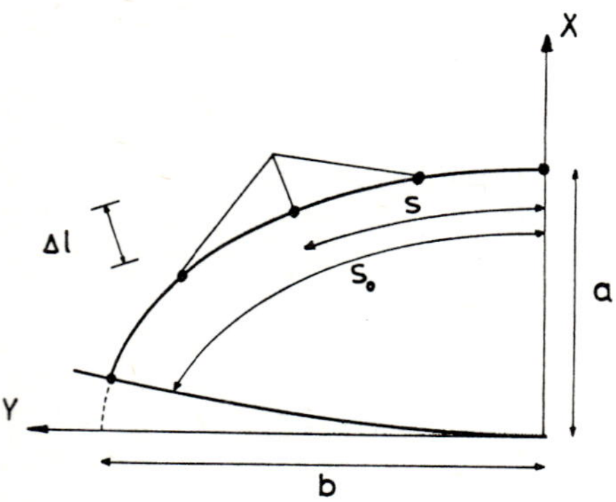

Figure 8
Crack advance for semi-elliptical border

The semi-elliptical crack is defined by the parameters a and b (see figure 8). Each of the nodes of the crack front has been translated along with the corresponding first ring of nodes in a direction perpendicular to the crack border. The value of Δl was chosen to be the same as in the previous cases.

The crack shapes which have been studied are defined on table 4 as well as the corresponding crack intensity factors as a function of the arc-length distance to the plane of symmetry. These crack shapes correspond to actual values obtained from experimental fatigue and stress corrosion tests. For computing K values, it has been supposed that a state of plane strain is present at every point of the crack border except at the node which is on the surface of the bar in which we suppose that we have a state of plane stress. This is an hypothesis which has not relevant influence on the final results as it has been shown earlier [14].

Diameter=7		$K/\sigma\sqrt{a}$				
a	b	S/S_o=0	S/S_o=1/4	S/S_o=1/2	S/S_o=3/4	S/S_o=1
0.4	0.267	1.02	1.06	1.15	1.24	1.32
0.4	0.500	1.36	1.35	1.31	1.27	1.24
0.4	0.615	1.49	1.47	1.41	1.29	1.23
0.4	0.800	1.63	1.61	1.53	1.34	1.21
0.7	0.467	1.08	1.13	1.21	1.31	1.38
0.7	0.875	1.36	1.35	1.31	1.25	1.30
0.7	1.077	1.48	1.47	1.40	1.26	1.32
0.7	1.400	1.60	1.59	1.51	1.29	1.33
1.0	0.667	1.14	1.19	1.30	1.41	1.48
1.0	1.250	1.38	1.36	1.31	1.22	1.39
1.0	1.538	1.49	1.47	1.40	1.21	1.42
1.0	2.000	1.60	1.59	1.49	1.23	1.43

Table - 4
Stress intensity factor for semi-elliptical cracks

The stress intensity factor values at the plane of symmetry have been represented on figure 9 as well as the results corresponding to the straight edge crack (a/b = 0). As it can be observed on this figure, for small cracks (a/D less than 0.15) the non dimensional stress intensity factor, $K/\sigma\sqrt{a}$, can be thought to be a function of a/b (this is only valid if the ellipse is "wide" = a/b less than 1).

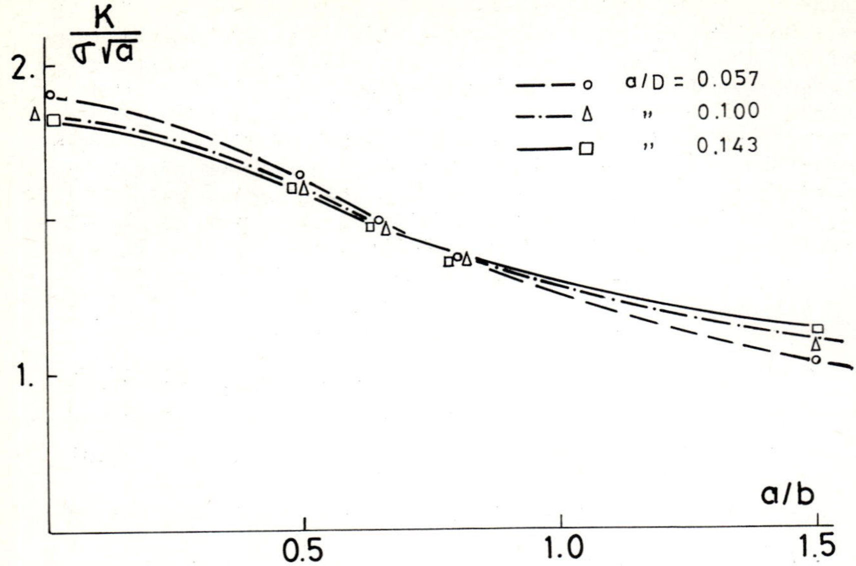

Figure 9
Stress intensity factor for semi-elliptical cracks.

Another important conclusion which can be drawn from the results of table 4 is that one can find crack shapes for which the maximum stress intensity factor along the crack front is not found at the center of the crack border. Furthermore it seems possible to determine through an iterative procedure the crack shapes which correspond to a constant stress intensity factor along the crack border. These crack shapes will be very useful to study fatigue crack propagation as they represent more realistic situations.

The stiffness derivative method has been used to derive some stress intensity factor values in two and three dimensional problems. This method is easy to implement in any finite element computer code. According to the results of this work, acceptable results can be found with coarse finite element meshes. This method is particularly recommended for studying three dimensional crack problems because it permits to calculate the crack intensity factor values at points of the crack border instead of having an averaged value along the crack.

REFERENCES

1. CHAN S.K., TUBA I.S. & WILSON W.K., On the Finite Element Method in Linear Fracture Mechanics, Eng. Fracture Mechanics, Vol. 2, pp. 1-17, 1970.

2. WILSON W.K., Finite Element Methods for Elastic Bodies Containing Cracks, Methods of Analysis and Solutions of Crack Problems, Ed. Sih G.C. Noordhoff, 1973

3. MORLEY L.S.D., A Modification of the Rayleigh-Ritz Method for Stress Concentration Problems in Elastostatics, J. of the Mechanics and Physics of Solids, Vol. 17, pp. 73, 1969.

4. WALSH P.F., The Computation of Stress Intensity Factors by a Special Finite Element Technique, Int. J. of Solids and Structures, Vol. 7, p. 1333, 1971.

5. TRACEY D.M., Finite Elements for Determination of Crack Tip Elastic Stress Intensity Factors, Eng. Fracture Mechanics, Vol. 3, pp. 255-265, 1971.

6. HILTON P.D. & SIH G.C., Application of the Finite Element Method to the Calculation of Stress Intensity Factors, Methods of Analysis and Solutions of Crack Problems, ed. Sih G.C. Noordhoff, 1973.

7. PAPAÏOANNOU S.G., HILTON P.D. & LUCAS R.A., A Finite Element Method for Calculating Stress Intensity Factors and its Application to Composites, Eng. Fracture Mechanics, Vol. 6, pp. 807-823, 1974.

8. MOWBRAY D.F., A Note on the Finite Element Method in Linear Fracture Mechanics, Eng. Fracture Mechanics, Vol.2, pp. 173-176, 1970.

9. HAYES D.J., Some Applications of Elastic-Plastic Analysis to Fracture Mechanics, Ph.D.Thesis, Imperial College of Science and Technology, London, 1970.

10. SUNDSTROM D., An Energy Condition for Initiation of Interfacial Microcracks at Inclusions, Eng. Fracture Mechanics, Vol. 6, pp. 483, 1974.

11. DAOUD O.E.K., CARTWRIGHT D.J. & CARNEY M., Strain Energy Release Rate for a Single-Edge-Cracked Circular Bar in Tension, J. of Strain Analysis, Vol. 13, pp. 83-89, 1978.

12. AAMODT B. & KLEM F., Application of Numerical Techniques in Practical Fracture Mechanics, Fracture Mechanics in Engineering Practice, ed. Stanley P., Applied Science Publishers, 1977.

13. PARKS D.M., A Stiffness Derivative Finite Element Technique for Determination of Crack Tip Stress Intensity Factors, Int. J. of Fracture, Vol. 10, pp. 487-502, 1974.

14. ASTIZ M.A., Ph.D.Thesis, Universidad Politécnica de Madrid, 1976.

15. ASTIZ M.A., ELICES M., SANCHEZ GALVEZ V., On Energy Release Rates in Axisymmetric Problems, Proc. 4th. Int. Conf. on Fracture, ed. Taplin D.M.R., University of Waterloo Press, Vol. 3 pp. 395-399, 1977.

16. BUECKNER H.F., Discussion, Fracture Toughness Testing, ASTM STP n°381, p. 82, 1964.

17. HARKEGAARD G., An Experimental Study of the Influence of Inclusions on the Fatigue Properties of Steel, Eng. Fracture Mechanics, Vol. 6 p. 795, 1974.

18. BENTHEM J.P. & KOITER W.T., Asymptotic Approximations to Crack Problems, Methods of Analysis and Solutions of -- Crack Problems, ed. Sih G.C., Noordhoff, 1973.

SOLUTION OF THREE DIMENSIONAL CRACK PROBLEMS USING THE BOUNDARY INTEGRAL EQUATION METHOD

J. HELIOT[1], R. LABBENS[1] and A. PELLISSIER-TANON[2]

(1) Creusot-Loire, Paris
(2) Framatome, Courbevoie

SUMMARY

The calculation of reliable stress intensity factors for three dimensional problems and any stress gradient is an industrial necessity. With the stress represented on the crack by a polynomial, influence functions depending only on the geometry, corresponding to each term of the polynomial are calculated.

The Boundary Integral Equation Method was used for the computation with an accuracy ± 5%.

The method was applied to semi elliptical cracks in cylinders and quarter circular cracks in nozzle corners. In this paper, the results are discussed and compared to the requirements of the ASME Code.

NOMENCLATURE

a, c	axes of an elliptical crack
b	radius of a circular crack
R_i	internal radius of a cylinder
t	reference length, thickness of a cylinder
x, y	coordinates in the plane of the crack
ϕ	angular parameter on an ellipse
θ	angle from the internal surface of the vessel
$K(\phi), K(\theta)$	stress intensity factor
$h_j(\phi), h_{ij}(\theta)$	polynomial influence functions
r	short distance along a normal to the crack front
E	Young's modulus
ν	Poisson's ratio
$E(k)$	complete elliptical integral of the second kind
Q	shape factor
p	internal pressure
σ_m, σ_b	membrane, bending stress
M_m, M_b	membrane, bending factor of the ASME Code

I - INTRODUCTION

Many industrial crack problems are three dimensional and the knowledge of the variation of the stress intensity factor along the front is necessary. Any stress intensity factor problem can be reduced to the equivalent problem of a loaded crack, but there is nothing such as a Saint Venant principle for reducing a given crack load to a simpler load ; the consequences of a modification of the load or the boundary conditions can difficultly be foreseen. A solution as exact as possible, with stresses variable on the crack area, is not an academic problem but an industrial necessity. This paper reports on calculation methods which were initiated for large vessels for chemical plants and developed mainly for nuclear vessels.

II - WEIGHT AND INFLUENCE FUNCTIONS

Closed form solutions for the stress intensity factor are known for embedded elliptical or circular cracks in an infinite solid [1,2]. Most three dimensional problems can have only a numerical solution.

The weight function concept, developed by Bueckner [3] and by Rice [4] was found of major interest since when it is known the calculation of the stress intensity factor is reduced to a simple integral. However the only closed form three dimensional weight functions are for a penny-shaped and a semi infinite crack in an infinite solid [5,2].

The finite element computation of these functions for two dimensional and axisymmetric problems raised no difficult problem[6,7]. For three dimensional problems, on a rather simple problem, the authors found that the computer time, in spite of possible improvements would be too long [8]. For this reason, the generality of the problem was reduced to influence functions related to a polynomial expression of the applied stress in one or two coordinates, and the possibilities of the Boundary Integral Equation (B.I.E.) method were investigated because this method reduces three dimensional elasticity problems to two dimensional problems on the boundary.

The method and the program were validated by a careful application to the algebraically solved problem of a penny shaped crack subjected to stresses in x^0, x, x^2, x^3, x^4, and applied to semi elliptical cracks in the meridional plane of a cylinder and quarter circular cracks in a nozzle corner (fig. 1, 2).

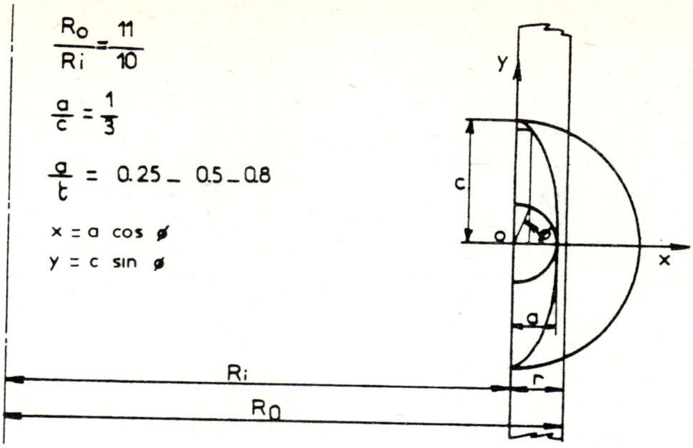

Fig. 1 - Semi elliptical crack in a cylinder

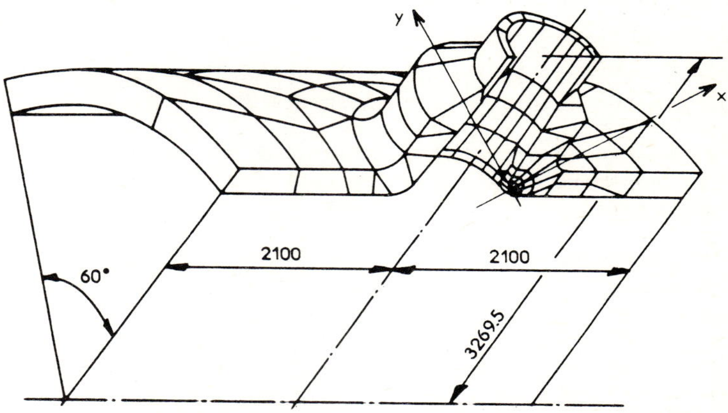

Fig. 2 - Quarter circular crack in a nozzle corner

For the first problem, the applied stress is defined as

$$\sigma\ (x) = \sigma_0 + \sigma_1 \frac{x}{t} + \sigma_2 \left(\frac{x}{t}\right)^2 + \sigma_3 \left(\frac{x}{t}\right)^3 + \sigma_4 \left(\frac{x}{t}\right)^4 \qquad (1)$$

and the polynomial influence functions are

$$h_j\ (\phi) = \frac{K_j\ (\phi)}{\sigma_j \left(\frac{a}{t}\right)^j \sqrt{\pi a}} \frac{E\ (k)}{\left(\cos^2 \phi + \frac{a^2}{c^2} \sin^2 \phi\right)^{\frac{1}{4}}} \qquad (2)$$

For the nozzle crack

$$\sigma\ (x,y) = \sum_{i,j} \sigma_{ij} \left(\frac{x}{t}\right)^i \left(\frac{y}{t}\right)^j \qquad (3)$$

and

$$h_{ij}\ (\theta) = \frac{\pi}{2} \frac{K_{ij}\ (\theta)}{\sigma_{ij} \left(\frac{b}{t}\right)^{i+j} \sqrt{\pi b}} \qquad (4)$$

III - B.I.E. METHOD - VALIDATION OF THE PROGRAM

The mathematics behind the B.I.E. method are not new ; they were rather recently well documented by Rizzo [9] and extended to crack problems by Cruse [10]. It is only recalled that the displacements and stresses are derived from two tensor kernels calculated on the boundary. Only the surface is meshed, and it is expected that although the matrix is complete and not symmetrical, the computer time can be significantly reduced with respect to finite elements.

The program Equations Integrales Tridimensionnelles (E.I.T.D.) was developed in France by the Centre d'Etudes Techniques des Industries de la Mécanique. It was described in [11,12] and presented in 1978 at the First International Conference on Numerical Methods in Fracture Mechanics [12].

Stress intensity factors were calculated by the extrapolation in a plane normal to the crack front of $\sqrt{\frac{\pi E}{1-\nu^2}}\,\frac{v}{2\sqrt{r}}$ on the crack surface. As the program can handle five loads it was convenient to calculate five influence functions in one run.

It was thought that for validating the program errors could be not more than a few per cent for a simple problem such as the embedded penny shaped crack.

The infinite solid was modelised by a cylinder with a radius four times the radius of the crack and a certain number of elements and grids were tested (fig. 3) with loads

$\sigma(x) = x^0, |x|, x^2, |x^3|, x^4$;

the influence functions are then

$$h_j(\phi) = \frac{\pi}{2}\,\frac{K'_j(\phi)}{a^j\sqrt{\pi a}} \qquad (5)$$

Using the weight function of the penny-shaped crack [5], these influence functions can easily be computed or algebraically calculated and compared with the results of the B.I.E. computation.

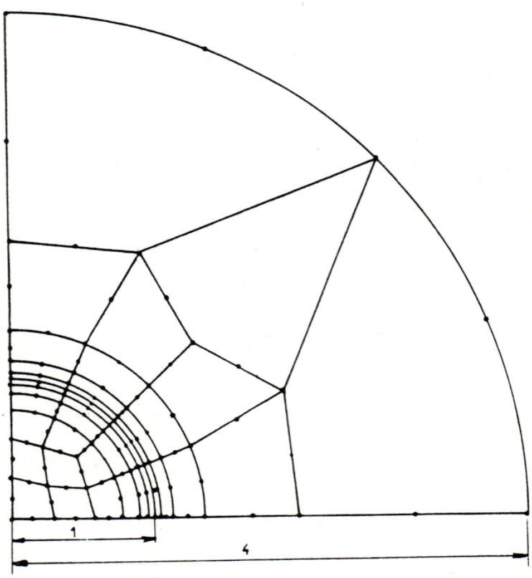

Fig. 3 - Grid of a penny-shaped crack

Isoparametric elements were used, with a parabolic interpolation of the stresses on the surface of the elements. The accuracy was satisfactory only when quarter point elements were used on the crack front, as they had already been used by other authors with finite element.

The accuracy was sensitive to the radial dimension of the crack front elements, or more precisely to relative load variation in these elements ; it did not depend much on the length of the front elements. A correlation was found between the error and the load variation in this element (fig. 4).

The depth of the front elements was limited so that the errors were not more than - 1 % for $h_0(\phi)$ and - 4 % for $h_4(\phi)$.

The results were later confirmed by tests on embedded elliptical cracks.

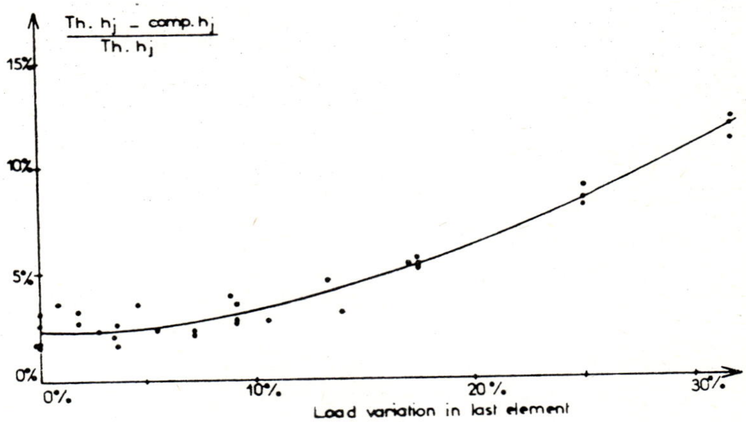

Fig. 4 - Curve of errors

IV - SEMI ELLIPTICAL CRACKS IN CYLINDERS

4,1 - This study was applied to nuclear vessels with $R_i/t = 10$ (fig. 1), axes ratio a/c = 1/3 (A.S.M.E. Code), and also a/c = 1/10, depth ratios a/t = 0.25 - 0.5 - 0.8.

The results are on fig. 5, 6, 7. They have already been published in [13,14]. The cracks a/c = 1/3 were studied by other authors and a very good coïncidence was found with [15,16,17].

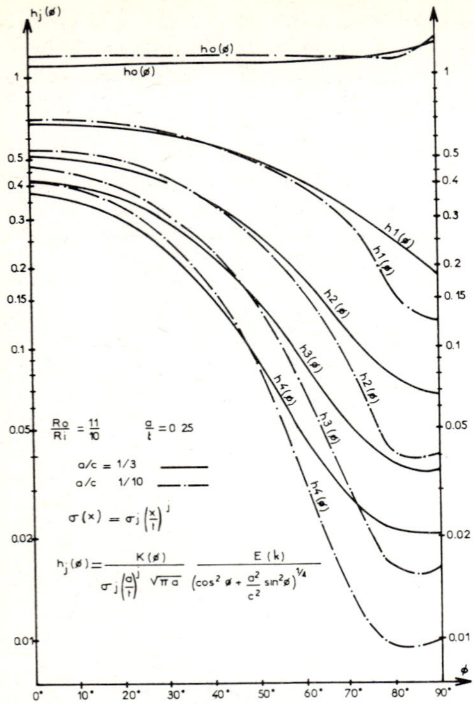

Fig. 5 - Influence fonctions

$\frac{R_i}{t} = 10 \; ; \; \frac{a}{c} = 0{,}25$

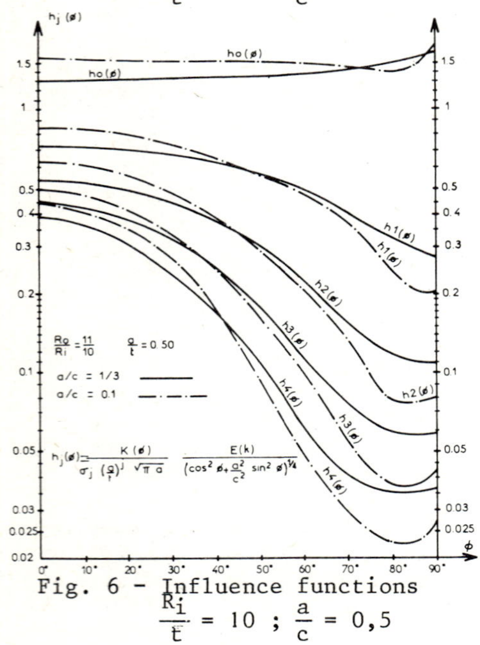

Fig. 6 - Influence functions

$\frac{R_i}{t} = 10 \; ; \; \frac{a}{c} = 0{,}5$

4,2 - It is important to see how the functions vary with a/c.

For the deep cracks (a/t = 0.8), with some uncertainty for h_1 and h_2 near the surface (ϕ=90°) the five functions are higher for a/c = 1/10 than for a/c = 1/3. For the shallow and the medium cracks they are higher near the deepest point (ϕ=0°) and lower near the surface.

All the functions are sensitive to the axes ratio, and the more the crack is deeper. Near the surface (ϕ=90°) the differences are not important for h0 (90°), but they are for the functions of higher order ; but as the values of these functions are low, and in K calculations are multiplied by $(a/t)^j$ these differences have no important consequences.

Let us consider two cracks of some depth with a/c = 1/3 and 1/10, subjected to uniform traction and bending (σ_m and σ_b > 0)

$$\sigma(x) = \sigma_m + \sigma_b \left(1 - \frac{2x}{t}\right)$$

$$= \sigma_m + \sigma_b - 2\sigma_b \frac{x}{t}$$

The stress intensity factor is given by

$$\frac{K(\phi)}{\sqrt{\pi a}} = \frac{\left(\cos^2 \phi + \frac{a^2}{c^2} \sin^2 \phi\right)^{\frac{1}{4}}}{E\ (k)} \left[(\sigma_m + \sigma_b)\ h_0\ (\phi) - 2\ \frac{a}{t}\ \sigma_b\ h_1\ (\phi)\right]$$

A numerical application for a/t = 0.5 yields

	$\frac{K(0)}{\sqrt{\pi a}}$	$\frac{K(90)}{\sqrt{\pi a}}$
a/c = 1/3	$1{,}15\ (\sigma_m+\sigma_b) - 0{,}66\ \sigma_b$ $= 1{,}15\ \sigma_m + 0{,}49\ \sigma_b$	$0{,}84\ (\sigma_m+\sigma_b) - 0{,}14\ \sigma_b$ $= 0{,}84\ \sigma_m + 0{,}7\ \sigma_b$
a/c = 1/10	$1{,}55\ (\sigma_m+\sigma_b) - 0{,}93\ \sigma_b$ $= 1{,}55\ \sigma_m + 0{,}62\ \sigma_b$	$0{,}53\ (\sigma_m+\sigma_b) - 0{,}06\ \sigma_b$ $= 0{,}53\ \sigma_m + 0{,}47\ \sigma_b$

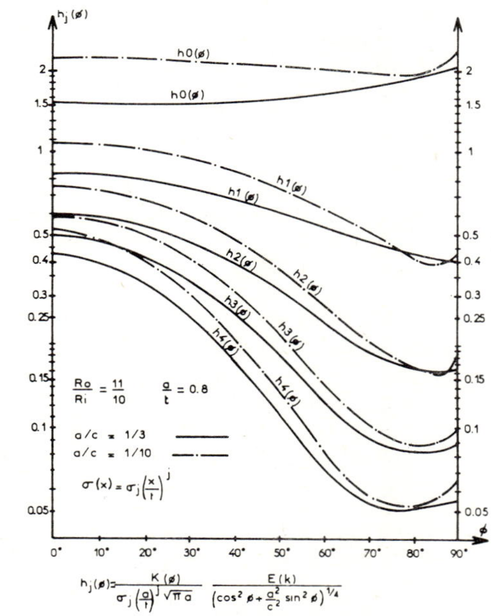

Fig. 7 - Influence functions

$$\frac{R_i}{t} = 10\ ;\ \frac{a}{c} = 0{,}8$$

If σ_b is sufficiently high, the crack a/c = 1/3 can be critical at the surface and not at the deepest point, and extend along the surface with a decrease of K at the surface and an increase at the deepest point. The crack may arrest as long as the stress intensity factor is not critical at the deepest point. This was confirmed by thermal shock experimentations on glass and steel cylinders [18,19].

4,3 - The influence functions for very long axial cracks (a/c = 0) are known [6]. The comparison for R_i/t = 10 at the deepest point is as follows

	$h_0(0)$	$h_1(0)$	$h_2(0)$	$h_3(0)$	$h_4(0)$
a/t=0.25					
a/c=0	1,451	0,814	0,600	0,491	0,423
a/c=1/10	1,179	0,705	0,548	0,467	0,418
a/c=1/3	1,112	0,681	0,524	0,436	0,377
a/t=0.5					
a/c=0	2,483	1,200	0,812	0,628	0,522
a/c=1/10	1,571	0,844	0,625	0,518	0,455
a/c=1/3	1,278	0,726	0,544	0,447	0,384
a/t=0.8					
a/c=0	5,749	2,372	1,436	1,024	0,798
a/c=1/10	2,269	1,092	0,763	0,608	0,519
a/c=1/3	1,516	0,839	0,615	0,498	0,425

Assuming that a slender semi elliptical crack (a/c = 1/10) is practically a long axial crack results in an overestimation, mainly for the lower orders and the deeper cracks, more than could be expected by the variation of E(k) which is respectively

$$E(1) = 1 \qquad E(0{,}995) = 1{,}016 \qquad E(0{,}943) = 1{,}113$$

4,4 - The influence function were also calculated for a semi elliptical crack (a/c=0.5) in a plate for the Benchmark Problem N° 1 [20].

Newman and Raju [16] published results for two opposed semi elliptical cracks in pressurized cylinders, with an approximate formula for all crack shapes $R_i/t > 4$, $0{,}2 < a/c < 1$, $a/t < 0.8$. Differences lower than ± 5 %, therefore not significant, were found for a/c = 1/3 ; but large differences were found for a/c = 0.1, which is beyong the scope of Newman's formulas.

The comparison of the results for the Benchmark Problem with this formula for R_i/t = 10 , a/c = 0.5, evidenced no difference higher than ± 5 % ; the conclusion was that these results could be used for the studied cylinders.

V - QUARTER CIRCULAR CRACKS IN A NOZZLE CORNER

5,1 - The Benchmark Problem N° 2 was studied before these cracks ; the results were compared to those of Palusamy and Raymund [21] and were found consistent.

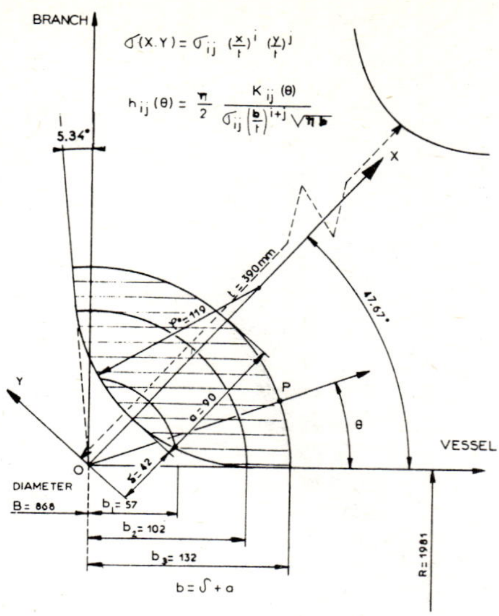

Fig. 8 - Nozzle Corner

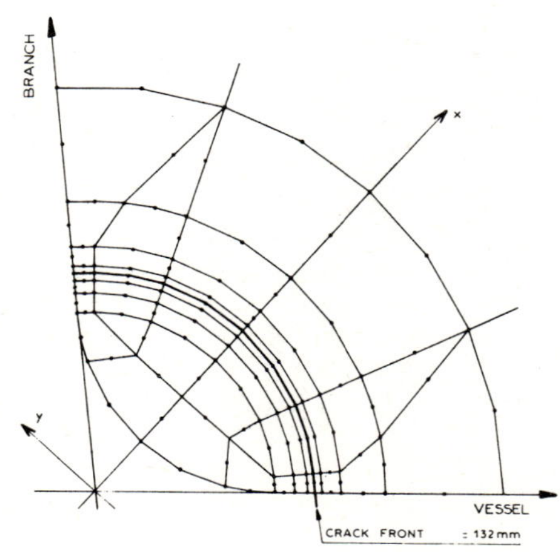

Fig. 9 - Mesh in the plane of the crack

The geometry of the nozzle, as on fig. 2, is detailed on fig. 8 and 9. Three radii were investigated b = 57, 102, 132 mm. The curves of the functions $h_{00}(\theta)$ and $h_{10}(\theta)$ are on fig. 10. The difference between the curve for the shallow crack and the others must be noted; it is surely due to the fact that the front of the shallow crack intersects the surface in the rounded part of the corner.

5,2 - The hoop stress resulting from pressure was represented on the area of the three cracks by a polynomial

$$\frac{\sigma_h(x,y)}{p} = \alpha_{00} + \alpha_{10}\,\frac{x}{t} + \alpha_{01}\,\frac{y}{t} + \alpha_{02}\left(\frac{x^2}{t}\right) + \alpha_{02}\left(\frac{y}{t}\right)^2$$

t is arbitrary and is chosen equal to the distance from the origin of the coordinates to the external surface on $0x$ (fig. 8), or 0,390 m. The following α_{ij} yield an approximation better than 2 % on the area of the three cracks

$$\alpha_{00} = 27,583$$
$$\alpha_{10} = 32,331$$
$$\alpha_{01} = 1,911$$
$$\alpha_{20} = 10,708$$
$$\alpha_{02} = 10,084$$

On account of the nearly symmetry α_{01} is small ; the higher order $h_{ij}(\theta)$ are always rather low and in the calculation of the stress intensity factor α_{20} and α_{02} are multiplied by $(b/t)^2$.

The two first terms were thought sufficient

$$\frac{K\ (\theta)}{\frac{2}{\pi}\ p\ \sqrt{\pi b}} = \alpha_{00}\ h_{00}\ (\theta) + \alpha_{10}\ \frac{b}{t}\ h_{10}\ (\theta)$$

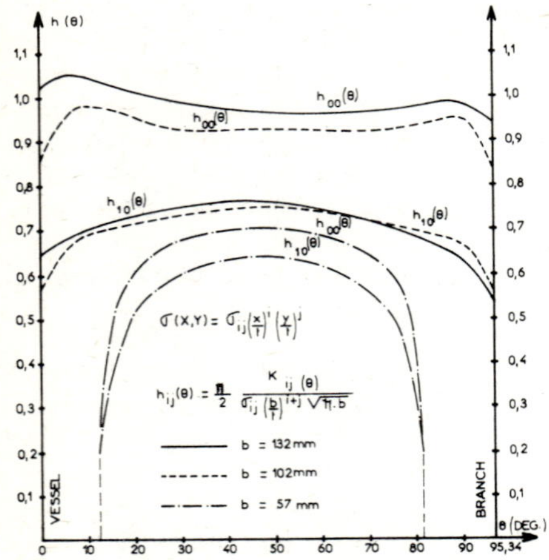

Fig. 10 - Influence functions in a nozzle corner

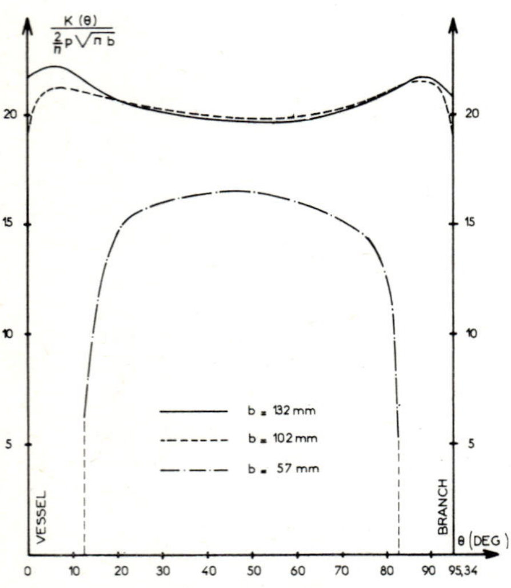

Fig. 11 - Non dimensional stress intensity factors for pressure

These non dimensional stress intensity factors are on fig. 11.

5,3 - The three curves of fig. 11 are nearly symmetrical.

The shallow crack exhibits a quick drop near the surface ; this is not surprising since the crack lies entirely in the rounded area.

With the two other cracks the non dimensional stress intensity factor is practically the same ; the differences near the surface have not a great practical importance, but they may be significant because the medium crack intersects the surface just at the limit of the rounded corner, and the deepest crack in the straight surfaces.

A tendency to a faster increase in the cylinder than in the branch appears, although only two calculations do not allow a firm conclusion. However, this is not unlikely since if the crack increases it will meet in the vessel a hoop stress near the membrane stress, while in the branch the reinforcement results in a lower stress.

Such an extension in the vessel wall was observed by Smith [22] in photoelasticimetric experimentations. A reasoning similar to that of par. 4,2 suggests that the crack may arrest since the extension results in a decrease of the stress intensity factor on the surface.

VI - COMPARISON WITH THE REQUIREMENTS OF THE A.S.M.E. CODE

The non mandatory requirements of the Appendix A of the Section XI of the A.S.M.E. Code result from considerations developed in the Welding Research Council Bulletin 1975 [23].

With the notations of the A.S.M.E. Code and of this study the stress intensity factors for an uniform tension

$$K_m\ (0) = \sigma_m \sqrt{\frac{\pi a}{Q}}\, M_m = \frac{\sigma_m \sqrt{\pi a}}{E(k)}\, h_0\ (0)$$

for a pure bending

$$K_b\ (0) = \sigma_b \sqrt{\frac{\pi a}{Q}}\, M_b = \frac{\sigma_b \sqrt{\pi a}}{E(k)} \left[h_0(0) - 2\,\frac{a}{t}\, h_1(0)\right]$$

The shape factor Q takes into account the plastic zone correction by

$$Q = E^2\ (k) - 0{,}21\ \left(\frac{\sigma_m}{\sigma_Y}\right)^2$$

σ_m/σ_Y is near 0,6 ; the difference between $\sqrt{Q}$ and $E(k)$ is about 4 % and will be neglected, so that

$$M_m = h_0\ (0)$$

$$M_b = h_0\ (0) - 2\,\frac{a}{t}\, h_1\ (0)$$

The comparison is on fig. 15 and 16.

It seems that M_m and M_b were derived from results obtained by Wilhem [24] and F.W. Smith [25] for long edge cracks in plates which were corrected for semi ellipses by $\sqrt{Q}$ or $E(k)$. It must be known that in 1972 no accurate computation had been performed on cylinders and ellipses, and the curves for M_m and M_b were very useful.

It appears now that two factors of overestimation exist in M_m and M_b, the calculation for a plate and not a cylinder which for t/R_i = 0.1 may have some importance only for deep cracks and the correction for a semi ellipse by $E(k)$ which as already seen in par. 4,3 is underestimated.

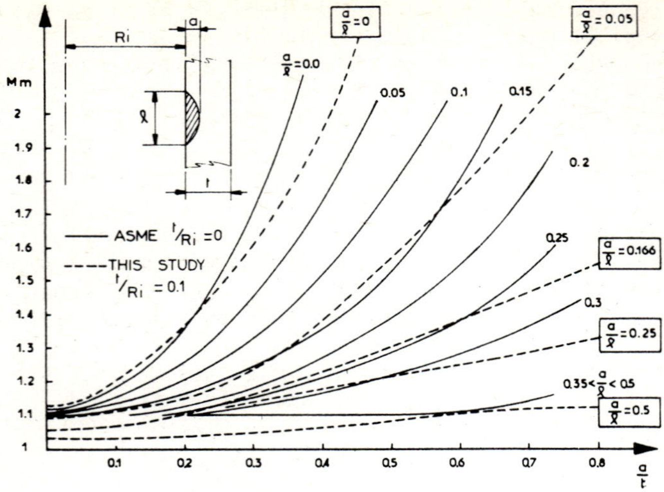

Fig. 12 - Comparison with the ASME Code - Membrane

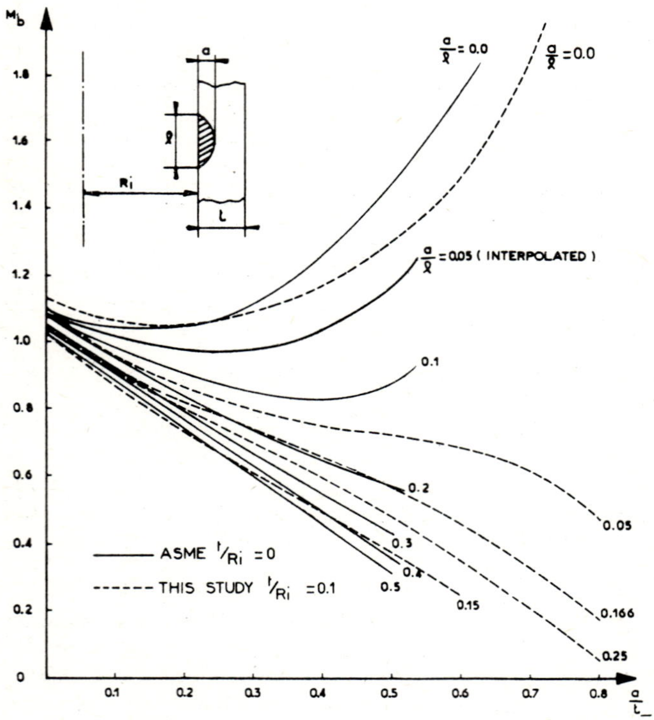

Fig. 13 - Comparison with the ASME Code - Bending

On fig. 12 and 13, it appears that the overestimation is never more than 10 % for a/t < 0.25, which is of no practical importance, but may be much more for deeper cracks.

In many problems, a two terms approximation is sufficient; the method developed in this study allow the solution of problems with more sophisticated gradients.

VII - CONCLUSION

A method has been developed for the industrial calculation of stress intensity factors in three dimensional problems. It was validated by an accurate comparison with algebraically calculated solutions and was applied to semi elliptical cracks in cylinders and quarter circular cracks in nozzle corners subjected to any stress gradient.

The expected accuracy ± 5 % was confirmed by comparison with other solutions.

A comparison with the Section XI of the A.S.M.E. Code evidences that the curves of the Code are overestimated by no more than 10 % for a/t ⩽ 0.25, but may be much more overestimated for deeper cracks.

REFERENCES

1. SHAH, R.C. and KOBAYASHI, A.S. - Stress Intensity Factor for an Elliptical Crack under Arbitrary Normal Loading. Engineering Fracture Mechanics, Vol. 3, pp. 71-96, 1971.

2. KASSIR, M.K. and SIH, G.C. - Three dimensional Crack Problems - Noordhoff, 1975.

3. BUECKNER, H.F. - Field Singularities and Related Integral Representations. in Methods of Analysis and Solutions of Crack Problems, ch. V, Noordhoff, 1973.

4. RICE, J.R. - Some Remarks on Elastic Crack Tip Stress Fields, International Journal of Solids and Structures, 1972, p. 751.

5. TADA, H., PARIS, P.C. and IRWIN, G.R. - The Stress Analysis of Cracks Handbook, Del Research Corporation, Hellertown, Pa, 1973.

6. LABBENS, R., PELLISSIER-TANON, A. and HELIOT, J. - Practical Method for Calculating Stress Intensity Factors through Weight Functions - Mechanics of Crack Growth, ASTM-STP 590, pp. 368,383, 1976

7. PARIS, P.C., Mc MEEKING, R.M. and TADA, H. - The Weight Function Method for Determining Stress Intensity Factors, Cracks and Fracture, ASTM-STP 601, pp. 471,483, 1976.

8. LABBENS, R., HELIOT, J. and PELLISSIER-TANON, A. - Weight Functions for Three Dimensional Symmetrical Crack Problems, Cracks and Fracture, ASMT-STP 601, pp. 448,467, 1976.

9. RIZZO, F.J. - An Integral Equation Approach to Boundary Value Problems of Classical Elastostatics, Quarterly of Applied Mathematics, pp. 83,95, April 1967.

10. CRUSE, T.A. - Application of the Boundary Integral Equation Method to Three Dimensional Stress Analysis, Computer and Structures, pp. 509,527, 1973.

11. LACHAT, J.C. and WATSON, J.O. - A Second Generation Boundary Integral Equation Program for Three Dimensional Elastic Analysis, Symposium on Boundary Integral Equations, American Society of Mechanical Engineers, June 1975.

12. LANGE, D.- 3.D Fracture Analysis using the Boundary Integral Equations Method - Numerical Methods in Fracture Mechanics, Jan. 1978.

13. HELIOT, J., LABBENS, R. and PELLISSIER-TANON, A. - Semi Elliptical Cracks in a Cylinder Subjected to Stress Gradients, in Fracture Mechanics, ASTM-STP 677, pp. 341, 364, 1979.

14. HELIOT, J., LABBENS, R. and PELLISSIER-TANON, A. - Application of the Boundary Integral Equation Method to Three Dimensional Crack Problems, ASME Pressure Vessel and Piping Conference, San Francisco, Aug. 1980.

15. Mc GOWAN, J.J. and RAYMUND, M. - Stress Intensity Factors Solutions for Internal Longitudinal Semi Elliptical Flaws in a Cylinder under Arbitrary Loadings, in Fracture Mechanics, ASTM-STP 677, pp. 365,380, 1979.

16. NEWMAN, J.C. and RAJU, I.S. - Stress Intensity Factors for Internal Surface Crack in Cylindrical Pressure Vessels, NASA Technical Memorandum 80073, 1979, Langley Research Center, Hampton, Va, 23665.

17. ATLURI, S.N. and KATHIRESAN, K. - Influence of Flaw Shapes on Stress Intensity Factors for Pressure Vessels Surface Flaws and Nozzle Corners, ASME Paper 79-PVP-65.

18. BLAUEL, J.G., KALTHOFF, J.K. and STAHN, D. - Model Experiments for Thermal Shocks Fracture Behavior, Journal of Engineering Materials and Technology, 1975.

19. CHEVERTON, R.D. - Pressure Vessel Fracture Studies Pertaining to a PWR LOCA-ECC Thermal Shock - Oak Ridge National Laboratory Report NUREG/TM-31, 1976.

20. HELIOT., LABBENS, R., and PELLISSIER-TANON, A. - Results for Benchamrk Problem N° 1 - International Journal of Fracture, pp. 197,202, Dec. 1979.

21. PALUSAMY, S.S., RAYMUND, M. and PETSCHE, J.F. - Stress Intensity Factors for a Corner Crack at the Edge of a Hole in a Plate, to be presented at the XIIIth National Symposium on Fracture Mechanics, Philadelphia, June 1980.

22. SMITH, C.W., JOLLES, M. and PETERS, W.H. - Stress Intensity Factors for Nozzle Cracks in Reactor Vessels, Virginia Polytechnic Institute and State University, Report VPI-E-76-25, Blacksburg, Va, 1976.

23. PVRC Recommendations on Toughness Requirements for Ferritic Materials, Welding Research Council Bulletin 175, Aug. 1972.

24. WILHEM, D.P. - Fracture Mechanics Guideline for Aircraft Structual Applications, AFFDL-Technical Report 69-111, Wright Patterson Air Force Base, Ohio.

25. SMITH, F.W. - Stress Intensity Factors for a Semi Elliptical Surface Flaw in a Plate, The Boeing Company, 1966.

APPLICATION OF THE BOUNDARY INTEGRAL EQUATION METHOD TO SOME FRACTURE PROBLEMS IN HIGH PRESSURE ENGINEERING

R.T. FENNER* and C.L. TAN†

SUMMARY

The boundary integral equation method for three-dimensional elastic stress analysis is applied to high pressure block fittings containing T-bores, both in the unflawed state and when cracks are present. Stress concentration factor and stress intensity factor solutions are obtained for both internal hydrostatic pressure acting in the bores, and the block subjected to external tension. The results presented should be of considerable practical use in the design of such high pressure engineering components. The paper also serves to illustrate the potential of the method of analysis and its advantages over other numerical techniques, such as the finite element method, for three-dimensional elastostatics. These advantages are particularly clear when the component concerned has a low surface-to-volume ratio and when large stress gradients are involved.

INTRODUCTION

The accurate prediction of stresses and strains in high pressure engineering components such as thick-walled cylinders with cross bores, and block fittings which are subjected to complex loading conditions, presents formidable analytical difficulties. The presence of cracks in these components makes the analytical treatment of the problems even more difficult, if not impossible. Nevertheless, the practical importance of obtaining accurate solutions in order to prevent catastrophic failure by brittle fracture is becoming increasingly evident with the scaling-up of plant, particularly

---

* Reader in Mechanical Engineering
† Lecturer in Mechanical Engineering
Department of Mechanical Engineering, Imperial College of Science & Technology, London SW7 2BX.

in the chemical industry. Numerical methods of analysis therefore have an increasingly important rôle to play. While the finite element (FE) method has been fairly successful as a tool for the fracture mechanics analysis of some high pressure engineering components, in many cases the size of the problem is too large for sufficiently accurate solutions to be obtained at acceptable cost, even when a sufficiently powerful computer is available. This is evidenced by the relatively few accurate three-dimensional FE stress intensity factor solutions available for cracked thick-walled cylinders, and the apparent lack of such solutions in the literature for cracks in thick-walled cylinders with cross bores, and in block fittings.

On the other hand, the boundary integral equation (BIE) method is now established as an accurate and economical numerical tool for elastic stress analysis. It is particularly suitable for problems which are three-dimensional with relatively low surface-to-volume ratios, and which involve high local stress gradients, such as those encountered in the vicinities of crack fronts. In using the BIE method, only the boundary of the solution domain needs to be discretised, as a result of which the dimensionality of the problem is effectively reduced by one: a three-dimensional volume problem is reduced to a two-dimensional surface one. Computer core storage requirements, program execution times and the amount of data preparation required are all reduced relative to the FE method for a given level of solution accuracy. Typically, computing costs are reduced by a factor of between five and ten.

The BIE method has been successfully applied by the present authors to the fracture mechanics analysis of surface cracks in thick-walled cylinders, both plain and with cross bores, subjected to internal pressure [1-3], and some comparisons with experimental results for fatigue crack growth showed very good agreement. This paper describes a further application of the BIE method, to corner-crack problems in block fittings. Some stress concentration results are presented for such components in the unflawed state, together with stress intensity factor solutions for cases when cracks are present. But first, a brief review of the BIE method for three-dimensional elastostatics is in order.

REVIEW OF THE BIE METHOD FOR THREE-DIMENSIONAL ELASTOSTATICS

The analytical basis of the BIE method is the transformation of the partial differential equations of elasticity (Navier's equations) applicable throughout the whole solution domain (Figure 1) into an integral identity for the displacements within the body, commonly known as Somigliana's identity

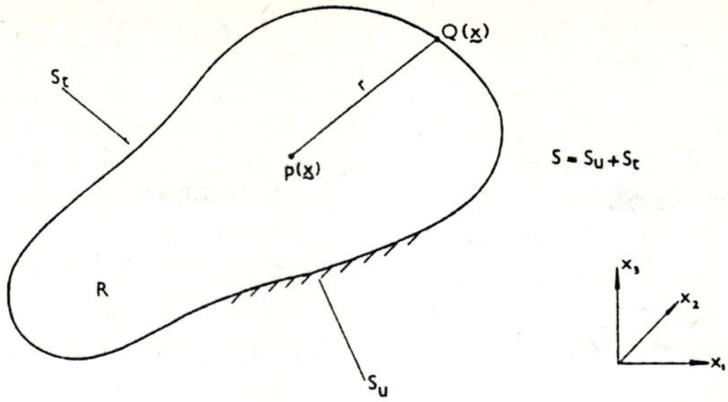

Figure 1: A boundary value problem in elastostatics

$$u_i(p) = \int_S t_j(Q)\, U_{ij}(p,Q)\, dS(Q) - \int_S u_j(Q)\, T_{ij}(p,Q)\, dS(Q) \quad (1)$$

where

$$U_{ij}(p,Q) = \frac{1}{16\pi\,\mu\,(1-\nu)} \left(\frac{1}{r}\right) [(3-4\nu)\, \delta_{ij} + r,_i\, r,_j] \quad (2)$$

$$T_{ij}(p,Q) = \frac{-(1-2\nu)}{8\pi\,(1-\nu)} \left(\frac{1}{r^2}\right) \left[\frac{\partial r}{\partial n} \left(\delta_{ij} + \frac{3}{(1-2\nu)}\, r,_i\, r,_j\right) - n_j\, r,_i + n_i\, r,_j\right] \quad (3)$$

In equation (1), $U_{ij}(p,Q)$ and $T_{ij}(p,Q)$ are the displacements and tractions, respectively, in the x_i directions at some surface point $Q(\underset{\sim}{x})$ due to orthogonal unit loads in the x_j directions at some internal point $p(\underset{\sim}{x})$; and $t_i(Q)$ and $u_i^j(Q)$ are the values of tractions and displacements, respectively, on the surface, S, at the point $Q(\underset{\sim}{x})$. Also, in (2) and (3), ν is Poisson's ratio, μ is the shear modulus, δ_{ij} is the Kronecker delta, r is the distance between p and Q, and the n_i are the components of the unit normal at Q.

Taking $p(\underset{\sim}{x})$ to a boundary point $P(\underset{\sim}{x})$ by the usual limiting process results in the BIE which relates the boundary tractions to boundary displacements

$$C_{ij}(P)\, u_j(P) + \int_S u_j(Q)\, T_{ij}(P,Q)\, dS = \int_S t_j(Q)\, U_{ij}(P,Q)\, dS \quad (4)$$

$$C_{ij}(P) = \lim_{\varepsilon\to 0} \int_{S_\varepsilon} T_{ij}(P,Q)\, dS \quad (5)$$

where S_ε is the surface of R_ε, which is the part of the solution domain, R, contained within a sphere of radius ε centred at P.

In the general problem illustrated in Figure 1, displacements are prescribed over part of the boundary surface, S_u, while tractions are prescribed over the remainder of the boundary, S_t. To solve (4) numerically for the unknown boundary tractions or displacements, the surface of the body is discretised into simple elements, over each of which the distributions of surface geometry and unknown tractions and displacements vary in some specified manner. In the present work, the modelling strategy employed is essentially that first proposed by Lachat & Watson [4,5], where the surface of the body is represented by eight-node isoparametric quadrilateral elements, and the geometry and distributions of the unknowns are expressed in terms of quadratic shape functions of the intrinsic coordinates, as are commonly used in FE methods. Thus

$$\left.\begin{aligned} x_i(\xi) &= N^c(\xi)\, x_i^c \\ u_i(\xi) &= N^c(\xi)\, u_i^c \\ t_i(\xi) &= N^c(\xi)\, t_i^c \end{aligned}\right\} \qquad (6)$$

where the $N^c(\xi)$ $(c = 1,2,\ldots,8)$ are the quadratic shape functions of intrinsic coordinates (ξ_1,ξ_2). If the surface of the body is represented by m such elements involving a total of q distinct nodes, substitution of (6) into (4) leads to

$$C_{ij}(P^a)\, u_j(P^a) + \sum_{b=1}^{m} \sum_{c=1}^{8} u_j(P^{d(b,c)}) \int_{S_b} T_{ij}(P^a,Q(\xi))\, N^c(\xi)\, J(\xi)\, d\xi$$
$$= \sum_{b=1}^{m} \sum_{c=1}^{8} t_j(P^{d(b,c)}) \int_{S_b} U_{ij}(P^a,Q(\xi))\, N^c(\xi)\, J(\xi)\, d\xi \qquad (7)$$

where P^a is the ath node, $d(b,c)$ is the number of the cth node of the bth element and $J(\xi)$ is the Jacobian of transformation. Equation (7) represents a set of $3q$ linear algebraic equations for the unknown values of displacements or tractions, fully populated with non-zero coefficients, the solution to which can be obtained by Gaussian elimination.

PROBLEM DEFINITION AND BOUNDARY DISCRETISATION

The block fitting analysed in the present study is represented by a cube, of side length $2H$, containing two cylindrical bores, each of radius R. As shown in Figure 2,

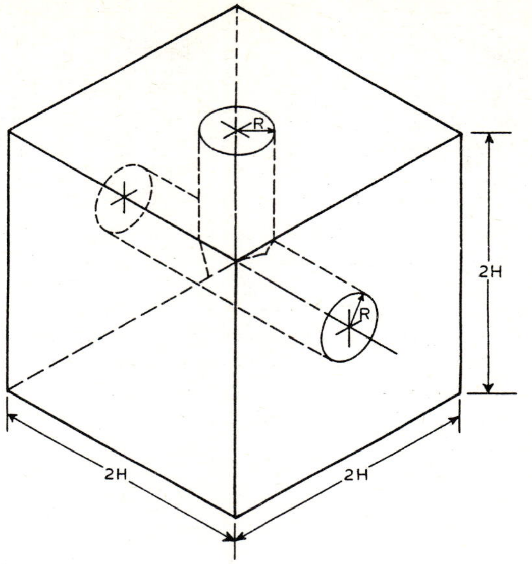

Figure 2: Block fitting containing T-bores

one of the bores runs completely through the block with its axis passing normally through the centres of two opposite faces. The axis of the other bore passes normally through the centre of one of the other faces and intersects the main bore at the centre of the block. Only the configuration where $R/H = 0.2$ is treated here, but two types of loading are considered - internal hydrostatic pressure, P, acting in the bores, and uniform tensile stress, P_o, applied to the faces of the block parallel to the T-bore axes. Under either of these forms of loading, the greatest stress concentrations occur at the two points (marked A in the diagram in Figure 5) which lie in the plane of the bore axes at the intersections of the sides of the holes. These points are also the usual sites of crack initiation, and fatigue cracks, once formed, grow in this plane through the bore axes.

For the fracture mechanics analyses, only the opening mode is considered. The crack fronts are assumed to be circular quadrants centred at the crack initiation points, A, and of radius a, as shown in Figure 7: two cracks of equal size are assumed to be present. These geometries provide reasonably good representations of some cracks observed in practice. Three different relative crack depths, $a/(H-R)$, are considered for each of the two different loading conditions, namely $a/(H-R) = 0.25$, 0.5 and 0.75. For the cases where internal hydrostatic pressure is applied to the bores, the same pressure is also assumed to act fully on the crack faces.

This represents the most severe situation for a given crack geometry and, particularly if the pressure is fluctuating, may somewhat over-estimate the effects of the actual loading [1]. In all cases, Poisson's ratio is taken to be 0.3.

Since there are two planes of symmetry for both forms of applied loading, only one quarter of the block fitting need be analysed. Figures 3 and 4 show typical boundary element mesh

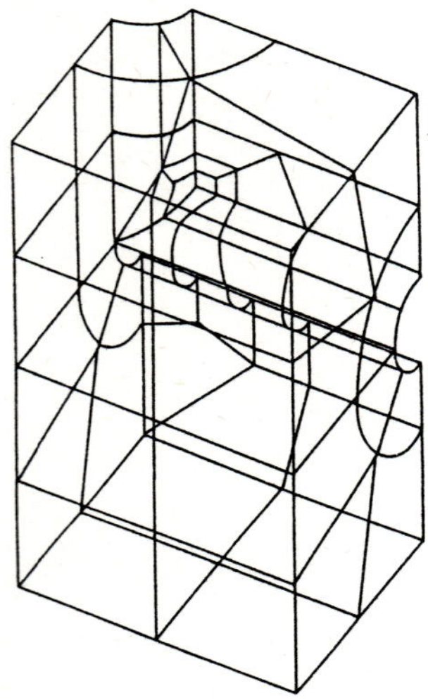

Figure 3: BIE boundary mesh for stress concentration problems

configurations used for cases without a crack and with a crack present, respectively. In each case, the same mesh is used for the two types of loading. For the block fitting in the unflawed state (Figure 3), 74 surface elements with 224 nodes are used to model the problem. With cracks present (Figure 4), up to 93 surface elements and 281 nodes are used, more elements being employed in the region of the crack front to ensure good accuracy. The meshes used are fine by BIE standards, and the results presented should be of adequate accuracy for engineering purposes.

NUMERICAL RESULTS AND DISCUSSION

Stress Concentrations

In the absence of cracks, stress concentrations in the block fitting are the main concern. Figure 5 shows the variation of the hoop stress along the wall of the main bore in the plane of the bore axes (between points A and B shown in the diagram) when the block is internally pressurised. The

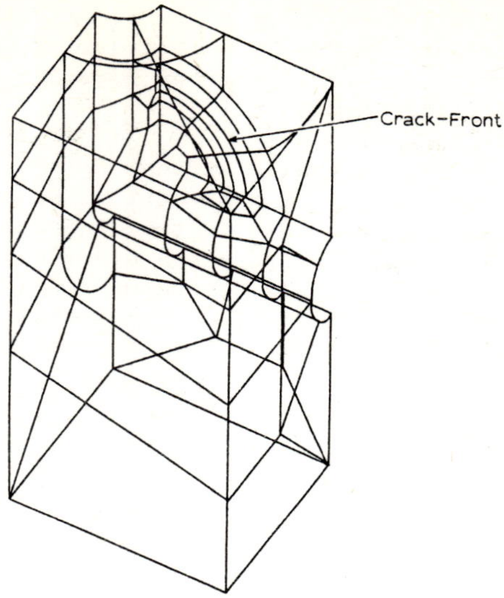

Figure 4: BIE boundary mesh for cracked block fitting with $a/(H-R) = 0.5$

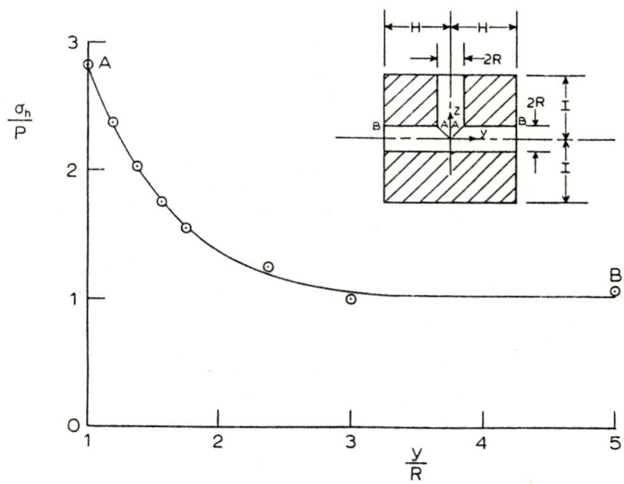

Figure 5: Hoop stress variation along the main bore of a block fitting with $R/H = 0.2$ and internal pressure P

hoop stress is greatest at the points A, where the magnitude is 2.82 P. Moving away from these points along the wall of the main bore, the hoop stress, as expected, decreases rapidly to a constant value of 1.08 P at points B on the free surface planes. This latter value compares well with the Lame solution for the maximum hoop stress in a thick-walled cylinder having a radius ratio of 5, which is also 1.08 P (to 3 significant figures).

When a uniform tensile stress, P_o, is applied to opposite faces of the block in the x-direction, perpendicular to the cross-sectional plane shown in the diagram in Figure 5, the stress concentrations at points A are much higher at $\sigma_x/P_o = 6.76$, and σ_x again decreases rapidly to a constant value of 3.00 P_o at points B. The variation of this direct stress is shown in Figure 6. A very similar variation of σ_x

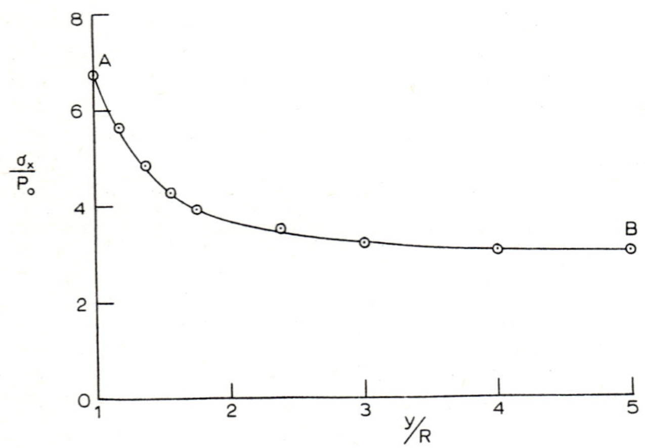

Figure 6: Variation of direct stress σ_x along the main bore of a block fitting with $R/H \cong 0.2$ and externally applied tensile stress P_o

is obtained along the wall of the cross bore in the same plane. The stress concentration factor of 3.00 at bore cross sections remote from the intersection is identical to the familiar result for a hole in an infinite body in simple tension, under conditions of either plane stress or plane strain.

Hooke & Demunshi [6], using the 'point matching' or 'boundary least squares' technique, have solved for the stress concentration factors around the intersection of two cylindrical holes in an infinite elastic body. Several ratios of bore radii were treated, including the present one of equal radii with one bore of infinite length and the other of semi-infinite length. Under internal hydrostatic pressure, the maximum stress concentration at points A was 2.68, about

5% lower than the value obtained from the present BIE analysis for a block fitting of finite size. For the case of a uniform tensile stress applied in the direction perpendicular to the plane of the bore axes at infinity, Hooke & Demunshi obtained a stress concentration factor of 5.60, some 17% lower than that obtained from the BIE analysis. While it may be possible to explain the relatively small difference between the results for internal pressure in terms of finite versus infinite body geometry, this is much less convincing for external tension where the difference is much greater. Bearing in mind that the boundary meshes used in the present work are already highly refined, any errors in the BIE results are much smaller than the 17% difference involved.

Stress Intensity Factors

In the presence of cracks, the stress intensity factor, K_I, is the characterising parameter for the stress field close to a crack front, and its value may vary along this front. In the present work, numerical values of K_I are obtained by the direct (extrapolation) method using the crack face displacement data [1]. Plane strain conditions are assumed to prevail at all points along the crack front except at the free surfaces, which are treated as under plane stress conditions. Also, the quarter-point node-shifting technique commonly used in FE fracture mechanics analyses (see, for example, Barsoum [7]) is applied to elements adjacent to a crack front, in order to introduce the expected $1/\sqrt{r}$ stress (strain) singularity there. In view of the modified order of singularity at points on the crack front very close to a free surface [1], numerical values of K_I obtained for such points should be treated with caution.

The computed values of normalised stress intensity factor, $K_I/P\sqrt{\pi a}$, around the crack fronts, for cases of cracked blocks subject to internal pressure, P, are shown in Figure 7. Position along a crack front is defined by the angular coordinate, θ, centred at A, and ranging from $\theta = 0$ at the wall of the cross bore to $\theta = 90°$ at the wall of the main bore. As θ increases from zero, the stress intensity factor decreases to a minimum at about $\theta = 45°$, and then increases again as θ approaches $90°$. While the value of K_I normalised with respect to the square root of the crack radius, a, passes through a minimum at a relative crack depth of about $a/(H-R) = 0.5$, the absolute value of K_I always increases with a.

Figure 8 shows the variations of the computed stress intensity factor, normalised by $P_Q\sqrt{\pi a}$, along the crack fronts for the three crack depths treated, when the block fitting is subjected to external tension. The general trends in these variations are similar to those for the internal pressure case, although the differences between the maximum and minimum values

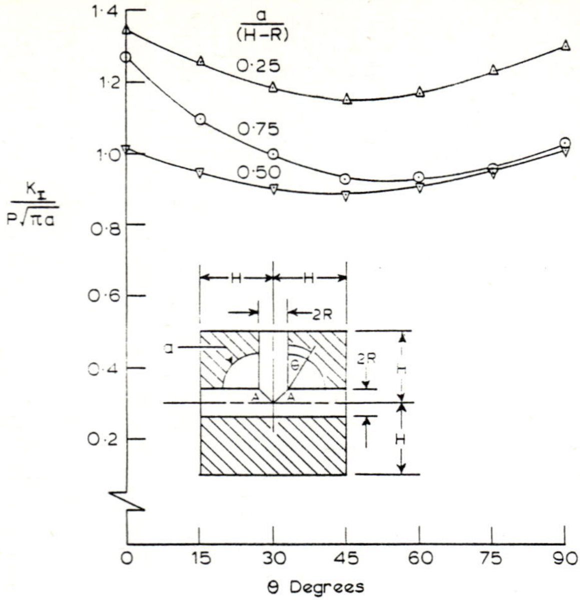

Figure 7: Variations of K_I along crack fronts in a block fitting with R/H = 0.2 and internal pressure P

of K_I around a crack front are more pronounced. Again, although normalised values of K_I decrease with increasing crack depth, absolute values increase.

The results presented should be useful in the design of high pressure block fittings and other similar components. While, for example, a bursting disc may be used to protect a block fitting against a pressure overload, this will not be sufficient to prevent brittle fracture if cracks are formed and grow in service. Knowledge of the critical size of crack which the block fitting can sustain makes it possible, for example, to provide a check port through which the pressurised fluid can leak when this size is reached.

CONCLUSIONS

A boundary integral equation method of three-dimensional elastic stress analysis has been successfully applied to some stress concentration and fracture mechanics analyses of square blocks containing circular T-bores subjected to internal hydrostatic pressure and external tension. The results should be of considerable practical use in the design of high pressure block fittings, and in predicting the safe life of such components known to be cracked to a certain depth. These

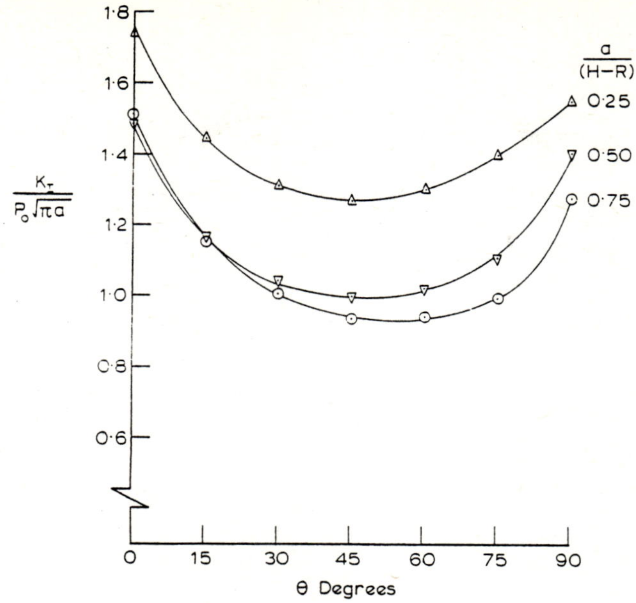

Figure 8: Variations of K_I along crack fronts in a block fitting with R/H = 0.2 and externally applied tensile stress P_o

problems do not appear to have been solved numerically before, possibly because of the very high cost of using other numerical techniques, such as the finite element method. As a matter of interest, all of the problems solved required less than 48000 words of core storage on either a CDC 6500 or CDC 7600 computer. The amount of central processor time used on the latter was up to 250 seconds.

REFERENCES

1. TAN, C.L., and FENNER, R.T. - Elastic Fracture Mechanics Analysis by the Boundary Integral Equation Method. Proc. R. Soc. Lond., Vol. A369, pp. 243-260, 1979.

2. TAN, C.L., and FENNER, R.T. - Stress Intensity Factors for Semi-Elliptical Surface Cracks in Pressurised Cylinders Using the Boundary Integral Equation Method. Int. J. Fracture, Vol. 16, No. 2, 1980 (in press).

3. TAN, C.L., and FENNER, R.T. - Boundary Integral Equation Stress Analysis of Some Problems in High Pressure Engineering. To be presented at the 2nd Int. Symp. on Innovative Numerical Analysis in Applied Engineering

Sciences, Montreal, 16-20 June 1980. Submitted for publication in the Proceedings of the Symposium.

4. LACHAT, J.C., and WATSON, J.O. - Effective Numerical Treatment of Boundary Integral Equations: a Formulation for Three-Dimensional Elastostatics. Int. J. Num. Methods Eng., Vol. 10, No. 5, pp. 991-1005, 1976.

5. LACHAT, J.C., and WATSON, J.O. - Progress in the Use of the Boundary Integral Equation Illustrated by Examples. Comp. Methods Appl. Mech. Eng., Vol. 10, pp. 273-289, 1977.

6. HOOKE, C.J., and DEMUNSHI, G. - Stress Concentration Factors Around the Intersection of Two Cylindrical Holes in an Infinite Elastic Body. J. Strain Analysis, Vol. 12, No. 3, pp. 217-222, 1977.

7. BARSOUM, R.S. - On the Use of Isoparametric Finite Elements in Linear Elastic Fracture Mechanics. Int. J. Num. Methods Eng., Vol. 10, No. 1, pp. 25-37, 1976.

STRESS INTENSITY FACTOR CALCULATION OF A MIXED MODE CRACK GROWTH PROBLEM

G. A. D. Briggs†, R. A. Smith*

† Department of Metallurgy, University of Oxford, Parks Road, Oxford.

* Cambridge University Engineering Dept., Trumpington Street, Cambridge.

SUMMARY

Mode I (tensile) and Mode II (in-plane sliding) stress intensity factors have been calculated for cracks growing from the points of maximum hoop stress on the boundary of an ellipse subjected to an angled compressive load. The use of Greens' functions for these calculations is discussed.

Some results of experimental fatigue tests on ellipses loaded in such a way are presented. The rate and direction of crack growth both change abruptly when the Mode I and II stress intensity factors are approximately equal.

INTRODUCTION

In certain engineering situations fatigue cracks may grow in components which experience cyclic compressive loads. In order to investigate the criteria controlling the rate and direction of crack growth some experiments have been conducted on aluminium plates containing a single hole.

The simplest case is a circular hole, Fig. 1. The hole acts as a stress concentrator, there is an elastic stress of some three times the applied stress at the sides of the hole at the ends of the diameter normal to the loading direction. If plastic yielding occurs at the first loading cycle then pure Mode I fatigue cracks may grow from these points in a direction perpendicular to the applied stress. There is also a tensile stress at the top and bottom of the hole, equal in magnitude to the applied compressive stress. If notches are made at these points, then fatigue cracks may grow parallel to the applied stress. Again these are pure Mode I cracks. Solutions exist for the stress intensity factors, K_I, Bowie

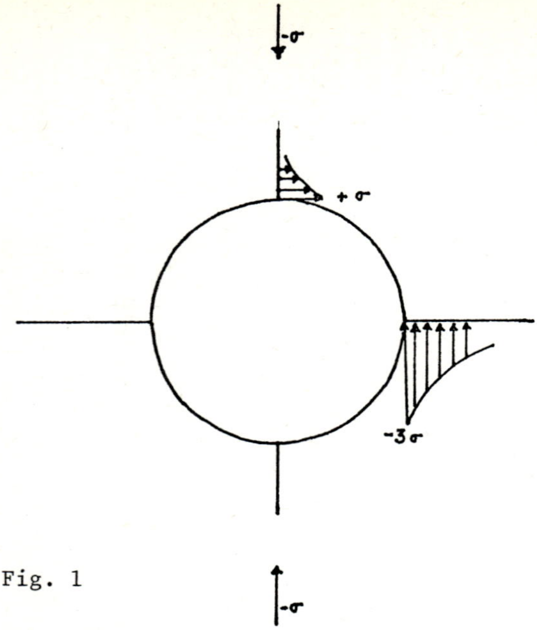

Fig. 1

(1), Newman (2), of such cracks. They first increase before decreasing as the crack grows further. For a given applied cyclical stress a minimum initial notch is required before crack growth can start and there is a maximum length, corresponding to the threshold, to which such cracks may grow.

The symmetry of the circular hole does not permit mixed Mode growth to occur. To investigate mixed Mode growth, angled fatigue cracks have been used by Smith (3), but problems arise because in general the sides of the crack will be in contact. Moreover if a crack does grow at some angle to the initially inclined fatigue crack, the detailed geometry may be modified by crack branching.

Some of these problems are absent when cracks are grown from an elliptical hole whose axes are inclined at some angle to the direction of the applied stress, Fig. 2. This has the advantage of being a precisely defined asymmetrical geometry. As for a circle, cracks may grow from regions of compression on the circumference, but these were largely ignored in the present study. Their influence on the stresses is omitted in the computations; it is recognised that this is an arguable source of error. The cracks of interest are those which may be grown from points of tension on the circumference of the ellipse.

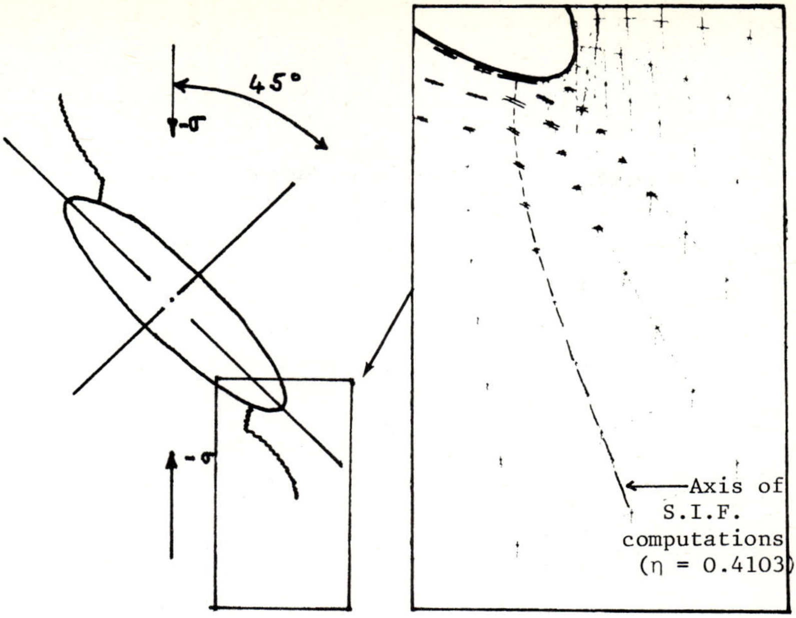

Fatigue crack growth ~~~~   Principal stresses: ═ Tension; -- Compression

Fig. 2

The points of maximum tension may be calculated from the expressions in Appendix A. Fine starter cracks were made in these positions in an ellipse with 24 mm major by 6 mm minor axes. This had been spark machined, with its major axis inclined at 45°, in a sheet, 75 mm square by 3.2 mm thick, of commercially pure aluminium. The plates were then subjected to a wholly compressive fatigue loading from -15.75 KN max to -2.50KN min.

Fatigue cracks were observed to propagate normal to the boundary of the ellipse. After they had grown on average about 2 mm, an abrupt change in growth direction was observed. This can be clearly seen in Fig. 2. This change in growth direction was accompanied by a several fold increase in crack growth rate, from about 5×10^{-6}mm/cycle to 10^{-4} mm/cycle.

COMPUTATION

In order to analyse this phenomenon it is desirable to know the stress intensity factors in Mode I (K_1) and Mode II (K_{11}) at the crack tip as it propagates. In principle it would be possible to determine these by boundary collocation methods or by finite element analysis. However, a much neater and more satisfying way to tackle the problem is to make use of Buekner's principle (4). The stress intensity factor for a crack in a body subject to external stresses is identical to that caused by stresses acting on the surface of a crack equal but opposite to those which would be present in the uncracked body under external load. The stress intensity factors may thus be computed by integrating, along the length, a, of the crack, the product of the stresses in an uncracked body σ_y, τ_{xy} and a Green's function, F(x), which indicates the stress intensity factor due to a point load, a technique discussed by Cartwright and Rooke (5),

$$K_1 = \frac{1}{\sqrt{\pi a}} \int_o^a \sigma_y(x)\ F(x)dx$$

$$K_{11} = \frac{1}{\sqrt{\pi a}} \int_o^a \tau_{xy}(x)\ F(x)dx$$

Because the initial crack growth is normal to the edge of the hole it is convenient to work in curvilinear coordinates, the crack growing approximately along a coordinate of constant η. The stresses for an inclined elliptical hole in an infinite plate subject to tensile stress have been found by a method developed by Stephenson (6), and are presented in Appendix A (for an excellent account see Timoshenko & Goodier, (7)). The principle stresses are plotted in Fig. 2, where it may be seen that the crack grows from a region of pure tension normal to the crack to one where the stress (before the arrival of the crack) is mainly shear. Since the Green's functions increase for stresses near the crack tip, it is to be expected that K_1 will initially increase proportional to the square root of the crack length and then decrease whereas K_{11} will increase more slowly, but continue to increase, as the crack becomes longer. The principle stresses also give an impression of the behaviour of the stress intensity factors after the change in direction of the crack growth, however these have not yet been computed.

Exact Green's functions do not exist for the present problem; to calculate them would be a major computational undertaking. For short cracks (as in this case), the simplest approximation is to use the Green's function for a loaded crack emanating from a free edge. Cartwright and Rooke (5) have compared various possible Green's functions for cracks from a circular hole. At the edge of the hole the different forms differ by at most 30%, but near the tip, where their value is largest (tending to infinity at the tip itself) they converge. Thus provided the stresses in the uncracked plate are known, and in the present case they are known exactly, even an approximate form of the Green's function will give a good indication of the stress intensity factor.

The Green's function used is derived from that presented by Buekner (4)

$$F(\frac{s}{a}) = (\frac{2}{s/a})^{\frac{1}{2}} \left[1 + 0.6147(\frac{s}{a}) + 0.2502(\frac{s}{a})^2\right]$$

where s is the distance measured along the crack from its tip.

Using the alternative form given by Hartrnaft and Sih (8):

$$F_e(x) = \frac{2}{\sqrt{1-x^2}} \left[1 + (1-x^2)\{ 0.2945 - 0.3912x^2 + 0.7685x^4 - 0.9942x^6 + 0.5094x^8\}\right]$$

where $x = 1 - s/a$
alters the results by less than 3%.

Even the use of the embedded crack function

$$F_c(x) = \frac{2}{\sqrt{1-x^2}}$$

reduces the calculated values of K by only about 10%.

Because of the coordinate system used for the stress field, the integration cannot be performed in closed form. It is of the form

$$K_1 = \frac{1}{\sqrt{\pi a}} \int_o^a \sigma_\eta(\xi)\ F\left[\frac{s}{a}\right] ds$$

where a is the crack length, and s the distance from the tip, along a coordinate of constant η. Appendix B shows how this may be turned into an integral over ξ, in the form:

$$K_1 = \frac{1}{\sqrt{\pi a}} \int_{\xi_1}^{\xi_o} \sigma_\eta(\xi)\ F\left[\frac{s(\xi)}{a}\right] H d\xi$$

where $H = ds/d\xi$.

The expression for K_{11} is identical except that $\tau_{\xi\eta}(\xi)$ is used instead of $\sigma_\eta(\xi)$.

This may be computed for various values of ξ_1, using a standard numerical analysis package. Within the integration a second numerical integration is called for to compute $s(\xi)$. This is well behaved (see Appendix B) and presents no problems.

However, the outer integration has a singularity at $\xi=\xi_1$ of the type $s^{-\frac{1}{2}}$. The package cannot cope with this, and the solution varies by a few percent depending on where the integration is truncated. Close to $\xi=\xi_1$ very little error is introduced by taking the Green's function as $\left[\frac{2a}{s}\right]^{\frac{1}{2}}$, and now an approximation for the stresses may be introduced, namely that their value is constant over this range and equal to their value at $\xi=\xi_1$, see Appendix 2. When the closed solution of the singularity is added to the numerical summation, the result is insensitive to the value of ξ at which the change-over took place.

RESULTS

The computed values of K_1 and K_{11} as a function of crack length are shown in Fig. 3. As suggested above, K_1 rises to a maximum and then decreases whereas K_2 increases slowly and steadily. It is particularly interesting to note where the two curves cross each other. This occurs at a crack length of about 2 mm, which is in the region in which the change of crack growth direction is observed experimentally.

Initially the crack grows normal to the free surface; there is little else it can do. As K_{11} grows it has initially little observable influence on the crack growth. When it becomes of about the same magnitude as K_1, it causes a sudden change in direction.

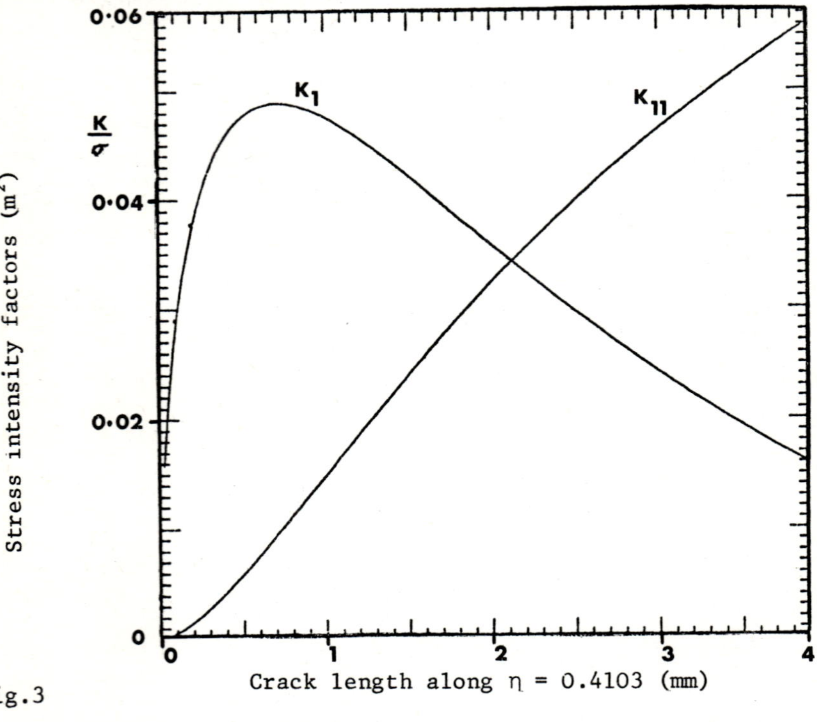

Fig.3

There is a remaining curiosity about this phenomenon. Before the change in direction the tip stresses due to K_{11} are compressive on the side further from the semi-minor axis and tensile on the side nearer: this can be seen from the principal stresses in Fig. 2. For brittle fracture it is generally assumed that the crack will prefer to grow in a direction where there is tension normal to its potential path (notably Erdogan and Sih, (9)). The question is well reviewed by Swedlow, (10). Criteria for the growth of fatigue cracks have yet to be established. (It is in this context that the present work was undertaken.) The next step must be to evaluate the reversed plastic strains around the crack tip. Knowledge of the stress intensity factors makes this possible.

REFERENCES

1. BOWIE, O. L. - Analysis of an infinite plate containing radial cracks originating at the boundary of an internal circular hole. J. Maths and Physics, Vol. 35, pp. 60-71, 1956.

2. NEWMAN, J. C. - An improved method of collocation for the stress analysis of cracked plates with various shaped boundaries. NASA Technical Note, NASA TN D-6376, 1971.

3. SMITH, R. A. - Interfaces of Wear and Fatigue, Proc. Int. Conf. Fundamentals of Lubrication, M.I.T., 1978 (in press).

4. BUEKNER, H. F. Weight Functions for the notched bar, ZAMM (Zeitschrift für Angewandte Mathematik und Mechanik) Vol. 51, pp. 97-109, 1971.

5. CARTWRIGHT, D. J., ROOKE, D. P. - Green's Functions in Fracture Mechanics, Fracture Mechanics - Current Status, Future Prospects, Ed. Smith, R. A., Pergamon, 1979.

6. STEVENSON, A. C. -Complex potentials in two-dimensional elasticity, Proc. R. Soc., Lond. A, Vol. 184, pp. 129-179, 1945.

7. TIMOSHENKO, S. P., GOODIER, J. N., Theory of Elasticity, McGraw Hill Kogakusha, 3rd Ed., 1970.

8. HARTRANFT, R. J., SIH, G. C. - Alternating method applied to edge and surface crack problems, Methods of analysis and solutions of crack problems, Ed. Sih, G. C., pp. 179-238, Noordhoff, Leyden, 1973.

9. ERDOGAN, F., SIH, G. C. - On the crack extension under plane loading and transverse shear. Trans ASME, Series D, J. Basic Engineering, Vol. 85, pp. 519-527, 1963.

10. SWEDLOW, J. L., - Criteria for growth of the angled crack, ASTM STP 601, pp. 506-521, 1976.

Appendix A

The elliptical coordinates are defined as

$$z = c \cosh\zeta \qquad \zeta = \xi + i\eta$$

giving $x = c \cos\xi\cos\eta \qquad y = c \sinh\xi\sin\eta$

$$\frac{dz}{d\xi} = c \sinh\zeta \qquad e^{2i\alpha} = \frac{\sinh\zeta}{\sinh\bar{\zeta}}$$

The boundary of the ellipse is defined by $\xi = \xi_o$. ξ_o and c may be determined from the semi-major and minor axes, corresponding to $\eta = 0, \pi/2$.

For an infinite plate with an elliptical hole whose semi-major axis is at an angle β to an applied stress σ at infinity, the complex elastic potentials are given by Timoshenko & Goodier (7) as:

$$4\psi(z) = Ac \cosh\zeta + Bc \sinh\zeta$$

$$4\chi(z) = Cc^2\zeta + Dc^2\cosh2\zeta + Ec^2\sinh2\zeta$$

where
$$A = Se^{2\zeta_o} \cos2\beta$$
$$B = S(1 - e^{2\zeta_o + 2i\beta})$$
$$C = -S(\cos2\zeta_o - \cos2\beta)$$
$$D = -\tfrac{1}{2}Se^{2\zeta_o}\cosh 2(\zeta_o + i\beta)$$
$$E = \tfrac{1}{2}Se^{2\zeta_o}\sinh 2(\zeta_o + i\beta)$$

The stresses are given by

$$\sigma_\xi = \sigma_\eta = 4\mathrm{Re}\psi'(z)$$

$$\sigma_\eta - \sigma_\xi + 2i\tau_{\xi\eta} = 2e^{2i\alpha}\{\bar{z}\psi''(z) + \chi''(z)\}$$

$$4\mathrm{Re}\psi'(z) = S.\mathrm{Re}(A + B \coth\xi)$$

$$2e^{2i\alpha} \left[z\psi''(z) + \chi''(z)\right] = S(-C \coth\zeta)$$

$$+ (4D = 2E \coth\zeta) \cos2\zeta$$

$$+ (4D - 2D \coth\zeta) \sinh2\zeta$$

$$-B\cosh\zeta \operatorname{csch}\zeta)/(\cosh2\xi - \cos2\eta)$$

at the boundary of the hole, $\sigma_\xi = 0$, so that

$$(\sigma_\eta)_{\xi=\xi_o} = S\left[\frac{\sinh 2\xi_o + \cos 2\beta - e^{2\xi_o}\cos 2(\beta - \eta)}{\cosh 2\xi_o - \cos 2\eta}\right]$$

giving maxima and minima at

$$n = \frac{1}{2}\sin^{-1}\left[\frac{-B\pm\sqrt{B^2 - AC}}{A}\right]$$

where

$$A = B_1^{\,2} + A_1^{\,2} \qquad A_1 = e^{2\xi_o}\cosh 2\xi_o \sin 2\beta$$

$$B = -B_1C_1 \qquad B_1 = -e^{2\xi_o}\cosh 2\xi_o \cos 2\beta + \sinh 2\xi_o + \cos 2\beta$$

$$C = C_1^{\,2} - A_1^{\,2} \qquad C_1 = e^{2\xi_o}\sin 2\beta$$

Appendix B

The integral to be evaluated is

$$K_1 = \frac{1}{\sqrt{\pi a}} \int_o^a \sigma_\eta(\xi) F\left[\frac{s}{a}\right] ds$$

It is more convenient to express this as an integral over ξ. For curvilinear coordinates, if s_ξ is length along a coordinate of constant η, then

$$ds = Hd\xi$$

where $$H^2 = \tfrac{1}{2}c^2(\cosh 2\xi - \cos 2\eta)$$

(a negative value of H should be taken for the present definition of s).

$$a = \int_{\xi_1}^{\xi_o} Hd\xi$$

and $$s(\xi) = \int_{\xi_1}^{\xi} Hd\xi$$

where ξ_1 describes the position of the crack tip. These integrals are well behaved and may be evaluated numerically at each stage in the outer integration, which becomes

$$K_1 = \frac{1}{\sqrt{\pi a}} \int_{\xi_1}^{\xi_o} \sigma_\eta(\xi) F\left[\frac{s}{a}(\xi)\right] Hd\xi$$

The expression for K_2 is identical except that $\tau_{\xi\eta}(\xi)$ is used instead of $\sigma_\eta(\xi)$.

The singularity is approximated to

$$K_1 = \frac{1}{\sqrt{\pi a}} \int_{\xi_1}^{\xi} \sigma_\eta(\xi_1)\ F\left[\frac{s}{a}\right] ds$$

$$= \frac{1}{\sqrt{\pi a}} \sigma_\eta(\xi_1) \int_o^{s(\xi)} (\frac{2a}{s})^{\frac{1}{2}} ds$$

$$= \frac{1}{\sqrt{\pi a}} \sigma_\eta(\xi_1) .\ \tfrac{1}{2}(2as(\xi))^{\frac{1}{2}}$$

$$= (\frac{s(\xi)}{2\pi})^{\frac{1}{2}}\ \sigma_\eta(\xi_1)$$

Analysis of mixed-mode stress intensity factors by body force method

Y.Murakami[(I)]

Summary

A method for the analysis of mixed-mode stress intensity factors for an arbitrarily shaped crack is described. The method is based on the body force method in which the solutions are obtained by the superpositions of stress field of a point force applied at an infinite plate.

The stress functions of a point force were used as a fundamental solution for superposition. The computer program developed was used to calculate mixed-mode stress intensity factors for various plane stress problems. The present method has two advantages as well as high accuracy, that is, (i) the stress intensity factors K_I and K_{II} can be obtained directly by the weighted integral of stress functions and (ii) the input data only related to the boundaries of the plate and the loading conditions are necessary.

The advantage (ii) is worth noticing because not only short CPU time but also saving the labor for the preparation of a great deal of input data is required.

1. INTRODUCTION

Recently the usefulness of stress intensity factor is generally accepted. A number of analyses for stress intensity factors in regard to fracture problems have been reported. These solutions which separately give the results for special problems are gathered in a few handbooks[1-3]. It is said that in spite of a great deal of solutions in the handbooks the person concerned can hardly find the very solution that he want. Although it may be possible to estimate approximate values from basic solutions in the handbooks, the real solutions are required. In order to satisfy the requirement, the development

(I) Associate professor, Department of Mechanical Engineering, Kyushu Institute of Technology, Tobata, Kitakyushu, 804 Japan of general computer program which covers broad extent of crack problems is necessary. Accordingly, such a method of analysis must be sufficiently versatile. The finite element method is versatile. However, it needs the labor for the preparation of input data and the accuracy in the analysis of crack problems should be improved.

In the present paper, described is a method to analyze accurately mixed-mode stress intensity factors in plane stress problems with short CPU time and less labor for preparation of input data. The basis is the body force method[4]. The method is versatile enough to solve crack problems under various boundary conditions. Stress intensity factors K_I and K_{II} can be directly calculated as output. The body force method was improved in several points so that the program developed could yield accurate results with less labor. The mixed-mode stress intensity factors in several simple problems and welded joints are calculated using the same computer program. Moreover, it is shown that the same program could be applied successfully to the prediction of crack propagation path[5,6]. The solutions presented in the present paper are only a part of many results obtained by the same computer program.

2. METHOD OF ANALYSIS

2.1 Basic stress functions

The method of analysis is the extended application of a separate paper[7] which is based on the body force method[4]. The resultant force boundary conditions are satisfied by the superposition of the fundamental solution of a point force which is applied at a point in the infinite plate with a stress-free crack (Fig.1).

The complex stress functions in this case are given as follows[7].

$$\Phi(\zeta) = M \cdot \log(\zeta - \zeta_0) - \bar{N} \cdot \log\left(\zeta - \frac{R^2}{\bar{\zeta}_0}\right) + \bar{N} \cdot \log\zeta + \bar{M} \cdot \frac{\left\{\left(\frac{R^2}{\bar{\zeta}_0} + \frac{\bar{\zeta}_0}{R^2}\right) - \left(\zeta_0 + \frac{1}{\zeta_0}\right)\right\}\frac{R^2}{\bar{\zeta}_0^{\,2}}}{\left(1 - \frac{1}{\bar{\zeta}_0^{\,2}}\right)\left(\zeta - \frac{R^2}{\bar{\zeta}_0}\right)} \tag{1}$$

$$\chi(\zeta) = N \cdot \log(\zeta - \zeta_0) - \bar{M} \cdot \log\left(\zeta - \frac{R^2}{\bar{\zeta}_0}\right) + \bar{M} \cdot \log\zeta - M \cdot \frac{\left(\bar{\zeta}_0 + \frac{1}{\bar{\zeta}_0}\right) - \left(\frac{R^2}{\zeta_0} + \frac{\zeta_0}{R^2}\right)}{\left(1 - \frac{1}{\zeta^2}\right)(\zeta - \zeta_0)} \tag{2}$$

$$\Psi^I(\zeta) = \chi(\zeta) - \bar{Z}\left(\frac{R^2}{\zeta}\right) \cdot \Phi^I(\zeta). \tag{3}$$

where, $z = Z(\zeta) = \frac{c}{2}\left(\zeta + \frac{1}{\zeta}\right)$, $M = -\frac{1+\nu}{8\pi}(X + iY)$, $N = \frac{3-\nu}{8\pi}(X - iY)$, $R^2 = 1$,

$z = x + iy$, (I: The derivative respect to z).

$\zeta = \xi + i\eta$,

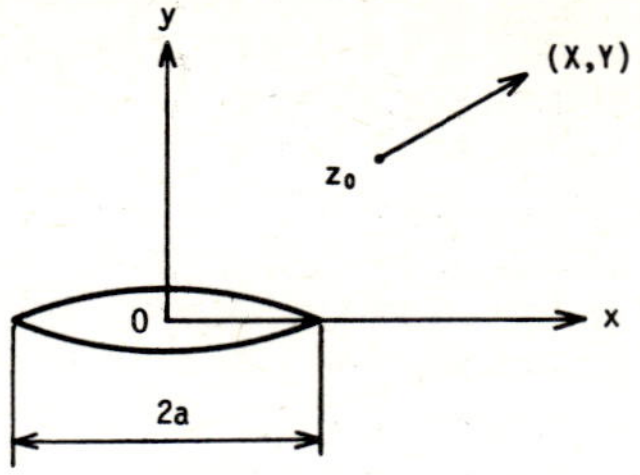

Fig.1 A point force applied at a point of an infinite plate with a stress-free crack.

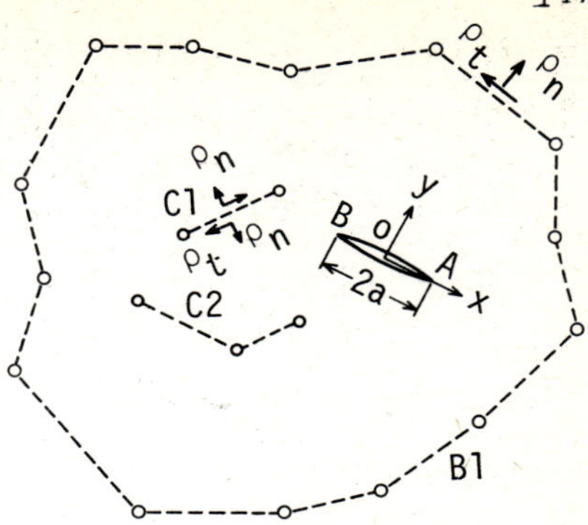

Fig.2 Distribution of body forces along boundaries and cracks.

This type of stress functions, including a traction-free crack or elliptical hole, have been applied successfully to various stress concentration problems[8-10].

The resultant force components P_x and P_y in x and y direction which act along an arbitrary path connecting two definite points are expressed by Eq.(4).

$$-P_y - iP_x = \bar{\Phi}(\zeta) + \{\bar{Z}(\bar{\zeta})\Phi'(\zeta) + \Psi'(\zeta)\} \qquad (4)$$

The basic equation for calculating stress intensity factors is

$$K_I - iK_{II} = 2\sqrt{\left(\frac{\pi}{a}\right)}\,\Phi'(\zeta)|_{\zeta=\pm 1} \qquad (5)$$

An example of the distribution of point forces along outer boundaries and other crack sites except for the stress-free crack are illustrated in Fig.2. Along other crack sites, point forces are distributed as a pair which consists of two forces with equal magnitude and opposite sign. Consequently, the stress field due to the pair of point forces is equivalent to derivative of the fundamental solution of Fig.1 (Eqs.(1)-(3)) respect to the coordinate normal to prospective crack surface. The fundamental solutions which express the stress field due to point forces distributed along the prospective crack sites are denoted by $\Phi^*(\zeta)$, $\chi^*(\zeta)$ and $\Psi^*(\zeta)$ in relation to $\Phi(\zeta)$, $\chi(\zeta)$ and $\Psi(\zeta)$. Thus, these are expressed as follows.

$$\Phi^*(\zeta) = \frac{\partial\Phi(\zeta)}{\partial y'} \ldots(6), \quad \chi^*(\zeta) = \frac{\partial\chi(\zeta)}{\partial y'} \ldots(7), \quad \Psi^*(\zeta) = \frac{\partial\Psi(\zeta)}{\partial y'} \ldots(8)$$

where, y' is the y'-coordinate of the application of point force in the physical plane when the origin is the center of stress-free crack and the x'-coordinate is taken in the parallel direction of each prospective crack (Fig.2).

2.2 Distribution of body forces and satisfaction of boundary conditions.

The solution is obtained if the densities of body force (consequently, distributed point forces) is determined so as to satisfy the resultant force condition of divisions along outer and crack boundaries. In the previous paper[7], the solution was obtained on the assumption that the body forces density was constant in each division. In the present paper, the distribution of body force that is linear within each division but discontinuous at both ends of it (Fig.3) was assumed. It is expected that linear distribution of body force will revise the accuracy remarkably[11]. The discontinuous distribution of body force will be inevitable at the corner of a finite plate, the end points of a division along which external forces (boundary conditions) are distributed and so on. So, the way of body force distribution requires that (a) the dividing points for the application of body forces and boundary conditions are jointly used, and (b) the body force densities at both ends of one division do not have the same values as those of neighboring divisions. If we consider the numbers of boundary conditions, the above described conditions (a) and (b) are related each other as follows. The unknowns of a problem are body force densities which should be distributed along boundary so as to satisfy the boundary conditions. The body force has two components, one is normal direction and the other is tangential which we distiguish by 'n' and 't' respectively. The normal component of body force which is constant along one division is denoted by ρ_n. The density $\rho_{n\ell}$ which is zero at the midpoint of the division varies linearly along the same division. Likewise, tangential body force densities are denoted by ρ_t and $\rho_{t\ell}$ respectively. Thus, we have the unknowns of four times of total division numbers N. In considering the numbers of boundary conditions, new points are added at the midpoint of each division in order to match the numbers of unknowns and the boundary conditions. Namely, the values of two components of resultant forces are satisfied along both halves of each divisions. Therefore, the numbers of the unknowns (body force densities) and boundary conditions are the same, that is, 4xN. As explained in the separate paper[7], two of boundary conditions (4xN) are not independent of others in a certain type of problems. In such cases, two conditions

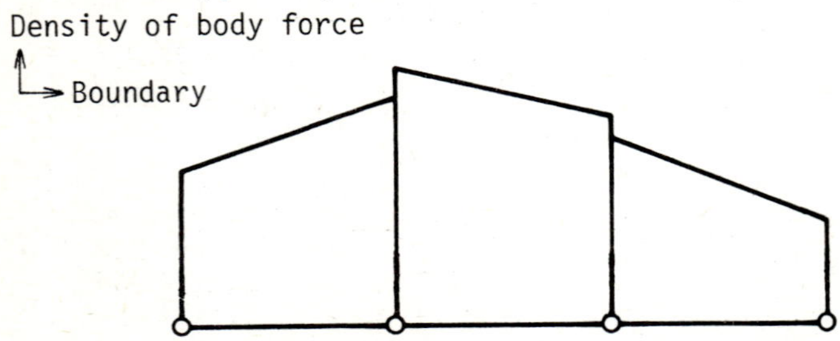

Fig.3 Linear and discontinuous distribution of body force.

must be replaced by the equilibrium conditions of total body forces distributed along the boundaries in x and y directions. After the density of point force distributed along boundaries is obtained, the stress intensity factors are calculated by integrating Eq.(5) numerically with weighted magnitude of body force (X,Y) along boundaries. The Gauss' numerical integral method was used for the calculation of the resultant forces (Eq.(4)) and the stress intensity factors.

2.3 Flow chart of the program

Figure 4 shows the flow chart of the calculation program 'MIXSIF'. The program details are summarized in Table 1. Most of running time in the present method is consumed for the numerical integral and the solution of simultaneous equations.

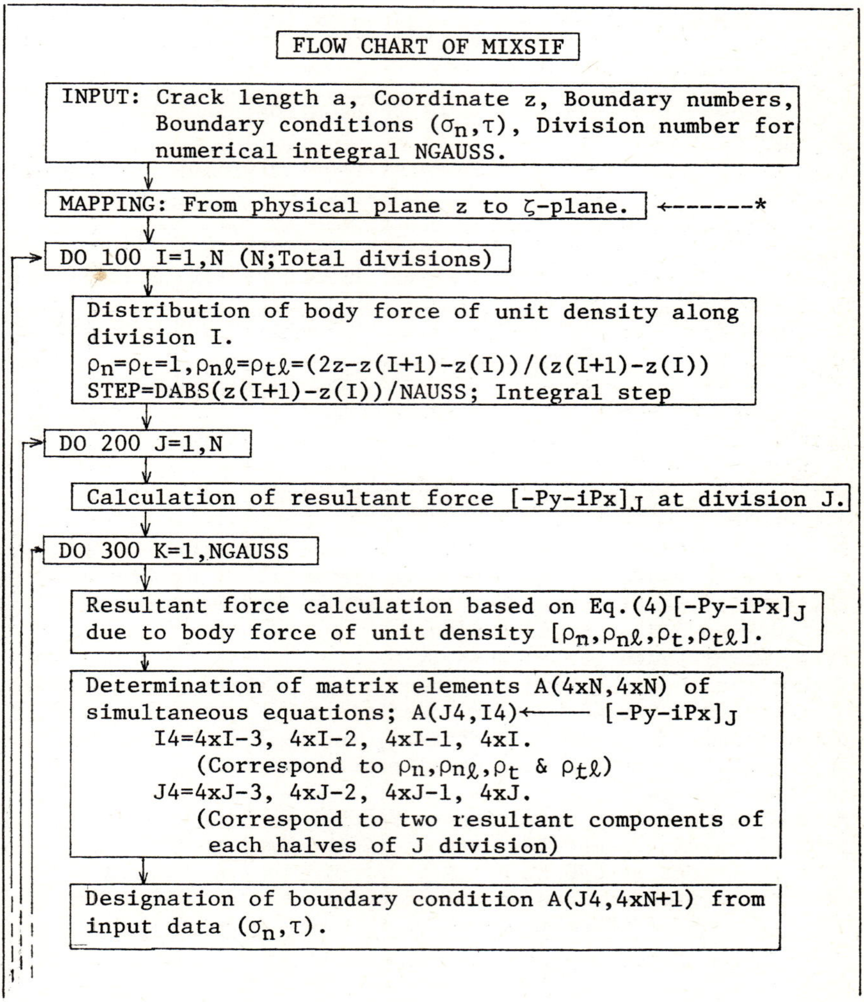

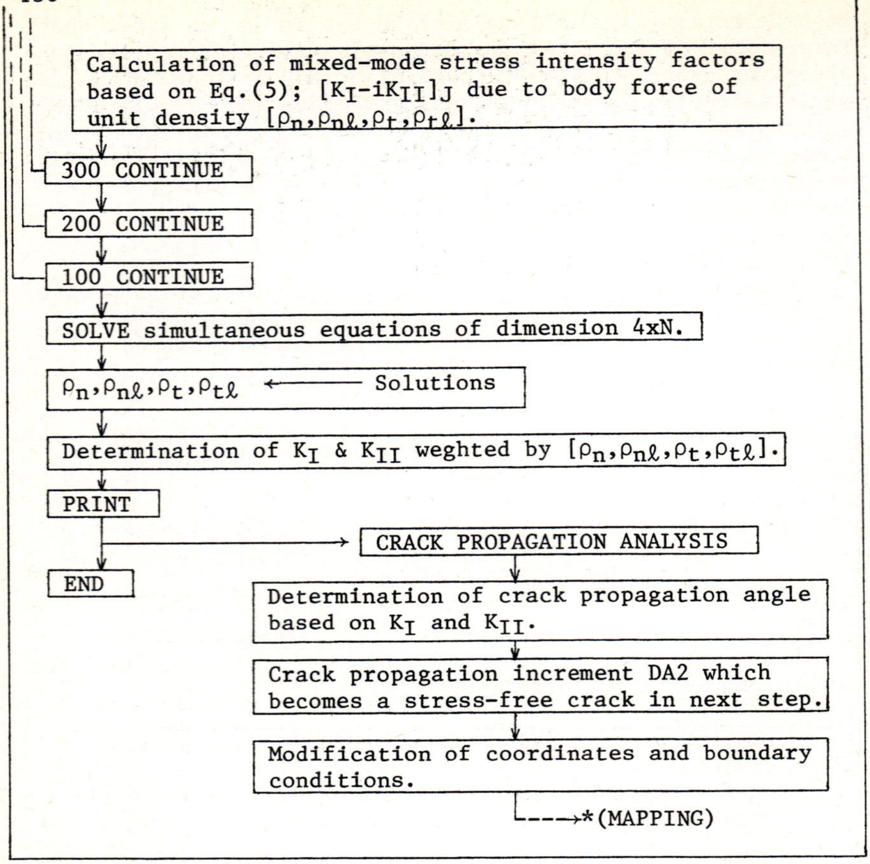

Fig.4

Table 1 PROGRAM DETAIL

Capability: Program yields stress intensity factors K_I and K_{II} for an inner or edge crack in the plate of arbitrary shape.

Limitation: Static analysis. Only stress boundary condition is designated at present time.

Programming Language: FORTRAN IV

Input: Nodal points of plate boundaries. Coordinates of nodal points. Stress boundary conditions.

Output: Stress intensity factors K_I and K_{II}. Crack propagation angle on maximum tangential stress criterion.

Software: Batch

Hardware: Currently operating on FACOM-190 (Kyushu Univ.)

Typical running time: The case of 12 boundary points required 20 second of CPU time. Most of simple two-dimensional crack problems can be solved accurately enough in 60 seconds of CPU time.

3. APPLICATIONS

3.1 Applications to simple problems.

Several applications of MIXSIF to simple problems are shown in Figs.5-8 and Tables 2-5. In some of following figures , the division points (may be called nodal points) marked with ∘ are remained in the figures. These marks will help to see the fineness of the divisions along boundaries. Details of the figures without division points are shown in the figures of Appendix.

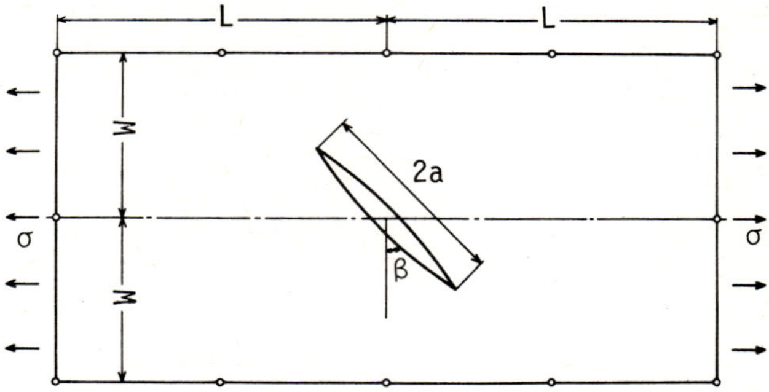

Fig.5

Table 2 Stress intensity factors for Fig.5 .
$F_I=K_I/\sigma\sqrt{\pi a}$, $F_{II}=K_{II}/\sigma\sqrt{\pi a}$, β=45deg. CPU time≃20sec.

$\frac{a}{W}$	F_I				F_{II}			
	Author L/W=2	[7] 2	[14] 2	[13] ∞	Author 2	[7] 2	[14] 2	[13] ∞
0.2	0.515	0.517	0.518	0.519	0.509	0.508	0.507	0.507
0.4	0.565	0.570	0.572	0.572	0.530	0.530	0.529	0.528
0.6	0.647	0.659	0.661	0.661	0.562	0.568	0.567	0.565

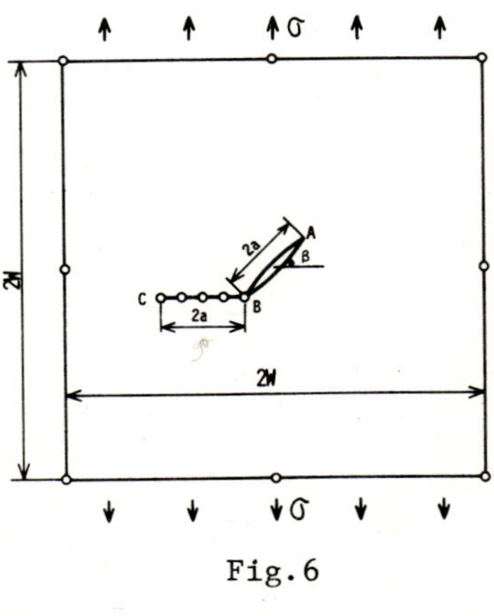

Fig.6

Table 3 Stress intensity factors for Fig.6 . β=45 deg.

$K_{IA}/\sigma\sqrt{2\pi a(1+\cos\beta)}$		$K_{IIA}/\sigma\sqrt{2\pi a(1+\cos\beta)}$	
Author W/a=50	[15] ∞	Author 50	[15] ∞
0.574	0.569	0.654	0.641

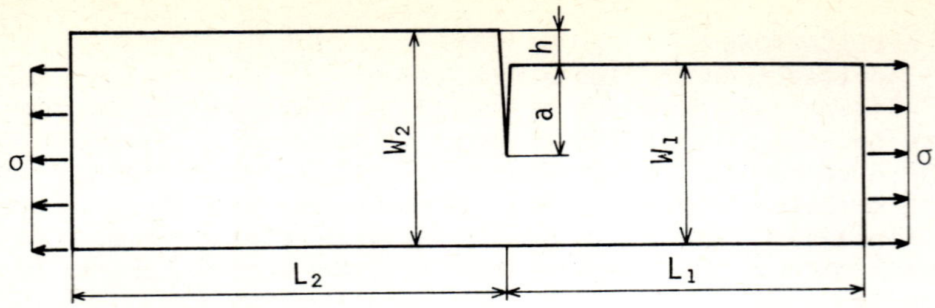

Fig.7 $L_1/W_1=L_2/W_2=2$

Table 4 F_I & F_{II} for Fig.7. $F_I=K_I/\sigma\sqrt{\pi a}$, $F_{II}=K_{II}/\sigma\sqrt{\pi a}$

h/W_1	a/W_1	0.1	0.2	0.3	0.4	0.5
0.2	F_I	1.27	1.40	1.64	2.03	2.61
	F_{II}	0.068	0.031	0.016	0.010	0.007
0.4	F_I	1.26	1.40	1.64	2.01	2.54
	F_{II}	0.066	0.030	0.015	0.008	0.006

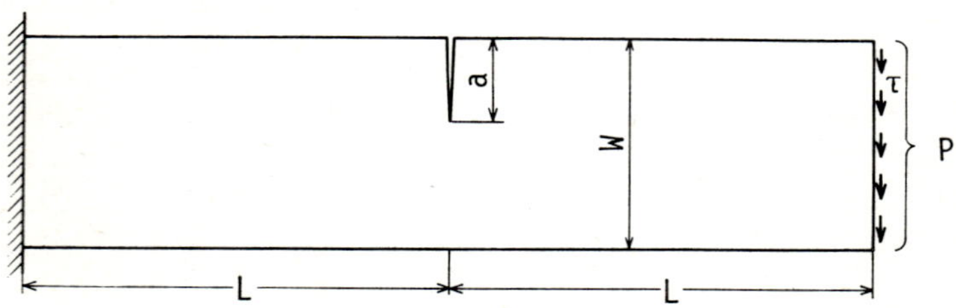

Fig.8 L/W=2, T:Plate thickness.

Table 5 Stress intensity factors for Fig.8.
$K_I=F_I\,\frac{6M}{W^2T}\sqrt{\pi a}$, $K_{II}=F_{II}\,\frac{P}{WT}\sqrt{\pi a}$, M=PL, P=τWT.

a/W	0.1	0.2	0.3	0.4	0.5
F_I	0.94	1.00	1.12	1.25	1.36
F_{II}	0.43	0.68	0.94	1.17	1.36

CPU time ≅43 sec.

3.2 Applications to welded joints.

Recent progress in welding engineering and its application are remarkable. Increase in reliability is contributing to applications. Although we have experienced many brittle fracture accidents in large scale structures and ships before the welding technology has accomplished the recent level, nowadays such a trouble is decreasing. However, from the viewpoints of stress analysis the welded structures are complicated and it prevents us from getting exact solutions even for a simple welded joint. This is the reason why the nominal stress at throat section is customarily used as design criterion. Although the design method based on the nominal stress at throat section is convenient for design simplification, the support by exact analysis should be made. Since most of welded joints have slits which may be regarded as cracks, the estimation of the safety from the viewpoint of stress intensity factor is necessary in addition to the rough estimation based on nominal stress.

Figures 9 and 10 show welded joints solved using the same program MIXSIF. The numerical results of these problems are shown in Table 6 and 7 respectively.

In lap joints (Fig.9 and Table 6), the stress intensity factor of mode II is important because the mode I stress intensity factor is negative.

In Tee joints (Fig.10 and Table 7), the stress intensity factors vary remarkably as the loading position changes under the condition of the constant nominal stress at throat section.

It is evident from these Tables that stress intensity factors have different values for the same value of nominal stress at throat. Thus, the current design method based on nominal stress is not reasonable. Welded joints may not necessarily fracture in the manner controlled by stress intensity factor. However, the safety evaluation on the basis of stress intensity factor as well as nominal stress at throat should be made.

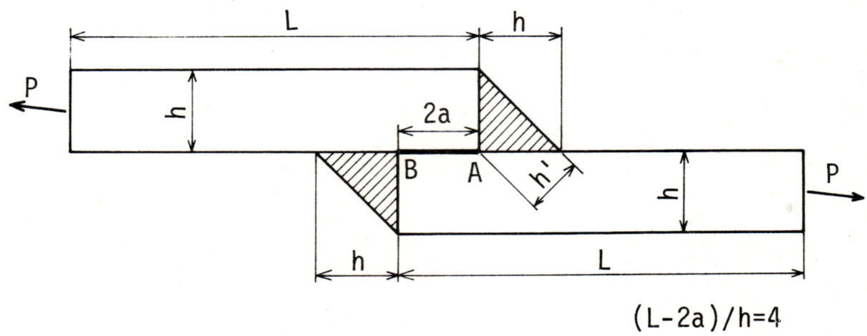

Fig.9 Lap joint

Table 6 Stress intensity factors for Fig.9 .

$$(K_I, K_{II}) = (F_I, F_{II}) \frac{P}{2bh'} \sqrt{\pi h'}$$

2a/h	0.5	1.0	2.0
F_I	-0.196	-0.243	-0.279
F_{II}	-0.106	-0.142	-0.203

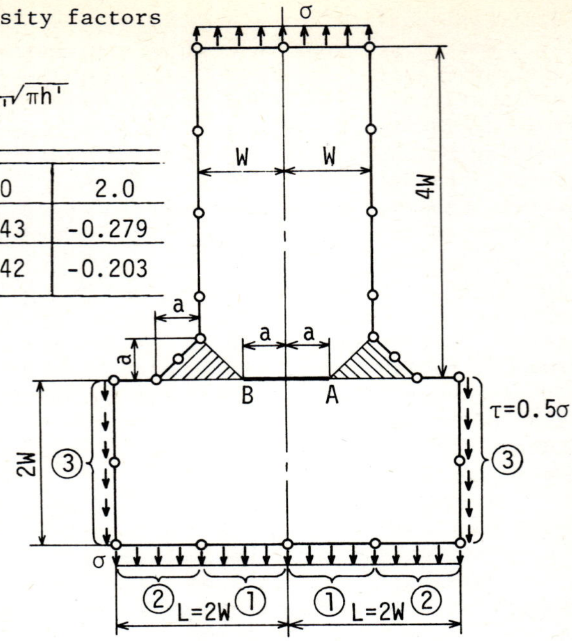

Fig.10(a)

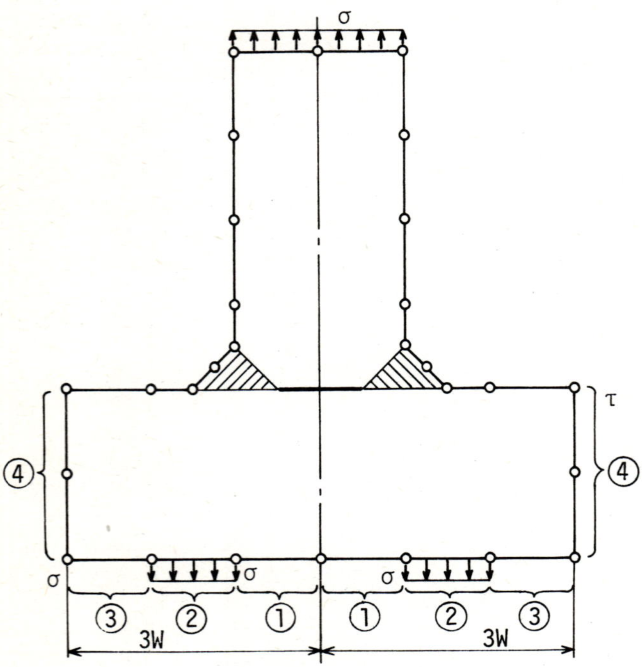

Fig.10(b)

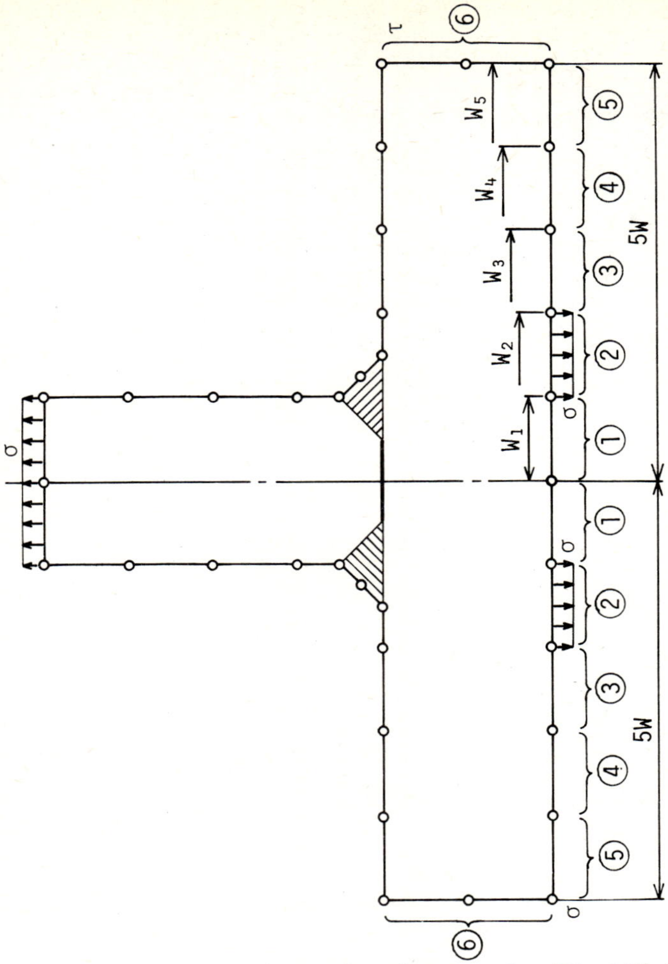

Fig.10(c)

Table 7 Stress intensity factors for Fig.10(a),(b) & (c).
$(K_I, K_{II}) = (F_I, F_{II})\sigma\sqrt{\pi a}$

Loading position	F_I			F_{II}		
	Fig.10 (a)	Fig.10 (b)	Fig.10 (c)	Fig.10 (a)	Fig.10 (b)	Fig.10 (c)
①	1.02	1.02	1.02	$\sim 2\times10^{-4}$	$\sim 2\times10^{-4}$	$\sim 9\times10^{-4}$
②	0.77	0.77	0.77	$\sim 6\times10^{-4}$	$\sim 1\times10^{-3}$	$\sim 2\times10^{-3}$
③	0.64	0.53	0.54	$\sim 7\times10^{-4}$	$\sim 2\times10^{-3}$	$\sim 4\times10^{-3}$
④	—	0.44	0.34	—	$\sim 2\times10^{-3}$	$\sim 5\times10^{-3}$
⑤	—	—	0.16	—	—	$\sim 6\times10^{-3}$
⑥	—	—	0.07	—	—	$\sim 7\times10^{-3}$

3.3 Application to the prediction of crack propagation path.

The prediction of crack propagation path becomes possible by successive calculation of mixed-mode stress intensity factors corresponding to increment of crack length. Details of analyses and experiments on crack propagation path are described in separate papers[5,6].

In the present paper, an example applied to the brittle fracture is shown in Fig.11. Figure 11 illustrates the path of brittle crack in the acryl plate with a square hole. The crack path was predicted on the Erdogan-Sih's criterion[12]. The crack was curved due to the presence of a square hole. The prediction is in good agreement with the experiment.

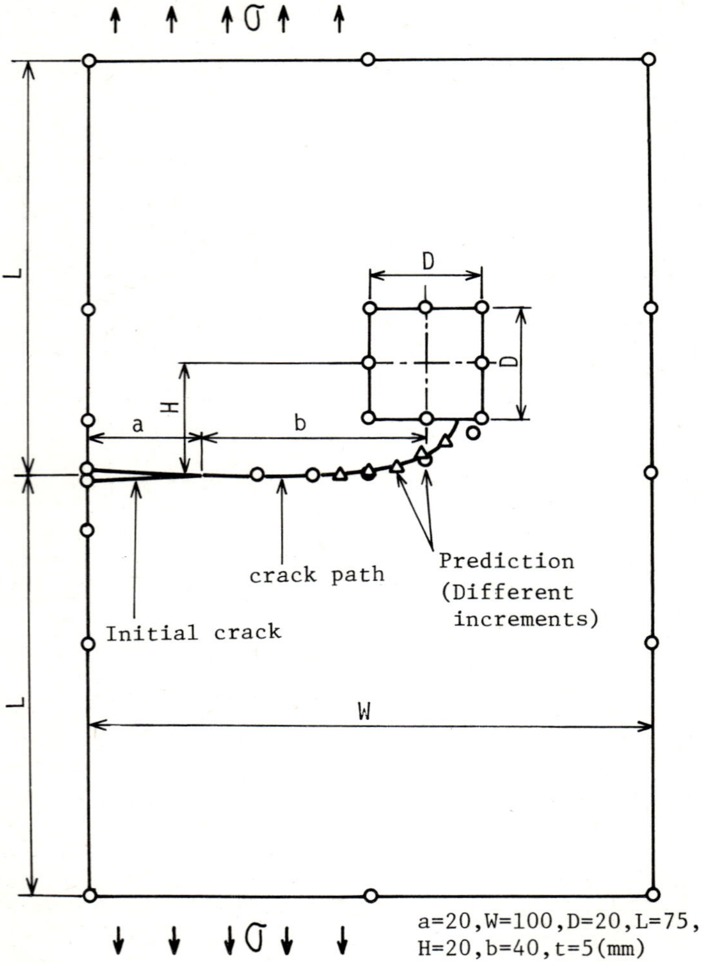

Fig.11 Path of crack propagation in acryl plate with a square hole.
(Solid curve:experiment, circular and triangular mark:prediction)

4. CONCLUSION

The method of analysis of mixed-mode stress intensity factors by the body force method was explained. The method is based on the stress function which expresses the stress field of a point force applied at a point of an infinite plate with a stress-free crack. These stress functions were superposed in order to satisfy the boundary conditions along the finite boundaries imagined in the infinite plate.

The choice of stress function and the distribution of point forces in an improved manner revised the accuracy and saved the labor for the preparation of input data. The point forces were distributed linearly within each division but discontinuously at both ends of them.

The calculation program developed was applied to,

(i) several simple problems with mixed-mode stress intensity factors,
(ii) welded joints, and
(iii) the prediction of crack propagation path.

REFERENCES

1. TADA,H., PARIS,P. and IRWIN,G. - The Stress Analysis of Handbook, Del Research Corporation, Hellertown, Pa., U.S.A., 1973.
2. SIH,G.C. - Handbook of Stress Intensity Factors, Institute of Fracture and Solid Mechanics, Lehigh Univ., Bethlehem, Pennsylvania, 1974.
3. ROOKE,D.P. and CARTWRIGHT,D.J. - Compendium of Stress Intensity Factors, Her Majesty's Stationary Office, London, 1974.
4. NISITANI,H. - Two-Dimensional Problems Solved Using a Digital Computer, Journal of the Japan Soc. Mech. Engrs., Vol.70, No.580, pp.627-635, 1967.
5. MURAKAMI,Y. - Prediction of Crack Propagation Path (Numerical Analyses and Experiments), In contribution to Int. Journ. Fracture.
6. MURAKAMI,Y. - Prediction of Crack Propagation Path (Numerical Analyses and Experiments), The paper to be presented at the Int. Conf. on Analytical and Experimental Fracture Mechanics, Rome, Italy, June 23-27, 1980.
7. MURAKAMI,Y. - Application of the Body Force Method to the Calculation of Stress Intensity Factors for a Crack in the Arbitrarily Shaped Plate, Engineering Fracture Mechanics, Vol.10, pp.497-513, 1978.
8. MURAKAMI,Y. - A Method of Stress Intensity Factor Calculation for the Crack Emanating from an Arbitrarily Shaped Hole or the Crack in the Vicinity of an Arbitrarily Shaped Hole, Trans. Japan Soc. Mech. Engrs., Vol.44, No.378, pp.423-432, 1978.
9. MURAKAMI,Y. - A Method of Stress Concentration Analysis for the Arbitrarily Shaped Plate with a Notch the Shape of which is expressed by a Part of an Ellipse, Trans. Japan Soc. Mech. Engrs., Vol.43, No.372, pp.2837-2844, 1977.

10. MURAKAMI,Y. and ARAKI,S. - Application of Body Force Method to the Analysis of Transverse Bending of an Arbitrarily Shaped Thin Finite Plate with an Elliptical Hole or Notch, Trans. Japan Soc. Mech. Engrs., Vol.45, No.397, pp.998-1006, 1979.
11. ISIDA,M. - Analysis of Stress Intensity Factors of Plates Containing an Arbitrary Array of Line and Branched Cracks, Trans. Japan Soc. Mech. Engrs., Vol.44, No.388, pp.1122-1132, 1978.
12. ERDOGAN,F. and SIH,G.C. - On the Crack Extension in Plates under Plane Loading and Transverse Shear, Journ. Basic Engng., Trans. Am. Soc. Mech. Engrs., Ser. D, Vol.85, No.4, pp.519-527, 1963.
13. TAMATE,O. and IWASAKA,N. - An Arbitrarily Oriented Crack in a Long Strip under Tension, Trans. Japan Soc. Mech. Engrs., Vol.42, No.356, pp.1054-1060, 1976.
14. KITAGAWA,H. and YUUKI,R. - Analysis of the Non-Linear Shaped Cracks in a Finite Plate by the Conformal Mapping Method, Trans. Japan Soc. Mech. Engrs., Vol.43, No.376, pp.4354-4362, 1977.
15. KITAGAWA,H. and YUUKI,R. - Stress Intensity Factors for Branched Cracks in an Infinite Body in the Two Dimensional Stress State, Trans. Japan Soc. Mech. Engrs., Vol.41, No. 346, pp.1641-1649, 1975.

APPENDIX (DIVISION PATTERNS)

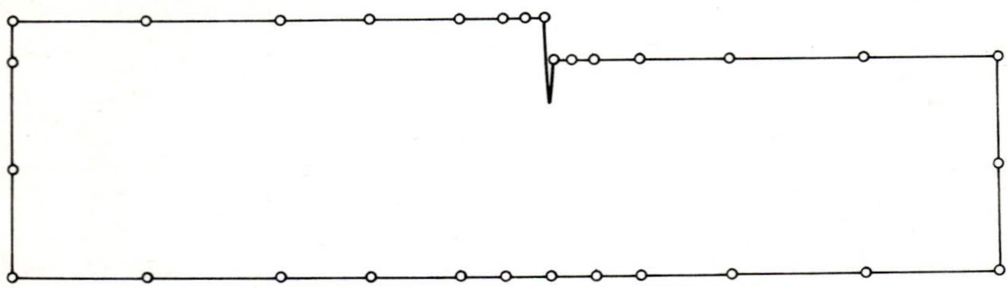

Fig.12

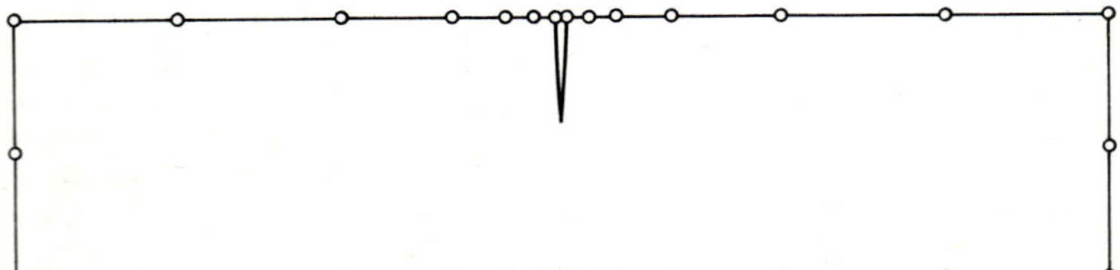

Fig.13

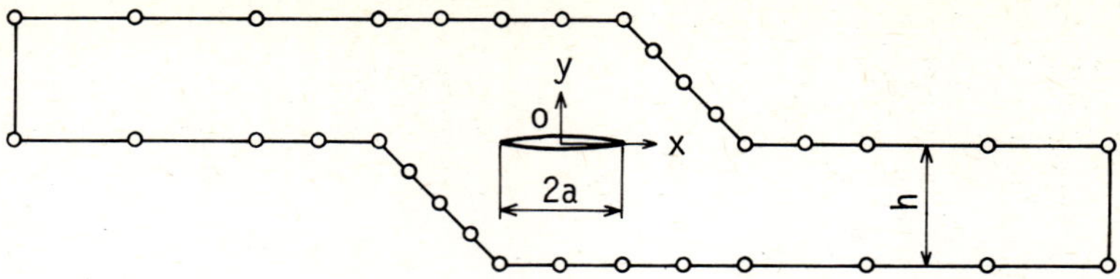

Fig.14

Contact Analysis in Crack Mechanics Using Penalty Function Method

G. Yagawa*, H. Hirayama**, Y. Ando§

SUMMARY

In improving safety and reliability of structural components, the study on crack under complex load is practically important. From the above-mentioned viewpoint, three typical examples of crack propagated from rivet hole, two-point supported specimen with an edge crack loaded by a rigid stamp and through-wall crack in a thick plate under bending load are studied in this paper. In the first and the second problems which involve the contact between different components, various assumptions of contact conditions are considered. In the third problem which has the contact between crack surfaces, the contact area on crack surfaces and the distribution of the stress intensity factor along the thickness direction are determined using the 3-D finite element method. In order to solve these contact problems a method based upon the penalty function is developed.

INTRODUCTION

In order to improve safety and reliability of structural components from the point of view of structural mechanics a number of researches have been made on various behaviors of cracks. Because cracks experience various types of loads in the actual structural components, the studies on cracks under not only simple tensile load but also complex one are important in practice.

The contact problems in crack mechanics are divided into two typical cases considering the relation between crack and contact surfaces.

*Associate Professor, **Graduate Student, §Professor
Department of Nuclear Engineering, University of Tokyo,
Hongo, Bunkyo-ku, Tokyo, Japan

One is the case in which cracked components are loaded through one or more components by contact. In this paper, a crack propagated from rivet hole and two-point supported specimen with an edge crack transversely loaded by a rigid stamp are analyzed. The former, which is important for safety of aircraft structures, has been taken by a few authors without considering friction [1]. In this paper, not only the effect of crack length but also the friction effect in the contact conditions are studied. About the latter, Nied and Erdogan [2] have studied for relatively short cracks. We analyze the same problems by a finite element method to obtain solutions for a long crack and for a large stamp. In this problem, it is necessary to treat a crack-contact problem, in which the distribution of the contact loads is not known apriori and dependent on the geometry of crack, specimen and stamp.

The other is the case involving contact problem between crack surfaces, which appears in the crack closure phenomenon. Two-dimensional crack closure problems have been studied on cyclic reversal bending of single cracked strip by Paris and Tada [3] and on Griffith crack under a general loading by Aksogan [4]. Rubayi and Ved [5] have studied a thick plate containing central crack under pure bending, by photoelastic analysis. Jones and Swedlow [6] and Yagawa and Nishioka [7] have solved the same problem theoretically. Jones and Swedlow [6] have considered the crack closure with Kirchhoff fourth order plate theory. In the present paper, through-wall crack in a thick plate under bending load is taken. The contact area on the crack surfaces and the distribution of the stress intensity factor along the thickness direction are determined using the three-dimensional finite element method.

In the present calculations, the crack tip singularity is taken into account by using the technique of moving the nodal points at the mid-edges of an element near the crack tip to quarter points [8-9].

In order to solve the contact problems, a method based upon the penalty function [10] is developed in this paper. The new method described here not only avoids the inconveniences as pointed out in References [11-16]: necessity of large transformation of global stiffness matrix, increase in degrees of freedom and repetitive calculation to solve the frictional condition, but also decreases the calculation time for repetitive procedure to solve the contact conditions. Concretely in this paper, the contact conditions are taken into account by inserting the special elements based on the penalty function method into the contact surfaces.

RELATION BETWEEN THE LAGRANGE MULTIPLIERS METHOD [17] AND THE PENALTY FUNCTION METHOD [10]

Before entering into the formulation for the contact

problems, consider the relation between the Lagrange multipliers method and the penalty function method by discussing the simple example.

We seek the differentiable continuous function $u=u(x)$ which makes the integral

$$I = \int_{x_1}^{x_2} f(x,u,u')\, dx \quad , \quad u' = \frac{du}{dx} \tag{1}$$

an absolute minimum with the condition

$$g(x,u) = 0 \tag{2}$$

where x and u are respectively an independent variable and a dependent variable. In Equation (1), x_1 and x_2 are constants, and $u(x_1)=u_1$ and $u(x_2)=u_2$ with u_1 and u_2 being constants. Functions f and g are assumed to have proper differentiability.

The first variation of the combination of Equations (1) and (2) using the Lagrange multiplier λ yields

$$\int_{x_1}^{x_2} \left(\frac{\partial f}{\partial u}\,\delta u + \frac{\partial f}{\partial u'}\,\delta u'\right)dx + \int_{x_1}^{x_2} \lambda \frac{\partial g}{\partial u}\,\delta u dx + \int_{x_1}^{x_2} g\delta\lambda dx = 0 \tag{3}$$

where 'δ' denotes variation. Integrating Equation (3) by parts considering $u(x_1)=u_1$ and $u(x_2)=u_2$, we can derive the Euler-Lagrange equation under the subsidiary condition as follows:

$$\frac{\partial f}{\partial u} - \frac{d}{dx}\left(\frac{\partial f}{\partial u'}\right) + \lambda\frac{\partial g}{\partial u} = 0 \quad \text{with} \quad g(x,u) = 0 \tag{4}$$

On the other hand, the first variation of the same problem with the penalty function method is

$$\int_{x_1}^{x_2} \left(\frac{\partial f}{\partial u}\,\delta u + \frac{\partial f}{\partial u'}\,\delta u'\right)dx + \int_{x_1}^{x_2} \alpha g\frac{\partial g}{\partial u}\,\delta u dx = 0 \tag{5}$$

where α is the penalty factor. If we integrate Equation (5) by parts considering $u(x_1)=u_1$ and $u(x_2)=u_2$, the Euler-Lagrange equation is obtained as

$$\frac{\partial f}{\partial u} - \frac{d}{dx}\left(\frac{\partial f}{\partial u'}\right) + \alpha g\,\frac{\partial g}{\partial u} = 0 \tag{6}$$

Equation (5) can be written as follows:

$$\int_{x_1}^{x_2} g\,\frac{\partial g}{\partial u}\,\delta u dx = -\left[\int_{x_1}^{x_2} \left(\frac{\partial f}{\partial u}\,\delta u + \frac{\partial f}{\partial u'}\,\delta u'\right)dx\right]/\alpha \tag{7}$$

Assuming $\frac{\partial g}{\partial u} \neq 0$, it is seen that g approaches zero in the limit when α approaches infinity. Using λ^* defined by

$$\lambda^* = \alpha g \tag{8}$$

Equation (6) is rewritten as follows:

$$\frac{\partial f}{\partial u} - \frac{d}{dx}\left(\frac{\partial f}{\partial u'}\right) + \lambda^* \frac{\partial g}{\partial u} = 0 \tag{9}$$

It can be seen here that λ^* is directly related to the Lagrange multiplier λ in Equation (4) in the limit when α approaches infinity, since Equation (9) has the additional condition g=0 in the limit.

CONTACT CONDITIONS

The contact problem is characterized by its proper nonlinearities due to the variation of contact area with deformation and the irreversibility of frictional effect. In order to treat the irreversibility with the incremental formulation, load is divided into incremental steps and the contact conditions are solved by repetitive calculation in each incremental step. To carry out this procedure, it is necessary that the contact conditions are classified into several cases and the equilibrium equations, the continuity conditions and the criteria are specified in each case.

Consider two bodies A and B as shown in Figure 1. Although we study the problem in the global coordinate system X, Y, Z, the local coordinate system $\bar{x}$, $\bar{y}$, $\bar{z}$ is employed at the contact surface. The latter coordinate system is fixed to body B and the $\bar{x}$-$\bar{y}$ plane is assumed to be a tangential plane of body B. Then, an outward normal vector from body B is the $\bar{z}$-axis. Let $\bar{u}_{\bar{i}}^{\beta}$ and $\bar{p}_{\bar{i}}$ indicate displacement of body β (β=A,B) and contact force applied to body B in the $\bar{i}$ direction ($\bar{i}=\bar{x},\bar{y},\bar{z}$), respectively. '$\Delta$', '$\circ$' and '-' denote respectively the incremental value, the initial value in each incremental step and the value prescribed in the local coordinate system. Distance $\bar{d}_{\circ}$ ($\bar{d}_{\circ} \geq 0$) from body A to body B is defined as shown in Figure 1.

Contact conditions are classified into following three cases:

1. Open condition on the area $\bar{C}_o$: In this condition gap remains open.
2. Sticking condition on the area $\bar{C}^k_{st}$ (k=x or y): In this condition gap remains closed and no sliding occurs in the $\bar{k}$ direction ($\bar{k}=\bar{x}$ or $\bar{y}$).
3. Sliding condition on the area $\bar{C}^k_s$ (k=x or y): In this condition gap remains closed and sliding occurs in the $\bar{k}$ direction ($\bar{k}=\bar{x}$ or $\bar{y}$).

Next, we consider the equilibrium equations, the continuity conditions and the criteria on each contact surface:

1. Open condition on $\bar{C}_o$: The equilibrium equations are

$$\bar{p}_{x\circ}+\Delta\bar{p}_x=0 \quad , \quad \bar{p}_{y\circ}+\Delta\bar{p}_y=0 \quad \text{and} \quad \bar{p}_{z\circ}+\Delta\bar{p}_z=0 \qquad (10a),(10b),(10b)$$

and the criteria are

$$\bar{p}_{z\circ}+\Delta\bar{p}_z\leqq 0 \quad \text{or} \quad \Delta\bar{u}_z^A-\Delta\bar{u}_z^B+\bar{d}_\circ\geqq 0 \qquad (11a),(11b)$$

2. Sticking condition on $\bar{C}_{st}^k$ (k=x or y): The continuity conditions are

$$\Delta\bar{u}_k^A-\Delta\bar{u}_k^B=0 \quad , \quad (k=x \text{ or } y) \quad \text{and} \quad \Delta\bar{u}_z^A-\Delta\bar{u}_z^B+\bar{d}_\circ=0 \qquad (12a),(12b)$$

and the criteria are

$$\Delta u_z^A-\Delta u_z^B+\bar{d}_\circ\leqq 0 \quad , \quad |(\bar{p}_{k\circ}+\Delta\bar{p}_k)|\leqq|\mu(\bar{p}_{z\circ}+\bar{p}_z)| \quad \text{and}$$

$$(\bar{p}_{k\circ}+\Delta\bar{p}_k)(\Delta\bar{u}_k^A-\Delta\bar{u}_k^B)\geqq 0 \quad , \quad (k=x \text{ or } y) \qquad (13a),(13b),(13c)$$

where μ denotes a friction co-efficient.

3. Sliding condition on $\bar{C}_s^k$ (k=x or y): The continuity condition and the equilibrium equation are respectively

$$\Delta\bar{u}_z^A-\Delta\bar{u}_z^B+\bar{d}_\circ=0 \quad \text{and} \quad \bar{p}_{k\circ}+\Delta\bar{p}_k=\pm\mu(\bar{p}_{z\circ}+\Delta\bar{p}_z) \quad , \quad (k=x \text{ or } y) \qquad (14a),(14b)$$

where a sign of μ depends on the sliding direction. The criteria are

$$\Delta\bar{u}_z^A-\Delta\bar{u}_z^B+\bar{d}_\circ\leqq 0 \quad , \quad |(\bar{p}_{k\circ}+\Delta\bar{p}_k)|\geqq|\mu(\bar{p}_{z\circ}+\Delta\bar{p}_z)| \quad \text{and}$$

$$(\bar{p}_{k\circ}+\Delta\bar{p}_k)(\Delta\bar{u}_k^A-\Delta\bar{u}_k^B)\leqq 0 \quad , \quad (k=x \text{ or } y) \qquad (15a),(15b),(15c)$$

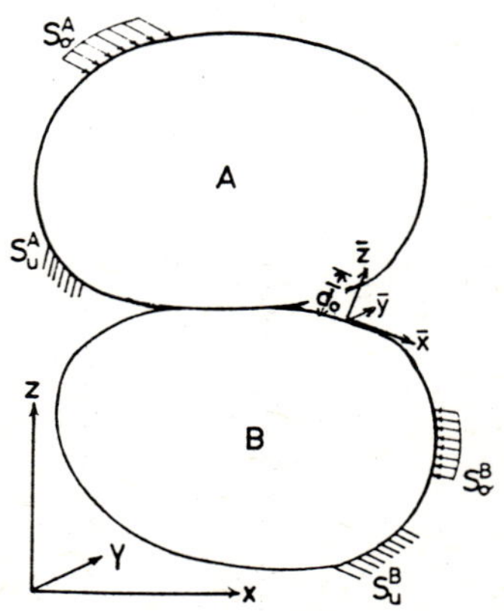

Figure 1. Coordinate systems of contact problem

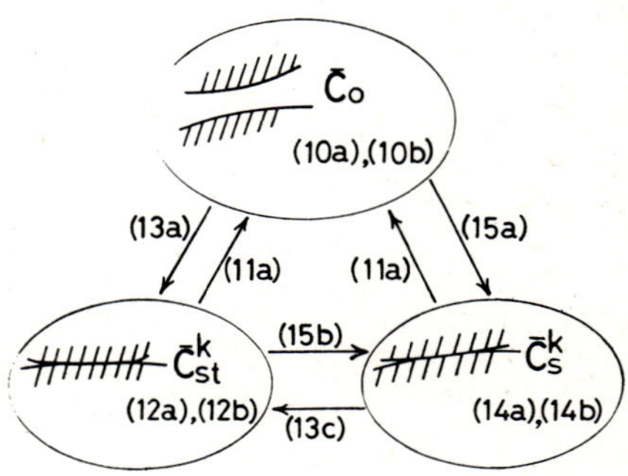

Figure 2. Relation among three contact conditions

Figure 2 shows the relations among the equilibrium equations, the continuity conditions and the criteria on the three different contact surfaces described above. In this figure, the criteria satisfied self-evidently are neglected. For example, in the process that the contact condition turns from $\overline{C}^k_{st}$ to $\overline{C}^k_s$ we need only Equation (15b), because Equations (15a) and (15c) are already satisfied with Equations (12a) and (12b).

THE PRINCIPLE OF VIRTUAL WORK

We assume that external forces f_i^β are applied on S_σ^β and displacements are prescribed on S_u^β, σ_{ij}^β, ε_{ij}^β, u_i^β, f_i^β and p_k denote stress, strain, displacement of body β (β= A or B), external surface force and contact force, respectively. The principle of incremental virtual work for the system consisting of bodies A and B in this problem is given by [17].

$$\begin{aligned}
\delta\Delta\pi = & \int_{V_A} \Delta\sigma_{ij}^A \delta\Delta\varepsilon_{ij}^A dV + \int_{V_B} \Delta\sigma_{ij}^B \delta\Delta\varepsilon_{ij}^B dV - \int_{S_\sigma^A} \Delta f_i^A \delta\Delta u_i^A dS - \int_{S_\sigma^B} \Delta f_i^B \delta\Delta u_i^B dS \\
& - \int_{\overline{C}^k_{st}} \delta\Delta\overline{p}_k^*(\Delta\overline{u}_k^A - \Delta\overline{u}_k^B) dS - \int_{\overline{C}^k_{st}+\overline{C}^k_s} \delta\Delta\overline{p}_z^*(\Delta\overline{u}_z^A - \Delta\overline{u}_z^B + \overline{d}_\circ) dS \\
& - \int_{\overline{C}^k_{st}} \Delta\overline{p}_k^*(\delta\Delta\overline{u}_k^A - \delta\Delta\overline{u}_k^B) dS - \int_{\overline{C}^k_s} \Delta\overline{p}_k(\delta\Delta\overline{u}_k^A - \delta\Delta\overline{u}_k^B) dS \\
& - \int_{\overline{C}^k_{st}+\overline{C}^k_s} \Delta\overline{p}_z^*(\delta\Delta\overline{u}_z^A - \delta\Delta\overline{u}_z^B) dS + \int_{\overline{C}_o} \overline{p}_{k\circ}(\delta\Delta\overline{u}_k^A - \delta\Delta\overline{u}_k^B) dS \\
& + \int_{\overline{C}_o} \overline{p}_{z\circ}(\delta\Delta\overline{u}_z^A - \delta\Delta\overline{u}_z^B) dS = 0 \quad , \quad (i=x,\ y,\ z;\ k=x,\ y) \qquad (16a)
\end{aligned}$$

with

$$\Delta\overline{p}_k = \pm\mu(\overline{p}_{z\circ} + \Delta\overline{p}_z) - \overline{p}_{k\circ} \quad \text{on } \overline{C}^k_s \quad , \ (k=x,\ y) \qquad (16b)$$

where $\Delta\overline{p}_x^*$, $\Delta\overline{p}_y^*$ and $\Delta\overline{p}_z^*$ are the Lagrange multipliers, whose physical meanings are the contact forces $\Delta\overline{p}_x$, $\Delta\overline{p}_y$ and $\Delta\overline{p}_z$, respectively.

Taking into account the relation between the Lagrange multipliers method and the penalty function method as discussed beforehand in this paper, Equations (16a) and (16b) are respectively rewritten as follows:

$$\delta\Delta\pi=\int_{V_A}\Delta\sigma^A_{ij}\delta\Delta\varepsilon^A_{ij}dV+\int_{V_B}\Delta\sigma^B_{ij}\delta\Delta\varepsilon^B_{ij}dV-\int_{S^A_\sigma}\Delta f^A_i\delta\Delta u^A_i dS-\int_{S^B_\sigma}\Delta f^B_i\delta\Delta u^B_i dS$$

$$-\int_{\overline{C}^k_{st}}\alpha(\Delta\overline{u}^A_k-\Delta\overline{u}^B_k)(\delta\Delta\overline{u}^A_k-\delta\Delta\overline{u}^B_k)dS-\int_{\overline{C}^k_s}\Delta\overline{p}_k(\delta\Delta\overline{u}^A_k-\delta\Delta\overline{u}^B_k)dS$$

$$-\int_{\overline{C}^k_{st}+\overline{C}^k_s}\alpha(\Delta\overline{u}^A_z-\Delta\overline{u}^B_z+\overline{d}_\circ)(\delta\Delta\overline{u}^A_z-\delta\Delta\overline{u}^B_z)dS+\int_{\overline{C}_0}\overline{p}_{k\circ}(\delta\Delta\overline{u}^A_k-\delta\Delta\overline{u}^B_k)dS$$

$$+\int_{\overline{C}_0}\overline{p}_{z\circ}(\delta\Delta\overline{u}^A_z-\delta\Delta\overline{u}^B_z)dS=0 \quad , \quad (i=x,\ y,\ z;\ k=x,\ y) \qquad (17a)$$

with

$$\Delta\overline{p}_k=\pm\mu\{\overline{p}_{z\circ}+\alpha(\Delta\overline{u}^A_z-\Delta\overline{u}^B_z+\overline{d}_\circ)\}-\overline{p}_{k\circ} \quad \text{on } \overline{C}^K_s \quad , \quad (k=x,\ y) \qquad (17b)$$

A sign of μ in Equation (17b) is determined by the sliding direction. It is noted here that $\Delta\overline{p}^*_x$, $\Delta\overline{p}^*_y$ and $\Delta\overline{p}^*_z$ are respectively written in terms of the displacements as follows:

$$\Delta\overline{p}^*_x=\alpha(\Delta\overline{u}^A_x-\Delta\overline{u}^B_x) \quad , \quad \Delta\overline{p}^*_y=\alpha(\Delta\overline{u}^A_y-\Delta\overline{u}^B_y) \quad , \quad \Delta\overline{p}^*_z=\alpha(\Delta\overline{u}^A_z-\Delta\overline{u}^B_z+\overline{d}_\circ)$$

(18a),(18b),(18c)

FINITE ELEMENT FORMULATION

Substituting Equation (17b) into Equation (17a), we obtain the finite element equation for this problem as follows:

$$\sum^e_{V_A}\delta\{\Delta U^A_e\}^t[K^A_e]\{\Delta U^A_e\}+\sum^e_{V_B}\delta\{\Delta U^B_e\}^t[K^B_e]\{\Delta U^B_e\}+\sum^e_{S^A_\sigma}\delta\{\Delta U^A_e\}^t\{\Delta F^A_e\}$$

$$+\sum^e_{S^B_\sigma}\delta\{\Delta U^B_e\}^t\{\Delta F^B_e\}-\sum^e_{\overline{C}_0+\overline{C}^k_{st}+\overline{C}^k_s}\begin{bmatrix}\delta\{\Delta U^A_e\}\\ \delta\{\Delta U^B_e\}\end{bmatrix}^t\left[[K^C_e]\begin{bmatrix}\{\Delta U^A_e\}\\ \{\Delta U^B_e\}\end{bmatrix}+\{R^C_e\}\right]=0 \quad ,$$

$$(k=x,\ y) \qquad (19)$$

where $\{\Delta U^\beta_e\}$ and $\{\Delta F^\beta_e\}$ (β=A or B) indicate element incremental displacement and element incremental force vector respectively, and $[K^\beta_e]$ denotes element stiffness matrix. Furthermore, $[K^C_e]$ and $\{R^C_e\}$ are respectively element stiffness matrix and element force vector of a special element inserted between contact surfaces in taking into account the contact conditions. Subscript e, superscripts β (β=A or B) and t indicate element value, discrimination between bodies A and B and transposition of matrix respectively.

Element incremental contact force vectors for each body are indicated by $\{\Delta P_e^A\}$ and $\{\Delta P_e^B\}$ in terms of displacements as follows:

$$\begin{bmatrix} \{\Delta P_e^A\} \\ \{\Delta P_e^B\} \end{bmatrix} = [K_e^C] \begin{bmatrix} \{\Delta U_e^A\} \\ \{\Delta U_e^B\} \end{bmatrix} + \{R_e^C\} \quad \text{on } \overline{C}_o + \overline{C}_{st}^k + \overline{C}_s^k \quad , \ (k=x,\ y) \tag{20}$$

While the first four terms of the right-hand-side of Equation (17a) are discretized with the ordinary elements, the other terms of Equation (17a) together with Equation (17b) are discretized with the special elements described above. The details of the formulations are shown for 3-D problems in the following.

THREE-DIMENSIONAL PROBLEMS

Let bodies A and B be discretized by the usual finite elements and the space between the contact surfaces be modeled by the 8-noded brick-type special elements as shown in Figure 3. The details of $\{\Delta P_e^A\}$, $\{\Delta P_e^B\}$, $\{\Delta U_e^A\}$ and $\{\Delta U_e^B\}$ in Equation (20) can be written for the special element as follows:

$$\{\Delta P^A\}^t = -\{\Delta P^B\}^t = \llcorner \Delta p_x^k, \Delta p_y^k, \Delta p_z^k, \Delta p_x^l, \Delta p_y^l, \Delta p_z^l, \Delta p_x^m, \Delta p_y^m, \Delta p_z^m, \Delta p_x^n, \Delta p_y^n, \Delta p_z^n \lrcorner \tag{21}$$

$$\{\Delta U^A\}^t = \llcorner \Delta u_x^{kA}, \Delta u_y^{kA}, \Delta u_z^{kA}, \Delta u_x^{lA}, \Delta u_y^{lA}, \Delta u_z^{lA}, \Delta u_x^{mA}, \Delta u_y^{mA}, \Delta u_z^{mA}, \Delta u_x^{nA}, \Delta u_y^{nA}, \Delta u_z^{nA} \lrcorner \tag{22}$$

$$\{\Delta U^B\}^t = \llcorner \Delta u_x^{kB}, \Delta u_y^{kB}, \Delta u_z^{kB}, \Delta u_x^{lB}, \Delta u_y^{lB}, \Delta u_z^{lB}, \Delta u_x^{mB}, \Delta u_y^{mB}, \Delta u_z^{mB}, \Delta u_x^{nB}, \Delta u_y^{nB}, \Delta u_z^{nB} \lrcorner \tag{23}$$

where for example Δp_x^k and Δu_x^{kA} denote the contributions from this special element to the nodal contact force in the x direction at the pair of nodes k_A and k_B and the nodal displacement in the x direction at the node k_A respectively.

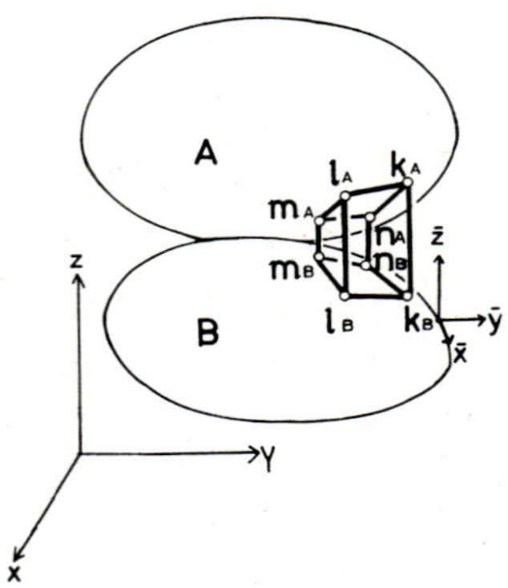

Figure 3. Special element between two bodies

Next, $[K_e^C]$ and $\{R_e^C\}$ in Equation (20) are obtained as follows:

$$[K_e^C]=[L_e]^t[\overline{K}_e^C][L_e] \quad , \quad \{R_e^C\}=[L_e]^t\{\overline{R}_e^C\} \qquad (24),(25)$$

where $[L_e]$ is the element direction cosine matrix from the global coordinate system to the local coordinate system, and $[\overline{K}_e^C]$ and $\{\overline{R}_e^C\}$ are defined as follows:

$$[\overline{K}_e^C]=\begin{bmatrix} \underset{\sim}{k}^c & -\underset{\sim}{k}^c \\ -\underset{\sim}{k}^c & \underset{\sim}{k}^c \end{bmatrix} \qquad \text{with} \qquad \underset{\sim}{k}^c=\begin{bmatrix} \underset{\sim}{k}^k & \underset{\sim}{0} & \underset{\sim}{0} & \underset{\sim}{0} \\ \underset{\sim}{0} & \underset{\sim}{k}^l & & \\ \underset{\sim}{0} & & \underset{\sim}{k}^m & \\ \underset{\sim}{0} & & & \underset{\sim}{k}^n \end{bmatrix} \qquad (26)$$

$$\{\overline{R}_e^C\}=\begin{Bmatrix} \underset{\sim}{r}^c \\ -\underset{\sim}{r}^c \end{Bmatrix} \qquad \text{with} \qquad \underset{\sim}{r}^c=\begin{Bmatrix} \underset{\sim}{r}^k \\ \underset{\sim}{r}^l \\ \underset{\sim}{r}^m \\ \underset{\sim}{r}^n \end{Bmatrix} \qquad (27)$$

Here,

$$\underset{\sim}{k}^h = \begin{cases} \begin{bmatrix} 0 & 0 & 0 \\ 0 & 0 & 0 \\ 0 & 0 & 0 \end{bmatrix} & \text{on } \overline{C}_o \\ \begin{bmatrix} \alpha & 0 & 0 \\ 0 & \alpha & 0 \\ 0 & 0 & \alpha \end{bmatrix} & \text{on } \overline{C}^x_{st}+\overline{C}^y_{st} \\ \begin{bmatrix} \alpha & 0 & 0 \\ 0 & 0 & \pm\mu\alpha \\ 0 & 0 & \alpha \end{bmatrix} & \text{on } \overline{C}^x_{st}+\overline{C}^y_{s} \\ \begin{bmatrix} 0 & 0 & \pm\mu\alpha \\ 0 & \alpha & 0 \\ 0 & 0 & \alpha \end{bmatrix} & \text{on } \overline{C}^x_{s}+\overline{C}^y_{st} \\ \begin{bmatrix} 0 & 0 & \pm\mu\alpha \\ 0 & 0 & \pm\mu\alpha \\ 0 & 0 & \alpha \end{bmatrix} & \text{on } \overline{C}^x_{s}+\overline{C}^y_{s} \end{cases} \quad , \quad \underset{\sim}{r}^h = \begin{cases} \begin{Bmatrix} -\overline{p}^h_{xo} \\ -\overline{p}^h_{yo} \\ -\overline{p}^h_{zo} \end{Bmatrix} & \text{on } \overline{C}_o \\ \begin{Bmatrix} 0 \\ 0 \\ \alpha\overline{d}^h_o \end{Bmatrix} & \text{on } \overline{C}^x_{st}+\overline{C}^y_{st} \\ \begin{Bmatrix} 0 \\ -\overline{p}^h_{yo}\pm\mu(\overline{p}^h_{zo}+\alpha\overline{d}^h_o) \\ \alpha\overline{d}^h_o \end{Bmatrix} & \text{on } \overline{C}^x_{st}+\overline{C}^y_{s} \\ \begin{Bmatrix} -\overline{p}^h_{xo}\pm\mu(\overline{p}^h_{zo}+\alpha\overline{d}^h_o) \\ 0 \\ \alpha\overline{d}^h_o \end{Bmatrix} & \text{on } \overline{C}^x_{s}+\overline{C}^y_{st} \\ \begin{Bmatrix} -\overline{p}^h_{xo}\pm\mu(\overline{p}^h_{zo}+\alpha\overline{d}^h_o) \\ -\overline{p}^h_{yo}\pm\mu(\overline{p}^h_{zo}+\alpha\overline{d}^h_o) \\ \alpha\overline{d}^h_o \end{Bmatrix} & \text{on } \overline{C}^x_{s}+\overline{C}^y_{s} \end{cases} \qquad (28)(29)$$

In Equations (28) and (29) the superscript h represents a pair of nodes k, l, m or n and $\overline{p}^h_{xo}$, $\overline{p}^h_{yo}$ and $\overline{p}^h_{zo}$ denote the contributions from this special element to the initial contact forces at the pair of nodes h_A and h_B. The initial value of the distance between nodes h_A and h_B is $\overline{d}^h_o$ in each incremental step. A sign of μ in Equations (28) and (29) is determined by the sliding direction.

RESULTS AND DISCUSSIONS

A CRACK PROPAGATED FROM RIVET HOLE

Consider a crack propagated from a rivet hole loaded through the contact between rivet and hole as shown in Figure 4a (Case 1). The same problem has been analyzed without the consideration of the friction in Reference [1]. The input data used for the numerical solution are shown in Table 1.

In this paper, we solve the same problem considering the friction effect between rivet and hole for various values of friction co-efficient μ from 0 through 0.5. Additional two other cases without the contact phenomenon as shown in Figures 4b and 4c are also analyzed for comparisons. In the former (Case 2) rivet and hole are continuous in the right hand half side of the hole, and in the latter (Case 3) hole without rivet is loaded by internal pressure in the right hand half side of the hole. In order to take into account the singularity at the crack tip, the interior nodes of the element near the crack tip are moved to quarter points [8-9] in the present calculations.

Nondimensional stress intensity factor K_I^* is determined in the plane stress condition from

$$K_I^* = \lim_{r\to 0} \sqrt{2}Eu(r)/8\sigma_o\sqrt{ar} \tag{30}$$

where r, u(r) and σ_o denote respectively the distance from the crack tip along the free surface of the crack (θ=90°), the crack opening displacement and the mean stress at the fixed edge of the plate (=5.236 kg/mm^2 in this calculation). To obtain better accuracy for K_I^* with Equation (30), the linear extrapolation to r=0 is made using several values of $K_I^*(r)$.

In the analysis of contact problems using the penalty function method it is necessary to determine the suitable value of the penalty factor α before calculation. Theoretically, the accuracy of the solution improves when the penalty factor α approaches infinity. In the numerical solution, however, unavoidable round-off error appears because of available word length of computers, so the suitable value for α may exist. It is thought that the best value of α is determined depending on the magnitude of Young's modulus, the size of the finite elements and the size of word of the computer used. Table 2 shows the relation obtained between K_I^* and α for Case 1 (a=7.5 mm and μ=0). From this table it is found that the penalty factor α in the range from 10^3 to 10^9 gives almost constant K_I^*. Hence, in the following calculation we use the penalty factor of $\alpha=10^6$.

Nondimensional stress intensity factors K_I^* for the three cases are compared in Table 3. Close agreement between the present result and the others in Reference [1] is seen from this table and it is also found that the friction co-efficient μ increasing from 0 to 0.5, K_I^* changes from the value for μ=0 to the value for the Case 2 and that the value for the Case 3 is smaller than the others.

The stresses σ_R and $\tau_{R\theta}$, which occur on the periphery of the hole in the normal and tangential directions respectively, are shown in Figures 5a and 5b for a=10 mm respectively. These figures show that the distributions of the stresses with increasing μ approach to those for the Case 2.

In Figure 6, the convergence characteristic of the non-dimensional stresses is shown for the Case 1 (a=7.5 mm and μ=0.5). In this problem we did not employ the incremental approach, since no irreversibility of the friction appears. From Figure 6 it is observed that the convergent solution is obtained in the 5-th stage of the calculation and the better assumption of the initial condition is thought to make the computer time shorter.

Table 1 The input data used for calculations (A crack propagated from a rivet hole)

Young's modulus	E	7,200 kgf/mm^2
Poisson's ratio	ν	0.3
	W	30 mm
	L	30 mm
	R	5 mm
	a	7.5 mm or 10 mm
F (Case 1 and Case 2)		314.16 kgf
p (Case 3)		31.416 kgf/mm^2

Table 2 Relation between K_I^* and α for Case 1 (a=7.5 mm and μ=0)

α	10^1	10^3	10^5	10^7	10^9	10^{11}
K_I^*	2.01	1.92	1.92	1.92	1.92	1.94

Table 3 Comparison of K_I^* among several finite element methods

Method	a = 7.5 mm Case 1 μ=0	a = 7.5 mm Case 1 μ=0.25	a = 7.5 mm Case 1 μ=0.5	a = 7.5 mm Case 2	a = 7.5 mm Case 3	a = 10 mm Case 1 μ=0	a = 10 mm Case 1 μ=0.25	a = 10 mm Case 1 μ=0.5	a = 10 mm Case 2	a = 10 mm Case 3
Present*	1.92	2.00	2.05	2.12	1.77	1.74	1.77	1.78	1.82	1.68
EPIC[1]*	1.89	—	—	2.13	1.75	—	—	—	—	—
CONTAC-P[1]*	1.91	—	—	2.18	1.77	—	—	—	—	—
CONTAC-P[1]**	1.92	—	—	2.14	1.79	—	—	—	—	—
MARC[1]**	1.93	—	—	2.09	1.71	—	—	—	—	—

*: Direct displacement method using Equation (30)

**: Direct stress method using the relation $K_I^* = \lim_{r\to 0} \frac{\sqrt{2r}}{\sigma_0\sqrt{a}}\sigma_x$

where σ_x is the stress in the perpendicular direction to crack prolongation line (θ=90°)

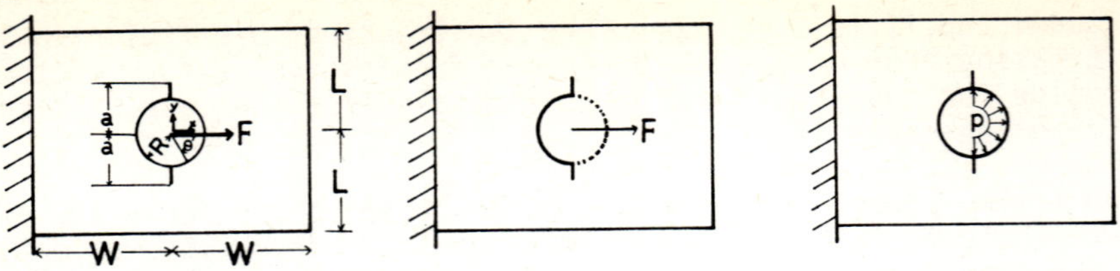

(a) Crack is loaded by contact between rivet and hole (Case 1)
(b) Rivet and hole are continuous in the right hand half side of the hole (Case 2)
(c) Hole is loaded by internal pressure in the right hand half side of the hole (Case 3)

Figure 4. Crack propagated from rivet hole

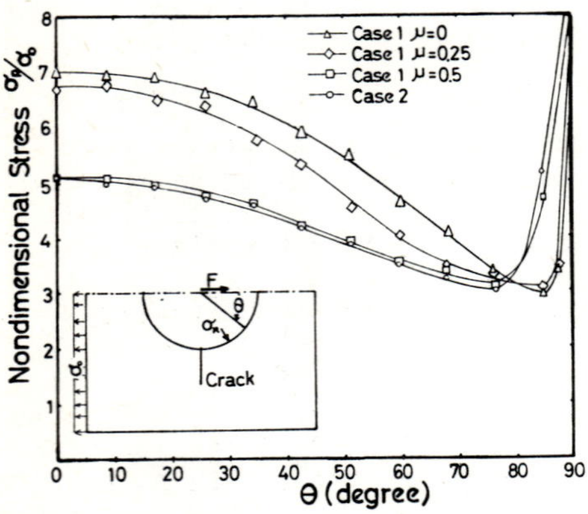

Figure 5a. Distributions of stress along the periphery of the hole

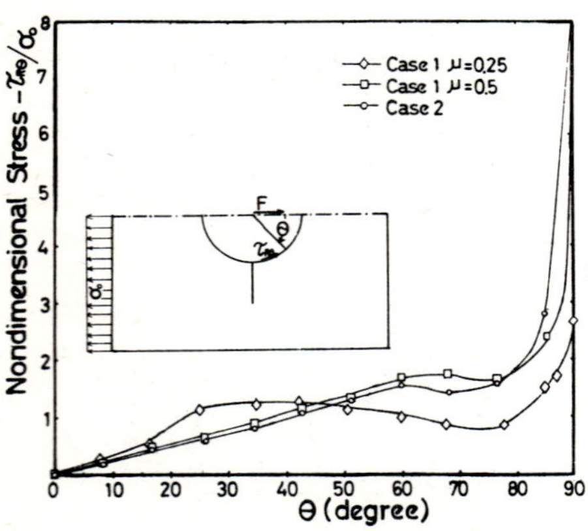

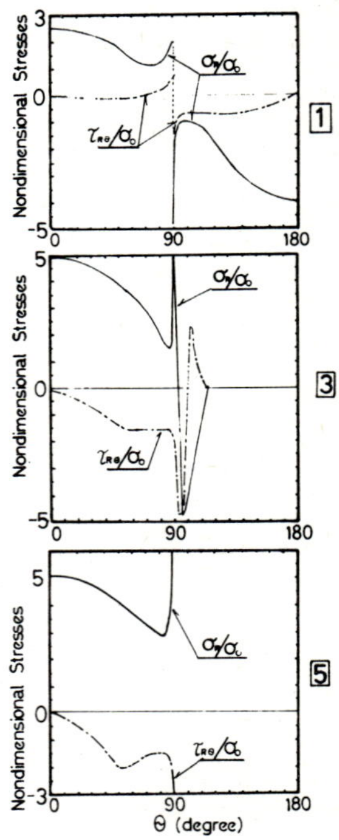

Figure 6. Convergence characteristics of the nondimensional stresses along the periphery of the hole

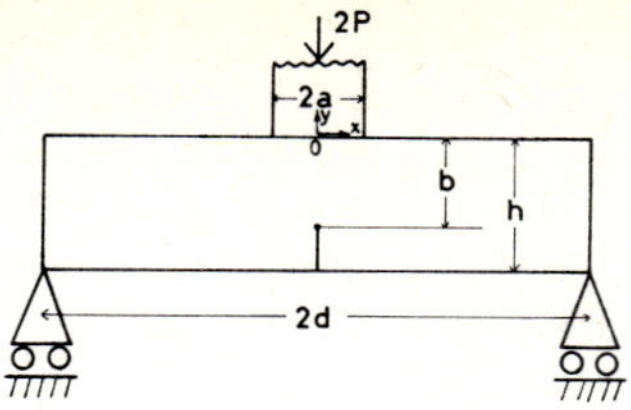

Figure 7a. Two-point supported specimen with an edge crack loaded by a rigid stamp

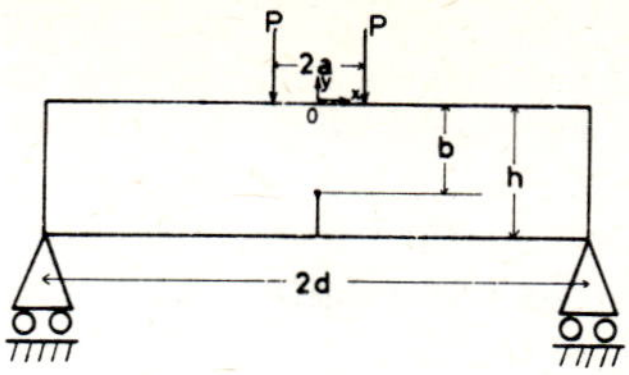

Figure 7b. Two-point supported specimen with an edge crack loaded in the four-point bend condition

TWO-POINT SUPPORTED SPECIMEN WITH AN EDGE CRACK LOADED BY RIGID STAMP

An analysis of two-point supported specimen with an edge crack loaded by a rigid stamp as shown in Figure 7a is carried out varying the crack length and the stamp geometry. The input data used for the numerical calculation are shown in Table 4. The Young's moduli ratio of 1000:1 between the stamp and the specimen are used for the approximation of the rigid condition of the stamp. The friction co-efficient between stamp and specimen is assumed to be zero. No incremental calculation was used here since the present problem involves no irreversibility of the friction.

The nondimensional stress intensity factor K_I^* is obtained by the extrapolation of the crack opening displacements along the free surface of the crack u(r) using the following equation in the plane strain condition.

$$K_I^* = \lim_{r\to 0} \sqrt{2}Eh^2u(r)/24(1-\nu^2)dP\sqrt{r(h-b)} \quad (31)$$

In order to obtain more accurate stress intensity factor using the extrapolation from the displacement, suitable finite element grids are necessary. Based on the preliminary calculations performed beforehand on the standard three point-bend specimen, we found that our finite element grids give only 1 percent deviation from the exact solution [18] for the stress intensity factor.

Table 4 The input data used for calculations (Two-point supported specimen with an edge crack)

Young's Modulus (specimen)	E	20 kgf/mm^2
Young's Modulus (stamp)	E'	20,000 kgf/mm^2
Poisson's Ratio (specimen and stamp)	ν	0.3
h		100 mm
d		200 mm
Friction Coefficient	μ	0

In Figure 8 the variations of the nondimensional stress intensity factors with the nondimensional stamp width are shown. Figure 9 shows the variations of the nondimensional stress intensity factors with the nondimensional distance between loading points in the four-point bend problem as given in Figure 7b. In Figure 9, the results under the stamp loading are also shown for comparison. Figure 10 shows the distributions of the normalized pressure under the stamp. In these figures the results by Nied and Erdogan [2] are also illustrated.

From Figure 8 it is found that the stress intensity factor for the constant crack length decreases as the width 2a of the stamp increases. It can be seen from Figure 9 that the stress intensity factor under the stamp loading is larger than the value in the four-point bend specimen. Furthermore, it is observed from Figure 10 that because of the bending of the specimen the contact pressure at x=0, y=0 decreases and the contact surfaces separate as the width of the stamp increases. From these observations it is confirmed that the stress intensity factor under the stamp loading approaches to the value in the four-point bend specimen when the width of the stamp increases.

It is concluded here that the contact problem between stamp and specimen with crack is distinct except when the width of the stamp is relatively large compared with the width of the specimen.

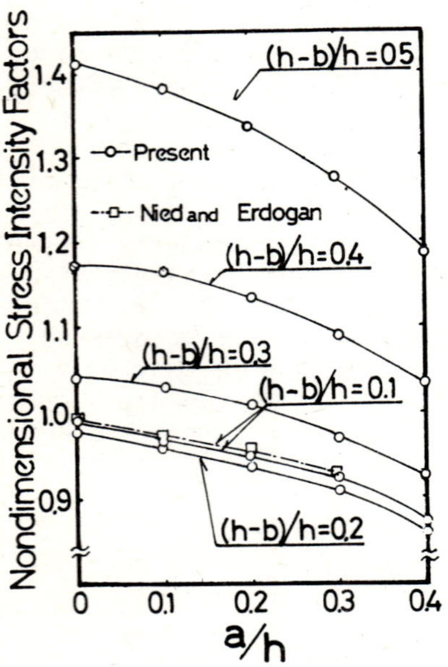

Figure 8. Variations of nondimensional stress intensity factor with crack length to specimen width ratio

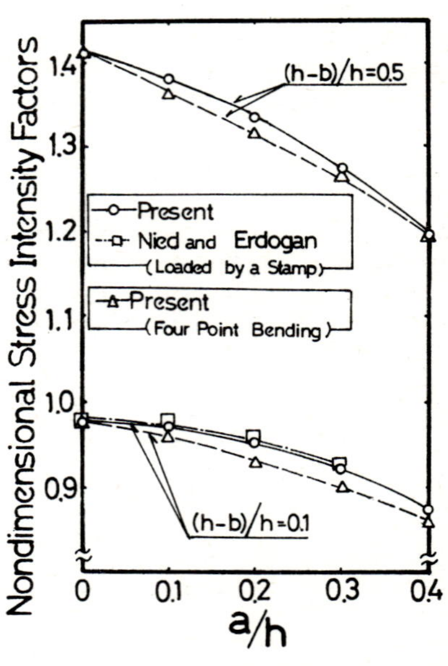

Figure 9. Nondimensional stress intensity factors under two different conditions

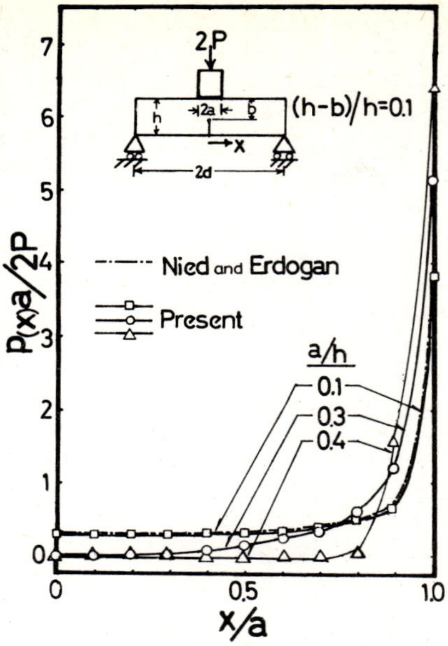

Figure 10. Normalized pressure distributions under a rigid stamp

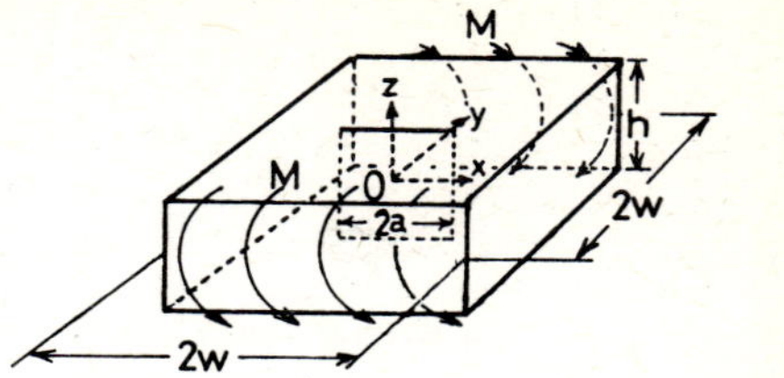

Figure 11a. Thick plate with a through-wall crack under bending load

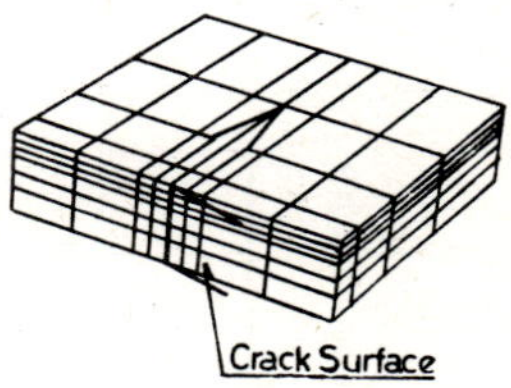

Figure 11b. Finite element grid

A THICK PLATE WITH A THROUGH-WALL CRACK UNDER BENDING

A thick plate with a through-wall crack under bending moment M as shown in Figure 11a is analyzed using the three-dimensional finite element method. The finite element grid is shown in Figure 11b. On the same problem Jones and Swedlow [6] have studied using Kirchhoff fourth-order plate theory without the three-dimensional consideration. Yagawa and Nishioka [7] have solved the same problem without the consideration of the crack closure. In the present paper, the cylindrical bend problem is studied with taking into account the phenomenon of the crack closure. Table 5 shows the input data used for the numerical calculation.

Figure 12 depicts the deformed shape of the cracked plate and the contact area at the crack surface obtained in this paper. In this figure the deformation is exaggerated for clarity.

Table 5 The input data used for calculations (A thick plate with a central through-wall crack)

Young's Modulus	E	20,000 kgf/mm^2
Poisson's Ratio	ν	0.3
w		100 mm
h		25 mm
a		50 mm

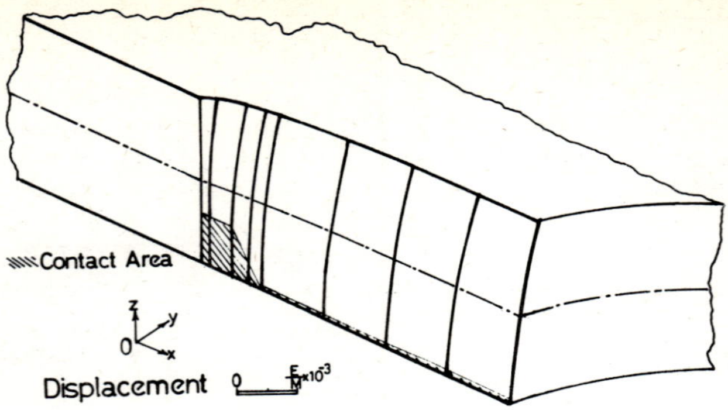

Figure 12. Deformed plate and contact area

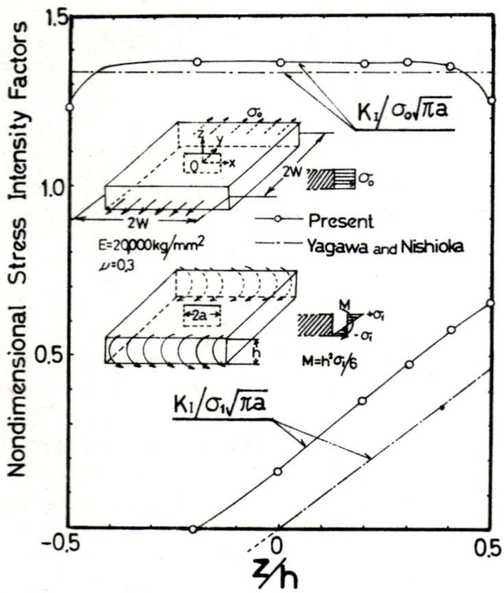

Figure 13. Distributions of nondimensional stress intensity factor along the thickness direction

The distributions of the elastic stress intensity factors are shown in Figure 13, where the nondimensional stress intensity factor is calculated using the relation:

$$K_I^*(z) = \lim_{r \to 0} \sqrt{2}E'u(z,r)/8\sigma'\sqrt{ar} \tag{32}$$

with

$$E' = E/(1-\nu^2) \quad \text{at } -h/2<z<h/2 \quad \text{and} \quad E' = E \quad \text{at } z=\pm h/2$$

where u(z,r) denotes the crack opening displacement and $\sigma'=\sigma_o$ for the plane extension problem and $\sigma'=\sigma_1$ for the bend problem as shown in Figure 13.

In order to check the suitability of the finite element grid the plane extension problem is analyzed and the result is shown in Figure 13. From this figure it is thought that the finite element grid used here is suitable. From the observation of Figure 13 it is clear that the nondimensional stress intensity factor with closure phenomenon obtained in this paper is larger than the value without it given in Reference [7].

CONCLUSIONS

The major conclusions obtained from this study are summarized below.

1. A method to solve contact problems is developed based on the penalty function method.
2. The two-dimensional analysis for the crack-contact problems is studied on a crack propagated from a rivet hole and two-point supported specimen with an edge crack loaded by a rigid stamp.
3. For the bending of a thick plate with a through-wall crack, the distribution of the stress intensity factor and the contact area are determined with the consideration of the crack closure.

REFERENCES

1. Second (Final) Report of Research Cooperation Subcommittee 46 on Research and Development of the Methods for Inelastic Structural Analysis, The Japan Society of Mechanical Engineers, 1979.

2. NIED, H. F. and ERDOGAN, F. - A Cracked Beam or Plate Transversely Loaded by a Stamp, Int. J. Solids Structures, Vol. 15, pp. 951-965, 1979.

3. PARIS, P. C. and TADA. H. - The Stress Intensity Factors for Cyclic Reversed Bending of a Single Edge Cracked Strip Including Crack Surface Interference, Int. Journ. of Fracture, Vol. 11, pp. 1070-1072, 1975.

4. AKSOGAN, O. - Partial Closure of a Griffith Crack under a General Loading, Int. Journ. of Fracture, Vol. 11, pp. 659-670, 1975.

5. RUBAYI, N. A. and VED, R. - Photoelastic Analysis of a Thick Square Plate Containing Central Crack and Loaded by Pure Bending, Int. Journ. of Fracture, Vol. 12, pp. 435-451, 1976.

6. JONES, D. P. and SWEDLOW, J. L. - The Influence of Crack Closure and Elasto-Plastic Flow on the Bending of a Cracked Plate, Int. Journ. of Fracture, Vol. 11, pp. 897-914, 1975.

7. YAGAWA, G. and NISHIOKA, T. - Finite Element Analysis of Stress Intensity Factors for Plane Extension and Plate Bending Problems, Int. J. Num. Meth. Engng., Vol. 14, pp. 727-740, 1979.

8. HENSHELL, R. D. and SHAW, K. G. - Crack Tip Finite Elements are Unnecessary, Int. J. Num. Meth. Engng., Vol. 9, pp. 495-507, 1975.

9. BARSOUM, R. S. - On the Use of Isoparametric Finite Elements in Linear Fracture Mechanics, Int. J. Num. Meth. Engng., Vol. 10, pp. 25-37, 1976.

10. ZIENKIEWICZ, O. C. - The Finite Element Method, McGraw-Hill, London, 1977.

11. WILSON, E. A. and PARSONS, B. - Finite Element Analysis of Elastic Contact Problems Using Differential Displacements, Int. J. Num. Meth. Engng., Vol. 2, pp. 387-395, 1970.

12. CHAN, S. K. and TUBA, I. S. - A Finite Element Method for Contact Problems of Solid Bodies-Part I. Theory and Validation, Int. J. Mech. Sci., Vol, 13, pp. 615-625, 1971.

13. TSUTA. T. and YAMAJI, S. - Finite Element Analysis of Contact Problem, Theory and Practice in Finite Element Structural Analysts, University of Tokyo Press, 1973.

14. FREDRIKSSON, B. - Finite Element Solution of Surface Nonlinearities in Structural Mechanics with Special Emphasis to Contact and Fracture Mechanics Problems, Comput. Structures, Vol. 6, pp. 281-290, 1976.

15. STADTER, J. T. and WEISS, R. O. - Analysis of Contact through Finite Element Gap, Comput. Structures, Vol. 10, pp. 867-873, 1979.

16. OKAMOTO, N. and NAKAZAWA, M. - Finite Element Incremental Contact Analysis with Various Frictional Conditions, Int. J. Num. Meth. Engng., Vol. 14, pp. 337-357, 1979.

17. WASHIZU, K. - Variational Methods in Elasticity and Plasticity, Second Edition, Pergamom Press, New York, 1975.

18. TADA, H., PARIS, P. C. and IRWIN, G. R. - The Stress Analysis of Cracks Handbook, Del Research Corporation, Hellertown, Pennsylvania, 1973.

FRACTURE ANALYSIS DATA FOR PARTIALLY OPENED CRACKS

by T H Richards[†] and P C Wood[††]

SUMMARY

When a structure is loaded in such a way that the faces of a pre-existing crack are pressed together over some part of their surface, a non-linear boundary value problem ensues. An application of, say, a finite element approach based simply on $\delta V = 0$ will yield solutions in which overlap of the crack faces is implied: when a crack tip actually lies in this overlap zone, a negative value is obtained for K_I. Here, we examine the above problem utilizing an iterative procedure in conjunction with a finite element formulation. Results are given for a number of cases.

NOTATION

[A]	Compatibility matrix.
[K]	Stiffness matrix.
K_I, K_{II}	Stress intensity factors.
[N]	Shape function matrix.
{Q}	Vector of generalized forces.
{q}	Vector of generalized co-ordinates.
{u}	Displacement field vector.
u, v	Displacement components.
V	Total potential energy.
{α}	Vector of core generalized co-ordinates incorporating stress intensity factors.
{λ}	Vector of Lagrange multipliers.
{ω}	Rotation.

† University of Aston in Birmingham.

†† Westland Helicopters, Yeovil.

INTRODUCTION

There is a considerable body of literature describing various analytical and numerical approaches for determining fracture mechanics data such as stress intensity factors. The vast majority of this work is concerned with situations in which the faces of pre-existing cracks are deemed either to separate (mode I behaviour) or to slide over each other (modes II and III). Very little attention seems to have been paid to situations wherein the crack faces are pressed together on some part of their surface whilst relative opening or sliding is still free to occur in the neighbourhood of the crack tip. A mode I example of such a partial closure situation can occur when, say, a normal edge crack exists in a plate subjected to in-plane combined bending and direct load: an inclined straight crack in a similar stress field or a curved crack in a more general field are situations wherein mixed mode I and II partial closure may occur. Solution of this problem by analytic means is very difficult with interesting work described by Bowie and Freese [1]. In the present work, a finite element treatment is presented.

In the displacement formulation of finite element stressing methods, certain rules must be obeyed in the selection of displacement models to ensure inter-element continuity. However, when physical boundaries of the solution domain are in close proximity, as is the case for the faces of a crack, the above criteria are not adequate to ensure that overlapping of the faces is not implied when stationarity is imposed on the potential energy functional: special precautions, requiring an iterative solution of the governing equations, are required for this non-linear problem.

In this paper, we describe an iterative approach used to predict the displaced profiles and stress intensity factors for a number of straight and curved partially opened cracks whose tips experience mode I and mode II loading.

FINITE ELEMENT ANALYSIS OF CRACK PROBLEMS

In the usual displacement finite element approach to stressing problems, a piecewise approximation to the structure's displacement field is assumed. On this basis, an expression for the total potential energy for the substitute discrete system so generated takes the form

$$V = \frac{1}{2}\{q\}^t [K]\{q\} - \{q\}^t\{Q\} \tag{1}$$

where the q's are the generalized co-ordinates associated with the displacement model. Application of the principle of stationary potential energy, $\delta V = 0$, yields the discrete governing equations of equilibrium [2]

$$[K]\{q\} = \{Q\} \tag{2}$$

The polynomial expansions used for "ordinary" stress analysis do not provide a good representation of the situation in the immediate neighbourhood of a crack tip so that special provision must be made for the singular conditions obtaining there. In recent years, a number of methods have been proposed and a useful assessment of most of them was offered in reference [3]. The approach adopted for the present work is one which has also been described previously [4]; a powerful attraction is that the stress intensity factors are determined directly and not inferred from any supplementary correlation calculations based on finite element nodal displacements or stresses as is required by some methods.

Delineating a small sub-region or "core" centred on the crack tip, we may use the classical near-tip singular solution of fracture mechanics, supplemented by rigid body translations and rotations, to express the displacement field within it in the form

$$\{u\}^c = [N(x_i)]\{\alpha\} \tag{3}$$

where $\{\alpha\}$ is a vector of core generalized co-ordinates which include the stress intensity factors explicitly. Embedding such a core within an ordinary finite element mesh, we may identify the generalized co-ordinates for nodes actually <u>on</u> the core/mesh interface as $\{q\}_1$. Then, Lagrange multipliers may be used to formulate a modified potential energy functional

$$\bar{V} = \frac{1}{2}\{\alpha\}[k]^c\{\alpha\} + \frac{1}{2}\{q\}^t[K]\{q\} - \{q\}^t\{Q\}$$
$$+ (\{q\}_1 - [A]\{\alpha\})^t\{\lambda\} \tag{4}$$

in which the core/mesh constraint is accommodated. Requiring $\delta\bar{V} = 0$ for equilibrium yields equations expressible in the form [4]

$$\begin{bmatrix} [K]^*_{11} & [K]^*_{12} \\ [K]^*_{21} & [K]_{22} \end{bmatrix} \begin{Bmatrix} \{q\}_2 \\ \{\alpha\} \end{Bmatrix} = \begin{Bmatrix} \{Q\}^*_1 \\ \{Q\}^*_2 \end{Bmatrix} \tag{5}$$

This approach has yielded values for the stress intensity factors comparing well with those generally accepted for standard cases.

Data preparation is minimized using a mesh generation scheme having a standard facility for treating the core/mesh zone [5].

A simple example of a partial crack closure situation is illustrated in Fig 1a. Here, a plate, containing a single edge crack, is subjected to in-plane combined tension and bending. Then, depending on the details of the loading, the faces of the crack may take up the form of Fig 1b. Under these circumstances, $\delta\bar{V} = 0$ from above would lead to a negative value for K_I and the physically unacceptable crack profile of Fig 1c. Evidently, special provision must be made for this eventuality and a negative K_I can provide the trigger for emplementing an iterative scheme for removing the overlap.

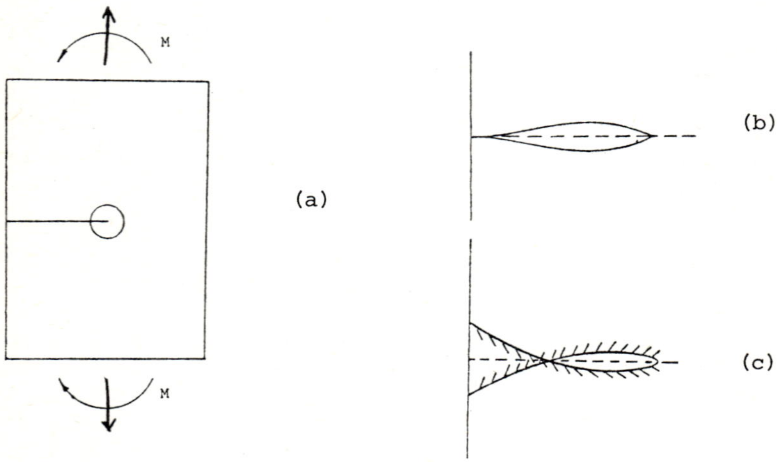

Figure 1
Illustration of partial closure

Three approaches were attempted and of these, the third proved to be the most useful. The average displacement scheme, which is only appropriate for mode I situations, proceeds essentially as follows: solve the problem in the ordinary way giving the overlap configuration, compute the average of the displacements at corresponding nodes on each crack face, re-solve using these average values as prescribed geometric boundary conditions over the overlap zone. An application of this approach was described in [6] and fresh results are compared with those abtained by the other two methods in Fig 2 of this report.

An alternative method is to proceed: (a) solve to give the overlap configuration as above: (b) attach a simple bar element to the pair of nodes corresponding to maximum overlap and solve this problem: (c) repeat the process until no overlap occurs. The difficulty with this approach is deciding on a suitable stiffness for this fictitious bar. Fig 3 illustrates the effect of varying "EA/L" on the partially closed

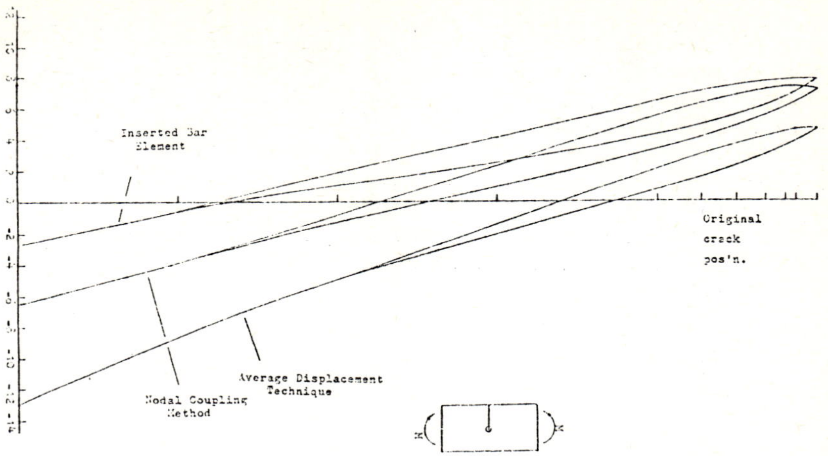

Figure 2
Crack profiles predicted by the three methods.

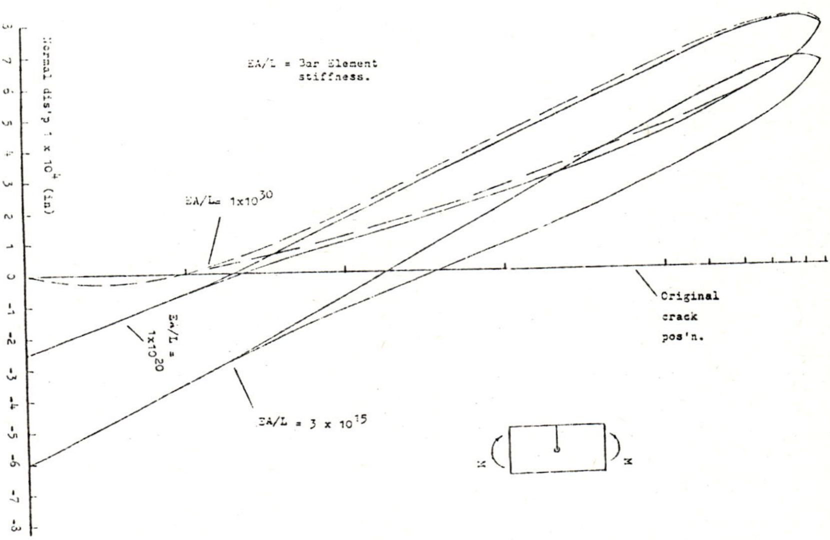

Figure 3
Influence of varying EA/L in fictitious bar method.

profile.

We have labelled the most useful approach as the nodal coupling technique and follows Cook [7]. In a mixed mode problem there will be relative displacements between material points normal and tangential to the crack faces. Friction effects will thus be called into play but in a practical situation, they will be rather indeterminate. The nodal coupling technique lends itself to examining two extremes providing solution bounds corresponding to no friction or no slip. In the former case, components of displacements normal to the crack faces in the overlap zone are matched at corresponding nodes whilst in the latter case, the nodes are completely coupled.

The method is iterative and proceeds essentially as follows:

(i) Solve $[K]^*\{q\}^* = \{Q\}^*$ in the ordinary way, a negative value for K_I signifies than an overlap as occurred.

(ii) Identify the point of maximum overlap, match the corresponding nodal displacements by modifying the stiffness equations.

(iii) Solve the modified equations.

(iv) Repeat the process until the overlap is removed.

NUMERICAL EXAMPLES

In the first example, a plate with a single edge crack is subjected to in-plane pure bending; the crack tip is in the tension part of the stress field. Table 1 compares values of crack tip displacements u, v, ω and stress intensity factors comprising the elements of $\{\alpha\}$, as predicted by the three approaches. Pound force and inch units are used. The non-zero values of K_{II} in this mode I situation provide a measure of the imperfection in the solutions.

Fig 2 shows the crack profiles corresponding to Table 1. Fig 3 shows the influence of varying "EA/L" in the inserted bar approach.

In the remaining examples, only the node coupling technique was used.

	Average Displ. 4 Iterations	Inserted bar 3 Iterations	Node coupling: No friction 3 Iterations	Node coupling: No Slip 3 Iterations
u	$-1.4778x10^{-3}$	$-1.2119x10^{-3}$	$-1.4307x10^{-3}$	$-1.4307x10^{-3}$
v	$4.574\ x10^{-4}$	$7.626\ x10^{-4}$	$7.2239x10^{-4}$	$7.2238x10^{-4}$
ω	$6.1152x10^{-4}$	$4.4712x10^{-4}$	$5.5087x10^{-4}$	$5.5087x10^{-4}$
K_I	$2.0499x10^{3}$	$2.7002x10^{3}$	$2.4403x10^{3}$	$2.4403x10^{3}$
K_{II}	$1.3728x10^{2}$	$-6.7916x10^{1}$	$-2.222\ x10^{0}$	$-2.304\ x10^{0}$

Table 1
Comparison of predictions using
the three approaches.

The study by Bowie and Freese [1] provides an opportunity for comparison between the present technique and an analytic approach. The structure comprises an infinite plate with a central crack; the loading is in-plane pure bending at infinity and the crack is at 45° to the overall uniaxial stress field: a mixed mode I and II situation thus obtains at the crack tips. See fig 4.

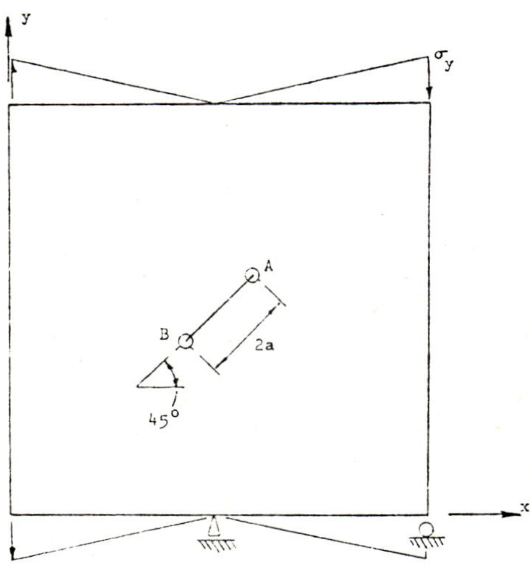

Figure 4
Model for comparing the present
approach with that of reference [1].

Table 2 compares stress intensity factor values for both crack tips. It will be seen that the Bowie and Freese values for K_I and K_{II} compare with those for the no-slip model tested here.

	Finite Element, 8 Iterations		B and F
	No Friction	No Slip	
u	3.4833×10^{-3}	3.4885×10^{-3}	
v	-5.1574×10^{-3}	-5.1424×10^{-4}	
ω	-1.1955×10^{-4}	-1.1391×10^{-3}	
K_I	0.0	0.0	0.0
K_{II}	-9.6642×10^{2}	0.0	0.0

Crack Tip A

	Finite Element, 8 Iterations		B and F
	No Friction	No Slip	
u	-3.5045×10^{-3}	-3.5097×10^{-3}	
v	2.1626×10^{-3}	2.1675×10^{-3}	
ω	-8.0094×10^{-4}	-8.0064×10^{-4}	
K_I	1.0466×10^{3}	1.0458×10^{3}	1.060×10^{3}
K_{II}	9.6444×10^{2}	1.0467×10^{3}	1.060×10^{3}

Crack tip B

Table 2
Fracture data for 45° centre crack
in "large" plate under in-plane bending.

Fig 5 shows the crack profile for the two finite element models.

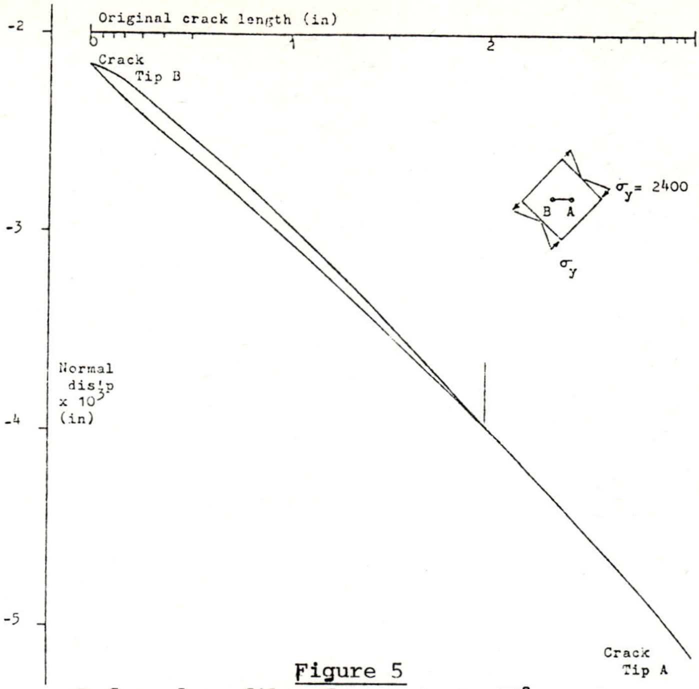

Figure 5
Deformed profile of a central 45° crack.

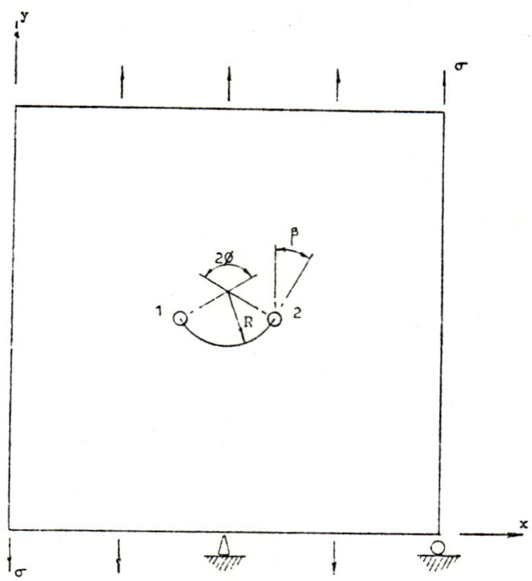

Figure 6
Circular arc crack in uniaxial stress field. $\phi = 60°$.

The case of a circular arc crack in a uniaxial overall stress field, Fig 6 has been treated in several ways. Rooke and Cartwright [8] show how K_I and K_{II} vary with orientation of the crack but accept that the negative values of K_I, which arise when partial closure occurs cannot be meaningful.

Table 3 shows that there is good agreement between the values of K_I and K_{II} quoted in reference [8] and those given by the first iteration in the present scheme. Figs 7 and 8 show the variation of K_I and K_{II} as a function of crack orientation and highlight the influence of partial closure. Figs 9 and 10 illustrate typical crack profiles.

Angle $\beta°$	90	60	30	0	-30	-60	-90
K_I*(R&C)	0.791	0.745	0.256	-0.163	-0.102	0.381	0.791
K_{II}*(R&C)	-0.116	0.303	0.624	0.488	0.056	-0.233	-0.116
K_I* (FE)	0.794	0.745	0.271	-0.153	-0.102	0.371	0.794
K_{II}*(FE)	-0.115	0.323	0.634	0.506	0.067	-0.243	-0.115

Table 3

Normalized K_I^* and K_{II}^* for circular arc crack. Comparison between first iteration FE solution and reference [8]

$$K_I^* = K_I\sqrt{\sin\phi}/K_o, \quad K_{II}^* = K_{II}\sqrt{\sin\phi}/K_o$$

$$K_o = \sigma\sqrt{\pi R\sin\phi}$$

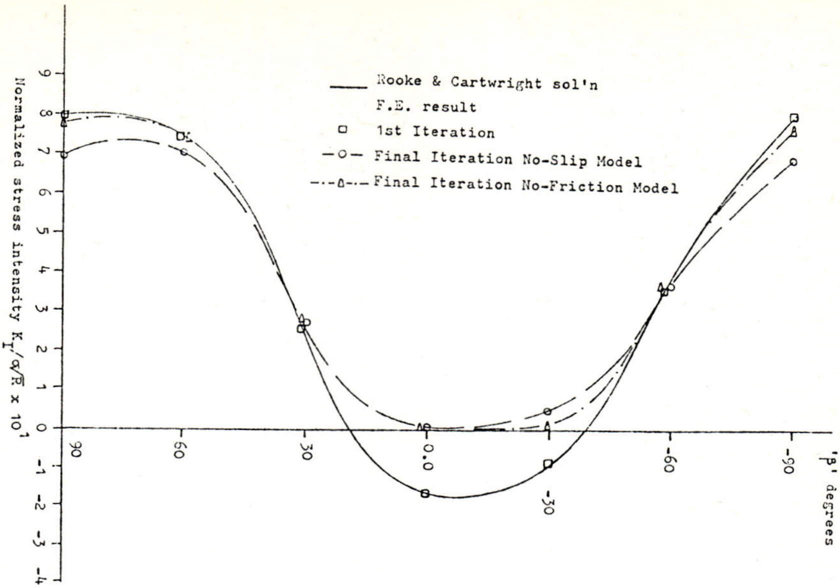

Figure 7
Variation of K_I with crack orientation.

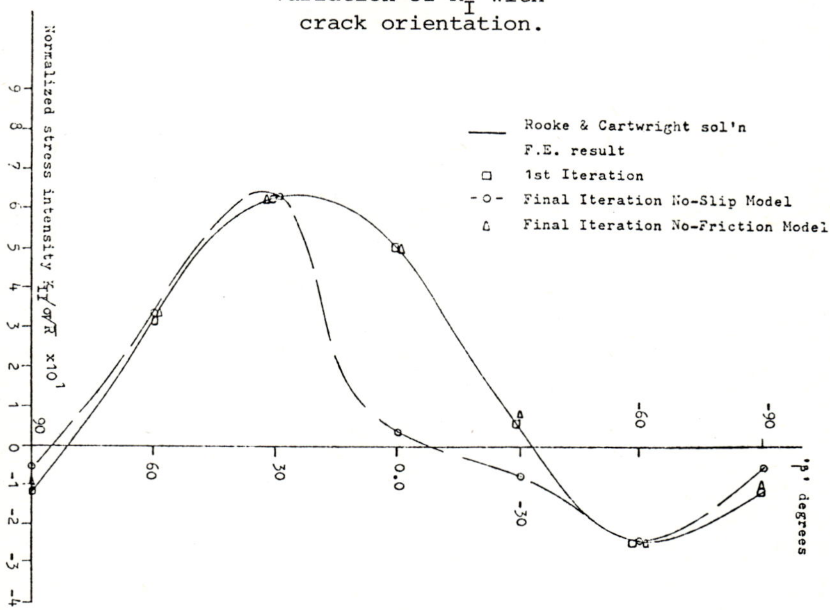

Figure 8
Variation of K_{II} with crack orientation

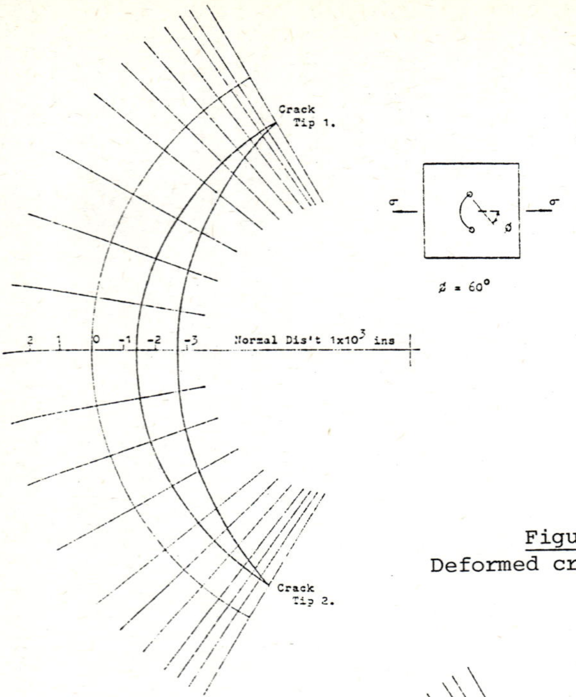

Figure 9
Deformed crack profile.

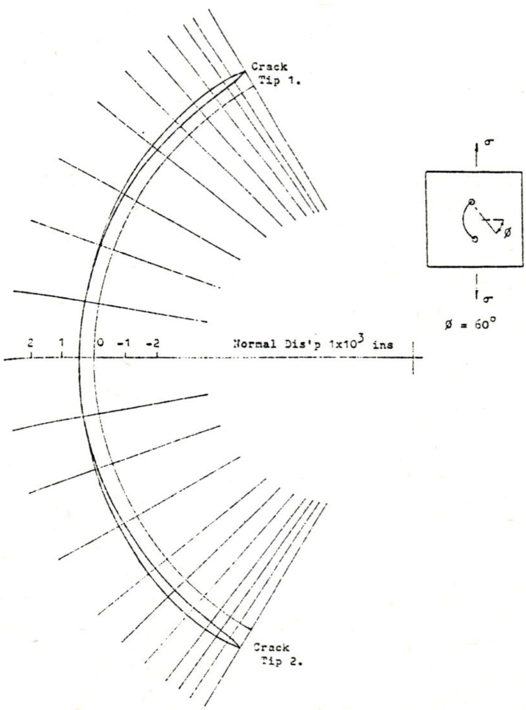

Figure 10
Deformed crack profile.

The final example is shown in Fig 11; it represents a crack near a hole. One iteration in the present approach gives a negative but small value for K_I at tip 1. After seven iterations, this K_I became zero whilst K_I for tip 2 changed only slightly in the third decimal place. The K_{II} values are zero as one would expect and these results compare well with those quoted in reference [8];see Table 4. Fig 12 gives a physical explanation for the small changes noted in Table 4: the overlap zone is a very small one.

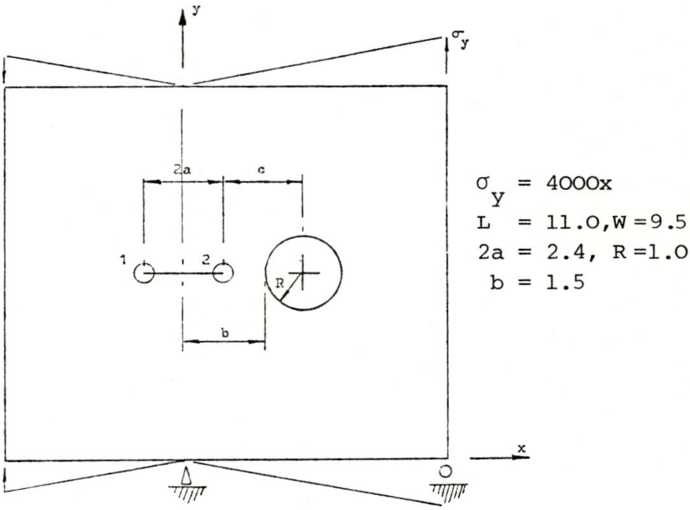

Figure 11
Crack near a drilled hole.

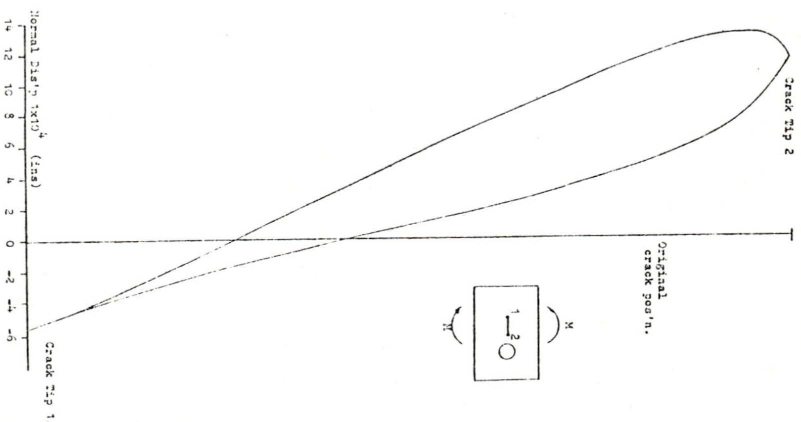

Figure 12
Deformed crack profile.

	Tip 1			Tip 2		
	R & C	Iter.1	Iter.7	R & C	Iter.1	Iter.2
K_I	0.00	-0.05	0.00	0.74	0.77	0.77
K_{II}	0.00	0.00	0.00	0.00	0.00	0.00

Table 4

CONCLUSION

A nodal coupling technique, used in conjunction with a crack tip singularity core, has provided an effective means of computing stress intensity factors for cases where partial crack closure occurs. In a case where a theoretical model has been offered, agreement is good; in the other examples offered, the results seem to be self consistent and give confidence in the adequacy of the approach.

REFERENCES

1. BOWIE, O L and FREESE, C E - On the overlapping problem in Crack Analysis. Eng. Frac. Mech., Vol 8, pp 373 - 379, 1976.

2. ZIENKIEWICZ, O C - The Finite Element Method. McGraw Hill (UK), 1977.

3. FAWKES, A J et al - An Assessment of Crack Tip Singularity Models for Use with Isoparametric Elements. Eng. Frac. Mech., Vol 11, pp 143 - 159, 1979.

4. RICHARDS, T H - On Using the Finite Element Method in Conjunction with Sub-Region Singular Solutions in Fracture Mechanics. Proc. 1st Int. Conf. on Numerical Methods in Fracture Mechanics, Eds. Owen, D R J and Luxmore, A R. 1978.

5. WOOD, P C - Application of the Finite Element Method to Problems in Fracture Mechanics, PhD Thesis, University of Aston in Birmingham, 1979.

6. RICHARDS, T H and ROBERTSON, A W - The Determination of Single and Mixed Mode Stress Intensity Factors for Engineering Components of Practical Interest, Fracture Mechanics in Engineering Practice, Ed. Stanley, P, Applied Science Publishers, 1977.

7. COOK, R D - Concepts and Applications of Finite Element Analysis, McGraw Hill, New York, 1973.

8. ROOKE, D P and CARTWRIGHT, D J - Compendium of Stress Intensity Factors, HMSO, 1977.

MULTILAYER-STRESS-HYBRID-FINITE-ELEMENT METHOD FOR FRACTURE ANALYSIS OF ANGLE-PLY LAMINATES.

T. Nishioka*, and S.N. Atluri*

Georgia Institute of Technology, Atlanta, Ga. 30332.

SUMMARY:

An efficient assumed stress hybrid finite element method, based on a complementary energy principle and a Lagrange-multiplier technique, is presented for fracture analysis of cracks and holes in anlge-ply laminates. A fully three-dimentional stress state in the laminate is accounted for. Asymptotic analytical solutions in the vicinity of cracks and holes are embedded in special crack and hole elements; inter-layer traction reciprocity condition is satisfied a priori; and cross-sectional warping of the laminate is allowed. The presented method leads to a direct computation of mixed mode stress-intensity factors near the crack-front or of stress-concentration factors near fastener holes in composites.

INTRODUCTION:

Fully three-dimensional stress analyses of angle-ply laminates with cracks and holes are necessary to understand (*i*) the mechanisms of non-self-similar crack growth in symmetric angle-ply laminates; (*ii*) the subcritical damage in the form of matrix crazing, splitting, and delamination that usually precedes final failure in a laminate; and (*iii*) the size-effects of holes.

Employing the multi-layer assumed-stress hybrid finite element method, as originally developed in [1], an analysis of a through-thickness crack in a $(\pm 45)_s$ laminate was performed in [2]. In the element development in [1], the following assumptions were made: (*i*) the transverse normal stress σ_{33} in the laminate is zero; (*ii*) the stress assumptions in the elements were of the usual polynomial nature, and hence cannot represent the mixed-mode singularities near the crack-

*Center for the Advancement of Computational Mechanics; School of Civil Engineering.

front; (*iii*) the stress assumption is independent for each lamina and thus the inter-laminar traction reciprocity condition was enforced through a Lagrange multiplier technique, which complicates the formulation and renders it prohibitively expensive for a large number of lamina. To obtain the desired accuracy, using the elements of [1], for a crack-problem, a very fine mesh was used in [2]. Also, eventhough one may obtain high stress-gradient solutions in the limit, it is often inconvenient to extract results for the mixed-mode K-factors near the crack-front. To study the effect of σ_{33}, the usual three-dimensional stress-hybrid finite element method, as originally proposed in [3], was used in [4] to study the problem of a through-crack in $(90^o/0^o)$s laminate. Once again, since stress singularities were not accounted for in the element development of [3], a very fine 3-D mesh was used in [4]. In fact, Ref. [4] employed a two-stage solution procedure, with a 9x18x2 mesh (with 1700 d.o.f.) and a second refined mesh of (11x9x4) 3-D elements (with 3600 d.o.f.) near the crack front.

As for the analysis of holes, a three dimensional complementary energy method was used in [5], wherein the equilibrated stress field was derived from 3 stress-functions. However, the stress-functions, and hence the stress-field, including the inplane stresses $\sigma_{\alpha\beta}$, were continuous at the interlaminar interfaces, in the assumptions in [5]. This is contrary to what one may expect in an angle-ply laminate, modeled as a multi-layer anisotropic medium (with layer-wise variation of anisotropic properties in global coordinates), wherein, under assumptions of perfect bonding, the inplane strains may be continuous at laminar interfaces, but not the inplane stresses. On the other hand, concepts of a "boundary-layer" (ie., a region of a laminate near the hole where the "plane-stress" solution is inadequate) were used in [6,7] in analyzing holes in several symmetric angle-ply laminates.

In the following we present the experiences of the authors' recent work [8,9] in the use of "multi-layer" assumed-stress "crack" and "hole" elements in analyzing cracks and holes in angle-ply laminates. In the interest of space, the mathematical details are omitted for the most part; some of these can be found in [8,9], and others will be published elsewhere.

BRIEF SYNOPSIS OF ANALYSIS PROCEDURE:

As elaborated upon in [8,9], the energy functional governing the modified complementary principle, and the associated development of a hybrid-stress finite element method for a multi-layer anisotropic laminate, can be written as:

$$\Pi_{HS}(\underset{\sim}{\sigma}^i;\underset{\sim}{u}^i) = \sum_n \left[\sum_i \left\{ \int_{V_n^i} (1/2)\underset{\sim}{\sigma}^{iT}\underset{\approx}{C}^i\underset{\sim}{\sigma}^i dv - \int_{\partial V_n^i} \underset{\sim}{T}^{iT}\underset{\sim}{u}^i ds + \int_{S_{\sigma n}^i} \bar{\underset{\sim}{T}}^{iT}\underset{\sim}{u}^i ds \right\} \right] \quad (1)$$

where: (*i*) $\underset{\sim}{\sigma}^i$ is the (6x1) vector of three-dimensional stress in the ith lamina which is assumed such that it satisfies equilibrium conditions within each lamina, as well as the interlaminar traction reciprocity conditions, a priori; ($\underset{\sim}{\sigma}^{iT}$ denotes the transpose of $\underset{\sim}{\sigma}^i$) (*ii*) $\underset{\approx}{C}^i$ is the matrix of elastic compliance coefficients for the ith lamina; (*iii*) $\underset{\sim}{T}^i$ are tractions, at the inter-element boundary for the ith lamina, which are derived from the assumed $\underset{\sim}{\sigma}^i$; (*iv*) $\underset{\sim}{u}^i$ are inter-element boundary displacements, which play the role of Lagrange multipliers to enforce inter-element traction reciprocity; and (*v*) $\bar{\underset{\sim}{T}}^i$ are prescribed tractions at $S_{\sigma n}^i$ of the ith lamina. In the above, $\underset{\sim}{u}^i$ are assumed such that cross-sectional warping of individual lamina are permitted, as elaborated upon on [8]. Further, we assume that elements n=1,2,...p are the special "crack-elements", or special "hole-elements" as the case may be, and that n>p are the "regular" elements" surrounding these special elements. For details of the development of these regular elements, Ref. [8] may be consulted.

In the special-element, we assume the stress-field to be,

$$\underset{\sim}{\sigma}^i = \underset{\approx}{P}^i\underset{\sim}{\beta} + \underset{\approx}{P}_s^i\underset{\sim}{\beta}_s^i \quad \text{in } V_n^i,\ n \leq P.$$

$$(i = 1,2,\ldots k \text{ Layers}) \quad (2)$$

where $\underset{\approx}{P}^i\underset{\sim}{\beta}$ are the usual polynomial functions, and $\underset{\approx}{P}_s^i\underset{\sim}{\beta}_s^i$ are special functions that account for the presence of a crack or a hole. Note that the parameters $\underset{\sim}{\beta}$ are the same throughout the thickness of the lamina, due to the rather novel way of assuming the (6x1) stress vector in each lamina, such that (*i*) the interlaminar traction reciprocity is satisfied a priori, and (*ii*) the inplane stresses $\sigma_{\alpha\beta}$ (α,β=1.2) are permitted to be discontinuous at the laminar interfaces, as detailed in [8].

In the following we give a very brief sketch of an interesting way of selecting special functions $\underset{\approx}{P}_s^i\underset{\sim}{\beta}_s^i$ for a crack or a hole, such that, once again, (*i*) the interlaminar traction reciprocity is satisfied a priori, and (*ii*) the inplane stresses $\sigma_{\alpha\beta}$ may be discontinuous at the layer interfaces. For further details the reader may consult Refs. [8,9].

We note that the special stress-field, in the vicinity of a crack or a hole, must satisfy the equilibrium equations, in the absence of body forces, as:

$$(\sigma^{is}_{mn})_{,n} = 0 \text{ and } \sigma^{is}_{mn} = \sigma^{is}_{nm}$$
$$(m,n=1,2,3;\ i=1,2,\ldots k) \quad (3)$$

In the case of a special multi-layer crack-element, we start by assuming σ^{is}_{33} in each layer (i=1,2,...k) as:

$$\sigma^{is}_{33} = \frac{K_4(x_3)}{\sqrt{2\pi r}} \cos(\theta/2) + \frac{K_5(x_3)}{\sqrt{2\pi r}} \sin(\theta/2) \quad (4)$$

where (r,θ,x_3) are cylindrical polar-coordinates centered at the crack front. In the case of a hole of radius "a", we assume σ^{is}_{33} as:

$$\sigma^{is}_{33} = \left\{\alpha_1(x_3)r^{-1} + \alpha_2(x_3)r^{-2} + \alpha_3(x_3)r^{-4}\right\} \cos 2\theta + \left\{\alpha_4(x_3)r^{-1} + \alpha_5(x_3)r^{-2} + \alpha_6(x_3)r^{-4}\right\} \sin 2\theta \quad (5)$$

where (r,θ,x_3) are cylindrical polar-coordinates centered at the center of a through-thickness hole in a laminate. In the above, $K_4(x_3)$, $K_5(x_3)\ldots\alpha_6(x_3)$ are assumed to be continuous polynomials through the laminate thickness; in specific, each of these is assumed to be a Hermitian interpolant in terms of $K_4(x_3^i)$ and $\partial K_4/\partial x_3$ at x_3^i, where x_3^i is the thickness coordinate of the interface between i^{th} and (i+1)th layers.

The third equilibrium equation, namely,

$$\sigma^{is}_{3\alpha,\alpha} = -\sigma^{is}_{33,3} \quad (\alpha=1,2) \quad (6)$$

is satisfied by setting $\sigma^{is}_{3\alpha} = \sigma^{ish}_{3\alpha} + \sigma^{isp}_{3\alpha}$ where, the additional subscripts h and p denote the homogeneous and particular solutions, respectively. The right-hand side of Eq. (6), for the crack and hole cases is taken from Eqs. (4) and (5), respectively. Thus, in each of these cases, the particular solution $\sigma^{isp}_{3\alpha}$ can be determined in terms of K_4, K_5, and $\alpha_1\ldots\alpha_6$ respectively. The homogenous solution, for the case of the crack element, is taken to be:

$$\sigma^{ish}_{3\alpha} = \frac{K_3^i(x_3)}{\sqrt{2\pi r}} F_\alpha(\theta, S_3^i) \quad (7)$$

where S_3^i is a complex number depending on the anisotropic elastic compliance coefficients, and $K_3^i(x_3)$ are once again taken to be continuous at the interlaminar interfaces. In the case of a hole, we assume that $\sigma^{ish}_{3\alpha}=0$.

The first two equilibrium equations viz.,

$$\sigma^{is}_{\alpha\beta,\beta} = -\sigma^{is}_{\alpha 3,3} \quad [\alpha,\beta=1,2] \tag{8}$$

are satisfied by setting $\sigma^{is}_{\alpha\beta} = \sigma^{ish}_{\alpha\beta} + \sigma^{isp}_{\alpha\beta}$. The right-hand-side of Eq. (8), for the crack and hole cases, can be substituted as $\sigma^{is}_{\alpha 3} = \sigma^{ish}_{\alpha 3} + \sigma^{ish}_{\alpha 3}$ where each of the terms is derived as sketched above. In the case of the crack, we take the homogenous solution to be:

$$\sigma^{ish}_{\alpha\beta} = \frac{K^i_4(x_3)}{\sqrt{2\pi r}} F_{1\alpha\beta}(\theta,S^i_1,S^i_2) + \frac{K^i_2(x_3)}{\sqrt{2\pi r}} F_{2\alpha\beta}(\theta,S^i_1,S^i_2) \tag{9}$$

where S^i_1 and S^i_2 are complex numbers defined in terms of the compliance properties of the ith lamina [8]. However, unlike the case of σ_{33} and $\sigma_{3\alpha}$, the parameters $K^i_1(x_3)$ and $K^i_2(x_3)$ are <u>not</u> required to be continuous at the inter-laminar interfaces, and thus are taken to be arbitrary for each lamina.

In the case of a hole, we take:

$$\sigma^{ish}_{\alpha\beta} = \alpha^i_7(x_3)\, F_{\alpha\beta}(r,\theta) \tag{10}$$

where $F_{\alpha\beta}(r,\theta)$ would involve terms of the type r^{-2}, r^{-4} etc. Thus Eq. (10) represents the two-dimensional analytical asymptotic solution for a hole in an anistropic plate, as may be found in [10]. Once again, the coefficients $\alpha^i_7(x_3)$ are permitted to be discontinuous at the interlaminar interfaces.

It may be noted that the parameters $K^i_1(x_3)$, $K^i_2(x_3)$, $K^i_3(x_3)$ etc. are related to the stress-intensity factors near the crack, while the parameters $\alpha^i_7(x_3)$ are related to the stress-concentration factors near holes.

Finally, we note that the element boundary-displacement, $\tilde{u}^i$ of Eq. (1), for the crack and hole elements, respectively, are chosen so as to reflect the known special asymptotic behavior. For further details, see [8,9].

We present below some representative results of computations based on the above formulations.

<u>THROUGH-THICKNESS EDGE CRACKS IN $(90^o/0^o)$ AND $(\pm 45^o)$s LAMINATES:</u>

The geometries of both the 4-ply laminates are shown in Fig. 1. In both the cases, (L/W)=1.0; (a/w)=0.2; $h=4h_o$, and w=1.5m. In the $(90^o/0^o)$s laminate, (h/W)=(1/300); while in the $(\underline{+}45^o)$s laminate, (h/W)=(1/100). The ply elastic constants, in the material principal directions are: for the $(90^o/0^o)$s laminate, $E_{11}=18.25\times10^6$psi $E_{22}=1.5\times10^6$psi; $E_{33}=1.5\times10^6$ psi; $G_{12}=G_{13}=G_{23}=0.95\times10^6$ psi; and $\nu_{12}=\nu_{23}=\nu_{13}=0.24$; while those for the $(\pm45^o)$s laminate are $E_{11}=20\times10^6$ psi; $E_{22}=2.1\times10^6$ psi; $E_{33}=2.1\times10^6$ psi; $G_{12}=G_{13}=G_{23}=0.85\times10^6$ psi, and $\nu_{12}=\nu_{13}=\nu_{23}=0.21$. These material properties are identical to those assumed to [2,4]. In both the problems, a uniform uniaxial stress is applied at the edges $X_2=\pm L_1$ and the top and bottom surfaces of the laminate as well as the crack-face, are assumed to be traction free. In the $(90^o/0^o)$s case only a quadrant of the specimen is modeled. The finite element mesh for the problems is depicted in Fig. 2, wherein, only the top half of the model is used for analysing the $(90^o/0^o)$s case.

In the $(90^o/0^o)$s case, Ref. [4] uses 2 meshes, the first a coarse mesh of (9x18x2) 3-D mesh with 1710 d.o.f. and the second, a fine inner mesh with (11x9x4) 3-D elements with 3600 d.o.f. In contrast, the present mesh consists of a (9x5x1) multi-layer hybrid element mesh with about 1100 total d.o.f.. The values of the normalized stress-intensity factors K_1^i (which are directly computed in the present procedure), and their variation with x_3 are shown in Fig. 3. As can be expected, there is a discontinuity in the K_1 value at the interface between the 0° and 90° plies, and morever, the K_1 value in the 90° ply is much higher than in the 0° ply. The stress-intensity factor in σ_{33}^i is shown in Fig. 4. Note that K_4 is zero at $x_3=0$, is continuous at the interlaminar interface, and its magnitude is much smaller than in K_1. The results for the stress-intensity factors are not given in [2], because the procedure used therein does not permit a direct extraction of these factors in a convenient manner. However, the present results for the inplane stress σ_{22}^i of each ply, along x_1 (near the crack front) are shown in Fig. 5, and are seen to correlate excellently with those in [2].

In the $(\pm45^o)$s case, the results in [2] were obtained by using the multi-layer hybrid elements of [1] by employing 2 meshes, the first, a coarse mesh with 1813 d.o.f., and the second, inner mesh with 1932 d.o.f. In contrast, the present solution is obtained by using a single mesh using the present multi-layer hybrid crack elements, with a total of about 2200 d.o.f. Fig 6 shows the directly computed mixed-mode stress-intensity factors K_I and K_{II} (in the in-plane stresses, $\sigma_{\alpha\beta}$, $\alpha,\beta=1,2$) in each lamina. Such results were not presented in [2]). However, a comparison of the detailed stress field (not included here for want of space) near the crack-tip indicates that the present results correlate excellently with those in [2]. From Fig. 6, it can be qualitatively understood that at the point of incipient crack-growth, if the direction of crack-growth in each lamina is determined by the magnitudes and signs of K_I and K_{II}, the crack growth will occur in different directions in each of the lamina. Any currently known growth-direction criterion such as: the maximum circumferential stress, strain energy density theory, or maximum strain energy release rate criterion may be used in determining the growth-direction in each lamina.

THROUGH-THICKNESS HOLE IN A $(90^o/0^o)s$ LAMINA:

We present briefly the results for the problem of a central hole in a 4-ply $(90^o/0^o)s$ boron/epoxy laminate whose properties are identical to those in [5,7], ie., $E_1=20\times10^6$ psi; $E_2=3\times10^6$ psi; $E_3=2\times10^6$ psi; $G_{12}=G_{13}=G_{23}=10^6$ psi; $\nu_{12}=\nu_{13}=\nu_{23}=$ 0.336. The finite element mesh used in analyzing this problem is shown in Fig. 7, with a 4x6 mesh consisting of the present multi-layer special hole elements near the hole. The results for the variation of the circumferential stress $\sigma_{\theta\theta}$ near the hole in the 0° and 90° lamina, respectively, are shown in Fig. 8. It is seen that all along the circumference the present results for both the lamina agree excellently with the boundary-layer analyses of [7], while in the 90° lamina, especially, considerable discrepancy exists between the present results and those of [5], with the results of [5] being higher. This discrepancy may be explained partially by the fact, as mentioned earlier, that the stress assumptions in [5] are such that even the inplane stresses $\sigma_{\theta\theta}$, $\sigma_{\theta r}$, and σ_{rr} are also continuous at the interlaminar interfaces. Further results for the problems of holes in laminates of different ply lay-ups will be presented elsewhere.

CLOSURE:

In the above, efficient, special, multi-layer-hybrid stress crack elements and hole elements are presented for the fracture analysis of cracks and holes in angle-ply laminates. Their versatility and accuracy have been demonstrated in a few test cases.

ACKNOWLEDGEMENTS:

This work was supported by U.S. AFOSR under contract No. 49620-78-C-0085, with Colonel J.D. Morgan, III as the program official. The authors gratefully acknowledge this support. They also thank Ms. Peggy Eiteman for her patient typing of this manuscript.

REFERENCES:

[1] Mau, S.T., Tong, P., and Pian, T.H.H., "Finite Element Solutions for Laminated Thick Plates" Journal of Composite Materials, vol 6, 1972, pp 304-311.

[2] Wang, S.S. Mandell, J.F., and McGarry, F.J., "A Multilayer Hybrid-Stress Finite Element Analysis of a Through-Thickness Edge Crack in a $\pm 45^{\circ}$ Laminate" Engineering Fracture Mechanics vol 9, 1977, pp217-238.

[3] Pian, T.H.H., "Derivation of Element Stiffness Matrices by Assumed Stress Distributions", AIAA Journal, vol 2, 1964, pp 1333-1336.

[4] Rybicki, E.F., and Hopper, A.T., "Analytical Investigation of Stress Concentrations Due to Holes in Fiber-Reinforced Plastic Laminated Plates: Three-Dimensional Models" AFML-TR-76-92, Battelle Columbus Labs, 1973.

[6] Tang, S., "Interlaminar Stresses Around Circular Cutouts in Composite Plates Under Tension" AIAA Journal, vol 15, 1977, pp 1631-1637.

[7] Tang, S., "A Variational Approach to Edge Stresses of Circular Cutouts in Composites", AIAA Paper 79-0802, 20th AIAA/ASME/ASCE/AHS SDM Conference, St. Louis, Mo, 1979, pp 326-332.

[8] Nishioka, T., and Atluri, S.N., "Fracture-Stress Analysis of Through-Cracks in Angle-Ply Laminates-An Efficient Assumed Stress Finite Element Approach-Part I", AIAA Paper No. 79-0801, 20th AIAA/ASME/ASCE/AHS SDM Conference, St. Louis, Mo, 1979, pp 315-325 (Also AIAA Journal, 1980, In Press).

[9] Nishioka, T., and Atluri, S.N., "Stress Analysis of Holes in Angle-Ply Laminates; An Efficient Assumed Stress" Special-Hole-Element "Approach" 21st AIAA/ASME/ASCE/AHS SDM Conference, Seattle, Wash. 1980 (to appear).

[10] Savin, G.N., Stress Concentration Around Holes, Bergamon Press, 1961.

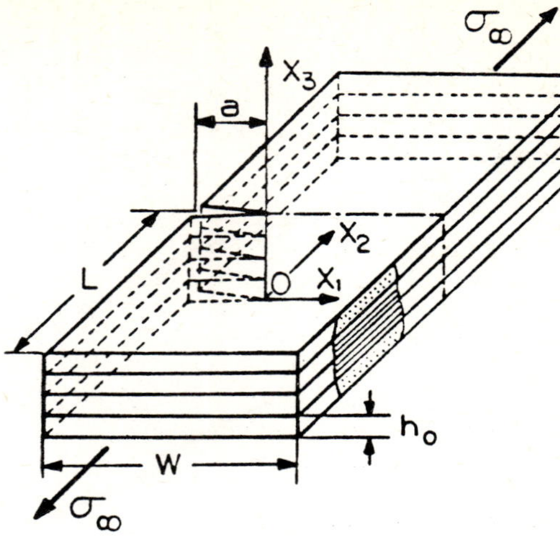

FIGURE 1

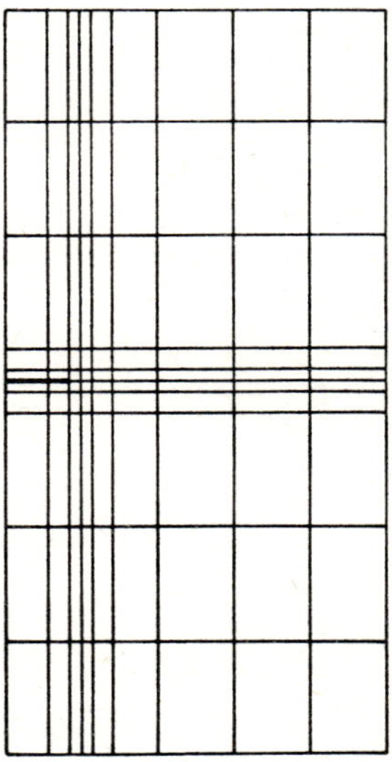

FIGURE 2

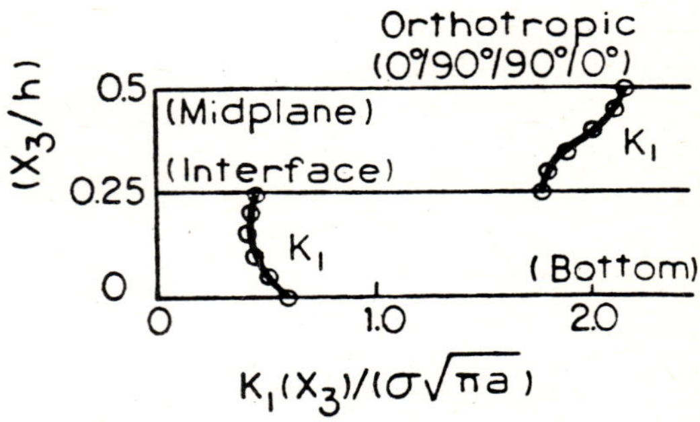

FIGURE 3

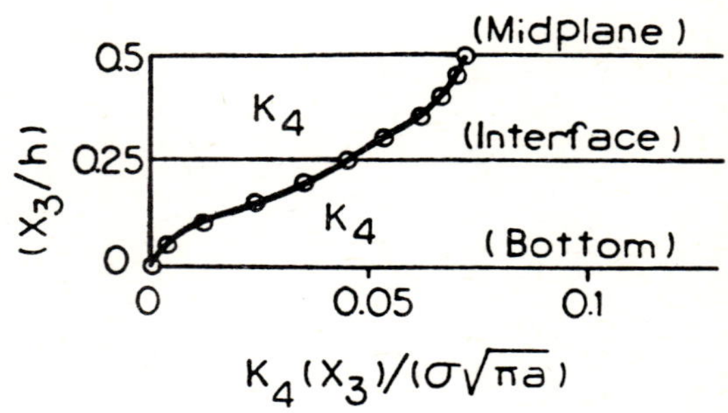

FIGURE 4

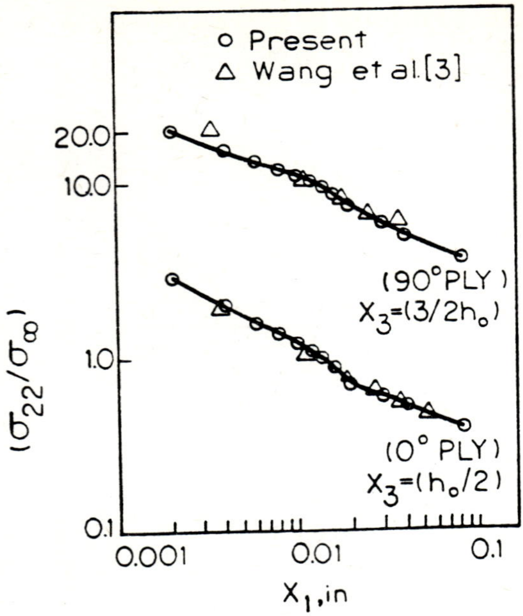

FIGURE 5

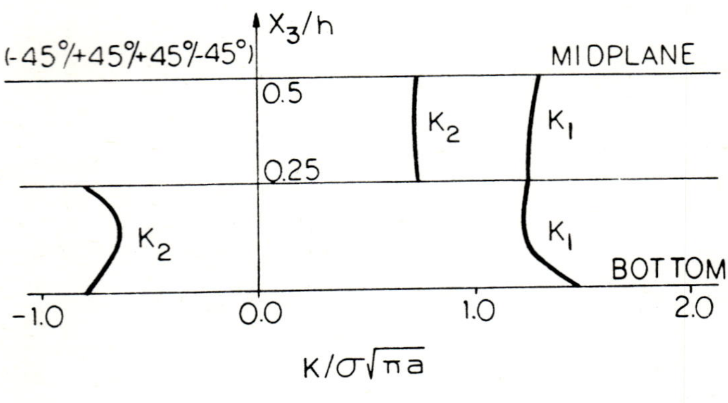

FIGURE 6

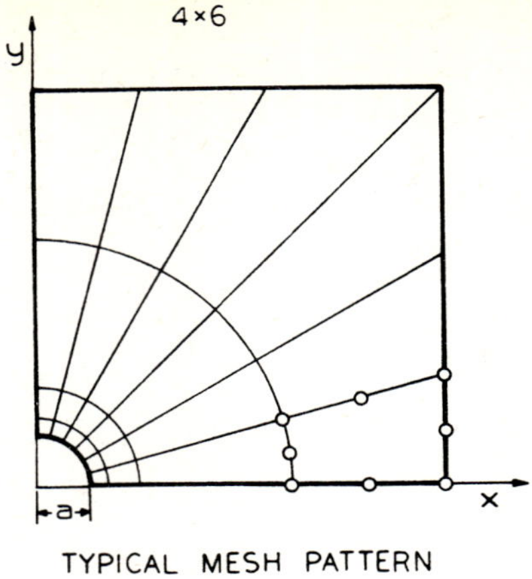

FIGURE 7

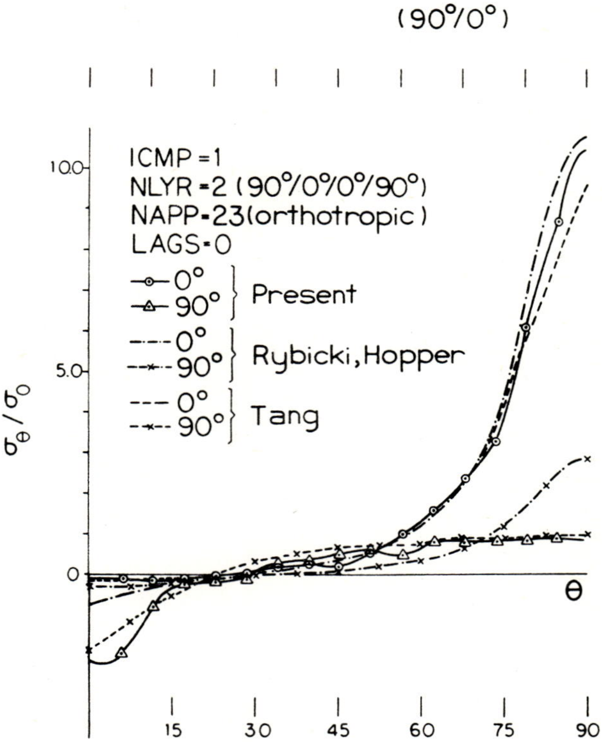

FIGURE 8

NUMERICAL CALCULATION OF STRAIN ENERGY RELEASE RATES FOR THERMAL CRACKS IN FIBRE-REINFORCED MATERIALS

H. Braun[1]
Daimler Benz AG Stuttgart

K. Herrmann[2]
FB 10 • Technische Mechanik
Universität - GH - Paderborn

SUMMARY

In this paper a quasistatic approach to thermal crack propagation in unidirectionally reinforced composite structures is given by means of a standard finite element program using constant strain elements. These calculations were performed in order to obtain micromechanical informations about the thermal shock resistance of such composite structures. Thus cracked unit cells of different shape of a two-phase composite as well as material sections enclosing several of such composite microcomponents are considered.

Firstly, using a cracked hexagonal unit cell of a Molybdenum fibre reinforced Aluminum matrix the crack surface energy as well as the strain energy release rate for the quasistatic extension of a thermal crack have been calculated. Secondly, considering a circular unit cell the energy release rates of cracks starting at the interface and running into the matrix material towards the external boundary were determined. Therein the transient thermal stress fields of this problem were caused by cooling the unit cell. Finally, assuming stationary temperature fields, the energy release rates of cracks extending between two fibers in an ensemble of composite microcomponents were calculated.

INTRODUCTION

Previous experiments show that different failure mechanisms dominate in the low- and high-fiber concentration ranges of unidirectionally reinforced composite materials. Thereby, for low-fiber concentration structures, cracks are generally first observed in the matrix material where the fibers act to

[1]Dr.-Ing., Research Associate
[2]Professor of Mechanics

restrain crack growth. To the contrary in the range of high-fiber concentration, debonding becomes the dominate failure mechanism, that means cracks often following the fiber-to-matrix interfaces. There exists a material dependent boundary between those two regions which is approximately given by a fiber volume fraction of 30 percent. Moreover, experimental investigations show a remarkable dependence of the fracture resistance of a fiber reinforced composite on the fiber volume fraction. An explanation of this phenomenon is given in a paper by Sih et al. [1].

Furthermore, crack propagation phenomena in fiber-reinforced materials are in general quite different from those in homogeneous continuum materials. Therefore, the direct application of fracture mechanics to cracked composite materials has to be made very carefully and requires the consideration of simpler models. Comprehensive surveys concerning the application of fracture mechanics to composite structures by considering the inherent difficulties were given in the papers by Cooper and Piggott [2], Brinson [3] and Sih [4]. Therein two standpoints can be taken

(a) micromechanical concept
(b) macromechanical theories

There exist essential differences between these two levels of investigation which are described in a paper by Smith [5]. The micromechanical aspect in thermal cracking of unidirectionally reinforced composite structures has been stressed recently in several papers by Herrmann et al. [6-10].

In this paper, using two different shaped unit cells of a unidirectional low-fiber concentration composite with circular fibers in a hexagonal array (cf. Fig. 1), the quasistatic thermal crack growth due to transient as well as stationary thermal stress fields has been studied by means of the finite element method. Fiber and matrix of these microcomponents consist of homogeneous, isotropic or anisotropic and linearly elastic materials where the temperature dependent thermoelastic material properties vary discontinuously at the fiber-matrix interface from the values $E_{ij}^{(f)}(T)$, $\nu_{ij}^{(f)}(T)$, $\alpha_i^{(f)}(T)$

of the fiber to the values $E_{ij}^{(m)}(T)$, $\nu_{ij}^{(m)}(T)$, $\alpha_i^{(m)}(T)$

of the matrix ($i,j = x,y$, E-Young's modulus, ν-Poisson's ratio, α-linear coefficient of thermal expansion). Besides, the conditions of perfect contact at the interface are assumed. Further, at time $t = t_o$ the cracked two-phase solid, having for $t < t_o$ the constant temperature T_o (temperature of the unstressed initial state) is subjected to the following temperature distributions in the cross section (cf. Figs. 2 and 3 for notation)

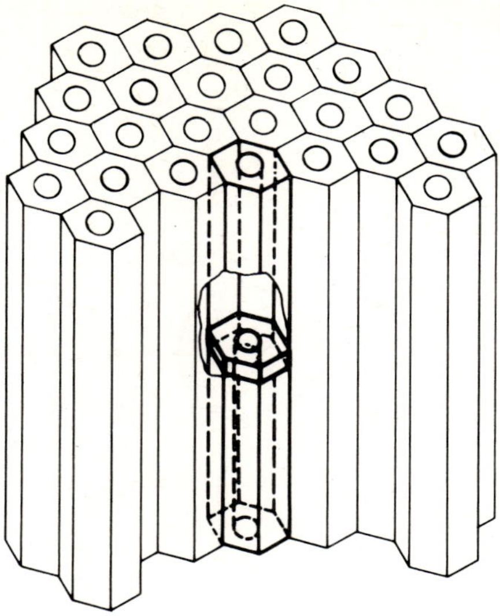

Fig.(1) - Fiber reinforced two-phase composite as a compound of hexagonal unit cells

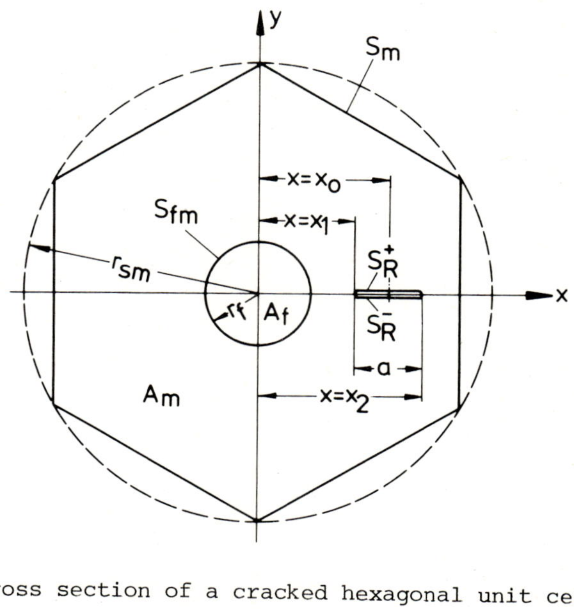

Fig.(2) - Cross section of a cracked hexagonal unit cell

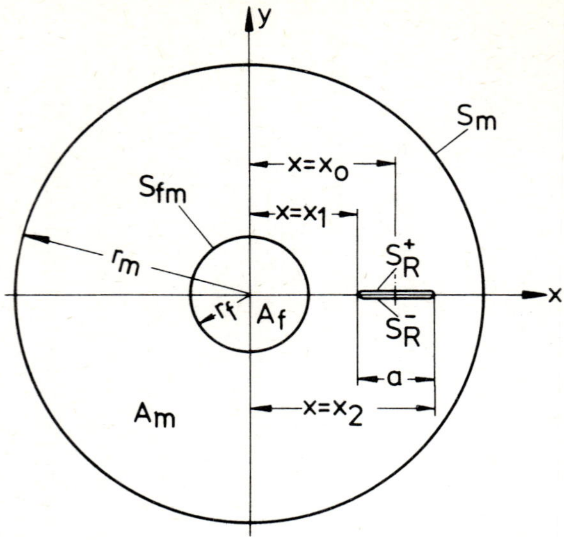

Fig.(3) - Cross section of a cracked circular unit cell

a) $$T(x,y,t) \begin{cases} T_f & \text{in } A_f \\ T_m & \text{in } A_m \end{cases} \qquad (1)$$

in which $T_f \neq T_m$ or $T_f = T_m \neq T_o$ and both temperatures are constant.

b) $T = T(x,y,t)$ is a solution of the following boundary-value problem of the heat conduction equation

$$\kappa_j T_{j,ii} - \dot{T}_j = 0 \quad ; \quad (j=f,m) \qquad (2)$$

Initial condition: $T_j(x,y,t_o) = T_o = \text{const}$ (3)

Boundary condition: $T_m(x,y,t) = T_e = \text{const on } S_m$ (4)

Continuity conditions:
$$\left.\begin{aligned} & T_f = T_m & (5) \\ & k_f \frac{\partial T_f}{\partial n} = k_m \frac{\partial T_m}{\partial n} & (6) \end{aligned}\right\} \text{ on } S_{fm}$$

with

k_j = thermal conductivity

ρ_j = density

c_j = specific heat

$\kappa_j = k_j(\rho_j c_j)^{-1}$, (j=f,m)

BASIC EQUATIONS AND FORMULATION OF THE BOUNDARY-VALUE PROBLEMS

The investigation of the self-similar quasi-static thermal crack growth in the inhomogeneous and linearly elastic microcomponents has been performed by using the concepts of linear elastic fracture mechanics. Applying the corresponding material law for an orthotropic medium

$$\varepsilon_{xx} = \sigma_{xx}(E_{xx}^{-1} - \nu_{xz}^{2}E_{zz}^{-1}) - \sigma_{yy}(\nu_{xy}E_{xx}^{-1} + \nu_{xz}^{2}E_{zz}^{-1}) + (\alpha_x + \nu_{xz}\cdot\alpha_z)\cdot\Delta T \tag{7}$$

$$\varepsilon_{yy} = -\sigma_{xx}(\nu_{xy}E_{xx}^{-1} + \nu_{xz}^{2}E_{zz}^{-1}) + \sigma_{yy}(E_{xx}^{-1} - \nu_{xz}^{2}E_{zz}^{-1}) + (\alpha_x + \nu_{xz}\cdot\alpha_z)\cdot\Delta T \tag{8}$$

$$\varepsilon_{xy} = \sigma_{xy}E_{xx}^{-1}(1+\nu_{xy}) \tag{9}$$

and by assuming the existence of a plane strain state the following mixed boundary-value problems of the plane thermoelasticity have to be solved

$$\sigma_{nn}(x,y) = \sigma_{nt}(x,y) = 0 \quad \text{on the external boundary } S_m \tag{10}$$

$$\sigma_{yy}(x,0) = \sigma_{xy}(x,0) = 0 \quad \text{on both crack surfaces } S_R^{+} \cup S_R^{-} \tag{11}$$

$$\left.\begin{array}{ll} \left[\sigma_{rr}(r,\phi)\right]_{r=r_f} = 0 & (12) \\ \left[\sigma_{r\phi}(r,\phi)\right]_{r=r_f} = 0 & (13) \\ \left[u_r(r,\phi)\right]_{r=r_f} = 0 & (14) \\ \left[u_\phi(r,\phi)\right]_{r=r_f} = 0 & (15) \end{array}\right\} \text{ at the fiber/matrix interface } S_{fm}$$

with the definition of the jump relations at the fiber/matrix interface S_{fm}

$$\left[\tau(r,\phi)\right]_{r=r_f} = \tau^{f}(r_f,\phi) - \tau^{m}(r_f,\phi) \tag{16}$$

The boundary-value problems (10)-(15) for the two composite structures have been solved under the assumption of plane strain conditions using the standard finite element computer program SAP IV together with the mesh-work generator program TOPONET [11]. Further, the finite element mesh-work for one-half of the symmetric specimens consist of about 500 to 670 triangular elements (cf. Figs. 4 and 8) focused essentially along the prospective crack line as well as along the interface and by using linear displacement functions. Because of the symmetry of the composite microcomponents with respect to the x-axis only the upper halves of the bimaterial specimens have to be considered. Thus, the following boundary conditions on the symmetry line have to be added

$$\sigma_{xy}(x,0) = 0 \quad , \quad u_y(x,0) = 0 \tag{17}$$

CALCULATION OF THE STRAIN ENERGY RELEASE RATE

A. Hexagonal Unit Cell

The determination of the strain energy release rate by use of standard finite element programs can be performed in different ways [12]. In this paper the desired crack extension force was calculated applying a principle originated by Bueckner [13] and proposed by Hayes [14] by use of an appropriate finite element formulation. This procedure bases essentially on the calculation of the crack surface energy. Assuming mode I-loading of a cracked specimen the crack closure from an original crack length $a + \Delta a$ to a shorter crack length a requires the energy

$$\Delta W_{cI} = -\Delta U_{cI} = \int_a^{a+\Delta a} \sigma_{yy}(x,a) \cdot u_y(x,a + \Delta a)\,dx \tag{18}$$

An analogous expression can be obtained for a mode II-loading of the cracked specimen

$$\Delta W_{cII} = -\Delta U_{cII} = \int_a^{a+\Delta a} \sigma_{xy}(x,a) \cdot u_x(x,a + \Delta a)\,dx \tag{19}$$

According to Irwin [15] the corresponding strain energy release rates have the form

$$G_j(a) = -\lim_{\Delta a \to 0} \frac{\Delta U_{cj}}{\Delta a} \quad ; \quad (j=I,II) \tag{20}$$

Further, an equivalent formulation with respect to the expressions (20) can be given by the definitions

$$G_I(a) = \lim_{\Delta a \to 0} \frac{1}{\Delta a} \{ \int_0^{a+\Delta a} \sigma_{yy}(x,0) u_y(x,a+\Delta a)\,dx - \int_0^{a} \sigma_{yy}(x,0) u_y(x,a)\,dx \} \tag{21}$$

and

$$G_{II}(a) = \lim_{\Delta a \to 0} \frac{1}{\Delta a} \{ \int_0^{a+\Delta a} \sigma_{xy}(x,0) u_x(x,a+\Delta a)\,dx - \int_0^{a} \sigma_{xy}(x,0) u_x(x,a)\,dx \} \tag{22}$$

Therein the stress $\sigma_{yy}(x,a=0)$ follows from the solution of the corresponding boundary-value problem for the uncracked specimen. On the contrary the crack surface displacement $u_y(x,a)$ can be obtained from a boundary-value problem for the cracked specimen only subjected to a load $-\sigma_{yy}(x,a=0)$ along the crack surfaces. Further, the approximate calculation of the limit values in the equations (21) and (22) by means of the finite element method corresponds to the procedure of Hayes mentioned above.

In case of the quasistatic extension of a thermal crack a calculation of the crack surface energies $U_{cI}(a)$ and $U_{cII}(a)$, respectively, in dependence on the crack length a is possible if the prospective crack line is known. Then the corresponding strain energy release rate can be obtained by a numerical differentiation of the crack surface energy with respect to the crack length a. Therein, in order to save computing time an application of the so-called superelement technique is often very helpful [16]. Applying a standard finite element program without a procedure for a partial matrix inversion computing time can only be saved by a reduction of the number of nodal points situated at the prospective crack line and used for an approximation of the function G (a). Thus, an appropriate selection of these nodal points is of great significance concerning a sufficient representation of the desired crack surface energy $-U_c$. As an example the quasi-static extension of a thermal crack is considered which starts at the circular interface of a self-stressed hexagonal unit cell of a $A\ell_2O_3$ matrix/Molybdenum fiber composite and runs towards the external stress-free boundary. Table 1 contains the temperature dependent thermoelastic material properties of the bimaterial specimen. Further, the fiber volume fraction is given by the reciprocal of the radius ratio $r_{sm}/r_f = 5.5$ which corresponds to a low fiber concentration composite. The thermal stress state in the cracked specimen has been caused by an isothermal cooling from the initial temperature $T_o = 1000^oC$ to the temperature $T_f = T_m = 0^oC$. The finite element mesh-work for one-half of the symmetric specimen consists of about 672 triangular elements with 374 nodes and is shown in Fig. 4. Further, Fig. 5 gives an enlarged section of this mesh-work in the vicinity of the straight crack line. Moreover, four different arrangements of nodal points for the approximate calculation of the crack surface energy $-U_c$ are shown. First of all the crack surface energy has been calculated for a maximum of nodal points that means for all feasible crack lengths. Secondly an approximation of the crack surface energy was performed using three different arrangements of nodal points. Therein the fourth approximation differs partly from the three others. Further, nearly one-half of the computing time could be saved by the transition from the first to the third approximation.

Al_2O_3

T [°C]	E [N/mm^2]	α [10^{-6}·K^{-1}]	ν
0	346700	7.8	0.260
200	340000	8.4	0.265
400	333400	9.0	0.270
600	324500	9.6	0.280
800	314900	10.2	0.310
1000	307700	10.8	0.365
1200	296500	11.4	0.450

molybdenum

T [°C]	E [N/mm^2]	α [10^{-6}·K^{-1}]	ν
0	324700	4.76	0.310
200	308000	5.05	0.313
400	288400	5.45	0.316
600	266800	5.92	0.319
800	240300	6.38	0.322
1000	212900	6.88	0.326
1200	168700	7.42	0.330

Table 1 - Thermoelastic material properties of the composite structure Al_2O_3 matrix/Molybdenum fiber in dependence on temperature

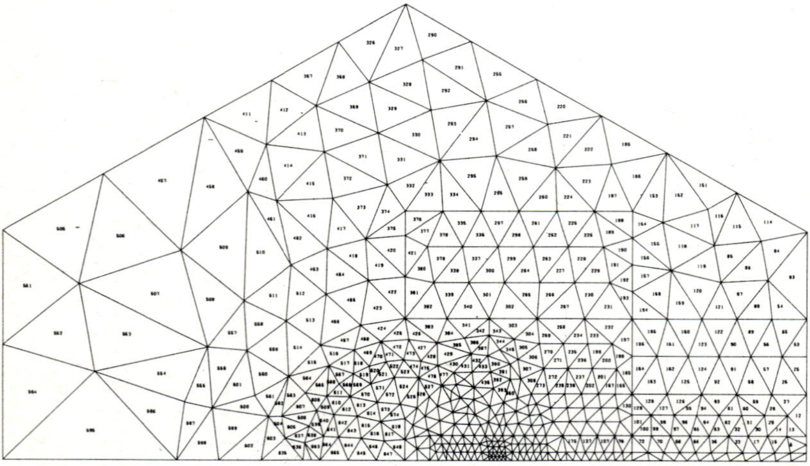

Fig. (4) - Finite element mesh-work for one-half of a hexagonal unit cell

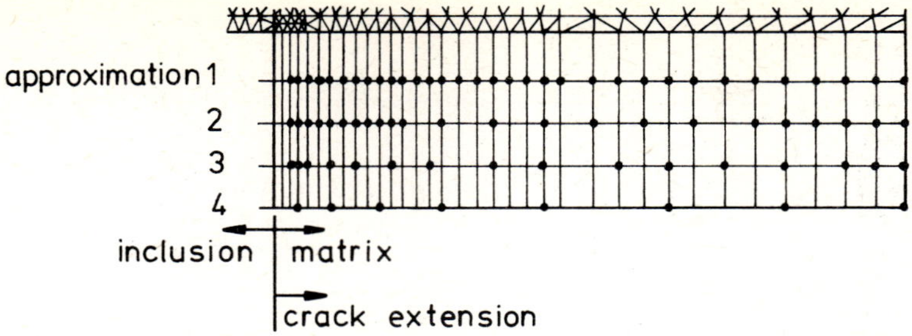

Fig.(5) - Different arrangements of crack border nodal points for the calculation of the crack surface energy

Moreover, the graph of the crack surface energy shown in Fig. 6 has been obtained using a chain of (q_a - 2) polynomials of the third order for the approximation of the intermediate values of the function $-U_c$. Therein q_a represents the number of nodal points along the crack line at the arrest crack length.

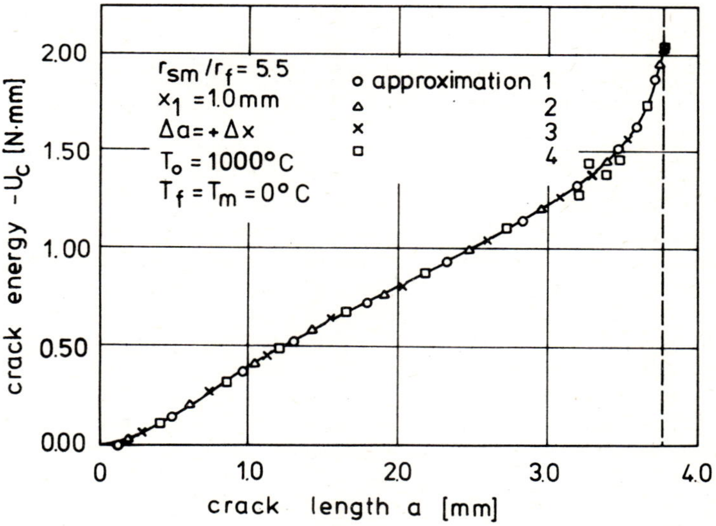

Fig.(6) - Crack surface energy of a cracked Al_2O_3 matrix/ Molybdenum fiber hexagonal composite microcomponent by using different arrangements of crack border nodal points

Finally, Fig. 7 shows the graph for the calculated strain energy release rate G_I of the propagating crack tip. The corresponding values of the stress intensity factor K_I can be obtained by using Irwin's formula [15]. Further, it should be noticed that all of the presented results have been related to a fiber radius of 1 mm. The treatment of geometrical similar specimens can be performed due to the existence of a similarity rule which could be derived using a dimension analysis [17].

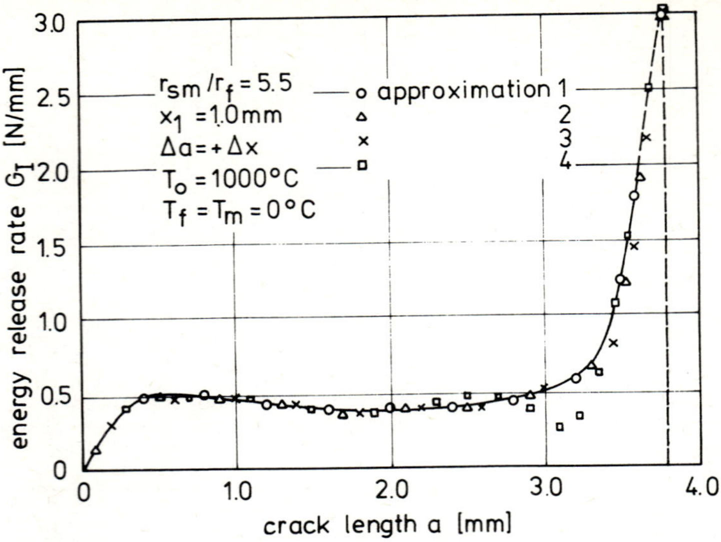

Fig.(7) - Strain energy release rate of a cracked $A\ell_2O_3$ matrix/Molybdenum fiber hexagonal unit cell

B. Circular Unit Cell

The foregoing considerations concerning the calculation of the strain energy release rate have been used also for the investigation of a quasistatic thermal crack growth in the circular unit cells of two different composite materials, namely in a $A\ell_2O_3$ matrix/Molybdenum fiber and a Aluminum matrix/Carbon fiber composite structure. The corresponding material properties are given in the tables 2 and 3 where in case of the thermophysical properties k and κ appropriate mean values have been used. Furthermore, Fig. 8 shows the finite element mesh-work for the circular unit cell which consists of about 502 triangular elements with 325 nodes. The thermal stress fields in the composite microcomponents were caused by well-defined instationary temperature fields which were obtained as solutions of the corresponding boundary-value problems (2)-(6) of the heat conduction equation [18].

Al_2O_3					
T [°C]	E [N/mm²]	α [1/10⁶·K]	ν [1]	k [N/sec·K]	ϰ [mm²/sec]
0	346700	7.8	0.260	28.7	8.22
200	340000	8.4	0.265		
400	333400	9.0	0.270		
molybdenum					
T [°C]	E [N/mm²]	α [1/10⁶·K]	ν [1]	k [N/sec·K]	ϰ [mm²/sec]
0	324700	4.76	0.310	133.0	51.1
200	308000	5.05	0.313		
400	288400	5.45	0.316		

Table 2 - Thermophysical material properties of the composite structure $A\ell_2O_3$ matrix/Molybdenum fiber

aluminum					
T [°C]	E [N/mm²]	α [1/10⁶·K]	ν [1]	k [N/sec·K]	ϰ [mm²/sec]
-100	74070	18.11	0.332	243.4	108.4
0	70390	22.50	0.339		
100	66710	24.60	0.346		

carbon fibre									
T	$E_{xx}=E_{yy}$	E_{zz}	G_{xy}	$\alpha_x=\alpha_y$	α_z	ν_{xy}	$\nu_{xz}=\nu_{yz}$	k	ϰ
[°C]	[N/mm²]			[1/10⁶·K]		[1]		[N/sec·K]	[mm²/sec]
-273	20000	225000	7042	4.5	0.0	0.42	0.3	122.7	110.6
0	20000	225000	7042	6.9	-0.5	0.42	0.3		
320	20000	225000	7042	8.7	-0.2	0.42	0.3		

Table 3 - Thermophysical material properties of the composite structure Aluminum matrix/Carbon fiber

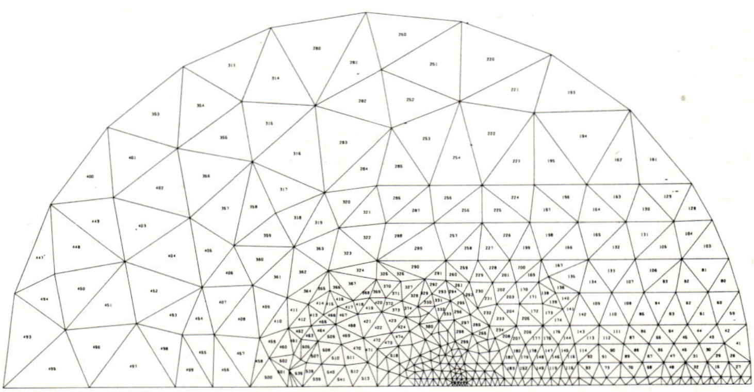

Fig.(8) - Finite element mesh-work for one-half of a circular unit cell

The radial temperature distributions for both bimaterial specimens are given in the figures 9 and 10.

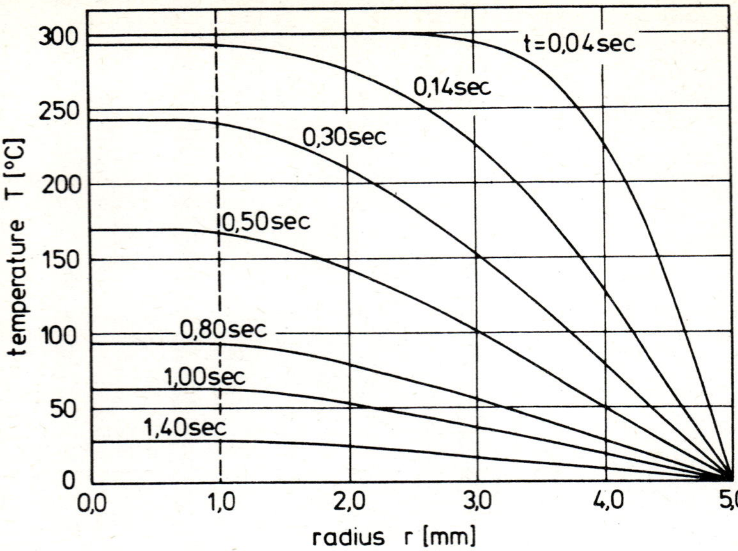

Fig. (9) - Radial temperature distribution in the cross section of a $A\ell_2O_3$ matrix/Molybdenum fiber circular unit cell

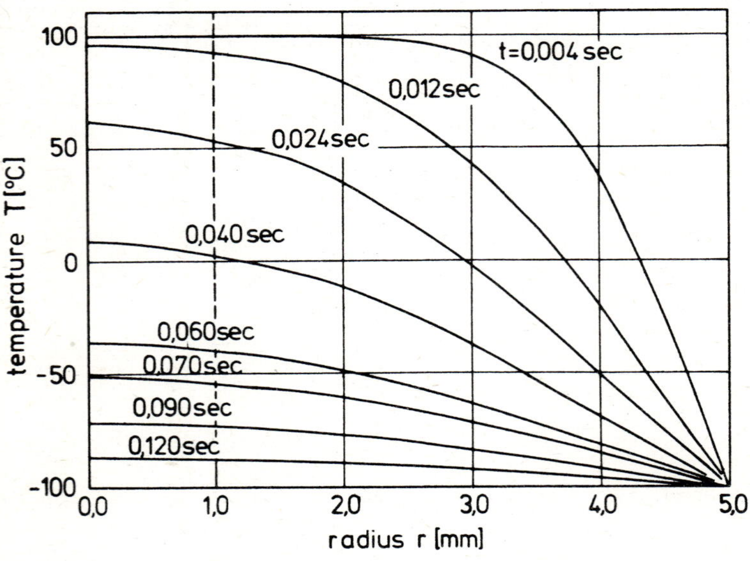

Fig. (10) - Radial temperature distribution in the cross section of a Aluminum matrix/Carbon fiber circular unit cell

It should be mentioned that for short time intervals immediately after the start of the heat conduction process crack closure can occur. Thus, in the figures 11 and 12 only those values of the strain energy release rate G_I are given where this effect cannot occur. Moreover, the corresponding values for $t \to \infty$ (stationary temperature distribution) as well as for the discontinuous temperature distribution $T_f = T_o \neq T_m = T_e$ have been calculated. Further, different initial temperatures were chosen, namely $T_o = 300^oC$ in case of the Al_2O_3/Molybdenum composite structure and $T_o = 100^oC$ in case of the Aluminum/Carbon unit cell. The corresponding temperatures of the environment were given by $T_e = 0^oC$ and $T_e = -100^oC$, respectively.

The graphs describing the crack extension force G_I show a remarkable behavior in so far as with the exception of the solid line belonging to the discontinuous temperature function all curves intersect in one point. Thus, the crack extension force G_I for cracks with a length $a < a_s$ (crack length at the intersection point) has an upper limit given by the corresponding G_I-values which belong to the stationary temperature distribution. On the contrary for crack lengths $a > a_s$ the upper limit for the crack extension force is marked by the solid curve belonging to the discontinuous temperature distribution. Therefore, the probability of a thermal crack growth initiated at the fiber/matrix interface and running into the matrix material increases with increasing time. Further, matrix cracks with a crack length $a = a_s$ or $a > a_s$, respectively, already existing in the composite microcomponent should grow rapidly after the application of a thermal shock.

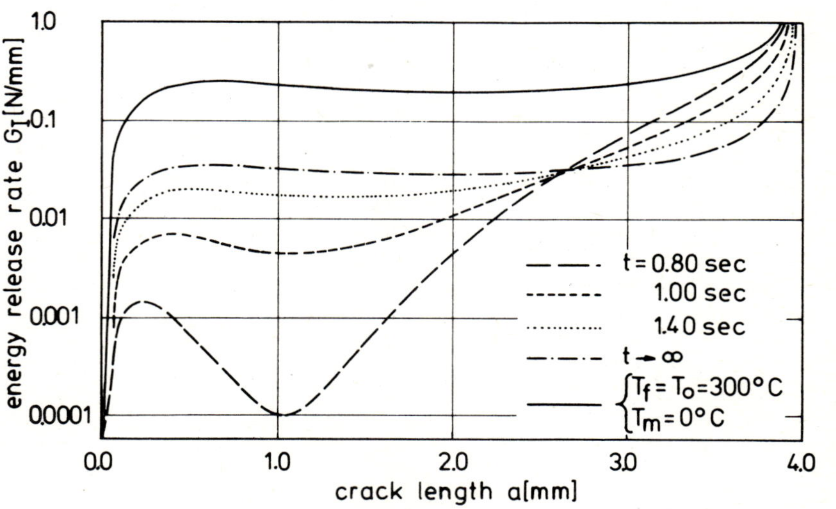

Fig. (11) - Crack extension force for different temperature distributions in the cross section of a Al_2O_3 matrix/ Molybdenum fiber circular unit cell

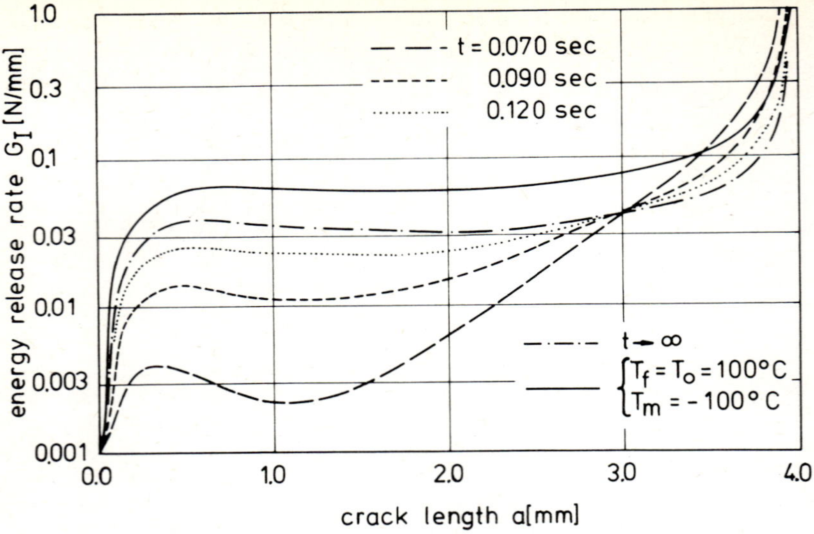

Fig.(12) - Crack extension force for different temperature distributions in the cross section of a Aluminum matrix/Carbon fiber circular unit cell

Finally, assuming stationary temperature fields, the strain energy release rates of cracks extending between two fibers in an ensemble of composite microcomponents were calculated. Further, numerical investigations concerning the interaction of collinear cracks in ensembles of microcomponents as well as crack-fiber interactions including the crack arrest phenomenon inside of single fibers have been performed [17,19].

CONCLUSION

Applying the definition of the crack closure work a calculation of the crack surface energy $-U_c$ using a standard finite element program has been presented. Therein the representation of the crack surface energy for the quasistatic extension of a thermal crack in a hexagonal unit cell of a fiber reinforced composite has been calculated for different arrangements of crack border nodal points.

Further, the knowledge of the crack surface energy allows the determination of the strain energy release rate G_I in case of the quasistatic extension of cracks originated at the fiber/matrix interfaces of circular unit cells which were subjected to well-defined instationary temperature distributions. Finally, it could be shown that the application of a discontinuous temperature distribution leads to a conservative estimation of the crack extension force belonging to the instationary temperature distribution.

REFERENCES

1. SIH, G. C., HILTON, P. D., BADALIANCE, R., SHENBERGER, P. S. and VILLARREAL, G. - Fracture Mechanics for Fibrous Composites, Analysis of Test Methods for High Modulus Fibers and Composites, Ed. Witney, J. J. ASTM, STP 521, ASTM Philadelphia, 1973.

2. COOPER, G. A. and PIGGOTT, M. R. - Cracking and Fracture in Composites, Fracture, Vol. 1, Ed. Taplin, D. M. R., University of Waterloo Press, 1977.

3. BRINSON, H. F. - The Methods of Experimental Mechanics Applied to Composite Materials, International Centre for Mechanical Sciences, Udine, Italy, 1978.

4. SIH, G. C. - Fracture Mechanics of Composite Materials, Fracture of Composite Materials, Ed. Sih, G. C. and Tamuzs, V. P., Sijthoff & Noordhoff, 1979.

5. SMITH, C. W. - Limitations of Fracture Mechanics as Applied to Composites, Inelastic Behaviour of Composite Materials, Ed. Herakovich, AMD, Vol. 13, ASME, New York, 1975.

6. HERRMANN, K. and FLECK, A. - Thermal Fracture in Compound Materials, Fracture, Vol. 3, Ed. Taplin, D. M. R., University of Waterloo Press, 1977.

7. HERRMANN, K. - Interaction of Cracks and Self-Stresses in a Composite Structure, Continuum Models of Discrete Systems, Ed. Provan, J. and Leipholz, H., SM Study No. 12, University of Waterloo Press, 1978.

8. BRAUN, H., FLECK, A. and HERRMANN, K. - Finite Element Analysis of a Quasistatic Crack Extension in a Unit Cell of a Fiber-Reinforced Material. Int. J. of Fracture, Vol. 14, R 3-6, 1978.

9. HERRMANN, K. and BRAUN, H. - Analysis of Cracks in Composite Structures Subjected by Thermal Loading, Fracture of Composite Materials, Ed. Sih, G. C. and Tamuzs, V. P., Sijthoff & Noordhoff, 1979.

10. HERRMANN, K., BRAUN, H. and KEMENY, P. - Comparison of Experimental and Numerical Investigations Concerning Thermal Cracking of Dissimilar Materials. Int. J. of Fracture, Vol. 15, R 187-190, 1979.

11. FÜHRING, H. - Finite Element Lösungen von Kerb- und Rißproblemen mit Hilfe automatischer Netzerzeugung. Report, Institut für Statik und Stahlbau, Techn. Hochschule Darmstadt, 1973.

12. HELLEN, T. K. and CESARI, F. - On the Solution of the Centre Cracked Plate with a Quadratic Thermal Gradient, Engng. Fract. Mechanics, Vol. 12, pp. 469 - 478, 1979.

13. BUECKNER, H. F. - The Propagation of Cracks and the Energy of Elastic Deformation, Trans. ASME, Vol. 80, pp. 1225 - 1229, 1958.

14. HAYES, D. J. - A Practical Application of BUECKNER's Formulation for Determining Stress Intensity Factors for Cracked Bodies, Int. J. of Fracture, Vol. 8, pp. 157 - 165, 1972.

15. IRWIN, G. R. - Fracture, Handbuch der Physik, Vol. 6, Ed. Flügge, S., Springer, Berlin, 1958

16. ZIENKIEWICZ, O. C. - The Finite Element Method in Engineering Science. Mc Graw - Hill, New York, 1971.

17. BRAUN, H. - Ph. D. Dissertation, University of Karlsruhe, 1979.

18. JAEGER, J. C. - Heat Conduction in Composite Circular Cylinders, Philosophical Magazine, Vol. 32, pp. 324 - 335, 1941.

19. BRAUN, H. and HERRMANN, K. - Quasistatische Rißausbreitung in der Einzeitszelle eines eigengespannten faserverstärkten Verbundwerkstoffes, GAMM-Tagung Wiesbaden, Germany, 1979
ZAMM, to be published

A HIGH ACCURACY FINITE ELEMENT ANALYSIS OF CRACKED RECTILINEAR ANISOTROPIC STRUCTURES SUBJECT TO BIAXIAL STRESS

GLENN R. HEPPLER[I] JORN S. HANSEN[II]

SUMMARY

A high order enriched element capable of allowing direct calculation of modes I and II stress intensity factors is utilized. This element provides a means of analysis which is exceptionally accurate (less than 5% error) and exhibits excellent economy. Both conforming and non-conforming formulations are examined and their relative influence on accuracy and economy are studied. The accuracy of the formulation is illustrated by comparing results for centre cracked specimens with other accepted references. The accuracy of the calculated stress intensity factors is found to vary with enriched element size, with crack length to width ratio, and to a lesser degree with the material properties. Results are presented for laminates of modern fibre composite materials subject to mode I and mixed mode I and II loading situations.

1.0 INTRODUCTION

The extensive introduction of brittle, fibre reinforced composite materials into aerospace structures over the last few years has led to attempts at predicting their behaviour when cracked. Because these structures are generally rectilinear anisotropic in nature and possess complicated boundary conditions, closed form analysis of their fractured state is not often possible being largely only amenable to numerical techniques.

I. Product Development Engineer, Boeing of Canada Ltd., Winnipeg, Manitoba, Canada.

II. Assistant Professor, University of Toronto, Institute for Aerospace Studies, Toronto, Ontario, Canada.

Within the context of a finite element formulation, several features are highly desirable. Some of these are [1]: minimal number of degrees of freedom necessary for an accurate solution, conforming elements to ensure monotone convergence, versatile element shape at the crack tip, and adaptability to handle material yielding. Further, most finite element procedures employ the method of Ritz and Strang and Fix [2] show that for this procedure to be valid trial functions must be sufficiently continuous across element boundaries (i.e., conforming), essential boundary conditions must be respected, and the energy functional must be computed exactly. Both Benzley [1] and Foschi and Barrett [3] have shown that conforming elements can produce accurate results but when these are compared to those obtained by Gifford and Hilton [4] they seem to require an excessive number of elements to achieve the same accuracy. The purpose of the work related here was to investigate conforming and nonconforming formulations. Particularly, it was hoped to develop a conforming element which exhibited comparable economy to the nonconforming case and would also guarantee monotone convergence of the solution.

2.0 FINITE ELEMENT FORMULATION

The present formulation is an extension of Gifford and Hilton's [4] twelve node element to rectilinear anisotropic materials. The element used by Gifford and Hilton [4] and developed by Zienkiewicz [5] is used for the conventional inplane load elements. A typical element is shown in Figure 1. The nodes are numbered as indicated and the basis functions are given by:

Corner Nodes:

$$N_i = \frac{1}{32}(1 + t_o)(1 + s_o)[-10 + 9(t^2 + s^2)]$$

Midside Nodes:

$$\left(t_i = \pm 1; \quad s_i = \pm \frac{1}{3}\right)$$

$$N_i = \frac{9}{32}(1 + t_o)(1 - s^2)(1 + 9s_o) \qquad (1)$$

(other midside node basis functions obtained by interchanging t and s). In the above expressions $t_o = tt_i$, $s_o = ss_i$ where t_i, s_i are the values of t, s at the i-th node. Here s, t are local coordinates s, t, $\in[-1, 1]$ and are defined by $s = x/a$, $t = y/b$. As usual, for s, t $\notin[-1, 1]$ it is understood that $N_i \equiv 0$. Thus the trial functions for conventional elements are expressed as

$$u_c = \sum_{i=1}^{12} N_i u_i, \qquad v_c = \sum_{i=1}^{12} N_i v_i \qquad (2)$$

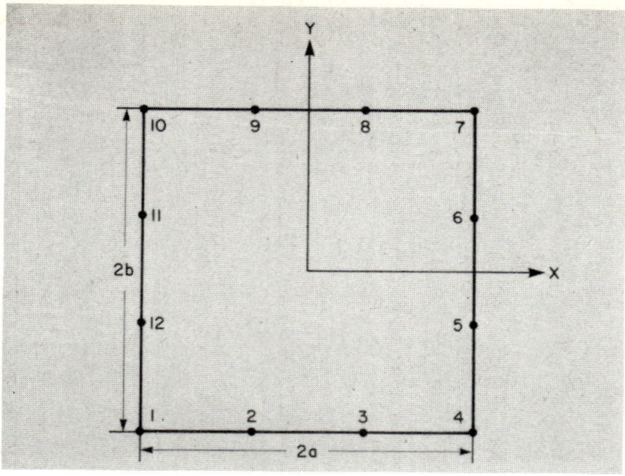

Figure 1 Numbering of a conventional 12 node element. Non-dimensional local coordinates are $s = x/a$, $t = y/b$.

where u_i, v_i are the nodal displacements in the x, y directions respectively. The singular portion of the trial functions u_s, v_s is given by

$$u_s = u_I + u_{II}, \qquad v_s = v_I + v_{II} \tag{3}$$

respectively. In turn [6] the singular components of the displacement field u_I, u_{II}, v_I, v_{II} are defined by

$$\begin{aligned} u_I &= K_I f_1(r, \theta), \qquad & u_{II} &= K_{II} f_2(r, \theta) \\ v_I &= K_I g_1(r, \theta), \qquad & v_{II} &= K_{II} g_2(r, \theta) \end{aligned} \tag{4}$$

where K_I, K_{II} are the mode one and two stress intensity factors and

$$f_1 = \sqrt{\frac{2r}{\pi}}\, \mathrm{Re}\left\{ \frac{1}{s_1 - s_2} \left[s_1 p_2 (\cos\theta + s_2 \sin\theta)^{1/2} \right.\right.$$

$$\left.\left. - s_2 p_1 (\cos\theta + s_1 \sin\theta)^{1/2} \right] \right\} \tag{5}$$

$$f_2 = \sqrt{\frac{2r}{\pi}}\, \mathrm{Re}\left\{ \frac{1}{s_1 - s_2} \left[p_2 (\cos\theta + s_2 \sin\theta)^{\frac{1}{2}} - p_1 (\cos\theta + s_1 \sin\theta)^{\frac{1}{2}} \right] \right\}$$

and where $g_{1,2}$ are obtained by substituting $q_{1,2}$ for $p_{1,2}$ in $f_{1,2}$ respectively. In the above, s_1, s_2 are the roots (generally complex) of

$$a_{11}s^4 + 2a_{16}s^3 + (a_{12} + a_{66})s^2 - 2a_{26}s + a_{22} = 0 \tag{6}$$

and can be expressed as $s_1 = \alpha_1 + i\beta_1$, $s_2 = \alpha_2 + i\beta_2$ where $\beta_1 > 0$, $\beta_2 > 0$. Furthermore, $p_{1,2}$, $q_{1,2}$ are given by

$$p_{1,2} = a_{11}s_{1,2}^2 + a_{12} - a_{16}s_{1,2};$$
$$q_{1,2} = \frac{a_{12}s_{1,2}^2 + a_{22} - a_{26}s_{12}}{s_{1,2}} \tag{7}$$

where a_{ij} are the elements of the material compliance matrix. Thus, the final expressions for the enriched trial functions are

$$u(s,t) = \sum_{i=1}^{12} N_i u_i + Z\left\{K_I\left[f_1 - \sum_{i=1}^{12} N_i f_{1i}\right] + K_{II}\left[f_2 - \sum_{i=1}^{12} N_i f_{2i}\right]\right\}$$
$$v(s,t) = \sum_{i=1}^{12} N_i v_i + Z\left\{K_I\left[g_1 - \sum_{i=1}^{12} N_i g_{1i}\right] + K_{II}\left[g_2 - \sum_{i=1}^{12} N_i g_{2i}\right]\right\} \tag{8}$$

where f_{1i}, ... represent the functions f_1, ... evaluated at the i-th node, and Z is the zeroing function in the conforming case (Section 2.1) and is unity in the nonconforming case.

The crack tip region is shown schematically in Figure 2 where a macro-element is formed completely surrounding the crack tip.

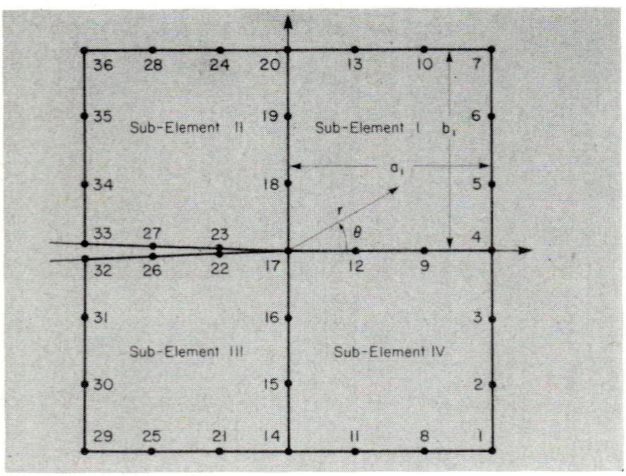

Figure 2 Macro-element modelling of the crack tip.

2.1 Zeroing Functions

Three different forms of the zeroing function were

chosen for study. As a baseline, the linear function used by Benzley [1] was considered. This function is of the form

$$
\begin{aligned}
Z_1 &= \tfrac{1}{4}(1 - s)(1 - t) \quad &\text{(sub-element I)}\\
Z_2 &= \tfrac{1}{4}(1 + s)(1 - t) \quad &\text{(sub-element II)}\\
Z_3 &= \tfrac{1}{4}(1 + s)(1 + t) \quad &\text{(sub-element III)}\\
Z_4 &= \tfrac{1}{4}(1 - s)(1 + t) \quad &\text{(sub-element IV)}
\end{aligned}
\qquad (9)
$$

The second was chosen in order to keep the zeroing function close to unity over the majority of the domain. It was chosen in the form

$$Z = \left[\tfrac{1}{4}(1 \pm s^5)(1 \pm t^5) \right]^{1/5} \qquad (10)$$

It was also argued that the zeroing function should be chosen so as to reinforce the desired singular behaviour of the trial function near the crack tip and hence should belong to the family of curves represented by the third curve in Figure 3 which is of the form

$$Z = \left[\tfrac{1}{4}(1 \pm s)(1 \pm t) \right]^5 \qquad (11)$$

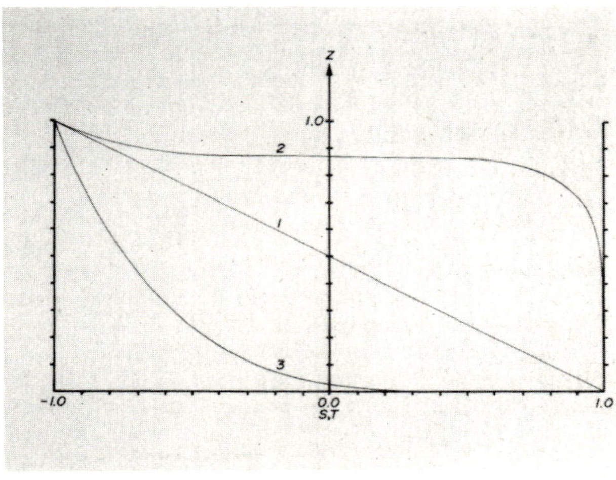

Figure 3 Zeroing functions.

3.0 TEST AND RESULTS

This section presents a series of results for cracked orthotropic specimens. The first set of results was obtained to assess the accuracy of the nonconforming formulation through comparisons to Bowie and Freese [7]. The second set considers mode I results for some typical composite laminates and the third set considers the same laminates in a mixed-mode loading situation. This is followed by an evaluation of the element in conforming formulations.

(i) Centre Cracked Specimens: Mode I - Comparisons with Bowie and Freese [7]

Bowie and Freese [7] have presented extensive data on centre-cracked orthotropic specimens for several different geometries and materials. Their work has been the standard for comparison due to its inherent accuracy. Their materials were defined in terms of only the imaginary part of the complex roots of the governing quartic equation (6). For the present comparison tests the relationship between the material parameters and β_1 and β_2 is governed by the equations:

$$\beta_1\beta_2 = \left(\frac{E_{11}}{E_{22}}\right)^{\frac{1}{2}}; \quad \beta_1 + \beta_2 = \sqrt{2}\left\{\left(\frac{E_{11}}{E_{22}}\right)^{\frac{1}{2}} + \frac{E_{11}}{2G_{12}} - \nu_{12}\right\}^{\frac{1}{2}} \qquad (12)$$

and tests were run for $\beta_2{}^2 = 2.5,\ 3.5,\ 4.5$ with $\beta_1 = 1$.

Crack length to specimen width ratios (L/W) of 0.1 to 0.8 were investigated for the $\beta_2{}^2 = 2.5$ and $\beta_2{}^2 = 3.5$ cases while L/W ratios of 0.3, 0.4, 0.5, 0.6 and 0.7 were treated for the $\beta_2{}^2 = 4.5$ case. All tests were performed on a specimen with a semi-height to width ratio of unity and of unit thickness. Bowie and Freese [7] define their stress intensity factor according to $K_I = \sigma\ \gamma\ \sqrt{L}$ where γ is the finite width correction factor such that its variation from unity is a measure of outer boundary influence. In order to be consistent with current conventions, K_I has been redefined as $K_I = \sigma\ \gamma\ \sqrt{\pi L}$.

The results from the above tests are presented in Figures 4 and 5. A number of features deserve mention. First, as may be seen in Figures 4a, b, c, there is no evidence of monotone convergence as the mesh is refined and the number of degrees of freedom is increased. Secondly, with reference to Figure 5, if the mesh is ignored and the error is considered to depend only on the enriched element ratio $(b_1/L)[(W-L)/a_1]$, then the magnitude of the error seems more predictable. For example, with enriched element ratios less than 3.0 and L/W > 0.2 the error in K_I never exceeds 5%. Thirdly, extensive tests have illustrated that the results are independent of the structural dimensions to an extraordinary degree. For example, an L/W = 0.5 crack modelled by the four basic elements, regardless of absolute dimensions (i.e., L = 2.5", 3" or 4") has a consistently reproducible error of 1.6%. Similarly if eight elements are used, the error is consistently less than 0.1%. It was also found that prescribing the ratio $(b_1/L)[(W-L)/a_1]$, of an enriched element resulted in a lack of sensitivity of the errors in K_I to the number of conventional elements used to complete the model. This insensitivity to the number of conventional elements allows one to solve problems with considerable economy as the accuracy is more dependent on macro-element size than on the mesh refinement.

It is to be noted that the aspect ratio of the enriched elements is crucial for accurate results and that only by proportioning all the sub-elements similarly can a satisfactory result be achieved. A possible interpretation of the mechanisms involved is as follows. Consider a threshold value of the macro-element ratio, $(b_1/L)[(W-L)/a_1]$, such that a macro-element described by this ratio will yield a minimum error in the stress intensity factor. Then, if an element ratio exceeds this value, the error will increase because the element trial function has not a sufficient flexibility to adequately cope with the larger domain. On the other hand, if the above element ratio is less than this threshold value, the singular elements do not cover enough of the crack tip domain and surrounding conventional elements are incapable of adequately modelling the local gradients. The point at which this transition occurs cannot be determined exactly as it will vary with crack geometry and the material properties. Also, it cannot be recognized due to the oscillatory behaviour of the nonconforming elements.

(ii) Centre Crack Specimens: Mode I - Typical Laminates

Due to the success enjoyed by the macro-element in determining the stress intensity factors for centre cracked specimens, three currently popular fibre reinforced epoxy materials

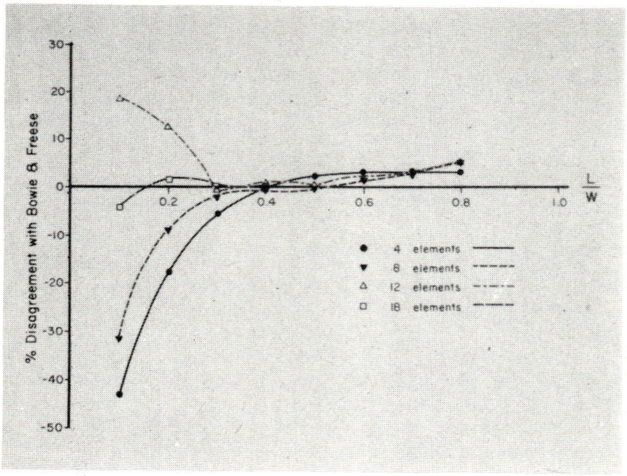

Fig. 4(a) Comparison with Bowie and Freese versus crack length to specimen width ratios (L/W) as total number of elements is varied; $\beta_2^2 = 2.5$.

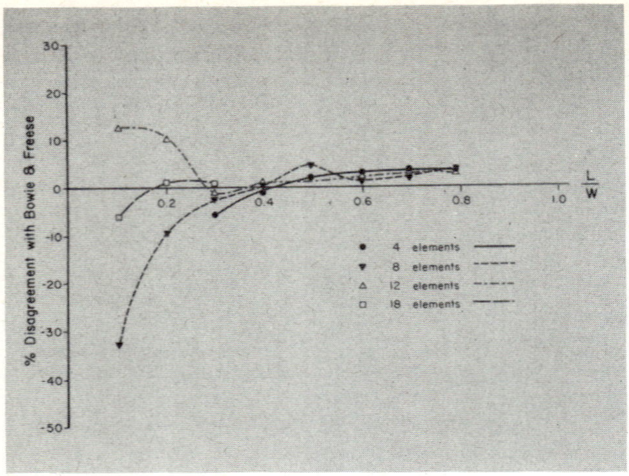

Fig. 4(b) Comparison with Bowie and Freese versus crack length to specimen width ratios (L/W) as total number of elements is varied; $\beta_2{}^2 = 3.5$.

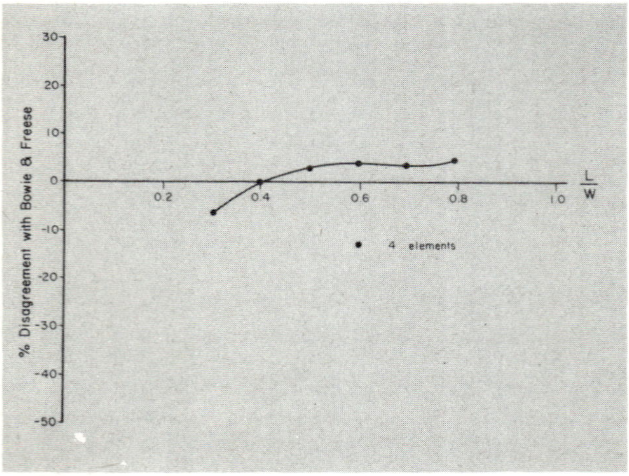

Fig. 4(c) Comparison with Bowie and Freese versus crack length to specimen width ratios (L/W) as total number of elements is varied; $\beta_2{}^2 = 4.5$.

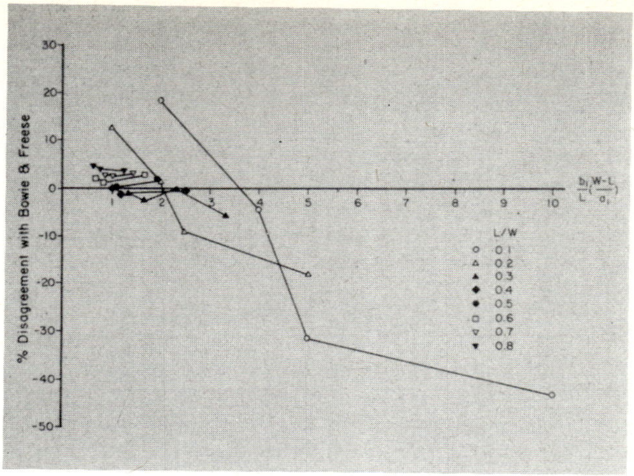

Fig. 5(a) Comparison with Bowie and Freese versus macro-element ratio for various crack length to specimen width (L/W) ratios; $\beta_2{}^2 = 2.5$.

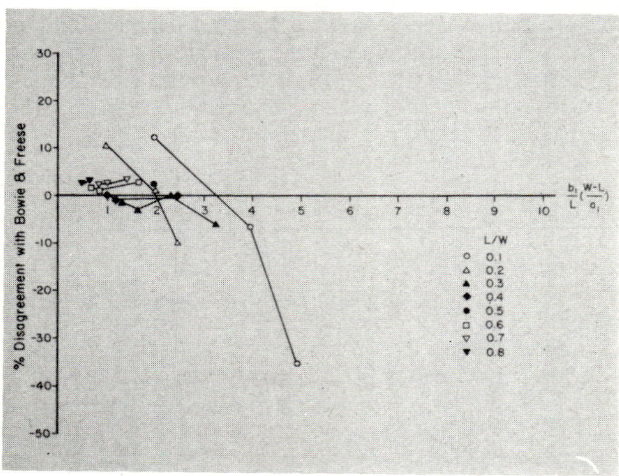

Fig. 5(b) Comparison with Bowie and Freese versus macro-element ratio for various crack length to specimen width (L/W) ratios; $\beta_2{}^2 = 3.5$.

were investigated for a number of centre crack geometries and fibre orientations. These materials were two graphite-epoxy composites (T300/SP286 and T300/NARMCO 5208) and one Kevlar-epoxy (Kevlar 49/CE3305) composite. The material parameters for the investigation are found in Table 1. For each material, three fibre geometries were considered: fibres perpendicular to the crack (0°), parallel to the crack (90°), and a four-ply symmetric balanced laminate which resulted in a zero coefficient of thermal expansion in the longitudinal direction. In addition, for each fibre geometry, three L/W ratios were investigated, these being 0.4, 0.5 and 0.6.

Table 1. Material Parameters for Mode I Tests

	Graphite-Epoxy T300/SP286	Graphite-Epoxy T300/NARMCO 5208	Kevlar-Epoxy Kevlar 49/ CE 3305
E_{11} psix10^6	20.5	26.3	12.2
E_{22} psix10^6	1.4	1.5	0.7
ν_{12}	0.26	0.28	0.41
ν_{21}	0.018	0.016	0.32
G_{12} psix10^6	0.59	1.0	0.02

The macro-element ratio was chosen to coincide with the value at which a minimum disagreement was obtained for the Bowie and Freese [7] comparison studies. These values are given in Table 2 along with the results of these investigations where it is observed that the stress intensity factor increases with increasing L/W as expected. Also, on comparing the results for these three cases, it was found that the 90° specimens show the largest K_I values, the zero CTE specimens the next largest and the 0° specimens had the smallest values. Based on the previous experience with these element configurations, it is expected that these results are within ~ 4% of their true values.

(iii) Edge Crack: Mixed Mode - Typical Laminates

To illustrate the mixed mode capability of this formulation, a cantilever beam specimen with a tension side edge crack was investigated for a uniform shear stress applied at the free end. The geometry of the test specimen is presented in Figure 6. The same materials, laminates, L/W ratios, and finite element ratios as were investigated in previous tests were considered. The results are presented in Table 3 and they show that mode I is by far the most significant for the situation under consideration. That is, K_I values are one order of magnitude greater than the corresponding K_{II} values. It is believed that these results are of an accuracy consistent with this formulation and should consequently be within ~ 5% of the actual values.

Table 2. K_I $(lb/in^{3/2})$ Values for the Typical Laminate Mode I Tests, W = 5.0"

	L/W	0.4	0.5	0.6
	$\frac{b}{L}, \frac{(W-L)}{(a)}$	1.0	1.0	0.83
Material	Case / NXxNY	2x6	2x4	2x4
T300/ SP286	0°	2.898	3.426	4.095
	90°	5.929	6.869	8.406
	0 CTE	3.095	3.797	4.903
T300/ 5208	0°	2.811	3.349	4.022
	90°	5.602	6.709	8.470
	0 CTE	3.007	3.650	4.659
K49/ CE3305	0°	2.829	3.362	4.032
	90°	5.769	6.826	8.528
	0 CTE	3.106	3.813	4.882

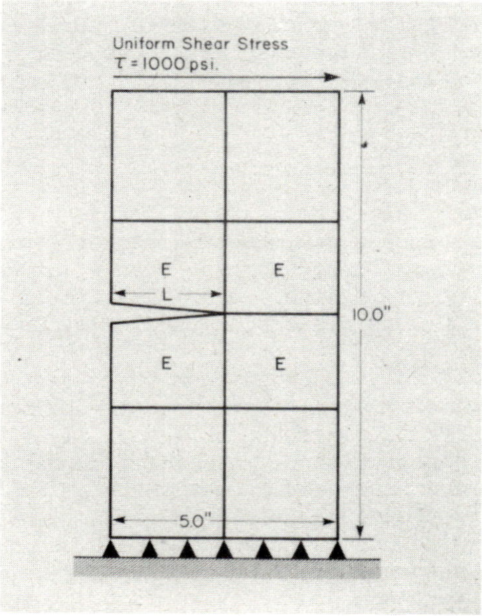

Fig. 6 Cantilever beam with uniform shear load for mixed mode K_I and K_{II} tests. Enriched elements are marked E.

Table 3. Mixed Mode Results for a Single Edge Cracked Cantilever Orthotropic Beam Subject to a Uniform Shear Stress ($lb/in^{3/2}x10^3$)

Material	L/W	0.4		0.5		0.6	
	Case	K_I	K_{II}	K_I	K_{II}	K_I	K_{II}
T300/ SP 286	0°	19.38	2.931	25.08	4.000	34.52	5.219
	90°	23.15	1.895	27.76	2.566	37.51	3.435
	0 CTE	22.49	3.318	28.46	3.743	40.77	4.232
T300/ 5208	0°	19.19	2.988	24.81	4.093	34.20	5.309
	90°	22.64	1.909	27.65	2.599	37.71	3.535
	0 CTE	21.90	3.178	28.13	3.901	40.16	4.580
K49/ CE 3305	0°	19.22	2.966	24.84	4.061	34.21	5.280
	90°	22.89	1.880	27.71	2.556	37.64	3.461
	0 CTE	22.31	3.221	28.33	3.733	40.72	4.386

(iv) Conforming Element Results

In order to determine what effect the introduction of conforming trial functions (via the zeroing function) would have on the accuracy of the enriched element, tests were conducted on the series of centre cracked specimens used by Bowie and Freese [7]. The $\beta_2^2 = 2.5$ case was chosen for comparison as this case was extensively treated in the nonconforming element studies.

Linear Zeroing Function Results

A crack length to specimen width ratio of 0.5 was used for comparison purposes because this geometry exhibited the best agreement in the nonconforming element studies. Seven different mesh configurations were investigated with the results from each test presented graphically in Figure 7.

It is immediately observed from these results that there is a substantial stiffening of the model upon the introduction of the zeroing function. The most favourable result obtained showed a disagreement with Bowie and Freese [7] of -61.7%. This was for the minimal model consisting of only the macro-element and no conventional elements. When conventional elements were introduced, the disagreement became more severe and it is felt that this was the result of several interacting factors which are discussed below. First, by forcing the enriched and conventional inter-element boundaries to be conforming, it is expected that the model will become stiffer due to the effective reduction in the degree of freedom enjoyed by the modelling functions. Secondly, insufficient con-

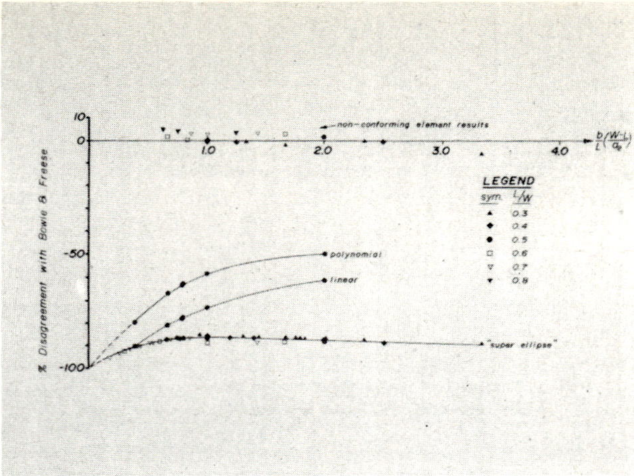

Fig. 7 Conforming element test results, different zeroing functions.

ventional elements were initially added to significantly change the stiffness of the model to a point that improved accuracy could be obtained. Thirdly, the introduction of the linear zeroing function interfered with the singular trial function's ability to model the stress field resulting in further effective stiffening of the model.

"Super-Ellipse" Type Zeroing Function Results

Due to the rather poor results obtained using a linear zeroing function, a nonlinear function based on the "super-ellipse" equation was introduced. This shape was chosen for the reasons previously mentioned (Section 2.1). In addition, a full range of L/W ratios was investigated in order to determine the effect of this parameter on the results. The results obtained are shown graphically in Figure 7.

It may be seen that the results obtained from this formulation were substantially poorer than those obtained in the previous case, indicating that this zeroing function did not have the desired effect but rather caused an increased stiffness.

The results of the L/W parameter study are shown in Figure 7. It is evident that these results are only slightly dependent on the L/W ratio with the smaller ratio values being marginally superior to the larger values. These differences in accuracy are so small that the formulation can be considered relatively insensitive to changes in the crack length to specimen width ratio. The least disagreement occurs for a macro-element aspect ratio of unity, indicating that for more refined finite element meshes this may be close to the optimum size for the macro-element.

"Polynomial" Type Zeroing Function Results

The third form of the zeroing function was used to investigate a series of element meshes that modelled an L/W = 0.5 specimen. This zeroing function returned a "best result" which under-estimated Bowie and Freese [7] by 50.2%.

The reasons for this disagreement are substantially the same as those given for the linear zeroing function. The improvement of the results is thought to be due to the fact that the slope of this function varies rapidly in a fashion similar to the stress field gradient, thus forcing conformity and reinforcing the trial function simultaneously.

It is thought that further improvements could be achieved by either refining the conventional element mesh or by refining the enriched element mesh. The latter of these possibilities seems the most likely to yield more accurate and economical results, particularly in view of Foschi and Barrett's [3] and Benzley's [1] results. Further, the trend indicates that increasing the order of this type of zeroing function may provide an improved result.

4.0 SUMMARY AND CONCLUSIONS

4.1 Nonconforming Case

The results presented have shown that excellent accuracy can be obtained from a nonconforming formulation with as few as eight elements. This is directly the result of the high order conventional and enriched elements being utilized which with the combination of accuracy and economy, makes them very attractive. A very significant finding is the effect that the macro-element's relative size, measured by the element ratio, has on the accuracy of the calculated K_I and K_{II} values. This has strong implications for the choice of the finite element mesh and is an area that needs thorough investigation. The only thing that detracts from the nonconforming formulation is the oscillatory nature of the stress intensity factors.

4.2 Conforming Case

The introduction of conforming trial functions resulted in very substantial stiffening of the finite element model that subsequently produced stress intensity factors that severely underpredicted the actual values. This resulted directly from the loss of flexibility suffered by themacro-element when the zeroing functions were introduced. It was also found that the choice of these zeroing functions has a marked influence on the quality of the results obtained and that they should be chosen to reinforce the expected singular behaviour near the crack tip. Improvements in the conforming formulation results can be expected if the enriched element mesh is refined to more than the four elements used here.

In conclusion, it is felt that the technique developed here is one of the best presented to date for the calculation of linear elastic planar stress intensity factors for rectilinear anisotropic materials subject to biaxial loading. However, as is often pointed out, real problems are seldom planar, never linear elastic, and always include effects not considered in this analysis. This should not detract from the use of this technique as, although it does not treat all aspects of the fracture problem, it does provide an accurate indication of the severity of the crack and a good basis for sound engineering judgements regarding cracked structures.

REFERENCES

1. BENZLEY, S. E. - Representation of Singularities with Isoparametric Finite Elements. Int. J. Num. Meth. Engng. 8, 537-545, 1974.

2. STRANG, G., FIX, G. J. - An Analysis of the Finite Element Method. Prentice-Hall, Englewood Cliffs, N.J., 1973.

3. FOSCHI, R. O., BARRETT, J. D. - Stress Intensity Factors in Anisotropic Plates Using Singular Isoparametric Elements. Int. J. Num. Meth. Engng., 10, 1281-1287, 1976.

4. GIFFORD, L. N., JR., HILTON, P. D. - Stress Intensity Factors by Enriched Finite Elements. Engng. Frac. Mech., 10, 485-496, 1978.

5. ZIENKIEWICZ, O. C. - The Finite Element Method in Engineering Science, 2nd Edn., McGraw-Hill, London, 107-109, 1971.

6. SIH, G. C., LIEBOWITZ, H. - Mathematical Theories of Brittle Fracture. In Fracture on Advanced Treatise, Vol. II, Ed. H. Liebowitz, 1968.

7. BOWIE, O. L., FREESE, C. E. - Central Crack in Plane Orthotropic Rectangular Sheet. Int. J. Frac. Mech., 8, No. 1, 49-58, 1972.

SECTION 2

NONLINEAR FRACTURE ANALYSIS

THE DOMINANCE OF THE CRACK TIP FIELDS OF INELASTIC CONTINUUM MECHANICS

D. M. Parks

Assistant Professor of Mechanical Engineering
Massachusetts Institute of Technology
Cambridge, Massachusetts 02139

"Fracture mechanics is, in a sense, the judicious interpretation of crack tip singular fields of continuum analysis"[1]

SUMMARY

Since the beginnings of fracture mechanics, a persistent theme (indeed, focus!) of the macroscopic description of crack growth phenomena has been a reliance on asymptotic continuum mechanics characterizations of the stress and deformation fields in the vicinity of a crack tip. In this paper, certain limits of suitability of "dominant singularity" approaches to inelastic crack behavior will be explored.

INTRODUCTION

The extreme complexity of the microscopic physical processes of deformation and fracture operating at the tip of a crack-like defect in a structural material have generally precluded the development of quantitative macroscopic models of crack growth phenomena which are explicitly based on the detailed spatial and temporal evolution of these fundamental processes. In this respect, it is indeed fortunate that considerable progress has been made in the fracture mechanics characterization of crack growth without requiring a complete and detailed modeling or understanding of the crack tip deformation mechanisms.

---

[1] D. C. Drucker and J. R. Rice, "Plastic Deformation in Brittle and Ductile Fracture", Engineering Fracture Mechanics, Vol. 1, 1970, pp. 577-602.

The principal means for achieving these successes has been asymptotic continuum mechanics characterization of the crack tip stress and deformation fields. For a given material model, the functional form of the local singular fields can often be determined to within a multiplicative scalar factor whose amplitude can be determined from a complete analysis of the overall loading and geometry considered. In certain cases, it can be argued that these asymptotic fields actually "dominate" the local crack tip vicinity over regions large by comparison to the scale of microstructural inhomogeneity of deformation and fracture process. In such instances, the amplitude of the field can be rationalized as a correlator of crack extension behavior.

The most well-known applications of asymptotic crack tip continuum analyses are the linear elastic singular fields associated with stationary crack tips [1-3]. For example, in the case of symmetric deformation in the crack tip region, the local stress field σ_{ij} associated with a linear elastic boundary value problem tends to:

$$\sigma_{ij}(r,\theta) \rightarrow \frac{K_I}{(2\pi r)^{1/2}} f_{ij}(\theta) \qquad (1)$$

as the local distance from the crack tip, r, tends to zero. The parameter θ is a polar angle measured from the plane ahead of the crack, and f_{ij} are known dimensionless functions of their argument. K_I is the stress intensity factor, which is undetermined from asymptotic analysis alone, but must be related to the overall geometry and loading of the region. Such relations have been compiled in handbook format for many loading and geometry combinations [4,5].

Once it became apparent that asymptotic results of the form of Eq. (1) suggested that the crack tip region was embedded in a field which was unique to within a multiplicative factor, K_I, then the amplitude of that parameter could be rationalized as a correlator of crack tip behavior. Material properties such as fracture toughness, K_{IC}, fatigue crack growth rate relations such as $da/dN = c(\Delta K)^m$, and other phenomena were well-described on the basis of the single-parameter relationship of Eq. (1).

Before pursuing other aspects of the approach, it is well to emphasize that Eq. (1) is an asymptotic result. At any finite distance from the crack tip, the complete solution to the boundary value problem posed will only be approximated by Eq. (1), but that approximation will become increasingly more accurate as the tip is approached. For example, in the Griffith problem of an embedded crack of length 2a in a large body subjected to a farfield uniform tensile stress σ^∞ perpendicular to the crack plane, the complete solution [6] for σ_{yy} on the plane ahead of the crack (y=0, $|x|>a$) gives

$$\sigma_{yy}(x,o) = \sigma^{\infty}|x|/(x^2-a^2)^{1/2} \qquad (2)$$

By comparison with Eq. (1), for $\theta=0$, $f_{yy}(0)=1$, $r=x-a$, and on using the stress intensity factor $K_I=\sigma^{\infty}(\pi a)^{1/2}$, we can see that in order for the dominant singularity approach to match the complete solution within 10%, it is necessary to be near enough to the crack tip such that r/a<.15, and to achieve 5% difference requires r/a<.07. Similar comparisons can be made for other field variables such as strain and displacement.

This comparison immediately suggests the existence of minimum, absolute crack size requirements in order to be able to expect that the K_I parameter be an appropriate correlator. For example, let the microstructural fracture process occur over some fixed material distance ρ^* around the crack tip. Clearly, unless the crack length a is at least several times ρ^*, there are not obvious grounds for correlations based on K_I. Size limitations of this type arise purely from the discrete nature of materials in contrast to the lack of a length scale in the continuum. Similar size comparisons between material fracture process zones and the spatial extents of regions dominated by the asymptotic singularities will be noted in the following sections for nonlinear constitutive models.

We have thus far left the precise meaning of the term "dominance" in the context of crack tip fields somewhat undefined. From the point of continuum mechanics, we can straightforwardly define a region of relative dominance by requiring that within the dominated region, the agreement between the complete and asymptotic solutions be within some specified tolerance. The second, and more difficult task is to "judiciously interpret" specific fracture phenomena in terms of the complete and asymptotic solutions of model boundary value problems. Results of this second step would presumably yield insight as to the "dominance" of crack tip fracture processes by the asymptotic fields. In this context, "dominance" can be ideally interpreted as the existence of a one-to-one relationship between crack tip phenomena (e.g., initiation, stable growth, etc.) and the amplitude of the singular field, but which is otherwise independent of loading and geometry. Of course, the ideal can be considerably complicated by material variability and by experimental, analytical, and judgemental limitations.

To further complicate the interpretation of fracture, we may note in the familiar linear elastic fracture mechanics an example of a different type of limitation to the dominance of crack behavior by singularity amplitude, as evidenced by the consideration of crack tip plastic zone size, r_y. Here, allowance is made for the inapplicability of a basic feature of the original continuum model (in this case, the linear constitutive law) within a boundary layer region at the crack tip. If the spatial extent of this region is determined *a posteriori* to be too large by comparison to the region of relative continuum dominance, then again we may not be able to expect correlations based on the singularity amplitude. Another example of this type of limitation to singularity dominance which can arise in fully plastic fracture mechanics is that the blunted region where finite deformations (which are generally neglected in the formulation of the boundary value and asymptotic analysis) occur is similarly large in comparison to the region of relative continuum dominance of the original formulation.

The subtle ambiguities inherent in this type of iteration of model, analysis, experimentation and interpretation can be quite formidable, but in the case of small-scale yielding in linear elastic plane strain fracture toughness testing, it was possible to define arbitrary, but useful limits to acceptable maximum plastic zone size while retaining a consistent and plausible interpretation of K_{IC}. [7]. These limits are expressed by:

$$15\, r_{y,c} \simeq 2.5 \left(\frac{K_{IC}}{\sigma_o}\right)^2 < a,\ B,\ w-a \qquad (3)$$

where a,B and w-a are crack length, specimen thickness, and uncracked ligament, respectively, σ_o is material flow stress, and $r_{y,c}$ is an approximate measure of plane strain plastic zone size at $K_I = K_{IC}$.

In fact, even the rather stringent requirements of ASTM Standard E-399 [7] do not provide fully unique characterizations of crack tip plastic fields in nominally elastic situations as has been noted by Larsson and Carlsson [8], Rice [9], and by Edmunds and Willis [10-13]. The source of the crack tip deviations is the existence in the asymptotic expansions of the elastic fields of higher order (nonsingular) terms that interact with the evolution of crack tip inelasticity. Such problems are treated with some generality in [10-13] in the spirit of matched asymptotic expansions. Salient features of these analyses can be obtained from the two term expansions considered in [8,9]. In particular, there exists a nonsingular nonvanishing stress component σ_{xx} parallel to the crack in linear elasticity so that Eq. (1) can be generalized [9] as:

$$\begin{bmatrix} \sigma_{xx} & \sigma_{xy} \\ \sigma_{yx} & \sigma_{yy} \end{bmatrix} \quad \frac{K_I}{\sqrt{2\pi r}} \begin{bmatrix} f_{xx} & f_{xy} \\ f_{yx} & f_{yy} \end{bmatrix} + \begin{bmatrix} T & 0 \\ 0 & 0 \end{bmatrix} \tag{4}$$

as $r \to 0$. Rather detailed finite element studies [8] indicate that the overall size and shape of plastic zone can vary considerably from one specimen type to another at nominally equal values of $K_I = 0.6\ \sigma_o a^{1/2}$. Rice's simple model [9] of the "T-effect" is in good qualitative agreement with the calculations of [8]. He finds, however, that the effects of this term on crack tip opening displacement δ_t and J-integral is much less than on global plastic zone size and shape. Depending on whether the microscale fracture mechanisms are stress induced (as cleavage microcracking) or deformation related (as in ductile void growth) the T-effect could manifest itself as a consistent bias in values of K_{IC} between different specimen types. Carlsson and Larsson's [8] results suggest that a normalized T given by $T\,a^{1/2}/K_I$ can range from -.59 for center-crack specimens to .03 for pure bending to 0.29 for the compact tension geometry. Similar results can be inferred from the early finite element calculations of Wilson [14]. As Rice [9] noted, specimens with negative T can be expected to increase the rapidity with which hydrostatically elevated crack tip stresses decrease with distance from the crack tip. To the extent that stress controlled fracture mechanisms must be triggered by a critical stress level acting over some material size scale [15], this tendency could lead to somewhat lower toughness values in compact geometries, as compared to center crack or bend specimens. On the other hand, an adaptation of Rice's [9] model with $\sigma_o \equiv \sqrt{3}\ \tau_o$ gives the influence of T on crack tip opening displacement as

$$\delta_t = .48 \frac{(1-\nu^2)K_I^{\,2}}{E\sigma_o} \quad [\ 1 - .54\ (T/\sigma_o) + 0(T/\sigma_o)^2\ ] \tag{5}$$

Thus we see that a negative T value serves to *increase* the crack tip deformation level as reflected in δ_t.

The examples which have been cited here for the cases of linear elastic or well-contained yielding serve to illustrate some of the considerations which must be taken into account if we wish to provide a rational framework upon which to base characterizations of crack extension behavior solely on the amplitude of a dominant continuum singularity.

Without intending to indicate a prescribed algorithmic approach to the assessment of applicability of any particular correlation between crack behavior and a continuum crack tip parameter, we note that generally the following types of inputs are synthesized:

(1) A continuum model is formulated, embodying constitutive and kinematical assumptions. Asymptotic analysis yields the nature of the near tip singular fields. The complete solution of a boundary value problem is obtained and regions of relative continuum dominance of the asymptotic solution are determined.

(2) Fractographic, metallurgical, and metallographic considerations suggest fracture process mechanisms and physical size scales over which they may act. These distances, normalized by actual component or specimen dimensions are compared with the regions of relative continuum dominance.

(3) Size scales are estimated for regions of uncertainty or inapplicability of basic features of the continuum model and compared with regions of relative continuum dominance.

(4) Raw data obtained from specific experiments is analyzed in light of (1-3) above, and judgements are made as to the suitability of correlations between features of the asymptotic fields and observed crack behavior.

In order to more firmly establish the soundness of the tentatively established correlation, it is also appropriate to examine data from fundamentally different crack configurations whose continuum models nonetheless admit similar asymptotic characterizations. Finally, it would seem appropriate to consider the sensitivity of the results of the continuum model to alternative, but equally plausible assumptions, especially as they relate to details of the constitutive idealization.

In view of the computational focus of this conference, it seems appropriate to review some of the studies which have been examined in the first point above the correspondences between complete numerical and asymptotic solutions in inelastic media. In the next section, attention is directed to fully plastic solutions of stationary cracks. A subsequent section considers problems of quasi-statically growing cracks in elastic-plastic materials, and finally some recent results in the analysis of creeping materials will be cited.

In several of these cases, it is found that, for many realistic absolute crack and fracture process zone sizes, there is insufficient leeway for uncertainty to assure that a unique correlation of crack growth behavior can be sensibly made on the basis of the dominant singularity alone. At this point we must recognize that we have sought to make such effectively empirical correlations largely in an effort to circumvent the need for formulating microstructural models of the fracture process whose dependencies on local crack tip stress and deformation histories would provide initiation and growth criteria. If these histories can be related through the asymptotic fields to a single parameter, then this can obviate the need for introducing the microstructural models of damage accumulation, etc. as a merely intermediate step in the correlation. But when in the more general case, single parameter descriptions fail to adequately characterize the local crack tip fields over microstructural size and material time scales, a predictive capability would seem to require the formulation and integration of such models through loading and crack growth histories. It may well be that in these non-singly-parameterized cases, the importance of the contributions of numerical and computational techniques to fracture analysis will surpass those which these disciplines have already made.

LARGE SCALE YIELDING AND FULLY PLASTIC PROBLEMS

For the case of power law hardening material behavior ($\sigma \alpha \varepsilon^N$ in the plastic range) the asymptotic forms of the stress and deformation fields for symmetric opening loads, collectively denoted HRR [16,17], have been determined. Again, in the limit as $r \to 0$, there obtains

$$\sigma_{ij} \to \left[\frac{EJ}{\sigma_o^2 r}\right]^{N/(1+N)} \sigma_o f_{ij}(\theta,N)$$

$$\varepsilon_{ij} \to \frac{\sigma_o}{E} \left[\frac{EJ}{\sigma_o^2 r}\right]^{1/(1+N)} g_{ij}(\theta,N) \qquad (6)$$

where σ_o is yield strength, E is Young's modulus, and J is the value of a an integral [18] which is precisely path-independent to nonlinear elastic (deformation theory plasticity) materials, and has been found to be substantially path-independent in numerical solutions of incrementally plastic materials as well. The functions g_{ij} and f_{ij} are again universal functions of their arguments, (though they differ depending on whether plain strain [16,17] or plane stress [19] is assumed). We see that again, there exists a single parameter, J, whose amplitude determines the asymptotic field, and on this basis we may presume to use it to correlate crack growth phenomena in such materials.

Before considering the issue of relative continuum dominance under tensile loading conditions, we may note that similar asymptotic expressions to Eqns. (16) exist for the case of antiplane (Mode III) deformation, and although this has been known for some years apparently only recently have the Mode III functions f_{ij} and g_{ij} been calculated [20].

The first indication of the strong dependence of the region of relative continuum dominance on strain hardening exponent N is found in Rice's [21] complete solutions for large scale, but contained, yielding in the edge-crack strip under longitudinal shear. Let $\gamma_o=\tau_o/G$ be shear yield strain. From the series representation used to express the complete field in the plastic zone, we can include additional terms in the expansion so that, for example, as $r\to 0$

$$\frac{\gamma_{yz}(r,\theta)}{\gamma_o} \to \left(\frac{J}{\tau_o\gamma_o r}\right)^{1/(1+N)} G_{yz}(\theta,N) + \left(\frac{A}{\tau_o\gamma_o r}\right)^{p} H_{yz}(\theta,N) + \left(\frac{B}{\tau_o\gamma_o r}\right)^{q} I_{yz}(\theta,N) + \ldots\ldots \tag{7}$$

where p and q are functions of N and $1/(1+N) > p > q > \ldots$, etc. Now, in the limit of low strain hardening as $N\to 0$, the term $1/1+N\to 1$ and so does p(N) and q(N): $p\to 1$ and $q\to 1$. In the limit, the higher order (less singular) terms with exponents p and q become as important (as singular) as the term governed by the "dominant" singularity whose amplitude is characterized by J. The low N limit of Eq. (7) then becomes:

$$\gamma_{yz}(r,\theta) \to \frac{1}{\tau_o r}\,[JG_{yz}(\theta,0) + AH_{yz}(\theta,0) + BI_{yz}(\theta,0) + \ldots] \tag{8}$$

so it is clear that there is no meaningful single-parameter representation of the crack tip fields, since different specimen geometries and load levels could plausibly yield the same J, but different A, B, etc. and hence different crack tip deformations.

The strong dependence of relative continuum dominance on N in such problems can be seen from complete solutions of pure power law materials obtained by Amazigo [22] who later [23] compared the one term (J-based) and complete solutions for γ_{yz} ahead of the crack ($\theta=0$) in the Mode III center-cracked strip. For the case of crack length a to strip width b given by $a/b = 0.75$, he compares the strain γ_{yz} on the remaining ligament, normalized by the minimum strain on this line, γ_G, at the (back) free surface. In assessing relative dominance, we can look for the relative radial distance at which the complete and asymptotic solutions give strains which agree to within a certain tolerance or, alternatively, can choose a strain level for which the relative radial distances, from complete and asymptotic solutions, required to achieve this strain level agree to within a given tolerance. We may note that for the moderate strain hardening of $N = .2$, it is required that $\gamma/\gamma_G > 6$ and $r/a < .02$ in order to obtain agreement to within 10% in r/a values for any fixed γ/γ_G. For the low strain hardening value $N = .1$, a 10% agreement is obtained only for $\gamma/\gamma_G > 25$ and $r/a < .003$. For $N = .05$, this agreement is secured only if $\gamma/\gamma_G > 300$ and $r/a < .0003$.

This work surely constitutes the most precise documentation of the dramatic reduction of region of dominance of the HRR type singularity as the non-hardening limit is approached. Of course, the nonuniqueness in plane strain crack tip states for the rigid/ideally plastic states had been known for some time, and McClintock [24] had commented on the possible consequences of such non-uniqueness on local crack tip fracture processes. For example, the Prandtl field associated with deep double edge cracked tension generates high triaxial tension ahead of the crack tip, with r^{-1} singular shear strains in a fan above and below the crack tip. In contrast, the tensile loading of the center cracked specimen or mid-ligament loading of the single edge-cracked specimen generates only modest triaxiality ahead of the crack, but intense plastic strain on 45° slip lines proceeding from the crack tip to the free surface.

The tendency of these rather different far field fully plastic flow fields to impinge on the (otherwise) J-controlled region at the crack in strain-hardening materials was investigated by McMeeking and Parks [25] and by Shih and German [26]. It was found in both investigations that the region of relative continuum dominance of the HRR singularity was considerably smaller in the large scale yielding regime in center and edge crack tension geometries than in the case of bending. As was noted in [25, 26], the practical importance of such studies is that considerably more stringent specimen size restriction must be imposed on J_{IC} testing in the unconstrained plane strain tension, as opposed to bending geometries. Suggested size limitations are $b\sigma_o/J_{IC} > M$, where b is remaining ligament size, with M = 25-50 for bending and M = 200 for tension.

Another practical application of this tendency could shape the numerical methods by which we obtain complete solutions or the strength of the asymptotic fields. It has been noted by Shih [27] that a possible cause of some of the discrepancies in J values obtained from numerical solutions of the fully plastic center crack geometry obtained in [28] and more recent results [29-31] is the fact that in [26], the Hilton-Hutchinson [32] element, which contained only the HRR field, was placed at the crack tip and given a radius roughly 2% of the crack length. If it turns out that the HRR field doesn't dominate over such relative distances, we would do well not to try to force it to do so in our singularity elements.

The evident relationship which would seem to exist between fully-plastic non-hardening slipline analysis of various notch/crack configurations and the extent of singular asymptotic dominance of crack fields in complete hardening solutions of these geometries has not been actively pursued. It would seem to be a topic worthy of further study.

There are a great deal of slip line fields for notches and cracks which have been compiled by McClintock [33]. A reasonable question would be: "What characteristics of a given slipline field affect the degree of 'impingement' which will occur at the crack tip?" Evidently, from the results of [25,26], the discrete shear on a single curved slipline in the Green and Hundy [34] slipline solution for a deep crack in bending can be readily accommodated by a local HRR field, while shear on a straight slipline cannot. Another interesting issue along this line of thought was raised in private communication by F. A. McClintock. It is perhaps tacitly assumed that the "worst" non-hardening fully plastic flow field (i.e., that field leading to least tough crack behavior) which could interact with a local HRR field is given by the Prandtl solution, and that other fully plastic fields are relatively more benign. But could a "worse" crack/notch configuration and corresponding slipline field, perhaps leading to both high triaxiality and discrete shearing on a single slipline, be found?

One slipline solution which is of practical relevance is that provided by Ewing [35] who followed on earlier work by Green [36] for bending of rather shallowly cracked specimens. In these cases, slipline fields can "break back" to the free surface from which the crack grew, with a resulting loss of triaxiality. This phenomenon coupled with a possible loss of relative continuum dominance, could perhaps contribute to the interpretation of experimental results such as those obtained by Muscati and Turner [37] on shallow cracks in three-point bending. Other modes of loss of constraint have been considered [38].

We close this section by noting that the question of single-parameter dominance of crack tip fields would seem to have important application in certain structural applications typified by a part-through surface crack in a plate or shell structure subject to predominantly membrane tensile loading. If the conditions are such that the structure is ductile and locally exhibits extensive plastic yielding near the surface crack, there would seem to be enough similarity of the plastic zone with the flow fields of the plane strain mid-ligament tension specimen that one could anticipate the possibility of a rather small region of HRR dominance at the tip. Indeed, in recent calculations for elastic-plastic surface cracks in plates and shells based upon the line-spring model of Rice [39,40], Parks [41] has found that the deformations and generalized loads along a surface crack tend, in the fully plastic regime, to the case of mid-ligament yielding. In fact, such deformations are consistent with the pressure vessel industry's time-honored principle of "leak before break" since, in leaking, the cracks invariably break through to the free surface by concentrated shearing on planes near 45° to the membrane tension.

ELASTIC-PLASTIC NONSTATIONARY CRACKS

The understanding of elastic-plastic fields associated with nonstationary cracks is considerably less developed than in the stationary case, and although some information concerning the limitations of applicability of macroscopic parameter/crack growth correlations has been obtained, much remains to be learned. Three areas of research in this field will be discussed.

The earliest approach to stable crack growth correlations typically in large plane stress components is based on Irwin's linear elastic R-curve concept; see for example, [42]. If the total extent of crack growth and time-integrated plastic zone is sufficiently small by comparison to the region of dominance of the elastic singularity, then there exists plausible grounds for correlation of applied K_I value with crack extension. In the elastic-plastic regime, Hutchinson and Paris [43] present a similarly conditional justification for a J-based resistance curve. They note that the incremental form of Eq. 6b for a cracked body of nonlinear elastic power hardening material which is simultaneously experiencing increments dJ of applied singularity strength and da of crack growth is:

$$d\varepsilon_{ij} \rightarrow \frac{\sigma_o}{E} \left(\frac{EJ}{\sigma_o^2 r}\right)^{1/(1+N)} \left\{\frac{1}{N+1} \frac{dJ}{J} g_{ij} + \frac{da}{r} h_{ij}\right\} \qquad (9)$$

where

$$h_{ij}(\theta) = \frac{1}{1+N} \cos\theta \, g_{ij}(\theta) + \sin\theta \frac{\partial}{\partial\theta} g_{ij}(\theta) \qquad (10)$$

They note that the first term in braces in Eq. (9) corresponds to proportional load increment while the second term is non-proportional. Since g_{ij} and h_{ij} are of comparable magnitude, the proportional straining term can overwhelm the other at radii r which satisfy $r >> (dJ/da)J$. If the region of relative continuum dominance of the HRR singularity exceeds radii which also greatly exceed the region of non-negligible non-proportional straining, Hutchinson and Paris argue that (a) deformation theory and incremental (flow) theory plasticity solutions of a crack growth increment would essentially coincide except in the zone of non-proportional deformation and (b) consequently there exists a rationale for the limited use of J vs. crack growth correlations. In some sense, this line of argument, which hinges critically on the rate, dJ/da, of increase of loading parameter to deformation parameter, is analogous to Shanley's [44] rationalization of the tangent modulus in plastic buckling problems.

The main questions which arise in the use of such a procedure are how much growth may be permitted and how large the slope dJ/da must be to justify the scheme. We may note that crack growth represents an inner impingement on a surrounding HRR field while the stationary crack far field flow fields can represent an outer impingement on the HRR fields which they in turn surround.

Numerical studies comparing incremental flow fields for the compact geometry with deformation and flow theories with different slopes dJ/da have been undertaken by Shih et al. [45] in an effort to answer some of the open questions. More calculations of this sort, with greater crack tip detail and details of constitutive modeling [46] would be of use in more precisely determining limitations of the approach.

The approach noted above has been based not on the asymptotic fields as $r \to 0$ associated with a power law material, but rather, on the stationary asymptotic field which surrounds the growing crack field. Indeed, no asymptotic solution for crack growth in a power hardening elastic-plastic material has been obtained. The asymptotic solution of Amazigo and Hutchinson [47] for steady growth in a bilinear elastic-plastic material is unlikely to provide a sufficiently accurate description of the near field in plane strain Mode I since the possibility of reversed plastic flow in a wake along the crack flank was neglected [55].

The only successful asymptotic analysis of elastic-plastic crack growth has been for the case of the ideal plasticity with a smooth yield surface [48-49]. It is found that the near tip stress field is similar to the Prandtl field, but with an elastic wedge of material somewhat above the crack flank. In contrast to the near equivalency of stationary and growing stress fields, however, it is found that a much weaker deformation singularity, $\varepsilon \sim \ln(1/r)$, results. Following Rice and Sorensen [50], Rice et. al. [48] develop from this asymptotic analysis a crack growth criterion based on a critical crack opening displacement at some fixed (microstructural) distance behind the growing crack tip, and rephrase the results of the model in terms of the two macroscopic parameters J and $(E/\sigma_o^2)\ dJ/da \equiv T$ [51] at initiation of crack growth. Some success has been obtained in correlating substantial amounts of crack growth in over a range of deformation from near small scale yielding through fully plastic behavior [52].

In view of the fully plastic nonuniqueness questions noted earlier for the non-hardening idealization, it is unlikely that complete numerical solutions could directly contribute to a complete assessment of the model, but certain parameters of the model which have been extracted from transient crack growth solutions in small scale yielding [53] could at least be extended to include T-effects as noted in the previous section for modest departures from ideal small-scale yielding conditions.

As a final example in this area, we cite numerical work on the idealized situation of steady crack growth in small scale yielding. The steady (with respect to a uniformly translating observer) condition obviates the need for incorporation of a growth criterion. The small scale yield condition conveniently affords a similarity length scale. Physically, such conditions might be approached in stress corrosion cracking.

Chitaley and McClintock [54] obtained an early such solution for nonhardening behavior in antiplane strain. Recently Dean and Hutchinson [55] and Parks et. al. [56] have investigated such problems in Mode I plane strain, including various hardening behaviors. It appears that strain hardening can substantially affect the levels of crack tip deformation as compared to ideal plasticity [55,56] and an uncritical application of the nonhardening growth model of [48,50] could substantially overpredict deformation levels sufficient to continue to sustain steady growth. Furthermore, it appears from [56] that the elastic wedge predicted theoretically [48,49] for nonhardening materials is even larger for steady growth in hardening materials, although this elastic unloading zone was not detected in the computations of ref. [55].

The question of the effect of constitutive idealization on the nature of near tip (or asymptotic) fields in cases like this of severely non-proportional deformation can be investigated. In ref. [55] a non-standard incremental plasticity model, based on corner theory [57], was used to examine antiplane deformations, in addition to standard smooth Mises yield surface theory. Little effect was noted. Similarly negligible effects of possible yield surface vertices were obtained in steady anti-plane growth from another idealization of a corner-type material [58]. These somewhat unexpected results can be rationalized by noting that a principal effect of a yield surface vertex is to soften the tangent modulus response in deviatoric directions orthogonal to the current stress state [59]. Unless material point stress histories contain a substantial component of such tangential loading, it is unlikely that much effect of this softened modulus will be seen. Indeed, the maximum angular measure of the distance between points on the yield surface where material points enter and leave the primary plastic zone in this problem is only a few degrees.

In sharp contrast, however, is the case in Mode I plane strain. If a complete Prandtl field were operative at the moving crack tip, then a material point would continuously move fully around the yield surface to a deviatoric stress point essentially the negative of that at which the plastic zone was first entered. There would then be ample opportunity for the soft tangential modulus to act. Even in the presence of an elastic unloading wedge [48, 49] a material point's trajectory during primary plastic deformation would span a considerable angular measure in stress space. In fact, a large effect in plane strain was noted in [58] on the basis of the highly idealized corner theory which gave essentially no effect on the anti-plane strain results. Further corner theory studies in Mode I plane strain are underway [56].

One final point on the 'judicious interpretation' of singularities should be made. In nonhardening growing crack problems, the form of the crack opening profile corresponds to the rather weak r•lnr form [48-54]. In small scale yielding, this term apparently affects the profile only over a very small fraction of the plastic zone [48, 53, 55, 56] and it could well turn out that, because the singularity is too weak, it fails to dominate fields over length scales comparable to process zone size. In a sense, this possible failure to dominate is entirely opposite to the situation encountered in the low hardening HRR fields, wherein the extreme strength of the deformation singularity causes it to die off rapidly.

CRACKS IN CREEPING MATERIALS

We close this paper with a very brief review of some recent asymptotic analyses of the fields associated with macroscopic crack tips in materials which exhibit time-dependent deformation.

Reidel and Rice [60] analyzed the fields near a stationary crack tip in an elastic/power law viscous material ($\dot{\varepsilon}_c \alpha \sigma^n$) subjected to a step loading. Instantaneously, elastic response is obtained and the initial solution is taken to be that of the K_I field. For time t>0, however, it is found that the crack tip stress fields are given by the HRR field based on a creep exponent n=1/N. In the steady state, the long term amplitude of this singular field is given by C*, which is a viscous analogue of the J integral It is found that the amplitude of this reduced singularity (as compared to $r^{-1/2}$ initial singularity) initially decays with time as t^{-1}. Approximate matching of the asymptotic short time and long time fields provides a rough indication of the time and deformation limits which mark the demarcation line between asymptotic fields which may be presumed to be K_I or C* "dominated". Reidel has recently extended the method to accommodate more general power law models of primary and tertiary creep deformation as well [61].

Again, since no sufficiently detailed complete solutions of such problems seem to yet exist, we may again anticipate questions of relative continuum dominance of such singular fields, although the interpretation in these cases becomes more difficult since both the strength and region of dominance of the local HRR-type fields are continuously changing with time. It could be that often generated shallow surface cracks in creep conditions may have to be rather large by comparison to common specimen geometries before their growth may be rationally correlated solely on the basis of these continuum singularities. Regarding the issue of influence of (or uncertainty in) constitutive model, Prof. J. L. Bassani has noted in private communication that, while a power law dependence of creep rate on stress level may well correlate data over limited ranges, the inferred power exponent can vary typically from 5 to 30, depending on stress level. Since a very broad spectrum of stress levels exists for these crack problems, attention should be given to this point.

Finally, Hui and Reidel [62] have obtained asymptotic results for steady quasi-static growth in an elastic/power law creeping material with small scale yielding. A new form of singularity denoted HR and dependent on creep exponent n, is obtained. The amplitude of this singular field can be related to crack velocity $\dot{a}$ and applied K_I level only by introducing a growth (fracture) criterion. Again, complete solutions of this problem are lacking, but we note that the near tip stress fields are singular as $r^{-1/(n-1)}$ for $n>3$ which is more singular than the HRR field which is as $r^{-1/(n+1)}$. Consequently, we may expect this strong singular field to die off rather rapidly, and it too may fail to dominate over physically relevant dimensions.

REFERENCES

1. Williams, M. L. -On the Stress Distribution at the Base of a Stationary Crack, J. Appl. Mech., Vol 24., pp. 109-114, 1957.

2. Irwin, G. R. -Fracture Mechanics, Structural Mechanics, Proceedings of the First Symposium on Naval Structural Mechanics, Ed. Goodier, J. and Hoff, N., Pergamon, 1960.

3. Rice, J. R. -Mathematical Analysis in the Mechanics of Fracture, Fracture, Vol. II, Ed. Leibowitz, H., Academic Press, 1968.

4. Tada, H., Paris, P. C. and Irwin, G. R. -The Stress Analysis of Cracks Handbook, Del Research, Hellertown, Pa., 1973.

5. Rooke, D. P. and Cartwright, D. J. -Compendium of Stress Intensity Factors, H. M. S. O., London, 1976.

6. Muskhelishvili, N. I. -Some Basic Problems in the Mathematical Theory of Elasticity (J. R. M. Radok, trans.), Noordhoff, Groningen, 1953.

7. Anon. -Plane Strain Fracture Toughness of Metallic Materials, ASTM Standard E399-78a, Annual Book of ASTM Standards, Philadelphia, 1979.

8. Larsson, S. G. and Carlsson, A. J. -Influence of Non-Singular Stress Terms and Specimen Geometry on Small-Scale Yielding at Crack Tips in Elastic-Plastic Materials, J. Mech. Phys. Solids, Vol. 21, pp. 263-277, 1973.

9. Rice, J. R. -Limits to the Small Scale Yielding Approximation for Crack Tip Plasticity, J. Mech. Phys. Solids, Vol. 22, pp. 17-26, 1974.

10. Edmunds, T. M. and Willis, J. R. -Matched Asymptotic Expansions in Nonlinear Fracture Mechanics, I, J. Mech. Phys. Solids, Vol. 24, pp. 204-224, 1976.

11. Edmunds, T. M. and Willis, J. R. -Matched Asymptotic Expansions in Nonlinear Fracture Mechanics II, J. Mech. Phys. Solids, Vol. 24, pp. 225-237, 1976.

12. Edmunds, T. M. and Willis, J. R. -Matched Asymptotic Expansions in Nonlinear Fracture Mechanics, III, J. Mech. Phys. Solids, Vol. 25, pp. 423-455, 1977.

13. Willis, J. R. -The Solution of Elastoplastic Fracture Problems by Matched Asymptotic Expansions, Fracture

Mechanics, Vol. XII, SIAM-AMS Proceedings, Ed. Burridge, R., American Mathematical Society, Providence, R.I., 1979.

14. Wilson, W. K. -Discussion in Plane Strain Crack Toughness Testing, ASTM STP 410, pp. 75-76, ASTM, Phil., 1966.

15. Ritchie, R. O., Knott, J. F., and Rice, J. R. -On the Relationship Between Critical Tensile Stress and Fracture Toughness in Mild Steel. J. Mech. Phys. Solids, Vol. 21, pp. 395-410, 1973.

16. Hutchinson, J. W. -Singular Behavior at the End of a Tensile Crack in a Hardening Material, J. Mech. Phys. Solids, Vol. 16, pp. 13-31, 1968.

17. Rice, J. R. and Rosengren, G. F. -Plane Strain Deformation Near a Crack Tip in a Power Law Hardening Material, J. Mech. Phys. Solids, Vol. 16, pp. 1-12, 1968.

18. Rice, J. R. -A Path Independent Integral and the Approximate Analysis of Strain Concentration by Notches and Cracks, J. Appl. Mech., Vol. 35, pp. 379-385, 1968.

19. Hutchinson, J. W. -Plastic Stress and Strain Fields at a Crack Tip, J. Mech. Phys. Solids, Vol. 16, pp. 337-347, 1968.

20. Riedel, H. -Cracks Loaded in Anti-Plane Shear Under Creep Conditions, Z. Metallkunde, Vol. 69, pp. 755-760, 1978.

21. Rice, J. R. -Stresses Due to a Sharp Notch in a Work-Hardening Elastic-Plastic Material Loaded by Longitudinal Shear, J. Appl. Mech., Vol. 34, pp. 287-298, 1967.

22. Amazigo, J. C. -Fully Plastic Center-Cracked Strip Anti-Plane Shear, Int. J. Solids Structures, Vol. 11, pp. 1291-1299, 1975.

23. Amazigo, J. C. -Some Mathematical Problems of Elastic-Plastic Crack Growth, Fracture Mechanics, Vol. XII, SIAM-AMS Proceedings, Ed. Burridge, R., American Mathematical Society, Providence, 1979.

24. McClintock, F. A. -Local Criteria for Ductile Fracture, Int. J. Fract. Mech., V. 4, pp. 101-130, 1968.

25. McMeeking, R. M. and Parks, D. M. -On Criteria for J-Dominance of Crack-Tip Fields in Large Scale Yielding, Elastic-Plastic Fracture, ASTM STP 668, Eds. Landes, J. D., Begley, J. A., and Clarke, G. A., American

Society for Testing and Materials, Philadelphia, 1979.

26. Shih, C. F. and German, M. D. -Requirements for a One-Parameter Characterization of Crack Tip Fields by the HRR Singularity, to appear, Int. J. Fract.

27. Shih, C. F. -Private Communication, 1978.

28. Goldman, N. L. and Hutchinson, J. W. -Fully Plastic Crack Problems: The Center-Cracked Strip under Plane Strain, Int. J. Solids Structures, Vol. 11, pp. 575-591, 1975.

29. Ranaweera, M. P. and Leckie, F. A. -Solution of Non-linear Fracture Problems by Direct Optimization, Numerical Methods in Fracture Mechanics, Eds. Luxmoore, A. R. and Owen, D. R. J., Swansea, 1978.

30. Parks, D. M. -Virtual Crack Extension: A General Finite Element Technique for J-Integral Evaluation, Numerical Methods in Fracture Mechanics, Eds. Luxmoore, A. R. and Owen, D. R. J., Swansea, 1978.

31. Hutchinson, J. W., Needleman, A. and Shih, C. F. - Fully Plastic Crack Problems in Bending and Tension, Fracture Mechanics, Ed. Perrone, N. J. et. al., University Press of Virginia, 1978.

32. Hilton, P. D. and Hutchinson, J. W. -Plastic Intensity Factors for Cracked Plates, Engr. Fract. Mech., Vol. 3, pp. 435-451, 1971.

33. McClintock, F. A. -Plasticity Aspects of Fracture, Fracture, Vol. III, Ed. Leibowitz, H., Academic Press, 1971.

34. Green, A. P. and Hundy, B. B. -Initial Plastic Yielding in Notch Bend Bars, J. Mech. Phys. Solids, Vol. 4, pp. 128-144, 1956.

35. Ewing, D. J. F. -Calculations on the Bending of Rigid/ Plastic Notched Bars, J. Mech. Phys. Solids, Vol. 16, pp. 205-213, 1968.

36. Green, A. P. -The Plastic Yielding of Shallow Notched Bars Due to Bending, J. Mech. Phys. Solids, Vol. 4, pp. 259-268, 1956.

37. Muscati, A. and Turner, C. E. -Post-Yield Behaviour of Shallow-Notched Alloy Steel Bars in Three Point Bending, Fracture Mechanics in Engineering Practice, Ed. Stanley, P. Applied Science Publishers, London, 1977.

38. Turner, C. E. -An Analysis of the Fracture Implications of Some Elastic Plastic Finite Element Studies, Numerical Methods in Fracture Mechanics, Eds. Luxmoore, A. R., and Owen, D. R. J., Swansea, 1978.

39. Rice, J. R. and Levy, N. -The Part-Through Surface Crack in an Elastic Plate, J. Appl. Mech., Vol. 39, pp. 185-194, 1972.

40. Rice, J. R. -The Line-Spring Model for Surface Flaws, The Surface Crack, Physical Problems and Computational Solutions, Ed. Swedlow, J. L., ASME, New York, 1972.

41. Parks, D. M. -The Inelastic Line-Spring: Estimates of Elastic-Plastic Fracture Mechanics Parameters for Surface-Cracked Plates and Shells, to be presented at ASME Pressure Vessel and Piping Conference, San Francisco, August, 1980.

42. Fracture Toughness Evaluation by R-Curve Methods, ASTM STP 527, 1974.

43. Hutchinson, J. W. and Paris, P. C. -Stability Analysis of J-Controlled Crack Growth, Elastic-Plastic Fracture, ASTM STP 668, Eds. Landes, J. D., Begley, J. A., and Clarke, G. A., ASTM, pp. 37-64, 1979.

44. Shanley, F. R. -Inelastic Column Theory, J. Aero. Sci., Vol. 14, 1946.

45. Shih, C. F., et. al. -Methodology for Plastic Fracture Combined Ninth and Tenth Quarterly Reports to EPRI, General Electric Corporate Research Publication, SRD-79-031, Jan. 31, 1979.

46. Rice, J. R. -Some Computational Problems in Elastic-Plastic Crack Mechanics, Numerical Methods in Fracture Mechanics, Eds. Luxmoore, A. R. and Owen, D. R. J., Swansea, 1978.

47. Amazigo, J. and Hutchinson, J. W. -Crack Tip Fields in Steady Crack-Growth with Linear Strain-Hardening, J. Mech. Phys. Solids, Vol. 25, pp. 81-97, 1977.

48. Rice, J. R., Drugan, W. J. and Sham, T. L. -Elastic Plastic Analysis of Growing Cracks, presented at ASTM 12 Annual Symposium on Fracture Mechanics, St. Louis, May, 1979, to appear in ASTM STP of Conference.

49. Slepyan, L. I. -Growing Crack During Plane Deformation of an Elastic-Plastic Body, Mekh. Tverdogo Tela, Vol. 9, pp. 57-67, 1974.

50. Rice, J. R. and Sorensen, E. P. -Continuing Crack Tip Deformation and Fracture for Plane Strain Crack Growth in Elastic-Plastic Solids, J. Mech. Phys. Solids, Vol.26 pp. 163-186, 1978.

51. Paris, P. C., Tada, H., Zahoor, A. and Ernst, H. -A Treatment of the Subject of Tearing Instability, U.S. Nuclear Regulatory Commission Report, NUREG-0311, August, 1977.

52. Hermann, L. and Rice, J. R. -Comparison of Experiment and Theory for Elastic-Plastic Plane Strain Crack Growth, presented a Conference on Micromechanisms of Crack Extension Churchill College, Cambridge, April 1980. To appear in special issue of Metal Science.

53. Sham, T. L. -A Finite Element Analysis of Quasi-Static Crack Growth in an Elastic-Perfectly Plastic Solid, ScM Thesis, Brown University, March, 1979.

54. Chitaley, A. D. and McClintock, F. A. -Elastic-Plastic Mechanics of Steady Crack Growth under Anti-Plane Shear, J. Mech. Phys. Solids, Vol. 19, pp. 147-163, 1971.

55. Dean, R. H. and Hutchinson, J. W. -Quasi-Static Steady Crack Growth in Small Scale Yielding, presented at ASTM 12 Annual Symposium on Fracture, St. Louis, May, 1979, to appear in ASTM STP of the Conference.

56. Parks, D. M., McMeeking, R. M., and Lam, P.S. -in preparation for publication, 1980.

57. Christoffersen, J. and Hutchinson, J. W. -A Class of Phenomenological Corner Theories of Plasticity, J. Mech. Phys. Solids, Vol. 27, pp. 465-488, 1979.

58. Parks, D. M. and Lam, P. S., -unpublished research, 1979.

59. Hutchinson, J. W. -Elastic-Plastic Behaviour of Polycrystalline Metals and Composites, Proc. Royal Soc. Lond. Ser. A., Vol. 319, pp. 247-272, 1970.

60. Riedel, H. and Rice, J. R. -Tensile Cracks in Creeping Solids, presented at ASTM 12th Annual Symposium on Fracture, St. Louis, May, 1979. To appear in ASTM STP of the Conference.

61. Riedel, H. -Creep Deformation at Crack Tips in Elastic-Viscoplastic Solids, Brown University Engineering Report MRL E-114, June, 1979.

62. Hui, C. Y. and Riedel, H. -The Asymptotic Stress and Strain Field Near the Tip of a Growing Crack under Creep Conditions, Brown University, Engineering Report MRL E-117, August 1979.

ELASTIC-PLASTIC ANALYSIS OF A CRACK BY USING CONTINUOUSLY DISTRIBUTED DISLOCATIONS

H. Miyamoto(I), H. Okamura(II), K. Kageyama(III) and H. Morita(III)

SUMMARY

A numerical method of elastic-plastic analysis of a cracked plate is proposed, based on the theory of continuously distributed dislocations. A plastic zone is approximated by the group of slip lines, and the effect of finite boundary is calculated. The plastic zones and the outer boundaries are divided into the linear segments, and the boundary conditions are satisfied by distributing dislocations on them. Some improvements in the numerical procedure make for higher accuracy and the reduction of computing time. The present method is applied to the elastic-plastic analysis of cracked plate and the estimation of the J-integral. The method is also extended to analysis of alternative shear rupture mechanism at crack tip and to computer simulation of fatigue crack propagation.

INTRODUCTION

In the present paper, the theory of continuously distributed dislocations is applied to the elastic-plastic analysis of a cracked plate. The outer boundaries and the silp lines are approximated by the groups of segments and the boundary conditions are satisfied by distributing dislocations. The fundamental solutions give the elastic field of an edge dislocation in an infinite cracked-plate. These solutions satisfy exactly the traction-free conditions on the crack surface, and contribute to the improvement of accuracy. Some other improvements in the numerical procedure make for higher accuracy and the reduction of computing time.

Interaction between a crack and the slip band in an infinite plate was studied by Yokobori et al. [1,2,3]. Plastic yielding on inclined slip-planes at a crack tip in an infinite plate was analysed numerically by Vitek[4] and Riedel[5]. McClintock proposed the dislocation

---

(I) Professor, Faculty of Science and Technology, Science University of Tokyo.
(II) Professor, Faculty of Engineering, University of Tokyo.
(III) Graduate Students, Faculty of Engineering, University of Tokyo.

superposition method, in which the outer boundary conditions are satisfied by distributing dislocations. This method is applied to elastic analysis of a cracked plate by Ogura[6]. Nishitani proposed the body force method[7]. Because of the close analogy between a dislocation and a point force in the form of stress functions, the present method has some similar points to the body force method. Especially, we emphasize the close relation between the present method and the method proposed by Murakami[8] for calculating the stress intensity factors of an arbitrarily shaped finite plate with a crack.

In order to examine the accuracy of the present method, some basic problems of Dugdale-BCS model and the elastic problems are analysed numerically by the method. Next, the inclined slip line model in the small scale yielding condition and in the rectangular cracked plate is analysed. Thirdly, the J-integral and CTOD of the rectangular center-cracked plate are calculated, and the relation between J and CTOD is examined. The method is applied to the more complicated problems; analysis of alternative shear rupture mechanism at a crack tip and computer simulation of fatigue crack propagation.

FUNDAMENTAL SOLUTIONS

1. Finite crack

Complex potential of an edge dislocation in an infinite plate with an elliptical hole can be obtained, according to the mirror image principle of Moriguchi[9] and to the similar procedure of Murakami's analysis[8]. The ellipse on the z-plane expressed by eq.(1) is mapped onto the circle ($|\zeta| = R$) on the ζ-plane by eq.(2).

$$x^2/a^2 + y^2/b^2 = 1 \quad , \quad z = x + iy \tag{1}$$

$$z = Z(\zeta) = \frac{C}{2}(\zeta + \frac{1}{\zeta}) \tag{2}$$

$$C = \sqrt{a^2-b^2} \qquad R = \sqrt{\frac{a+b}{a-b}} \tag{3}$$

The position of a dislocation is $z_0 = x_0 + iy_0 = Z(\zeta_0)$. The x and y components of the Burgers vector are b_x and b_y respectively. The complex potentials of an edge dislocation and an elliptical hole $\Phi(\zeta)$ and $\Psi(\zeta)$ are given by eqs.(4) - (7).

$$\Phi(\zeta)=B\log(\zeta-\zeta_0)-B\log(\zeta-R^2/\bar{\zeta}_0)+(B+D)\log\zeta-\bar{B}\frac{R^2[z_0-Z(R^2/\bar{\zeta}_0)]}{\frac{C}{2}(\bar{\zeta}_0^2-1)(\zeta-R^2/\bar{\zeta}_0)} \tag{4}$$

$$\Psi^I(\zeta) = \chi(\zeta) - \bar{Z}(R^2/\zeta)\Phi^I(\zeta) \tag{5}$$

$$\chi(\zeta)=\bar{B}\log(\zeta-\zeta_0)-\bar{B}\log(\zeta-R^2/\bar{\zeta}_0)+(\bar{B}+\bar{D})\log\zeta-B\frac{\zeta_0^2[\bar{z}_0-Z(R^2/\zeta_0)]}{\frac{C}{2}(\zeta_0^2-1)(\zeta-\zeta_0)} \tag{6}$$

$$B = \frac{G}{\pi(\kappa+1)}(b_y-ib_x) \qquad \kappa = \begin{cases} (3-\nu)/(1+\nu) & \text{plane stress} \\ 3-4\nu & \text{plane strain} \end{cases} \tag{7}$$

Suffix I in eq.(5) indicates the differential operation with respect to z, that is, $\Phi^{I}(\zeta) = (\frac{d\Phi}{d\zeta})/(\frac{dz}{dz})$. G and ν are the elastic shear modulus and Poisson's ratio respectively. As we deal with an ideal crack ($b \to 0$) in the present paper, $C = a$ and $R = 1$. D is related to the many-valuedness of displacement around a crack or an elliptical hole, u_c and v_c [10].

$$D = \frac{G}{\pi(\kappa+1)}(v_c - iu_c) \tag{8}$$

B+D is equal to zero in the case that the single-valuedness of displacement is satisfied at infinity. Vitek[11] gave a similar solution, but his solution is confined to the case that D equals zero.

The components of total force per unit thickness P_x and P_y are obtained by eq.(9).

$$-P_y - iP_x = \overline{\Phi}(\overline{\zeta}) + \overline{Z}(\overline{\zeta})\Phi^{I}(\zeta) + \Psi^{I}(\zeta) \tag{9}$$

Stress singularity at a crack tip results from the interaction between a dislocation and a crack. The stress intensity factor at a crack tip on the right side is given by eq.(10).

$$K_I - iK_{II} = \frac{G}{(\kappa+1)\sqrt{\pi a}}\Big[(b_y - ib_x)\Big(2 - \sqrt{\frac{z_0+a}{z_0-a}} - \sqrt{\frac{\overline{z}_0+a}{\overline{z}_0-a}}\Big)$$
$$-(b_y+ib_x)(z_0-\overline{z}_0)\Big\{\frac{1}{\sqrt{\overline{z}_0^2-a^2}} + \frac{1}{\overline{z}_0-a}\Big(\frac{a}{\sqrt{\overline{z}_0^2-a^2}} - \sqrt{\frac{\overline{z}_0+a}{\overline{z}_0-a}}\Big)\Big\}$$
$$+2(v_c - iu_c)\Big] \tag{10}$$

2. Semi-infinite crack

For the analysis in small scale yielding, the fundamental solution is extended to the case of a semi-infinite crack. The semi-infinite crack lies on the negative part of x-coordinate axis ($y=0$ and $x<0$), and the crack tip on the right side is situated at the origin. The total forces P_x and P_y at a point $z = x+iy$ are given by eq.(11).

$$-P_y - iP_x = 2\overline{B}\log\left|\frac{\sqrt{z}-\sqrt{z_0}}{\sqrt{z}+\sqrt{z_0}}\right| + B\,\frac{z_0-\overline{z}_0}{2\sqrt{z_0}}\Big(\frac{1}{\sqrt{\overline{z}}-\sqrt{\overline{z}_0}} + \frac{1}{\sqrt{\overline{z}}+\sqrt{\overline{z}_0}}\Big)$$
$$-\frac{z-\overline{z}}{2\sqrt{z}}\Big[B\Big\{\frac{1}{\sqrt{z}-\sqrt{z_0}} - \frac{1}{\sqrt{z}+\sqrt{\overline{z}_0}}\Big\} + \overline{B}\,\frac{z_0-\overline{z}_0}{2\sqrt{\overline{z}_0}(\sqrt{z}+\sqrt{\overline{z}_0})^2}\Big] \tag{11}$$

PROCEDURE OF ANALYSIS

In the present method , the outer boundaries and the plastic zones are divided into the segments and approximated by the groups of linear segments. The boundary conditions are satisfied by distributing dislocations on the segments. The dislocation density is assumed to be constant and continuous within each segment, and the total force conditions are satisfied on the boundaries. The total force method is more preferable than the stress method , because the average quantities of boundary conditions along the segments are always satisfied. In order to obtain the solution of continuously distributed dislocations, integration of eq.(9) or (11) needs to be carried out on each segment. We use 3 Gauss points and obtain a summation expression.

1. Elastic analysis

The present method is applied to the numerical analysis of the stress intensity factor K_I of a center-cracked plate with width 2W and length 2L, subjected to a uniform stress σ. The configuration of the specimen is shown in Fig.1. The outer boundary of specimen is divided into linear segments as shown in Fig.3. Taking advantage of symmetry with respect to the x and y axes, we may analyse quarter part of a plate. The stress intensity factor is calculated by integrating eq.(10) numerically along the boundary.

The effect of the number of segments M is examined. Fig.2 shows the results for L/W of 2 and a/W of 0.3. A linear relation is obtained between the numerical results and 1/M . We can calculate K-values more accuratelly by using extrapolation with respect to two values of 1/M. The practically exact solutions are obtained with small number of segments. This property on numerical error is an advantage of the present method, as in the case of the body force method. Table 1 shows the numerical results obtained by the present method, compared with Isida's exact solutions[12]. The results of extrapolation have accuracy better than 0.5% for L/W of 1 and 0,1% for L/W of 2. Without extrapolation, accuracy is obtained better than 1% for M of 20 in the case of L/W=2.

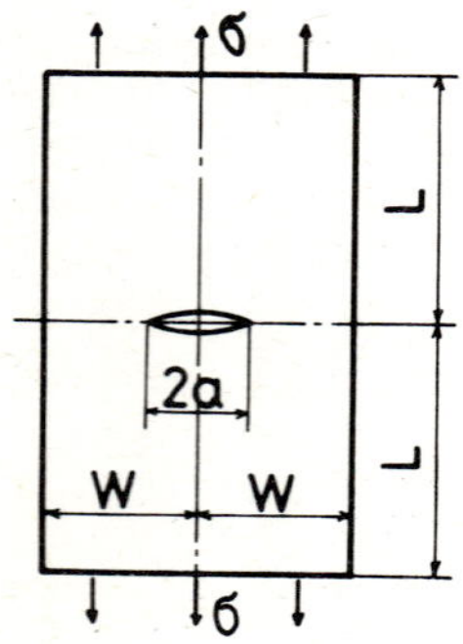

Fig.1 Rectangular center-cracked plate

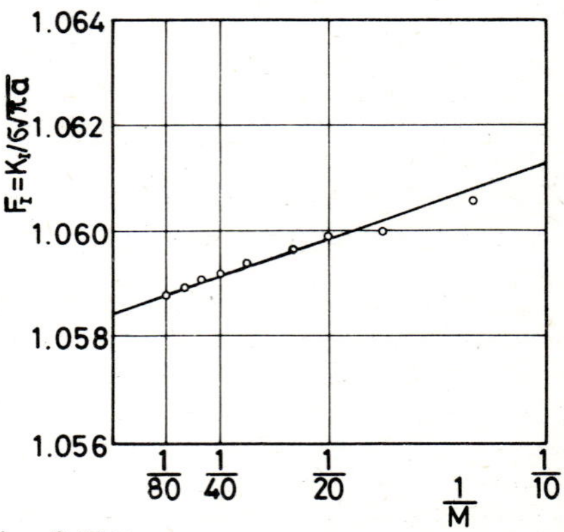

Fig.2 Effect of number of segment on results of normalised stress intensity factor

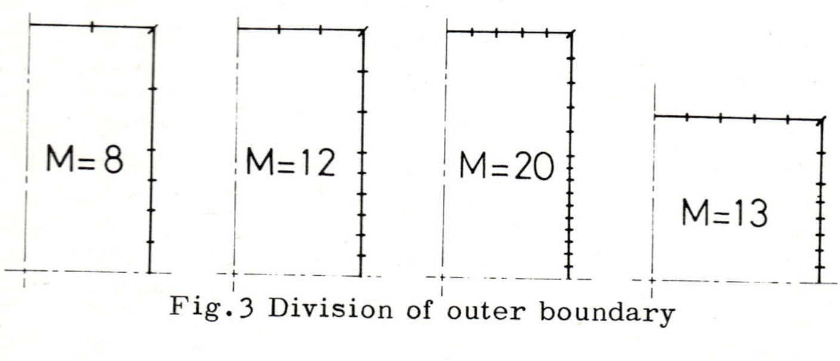

Fig.3 Division of outer boundary

Table 1 Stress intensity factors of rectangular center-cracked plate

L/W	M1,M2	a/W	$F_I = K_I/\sigma\sqrt{\pi a}$			
			The Present Method			Isida
			M1	M2	Extrapolation	
1.0	13,26	0.2	1.0762	1.0631	1.050	1.055
		0.4	1.2382	1.2248	1.211	1.216
		0.6	1.5027	1.4900	1.477	1.481
2.0	20,40	0.1	1.0073	1.0070	1.007	1.006
		0.3	1.0599	1.0592	1.058	1.058
		0.5	1.1903	1.1886	1.187	1.187
		0.7	1.4942	1.4911	1.488	1.488
		0.8	1.8233	1.8206	1.818	1.816

2. Dugdale-BCS model[13,14] in a finite plate

Fig.4 shows schematically the configuration of specimen. The outer boundary and the plastic zones are divided into M and N pieces of segment respectively. Since the stress and the dislocation density are to be finite at the tip of plastic zone, the applied stress σ and the size of plastic zone r_Y cannot be chosen independently. Assuming proportional loading, the intensity of applied stress can be determined for a given r_Y. It is required on the segment at the tip of plastic zone that the dislocation density is zero and that the yield condition is satisfied.

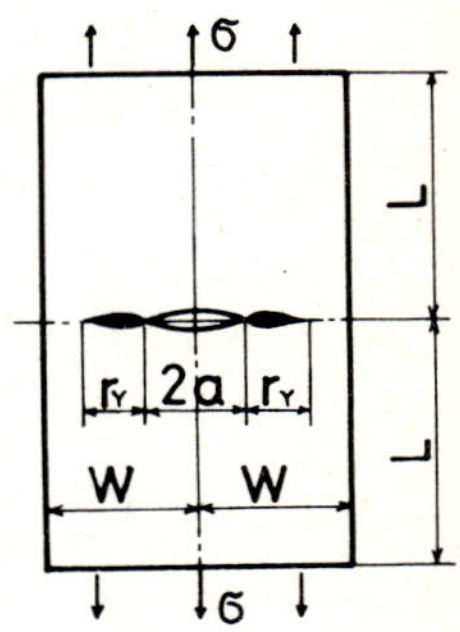

Fig.4 Dugdale-BCS model in a rectangular plate

Table 2 shows the results of the normalized applied stress σ/σ_Y for r_Y/W of 0.1, L/W of 2 and a/W of 0.3. Difference in the results for M of 20 and 40 is less than 0.1%. Therefore, extrapolation is carried out with respect to 1/N only, assuming M is 20 in the problem. The practically exact results are obtained in the case of $(a+r_Y) \leqq 0.8W$. The relation between r_Y/W and σ/σ_Y are shown in Fig.5. The results of the present method agree quite well with Isida's results[15]. Fig.6 shows the results of normalized CTOD,

Table 2 Effect of number of segments M and N on results of applied stress (L/W=2, a/W=0.3 and r_Y/W=0.1)

N / M	10	20	40	Extrapolation	
				10-20	20-40
20	0.4263	0.4273	0.4278	0.4283	0.4282
40	0.4267	0.4277	0.4281	0.4287	0.4286
Extrapolation	0.4271	0.4281	0.4286	0.4291	0.4290

compared with a curve for an infinite plate. As Isida has already suggested[15], the effect of finite boundary can be neglected in the relation between CTOD and r_Y/a.

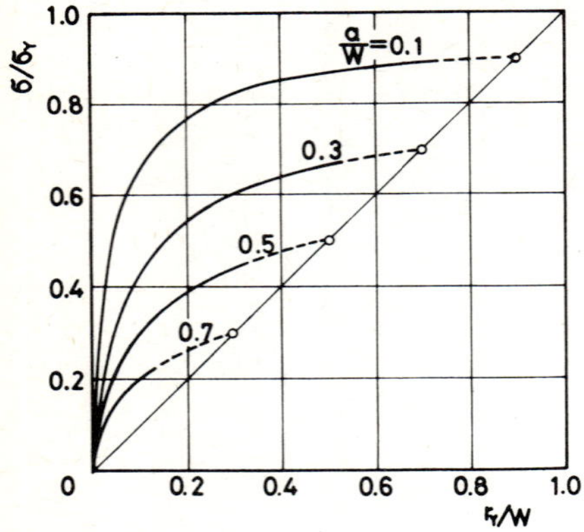

Fig.5 Relation between σ/σ_Y and r_Y/W for the Dugdale-BCS model in a rectangular plate

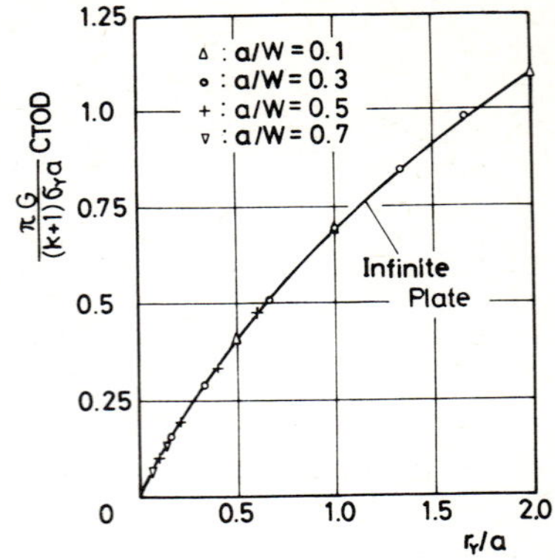

Fig.6 Relation between CTOD and r_Y/a for the Dugdale-BCS model in a rectangular plate

INCLINED SLIP LINE MODEL

1. Inclined slip line in small scale yielding

The length of slip line is small enouth, compared with the crack length and other dimensions. Assuming small scale yielding, the shear deformation on the inclined slip line at crack tip is analysed numerically for two types of model of (a) single slip and (b) symmetrical double slips, as schematically shown in Fig.7. The fundamental solution for small scale yielding is given by eq.(11).

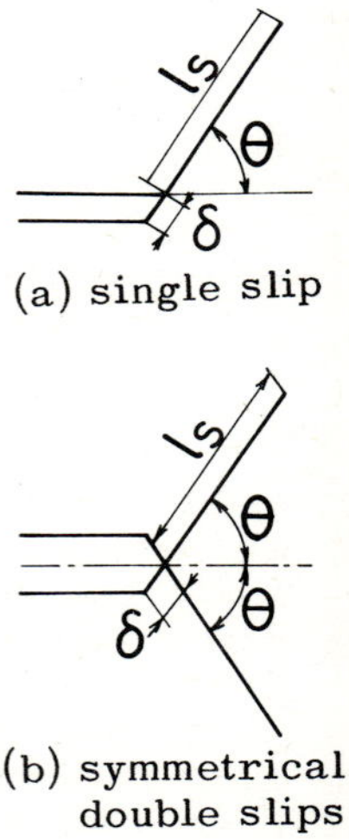

Fig.7

The inclined slip line with length l_s emanates from the tip of semi-infinite crack, subjected to the applied stress intensity factor K_I. The amount of shear slip at a crack tip δ, as defined in Fig.7, and the stress singularity in mode II at the tip of slip line $\overline{K}_{II}$ is calculated numerically, assuming that the critical shear stress τ_c is constant on the slip line. This solution can be obtained by superposing the solution of the equilibrium slip line with the shear stress τ_c on the solution of the blocked slip line with no frictional stress (see. Fig.8). For the both models of single slip and symmetrical double slips, the results are expressed as eqs. (12) and (13).

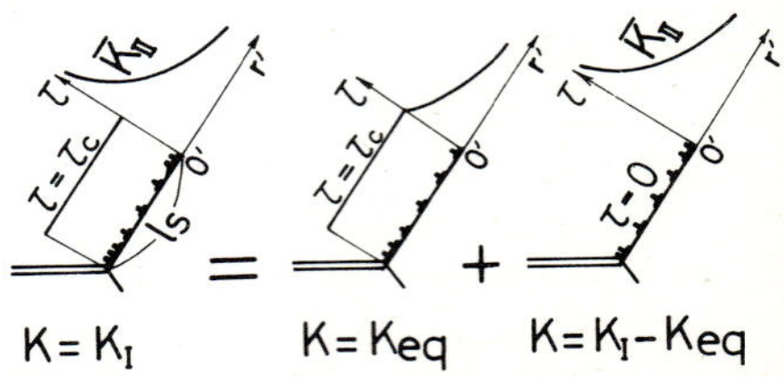

Fig.8 Superposition of solutions

$$\delta/l_s = 4f_1(\theta)\left[\frac{K_I}{E'\sqrt{2\pi l_s}}\cos\frac{\theta}{2}\sin\theta - \frac{4\tau_c}{\pi E'}f_2(\theta)\right] + \frac{8\tau_c}{\pi E'}f_3(\theta) \qquad (12)$$

$$\overline{K}_{II} = \frac{1}{2}g(\theta)\left[K_I\cos\frac{\theta}{2}\sin\theta - \frac{4}{\pi}\tau_c\sqrt{2\pi l_s}\,f_2(\theta)\right] \qquad (13)$$

where, $E'=E$ for plane stress or $E'=E/(1-\nu^2)$ for plane strain. $f_1(\theta)$–$f_3(\theta)$ and $g(\theta)$ are the functions of the slip line angle θ and determined numerically. Results of $f_1(\theta)$–$f_3(\theta)$ and $g(\theta)$ are shown in Fig.9 (a) and (b).

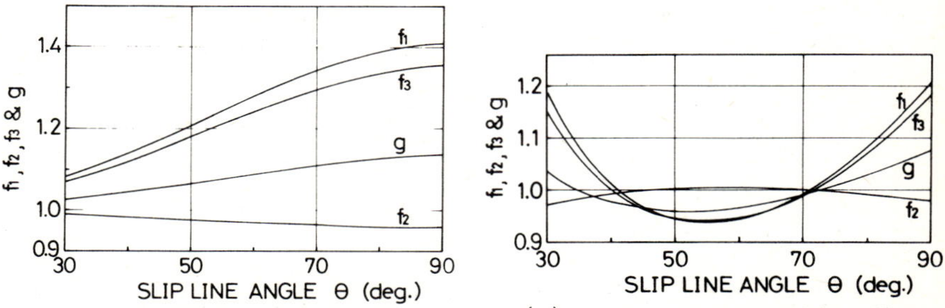

(a) single slip (b) symmetrical double slips

Fig.9 $f_1(\theta) - f_3(\theta)$ and $g(\theta)$ in eqs.(12) and (13)

2. Inclined slip line in a rectangular center-cracked plate

The configuration of the model is schematically shown in Fig.10. The number of outer boundary segments M is chosen to be 20. The aspect ratio L/W is restricted to 2 in the present analysis. The slip line is considered to be in the equilibrium state. The stress and the dislocation density are to be finite at the tip of slip line.

A suitable direction of slip line is discussed. We assume that slip occurs to the direction which makes r_Y maximum under a given σ. In the present method, σ is determined for a given r_Y, and the assumption can be rewritten that slip occures to the direction which makes σ minimum for the constant r_Y. Fig.11 shows the relation between the slip line angle θ and the normalized applied stress σ/τ_c for the several ratios of r_Y/a. σ/τ_c has a minimum value for a certain θ. The minimum points are jointed by a dotted

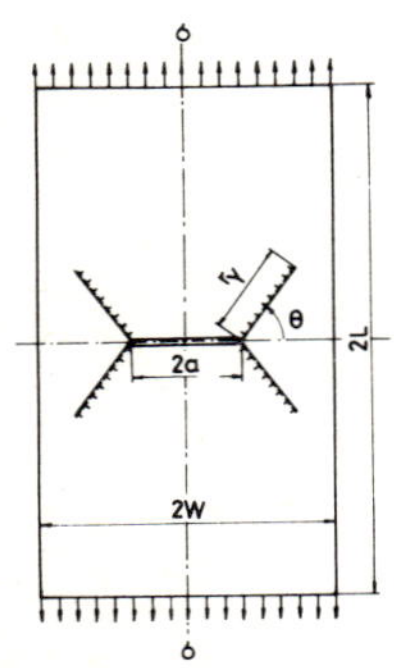

Fig.10 Inclined slip line model in a center-cracked plate

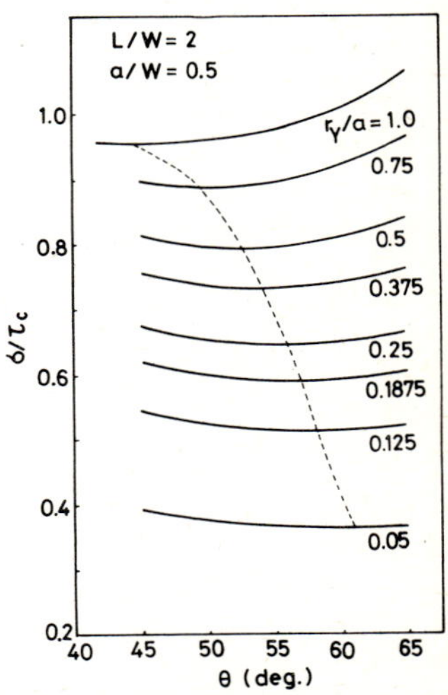

Fig.11 Relation between applied stress and slip angle for a given r_Y

curve by using Lagrange's interpolation formula. The dotted curve represents the relation between the applied stress and a suitable slip angle. The suitable slip angle is about 60° for the small value of σ/τ_c, and decreases with increasing the applied stress. Near general yielding, the suitable slip angle reaches 45°.

Fig.12 shows the results of the normalized CTOD and stretched zone width (SZW) for a/W of 0.2 and 0.5, in which θ is a parameter.

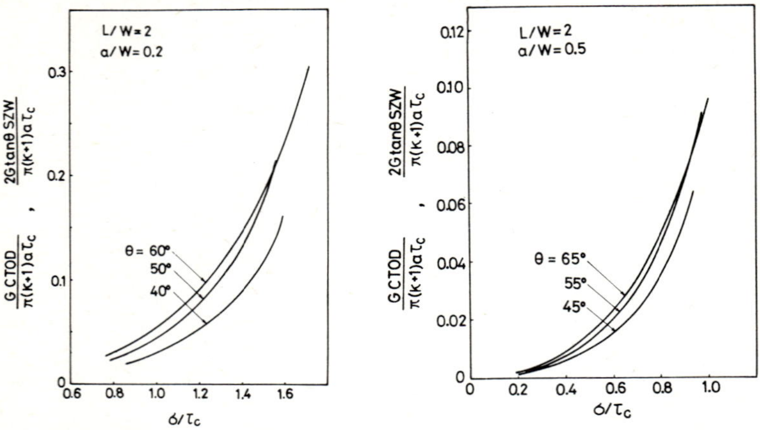

Fig.12 Results of CTOD and SZW for the inclined slip line model in a rectangular center-cracked plate

ESTIMATION OF THE J-INTEGRAL

The J-integral proposed by Rice[16] is defined as

$$J = \int_{\Gamma}(W^{tot}dx_2 - T_i\frac{\partial u_i}{\partial x_1}ds) \tag{14}$$

where, W^{tot} is the total strain energy density, T_i and u_i are the x_i components of traction vector and displacement respectively. Γ is the contour chosen to encircle the crack tip. The crack is in the x_1 direction and the summation convention is used.

Within the framework of the theory of continuously distributed dislocations, the J-integral can be interpreted as the total sum of the generalized forces acting on both crack tip and continuously distributed dislocations in the plastic zone[17,18]. When the contour is chosen to encircle the plastic zone, J can be written as

$$J = J(\text{crack}) + J(\text{dis.}) \tag{15}$$

J(crack) is the J-value due to a crack tip. J(dis.) is equal to the Peach-Koehler force acting on the continuously distributed dislocations in the plastic zone Ω_p.

$$J(\text{crack}) = \frac{(\kappa+1)}{8G}(K_I^{ap} + K_I^{in})^2 \tag{16}$$

$$J(\text{dis.}) = \int_{\Omega_p}(\sigma_{i2}\alpha_{3i} - \sigma_{i3}\alpha_{2i})dx_1dx_2 \tag{17}$$

K_I^{ap} and K_I^{in} are the stress intensity factors due to the applied stress and the internal stress respectively. α_{ij} is the dislocation density tensor defined by Kröner [19]. J(crack) and J(dis.) can be easily estimated by the present method.

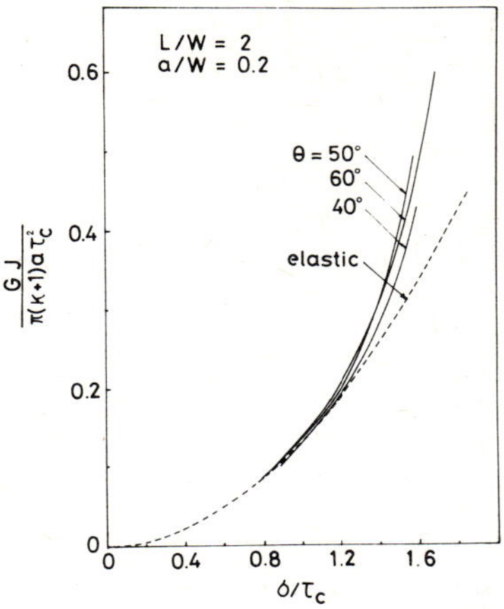

(a) a/W = 0.2

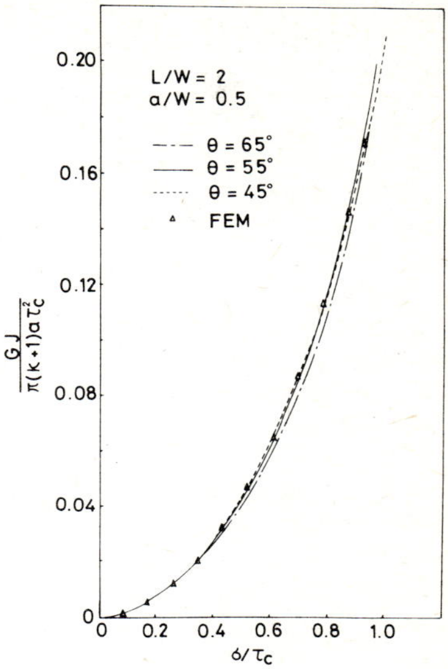

(b) a/W = 0.5

Fig.13 J-value for a rectangular center-cracked plate

This method is applied to estimation of J for the rectangular center-cracked plate with the inclined slip lines, as shown in Fig. 10. Fig.13(a) and (b) show the relation between the applied stress and the J-value for a/W of 0.2 and 0.5. Normalizing J and σ by use of τ_c, we can make J-σ curve nearly independent of θ. The results of the finite element method are also shown in Fig.13 (b). The close agreement is obtained between the two methods.

The relation between J and CTOD is examined. A parameter λ is defined as

$$\lambda = J/(\sigma_Y \mathrm{CTOD}) \qquad (18)$$

$\lambda = 1$ (constant) for the Dugdale-BCS model. In the inclined slip line model, λ varies with plastic deformation. Fig.14 shows the relation between λ and the normalized mean stress of ligament. λ decreases as plastic zone size increases. λ depends largely on the slip angle. The dash-dot curve corresponds to the suitable slip angle mentioned in the previous chapter. This curve shows that λ is about 1.8 for small scale yielding and that λ decreases to 1.2 near general yielding.

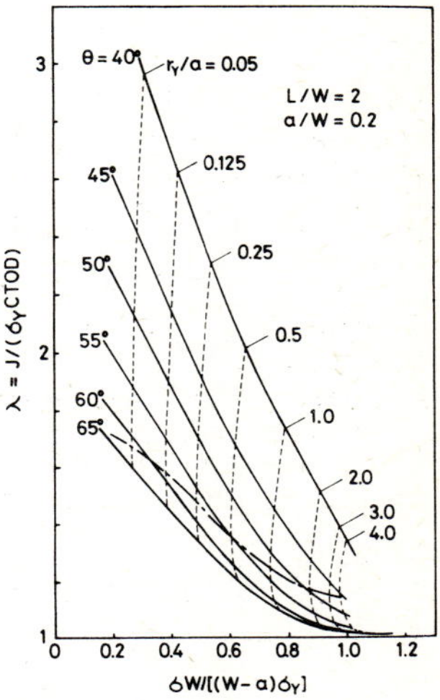

Fig.14 Change of λ for the inclined slip line model

The slip band is composed of the group of parallel slip lines and the alternative shear rupture mechanism at a crack tip, as shown in Fig.15, is analysed numerically by the present method. This mechanism of crack tip blunting is experimentally observed in the single crystal[20]. Kuo and Liu[21] called the model as the unzipping model and analysed it by using the finite element method. But in their analysis, the effect of formation of slip band by many slip lines seems to be neglected. The present method can be applied to the problem. The crack tip bluntind model by the alternative shear slip is analysed numerically. The slip angle is chosen to be 60° and two types of slip band model, that is, the blocked slip band model and the equilibrium slip band model are analysed.

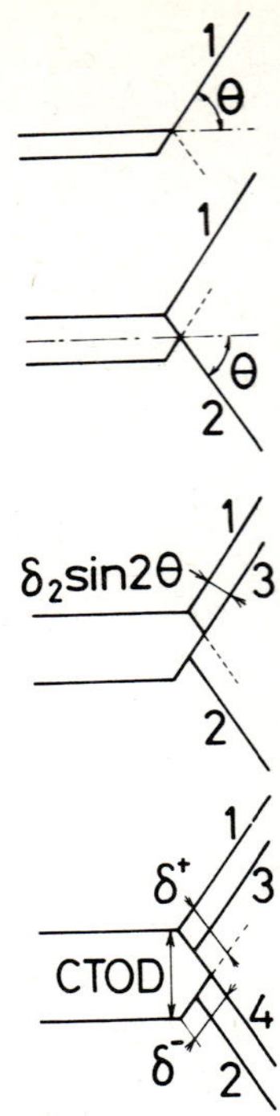

Fig.15 Alternative shear rupture model

1. Blocked slip band model

The slip band is blocked by the grain boundary or other barriers. The slip band length l_s is given and the stress has a singularity of $\overline{K}_{II}$ at the tip of slip line. The frictional shear stress on the slip lines is assumed to be zero. Increment of slip $\Delta\delta/l_s$ is chosen to be 0.001, 0.01 and 0.1. The total amount of slip during k steps is $0.5k\Delta\delta$ (see Fig.15). Fig.16 shows the relation between the step number and the applied stress intensity factor normalized by using $\Delta\delta$. A dash-dot line represents the results of the symmetrical double slip lines. The effect of slip band width is negligible if the width is less than one-tenth of the slip band length.

2. Equilibrium slip band model

The length of slip band is larger than the grain size, and the stress and the dislocation density are to be finite at the tip of slip lines. The slip band length increases as the applied stress intensity factor increases. In the present analysis, we assumes that the increment of slip band length by a step is constant. The shear yield stress τ_c/E' is chosen to be 2.5×10^{-4}, 10^{-3} and 4×10^{-3}. The range of the tensile yield stress corresponds to 11 - 170 kg/mm^2 for steel.

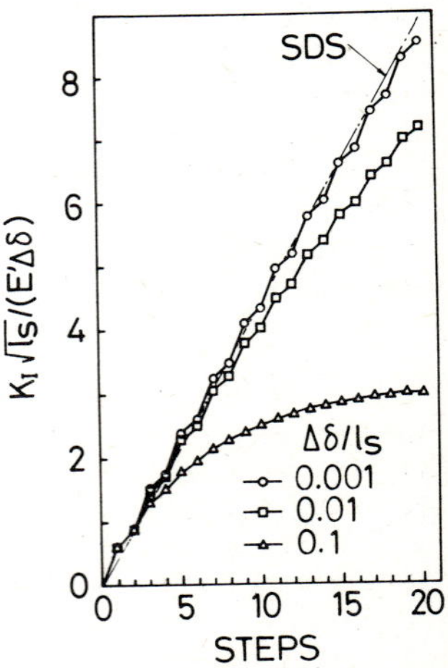

Fig.16 Change of K_I with crack tip blunting for blocked slip band model

Fig.17 shows the relation between the normalized total amount of slip and the number of step. For the small value of τ_c/E', δ/l_s approaches asymptotically to the value of symmetrical double slip line model. Fig.18 shows the change of $(K_I^2/E')/(2\tau_c CTOD)$. K_I^2/E' is the J-value for small scale yielding. The ratios become nearly constant after 10 steps. The stationary values are in the range between 1.9 - 2.2.

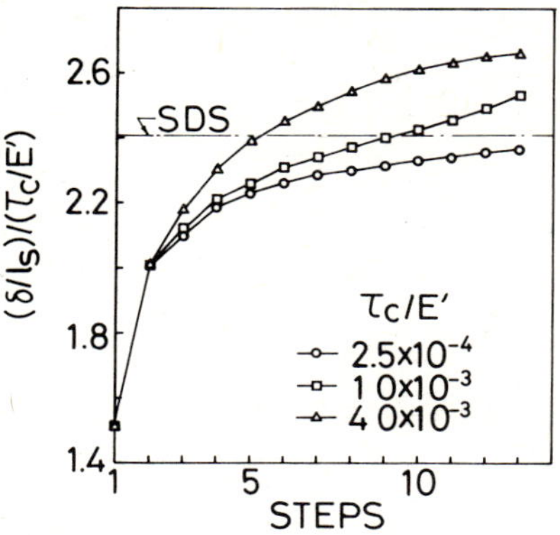

Fig.17 Change of δ/l_s for equilibrium slip band model

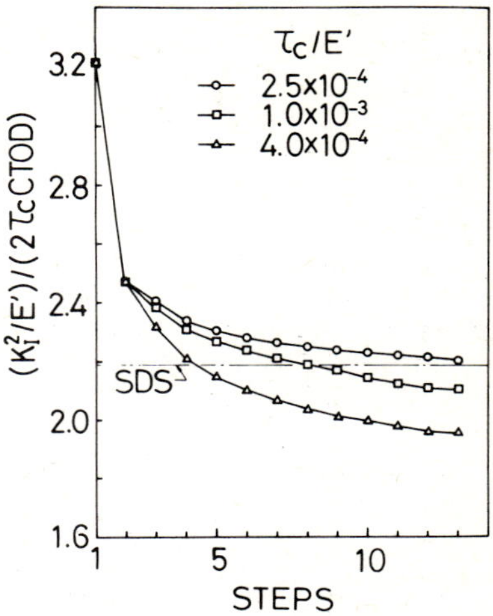

Fig.18 Chande of $(K_I^2/E')/(2\tau_c CTOD)$ for equilibrium slip band model

SIMULATION OF FATIGUE CRACK PROPAGATION

The shear rupture deformation at a crack tip is approximated with the inclined slip lines. The fatigue crack propagation, shown schematically in Fig.19, is simulated by using the present method. Striation spacing S and CTOD are calculated. The crack growth increment during a cycle is chosen to be equal to S and the crack propagates under cyclic loading. The relation between S and ΔK is examined for an infinite plate with a crack of length 2a under R=0 loading. The effect of work hardening is also considered, but the effect of crack closure is neglected in the present paper.

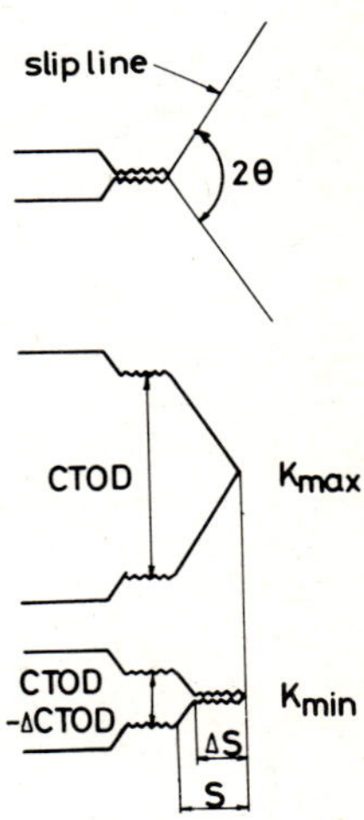

Fig.19 Model of fatigue crack propagation

1. Results for elastic perfectly-plastic body

First, the frictional shear stress τ_c is assumed to be constant on the slip lines. The mechanical properties is chosen to be those of aluminum alloy with E=7500 kg/mm^2 , ν = 0.3 and τ_c = 14kg/mm^2 . The length of an initial crack is 10 mm.

Change of S and ΔS during crack propagation is shown in Fig.20. The striation spacing and the plastic zone size become nearly constant after 5 cycles, so the value of 5-th cycle is considered

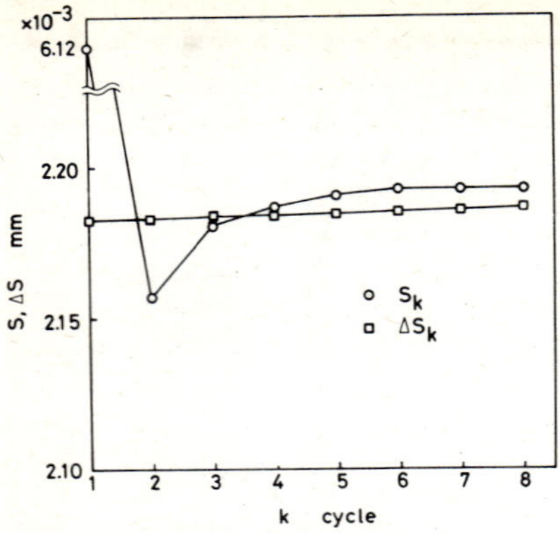

Fig.20 Change of S and ΔS during fatigue crack growth (θ = 60° and Δσ= 14kg/mm^2)

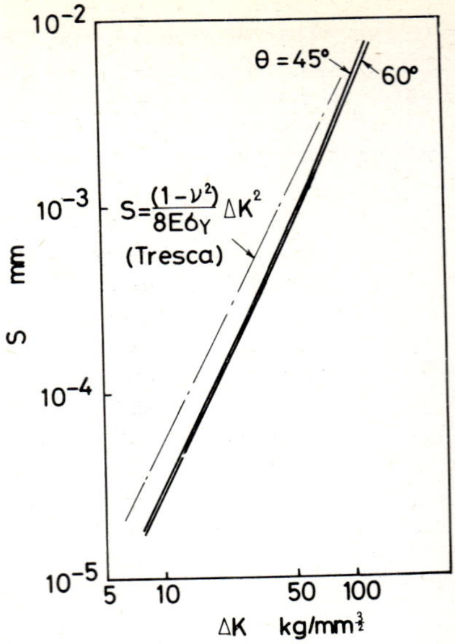

Fig.21 Relation between S and ΔK

as the stationary value. Fig.21 shows the relation between the stationary value of S and ΔK with respect to θ of 45° and 60°. If we represent the relation as $S = C(\Delta K)^m$, m is equal to 2 for small scale yielding, and m increases to 2.5 as the plastic zone size increases.

2. Effect of work hardening

The kinematic hardening law , as shown in Fig.22, is introduced to the inclined slip line model. The plastic shear strain γ_p and $\Delta\gamma_p$ are defined as the average value with respect to the width of crack tip slip band t_Y, as shown in Fig.23. Fig.24 shows the

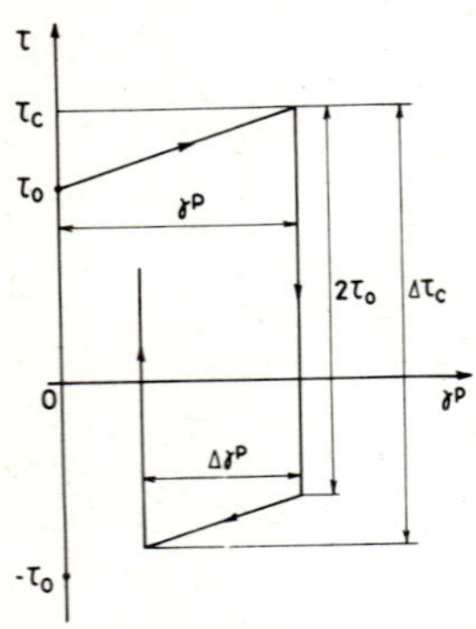

Fig.22 Kinematic hardening law

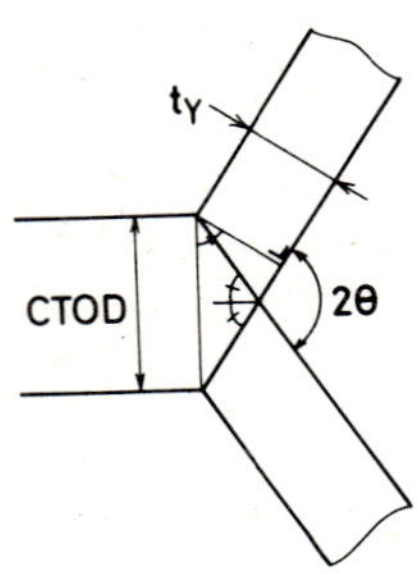

Fig.23 Width of crack-tip slip band t_Y

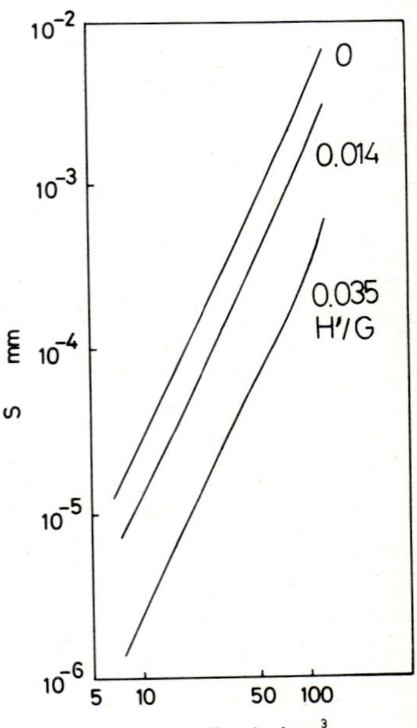

Fig.24 Effect of H' on the relation between S and ΔK

effect of strain hardening rate H' on the relation between S and ΔK. $H' = d\tau/d\gamma_p$ and H'/G is chosen to be 0, 0.014 and 0.035. The results shows that the strain hardening rate has significant influence upon the relation between S and ΔK for the present model.

CONCLUSIONS

A numerical method of elastic-plastic analysis of a cracked plate is proposed, based on the theory of continuously distributed dislocations. The present method is applied to some basic problems. The numerical results are obtained with very high accuracy. The method is applied to analysis of plastic deformation on the inclined slip line for both case of an infinite plate and a finite plate.
A procedure of estimation of the J-integral is proposed, based on the theory of continuously distributed dislocations. The method is applied to estimation of J for a rectangular center-cracked plate, and the relation between J and CTOD is examined. The present method is also applied to analysis of alternative shear rupture deformation at a crack tip under monotonic loading and to simulation of fatigue crack propagation.

ACKNOWLEDGMENTS

The developement of the present method was initiated by the achievement of the extended body force method proposed by Murakami [8]. The fundamental consideration in his paper is also helpful for the present method. The authors wish to thank Dr. Y. Murakami, Associate Professor of Kyushu Institute of Technology. The authors are also grateful to Dr. H. Nishitani, Professor of Kyushu University, for his comments on accuracy of the present method.

REFERENCES

1. YOKOBORI, T. and KAMEI, A.- A Criterion for Low-stress Brittle Fracture of Materials with Crack or Sharp Notch Based on Combined Micro-and-macro Fracture Mechanics. Eng. Fract. Mech.,vol.5, pp.819-836, 1973.

2. YOKOBORI, T.- Criteria for Nearly Brittle Fracture. Int. J. Fract. Mech., vol.4 no.2, pp.179-208, 1968.

3. KAMEI, A. and YOKOBORI, T. - Some Results on Stress Intensity Factors of the Cracks and/or Slip Bands System. Rep. Res. Inst. Strength and Fract. Mat. Tohoku Univ., vol.10 no.2, pp.29-93, 1974.

4. VITEK, V.- Yielding on Inclined Planes at the Tip of a Crack Loaded in Uniform Tension. J. Mech. Phys. Solids, vol.24, pp.263-275, 1976.

5. RIEDEL, H.- Plastic Yielding on Inclined Slip-planes at a Crack Tip. J. Mech. Phys. Solids, vol.24, pp.277-289, 1976.

6. OGURA, K.- Analysis of Stress Intensity Factors by Dislocation Superposition Method. Preprint of JSME, no.710-10, pp.33-36, 1971 (in Japanese).

7. NISHITANI, H.- Solutions of Notch Problems by Body Force Method, Mechanics of Fracture, vol.5, Ed. Sih, G. C., Noodhoff International Publishing, 1978.

8. MURAKAMI, Y. - Application of the Body Force Method to the Calculation of Stress Intensity Factors for a Crack in the Arbitrarily Shaped Plate. Eng. Fract. Mech., vol.10, pp.497-513, 1978.

9. MORIGUCHI, S.- Theory of Two-dimensional Elasticity, Iwanami, 1957 (in Japanese).

10. MIYAMOTO, H. and KAGEYAMA, K.- Numerical Analysis of the Interaction between a Crack and the Arbitrarily Distributed Dislocations. Proc. Fract. Mech. Tech., vol.2, Ed. Sih, G C., Sijthoff and Noordhoff International Publisher, pp.957-964, 1977.

11. VITEK, V.- Yielding from a Crack with Finite Root-radius Loaded in Uniform Tension. J. Mech. Phys. Solids, vol.24, pp.67-76, 1976.

12. ISIDA, M.- Effect of Width and Length on Stress Intensity Factors of Internally Cracked Plate under Various Boundary Conditions. Int. J. Fract. Mech., vol.7 no.3, pp.301-316, 1971.

13. DUGDALE, D. S.- Yielding of Steel Sheets Containing Slits. J. Mech. Phys. Solids, vol.8 no.2, pp.100-104, 1960.

14. BILBY, B. A. COTTRELL, A. H. and SWINDEN, K. H.- The Spread of Plastic Yield from a Notch. Proc. Roy. Soc. London, A272, pp.304-314, 1963.

15. ISIDA, M.- Stress Intensity Factors and Yielding on Dugdale-type Plastic Zone in an Internally Cracked Finite Plate. Preprint of JSME, no.710-7, pp.119-126, 1971 (in Japanese).

16. RICE, J. R.- A Path Independent Integral and the Approximate Analysis of Strain Concentration by Notches and Cracks. Trans. ASME, J. Appl. Mech., vol.35 no.2, pp379-386,1968.

17. MIYAMOTO, H. and KAGEYAMA, K.- Extension of J-integral to the General Elasto-plastic Problem and Suggestion of a New Method for Its Evaluation. Numerical Methods in Fracture Mechanics, Ed. Luxmoore, A. R. and Owen, D. R. J., pp.479-486, 1978.

18. MURA, T. - On the Extended J-integral. Recent Research on Mechanical Behavior of Solids, Ed. Miyamoto, H., pp.237-243, University of Tokyo Press, 1979.

19. KRÖNER, E.- Kontinuums Theorie der Versetzungen und Eigenspannungen, Springer Verlag, 1958.

20. NEUMANN, P., VEHOFF, H. and FUHLROTT, H.- On the Mechanics of Fatigue Crack Growth. Fracture 1977, vol.2, Ed. Taplin, D. M. R., pp.1313-1324, University of Waterloo Press, 1977.

21. KUO, A. S. and LIU, H. W.- An Analysis of Unzipping Model for Fatigue Crack Growth. Scripta Metallurgica, vol.10, pp.723-728, 1976.

GEOMETRIC AND MATERIAL NONLINEARITY PROBLEMS "LAGRANGIAN AND UPDATED LAGRANGIAN FORMULATIONS"

M.S. Gadala*, M.A. Dokainish** and G. AE. Oravas**
*Atomic Energy of Canada Ltd., Mississauga, Ontario, Canada
**McMaster University, Hamilton, Ontario, Canada L8S 4L7

SUMMARY

A concise survey of formulation methods of geometric and material nonlinearity problems is given. Differences between existing formulations in the literature are discussed. A consistent Lagrangian and updated Lagrangian formulations are derived from the energy balance equation transformed to the proper reference configuration. Differences between the existing formulations and similar ones in the literature are found to be in specific geometric nonlinear terms in the final incremental equation as well as in the definition of the load increment vector.

PART 1

SURVEY OF FORMULATION METHODS

1.1 INTRODUCTION

In describing the motion of a body, there are four methods of formulations; Truesdell [1] called them:

1. Material description; independent variables are the particle or the body-point X, and the time t. The description is the most natural one in concept and is the only one used in analytical dynamics.

2. Referential description; independent variables are the position $\bar{X}$ of a particle X in an arbitrary chosen reference configuration, and the time t. It is important to notice that the choice of the reference configuration is arbitrary and essentially would not affect the results. A particular referential description was introduced by Euler where the position $\bar{X}$ of the body-point X at the particular time t = 0 is used to describe the motion. This particular description is often called Lagrangian formulation in the literature.

3. Spatial description; independent variables are the current position $\bar{x}$ of a particle X, and the time t. This description was introduced by Bernoulli and D'Alembert but it is usually called Eulerian description. It is important to emphasize the fact that the independent variable $\bar{x}$ is a function of the Lagrangian position $\bar{X}$ and the time t, i.e. $\bar{x} = \bar{x}(\bar{X},t)$. Hence all material derivatives will be more difficult to handle in spatial description.

4. Relative description; independent variables are the position $\bar{x}$ in a current configuration and a variable time τ. In relative description the reference configuration varies with time. The current configuration is taken as the reference configuration and the past and future configurations are described relative to the current one. The variable time τ is the time when the particle X occupied a position $\bar{\xi}$, where $\bar{\xi} = \bar{\xi}_t(\bar{x},\tau)$. It is important to realize that the relative description is referential or Lagrangian in nature, in the sense that the reference position is now denoted by $\bar{x}$ at time t rather than $\bar{X}$ at time t = 0. This will justify the classification of updated Lagrangian formulation as a special case of the relative description as will be discussed later.

In finite element application to continuum mechanics problems, only the last three descriptions are used. In what follows, we will try to assess different finite element formulations, existing in the literature, within the frame work of the definitions given above.

1.2 LAGRANGIAN (REFERENTIAL) FORMULATION

Lagrangian formulation is by far the most widely used formulation in finite element applications to nonlinear structural problems [2-11]. While most of the Lagrangian finite element formulations utilize incremental method with linear increments, the differences basically lie in the simplifying assumptions of the kinematic relations within the increment and in the interpretation of the energy balance equation in the undeformed Lagrangian configuration. Within the same simplifying assumptions, different forms of the final nonlinear equilibrium equations are owing to different choices of stress measures and their conjugate strain definitions.

Many of the early incremental finite element formulations are incomplete in that the variations in loading and rotations of the element are not properly represented [9]. The most general approach to incremental Lagrangian formulation is to consider a fully nonlinear kinematic relations within a linear increment [9-11]. When fully nonlinear kinematic relations are used within a linear increment, the number of stiffness contributions to the nonlinear element stiffness matrix are found to be three matrices [9-11]. These are: the usual small displacement or the incremental stiffness matrix, the initial stress or the geometric or the tangent stiffness matrix and, finally, the initial displacement or the initial rotation stiffness matrix. A fourth stiffness matrix: the initial load stiffness matrix or the load correction matrix will be discussed when considering different implementations of the energy balance equation.

Returning back to the original point of the simplifying kinematic assumptions, Wunderlich [5] has introduced a systematically consistent development of an incremental Lagrangian formulation on the basis of the equations of continuum mechanics. However, he omits specific nonlinear terms for the intended linearization process within the increment, when they appear with other incremental quantities. In references [2,3] a similar simplified approach is applied to plate and shell problems. Fully nonlinear kinematic relations are used by Oden [9], Hibbit et al. [10] and followed by many other authors. Needleman [4] introduces a Lagrangian scheme which is derived from a variational principle due to Hill [13,14]. Bathe et al. [7-8] also have followed a general approach but they present an implicit final form of the nonlinear equilibrium equations where all the nonlinear stiffness contributions are combined in one stiffness term. We may conclude this particular discussion by saying that none of the stiffness matrices contributing to the total stiffness of the element should in general be omitted. In specific applications, however, certain simplifications of these stiffness expressions may be achieved.

Another point worthy of discussion is the energy balance equation used by different authors. If we consider three different configurations for the body; the original configuration C_o at time $t = 0$, a current configuration C_1 at time $t = \tau$ and a neighbourhood configuration C_2 at time $t = \tau + \Delta t$, then we may write the energy balance equation in configuration C_2 as

$$\frac{d}{dt}\,{}^2_2K = \int_{{}^2V} {}^2\rho\, {}^2\bar{b} \cdot {}^2\bar{v}\, d\,{}^2V + \int_{{}^2A} d\,{}^2\bar{A} \cdot {}^2\bar{\bar{\sigma}} \cdot {}^2\bar{v}$$

$$- \int_{{}^2V} {}^2\bar{\bar{\sigma}} : {}^2\bar{\bar{D}}\, d\,{}^2V \qquad (1)$$

where the left superscripts indicate the configuration at which the quantity is measured whereas the left subscripts indicate the configuration to which the quantity is referred, and ${}^{2}_{2}K$ is the kinetic energy, ${}^{2}\rho$ is the material density, ${}^{2}\bar{b}$ is the body force per unit of current volume, ${}^{2}V$ is the current volume, ${}^{2}\bar{v}$ is the velocity, ${}^{2}\bar{A}$ is the area of the current loading surface, ${}^{2}\bar{\bar{\sigma}}$ is the Cauchy stress tensor, ${}^{2}\bar{\bar{D}}$ is the rate of deformation tensor and, finally, ${}^{2}\bar{\bar{\sigma}} : {}^{2}\bar{\bar{D}}$ indicates the double dot product of the two second order tensors ${}^{2}\bar{\bar{\sigma}}$ and ${}^{2}\bar{\bar{D}}$, i.e. ${}^{2}\bar{\bar{\sigma}} : {}^{2}\bar{\bar{D}} = {}^{2}\sigma_{ij}\ {}^{2}D_{ij}$.

A consistent transformation of eqn. (1) from the current deformed configuration to the Lagrangian undeformed configuration may give [9,11]

$$\frac{d}{dt}\,{}^{2}_{o}K = \int_{{}^{o}V} {}^{o}\rho\ {}^{2}\bar{b} \cdot {}^{2}\bar{v}\ d\,{}^{o}V + \int_{{}^{o}A} d\,{}^{o}\bar{A} \cdot {}^{2}_{o}\bar{\bar{S}} \cdot {}^{2}_{o}\bar{\bar{F}}^{T} \cdot {}^{2}\bar{v} - \int_{{}^{o}V} {}^{2}_{o}\bar{\bar{S}} : \frac{d\,{}^{2}_{o}\bar{\bar{E}}}{dt}\, d\,{}^{o}V \qquad (2)$$

where ${}^{2}_{o}K$ is the kinetic energy in the undeformed configuration, ${}^{o}\rho$ is the material density in the undeformed configuration, ${}^{2}_{o}\bar{\bar{S}}$ is the second Piola-Kirchoff stress tensor, ${}^{2}_{o}\bar{\bar{F}}^{T}$ is the transpose of the deformation gradient tensor and ${}^{2}_{o}\bar{\bar{E}}$ is the Lagrangian strain tensor.

Many authors, however, differ in considering the surface traction term in eqn. (2) by dropping ${}^{2}_{o}\bar{\bar{F}}^{T}$ from its expression [7,8,10]. The expression of the surface traction term presented here leads to a consistent development of the load correction matrix or the initial load stiffness matrix [9, 11]. Hibbit et al. [10] introduce a rather artificial argument as $\overline{\Delta F} = \Delta_{load}\,\bar{F} + \Delta_{geom}\,\bar{F}$, where $\bar{F}$ is the surface traction vector or the body force vector, to obtain the form of the load correction vector. The term obtained by Hibbit et al., will have a zero value in some loading cases, while the term derived from the expression presented here will always possess some value in general. In [7,8], Bathe et al. consider a similar treatment to that given by Hibbit et al. for the case of loads that are independent of the configuration of the body, however, for deformation-dependent loading they consider a similar approach to the one given here. It is our opinion that this term is obtained by consistent transformation of the energy balance equation with no reference to the type of loading, hence it should be applied to all types of loadings. The importance of including the load correction term in incremental formulations was pointed out by Oden [15] and Oden and Key [16].

The final point of discussion under the scheme of Lagrangian formulation is the choice of compatible stress and strain measures. To provide a consistent discussion, we will first recall the following definitions

$$\bar{\bar{T}} = |\bar{\bar{F}}|\ \bar{\bar{F}}^{-1} \cdot \bar{\bar{\sigma}} \qquad (3a)$$

where $\bar{\bar{F}}^{-1}$ is the inverse of $\bar{\bar{F}}$, and $|\bar{\bar{F}}| = \bar{\bar{F}} \, {}^{x}_{x} \, \bar{\bar{F}} : \bar{\bar{F}}/3!$

: Unsymmetric first Piola-Kirchhoff stress tensor or the nominal stress tensor or Piola-Lagrange stress tensor.

$$\bar{\bar{S}} = |\bar{\bar{F}}|\ \bar{\bar{F}}^{-1} \cdot \bar{\bar{\sigma}} \cdot (\bar{\bar{F}}^{-1})^{T} = \bar{\bar{T}} \cdot (\bar{\bar{F}}^{-1})^{T} \qquad (3b)$$

: Symmetric second Piola-Kirchhoff stress tensor or Kirchhoff-Treftz stress tensor.

$$\bar{\bar{F}} = \bar{\bar{R}} \cdot \bar{\bar{U}} = \bar{\bar{V}} \cdot \bar{\bar{R}} \qquad (3c)$$

: The deformation gradient tensor, whereas $\bar{\bar{R}}$ is the rotation tensor, $\bar{\bar{U}}$ and $\bar{\bar{V}}$ are the right and left stretch tensors respectively.

$$\bar{\bar{E}} = \frac{1}{2}\,[\bar{\bar{F}}^{T} \cdot \bar{\bar{F}} - \bar{\bar{1}}] \qquad (3d)$$

: Euler-Lagrange or Lagrange or Green strain tensor.

$$\bar{\bar{\varepsilon}} = \frac{1}{2}\,[\bar{\bar{1}} - (\bar{\bar{F}}^{-1})^{T} \cdot \bar{\bar{F}}^{-1}] \qquad (3e)$$

: Eulerian or Almansi strain tensor.

$$\bar{\bar{L}} = \frac{\bar{v}\ \partial}{\partial \bar{x}} = \bar{\bar{D}} + \bar{\bar{W}} \qquad (3f)$$

: The velocity gradient tensor, whereas $\bar{\bar{D}}$ is the rate of deformation tensor and $\bar{\bar{W}}$ is the spin or vorticity tensor.

Based on the definition of the rate of work per unit final volume expressed as $\bar{\bar{\sigma}} : \bar{\bar{D}}$, Hill [17] defines particular conjugate pairs of stress and strain variables:

$$\{\bar{\bar{\sigma}}, \bar{\bar{L}}\},\ \{\bar{\bar{S}}, \dot{\bar{\bar{E}}}\},\ \{|\bar{\bar{F}}|\ \bar{\bar{F}}^{T} \cdot \bar{\bar{\sigma}} \cdot \bar{\bar{F}}, \dot{\bar{\bar{\varepsilon}}}\},\ \{\bar{\bar{T}} \cdot \bar{\bar{R}}, \dot{\bar{\bar{U}}}\} \text{ and } \{\bar{\bar{T}}, \dot{\bar{\bar{F}}}\},$$

where $\dot{(\ )} \equiv d(\ )/dt$. In finite element formulations, however, the most widely used stress measures are the second and the first Piola-Kirchhoff stress tensors $\bar{\bar{S}}$ and $\bar{\bar{T}}$.

The first conjugate pair $\{\bar{\bar{S}}, \bar{\bar{E}}\}$ is used by many analysts [7,8,10,11] especially in the case of both geometric and material nonlinear problems. The transformation of the energy balance equation from the current deformed configuration to the Lagrangian undeformed configuration lends itself to an expression in terms of $\bar{\bar{S}}$ and $\bar{\bar{E}}$. Based on the symmetry of the second Piola-Kirchhoff stress tensor $\bar{\bar{S}}$, the formulation identically satisfies the moment of momentum or the rotational equilibrium equations.

The use of the second conjugate pair $\{\bar{\bar{T}}, \bar{\bar{F}}\}$ usually arise from formulating the problem based on a complementary energy principle [6,18-22]. Professor G. AE. Oravas has given a comprehensive account of the history of the principle of minimum complementary energy in the linear theory of elasticity in his introduction and biographical portrait section to the English translation of Castigliano's pioneering book [23]. If the finite element formulation is based on a complementary energy principle, then unlike the linear theory of elasticity, it is not possible to derive a single field complementary energy principle for finite elasticity involving $\bar{\bar{S}}$ alone as a field variable [6,24]. Also, it is not a straightforward task to satisfy the linear momentum balance equation when it is expresssed in the undeformed coordinates. This, in fact, has motivated the study of constructing alternate complementary energy principles for finite elasticity.

1.3 UPDATED LAGRANGIAN OR RELATIVE FORMULATION

As discussed above, the referential description employs some assigned reference configuration. If the reference configuration is taken at a variable time t, then we have a relative description, which is referential in nature. In a general relative description we describe the past and future relative to the present. Hence, we may be marching backward or forward with respect to the current configuration. In updated Lagrangian formulation, we describe the future with respect to the present, i.e. we march backward, which is a special case of the relative description. In updated Lagrangian or relative formulation, the independent variables are $\bar{x}$ and τ, where $\bar{x}$ is the position occupied by the material point X at time t. This indicates that $\bar{x}$ is independent of the variable time τ. A feature which justifies the referential nature of the relative description and its distinction from spatial or Eulerian description where the two variables $\bar{x}$ and t used to describe the motion are dependent as $\bar{x} = \bar{x}\ (\bar{X},t)$. For the same reason, material derivatives and other relations which are obvious and easy to derive in material or referential description seem to be more difficult to handle in spatial description.

It is unfortunate that facts in the above discussion are ignored in finite element literature and many analysts describe relative or updated Lagrangian formulation under the name of Eulerian formulation. Moreover, little effort has been made in the development of a consistent and detailed spatial or Eulerian description.

PART 2
METHODS OF FORMULATION

2.1 LAGRANGIAN (REFERENTIAL) FORMULATION

(i) Comment on Notations

In all formulations to follow we adopt both the direct and the indicial tensor notations with the summation convention. Bold face letters are used to indicate tensor quantities, lower case letters for vectors and upper case letters for higher order tensors (in typewritten manuscript, tensors are identified by a number of bars equal to the order of the tensor on the top of the letter). All repeated indices are to be summed throughout their admissible range unless they are enclosed by parentheses. Right subscripts in upper case refer to nodal points, while lower case right Greek scripts refer to generalized displacements. Left superscripts (numbers) indicate the configuration at which the quantity is measured whereas left subscripts (numbers) indicate the relative configuration to which the quantity is referred. All quantities associated with nodal points are underscored.

(ii) Basic Equations and Assumptions

General Assumptions

A general body which occupies a finite region of Euclidean space is considered. Subjected to prescribed body forces and surface tractions the body undergoes the motion $\bar{x} = \bar{x}(\bar{X},t)$. Three configurations of the body are considered. For simplicity, without any loss of generality, a fixed rectangular frame with cartesian coordinate system in three dimensional space is established to describe the motion. Only perfect mechanical systems which take no account of thermodynamical effects will be considered. Complete nonlinear kinematic relations within a linear increment will be used throughout the formulation.

Further simplifying assumptions will be mentioned where introduced.

Deformation Gradients

Referring to Fig. (1), we may write

$$ {}^{1}\bar{x} = {}^{o}\bar{X} + {}^{1}\bar{u} \qquad , \qquad {}^{2}\bar{x} = {}^{o}\bar{X} + {}^{2}\bar{u} = {}^{1}\bar{x} + {}^{2}_{1}\bar{u} \qquad (4a,b) $$

The deformation gradient at configuration C_1 related to configuration C_o may be given by

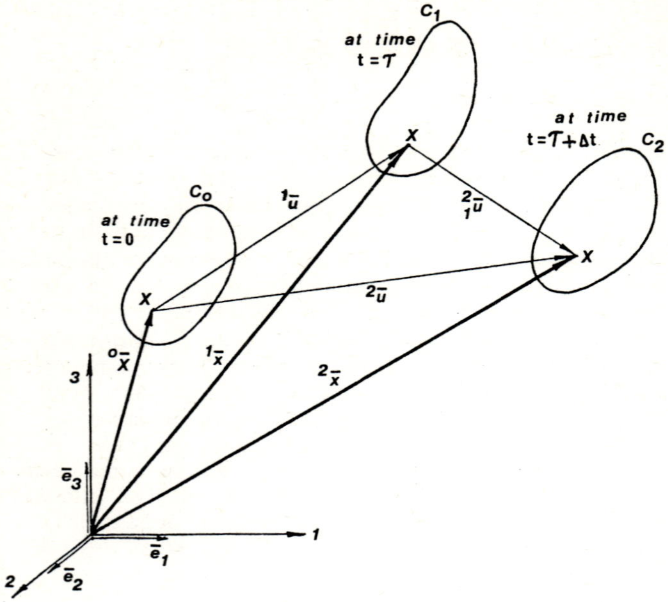

Figure (1) Geometry of motion from configuration C_o to C_1 to C_2 .

$$ {}^{1}_{o}\bar{\bar{F}} = \frac{{}^{1}\bar{x}\;\partial}{\partial\;{}^{o}\bar{X}} = \bar{\bar{1}} + \frac{{}^{1}\bar{u}\;\partial}{\partial\;{}^{o}\bar{X}} \tag{5a} $$

In a similar manner, other deformation gradients may be formed

$$ {}^{2}_{o}\bar{\bar{F}} = \frac{{}^{2}\bar{x}\;\partial}{\partial\;{}^{o}\bar{X}} = \bar{\bar{1}} + \frac{{}^{1}\bar{u}\;\partial}{\partial\;{}^{o}\bar{X}} + \frac{{}^{2}_{1}\bar{u}\;\partial}{\partial\;{}^{o}\bar{X}} = {}^{1}_{o}\bar{\bar{F}} + \frac{{}^{2}_{1}\bar{u}\;\partial}{\partial\;{}^{o}\bar{X}} \tag{5b} $$

and

$$ {}^{2}_{1}\bar{\bar{F}} = \frac{{}^{2}\bar{x}\;\partial}{\partial\;{}^{1}\bar{x}} = \bar{\bar{1}} + \frac{{}^{2}_{1}\bar{u}\;\partial}{\partial\;{}^{1}\bar{x}} \tag{5c} $$

Strain Tensors

We will be mainly dealing with Green-Lagrange strain tensor, which at different configurations may be given by

$$ {}^2_0\bar{\bar{E}} = \frac{1}{2}\,[{}^2_0\bar{\bar{F}}^T \cdot {}^2_0\bar{\bar{F}} - \bar{\bar{1}}] = \frac{1}{2}\,[\frac{\partial\,{}^2\bar{u}}{\partial\,{}^0\bar{X}} + \frac{{}^2\bar{u}\,\partial}{\partial\,{}^0\bar{X}} + \frac{\partial\,{}^2\bar{u}}{\partial\,{}^0\bar{X}} \cdot \frac{{}^2\bar{u}\,\partial}{\partial\,{}^0\bar{X}}] \qquad (6) $$

Substituting ${}^2\bar{u} = {}^1\bar{u} + {}^2_1\bar{u}$, eqn. (6) may be written in the form

$$ {}^2_0\bar{\bar{E}} = {}^1_0\bar{\bar{E}} + (\Delta\,{}^1_0\bar{\bar{E}}^{linear} + \Delta\,{}^1_0\bar{\bar{E}}^{nonlinear}) \qquad (7a) $$

where

$$ {}^1_0\bar{\bar{E}} = \frac{1}{2}\,[\frac{\partial\,{}^1\bar{u}}{\partial\,{}^0\bar{X}} + \frac{{}^1\bar{u}\,\partial}{\partial\,{}^0\bar{X}} + \frac{\partial\,{}^1\bar{u}}{\partial\,{}^0\bar{X}} \cdot \frac{{}^1\bar{u}\,\partial}{\partial\,{}^0\bar{X}}] \qquad (7b) $$

$$ \Delta\,({}^1_0\bar{\bar{E}}^{linear}) = \frac{1}{2}\,[\frac{\partial\,{}^2_1\bar{u}}{\partial\,{}^0\bar{X}} + \frac{{}^2_1\bar{u}\,\partial}{\partial\,{}^0\bar{X}} + \frac{\partial\,{}^1\bar{u}}{\partial\,{}^0\bar{X}} \cdot \frac{{}^2_1\bar{u}\,\partial}{\partial\,{}^0\bar{X}} + \frac{\partial\,{}^2_1\bar{u}}{\partial\,{}^0\bar{X}} \cdot \frac{{}^1\bar{u}\,\partial}{\partial\,{}^0\bar{X}}] \qquad (7c) $$

and

$$ \Delta\,({}^1_0\bar{\bar{E}}^{nonlinear}) = \frac{1}{2}\,[\frac{\partial\,{}^2_1\bar{u}}{\partial\,{}^0\bar{X}} \cdot \frac{{}^2_1\bar{u}\,\partial}{\partial\,{}^0\bar{X}}] \qquad (7d) $$

We also have

$$ {}^1_1\bar{\bar{E}} = \frac{1}{2}\,[\frac{\partial\,{}^1\bar{u}}{\partial\,{}^1\bar{x}} + \frac{{}^1\bar{u}\,\partial}{\partial\,{}^1\bar{x}} + \frac{\partial\,{}^1\bar{u}}{\partial\,{}^1\bar{x}} \cdot \frac{{}^1\bar{u}\,\partial}{\partial\,{}^1\bar{x}}] \qquad (8a) $$

$$ \Delta\,({}^1_1\bar{\bar{E}}) = \frac{1}{2}\,[\frac{\partial\,(\Delta\,{}^1\bar{u})}{\partial\,{}^1\bar{x}} + \frac{(\Delta\,{}^1\bar{u})\,\partial}{\partial\,{}^1\bar{x}} + (\frac{\partial\,(\Delta\,{}^1\bar{u})}{\partial\,{}^1\bar{x}} \cdot \frac{{}^1\bar{u}\,\partial}{\partial\,{}^1\bar{x}} + \frac{\partial\,{}^1\bar{u}}{\partial\,{}^1\bar{x}} \cdot \frac{(\Delta\,{}^1\bar{u})\,\partial}{\partial\,{}^1\bar{x}})] \qquad (8b) $$

and

$$ {}^2_1\bar{\bar{E}} = {}^1_1\bar{\bar{E}} + \Delta\,({}^1_1\bar{\bar{E}}) \qquad (8c) $$

Displacement Assumption

For Lagrangian (or total Lagrangian) formulation, we assume the displacement shape function in the form

$$ {}^2_1u_i = \psi_{i\alpha}\,({}^0\bar{X})\;{}^2_1\underline{u}_\alpha \quad \text{and} \quad {}^1u_i = \psi_{i\alpha}\,({}^0\bar{X})\;{}^1\underline{u}_\alpha \qquad (9a,b) $$

whereas for the updated Lagrangian formulation, we have

$$ {}_1^2u_i = \psi_{i\alpha}\,({}^1\bar{x})\;{}_1^2\underline{u}_\alpha \text{ and } {}^1u_i = \psi_{i\alpha}\,({}^1\bar{x})\;{}^1\underline{u}_\alpha \qquad (9c,d)$$

where $\underline{u}_\alpha$ is the vector of nodal displacements and $\psi_{i\alpha}$ is the usual finite element shape function. Later on, the implicit dependence of ψ on either $\bar{X}$ or $\bar{x}$ will be understood.

Incremental and Total Strain Rates

Substituting the displacement assumptions, eqns. (9), into the incremental Lagrangian strain, eqns. (7), and utilizing the symmetry of the strain tensor, then for a linear increment we may obtain

$$\Delta\,({}_o^1\bar{\bar{E}}) = \{[\frac{\partial\,\psi_{i\alpha}}{\partial\,{}^oX_j} + \frac{\partial\,\psi_{k\beta}}{\partial\,{}^oX_i}\,\frac{\partial\,\psi_{k\alpha}}{\partial\,{}^oX_j}({}^1\underline{u}_\beta)]\;{}_1^2\underline{u}_\alpha\}\;\bar{e}_i\;\bar{e}_j \qquad (10)$$

The material time derivative of eqn. (10) may be given by replacing ${}_1^2\underline{u}_\alpha$ with ${}_1^2\underline{v}_\alpha$ where ${}_1^2\underline{v}_\alpha$ are the incremental nodal velocities from configuration C_1 to C_2. It should be noted that in eqn. (10b) the displacement ${}_1^2\bar{u}$ is a function of the relative time τ at configuration C_1 whereas the displacement ${}_1^2\bar{u}$ is a function of the actual time t at configuration C_2. This, as discussed above, indicates the referential nature of the analysis.

In a similar way, starting from eqn. (8b) and utilizing the relation $\Delta\;{}^1\bar{u} = {}_1^2\bar{u}$, then we may write

$$\Delta\,({}_1^1\bar{\bar{E}}) = \{[\frac{\partial\,\psi_{i\alpha}}{\partial\,{}^1x_j} + \frac{\partial\,\psi_{k\beta}}{\partial\,{}^1x_i}\,\frac{\partial\,\psi_{k\alpha}}{\partial\,{}^1x_j}\,({}^1\underline{u}_\beta)]\;{}_1^2\underline{u}_\alpha\}\;\bar{e}_i\;\bar{e}_j \qquad (11a)$$

which yields

$$\frac{d}{dt}\,(\Delta\;{}_1^1\bar{\bar{E}}) = \{[\frac{\partial\,\psi_{i\alpha}}{\partial\,{}^1x_j} + \frac{\partial\,\psi_{k\beta}}{\partial\,{}^1x_i}\,\frac{\partial\,\psi_{k\alpha}}{\partial\,{}^1x_j}\,({}^1\underline{u}_\beta)]\;{}_1^2\underline{v}_\alpha\}\;\bar{e}_i\;\bar{e}_j \qquad (11b)$$

Once again, the referential nature of the analysis is implied when obtaining eqn. (11b) from (11a) by the assumption ${}^1\bar{x} = {}^1\bar{x}\,({}^o\bar{X},\tau)$. Analogously, we may write

$$\frac{d}{dt}\,({}_o^2\bar{\bar{E}}) = \frac{d}{dt}\,({}_o^1\bar{\bar{E}} + \Delta\;{}_o^1\bar{\bar{E}}) = \frac{d}{dt}\,(\Delta\;{}_o^1\bar{\bar{E}}) \qquad (12a)$$

$$\frac{d}{dt}\,({}_1^2\bar{\bar{E}}) = \frac{d}{dt}\,({}_1^1\bar{\bar{E}} + \Delta\;{}_1^1\bar{\bar{E}}) = \frac{d}{dt}\,(\Delta\;{}_1^1\bar{\bar{E}}) \qquad (12b)$$

Utilizing the deformation gradients given by eqns. (5) together with the displacement assumptions, eqns. (9), and the general assumptions discussed above, we may write

$$ {}^{1}_{o}F_{ij} = \frac{\partial\, {}^{1}x_i}{\partial\, {}^{o}X_j} = \delta_{ij} + \frac{\partial\, \psi_{i\alpha}}{\partial\, {}^{o}X_j}\, {}^{1}\underline{u}_{\alpha} \qquad (13a) $$

$$ {}^{2}_{1}F_{ij} = \frac{\partial\, {}^{2}x_i}{\partial\, {}^{1}x_j} = \delta_{ij} + \frac{\partial\, \psi_{i\alpha}}{\partial\, {}^{1}x_j}\, {}^{2}_{1}\underline{u}_{\alpha} \qquad (13b) $$

(iii) Incremental Formulation

We base the formulation on the mechanical energy balance equation , eqn. (2) [9,11]. For the intended incrementation procedure, it is convenient to focus attention on a single element with the understanding that equations for an individual element are only valid when contributions from all elements sharing a particular node are added together. With this in mind, then substituting the displacement assumptions, eqns. (9), into the mechanical energy balance equation in the undeformed state, eqn. (2), assuming an arbitrary virtual velocities and, finally, considering an incremental formulation, the final equilibrium equations may take the form (for details of the analysis, see reference [11])

$$ \Delta\, {}^{1}_{o}\bar{R}_{\alpha} = [{}^{1}_{o}K^{(1)}_{\alpha\beta} + {}^{1}_{o}K^{(2)}_{\alpha\beta} + {}^{1}_{o}K^{(3)}_{\alpha\beta} - {}^{1}_{o}K^{(4)}_{\alpha\beta}]\, \{\Delta\, {}^{1}\underline{u}_{\beta}\} $$

$$ + [{}^{1}_{o}M_{ab}]\, \{W\ {}^{1}\ddot{u}_{b}\} \qquad (14) $$

$$ \Delta\, {}^{1}_{o}\bar{R}_{\alpha} = \{\int_{{}^{o}V} \Delta\, ({}^{o}\rho\ {}^{1}b_i)\ \psi_{i\alpha}\, d\, {}^{o}V + \int_{{}^{o}A} (\Delta\, {}^{1}_{o}S_{n_i})\, [\psi_{i\alpha} $$

$$ + \frac{\partial\, \psi_{i\beta}}{\partial\, {}^{o}X_j}\, \psi_{j\alpha}\, {}^{1}\underline{u}_{\beta}]\, d\, {}^{o}A \qquad (15) $$

is the load increment vector;

$$ {}^{1}_{o}K^{(1)}_{\alpha\beta} = \int_{{}^{o}V} \frac{\partial\, \psi_{i\alpha}}{\partial\, {}^{o}X_j}\, {}^{1}_{o}D_{ijkl}\, \frac{\partial\, \psi_{k\beta}}{\partial\, {}^{o}X_l}\, d\, {}^{o}V \qquad (16a) $$

is the usual small displacement or the icnremental stiffness matrix;

$$ {}^{1}_{o}K^{(2)}_{\alpha\beta} = \int_{{}^{o}V} {}^{1}_{o}S_{ij}\, \frac{\partial\, \psi_{k\beta}}{\partial\, {}^{o}X_i}\, \frac{\partial\, \psi_{k\alpha}}{\partial\, {}^{o}X_j}\, d\, {}^{o}V \qquad (16b) $$

is the inital stress or the geometric or the tangent stiffness matrix;

$$ {}^{1}_{o}K^{(3)}_{\alpha\beta} = \int_{{}^{o}V} {}^{1}_{o}D_{ijkl} \{ (\frac{\partial \psi_{r\gamma}}{\partial\, {}^{o}X_i} \frac{\partial \psi_{r\alpha}}{\partial\, {}^{o}X_j} \frac{\partial \psi_{k\beta}}{\partial\, {}^{o}X_l} + \frac{\partial \psi_{i\alpha}}{\partial\, {}^{o}X_j} \frac{\partial \psi_{w\gamma}}{\partial\, {}^{o}X_k} \frac{\partial \psi_{w\beta}}{\partial\, {}^{o}X_l}) $$

$$ {}^{1}\underline{u}_{\gamma} + (\frac{\partial \psi_{r\gamma}}{\partial\, {}^{o}X_i} \frac{\partial \psi_{r\alpha}}{\partial\, {}^{o}X_j} \frac{\partial \psi_{w\eta}}{\partial\, {}^{o}X_k} \frac{\partial \psi_{w\beta}}{\partial\, {}^{o}X_l} \, {}^{1}\underline{u}_{\gamma} \, {}^{1}\underline{u}_{\eta}) \} \, d\, {}^{o}V \qquad (16c) $$

is the initial displacement or the initial rotation stiffness matrix;

$$ {}^{1}_{o}K^{(4)}_{\alpha\beta} = \int_{{}^{o}A} {}^{1}_{o}S_{n_i} \frac{\partial \psi_{i\beta}}{\partial\, {}^{o}X_j} \psi_{j\alpha} \, d\, {}^{o}A \qquad (17) $$

is the initial load or the load correction matrix, and finally

$$ {}^{1}_{o}M_{\alpha\beta} = \int_{{}^{o}V} {}^{o}\rho \, (\psi_{i\alpha} \psi_{i\beta}) \, d\, {}^{o}V \qquad (18) $$

is the consistent mass matrix.

In the above analysis, we assumed a stress strain relation of the form

$$ \Delta ({}^{1}_{o}S_{ij}) = {}^{1}_{o}D_{ijkl} \, \Delta ({}^{1}_{o}E_{kl}) $$

where possible forms of D_{ijkl} for specific applications may be discussed independently [11].

So far, the results presented here differ from similar ones in the literature [7,10] in the definition of the load correction matrix, eqn. (17), and the definition of the load increment, eqn. (15). In [10], Hibbit et al. introduce a rather artificial argument as $\overline{\Delta F} = \Delta_{load}\bar{F} + \Delta_{geom}\bar{F}$, where $\bar{F}$ is the surface traction vector or the body force vector, to obtain the form of the load correction vector. The term obtained by Hibbit et al. will have a zero value in some loading cases, while the term presented here will always possess some value in general. In [7] Bathe et al. introduce similar arguments to the ones presented here but for deformation dependent loadings. The formulation presented here coincides with the procedure given by Oden [9].

2.2 UPDATED LAGRANGIAN OR RELATIVE FORMULATION

As discussed above, in updated Lagrangian formulation we describe the future in terms of the current configuration.

Referring to Fig. 1, the independent variables will be the position ${}^{2}\bar{x}$ and the variable time τ, where ${}^{2}\bar{x}$ (${}^{o}\bar{X}$,t), i.e. ${}^{2}\bar{x}$ is independent of τ which indicates the referential nature of the analysis (compare to independent variables of Lagrangian formulation; ${}^{o}\bar{X}$ and t which have the same character). The energy balance equation referred to the updated or variable Lagrangian reference configuration C_1 may be written in the form

$$\frac{d}{dt}\,{}^{2}_{1}K = \int_{{}^{1}V} {}^{1}\rho\, {}^{2}\bar{b} \cdot \frac{d\,{}^{2}\bar{u}}{dt}\, d\,{}^{1}V + \int_{{}^{1}A} d\,{}^{1}\bar{A} \cdot {}^{2}_{1}\bar{\bar{S}} \cdot {}^{2}_{1}\bar{\bar{F}}^{T} \cdot \frac{d\,{}^{2}\bar{u}}{dt}$$

$$- \int_{{}^{1}V} {}^{2}_{1}\bar{\bar{S}} : \frac{d\,{}^{2}_{1}\bar{\bar{E}}}{dt}\, d\,{}^{1}V \qquad (19)$$

Considering eqn. (19) term by term, we have

Kinetic energy term

$$J_1 \Rightarrow \frac{d}{dt}\,{}^{2}_{1}K = \frac{d}{dt}\int_{{}^{1}V} \frac{1}{2}\,{}^{1}\rho \left(\frac{d\,{}^{2}\bar{u}}{dt} \cdot \frac{d\,{}^{2}\bar{u}}{dt}\right) d\,{}^{1}V$$

following the same procedure of developing eqns. (18), we may write

$$\Delta J_1 = [{}^{1}_{1}M_{\alpha\beta}]\ \{\Delta\,{}^{1}\ddot{\underline{u}}_{\beta}\} \qquad (20a)$$

where

$${}^{1}_{1}M_{\alpha\beta} = \int_{{}^{1}V} {}^{1}\rho\, (\psi_{i\alpha}\, \psi_{i\beta})\, d\,{}^{1}V \qquad [18]\ (20b)$$

is the consistent mass matrix, and for the sake of comparison, the numbers of corresponding Lagrangian expression are indicated in square brackets.

Load vector term

$$J_2 \Rightarrow \int_{{}^{1}V} {}^{1}\rho\, {}^{2}\bar{b} \cdot \frac{d\,{}^{2}\bar{u}}{dt}\, d\,{}^{1}V + \int_{{}^{1}A} d\,{}^{1}\bar{A} \cdot {}^{2}_{1}\bar{\bar{S}} \cdot {}^{2}_{1}\bar{\bar{F}}^{T} \cdot \frac{d\,{}^{2}\bar{u}}{dt} \Rightarrow {}^{2}_{1}R$$

Substituting expressions for the displacement assumption, the deformation gradient and, finally, considering an arbitrary virtual velocities ${}^{2}_{1}\underline{v}_{\alpha}$, we may have

$$J_2 = \int_{{}^{1}V} {}^{1}\rho\, {}^{2}b_i\, \psi_{i\alpha}\, d\,{}^{1}V + \int_{{}^{1}A} {}^{2}_{1}S\, n_i \left(\psi_{i\alpha} + \frac{\partial\,\psi_{i\beta}}{\partial\,{}^{1}x_j}\, \psi_{j\alpha}\, {}^{2}_{1}\underline{u}_{\beta}\right) d\,{}^{1}A$$

Increments of the above equation give

$$\Delta J_2 = [\int_{^1V} \Delta ({}^1\rho\ {}^1b_i)\ \psi_{i\alpha}\ d\ {}^1V] + [\int_{^1A} (\Delta\ {}^2_1S_{n_i})\ (\psi_{i\alpha}$$

$$+ \frac{\partial\ \psi_{i\beta}}{\partial\ {}^1x_j}\ \psi_{j\alpha}\ {}^2_1\underline{u}_\beta)\ d\ {}^1A + \int_{^1A} {}^2_1S_{n_i}\ \{\frac{\partial\ \psi_{i\alpha}}{\partial\ {}^1x_j}\ \psi_{j\alpha}\ (\Delta\ {}^2_1\underline{u}_\beta)\}\ d\ {}^1A]$$

substituting the assumptions

$$\Delta\ {}^2_1S_{n_i} = \Delta\ ({}^1_1S_{n_i} + \Delta\ {}^1_1S_{n_i}) = \Delta\ {}^1\sigma_{n_i} + O[\Delta(\Delta\ {}^1_1S_{n_i})]$$

where ${}^1_1\overline{\overline{S}} = {}^1\overline{\overline{\sigma}}$, and $\Delta\ {}^2_1\underline{u}_\beta = \Delta\ (\Delta\ {}^1\underline{u}_\beta) \simeq O[\Delta(\Delta\ {}^1\underline{u}_\beta)]$

we obtain

$$\Delta J_2 = \overline{\Delta\ {}^1_1R} = \Delta\ {}^1_1\overline{R} + [{}^1_1K^{(4)}_{\alpha\beta}]\ \{\Delta\ {}^1\underline{u}_\beta\} \tag{21a}$$

where

$$\Delta\ {}^1_1\overline{R} = \{\int_{^1V} \Delta\ ({}^1\rho\ {}^1b_i)\ \psi_{i\alpha}\ d\ {}^1V + \int_{^1A} (\Delta\ {}^1\sigma_{n_i})\ \psi_{i\alpha}\ d\ {}^1A\} \qquad [15]\ (21b)$$

is the load increment vector, and

$${}^1_1K^{(4)}_{\alpha\beta} = \int_{^1A} \Delta\ {}^1\sigma_{n_i}\ \frac{\partial\ \psi_{i\beta}}{\partial\ {}^1x_j}\ \psi_{j\alpha}\ d\ {}^1A \qquad [17]\ (21c)$$

is the initial load or the load correction matrix.

Strain energy rate term

$$J_3 \Rightarrow \int_{^1V} {}^2_1\overline{\overline{S}} : \frac{d}{dt}\ {}^2_1\overline{\overline{E}}\ d\ {}^1V$$

Once again, following the same procedures of developing eqns. (16), we may obtain

$$\Delta J_3 = [{}^1_1K^{(1)}_{\alpha\beta} + {}^1_1K^{(2)}_{\alpha\beta} + {}^1_1K^{(3)}_{\alpha\beta}]\ \{\Delta\ {}^1\underline{u}_\beta\} \tag{22a}$$

where

$${}^1_1K^{(1)}_{\alpha\beta} = \int_{^1V} \frac{\partial\ \psi_{i\alpha}}{\partial\ {}^1x_j}\ {}^1_1D_{ijkl}\ \frac{\partial\ \psi_{k\beta}}{\partial\ {}^1x_l}\ d\ {}^1V \qquad [16a]\ (22b)$$

is the usual small displacement or the incremental stiffness matrix;

$$ {}_{1}^{1}K_{\alpha\beta}^{(2)} = \int_{{}^{1}V} {}^{1}\sigma_{ij} \frac{\partial \psi_{k\beta}}{\partial\, {}^{1}x_i} \frac{\partial \psi_{k\alpha}}{\partial\, {}^{1}x_j} d\, {}^{1}V \qquad [16b]\ (22c) $$

is the initial stress or the geometric or the tangent stiffness matrix;

$$ {}_{1}^{1}K_{\alpha\beta}^{(3)} = \int_{{}^{1}V} {}_{1}^{1}D_{ijkl} \{ (\frac{\partial \psi_{r\gamma}}{\partial\, {}^{1}x_i} \frac{\partial \psi_{r\alpha}}{\partial\, {}^{1}x_j} \frac{\partial \psi_{k\beta}}{\partial\, {}^{1}x_l} + \frac{\partial \psi_{i\alpha}}{\partial\, {}^{1}x_j} \frac{\partial \psi_{w\gamma}}{\partial\, {}^{1}x_k} \frac{\partial \psi_{w\beta}}{\partial\, {}^{1}x_l}) \, {}^{1}\underline{u}_{\gamma} $$

$$ + (\frac{\partial \psi_{r\gamma}}{\partial\, {}^{1}x_i} \frac{\partial \psi_{r\alpha}}{\partial\, {}^{1}x_j} \frac{\partial \psi_{wn}}{\partial\, {}^{1}x_k} \frac{\partial \psi_{w\beta}}{\partial\, {}^{1}x_l} \, {}^{1}\underline{u}_{\gamma} \, {}^{1}\underline{u}_{n}) \} \, d\, {}^{1}V \qquad [16c]\ (22d) $$

is the initial displacement or the initial rotation stiffness matrix.

Collecting results of eqns. (19-22), the final incremental equilibrium equations take the form

$$ \Delta\, {}_{1}^{1}\bar{R}_{\alpha} = [{}_{1}^{1}K_{\alpha\beta}^{(1)} + {}_{1}^{1}K_{\alpha\beta}^{(2)} + {}_{1}^{1}K_{\alpha\beta}^{(3)} - {}_{1}^{1}K_{\alpha\beta}^{(4)}] \{ \Delta\, {}^{1}\underline{u}_{\beta} \} + [{}_{1}^{1}M_{\alpha\beta}] \{ \Delta\, {}^{1}\ddot{\underline{u}}_{\beta} \} \qquad [14]\ (23) $$

where all individual terms are defined above.

The formulation developed here differs from similar ones in the literature in the development of the load correction matrix, the definition of the load increment, the expressions and the number of stiffness matrices contributing to the total incremental nonlinear stiffness matrix. In [7,8], Bathe et al. introduce the same load increments for both Lagrangian and updated Lagrangian formulations. In the formulation presented here, there is a distinct difference between the two expressions, eqns. (15, 21b). The dependence of the load increment vector on the deformation gradient in only the Lagrangian formulation and not in the updated Lagrangian version may be visualized from the nature of the reference configuration in each case. For the same reason, it would be expected that the load correction matrix in updated Lagrangian case is dependent on $\Delta\, {}^{1}\sigma_n$, the increment of the surface traction vector from configuration C_1 to C_2, with comparison to its dependence on ${}_{o}^{2}S_n$, the surface traction vector from configuration C_o to C_2 in the Lagrangian case. In [5] Wunderlich introduces kinematic simplifications which eliminate the initial displacement stiffness matrix in the updated Lagrangian formulation. The above analysis shows that the number of stiffness matrices contributing to the total nonlinear stiffness matrix of the element are the same in both Lagrangian and updated Lagrangian formulations and only simplifying kinematic assumptions would eliminate some of these matrices. It is this reasons which lends similar nature to both the Lagrangian and the updated Lagrangian formulations and makes them distinct from the features of an Eulerian formulation. It should be noticed that the definition of the material property tensor is different in each formulation, however, proper transformation should make it possible to change one into the other.

ACKNOWLEDGEMENT

The authors would like to thank Miss P. Dillon for her patience and understanding in typing this manuscript.

REFERENCES

[1] TRUESDELL, C. - A First Course in Rational Continuum Mechanics, Volume 1: General Concepts, Academic Press, New York, 1977.

[2] KAWAI, T. and YOSHIMURA, N. - Analysis of Large Deflection of Plates by the Finite Element Method, Int. J. Num. Meth. Engng., Vol. 1, pp. 123-133, 1969.

[3] BERGAN, P.G. and CLOUGH, R.W. - Large Deflection Analysis of Plates and Shallow Shells Using the Finite Element Method, Int. J. Num. Meth. Engng., Vol. 5, pp. 543-556, 1973.

[4] NEEDLEMAN, A. - A Numerical Study of Necking in Circular Cylindrical Bars, J. Mech. Phys. Solids, Vol. 20, pp. 111-127, 1972.

[5] WUNDERLICH, W. - Incremental Formulations for Geometrically Nonlinear Problems, In Formulations and Computational Algorithms in Finite Element Analysis: U.S.-Germany Symposium, Ed. K.J. Bathe et al., pp. 193-240, 1976.

[6] MURAKAWA, H. and ATLURI, S.N. - Finite Elasticity Solutions using Hybrid Finite Elements Based on a Complementary Energy Principle, J. of Appl. Mech., ASME Trans., Vol. 45, pp. 539-547, 1978.

[7] BATHE, K.J. and OZDEMIR, H. - Elastic-Plastic Large Deformation Static and Dynamic Analysis, Computers & Structures, Vol. 6, pp. 81-92, 1976.

[8] BATHE, K.J. and BOLOURCHI, S. - Large Displacement Analysis of Three-Dimensional Beam Structures, Int. J. Num. Meth. Engng., Vol. 14, pp. 961-986, 1979.

[9] ODEN, J.T. - Finite Elements of Nonlinear Continua, McGraw Hill, New York, 1972.

[10] HIBBIT, H.D., MARCAL, P.V. and RICE, J.R. - A Finite Element Formulation for Problems of Large Strain and Large Displacements, Int. J. Solids & Structures, Vol. 6, pp. 1069-1086, 1970.

[11] GADALA, M.S., ORAVAS, G. AE. and DOKAINISH, M.A. - Geometric and Material Nonlinearity Problems - Formulation Aspects, submitted for presentation in the First International Conf. on Numerical Methods for Non-Linear Problems, Swanswea-England (1980).

[12] MURRAY, D.W. and WILSON, E.L. - Finite-Element Large Deflection Analysis of Plates, J. Engng. Mech. Div., ASCE Proc., Vol. 95, pp. 143-165, 1969.

[13] HILL, R. - A General Theory of Uniqueness and Stability in Elastic-Plastic Solids, J. Mech. Phys. Solids, Vol. 6, pp. 236-249, 1958.

[14] HILL, R. - Some Basic Principles in the Mechanics of Solids Without a Natural Time, J. Mech. Phys. Solids, Vol. 7, pp. 209-225, 1959.

[15] ODEN, J.T. - Discussion of: Finite Element Analysis of Nonlinear Structures, J. Struct. Div., ASCE Proc., Vol. 95, pp. 1379-1381, 1969.

[16] ODEN, J.T. and KEY, J.E. - Numerical Analysis of Finite Axisymmetric Deformations of Incompressible Elastic Solids of Revolution, Int. J. Solids & Structures, Vol. 6, pp. 697-518, 1970.

[17] HILL, R. - On Constitutive Inequalities for Simple Materials - I, J. Mech. Phys. Solids, Vol. 16, pp. 229-242, 1968.

[18] DE VEUBEKE, B.F. - A New Variational Principle for Finite Elastic Displacements, Int. J. Engng. Sci., Vol. 10, pp. 745-763, 1972.

[19] ATLURI, S.N. - On Hybrid Finite Element Methods in Solid Mechanics, In Advances in Computer Methods for Partial Differential Equations, Ed. Vishnevetsky, AICA, Rutgers University, pp. 346-356, 1975.

[20] ATLURI, S.N. and MURAKAWA, H. - On Hybrid Finite Element Models in Nonlinear Solid Mechanics, Proc. of the Int. Conf. on Finite Elements in Nonlinear Solid and Struct. Mechs., Geilo-Norway, pp. 3-40, Aug. 29-Sept. 1, 1977.

[21] KOITER, W.T. - On The Principle of Stationary Complementary Energy in The Nonlinear Theory of Elasticity, SIAM J. of Appl. Math., Vol. 25, pp. 424-434, 1973.

[22] ZUBOV, L.M. - The Stationary Principle of Complementary Work in Nonlinear Theory of Elasticity, J. Appl. Math. Mech., Vol. 34, pp. 228-232, 1970.

[23] CASTIGLIANO, C.A.P. - The Theory of Equilibrium of Elastic Systems and its Applications, Dover Public., New York, 1966.

[24] WASHIZU, K. - Variational Methods in Elasticity and Plasticity, 2nd edition, Pergamon Press, 1975.

Two- and Three-dimensional elasto-plastic FEM-calculation of CT-specimens

by Joachim Redmer and Winfried Dahl

Institut für Eisenhüttenkunde der RWTH Aachen
Intzestr. 1
5100 Aachen

Summary

The 3-dimensional stress state is calculated, using a 3-dimensional elasto-plastic Finite Element Method in displacement formulation. The important stress distributions and the plastic zones are plotted and kontinuumsmechanical interpretations of the results are presented. It is shown, that the 2-dimensional models are unable to take a few of the most important effects into account. So the 2- and 3-dimensional discriptions end up to be contradictionary in some points.

Introduction

The Finite Element Method (FEM) became a common instrument in structural analysis and it is by now also well established in fracture mechanics. This work wants to show some basic results of the 3-dimensional elasto-plastic calculation of CT-specimen. The specimens were experimentally investigated at the "Institut für Eisenhüttenkunde".

The FEM program used, works with 20-node isoparametric brick elements (1), and a frontal solution technique (2). The work expressions are formulated in displacements and it is assumed, that the strains and the displacements are small enough to use first order geometrical theory. So we solve a geometrically linear, but materially non-linear problem. This seems to be justified, because the strains in the examples discussed are always less than 8 %.

The plasticity is solved by an incremental method, using initial stresses. The stiffness is kept constant, so that the linear equation system has to be solved only once, namely during the first load increment. For the following load steps only the residual forces have to be calculated and the new right hand side has to be resolved. To accelerate the incremental process, the "alpha-constant-stiffness method" is used, as it was introduced by Nayak (3).

The FEM programm used was originally written by Nayak within his thesis at Swansea. It was however largely modified by us. Two of the most important modifications are, that it is now at least five times faster than in its original version, and that it is able to model stable crack growth in 3-dimensional structures. The modelling of stable crack growth will however not be a subject of this paper.

Results

Fig. (1):

Typical stress-strain curve

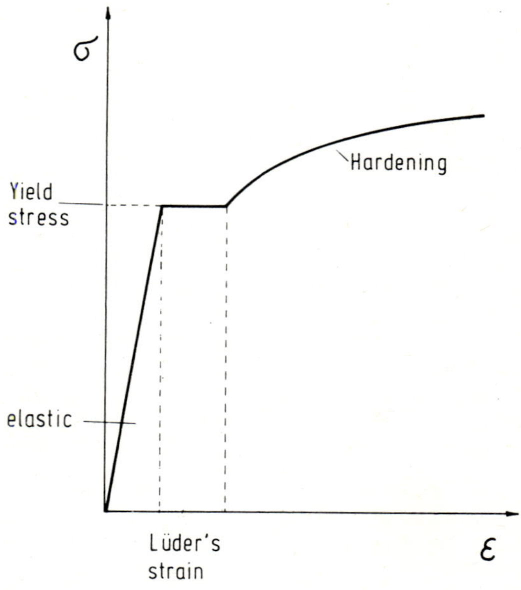

Fig (1) shows a typical uniaxial stress-strain curve of the investigated steels. This curve is taken into account in the calculation via the Von-Mises yield criterion. It has to be noted, that the curve has a Lueders-strain, which is equivalent to a zone of ideal plasticity.

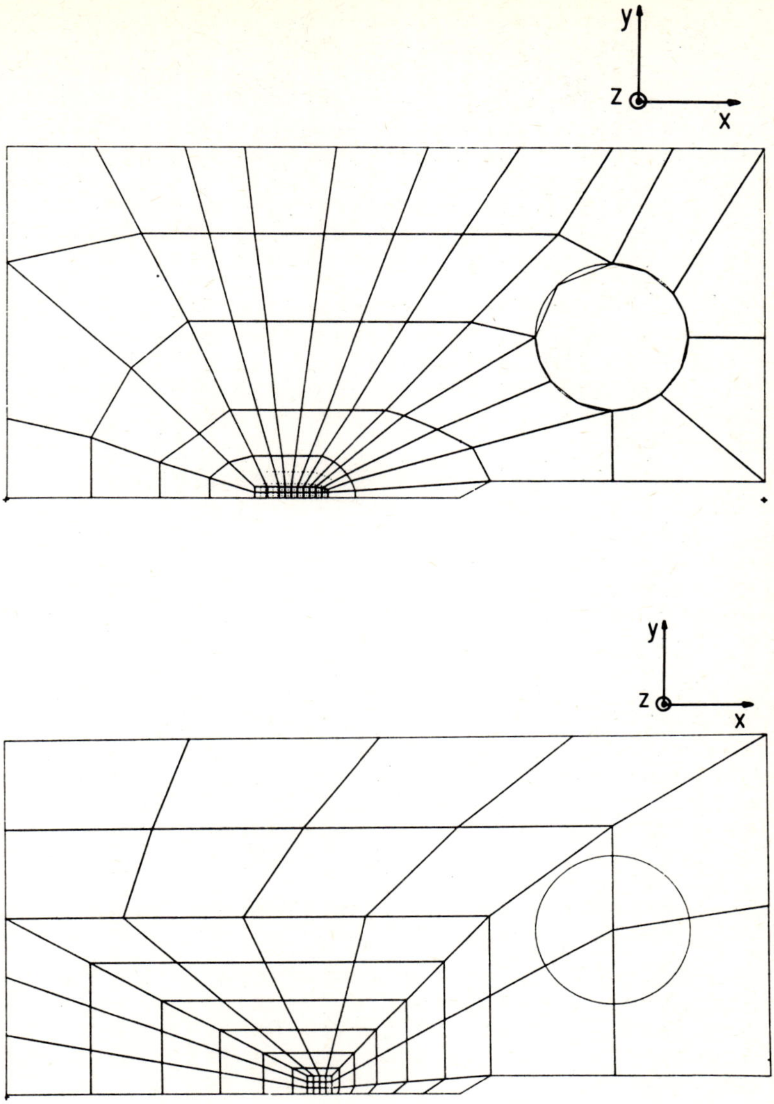

Figs.(2a, 2b): Typical FEM meshes, as used in the calculations

In figs. (2a, 2b) two typical meshes, which were used, are drawn. (The figs. also define the coordinate system we are referring to in this paper). In this figs. of course only the 2-dimensional basic meshes are plotted. For the discussed 3-d calculations we put three 20-node isoparametric brick elements in z-direction. For there is a symmetrie plane in the z-direction, only one half of the specimen thickness has to be calculated. This means, that in fact 6 quadratic elements describe the total thickness of the specimen, which should be enough. The second symmetrie plane (in y-direction) is well known from all 2-d calculation, so that at the end one quarter of the total specimen has to be modelled. In the 3-d calculations we used about 250 elements with about 4500 variables to modell this quarter. About 110 load increments are necessary to reach the maximum load.

In figs. 2a and 2b two different methods of constructing a mesh for a CT-specimen are shown: with and without modelling of the whole. Both methods have their disadvantages. If the whole is modelled, the circle is deformed into an ellypsis, because it is not taken into account that the load pin has an supporting effect in y-direction. On the other hand, if the whole is not modelled, the load-pin now is assumed to have a rigid link to the CT-specimen. This effect is responsible for a force transmission on the lower half of the circle, which is in contradiction to the nature of the experiment. The results of both methods do not differ in the region of the crack-tip. In this paper calculations with and without modelling the whole are presented. Anyhow it seems to be better to us not to modell the whole in future, because the elements in the region of the whole are wasted.

Fig. (3) shows the comparison of 2- and 3-dimensional calculated load-deflection curves with the experimentally measured curve. One can see that the plane-strain curve is much over-stiff. The plane stress curve is a little bit too flexible in the plastic region, but modells the experiment quite well. The 3-dimensional curve is slightly too stiff in the plastic part and so it seems, that the plane stress modell describes the experiment best. This is of course not possible. In fact 2 errors are compensating one another. First a plane stress calculation must be too flexible. This effect is compensated by the fact, that a FEM in displacement formulation produces an over-stiff solution (specially in the plastic part).Generally it can be said,that the 3-dimensionally calculated curve is in reasonable agreement with the experiment.

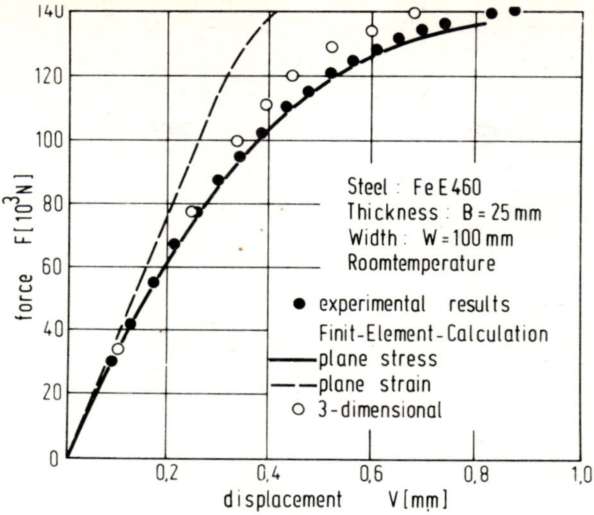

Fig. (3): Comparison between calculated and experimental load deflection curves

We now come to the discussion of the stress distribution. Referring to the coordinate-system defined in figs. (2) we look at the variation of the stresses in z-direction at a point in the x-y plane.

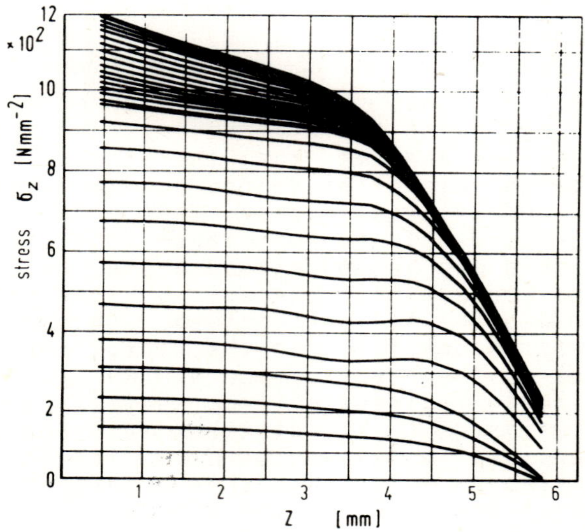

Fig. (4): Variation of σ_z from mid-plane to surface

Fig. (4) shows the variation of σ_z from the mid-plane of the specimen (left) to the surface-plane. The different curves refer to different load steps. As it had to be expected, the stress goes to zero at the surface of the specimen. This is in accordance with the two dimensional models , which assume a transition from the plane strain situation in the mid of the specimen to the plane stress situation on the surface.

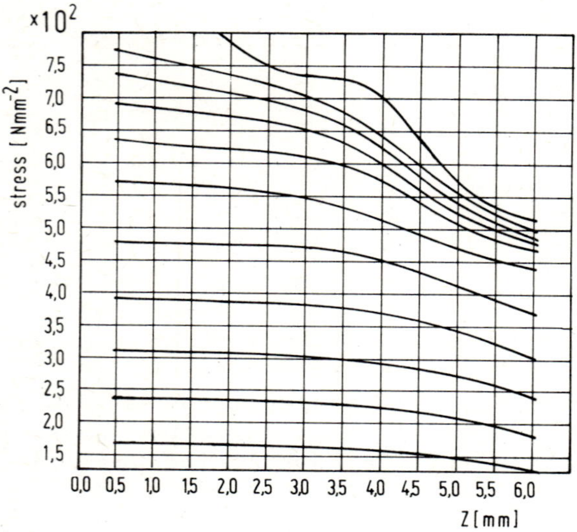

Fig. (5): Variation of σ_y from mid-plane to surface

Fig. (5) shows the distribution of the σ_y stress. The lowest curve represents the purely elastic case. One can see, that the stress is decreasing from the mid-plane to the surface plane even for this purely elastic case. This effect can be more clearly mentioned for higher load steps. Mechanically this can be understood quite easily by taking into account, that there is a higher constraint in the mid-plane of the specimen. The same effect is by the way responsible for the fact, that the y-displacement on the free surface of the crack-plane is increasing in z-direction from surface to mid-plane. The variation of σ_y is not taken into account in the two dimensional models , therefore its influence on to the all-over solution has to be discussed. It is definitely true, that, as the 2-d modells assume, we have an approximate plane strain situation on the infinitesimally thin mid-plane of the specimen, and that we have a plane stress situation on the surface-plane of the same infintesimal size. But because of the variation of σ_y this two extreme situations are not comparable to one another, for they are both referring to different maximal principal stresses.

In the 3-dimensional model we have three shear stresses: τ_{xy}, τ_{xz} and τ_{yz}. Only one of them, τ_{xy}, can be taken into account in a 2-d model. It is well known, that the two additional 3-d shear stresses have a strong influence on to the Von Mises effective stress which is defined by:

$$\sigma_{ef} = \frac{1}{\sqrt{2}} \sqrt{(\sigma_y - \sigma_x)^2 + (\sigma_x - \sigma_z)^2 + (\sigma_z - \sigma_y)^2 + 6(\tau_{xy}^2 + \tau_{yz}^2 + \tau_{zx}^2)} \quad (1)$$

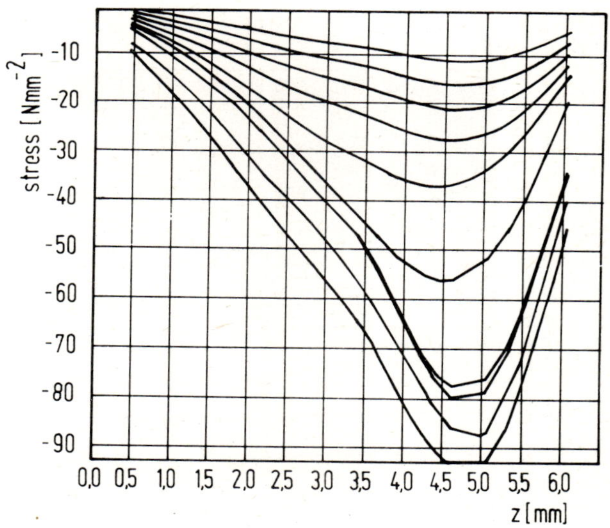

Fig. (6): Variation of shear stress τ_{xz} from mid-plane to surface

Fig. (6) shows the behaviour of the stress τ_{xz} (τ_{yz} gives a similar impression). The symmetrie condition on the mid-plane, respectively the equilibrium condition on the surface-plane (traction-free surface) requiresthe two shear stresses to be zero at these two planes. Fig. (6) shows, that the FEM solution satisfies this requirement. It has to be noted now, that the kinematic link between the flexible plane stress state at the surface and the stiff approximate plane strain state at the mid-plane can only exist,because of the presence of τ_{xz} and τ_{yz}. Without the ability of transmittling shear in z-direction, the material would behave in this direction like a liquid without kinematic link and the differently stiff states would exist independent from one another, as the 2-d model strictly speaking assumes. Thus the two shear stresses cannot be zero all over the thickness from the mechanical point of view. The 3-d FEM solution verifies this qualitive consideration. Fig. (6) shows that τ_{xz} and τ_{yz} have extremities between the surface and the mid-plane.

Knowing about the importance of the shear transmission for a correct discription of the real physical behaviour, and knowing about its influence on to the yield-function as given in equation (1), one can no more expect a 2-d modell to give the correct answers about the size of the plastic zones. The above mentioned decrease of σ_y from mid-plane to the surface which also influences the yield-function supports this effect.

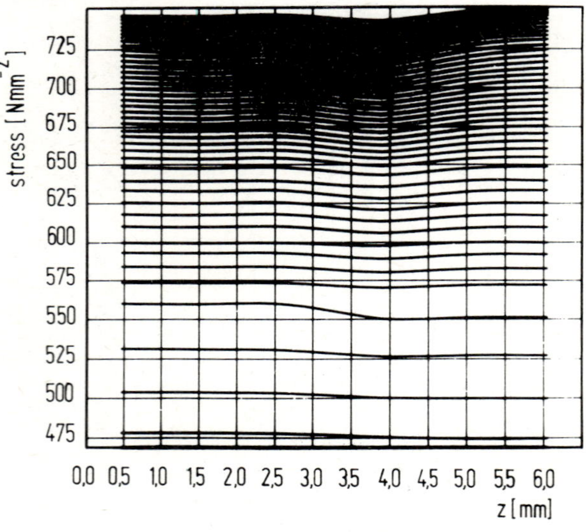

Fig. (7): Variation of the effective stress from mid-plane to surface

Fig. (7) shows the Von Mises effective stress. The 2-d modell would expect this stress to increase distinctly in vicinity to the surface (plane stress situation). After the discussion in the preceding paragraphs no doubt arises, that the 3-d calculation will not verify this presumption. The Von Mises stress is rather constant from the mid-plane to the surface, as can be seen in fig. (7). It has to be mentioned,that this general development of the effective stress is not fully independent of the point in the x-y plane at which it is plotted.It may have a small maximum in front of the surface which is probably effected by the influence of the shear stresses τ_{xz} and τ_{yz}.

We now come to perspective graphs of the effective stress presented in figs. (8) to (10) (the applied load is increasing from fig. (8) to fig. (10).

$$\sigma_V = \frac{1}{\sqrt{2}}\sqrt{(\sigma_Y-\sigma_X)^2+(\sigma_X-\sigma_Z)^2+(\sigma_Z-\sigma_Y)^2+6(\tau_{XY}^2+\tau_{YZ}^2+\tau_{ZX}^2)}$$

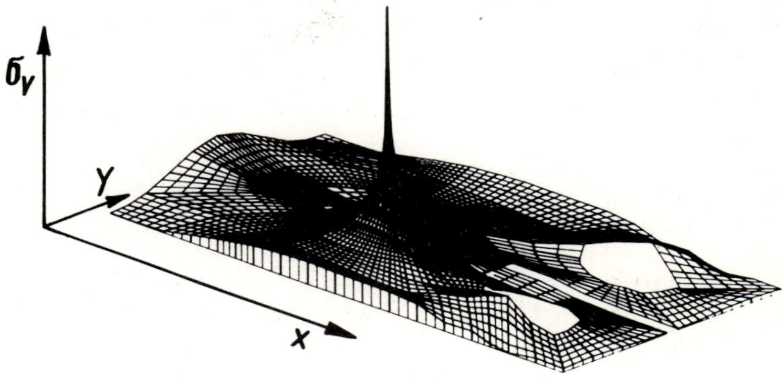

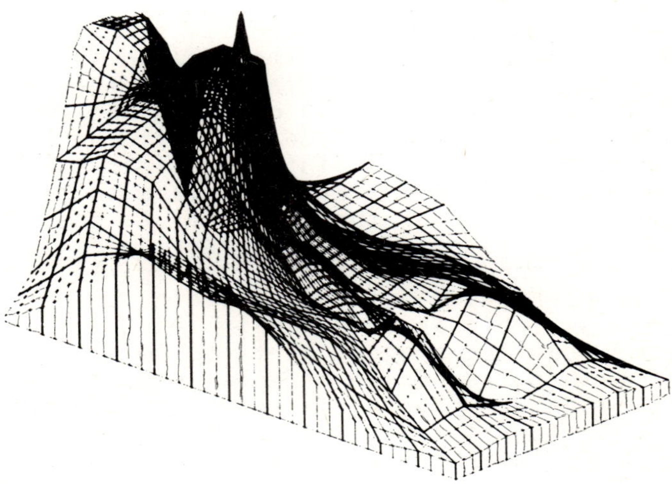

Figs. (8) and (9): Perspective graph of the effective stress over the x-y-plane near to the surface

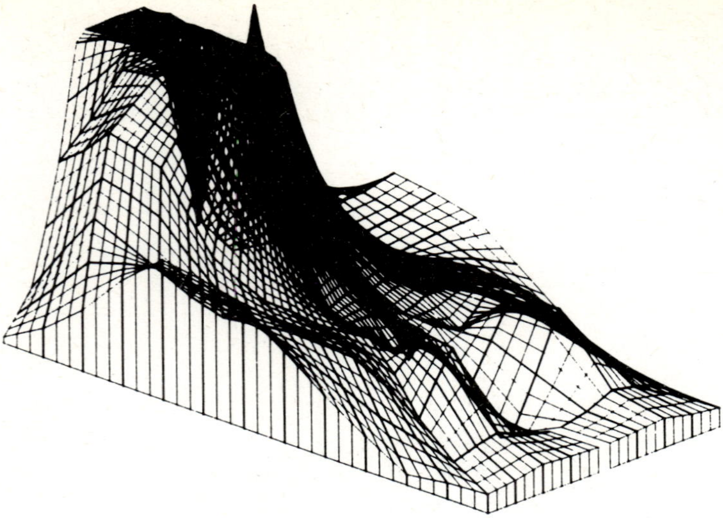

Fig. (10): Perspective graph of the effective stress over the x-y-plane near to the surface

The graphs show the development of the Von Mises stress in an x-y-plane near to the surface (last Gauss-point layer in front of the surface). Fig. (8) represents the purely elastic case. One can see the influence of the singularity in the region of the crack-tip and the material strength in the 2 pressure zones of the specimen.In fig.(9) we already have plasticity in the vicinity of the crack-tip. As the uniaxial stress-strain curve of the material shows a small region of ideal-plastic behaviour,the Von Mises stress levels into the yield-stress of the material. In fig.(9) plasticity is also present in the pressure zone of the specimen. At the crack-tip the effective stress left the ideal-plastic plateau, and strain hardening started. In fig.(1o) the two plastic zones have joined one another. Thus the state of general-yield is obtained in this x-y plane. One can see,that the biggest part of the plastic zone is still within the region of ideal plasticity (Lueders Band). Strain hardening is restricted to a relatively small vicinity of the crack-tip.This means,that it is very important to model the Lueders-Strain in the uniaxial stress-strain curve. Without modelling this effect the hardening would necessarily have to start already at the boundary of the plastic zone. Regarding the relative large region of ideal plasticity this neglect would definitely overstiff the solution. It is also interesting,that in the region of the centre of rotation of the specimen one can still find a small "valley" of relatively low material strength.

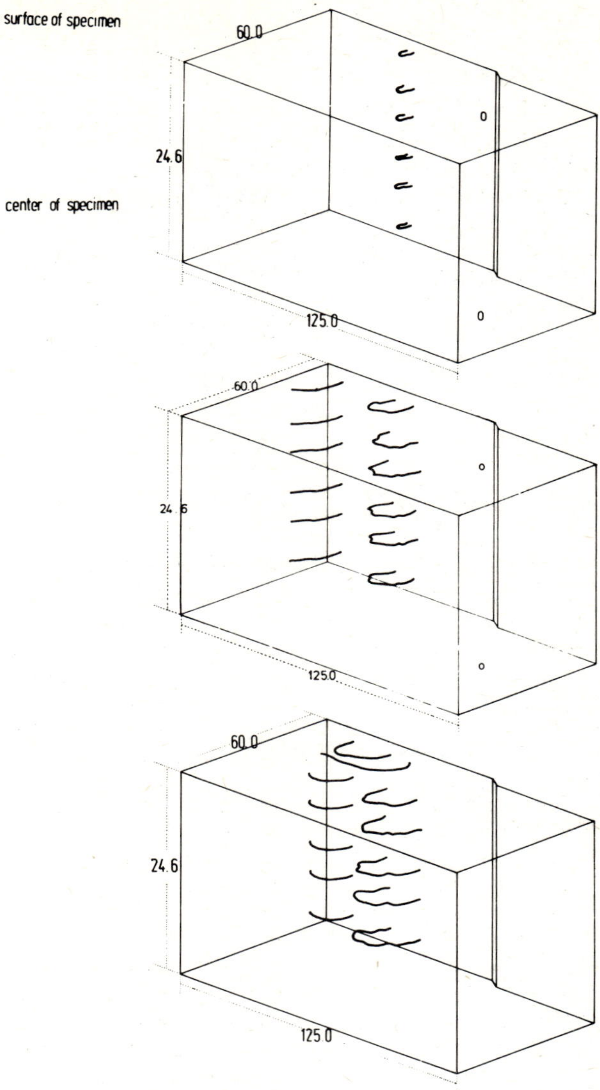

Figs. (11) to (13): Development of the plastic zone with increasing load

Now we want to discuss the development of the plastic zones. Figs. (11) to (13) show perspective graphs of the plastic zones in one quarter of the investigated CT-specimen. The load is increasing from fig. (11) to fig. (13). The z-axis scale is distorted relatively to the scale of the x-y plane in order to arrange a proper impression of the changes in plastic zone size from the mid-plane to the surface plane. The mid-plane is on bottom of the graphs and the surface plane is on top. The plastic zones are always plotted in the Gauss-point layers.

In fig. (11) plasticity has already occurred in the region of the crack tip. One can see that there is no significant difference between the size of the plastic zone from the mid-plane to the surface plane. In fig. (12) the pressure region of the specimen already started to plastify. Still the plastic zones show no variation in z-direction which is in contradiction to the two dimensional "dog-bone" model. Having in mind the results of the preceding paragraphs one can no more expect the plastic zones to verify this two dimensional prediction. In fig. (13) finally a difference between the mid-plane and the surface appears. We still have two plastic zones at the mid-plane whereas the plastic zones at the surface have joined one another thus establishing the state of general yield at the surface. In the vicinity of the specimen s centre of rotation one still can see a purely elastic region. This corresponds to the "valley" of small material strength in fig. (10).

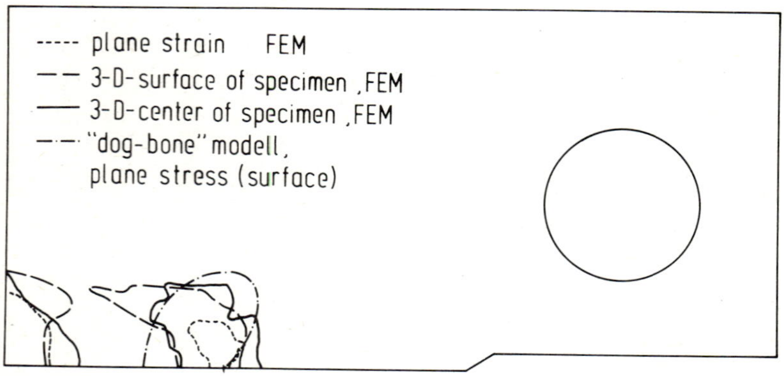

Fig. (14): Comparison of plastic zones calculated with different methods

In fig. (14) the plastic zones investigated by use of different methods are plotted shortly before reaching the state of general yield at the surface. The plastic zones of the 3-d FEM calculation at the surface are just on their way of joining one another. The plastic zone of the FEM plane strain calculation is even smaller than the 3-d FEM plastic zone at the mid-plane. For this circumstance two reasons are responsible. The "plane strain" situation in the mid-plane of the specimen is only an approximate one and the σ_y-stress in the mid-plane of the 3-d model must be higher than the averaged σ_y-stress of the plane strain model in order to satisfy the equilibrium condition. The plastic zone calculated by use of the "dog-bone" model for the plane stress situation at the surface clearly overestimates the size of the more realistic 3-d FEM plastic zone and produces a completely wrong shape.

Discussion

It was shown, that the 3-d calculation and the 2-d model produces different answers about the development of the plastic zones. The 2-d model neglects some of the most important effects thus becoming insufficient. The three dimensional shear stresses τ_{xz} and τ_{yz} may not be neglected because they are responsible for the kinematic link between surface and centre of the specimen. It is definitely true that we have an approximate plane strain situation in the centre of the specimen and that we have a plane stress situation on the surface but they may not be treated independently from one another. The decrease of σ_y, for which a qualitative mechanical proof was presented, is also of high importance.

Some further investigations, which have not been subject of this paper, show that the all over plastic zone decreases with increasing thickness.

Finally it is necessary to mention that the presented results are valid for a straight crack front. If the crack propagates in the centre of the specimen in form of the well-known "thumb-nail" the plastic zone at the surface becomes bigger than the plastic zone in the centre. This has of course nothing to do with the "dog-bone" model because the "thumb-nail" is a three dimensional effect itself and was never included in the two dimensional model. If the "dog-bone" model is used and a "thumb-nail" effect is present the correct answer about the plastic zone development might be carried out. But if such a case happens it only can happen because two errors are compensating one another just by chance.

References

1. ZIENKIEWICZ,O.C.-The Finite Element Method in Engineering Science, McGraw Hill,London (1971)

2. IRONS,M.B.-A Frontal Solution Proram for Finite Element Analysis.Int.J.Num.Meth.Eng.Sc., Vol.2,No.1,(1970)

3. NAYAK,G.C.,O.C.ZIENKIEWICZ,-Elasto-Plastic Stress Analysis.A Generalisation for Various Constitutive Relations Including Strain Softening. Int.J.Num.Meth.Eng.Sc., Vol.5,113-135 (1972)

THE EFFECT OF DILATING PLASTICITY ON SOME ELASTIC-PLASTIC STRESS AND STRAIN CONCENTRATION PROBLEMS RELEVENT TO FRACTURE

D.K. Brown,* J.W. Hancock,* R. Thomson* and D.M. Parks.†

SUMMARY Finite elastic-plastic deformation analysis of some circumferentially notched tensile specimens used to develope triaxial stress fields are presented. As a basis for comparison the results from convential elastic-plastic finite element analysis using the Prandtl - Reuss equations and the von Mises yield criterion are first given. These are compared with the results obtained using Gurson's dilating plastic constitutive equations with varying amounts of initial porosity. The potency of microvoid growth and plastic dilation in changing stress and strain states is discussed. An alternative approach to determining porosity using a discrete void and McClintock's model along with finite element stress and strain data, is described and resulting porosity values compared.

INTRODUCTION

The ability to deform plastically is an important property of structural materials both from the viewpoint of fabrication by a metal forming process and for their subsequent use in structures. In a structure plastic deformation limits the elastic stress concentrations that arise but only at the expense of replacing the elastic stress concentration by a plastic strain concentration. The limits to ductility therefore play in important role in forming processes and in structural failure, and in both cases the associated problems of analysis involve multi-axial states of stress and finite deformations. During finite plastic deformation many studies have established that voids nucleate at inclusions and second phase particles and subsequently grow in the plastic strain field. The growth of voids in a viscous material was studied initially

* University of Glasgow, Scotland.
† Massachusetts Institute of Technology, U.S.A.

by Berg [1] who showed than in plane strain the way in which the mean radius R_v of the void grew with remotely applied effective strain $\bar{\varepsilon}$ could be expressed as

$$\frac{dR_v}{d\bar{\varepsilon}} = \frac{3}{2} \left(\frac{\sigma_m}{\bar{\sigma}}\right) R_v$$

In this equation σ_m is the hydrostatic component of the remotely applied stress state and $\bar{\sigma}$ is the equivalent stress. The non-dimensional parameter $\left(\frac{\sigma_m}{\bar{\sigma}}\right)$ also features in the analysis of the growth of spherical voids during plastic deformation by Rice & Tracey [2] who obtained expressions of the form

$$\frac{dR_v}{d\bar{\varepsilon}} = 0.28 \exp \left(\frac{3}{2} \frac{\sigma_m}{\bar{\sigma}}\right) . R_v$$

The same non-dimensional stress state parameter $\left(\frac{\sigma_m}{\bar{\sigma}}\right)$ has been used by Hancock & Mackenzie [3], Mackenzie Hancock & Brown [4], and Hancock, Brown & Cowling [5] to present ductile failure loci in which the strain to initiate failure is presented as a function of the state of stress $(\sigma_m/\bar{\sigma})$. The essential result being the failure strain decreases markedly with increasing triaxiality in accord with the general form of the hole growth equations.

The void growth equations essentially apply to a single void in a infinite matrix but are widely applied to the growth of arrays of holes in homogeneous deformation. Homogeneous deformation is however terminated by flow localisation which is widely accepted to define the limits of macroscopic ductility. Void coalescence and failure occur within such localised bands with only negligible additional remote strains.

There are however a range of possible viewpoints on flow localisation and void coalescence. The work of Rudnicki & Rice [6] is motivated to the view that localisation may be viewed as a bifurcation in a homogeneous continuum flow field due to a change in the nature of the equations of equilibrium expressed in rate form. Although the localisation condition has often been associated with traction maxima, Rudnicki & Rice show that the flow localisation conditions depend sensitively on the nature of the constitutive equations & that deviations from normality allow localisation at a positive hardening rate. At a macroscopic level the presence of voids growing in the pre-localised deformation field can significantly alter the bulk constitutive equation of what can be regarded as a plastically dilating continuum which allows the possibility of strain softening. The general form of yield criteria and flow rules for such porous ductile aggregates has been studied by Gurson [7] & Tvergaard [8]. Using a range of simple cell models containing single spherical or cylindrical voids, Gurson [7] considered approximate velocity fields to determine the form of the upper bound yield functions for a non-hardening matrix. These approximate yield functions show the important effects of porosity and triaxiality in producing plastic dilation and strain softening. The yield condition proposed by Gurson takes the form

$$\phi = \left(\frac{\overline{\Sigma}}{\overline{\sigma}}\right)^2 + 2fq_1 \cosh\left(\frac{q_2}{2}\frac{\Sigma_m}{\overline{\sigma}}\right) - (1 + q_3 f^2) = 0$$

with the constants $q_1 = q_2 = q_3 = 1$. In this equation f is the volume fraction of holes, Σ_m and $\overline{\Sigma}$ are the mean and effective stresses of the porous aggregate and $\overline{\sigma}$ is the corresponding equivalent flow stress of the incompressible matrix. However, Tvergaard [8] on the basis of a numerical analysis of the growth of a double periodic array of circular cylindrical voids, introduced and adjusted the q constants as a result of finding better agreement with $q_1 = 1.5$, $q_2 = 1$ and $q_3 = q_1^2$ for a power hardening exponent $n = 0.1$ and also $q_1 = 1.8$ for $n = 0.2$. Nevertheless the pressure sensitive form of the yield locus and flow rules is retained and in view of the approximate nature of the yield functions Gurson's form has initially been used in the current work.

The work presented here was motivated to investigating the effects of plastic dilation and strain softening on the flow fields of axisymmetric notched tensile specimens shown in Figure 1. In these specimens the ratio of the external radius of the notch R, to the radius of the minimum cross section a controls the triaxiality $(\sigma_m/\overline{\sigma})$, and for this reason such specimens have been widely used to study the effect of stress state on ductile failure processes ([3], [4], [5], [9], [10].) However previous work has ignored the effect of porosity and concentrated on convential elastic-plastic analysis with plastic incompressibility. In this work the effect of porosity and void growth on the flow field has been a main theme, and results are presented for two initial porosities fo = 0.01 and fo = 0.1, the former being approximately representative of structural materials and the latter being used to assess the sensitivity of the results to fo, both being compared with the results of a convential finite deformation [11] elastic-plastic analysis.

Rather than attempt to present full field solutions the stresses have been presented in terms of the non-dimensional parameter $(\sigma_m/\overline{\sigma})$ for the porous aggregate in view of its important role in both the void growth and porous yield equations. This allows a more realistic assessment of the macroscopic stress and strain parameters at failure initiation and flow localisation, in the context of aggregate continuum theories of localisation such as those developed by Rudnicki

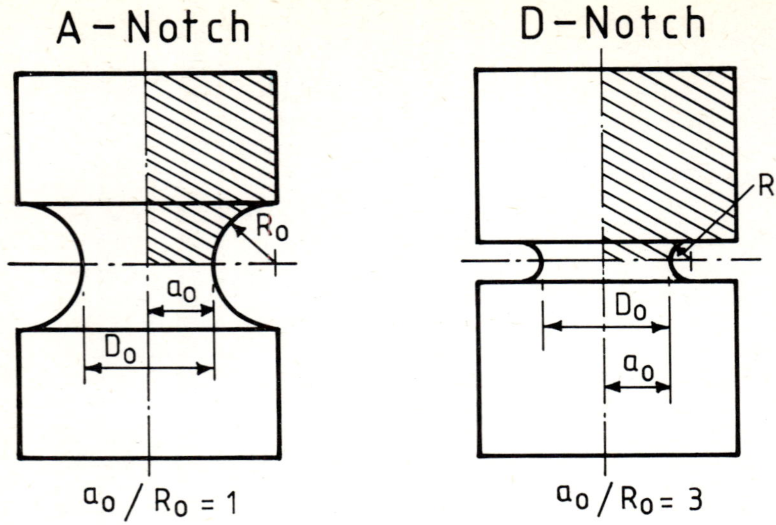
A-Notch
R₀
a₀
D₀
a₀ / R₀ = 1
D-Notch
R₀
D₀
a₀
a₀ / R₀ = 3

Fig. 1.

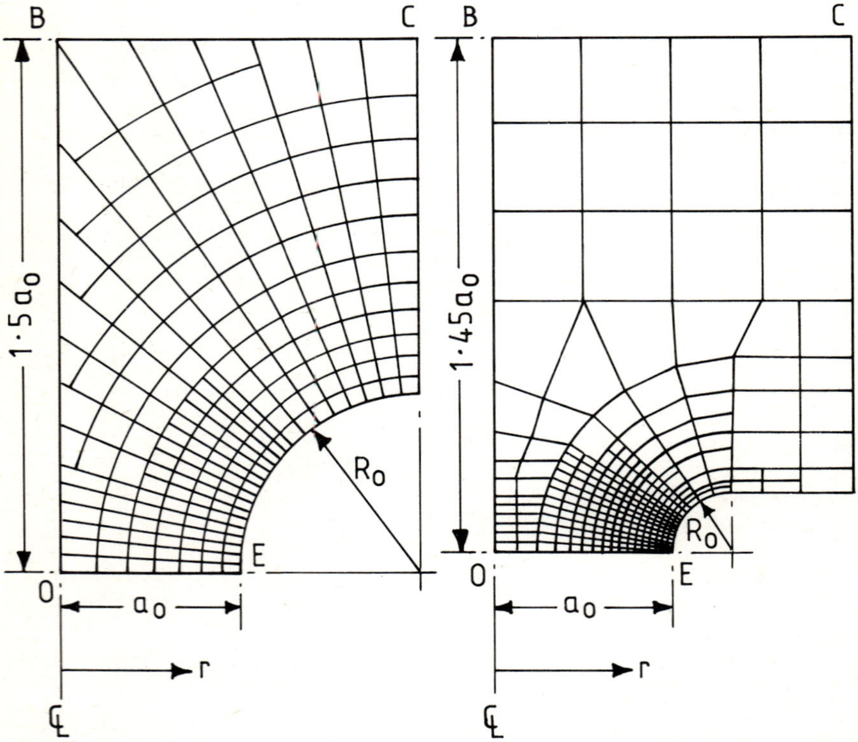
B
C
1·5a₀
R₀
E
O
a₀
r
℄
B
C
1·45a₀
R₀
O
E
a₀
r
℄

Fig. 2.

& Rice [6]. A central feature of the Rudnicki & Rice analysis is the effect of strain state, since such an effect was not found by Hancock, Brown and Cowling [5] who incline to the view that at least in the materials that they studied the local details of void geometry and interaction can not be ignored. Nevertheless even when failure is determined by the details of the local void-void interaction the remote homogeneous flow field in which such voids initially grow is that of a random array of holes which is appropriately described as dilating porous continuum. Thus whether or not the local details of void interaction are important in localisation failure, pressure sensitive yield and plastic flow analysis are appropriate for determining the failure conditions.

2. SOLUTIONS USING MISES YIELD SURFACE & ASSOCIATED FLOW RULES

The finite element program used for the solutions, was developed at Brown University and the elements used here were isoparametric quadrilaterals over which uniform dilation was enforced [12], The program has finite strain capability [13] and used tangent modulus determination of plastic strains [10] The power low stress-strain curve used in the program is given by

$$\left(\frac{\bar{\sigma}}{\sigma_y}\right)^{\frac{1}{n}} - \left(\frac{\bar{\sigma}}{\sigma_y}\right) = \frac{3G\bar{\varepsilon}^p}{\sigma_y} \quad \text{for } \frac{\bar{\sigma}}{\sigma_y} \geq 1$$

-the choice of index n depending on the material response. For all results, to enable easier comparison a value of n = 0.2 was used, this being characteristic of steels such as 50D and EN8.

Two particular axisymmetric geometries were considered as shown in Figure 1, these being the notched bar geometries designated A- and D- notch with 203 and 282 elements and 240 and 323 nodes respectively. The discretazations of the symmetrical quarter are shown in Figure 2. The grid was 'loaded' by applying a uniform displacement along boundary BC, the axial displacement to cause initial yielding at E being denoted by a loading of 100%.

Distributions of $\sigma_m/\bar{\sigma}$ and $\bar{\varepsilon}^p$ at 4 sample loadings (%) are shown in Figure 3. The results from Figure 3 are used in section 4.

Full results giving distributions of stresses and strains for the A- notch (perfectly plastic, n = 0.1, 0.2) and D-notch (n = 0.1, 0.2) have also been compiled and are shortly to be published, in addition to the results for another notch (n = 0.1, and 0.2), which has the same a_o/R_o ratio of the A- notch but with a flank angle of 90°.

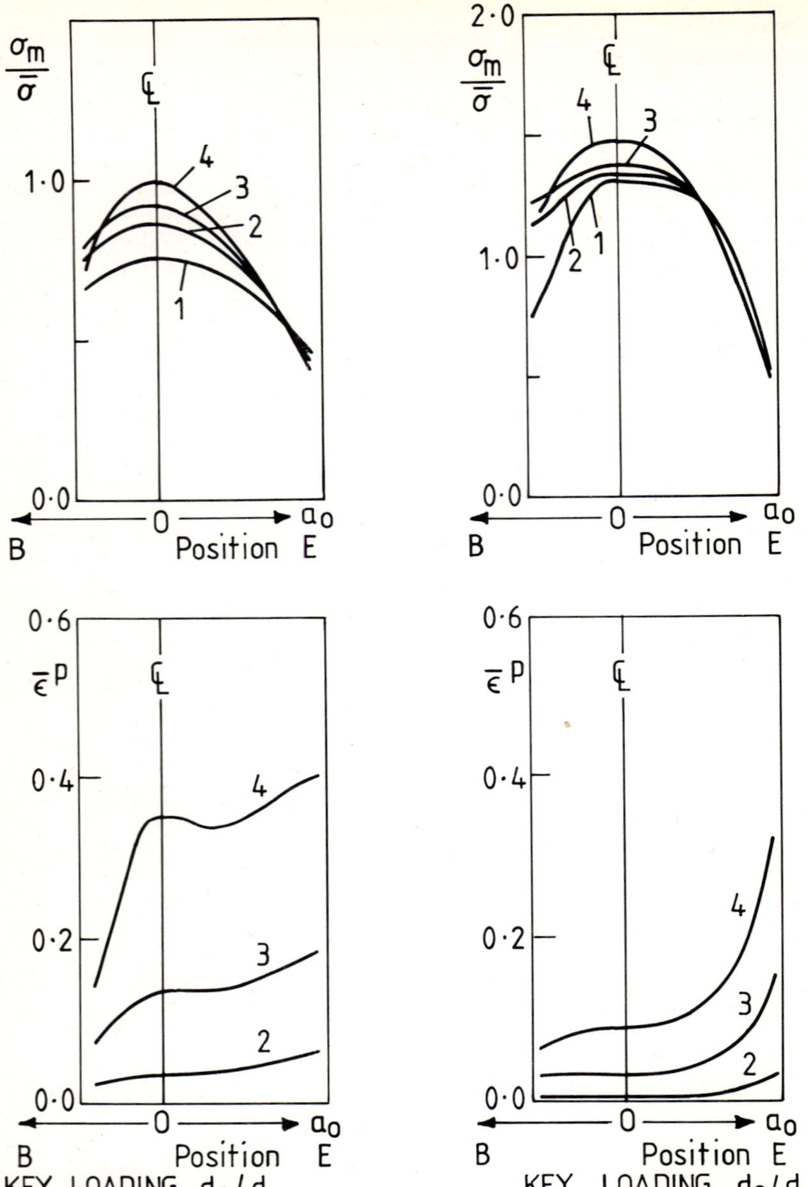

KEY	LOADING	d_0/d
1.	100%	1·001
2.	390%	1·027
3.	830%	1·087
4.	1550%	1·212

A–Notch

KEY	LOADING	d_0/d
1.	100%	1·000
2.	1105%	1·006
3.	4465%	1·031
4.	8065%	1·067

D–Notch

Fig. 3.

3. SOLUTIONS USING GURSON's YIELD SURFACE & FLOW RULES FOR POROUS DUCTILE MEDIA

In order to examine the effects of void growth & the development of porosity on flow fields which feature tensile hydrostatic stress component the two notch geometries of Figure 1 have been analysed using Gurson's approximate yield function based on the spherically symmetric deformation of a voided cell ($q_1 = q_2 = q_3 = 1$). The results of adjusting the q constants in accord with Tvergaard's suggestion is to be reported later.

Although hole nucleation is a significant metallurgical feature it introduces a further variable so that for simplicity this work has presumed the pre-existance of holes, despite the program having a capability for dealing with nucleation.

The analysis of porous plasticity has been terminated at the onset of a negative strain hardening rate in the central elements of the notch. The salient features of the porous flow fields are summarised in Figures 4 and 5. Comparisons between these and the incompressible plastic results maybe made in several ways but here comparison is in terms of the diameter of the minimum cross section D normalised with respect to its initial value Do. Comparisons in terms of the axial displacement on the loaded boundary BC would produce different numerical results but exhibit similar trends. Figures 4a and 5a show the axial load normalised with respect to the load to cause initial yield (designated 100%) as a function of (Do/D). In figures 4b, c, d and 5b, c, d. the development of $\Sigma_m/\bar{\Sigma}$ and porosity f, at the ℄O and $\bar{\varepsilon}^p$ at the ℄O and surface E, are shown for the two notches. Figure 6 shows the development of axial load with axial displacement loading (%) of the upper boundary BC.

4. DISCUSSION

A central role of void growth is to reduce the aggregate flow stress of a porous medium. This is reflected in the flow stress shown in Tables 1 and 2 where after comparable strains (Do/D) the flow stress $\bar{\Sigma}$ (normalised with respect to the initial matrix yield stress σ_y), is markedly reduced by increasing initial porosity, an effect which becomes more pronounced in highly triaxial states of stress due to the exponential amplification of the stress state in the hole growth equations. Similarly the mean stress Σm, again normalised with respect to the initial yield stress shows the same trends so that void growth is a potent way of relieving the hydrostatic stress and lowering all the stress levels. See Tables 1 and 2.

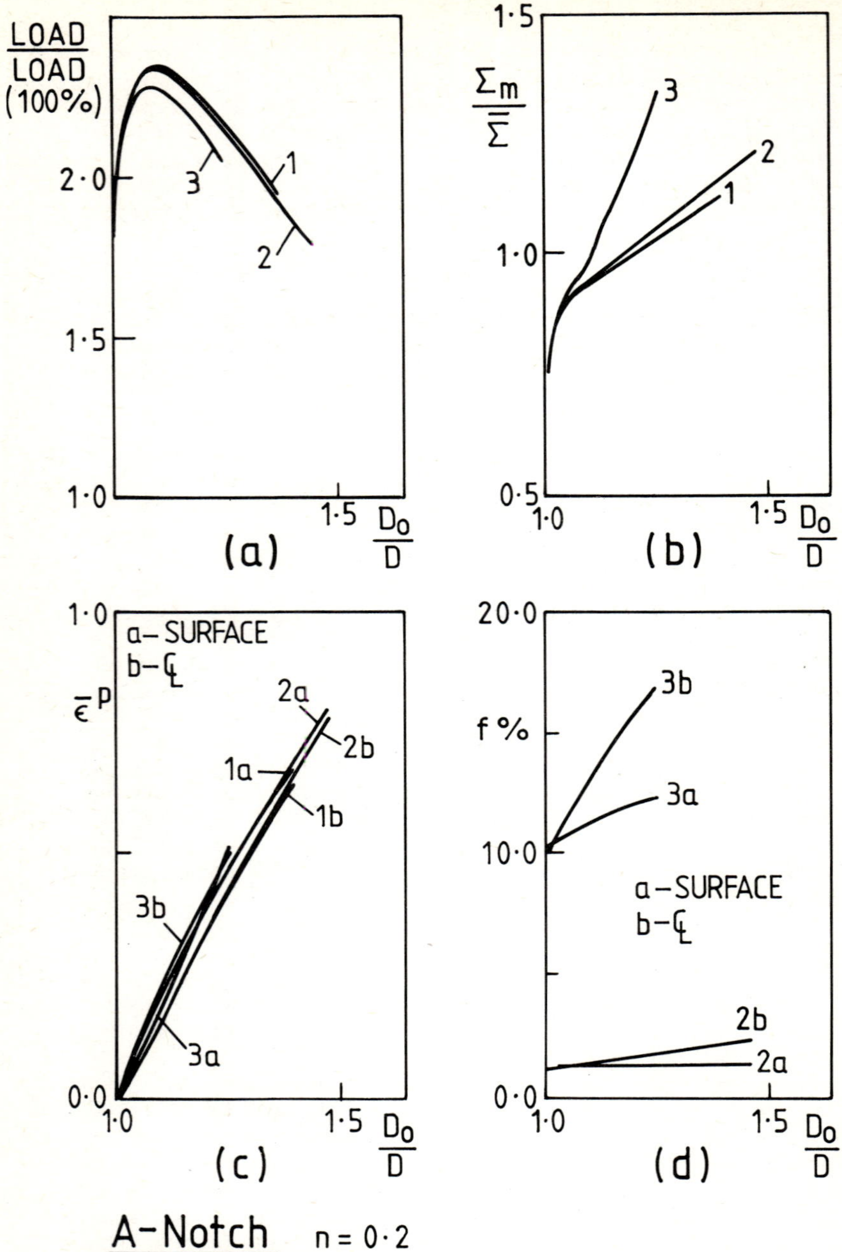

A-Notch $n = 0 \cdot 2$

KEY :- 1. NON-POROUS
2. f_o = 1 %
3. f_o = 10 %

Fig. 4.

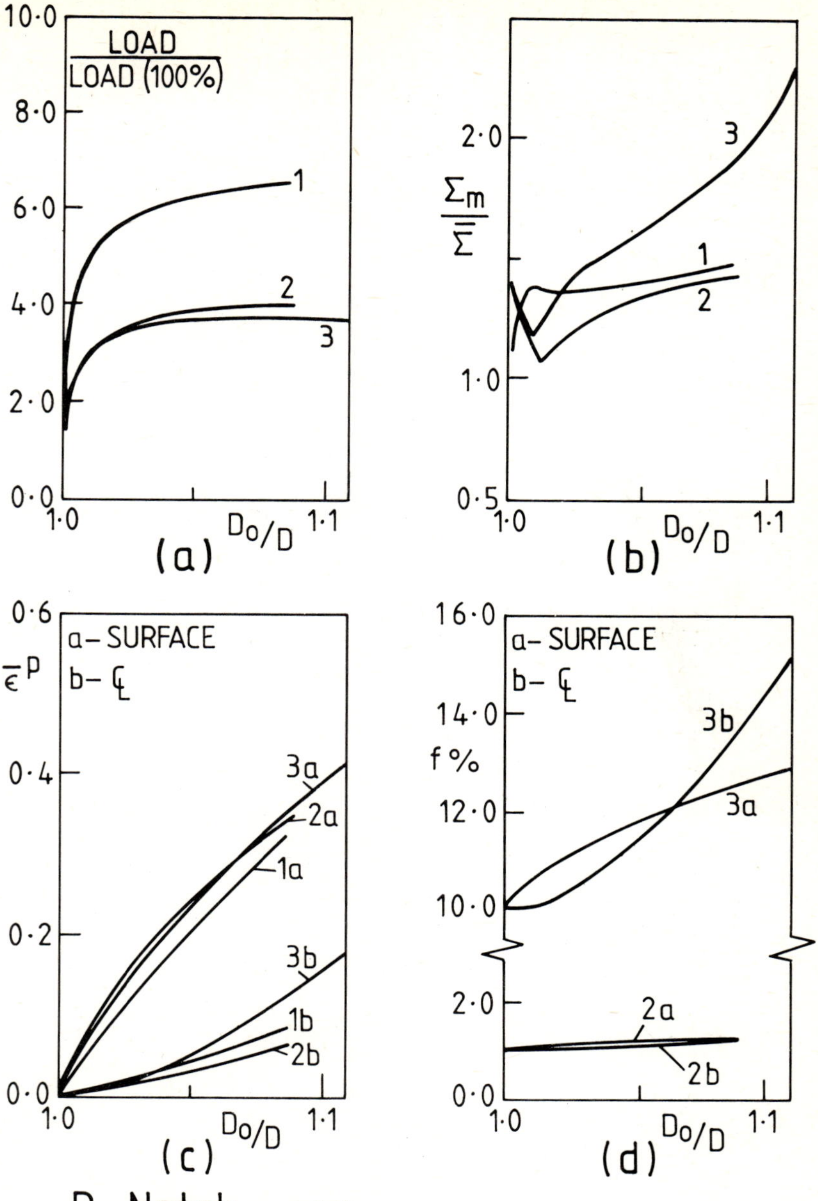

D–Notch $n = 0{\cdot}2$

KEY :– 1. NON–POROUS
2. f_o = 1%
3. f_o = 10%

Fig. 5.

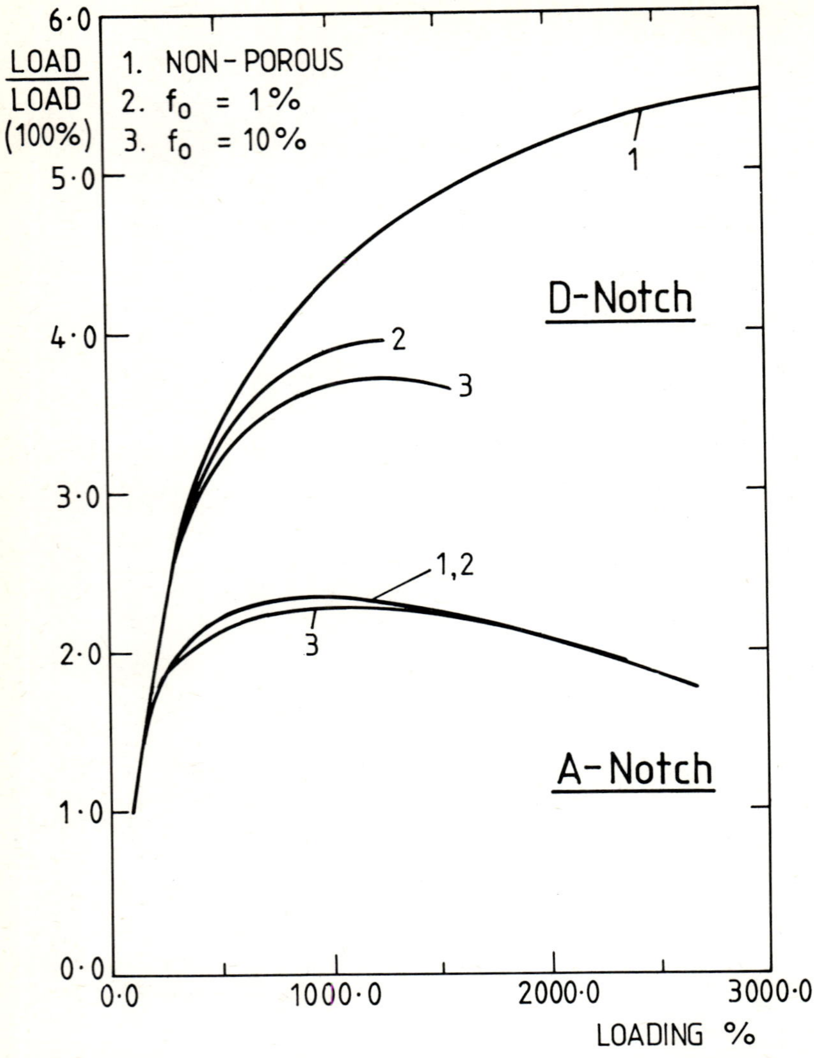

Graphs of Load Against Axial Displacement for A-and D-Notches and Different Porosities

Fig. 6.

Table 1 stresses at ℄0 of D- notch at $D_o/D = 1.06$

Stress / Solution	Σ_m/σ_y	$\overline{\Sigma}/\sigma_y$	$\Sigma_m/\overline{\Sigma}$
Non-porous	3.27	2.30	1.42
1% initial porosity	1.87	1.41	1.33
10% initial porosity	1.67	1.0	1.67

Table 2 Stresses at ℄0 of A- notch at $D_o/D = 1.15$

Stress / Solution	Σ_m/σ_y	$\overline{\Sigma}/\sigma_y$	$\Sigma_m/\overline{\Sigma}$
Non-porous	1.91	1.99	0.96
1% initial porosity	1.87	1.92	0.97
10% initial porosity	1.60	1.48	1.08

The marked effect observed on the stress levels at the centre of the notches is however less pronounced in the less triaxial stress states observed at the surface as expected. This reduction in stress levels is of course reflected in the axial load bearing capacity of the notches shown in Figures 4a and 5a.

Turning now to the non-dimensional grouping ($\Sigma_m/\overline{\Sigma}$) which appears in the hole growth equations and the ductile loci it is interesting to note that realistic initial levels of porosity ($\approx$.01) do not markedly effect this parameter, so that sensible estimates may be made from incompressible plastic analysis. However as void growth has a larger effect in relieving Σ_m than in reducing $\overline{\Sigma}$, at higher levels of porosity $\Sigma_m/\overline{\Sigma}$ is elevated over that of the incompressible plastic solution as shown in Figures 4b and 5b.

Although the flow rules allow a plastic dilation this does not appear to have a marked effect on the development of the equivalent plastic strain $\overline{\varepsilon}^p$ so that in Figures 4c and 5c the curves are very similar. The main effect of the notch geometry on the incompressible plastic strain distribution is retained, with the sharper notch exhibiting a marked strain gradient from the high strain concentrations at the notch surface and the late development of plasticity in the central elements. In contrast the blunt notch shows an almost uniform flow field independent of plastic dilation, so that in both cases good approximations to the matrix strain may be taken from an incompressible solution at quite high levels of porosity & void growth. The void growth which produces the corresponding plastic dilation is shown in

Figures 4d and 5d and as required by the void growth equation this occurs most rapidly in triaxial states of stress as can be seen by comparing the development of f in the low triaxiality at the surface of the notch with the high triaxiality of the central regions. This feature is also shown in comparing the plastic dilation in the centre of the A- & D- notches although in the latter the comparison is complicated by the late development of plasticity & the marked strain gradient. Figure 7, shows the contours of porosity in the D- notch with fo = 10% at Do/D = 1.111 and loading 1570%.

It is interesting to compare the porosity & void growth calculated through a dilating plastic approach with that obtained using the strain increments and stress states obtaining to the incompressible plastic analysis. As usual in such calculations [15] the stresses and strains at the point of interest are applied in the void growth equations as if they were remotely applied to a void and that the growing void does not affect them in any way. Here McClintock's [16] equations are used for an ellipse with semi-axes c, d. The hole growth equations are presented in terms of the mean radius $R_V = \frac{1}{2}(c + d)$ and the eccentricity $M_V = (\frac{c - d}{c + d})$. During an increment of plastic strain the initial values of R_V & M_V, denoted R_{vo} & M_{vo} change according to

$$R_v = R_{vo} \exp\{\tfrac{1}{2} \alpha\beta + \tfrac{1}{4} d\bar{\varepsilon}^p\}$$

$$M_v = \frac{1}{\gamma} + \{M_{vo} - \frac{1}{\gamma}\}\exp(-\alpha\beta)\}$$

where

$$\alpha = \sqrt{3}\, d\bar{\varepsilon}^p / (1 - m), \quad \beta = \sinh\{\frac{\sqrt{3}}{2} \cdot (1 - m)\gamma\}$$

$$\gamma = 2^{\sigma}m/\bar{\sigma} + \frac{1}{3} \quad \text{and } m \text{ from } \bar{\sigma} = \bar{\sigma}_o (\bar{\varepsilon}^p)^m$$

Initially a discrete spherical void of unit radius is assumed to be located at the ℄0, surrounded by a certain volume of matrix corresponding to the initial porosity. The record of development of $\sigma_m/\bar{\sigma}$ and $\bar{\varepsilon}^p$ at that location for A- and D- notches with $n = 0.2$ and no porosity is obtained from graphs such as Figure 3 at different load steps. The void is then subjected to increments of $d\bar{\varepsilon}^p$ at the corresponding stress state $\sigma_m/\bar{\sigma}$. The initial sphere changes into an ellipsoid of increasing eccentricity and, assuming there is no matrix volume change, an new porosity can be calculated. Alternatively the two dimensionally equivalent cylindrical void growing into a prismatic elliptical void can be considered. The results of both approaches and a comparison with the finite element results are shown below:-

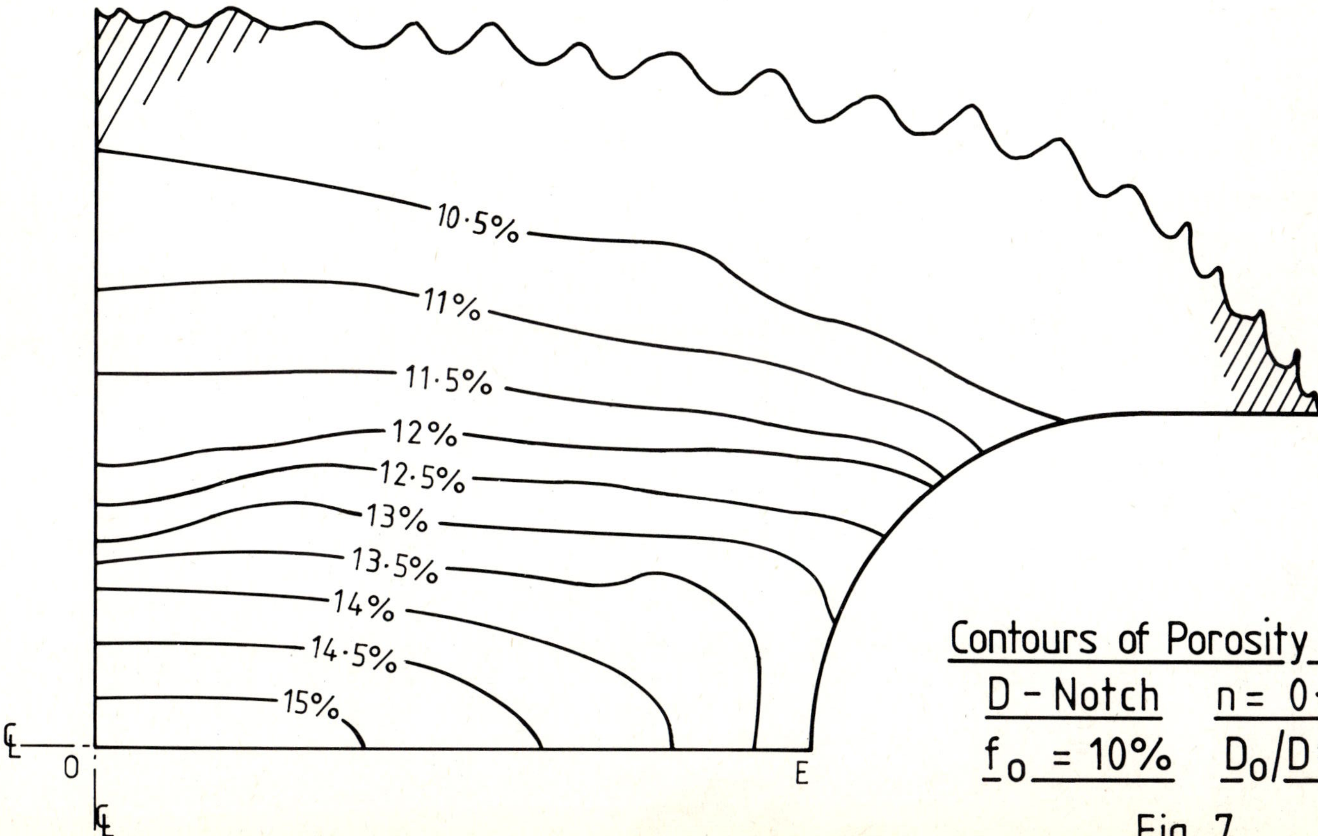

Contours of Porosity f

D - Notch $n = 0{\cdot}2$

$f_o = 10\%$ $D_o/D = 1{\cdot}111$

Fig. 7.

Table 3

Notch	Initial Porosity	Solution	$\bar{\varepsilon}^p_{ave}$	f_ε %	
A	1%	(1) (2)	23.1% 23%	1.36 2.75	 (2.36)
A	10%	(1) (2)	19.33% 19.0%	12.94 19.8	 (18.1)
D	1%	(1) (2)	16.8% 17.0%	1.21 2.54	 (2.08)
D	10%	(1) (2)	21% 21%	15.37 29.8	 (23.66)

where $\bar{\varepsilon}^p_{ave} = 2 \log_e {}^{Do}/D$, (1) F.E. Solution - porous

(2) Ellipsoid solution (Prismatic Ellipse Solution)

Due to the approximate nature of the yield criteria & flow rules for dilating plasticity & in particular their dependence on the indealised symmetrical geometry of the voided cells considered, the discrepancy between discrete void and dilating plasticity porosities should not be overemphasised. The void growth equations must be expected to give the better representation of the geometry change of a discrete void at low volume fractions but cannot show the effect of dilation & strain softening on the complete flow fields. This can only be achieved by a dilating plastic approach. The best compromise at low volume fractions is thus to use the void growth equations in conjunction with the stresses & strains from dilating plastic analysis which essentially forms the remote flow field for a single void.

Analysis on this basis provides information both about detailed void geometry and the aggregate properties at failure.

5. CONCLUSIONS

The dilating plasticity analysis has shown that even small initial porosities ($\approx .01$) are potent in relieving the hydrostatic stress component and causing strain softening of the deviatoric components. However the non-dimensional grouping ($^{\sigma m}/_{\bar{\sigma}}$) which appears in both the void growth & dilating yield equations is relatively insensitive to these effects at the porosities representative of the inclusion content of structural materials. The use of the void growth equations coupled with incompressible plastic analysis is thus justified in these cases, particularly as the matrix strain distributions are insensitive to dilating plasticity.

ACKNOWLEDGEMENTS

Most of the work was carried out under the Science Research Council's Grant GR/B/21843 within the Marine Technology Programme. The SRC sponsored Prof Parks as a Senior Visiting Research Fellow in the University of Glasgow during the summer of 1979.

Gratitude is also expressed to Prof. J.R. Rice for use of the finite element program.

REFERENCES.

1. BERG, C.A. - The motion of cracks in plane viscous deformation. Proc. 4th U.S. Nat. Cong. Appl. Mech., Vol. 2, P. 885, 1962.

2. RICE, J.R. and TRACEY D.M. - On the ductile enlargement of voids in triaxial stress fields. J. Mech. Phys. Solids., Vol. 17, P. 201, 1969.

3. HANCOCK, J.W. and MACKENZIE A.C. - On the mechanisms of ductile failure in high strength steels subject to multi-axial states of stress. J. Mech. Phys. Solids., Vol. 24 P. 147, 1976.

4. MACKENZIE, A.C. HANCOCK, J.W. and BROWN, D.K. - On the influence of state of stress on ductile failure initiation in high strength steels. Engng. Fracture Mech., Vol. 9 P. 167, 1976.

5. HANCOCK, J.W. BROWN D.K. and COWLING, M.J. - On the role of strain and stress state in ductile failure. To be published.

6. RUDNICKI, J.W. and RICE, J.R. - Conditions for flow localization of deformation in pressure sensitive dilatant materials. J. Mech. Phys. Solids, Vol. 23, P. 371, 1975.

7. GURSON, A.L. - Continium theory of ductile rupture by void nucleation and growth: Part 1 - Yield criteria and flow rules for porous ductile media. J. Engng. Mat. Tech., Trans. A.S.M.E., Vol. 99, P.2, 1977

8. TVERGAARD, V. - Influence of voids on shear band instabilities under plane strain conditions. To be published.

9. McCLINTOCK, F.A. - Plasticity aspects of fracture. Fracture: An Advanced Treatise. Ed. Vol. III, P.47, Academic Press, New York, 1971.

10. MACKENZIE, A.C. - Crack initiation in circumferentially notched bars in Torsion. S.M. Thesis, Dept. of Mech. Engng., M.I.T, 1958.

11. BROWN D.K. and McMEEKING, R.M. - A finite element analysis of a circumferentially notched tensile specimen. Fracture 1977, Vol. 3, P. 507, University of Waterloo Press, Waterloo, 1977.

12. NAGTEGAAL, J.C. PARKS, D.M. and RICE, J.R. - On numerically accurate finite element solutions in the fully plastic range. Comp. Meth. in Appl. Mech. and Engng., Vol.4 P.153 1974.

13. McMEEKING, R.M. and RICE, J.R. - Finite element formulations for problems of large elastic - plastic deformation. Int. J. Solids Struc., Vol. 11, P.601, 1975.

14. RICE, J.R. and TRACEY, D.M. - Computational fracture mechanics. Numerical and Computer Methods in Structural Mechanics., Eds. S.J. Fenves et al., Academic Press New York, 1973.

15. RICE, J.R. and JOHNSON, M.A. - The role of large crack tip geometry changes in plane strain fracture. Inelastic Behaviour of Solids., (Eds., M.F. Kanninen et al), McGraw - Hill, New York, 1970.

16. McCLINTOCK, F.A. - A criterion for ductile fracture by the growth of holes. J. Appl. Mech. Trans. A.S.M.E., Vol. 35, P.363, 1968.

ELASTIC-PLASTIC ANALYSIS OF CRACK OPENING AND CRACK CLOSURE [1]

by

D. BERGEZ[2] ; Q.S. NGUYEN[3] ; D. RADENKOVIC[3]

SUMMARY

A short review of some papers concerning elastic-plastic analysis of crack growth shows that near tip solutions for particular situations are now well known. The analyses indicate that the crack opening and the extent of the plastic zone depend on loading conditions. The energy dissipated during cracking corresponds to a volumetric distribution with a zero crack driving force.

The appropriateness of the classical elastic-plastic model to give relevant informations about crack growth criteria, without taking into account the particular behaviour of the process zone is discussed.

INTRODUCTION

The use of FEM analysis in the study of elastic-plastic fracture started about ten years ago. A considerable number of papers concerning this subject from different points of view have been produced during this short lapse of time. Many different geometries have been analysed (three or four point bent specimen, CT specimen, wide plate with central crack, etc) ; in most of the cases plane, resp. anti-plane, problem of a crack growing in its own plane was treated. A systematic comparison of different results obtained is hardly possible. Only selected data, concerning one or the other hypothesis under discussion, are given in the different papers, as FEM analysis

---

1 Study supported partly by Research Division of Elf-Aquitaine.

2 S.N.E.A. (P), 64001 Pau, France.

3 LABORATOIRE DE MECANIQUE DES SOLIDES, ECOLE POLYTECHNIQUE, 91128 Palaiseau, France.

yield too many useful informations to be easily presented.

The following short review of some papers (resulting partly from our own numerical experiences) tends to recognize that near tip solutions for many particular situations are now well understood. At the same time it is shown that crack tip opening is not directly linked to the amount of energy supplied to a restricted process zone (as it happens in the elastic case) but that it depends on the evolution of the whole plastic zone, i.e. on loading conditions. From the energy considerations it is seen that a pure elastic-plastic analysis does not allow us to distinguish the seperation work from the energy dissipated in the plastic deformation, which depends as a whole on the geometry of the cracked body.

1.- ELASTIC-PLASTIC SOLUTIONS FOR STATIONARY AND PROPAGATING CRACKS

Many different numerical solutions of elastic-plastic crack problems, partly backed up by the analysis of the near tip asymptotic behaviour, concern, roughly speaking, following situations :

i - stationary crack response under increasing loading ;

ii- quasistatic steady crack growth ;

iii- transient crack growth simulated by node release ;

iv- fatigue crack growth simulation.

1.1- Stationary crack response

Closed form solution for the elastic-plastic near field for a stationary crack under anti-plane shear was given by Hult & Mc Clintock [8]. Classical results for the plane problem are due to Rice & Rosengren [15] and to Hutchinson [9].

In the case of isotropic hardening of the type $\sigma = \sigma_o\ (\gamma/\gamma_o)^N$, where σ is shearing stress (yield point being σ_o) and γ the corresponding plastic strain, the displacement discontinuity $[u]$ near the crack tip is given by

$$[u] \sim r^{\frac{N}{1+N}}$$

extremal values of N being N = 0 for perfectly plastic and N = 1 for perfectly elastic material.

For a perfectly plastic material the FEM calculations reported by different authors are in good agreement with Prandtl's scheme, where the order of the strain singularity is $\gamma \sim r^{-1}$. The so called special finite elements embody sometimes this type of singularity.

As for the physical relevance of these results in crack mechanics, the problem of the appropriate constitutive equations to be used in the highly strained zone has been raised by Rice [17] . Also, the role of the level of triaxiality in the process zone is not clearly assessed although it has been often mentionned.

The influence of the far off conditions on the near crack field has been discussed by Mc Meeking & Parks [11], who take into account the finite deformation and a particular initial geometry (blunted crack).

It is obvious, but often forgotten in the applications, that the crack opening depends on the presence of residual stresses which can not be ignored in the evaluation of the toughness near the welds (in practical situations initial cracks occur most often in these regions). By generating stochastic distribution of residual stresses, as their real distribution is usually unknown, Bergez [3] has examined their influence on the crack opening.

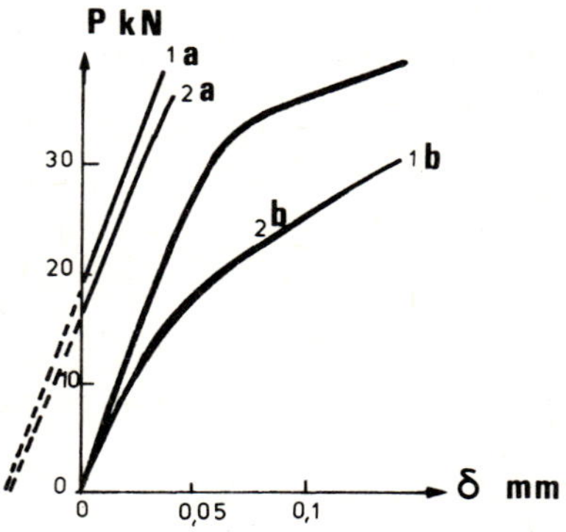

Fig.1- Influence of residual stresses on the crack opening displacements.

The evolution of the near tip displacement (node next to the crack tip) with the increasing loading in a CT specimen is shown on the fig. 1 for four different residual stress distributions, either tensile or compressive on the line ahead of the crack ; the tick line gives the corresponding history history without residual stresses.

Let us finally mention one of the first attempts to analyse the three-dimensional stress-distribution in an edge cracked bent specimen, due to Bui & Nguyen [5] . The expected difference between the "plane strain" in the central plane and the "plane stress" (with markedly larger plastic zone) in the lateral planes was found.

1.2- Quasistatic steady crack growth

The crack tip opening of a growing crack is considerably reduced in comparison with the response of a stationary crack, due to the continuous plastic deformation of the surrounding material.

First solution confirming this effects, due to Chitaley & Mc Clintock [6], concerns steady crack growth under anti-plane shear. Plane problem was analysed by Slepyan [20] who has demonstrated the presence of three zones around the crack tip (plastic loading, elastic unloading, reloading due to residual plastic deformation). Analogous solutions have been discussed by Amazigo & Hutchinson [1] for hardening materials and by Rice, Drugan & Sham [16] for the case of ideal plasticity. They find crack tip opening displacement in the form

$$[u] \sim r\, C_1(r) - C_2\, r \log r$$

where $C_2 > 0$ and $C_1(r)$ is a regular function. If $r \to 0$ the opening is smaller than in the elastic case, but at $r = 0$ the tangent remains vertical.

A recent numerical approach concerning steady crack growth is due to Nguyen & Rahimian [14] ; the steady response was obtained as a solution of an equivalent stationary problem with non-local constitutive equations. Their results are in good accordance with the forementioned theoretical studies.

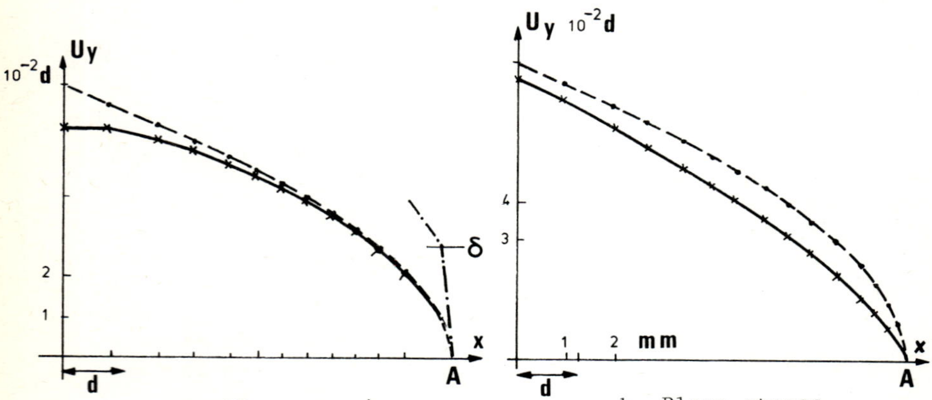

a. Plane strain b. Plane stress

Fig.2- Steady crack growth : opening profiles
Elastic analysis : - - - - - -
Plastic analysis : Steady growth ———
Stationary —·—·—·—

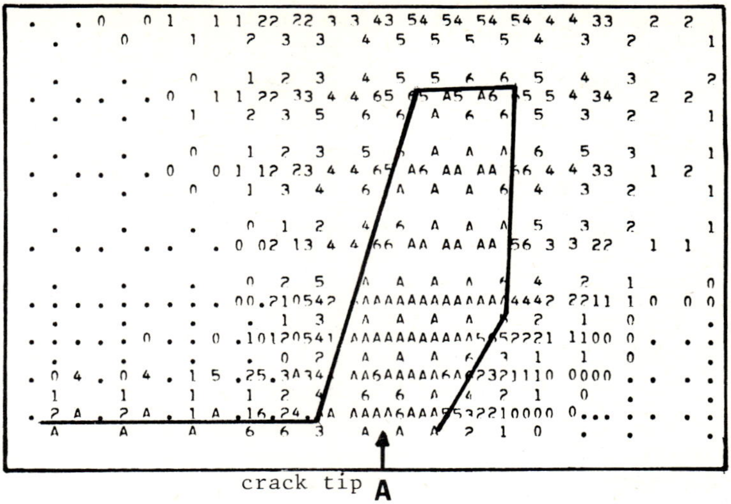

a. Plane strain

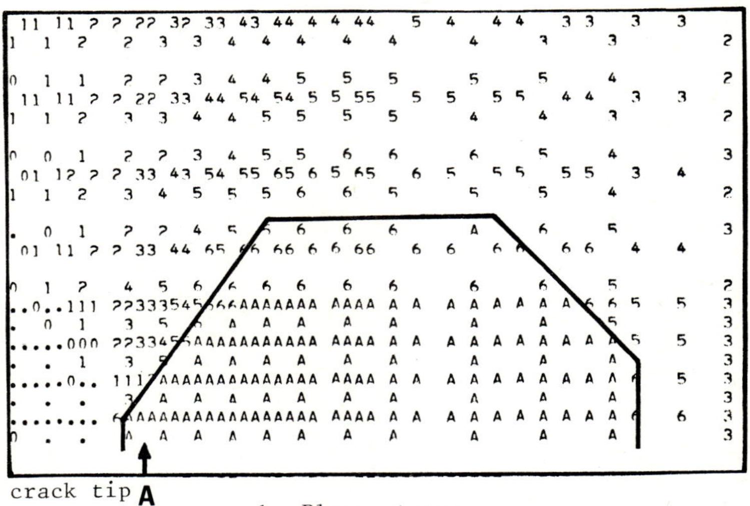

b. Plane stress

Fig.3- Steady crack growth : spread of the plastic zone

The crack opening profiles are shown in the fig. 2 ; the steady growth opening is smaller than in the elastic case for plane strain and still smaller for plane stress ; on the contrary, the stationary crack opening is much larger than in the elastic case.

Figure 3 showsthe spread of the plastic zone for plane strain and plane stress. In the first case a wake, corresponding to secondary plastic zone along the crack tip , can be noticed ; the numerical treatment is not accurate enough to distinguish the intermediate elastic zone. Plastic zone is much wider (practicaly circular) for plane stress, which corresponds to the reduced intensity of the opening ; in this case no reloading plastic wake is apparent although it probably exists.

1.3- Crack growth simulated by node release

Many numerical studies are concerned with crack growth simulated by node release ; starting with the work of Anderson [2] a number of papers appear regularly, cf. e.g. in *FRACTURE* (ICF 4, June 1977) ; Kfouri & Rice [10], Rousselier [18], etc, or in *ELASTIC-PLASTIC FRACTURE* (ASTM, STP 668, Nov. 1977) : Sorensen [21] , d'Escatha & Devaux [7] and others.

Different specimen geometries are treated and different node release conditions are adopted. The obtained crack opening profiles are more or less regular, depending on the mesh size and on the nodal release technique. A systematic comparison of results is not possible, but a general trend is clear. After initial blunting corresponding to a stationary crack, the crack tip penetrates into the plastic zone forming a new profile with reduced opening which might tend to the form of a steady growing crack.

Figure 4 shows a typical result obtained by Rousselier [19] . The solutions are influenced by the idealization of the process zone (resp. node release conditions) to such extent that careful comparison with experimental results (if available !) is not always a simple matter [19] .

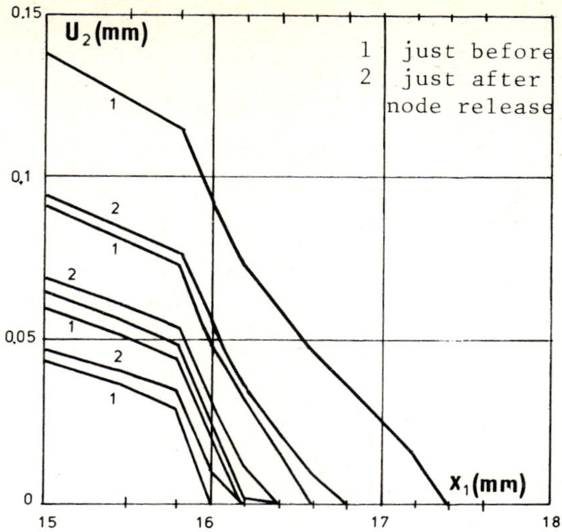

Fig.4- Evolution of crack opening by nodal release

1.4- Fatigue crack growth simulation

The reduction of the crack opening of a propagating crack plays an important role when the phenomenon of closure of fatigue cracks is studied.

In fact, the crack opening in a tension cycled specimen at F max is of the some order of magnitude as in steady growth i.e. less than in the elastic case. The cycling response being practicaly elastic, a partial closure of the crack lips occurs before F min is reached. This effect is particularly pronounced in plane stress conditions.

Fatigue cracks grow progresssively, at a very small rate (e.g. 10^{-3} mm per cycle), so that the numerical simulation, in order to be practicaly possible, entails a number of assumptions : one numerical release step (say 10^{-1} mm) summarizes hundreds of cycles during which, in principle, a steady regime of residual stresses is established.

Interesting numerical studies are due to Newmann [12]. Some results obtained by Bozec [4] are presented here.

Crack opening profiles during the propagation are shown in the fig. 5.

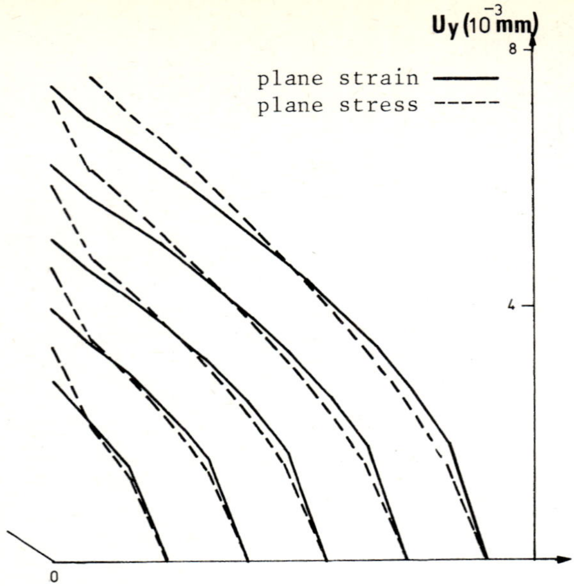

Fig.5- Fatigue crack growth

The difference in the progressive crack closure with decreasing loading between plane stress (a) and plane strain (b) conditions can be seen in the fig. 6.

2.- ENERGY CONSIDERATIONS

In the classical LEFM a "crack driving force" is easily defined, corresponding either to the energy release rate G or to the J integral. It was natural to look after equivalent energy concepts in the case of elastic-plastic fracture. Unfortunately such simple unambiguous notions do not exist here.

A thorough energy analysis (including dynamic and thermal effects) is due to Q.S. Nguyen [13],[14] ; a simplified version of this analysis is presented in the following.

2.1- Energy balance during crack growth

For simplicity only two-dimensional cases are considered in the following. Let us denote by Ω the domain occupied by the cross-section of the body, its boundary by $\partial\Omega$ and the actual length of the crack by a. The crack growth rate $\dot{a}=da/dt$ corresponds to the action of external forces F(t), variable in time. Let small displacements be denoted by u, the corresponding strains not necessarily elastic by ε and the stresses by σ.

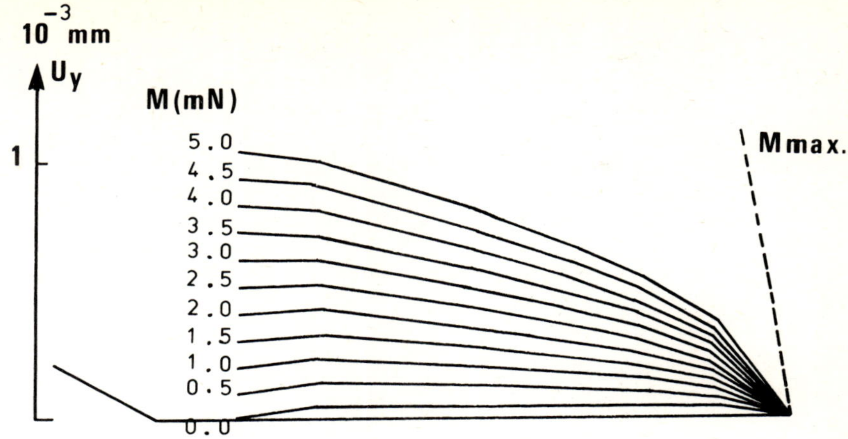

a. Plane strain

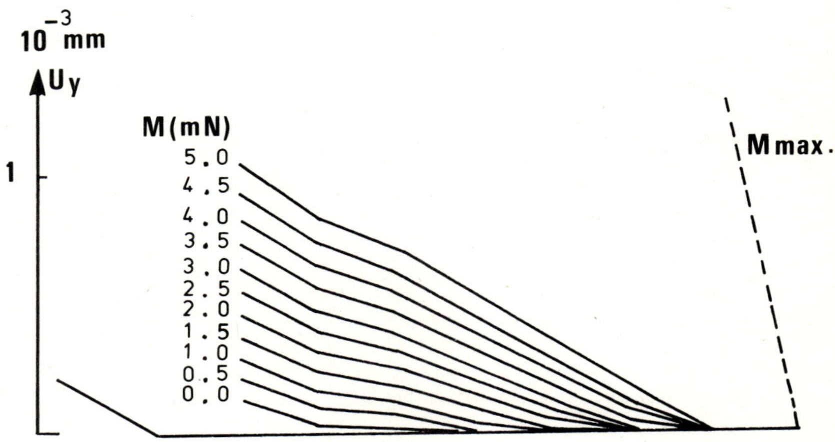

b. Plane stress

Fig.6- Crack closure profiles with decreasing loading

The rate of dissipation of the energy is given by

$$D = P_e - d/dt\ \phi \qquad (1)$$

where $P_e = \int_{\partial\Omega} F\dot{u}dS$ represents the rate of work of external forces and $\phi = \int_{\Omega} Wd\Omega$ the recoverable energy. In order to calculate the rate $d/dt\ \phi$, it is necessary to isolate eventual singularities at the crack tip by a small domain Σ moving with the crack tip, whose external normal is n and the maximal dimension R, and to let subsequently R tend to zero.

The principle of virtual work gives immediately

$$\int_{\partial\Omega} F\dot{u}dS + \int_{\partial\Sigma} \sigma n\dot{u}dS = \int_{\Omega-\Sigma} \sigma\dot{\varepsilon}d\Omega$$

Remembering that the specific dissipation is given by $\mathcal{D} = \sigma\ \dot{\varepsilon} - \dot{W}$ we find

$$P_e = \int_{\Omega-\Sigma} \dot{W}\ d\Omega + \int_{\Omega-\Sigma} \mathcal{D}\ d\Omega + \int_{\partial\Sigma} \sigma n\dot{u}dS \qquad (2)$$

Let us introduce the auxiliary quantity $\phi^{\Sigma} = \int_{\Omega-\Sigma} W\ d\Omega$ whose time derivative is easily calculated taking into account that Σ is a domain on translation

$$d/dt\ \phi^{\Sigma} = \int_{\Omega-\Sigma} \dot{W}\ d\Omega - \int_{\partial\Sigma} \dot{a}\ Wn_1\ dS \qquad (3)$$

By eliminating between (2) and (3), we obtain :

$$P_e - d/dt\ \phi^{\Sigma} = \int_{\Omega-\Sigma} \mathcal{D}\ d\Omega + \dot{a} \int_{\partial\Sigma} (Wn_1 + n.\sigma.\dot{u}/\dot{a})\ dS \qquad (4)$$

Now, if W satisfies the condition of the singularity transport

$$\dot{W} = -\ \dot{a}\ W_{,1} + \text{integrable function} \qquad (5)$$

if can be shown [13] that

$$d/dt\ \phi = \lim_{R\to o} d/dt\ \phi^{\Sigma} \qquad (6)$$

In this case the left hand side of (4) tends in the limite, with $R \to o$, to the value of the right hand side of (1). If the near tip velocity satisfies a singularity transport condition analogous to (5), the integral of the second term at the right hand side of (4) can be written in the form

$$G = \lim_{R\to o} \int_{\partial\Sigma} (Wn_1 - n.\sigma.u_{,1})\ dS \qquad (7)$$

With this the rate of dissipation (1) is given by

$$D = \int_{\Omega} \mathcal{D}\ d\Omega + G\ \dot{a} \qquad (8)$$

Remember that the expression (7) and (8) have been deduced without any particular assumption on the behaviour of the material, so that they are valid for a general elastic-visco-plastic body, in so far the corresponding near tip solution satisfies the condition (5).

2.2- Dissipation analysis

For a perfectly elastic material we have $D = 0$, so that (8) yields

$$D = G\,\dot{a}$$

where G is Irwin's crack driving force. As the integral (7) is path independent in this case, the domain Σ can be arbitrarily chosen ; this gives the energy interpretation to Rice-Eshelby's integral J.

In the case of viscoelastic materials both terms in (8) are non-zero.

A detailed analysis of near tip fields for elastic-plastic materials, with or without hardening, cf. e.g. [1], [16] shows that the transport condition (5) is satisfied so that (8) is meaningful. However it can be shown [14] that in all these cases $G = 0$, as Rice conjectured since 1968, so that here

$$D = \int_{\Omega} D \, d\Omega$$

This would mean that when the crack propagates through a plastic material, the supplied energy is dissipated in plastic strain and the separation occurs without any supplementary loss of energy.

Before commenting this result, let us note that the expression (4) for an arbitrary contour $\partial\Sigma$ does not yield a direct energy interpretation of the J integral. Yet, it can be shown, cf. [14], that for steady growth we have

$$J_{\Sigma} = \int_{\partial\Sigma} (W n_1 - n.\sigma.u_{,1}) \, dS = 1/\dot{a} \int_{\Sigma} D \, d\Omega$$

so that $J_{\Sigma}\,\dot{a}$ represents the plastic dissipation in the domain Σ, which with decreasing R tends to zero. Numerical computations [14] are in good agreement with this conclusion.

2.3- Comments on the elastic-plastic analysis of fracture

The obtained results are in a certain sense paradoxical. The crack driving force is well defined only in the case of illimited elasticity, i.e. under physically inacceptable assumption that the plastic zone is reduced to zero. As soon as the existence of such a zone is taken into account the notion of a work of separation in the cracking process does not appear clearly, since it cannot be distinguished from the dissipation in the whole system. The creation of a material discontinuity by local shearing is of course possible, but can not in itself represent the mechanism of tearing.

In fact, in the small scale yielding conditions the dissipated energy is concentrated near the crack tip and correspond to separation work. In such situations, the "far off" elastic field being not essentially perturbed (St Venants's principle) J resp. G, computed by the linear elastic theory, give an adequate measure of the crack driving force.

The situation is completely different when the plastic zone is sufficiently large so that the dissipated energy corresponds to a volumic distribution : no relevant crack tip parameter is obtained in this case. On the one hand, it can be expected that the process of separation in the zone submitted to gradually increasing strain is in itself different. On the other hand, the physical meaning of the asymptotic solutions near the tip of a growing crack can be questioned. In a hyperbolic problem eventual changes of the local boundary conditions, due to the presence of the process zone, are transfered along the characteristics into the plastic domain.

In conclusion, we should say that the classical elastic-plastic model is perhaps inadequate in the context of rupture mechanics <u>if not implemented</u> by appropriate idealization of the process zone.

CONCLUDING REMARKS

The final aim of all elastic-plastic analyses of cracking is to satisfy an essential practical need : to establish <u>fracture criteria</u> which would allow the engineer to decide whether a structure is safe against steady or unsteady crack growth in given conditions.

Two classical approaches in this line are :

i - a purely mechanical approach : the estimation of a critical crack opening displacement (COD) or of a critical maximal strain, with various definitions depending on the authors ;

ii- energy approach by means of a J integral or other more or less clearly defined energy criteria, more or less clearly linked to this integral.

The foregoing analyses show that the doubts can be raised concerning the justification by the pure elastic-plastic analysis, of many speculations concerning crack-growth criteria based either on a characteristic deformation, or on a specified energy intensity.

Rice stated two years ago, here in Swansea [17] , that it might be argued that the now voluminous computational work in elastic-plastic crack mechanics has contributed to the development of numerical techniques, but that its impact on the understanding of fracture is, with some important exceptions, less significant. In our opinion the important achievement is the precise knowledge of the elastic-plastic near crack fields from which, in turn, a clear understanding of the limitations of the adopted model can be deduced. We may reasonably expect that, with sophisticated numerical techniques now at our disposal, a better modeling of the process zone behaviour will help the research workers to give acceptable answers to many actually open questions.

REFERENCES

1 AMAZIGO J.C. and HUTCHINSON J.W. - "Crack-tip fields in steady crack-growth with linear hardening" - J. Mech. Phys. of Solids, vol. 25, n° 2, pp. 87-97, 1977.

2 ANDERSON H. - "A finite element representation of stable crack growth" - J. Mech. Phys. of Solids, vol. 21, n° 5, pp. 337-356, 1973.

3 BERGEZ D. - "Influence des contraintes résiduelles sur la mesure du COD"- Practical Application of Fracture Mechanics to the prevention of Failure of Welded Structures, IIW, Colloquium, Bratislava, 1979.

4 BOZEC J. - "Propagation d'une fissure de fatigue dans une éprouvette de flexion" - Thesis (Dr. Ing), Bordeaux, 1977.

5 BUI H.D. and NGUYEN Q.S. - "Plastic bending of an edge-cracked specimen" - Engineering Fract. Mech., vol. 5, n°2 pp. 509-511, 1973.

6 CHITALEY A.D. and Mc CLINTOCK F.A. - "Elastic-Plastic mechanics of steady crack growth under anti-plane shear" - Jour. Mech. Phys. of Solids, vol. 19, n° 3, pp. 147-163, 1971.

7 d'ESCATHA Y. and DEVAUX J.C. - "Numerical study of initiation, stable crack growth, and maximum load, with a ductile fracture criterion based on the growth of Holes" - Elastic Plastic Fracture, STP 668, Ed. Landes/Begley/Clarke, ASTM, 1977.

8 HULT J.A.H. and Mc CLINTOCK F.A. - "Elastic plastic stress and strain distribution ground sharp notches under repeated shear" - Ninth Int. Cong. Appl. Mech., vol. 8, Free University of Brussels, 1956.

9 HUTCHINSON J.W. - "Singular behaviour at the end of a tensile crack in a hardening material" - Jour. Mech. Phys. of Solids, vol. 16, n° 3, pp. 13-31, 1968.

10 KFOURI A.P. and RICE J.R. - "Elastic plastic separation energy rate for crack advance in finite steps" - Fracture, ICF 4, Vol. 1, Waterloo, Canada, 1977.

11 Mc MEEKING R.M. and PARKS D.M. - "On criteria for J-dominance of crack tip fields in large scale yielding" - Elastic Plastic Fracture, STP 668, Ed. Landes/Begley/Clarke, ASTM, 1977.

12 NEWMAN J.C. - "A finite element analysis of fatigue crack closure"- Mechanics of crack growth, STP 590, Ed. Rice-Paris, ASTM 1974.

13 NGUYEN Q.S. - "Méthodes énergétiques en Mécanique de la Rupture" - Journal de Mécanique, vol. 19, n° 2, (to be published), 1980.

14 NGUYEN Q.S. and RAHIMIAN M. - "Mouvement permanent d'une fissure en milieu élastoplastique", Journal de Mécanique Appliquée (to be published).

15 RICE J.R. and ROSENGREN G.F. - "Plane strain deformation near a crack tip in a power-law hardening material" - Journ. Mech. Phys. of Solids, vol. 16, n° 3, pp. 1-12, 1968.

16 RICE J.R., DRUGAN W.J., SHAM T.L. - "Elastic plastic analysis of growing cracks" - ASTM 12th Annual Symposium on Fracture Mechanics, Washington University, May, 1979.

17 RICE J.R. - "Some computational problems in elastic-plastic crack mechanics" - Numerical Methods in Fracture Mechanics, Ed. Luxmoore A.R. and Owen D.R. J., University College, Swansea, 1978.

18 ROUSSELIER G. - "A numerical approach for stable crack growth and fracture criteria" - Fracture, ICF 4, Vol. 3, Waterloo, Canada, 1977.

19 ROUSSELIER G. - "Contribution à l'étude de la rupture des métaux dans le domaine de l'élastoplasticité" - Thesis (Dr. Ing.), Paris VI, 1979.

20 SLEPYAN L.I. - "Growing crack during plane deformation of an elastic-plastic body" - Isv. ANSSSR. Mekh. Tver. Tela, Vol. 9, n° 1, pp. 57-67, 1974.

21 SORENSEN E.P. - "A numerical investigation of plane strain stable crack growth under small-scale yielding conditions" Elastic-Plastic Fracture, STP 668, Ed. Landes/Begley/ Clarke, ASTM, 1977.

DEFORMATION THEORY OF PLASTICITY AND ITS INSTALLATION IN THE FINITE ELEMENT ANALYSIS ROUTINE

Y. Yamada*, Y. Huang** and I. Nishiguchi**

Institute of Industrial Science, University of Tokyo
7-22-1, Roppongi, Minato-ku, Tokyo, 106 Japan

SUMMARY

An incremental expression for the Hencky consitutive equation in deformation theory of plasticity is obtained. It connects the stress-rate with the strain-rate and its similarity to the Prandtl-Reuss equation in incremental theory facilitates its installation in existing finite element elastic-plastic analysis routine. The Hencky equation, when applied in its primary form, requires the iterative solution procedure in one way or the other and this means a departure from the standard finite element analysis routine. For comparisons of both theories of plasticity, two numerical examples are worked out. They are combined tension-torsion of thin circular tube and tensile loading of the center crack specimen under plane strain condition. In the latter case, the J-integrals are found almost path-independent for both theories in the regime of medium as well as small scale yeilding for a material having bilinear work-hardening property.

INTRODUCTION

The deformation theory of plasticity dates back to Hencky [1] and Nadai [2]. Although the physical soundness of the theory was criticized by Prager [3] and Hill [4], we can notice in the literature of plasticity theory that reconsiderations or reassessments were given repeatedly to it in later times. Among them are the works of Winzer and Prager [5] and Edelman [6] who examined and/or generalized the heuristic theorem of Ilyushin [7,8] on the coincidence of plasticity solutions obtained with incremental and deformation theories. An important conclusion to note from these attainments is for a large class of loading criteria that 'if the boundary loads on a body increase in proportion to one another and if the stress-strain relation used is of the power-term type, then the inte-

* Professor
** Graduate Student

rior state of stress is of the form $\sigma_{ij}=K_o\sigma_{ij}^{o}$, where K_o, a constant over the volume, is the proportionality factor of the boundary loads.'(I) In the meantime, Rice advocated the path independence of J-intergral for the material obeying the deformation theory of plasticity [9] and the crack tip stress and strain singularity solutions were obtained for a power-law hardening material [10-13]. These renewed the attention to the deformation theory and also were the stimuli of our performing the present work.

In this paper, an incremental expression for the Hencky constitutive equation is derived first for the general three dimensional case. It assumes a similar form to the Prandtl-Reuss equation and can be incorporated in the standard finite element analysis routine with minor modifications of the program. Incremental relations are also given in the form specialized for the application to plane stress problem and combined tension-torsion of a thin circular tube. It is shown that the solutions obtained with the deformation theory deviate from those of the incremental theory for the stress or strain paths with sharp bends. However, in case of the center crack specimen under remote uniform tensile stress, near tip loading condition is approximately a proportional one. Therefore, the solutions of both plasticity theories are practically identical and so the path-independent property of the J-integral is retained as well in the solution with the incremental theory.

INCREMENTAL EXPRESSION OF HENCKY EQUATION

In the deformation theory, the total strain $\{\varepsilon\}$ comprises the elastic part $\{\varepsilon^e\}$ and plastic part $\{\varepsilon^p\}$. Thus

$$\{\varepsilon\} = \{\varepsilon^e\} + \{\varepsilon^p\} \tag{1}$$

Usually, the following law is applied to $\{\varepsilon^p\}$

$$\{\varepsilon^p\} = \lambda_s\{\partial f/\partial\sigma\} \tag{2}$$

where f represents the yield condition or the plastic potential. To obtain an incremental expression, we differentiate eq.(2) to have

$$\{d\varepsilon^p\} = d\lambda_s\{\partial f/\partial\sigma\} + \lambda_s d\{\partial f/\partial\sigma\} \tag{3}$$

In the Hencky constitutive equation, the von Mises yield condition is assumed such that

$$f^2 = \frac{1}{2}[(\sigma_y-\sigma_z)^2+(\sigma_z-\sigma_x)^2+(\sigma_x-\sigma_y)^2+6(\tau_{yz}^2+\tau_{zx}^2+\tau_{xy}^2)] \\ =\frac{3}{2}\sigma_{ij}'\sigma_{ij}'=\bar{\sigma}^2 \tag{4}$$

(I) From the conclusions of Edelman [6].

Therefore

$$\{\partial f/\partial\sigma\} = \frac{3}{2\bar{\sigma}}\{\sigma'\} \tag{5}$$

and

$$d\{\partial f/\partial\sigma\} = \frac{3}{2\bar{\sigma}}(\{d\sigma'\}-\frac{\{\sigma'\}}{\bar{\sigma}}d\bar{\sigma}) \tag{6}$$

In the present context, the tensor definition is applied to strain component. $\bar{\sigma}$ is often called the equivalent stress and $\{\sigma'\}$ denotes the deviatoric stress

$$\lfloor\sigma'\rfloor = \lfloor\sigma_x{}' \ \sigma_y{}' \ \sigma_z{}' \ \tau_{yz} \ \tau_{zx} \ \tau_{xy}\rfloor \tag{7}$$

Introducing eqs.(5) and (6) to eq.(3), we have for a generic component of plastic strain increment

$$d\varepsilon_{ij}{}^p = \frac{3}{2\bar{\sigma}}(\sigma_{ij}{}'d\lambda_s+\lambda_s d\sigma_{ij}{}'-\lambda_s\frac{\sigma_{ij}{}'}{\bar{\sigma}}d\bar{\sigma}) \tag{8}$$

The deviatoric strain increment for the isotropic material is then given by

$$d\varepsilon_{ij}{}' = \frac{1}{2G}d\sigma_{ij}{}' + \frac{3}{2\bar{\sigma}}(\sigma_{ij}{}'d\lambda_s +\lambda_s d\sigma_{ij}{}'-\lambda_s\frac{\sigma_{ij}{}'}{\bar{\sigma}}d\bar{\sigma}) \tag{9}$$

where G is the modulus of rigidity.

We define here the equivalent plastic strain increment $\overline{d\varepsilon^p}$, so that it affords the plastic work increment, when combined with $\bar{\sigma}$, as

$$dW^p = \sigma_{ij}d\varepsilon^p{}_{ij} = \sigma_{ij}{}'d\varepsilon_{ij}{}^p = \bar{\sigma}\,\overline{d\varepsilon^p} \tag{10}$$

Substituting $d\varepsilon_{ij}{}^p$ from eq.(8) into eq.(10)

$$\frac{3}{2\bar{\sigma}}\sigma_{ij}(\sigma_{ij}{}'d\lambda_s+\lambda_s d\sigma_{ij}{}'-\lambda_s\frac{\sigma_{ij}{}'}{\bar{\sigma}}d\bar{\sigma}) = \bar{\sigma}\,\overline{d\varepsilon^p} \tag{11}$$

In view of the identity $\sigma_{ij}d\sigma_{ij}{}' = \sigma_{ij}{}'d\sigma_{ij}{}' = \frac{2}{3}\bar{\sigma}d\bar{\sigma}$, the second and third terms in eq.(11) are cancelled out. Then, considering the identity $\sigma_{ij}\sigma_{ij}{}' = \sigma_{ij}{}'\sigma_{ij}{}' = \frac{2}{3}\bar{\sigma}^2$, we have

$$d\lambda_s = \overline{d\varepsilon^p} \tag{12}$$

or

$$d\lambda_s = \frac{d\bar{\sigma}}{H'} = \frac{3}{2}\,\frac{\sigma_{ij}{}'d\sigma_{ij}{}'}{\bar{\sigma}H'} \tag{13}$$

where H' represents the strain-hardening rate or the slope of the $\bar{\sigma}-\int\overline{d\varepsilon^p}$ curve. Integration of eq.(13) yields

$$\lambda_s = \overline{\varepsilon}^p = \overline{\sigma}/H_s \tag{14}$$

H_s is the secant modulus of the $\overline{\sigma}-\int\overline{d\varepsilon^p}$ curve.

In order to obtain a formula for $d\lambda_s$ which is expressed in terms of the strain increment $d\varepsilon_{ij}$ in place of $d\sigma_{ij}'$, we solve eq.(9) for $d\sigma_{ij}'$ to have

$$d\sigma_{ij}' = \frac{2G\overline{\sigma}}{\overline{\sigma}+3G\lambda_s}[d\varepsilon_{ij}' - \frac{3}{2\overline{\sigma}}(1-\frac{H'\lambda_s}{\overline{\sigma}})\sigma_{ij}'d\lambda_s] \tag{15}$$

Substituting eq.(15) into eq.(13) and solving for $d\lambda_s$, we get after some manipulation of relevant terms

$$d\lambda_s = \frac{3G}{(H'+3G)\overline{\sigma}}\sigma_{ij}'d\varepsilon_{ij} \tag{16}$$

Substituting $d\lambda_s$ from eq.(16) back to eq.(15) and using eq.(14)

$$d\sigma_{ij}' = \frac{2G}{1+3G/H_s}[d\varepsilon_{ij}' - \frac{9G(1-H'/H_s)\sigma_{ij}'\sigma_{k\ell}'}{2(H'+3G)\overline{\sigma}^2}d\varepsilon_{k\ell}] \tag{17}$$

The stress increment $d\sigma_{ij}$ obtainable from the deviatoric part $d\sigma_{ij}'$ as given above is for the isotropic Hencky material

$$d\sigma_{ij} = \frac{2G}{1+3G/H_s}[d\varepsilon_{ij} + \frac{\nu+(1+\nu)G/H_s}{1-2\nu}\delta_{ij}d\varepsilon_{ii} - \frac{9G(1-H'/H_s)\sigma_{ij}'\sigma_{k\ell}'}{2(H'+3G)\overline{\sigma}^2}d\varepsilon_{k\ell}] \tag{18}$$

where δ_{ij} represents the Kronecker delta.

The corresponding constitutive equation in incremental theory, i.e. the Prandtl-Reuss equation, is known to be expressible as [14]

$$d\sigma_{ij} = 2G[d\varepsilon_{ij} + \frac{\nu}{1-2\nu}\delta_{ij}d\varepsilon_{ii} - \frac{9G\sigma_{ij}'\sigma_{k\ell}'}{2(H'+3G)\overline{\sigma}^2}d\varepsilon_{k\ell}] \tag{19}$$

Comparing eqs.(18) and (19), it can be seen that both theories differ each other only by the term H_s, which is naturally a variable depending on the stress and/or strain values at each stage of deformation. An important conclusion here is that by virtue of the incremental expression worked out as eq.(18), the solution with the deformation theory is executable on the standard elastic-plastic solution routine with minor modifications of the computer program.

Moreover, it is worth noting that expressions for $d\lambda_s$ given either by eq.(13) or eq.(16) are identical in both theories, more specifically in the Hencky equation and the Prandtl-Reuss equation.

In the primary application of the deformation theory, the Hencky equation is often expressed in the following way

$$\varepsilon_{ij}' = (1/2G+\phi)\sigma_{ij}' \qquad (20)$$

It can be easily seen from eqs.(2) and (5) that parameter ϕ is connected with λ_s and H_s of the present paper as

$$\phi = \frac{3\lambda_s}{2\bar{\sigma}} = \frac{3}{2H_s} \qquad (21)$$

Then

$$d\phi = \frac{3}{2\bar{\sigma}}(d\lambda_s - \frac{\lambda_s}{\bar{\sigma}}d\bar{\sigma}) = \frac{3}{2\bar{\sigma}}(d\lambda_s - \frac{2}{3}\phi d\bar{\sigma}) \qquad (22)$$

Using the relation $d\bar{\sigma} = H'd\lambda_s$ from eq.(13), we have

$$d\phi = \frac{3}{2\bar{\sigma}}(1 - \frac{2}{3}\phi H')d\lambda_s = \frac{3}{2\bar{\sigma}}(1-H'/H_s)d\lambda_s \qquad (23)$$

Eq.(23) is useful to obtain the formula for $d\phi$ from $d\lambda_s$ which is given by eq.(13) or (16).

SPECIFIC FORMS FOR PLANE STRESS AND COMBINED TENSION-TORSION PROBLEMS

Eq.(18) for the deformation theory and eq.(19) for the incremental theory can be written in matrix form respectively as

$$\{d\sigma\} = [D^{ep}]_{tot}\{d\varepsilon\} \qquad (24)$$

and

$$\{d\sigma\} = [D^{ep}]_{inc}\{d\varepsilon\} \qquad (25)$$

These equations are directly applicable to the three dimensional and plane strain problems, and so forth. They should be modified, however, for the cases wherein the stress conditions are specified. The plane stress field and the combined torsion-tension of thin tube are such examples. In practice, the modification can be carried out numerically in the finite element analysis program. But it would be useful to include here the explicit expressions of $[D^{ep}]_{tot}$ for these two cases.

The elastic-plastic stress-strain matrix for plane stress case of the Hencky material is given as

$$\begin{Bmatrix} d\sigma_x \\ d\sigma_y \\ d\tau_{xy} \end{Bmatrix} = [D^{ep}]_{tot} \begin{Bmatrix} d\varepsilon_x \\ d\varepsilon_y \\ d\gamma_{xy} \end{Bmatrix} \qquad (26)$$

where $d\gamma_{xy}$ represents the engineering shear strain and $[D^{ep}]$ is

$$[D^{ep}]_{tot} = \frac{E}{\beta\delta}\begin{bmatrix} \gamma+\delta & \beta-3\zeta & 0 \\ & \gamma+\delta & 0 \\ \text{sym.} & & \beta/2 \end{bmatrix}$$

$$-\frac{EF}{\beta\delta}\begin{bmatrix} (\gamma\sigma_x-\zeta\sigma_y)^2 & (\gamma^2+\zeta^2)\sigma_x\sigma_y-\gamma\zeta(\sigma_x{}^2+\sigma_y{}^2) & \beta\tau_{xy}(\gamma\sigma_x-\zeta\sigma_y) \\ & (\zeta\sigma_x-\gamma\sigma_y)^2 & \beta\tau_{xy}(\gamma\sigma_y-\zeta\sigma_x) \\ \text{sym.} & & \beta^2\tau_{xy}{}^2 \end{bmatrix} \qquad (27)$$

where E denotes Young's modulus and F is

$$F = \frac{3\alpha}{\beta-\alpha\zeta(\sigma_x+\sigma_y)^2} \qquad (28)$$

with

$$\alpha = \frac{3G(1-H'/H_s)}{2(H'+3G)\bar{\sigma}^2} \qquad (29)$$

$$\beta = 3(1-\nu) + \frac{3E}{2H_s}, \quad \gamma = 2 - \nu + \frac{3E}{2H_s}, \quad \delta = 1 + \nu + \frac{3E}{2H_s}, \quad \zeta = 1-2\nu \qquad (30)$$

Expression for $d\lambda_s$ corresponding to eq.(16) is

$$d\lambda_s = \frac{2F\bar{\sigma}}{3(1-H'/H_s)}[(\gamma\sigma_x-\zeta\sigma_y)d\varepsilon_x+(\gamma\sigma_y-\zeta\sigma_x)d\varepsilon_y+\beta\tau_{xy}d\gamma_{xy}] \qquad (31)$$

For the combined tension-torsion of thin circular tube

$$\begin{Bmatrix} d\sigma \\ d\tau \end{Bmatrix} = [D^{ep}]_{tot}\begin{Bmatrix} d\varepsilon \\ d\gamma \end{Bmatrix} \qquad (32)$$

where $d\gamma$ represents the engineering shear strain and

$$[D^{ep}]_{tot} = \begin{bmatrix} \dfrac{E}{1+E/H_s} - \dfrac{(1-H'/H_s)(1+3G/H_s)E^2\sigma^2}{(1+E/H_s)S'} & \dfrac{-3(1-H'/H_s)EG\sigma\tau}{S'} \\ \text{sym.} & \dfrac{G}{1+3G/H_s} - \dfrac{9(1-H'/H_s)(1+E/H_s)G^2\tau^2}{(1+3G/H_s)S'} \end{bmatrix} \qquad (33)$$

where

$$S' = (H' + \frac{3EG}{H_s})\bar{\sigma}^2 + (E + \frac{3GH'}{H_s})\sigma^2 + 3(3G + \frac{EH'}{H_s})\tau^2 \qquad (34)$$

Expression for $d\lambda_s$ in this case is

$$d\lambda_s = \frac{\bar{\sigma}}{S'}[(1+3G/H_s)E\sigma d\varepsilon + 3(1+E/H_s)G\tau d\gamma] \qquad (35)$$

NUMERICAL EXAMPLES

Incremental relations derived in the preceding section are incorporated in the existing standard elastic-plastic analysis routine and the solutions with deformation and incremental theories are obtained side by side. Examples studied are combined tension-torsion of a thin circular tube and tensile loading of the center crack specimen.

A program [15] specially developed for the analysis of combined loadings of uniform cross-section bar and beam elements is utilized in the solution of the first example. Fig.1 shows the finite element idealization. By the axisymmetric nature of the problem, a sector of appropriate including angle, 2.8° in the present example, is chosen and divided into two triangular prism elements. Each nodal point is endowed with three degrees of freedom (DOF), i.e. u,v in the cross-section and w in the axial direction, so that totally nine DOF are associated with the individual element. Oblique boundary conditions are imposed along two radial planes bounding the sector. Dimensions and material data are as follows. Inner radius r_i=20mm, outer radius r_o=21mm Young's modulus E=19×10^{3}kgf/mm^2(186×10^{3}MPa), strain-hardening rate H'=380 kgf/mm^2 (3.72×10^{3}MPa), yield stress σ_y=26.2 kgf/mm^2 (0.257×10^{3}MPa) and Poisson's ratio ν=0.3. Bilinear strain-hardening property is assumed.

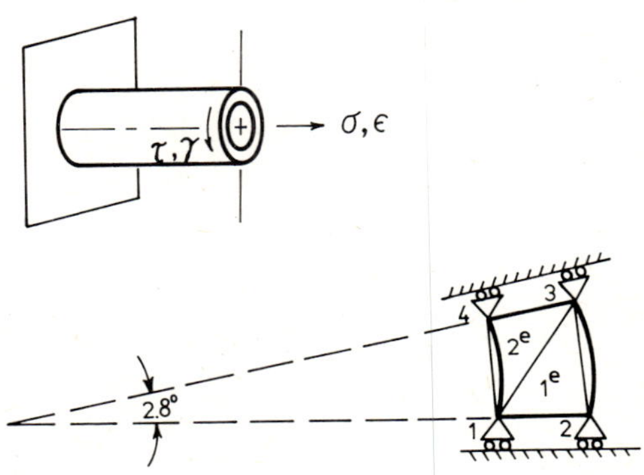

Fig.1 Tension-torsion of thin circular tube

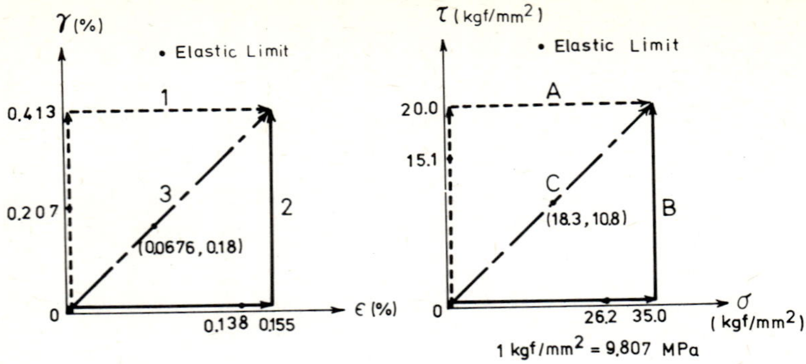

Fig.2 Strain and stress paths of combined tension-torsion

Strain and stress paths assumed in the numerical computation are depicted in Fig.2. Besides two paths with a 90° bend, included is the radial path of the proportional loading. Table 1 compares the final state of stress for the case in which the strain path is prescribed. It is seen that the final stress state in the deformation theory is practically independent on the strain path, in contrast to the strong dependency in case of the incremental theory. This verifies that the incremental relations as well as the computer programs are precisely developed in the present work. The same conclusions apply to the final strain state of Table 2, the result obtained under the prescribed stress path conditions.

	strain path number	strain(%) γ	ε	stress(kgf/mm^2) τ	σ
deformation theory	1			13.28	14.00
	2	0.413	0.155	13.84	14.52
	3			13.20	14.00
incremental theory	1			9.12	22.25
	2	0.413	0.155	15.74	8.93
	3			13.13	14.18

Table 1 Comparison of final stress state under prescribed γ-ε strain path, 1 kgf/mm^2=9.807MPa

	stress path identification	stress(kgf/mm^2) τ	σ	strain(%) γ	ε
deformation theory	A			7.63	4.34
	B	20.0	35.0	7.68	4.48
	C			7.85	4.44
incremental theory	A			9.50	2.04
	B	20.0	35.0	3.58	5.61
	C			7.79	4.43

Table 2 Comparison of final strain state under prescribed τ-σ stress path, 1 kgf/mm^2=9.807MPa

The center crack specimen analyzed is shown in Fig.3, together with the material data assuming the bilinear hardening behavior. The path independent property of the so-called J-integral is known to hold [10-13] for the material obeying the

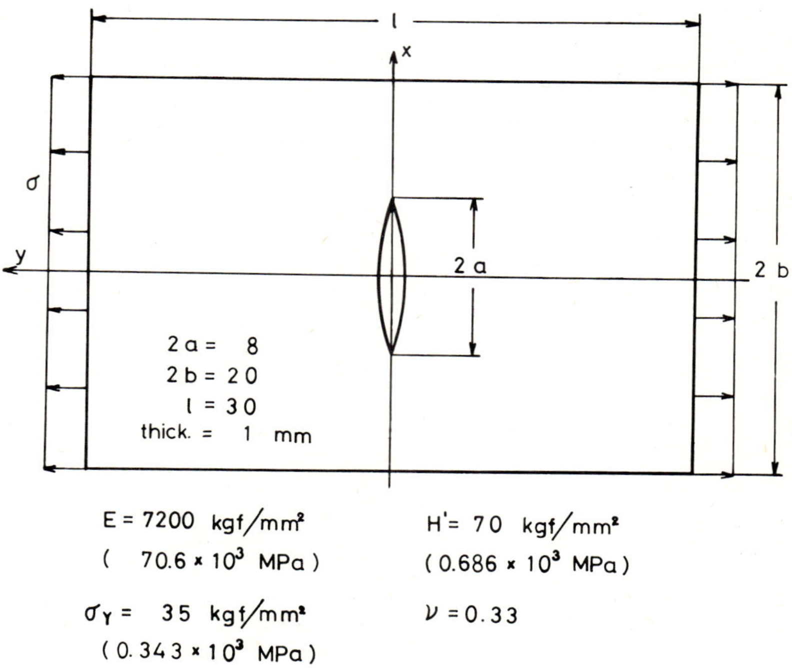

Fig.3 Central crack specimen under uniform remote tensile stress

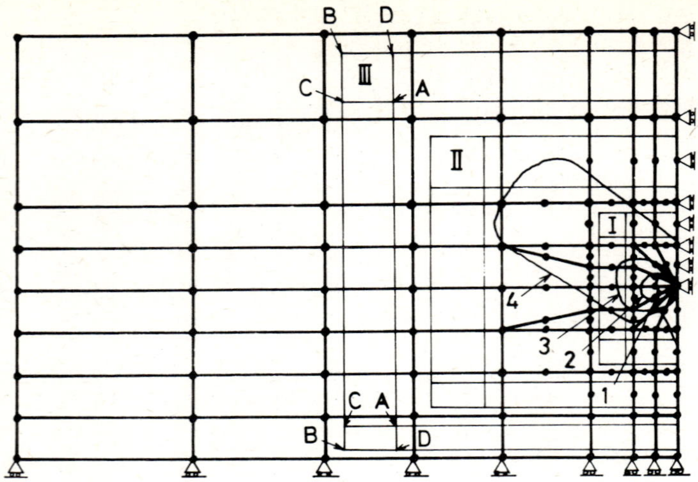

stress level σ/σ_Y

1. 0.323   2. 0.412   3. 0.526   4. 0.699

Fig.4 Mesh division, paths for J-integral and development of plastic enclave in center crack specimen

deformation theory of plasticity as well as the linear elastic law. Moreover, the stress intensity factor K_I is obtainable from the J-integral for the elastic and the small yielding cases, e.g. under the plane strain condition

$$K_I = \sqrt{EJ/(1-\nu^2)} \tag{36}$$

J-integral kgf/mm (kN/m in the parenthesis)

class of path	path identification in each class				mean
	A	B	C	D	
I	0.1867 (1.831)	0.1852 (1.816)	0.1859 (1.823)	0.1854 (1.818)	0.1858 (1.822)
II	0.1858 (1.822)	0.1917 (1.880)	0.1918 (1.881)	0.1847 (1.811)	0.1885 (1.849)
III	0.1858 (1.822)	0.1894 (1.857)	0.1885 (1.849)	0.1860 (1.824)	0.1874 (1.838)

total mean J=0.1872 (1.836)

Table.3 J-integral estimation along twelve paths of Fig.4

The J-integral is estimated along three classes of paths I to III from inside to outside of the crack tip as shown in Fig.4. As can be seen from the figure each class of path is divided further into four depending on the Gauss integration points being passed through. Variable number nodes elements are used and a quarter of the specimen is analyzed due to the symmetric configuration and the loading condition.

Table 3 compares the values of the J-integral obtained along twelve different paths at the remote uniform stress level of 10 kgf/mm^2 (98.07MPa). The elastic stress and strain state is assumed in this example. It is seen that the path-independent property of the J-integral is assured numerically.

Table 4 summarizes the stress intensity factor from J-integral of Table 3. The deviation of total mean value K_I=38.90 $kgf \cdot mm^{-3/2}$ ($12.06 \times 10^3 kN \cdot m^{-3/2}$) from the theoretical prediction K_I=39.30 $kgf \cdot mm^{-3/2}$ ($12.19 \times 10^3 kN \cdot m^{-3/2}$)[16] is only about 1%. Even the maximum deviation between the paths is 2.4% and the difference between the mean values in each class of integration paths is below 1%. In contrast, the K_I-value estimated from the near tip crack opening displacement showed an error of order 10%. Although the incorporation of singularity element [17] may be useful for improving the estimate, the present analysis is based exculsively on the ordinary 4 to 8 variable number nodes parametric element.

K_I $kgf \cdot mm^{-3/2}$ ($kN \cdot m^{-3/2}$ in the parenthesis)

class of path	path identification in each class				mean
	A	B	C	D	
I	38.84 (12.05×10^3)	38.68 (12.00×10^3)	38.76 (12.02×10^3)	38.70 (12.00×10^3)	38.75 (12.02×10^3)
II	38.75 (12.02×10^3)	39.36 (12.21×10^3)	39.37 (12.21×10^3)	38.63 (11.98×10^3)	39.03 (12.10×10^3)
III	38.75 (12.02×10^3)	39.12 (12.13×10^3)	39.03 (12.10×10^3)	38.77 (12.02×10^3)	38.91 (12.07×10^3)

total mean K_I=38.90(12.06×10^3)

Table 4 Stress intensity factor K_I from J-integral computed by eq.(36)

In the following, some features of the solution of the center crack specimen are summarized. The plastic enclaves obtained with the deformation theory, on the Hencky equation, at four different levels of remote stress σ/σ_y are depicted in Fig.4. No essential difference is found in the plastic enclaves at the same or almost the same level of stress in the solution with the incremental theory. ϕ in Table 5 is the strain-stress ratio calculated by eq.(21). Similarly ϕ_x, ϕ_y ϕ_{xy} are the ratios defined as follows at the same point, i.e. a Gauss point near the crack tip

$$\phi_x = \varepsilon_x{}^p/\sigma_x{}', \quad \phi_y = \varepsilon_y{}^p/\sigma_y{}', \quad \phi_{xy} = \gamma_{xy}{}^p/(2\tau_{xy}) \tag{37}$$

and ϕ_I is estimated by an integration of $d\phi_I$ which is expressible as the following by eqs.(16) and (23)

$$d\phi = \frac{3}{2\bar{\sigma}}(1-H'/H_s)d\lambda_s = \frac{9G(1-H'/H_s)}{2(H'+3G)\bar{\sigma}^2}\sigma_{ij}{}'d\varepsilon_{ij} \tag{38}$$

or by the relation of eq.(21)

$$d\phi = \frac{3G(3/2-\phi H')}{(H'+3G)\bar{\sigma}^2}\sigma_{ij}{}'d\varepsilon_{ij} \tag{39}$$

stress level σ/σ_y	0.323	0.412	0.526	0.699
ϕ_x × 10^6	8.208	87.87	233.6	657.5
ϕ_y	8.265	88.42	235.2	660.9
ϕ_{xy}	8.293	88.71	232.2	654.3
ϕ	8.246	88.38	234.1	657.9
ϕ_I	8.250	88.44	234.4	658.8
J_I	0.2372	0.3907	0.6633	1.450
J_{II}	0.2406	0.3966	0.6760	1.491
J_{III}	0.2393	0.3945	0.6728	1.481
J_{mean}	0.2390	0.3939	0.6707	1.474

ϕ:mm^2/kgf, J:kgf/mm

Table 5 Strain-stress ratio ϕ and J-integral, from the solution by deformation theory.
1 mm^2/kgf=0.10197MPa^{-1}, 1 kgf/mm=9.807kN/m

It can be seen that all ϕ's are almost equal and the prerequisite of the deformation theory is satisfied by the solution of the present paper based on the incremental expression of the theory. Furthermore, the J-integrals in Table 5 are almost path-independent, in accordance with the theoretical predictions [10-13].

stress level σ/σ_y	0.323	0.412	0.523	0.683
$\phi_x \times 10^6$ ϕ_y ϕ_{xy}	8.218 8.267 8.314	86.69 88.15 88.28	222.3 237.2 204.6	582.7 635.0 529.0
$\phi \times 10^6$ ϕ_I	8.243 8.243	87.87 87.94	225.7 226.0	588.5 589.4
J_I J_{II} J_{III}	0.2372 0.2406 0.2393	0.3909 0.3969 0.3947	0.6539 0.6663 0.6632	1.335 1.376 1.367
J_{mean}	0.2390	0.3942	0.6611	1.359

Table 6 Strain-stress ratio ϕ and J-integral, from the solution by incremental theory. 1 $mm^2/kgf=0.10197MPa^{-1}$, 1 kgf/mm=9.807kN/m

Table 6 summarizes the corresponding solutions obtained with the incremental theory on the Prandtl-Reuss equation. As the natural consequence, the coincidence of the ϕ's estimated by different formulas can not be expected in this case. However, it should be emphasized that the path-independence of the J-integral is preserved approximately in the solution, which in turn shows that the near tip stress and strain state is nearly a proportional one.

CONCLUDING REMARKS

By an exploitation of the incremental representation of the constitutive equation of deformation theory of plasticity, i.e. the Hencky equation, it becomes possible to make direct comparison of different plasticity theories on the standard elastic-plastic analysis routine. By applying the present formulation, the difference and/or the coincidence of the solutions of the deformation and incremental theories are elucidated in two examples, the combined tension-torsion of a thin circular tube and the central crack specimen. It should be pointed out that the paper exemplifies again that the numerical solutions on proper formulation can give clear understanding of seemingly difficult problems. When viewed from only empirical basis without resorting to and/or acknowledging the latest finite element analysis capabilities, the problems will never have been solved.

REFERENCES

1. HENCKY, H. - Zur Theorie plastischer Deformationen und der hierdurch im Material hervorgerufenen Nachspannungen, Z. für angewandte Math. Mech., Vol. 4, pp. 323-334, 1924.
2. NADAI, A. - Plasticity, A Mechanics of the Plastic State of Matter, McGraw-Hill, New York, 1931.
3. PRAGER, W. - Theory of Plastic Flow versus Theory of Plastic Deformation, J. Appl. Phys., Vol. 19, No. 6, pp. 540-543, 1948.
4. HILL, R. - The Mathematical Theory of Plasticity, Oxford, Clarendon Press, 1950.
5. WINZER, A and PRAGER. W. - On the Use of Power Laws in Stress Analysis Beyond the Elastic Range, J. Appl. Mech., Trans. ASME, Vol. 14, No. 4, pp. A281-A284, 1947.
6. EDELMAN, F. - On the Coincidence of Plasticity Solutions Obtained with Incremental and Deformation Theories, Proc. 1st Nat. Congr. Appl. Mech., ASME, pp.493-498, 1951.
7. ILYUSHIN, A. A. - Relation between the Theory of St. Venant-Levy-Mises and the Theory of Small Elastic-Plastic Deformations, Prikl. Mat. Mekh., Vol. 9, pp. 207-218, 1945, Translation by the Appl. Math. Group, Brown University, Providence, R. I.
8. ILYUSHIN, A. A. - The Theory for Small Elastic-Plastic Deformations, Prikl. Mat. Mekh., Vol. 10, pp. 347-356, 1946, Translation by the Grad. Div. of Appl. Math., Brown University, Providence, R. I.
9. RICE, J. R. - A Path-Independent Integral and the Approximate Analysis of Strain Concentration by Notches and Cracks, J. Appl. Mech, Vol. 35, pp.379-386, 1968.
10. RICE, J. R. and ROSENGREN, G. F., - Plane Strain Deformation Near a Crack Tip in a Power-law Hardening Material, J. Mech. Phys. Solids, Vol. 16, pp. 1-12, 1968.
11. Hutchinson, J. W. - Singular Behavior at the End of a Tensile Crack in a Hardening Material, J. Mech. Phys. Solids, Vol. 16, pp. 13-31, 1968.
12. RICE, J. R. - Elastic-Plastic Fracture Mechanics, The Mechanics of Fracture, Ed. Erdogen, F., AMD-Vol. 19, ASME, pp. 23-53, 1976.
13. RICE, J. R., MCMEEKING, R. M., PARKS, D. M. and SORENSEN, E. P. - Recent Finite Element Studies in Plasticity and Fracture Mechanics, Computer Methods in Appl. Mech. and Engng., Vol. 17/18, pp. 411-442, 1979.
14. YAMADA, Y., YOSHIMURA, N. and SAKURAI, T., - Plastic Stress-Strain Matrix and Its Application for the Solution of Elastic-Plastic Problems by the Finite Element Method, Int. J. Mech. Sci., Vol. 10, pp. 343-354, 1968.
15. YAMADA, Y. and TAKATSUKA, K., - Elastic Analysis of Beams with Uniform Cross-Section under Combined Loadings, Seisan Kenkyu, Monthly J. of Institute of Industrial Science, Vol. 23, pp. 531-535, 1971.

16. TADA, H., PARIS, P. and IRWIN, G., - The Stress Analysis of Cracks Handbook, Del Research Corporation, Heller Town, Pa. USA, 1973.

17. YAMADA, Y., EZAWA, Y., NISHIGUCHI, I. and OKABE, M., - Reconsiderations on Singularity or Crack Tip Element, Int. J. Num. Meth. Engng., Vol. 14, pp. 1525-1544, 1979.

EVALUATION OF J_K INTEGRALS FOR A CRACK IN TWO-PHASE MATERIALS.

Hiroshi Miyamoto and Masanori Kikuchi

Summary:

In this study, the J integral is extended to the problem of a crack in an inhomogeneous material. When a crack exists in the inhomogeneous material, the path independency of the J integral becomes invalid, and it is difficult to evaluate the J integral.

Here the definition of the J integral is extended for these problems. This extension is accomplished by introducing a new term in the J integral definition. This new J integral has the path independent characteristics even when the crack exists in the inhomogeneous material.

J_y integral is also introduced to evaluate the asymmetricity of the boundary conditions.

Several examples are analysed by using these new J integrals and the effect of phase boundary on crack extension is discussed.

1. Introduction

The J integral is considered to be one of a most effective parameter of fracture in elasto-plastic state. The availability is due to the fact that the J integral holds the path independency and so that its analytical and experimental evaluations are very easy. But when a crack exists in multiphase materials, for example, composite materials or a welded structure, path independency of the J integral becomes invalid, and the evaluation of the J integral value is difficult. This is because the discontinuities of stresses and strains occur at the phase boundaries. Moreover, the boundary con-

Department of Mechanical Engineering, Faculty of Science and Technology, Science University of Tokyo, Yamazaki, Noda, Chiba (278), Japan ditions become to be asymmetric in these cases, so a single component of J integral is not enough to evaluate the crack extension.

In this paper, first the J_x and J_y integrals are introduced. For the asymmetric problem, for example, a crack exists on a phase boundary, only the J integral proposed by Rice [1] is insufficient and one more component of J vector may be necessary. The J integral defined by Rice is considered to be the force acting on the crack tip parallel to the crack, say J_x. One more J component, J_y, is the force acting on the crack tip perpendicular to the crack. These two values have path independent characteristics.

Secondly, the definitions of these J integrals are extended to the multi-phase materials. This extension is carried out by an introduction of an additional term in the J integral definition by Rice. The additional term concerns with the integral along the phase boundary. The new definition of the J integral preserves the path independency as long as the material is in elastic or non-linear elastic state. The physical meaning of this new J integral does not alter by this modification and it means the energy release rate per unit elongation of a crack. By using this, the crack behaviours in multi-phase materials may be discussed well.

Several examples are analysed by using finite element analyses. One is when a crack exsits along the phase boundary. In this case, the additional term in J_x integral becomes zero. The other is when a crack exists perpendicular to the phase boundary. In this case, the additional term to J_y integral becomes to zero. In both cases, J_x and J_y integrals are verified to be path independent values by considering the additional terms. Then by changing the crack length and the elastic modulus, the behaviors of J value are discussed Elasto-plastic analysis is also carried out by using three point bend specimen composed of two phase materials. The J integral value is evaluated and compared with the Rice's conventional J values.

2. J_x and J_y integrals.

J_x and J_y integrals are defined as follows:

$$J_x = \int_\Gamma (W\delta_{jx} - \sigma_{ij}U_{i,x})dS_j \qquad (1)$$

$$J_y = \int_\Gamma (W\delta_{jy} - \sigma_{ij}U_{i,y})dS_j \qquad (2)$$

where Γ denotes the path from point A, in Fig. 1, to point B surrounding the crack tip.

J_x and J_y are considered to be the component of the force vector acting on the crack tip [2]. J_x is the component parallel to the crack surface and J_y is perpendicular to it. When the boundary conditions are symmetric with the crack surface, the deformations are symmetric and J_y becomes zero. Then only J_x value must be considered. But when a crack exists in a multiphase material, the boundary conditions are not symmetric with the crack surface, so J_y exists as non-zero value.

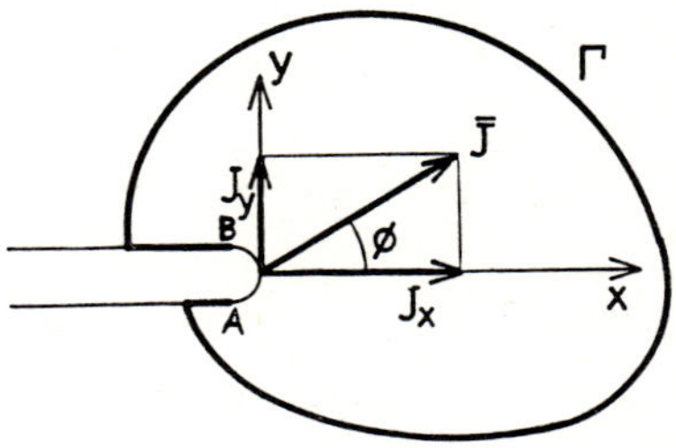

Fig.1 J_x and J_y of a crack.

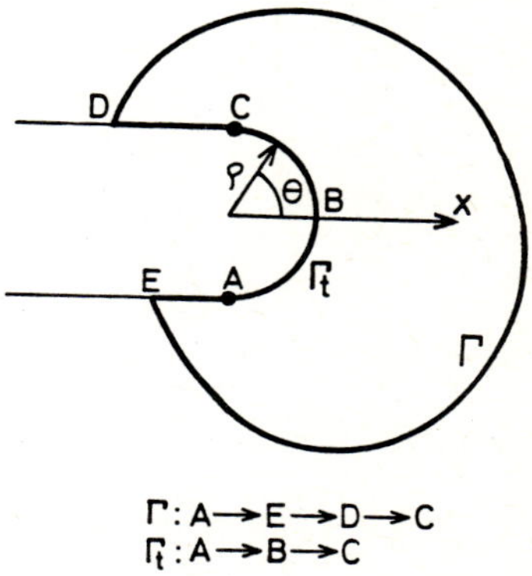

Fig.2 Γ and Γ_t of a crack tip.

When the crack tip has a radius ρ, as in Fig.2, J_x and J_y are equal to the following equations;

$$J_x = \int_{\Gamma_t} W dy \tag{3}$$

$$J_y = \int_{\Gamma_t} W dx \tag{4}$$

because of the path independency of J intergrals. As $x=\rho\cos\theta$ and $y=\rho\sin\theta$, so eqs. (3) and (4) are rewritten as follows;

$$J_x = \int_{-\pi/2}^{\pi/2} \rho W \cos\theta d\theta \tag{5}$$

$$J_y = \int_{-\pi/2}^{\pi/2} \rho W \sin\theta d\theta \tag{6}$$

By Eqs. (5) and (6), it is noticed that J_x and J_y are the measure of the strain energy stored on the crack surface. So it is estimated that these two values may be related to the condition of crack extension, not only by J_x [3]. Hereafter, these two values are considered together.

3. Definition of the J integrals of a crack in multi-phase materials.

In Fig.3, a crack exists near the phase boundary. In this case, J_x and J_y integrals are defined as follows;

$$J_x = \int_{\Gamma_1} (W\delta_{jx} - \sigma_{ij} U_{i,x}) ds_j - \int_{\Gamma_2} (W\delta_{jx} - \sigma_{ij} U_{i,x}) ds_j \quad (7)$$

$$J_y = \int_{\Gamma_1} (W\delta_{jy} - \sigma_{ij} U_{i,y}) ds_j - \int_{\Gamma_2} (W\delta_{jy} - \sigma_{ij} U_{i,y}) ds_j \quad (8)$$

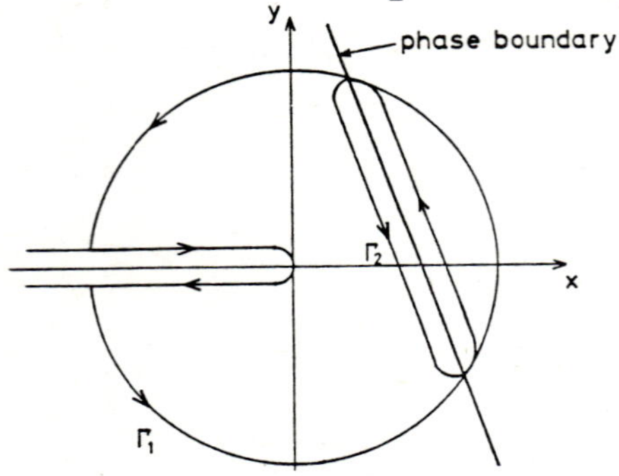

Fig.3 A crack and a phase boundary.

where Γ_2 means the closed path in Γ_1 [4].

The physical meaning of eqs. (7) and (8) are as follows. On the phase-boundary, the discontinuities of stresses and strains occur and the first terms of eqs. (7) and (8) mean the force acting on both the crack tip and these discontinuities. Then the second terms eliminate the contributions of the phase boundary. So eqs. (7) and (8) mean the force vector acting on only the crack tip. This is equal to the energy release rate per unit elongation of the crack. In the special cases, as shown in Fig.4(a) and (b), when a crack exists parallel or perpendicular to the phase boundary, eqs. (7) and (8) can be simplified by considering the continuities of stresses and stains at the phase boundary.

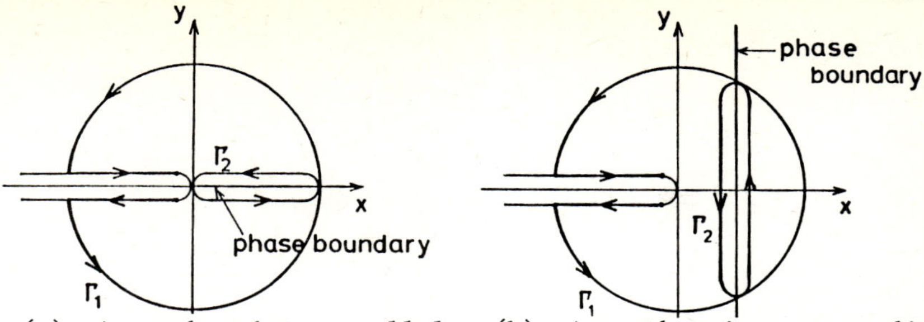

(a) A crack exists parallel to the phase boundary.

(b) A crack exists perpendicular to the phase boundary.

Fig. 4.

For the Fig.4(a), J_x and J_y are expressed as;

$$J_x = \int_{\Gamma_1} \left\{ (W - \sigma_x \frac{\partial u}{\partial x} - \tau_{xy} \frac{\partial v}{\partial x}) dy + (\tau_{xy} \frac{\partial u}{\partial x} + \sigma_y \frac{\partial v}{\partial x}) dx \right\} \tag{9}$$

$$J_y = -\int_{\Gamma_1} \left\{ (W - \tau_{xy} \frac{\partial u}{\partial y} - \sigma_y \frac{\partial v}{\partial y}) dx + (\sigma_x \frac{\partial u}{\partial y} + \tau_{xy} \frac{\partial v}{\partial y}) dy \right\} + \int_{\Gamma_2} (-W + \sigma_x \frac{\partial u}{\partial x}) dx \tag{10}$$

and for Fig.4(b), they are expressed as;

$$J_x = \int_{\Gamma_1} \left\{ (W - \sigma_x \frac{\partial u}{\partial x} \; \tau_{xy} \frac{\partial v}{\partial x}) dy + (\tau_{xy} \frac{\partial u}{\partial x} + \sigma_y \frac{\partial v}{\partial x}) dx \right\} - \int_{\Gamma_2} (-W + \sigma_y \frac{\partial v}{\partial y}) dy \tag{11}$$

$$J_y = -\int_{\Gamma_2} \left\{ (W - \tau_{xy} \frac{\partial u}{\partial y} - \sigma_y \frac{\partial v}{\partial y}) dx + (\sigma_x \frac{\partial u}{\partial y} + \tau_{xy} \frac{\partial v}{\partial y}) dy \right\} \tag{12}$$

eqs.(9) and (12) don't contain the additional integral terms along Γ_2 because $C_{ijkl,x}=0$ in Fig.4(a) and $C_{ijkl,y}=0$ in Fig. 4(b) where C_{ijkl} means elastic constants.

4. The verification of the path independency of the J_x and J_y integrals.

By using Gauss's theorem, it is easy to verify the path independency of J_x and J_y integrals defined by eqs.(7) and (8) as follows.

Consider the two J_ℓ values, J_ℓ^1 and J_ℓ^2, defined by two

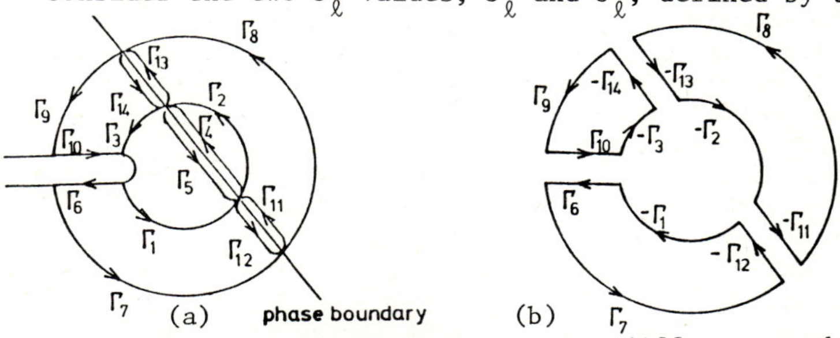

Fig.5 Two J values defined by two different paths.

different paths as follows(Fig.5(a);

$$J_{\ell}^{1} = \int_{\Gamma_1+\Gamma_2+\Gamma_3} (W\delta_{j\ell}-\sigma_{ij}U_{i,\ell})ds_j - \int_{\Gamma_4+\Gamma_5} (W\delta_{j\ell}-\sigma_{ij}U_{i,\ell})ds_j \qquad (13)$$

$$J_{\ell}^{2} = \int_{\Gamma_6+\Gamma_7+\Gamma_8+\Gamma_9+\Gamma_{10}} (W\delta_{j\ell}-\sigma_{ij}U_{i,\ell})ds_j - \int_{\Gamma_{11}+\Gamma_4+\Gamma_{13}+\Gamma_{14}+\Gamma_5+\Gamma_{12}} (W\delta_{j\ell}-\sigma_{ij}U_{i,\ell})ds_j \qquad (14)$$

where Γ_1, Γ_2,...Γ_{14} are each part of paths.

$$J_{\ell}^{2}-J_{\ell}^{1} = \int_{\Gamma_6+\Gamma_7-\Gamma_{12}-\Gamma_1} (W\delta_{j\ell}-\sigma_{ij}U_{i,\ell})ds_j + \int_{\Gamma_8-\Gamma_{13}-\Gamma_2-\Gamma_{11}} (W\delta_{j\ell}-\sigma_{ij}U_{i,\ell})ds_j$$
$$\int_{\Gamma_9+\Gamma_{10}-\Gamma_3-\Gamma_{14}} (W\delta_{j\ell}-\sigma_{ij}U_{i,\ell})ds_j \qquad (15)$$

As shown in Fig.5(b), each term of integration becomes zero because the each path closes excluding any singularities, and by using Gauss's theorem. Then $J_{\ell}^{2}=J_{\ell}^{1}$. The path independency of J_{ℓ} is verified.

Then the path independency of these is checked by the example analyses by using finite element method.

Fig.6(a) and (b) show the examples. In Fig.6(a), a crack exists perpendicular to the phase boundary and in Fig.6(b), a crack exists parallel to the phase boundary. Elastic analyses are carried out. The phase where the crack exists is called material 1 and the elastic constants are denoted as E_1 and ν_1. The adjacent phase is called material 2 and the elastic constants are denoted as E_2 and ν_2. E_1 is chosen to 21000.0kgf/mm^2 and ν_1 is 0.3. Two cases where E_2/E_1=0.5 and 2.0 are analysed. In Fig.6(b), E_2/E_1=0.1 is analysed. In all cases ν_2 is considered to be equal to ν_1.

Results are shown in Fig.7(a) and (b). In these figures, the x mark means the J integral values where the second terms of eqs.(7) and (8) are neglected and the • mark means where they are considered. In Fig.7(a), both values coincide well when the integral paths are in the materal 1, but when the paths cross the phase boundary, they separate. In Fig.7(b), they all separate because all paths cross the phase boundary. By these results, it is verified that these J integrals have path independent characteristics and it is necessary to consider the second terms of eqs.(7) and (8) to evaluate the J values of a crack in multi-phase materials.

5. Evaluation of J_x integral in three storied lamella.

In this section, the effect of the phase boundary on the crack growth is discussed. J_x values of a crack in a three storied lamella, as shown in Fig.6(a), are analysed by

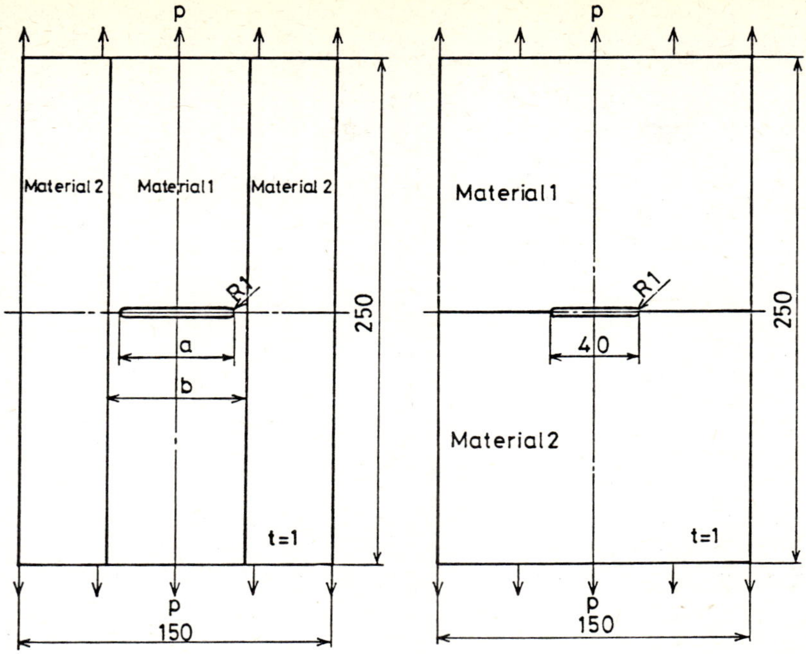

(a) A crack exists parallel to the phase boundary.

(b) A crack exists perpendicular to the phase boundary.

Fig.6

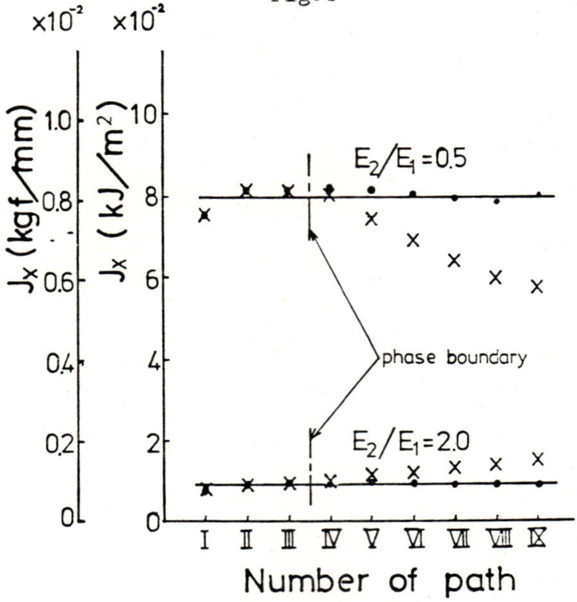

Fig.7(a) J_x values of Fig.6(a).

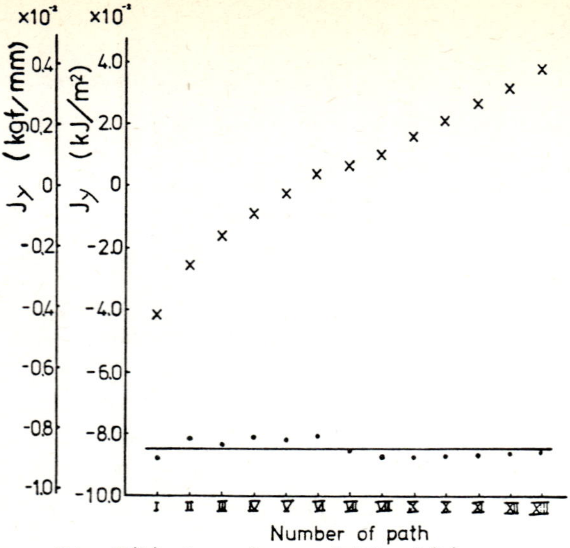

Fig.7(b) J_y values of Fig.6(b).

changing the crack length. Three cases, E_2/E_1=0.5, 2.0 and 10.0, are analysed by changing a/b value. The homogenous material, E_2/E_1=1.0, is also analysed for the comparison. Elastic analyses are carried out for all cases. Fig.8 shows one example of mesh division for the analyses. Seven paths are used to evaluate J_x values and the average of them are used.

Results are shown in Fig. 9. When the crack exists in the homogeneous material, J_x value increases slowly as a/b value increases. When the crack exists in the hard material surrounded by the soft material, E_2/E_1=0.5, J_x value increases rapidly as a/b increases. This result means that in this case, the crack tends to grow as it grows. But when the crack extends out of the phase boundary, it decreases suddenly and then increases slowly. Contrary to this result, when the crack exists in the soft material surrounded by the hard material, E_2/E_1=2.0 and 10.0, J_x value decreases as a/b increases, and at the phase boundary it decreases suddenly. This means that in this case the crack tends to stop as it grows. But when the crack extends out of the phase boundary, it increases suddenly and then increases slowly.

By these results, it is shown that the phase boundary behaves as the barrier or accelarator of the crack growth.

6. Evaluation of J_x integral of the three point bend specimen.

Up to the present, all analyses are carried out in elastic states. In this section, the elasto-plastic analyses are carried out for three point bend specimen composed of two materials.

Fig.8 Mesh division of three storied lamella.

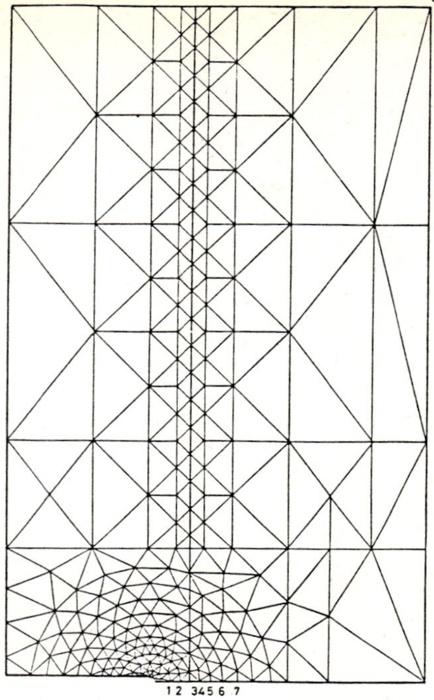

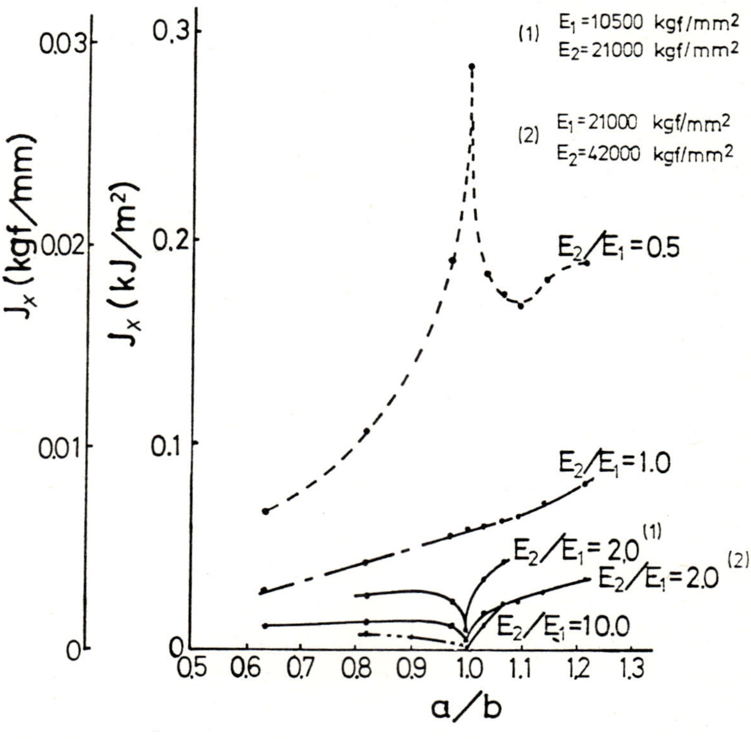

Fig.9 J_x value of the three storied lamella.

Fig.10 shows the three point bend specimen. This is the bench mark problem of this conference. The differences of the weld material and the parent material are the yield stresses and the hardening ratioes. Elastic constants are equal between them. The mechanical properties of the two materials are shown in Table 1. The ligament of the crack tip is composed of one material, weld metal. The crack length, a, is 25.5mm.

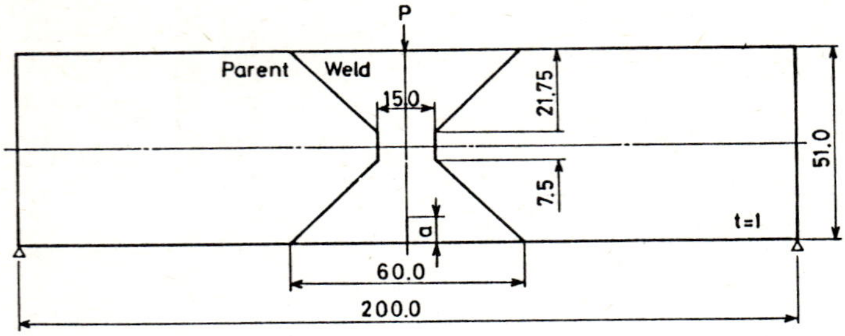

Fig.10 The three point bend specimen configuration.

Fig.11 shows the J_x values for each path. In these figures, x marks mean that the J_x values are evaluated by using Rice's equation and • marks mean that eq.(7) is used. When the deformation is yet small, both values coincide well. This is because the elastic constants are equal between them. But as the deformation proceeds, they separate gradually, and the J_x values of Rice loses its path independency. The new definition of the J_x has the path independency even is these states. By these results, eqs.(7) and (8) are applicable to the crack problem in the multi-phase materials for both elastic and elasto-plastic states.

	weld	parent
E (kN/mm^2)	210.0	210.0
ν	0.3	0.3
σ_Y (kN/mm^2)	0.78	0.60
H' (kN/mm^2)	0.61	0.47

Table 1. Mechanical properties of three point bend specimen.

Then the results are compared with the Rice's conventional J value [5]. It is expressed by the next equation.

$$J = \frac{2}{b}\int_{o}^{\delta}\left(\frac{P}{B}\right)d\delta \tag{16}$$

The J_x value obtained by this equation is called J_R, and the J values of present analysis are called J_I. J_I value is the average value of all paths. These two values are compared with each other in Fig.12. They agree very well. This is because eq.(16) assumes that the deformation of the bend bar specimen is the function of bending moment, M, and the length of the ligament length, b, as the next equation.

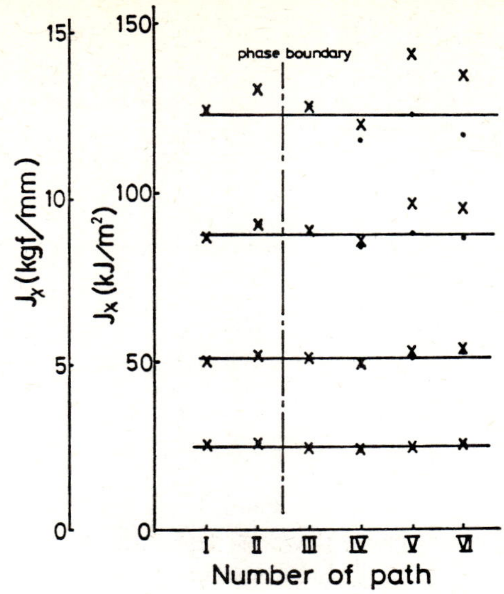

Fig.11 J_x values for each path.

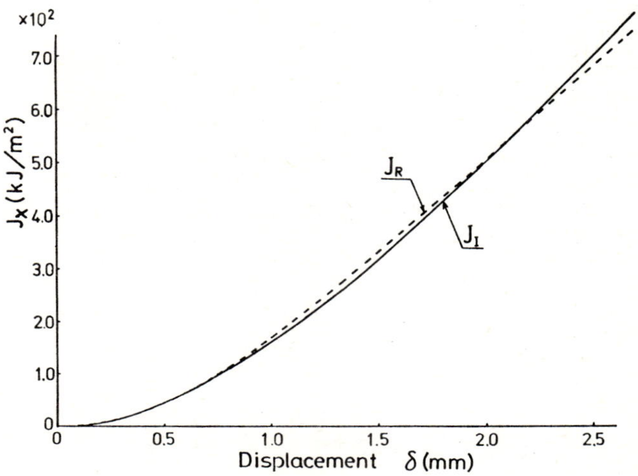

Fig.12 J vs. displacement curve.
(J_I: present analysis,
J_R: Rice's conventional J value)

$$\theta crack = h(\frac{M}{b^2}) \tag{17}$$

As in this case, the ligament of the crack tip is composed of only one material, so in this case, eq.(17) may be valid approximately. But in other cases, where the crack tip ligament is composed of more than two materials, this may be valid.

Conclusion:

The results are summarized as follows:

(1) A new J ingegral is introduced for a crack in a multi-phase material. This new J integral has the path independency even in the multi-phase material.

J_y integral is also extended to the multi-phase problem.

(2) The effect of the phase boundary on the crack extension is discussed. The phase boundary behaves as a barrier to elasto-plastic problems.

(3) Three point bend specimen is analysed in elasto-plastic state. It is shown that the new J integral is applicable to elasto-plastic problems.

Acknowledgements:

Authors wish to express their appreciation to Professor S.N. Atluri, Georgia Institute of Technology, for his helpful comments and to Mr. T. Kimura and Mr. T. Ueno, students of Science University of Tokyo, for their support and helpful discussion.

References:

[1] Rice, J.R.A. Path Independent Integral and Approximate Analysis of Strain Concentration by Notches and Cracks. Trans. ASME, J. Appl. Mech., vol 25, pp 379, 968.

[2] Eshelby, J.D., The Continuum Theory of Lattice Defects. Prog. Solid State Physics, vol 3, pp 79, 1956.

[3] Tirosh, J., Incipient Fracture Angle, Fracture Loci and Critical Stress for Mixed Mode Loading, Engg. Fracture Mech., vol 9, pp 607, 1977.

[4] Chen, W.H., On the J-Integral for Nonhomogeneous Cracked Composites, Trans. 5th Int. Conf. SMiRT, vol M, 1979.

[5] Rice, J.R. et al., Some Further Results of J-Integral Analysis and Estimates. ASTM STP536, pp 231, 1973.

CONTOUR INTEGRALS FOR SMALL CRACKS AT SHARP CORNERS IN NON WORK HARDENING ELASTIC PLASTIC MATERIALS

W S BLACKBURN AND J N SWINGLER

SUMMARY

The value of a contour integral around a crack tip is a parameter used to assess whether or not a crack will propagate in a non linear material. For a linear elastic material the value tends to zero for small cracks. For anti-plane strain the value is asymptotically proportional to the $\beta/(2\pi - \beta)$ power of the crack size for small cracks at a sharp corner containing an angle β. For a non work hardening elastic plastic material the integral is finite when the crack size is zero. Thus for sufficiently large dimensions of the structure a large value of the contour integral may occur even for very small cracks.

Finite element methods have been used here to investigate this further in the case of angles of 45° and 90°.

The numerical results for plane strain have shown that the integral was approximately proportional to stress squared and crack size to a power of about 0.25 for a 90° step and 0.055 for a 45° step.

1. INTRODUCTION

To assess the possibility of fracture arising from a crack like defect in a component in service, it is necessary to compare the values of a parameter associated with the crack in service and in test conditions and to ensure that the value of the parameter in service conditions is less than the value which arose when the crack began to propagate in the test conditions. Where linear elastic conditions may be assumed, such a parameter would be the stress intensity factor, while for more general situations, the path independent integrals J or J* may be used {1}.

CEGB SE-SSD Canal Road, Gravesend, Kent.

Handbooks of stress intensity factors (e.g {2}) exist containing values for many geometries. However, they do not contain results for cracks arising at sharp corners, locations where cracks are particularly likely to occur, e.g. where two components are forced together by centrifugal loading or where an attachment has been welded to a structure.

For a linear elastic material, the asymptotic behaviour of the stress near a sharp corner of depth a is known {3,4} to be proportional to $(r/a)^{\lambda}$ at a distance r from the corner, where for out of plane deformation $1 + \lambda = \frac{1}{2}\pi/(\pi-\frac{1}{2}\beta)$ and for in plane deformation $(1 + \lambda) \sin \beta = \pm \sin((2\pi-\beta)/(1+\lambda))$, where β is the angle at the corner. For a linear material, dimensional analysis implies that the stress intensity factor for a very small crack of length c must be proportional to $c^{\frac{1}{2}+\lambda}a^{-\lambda}$. For anti-plane strain $\lambda = -1/3$ or $-3/7$, whereas for plane deformation $\lambda = -.456$ or $-.495$ for angles of 90° and 45° respectively. Hence for these angles the stress intensity factor is proportional to $c^{1/6}$ and to $c^{1/14}$ respectively for out of plane deformation, while for in plane deformation dependence on c is even less. Since the investigation commenced, Hasebe and Iida {5} have determined the stress intensity factor for a crack at the apex of a sharp notch in a half plane under tension. They confirmed that for notch angles up to 40° it is adequate to use the stress intensity factor for a crack of depth equal to the sum of the depths of the crack and the notch.

For anti-plane strain the solution for a sharp notch in a half space has been extended to a non work hardening elastic plastic material by Hult and McClintock {6}, Koskinin {7}, Rice {8} and Arthur and Blackburn {9}, while (under the assumption that slip is confined to a single plane) a corresponding solution has been obtained by Smith, both for the above geometry {10}, and {11} for a 90° step on the surface of a half space which is uniformly stressed away from the step. It is to be noted that the contour integrals J and J* have a non zero value even in the absence of a crack and this will be true for in-plane deformation also.

The present investigation falls into two parts, an extension of the application of the analytic methods for anti-plane geometries (reported in the appendix), and a numerical investigation of the values of the contour integrals around the tip of a small crack at the root of a 45° or 90° step on the surface of a thick plate under a uniform tension away from the step. A non work hardening elastic plastic material is considered. In particular, it was intended that the dependence of the contour integrals on stress and crack size should be determined, and the differences between in-plane and out-of-plane behaviour should be noted, and the relative crack size at which crack size effects become important should be discovered.

2. NUMERICAL METHOD

The geometries considered are doubly symmetric quarters of the body whose cross sections are shown below. Uniform tensile loads were applied on the ends.

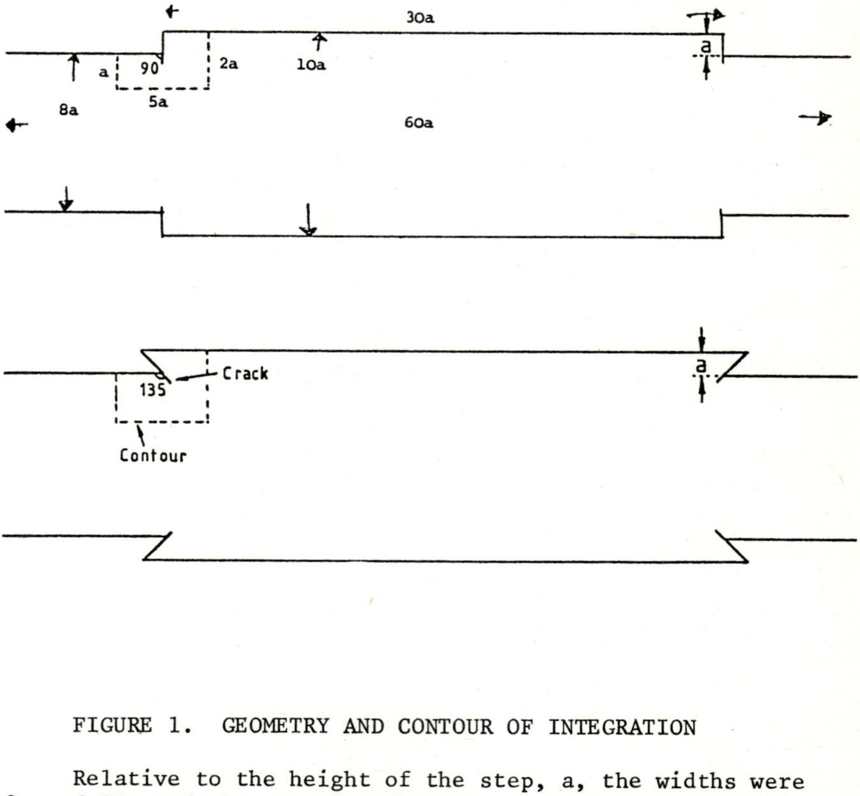

FIGURE 1. GEOMETRY AND CONTOUR OF INTEGRATION

Relative to the height of the step, a, the widths were 8a and 10a and the step length and overall length were 30a and 60a. The stress, σ, was applied in plane strain with Poisson's ratio 0.33. The magnitude was increased to 90% of the yield stress Y which was 0.1% of Young's modulus E.

Basic finite element meshes were generated with 140 or 147 quadrilateral elements with vertex and mid edge nodes.

A crack whose projected length normal to the applied stress was either 4%, ½% or 1/16% of the step height was then introduced at the reentrant corner and the four elements surrounding the tip replaced by 72 triangular elements, of which the eight around the tip were special elements which allow for the displacement to vary partly in proportion to the square root of the distance from the tip {12}. The size of the special elements was one sixteenth the size of the crack.

The displacements and stresses were calculated (with a tolerance of 5% overall) and the contour integrals around the tip were determined by the BERSAFE finite element programs {13} on the CEGB IBM 370/165 computer. The contour for which the results are presented was from the 90° or 135° corner at the crack mouth to the opposite side of the crack mouth via the surface and the dotted line (indicated in the figure) whose dimensions are a, 5a and 2a respectively.

3. RESULTS

Dimensionless values $EJ_1/(aY^2)$, $EJ_2/(aY^2)$, $EJ_1^*/(aY^2)$ and $EJ_2^*/(aY^2)$ were calculated for the configurations with 45° and 90° notch angles for the cases of crack depths of 0, 0.000625a, 0.005a and 0.04a. The subscripts $_1$ and $_2$ refer to the directions parallel and perpendicular to the crack. There was little difference between the values of J_1 and J_1^* except for the largest cracks. The values of $EJ_1/(aY^2)$ and $EJ_1^*/(aY^2)$ are presented below as functions of σ/Y. The behaviour of J_2 and J_2^* is similar.

For the smaller values of σ/Y (in most cases until σ/Y exceeds ½) the results are approximately proportional to σ^2 and for the larger values, when the plastic zone size is larger than the crack size, the results become approximately independent of crack size.

Until plasticity occurs at any of the Gauss points, linear elastic conditions are being represented. Hence when there is no crack, the values computed for the integrals should represent numerical errors. When there is a crack the value of the integral may be obtained by displacement substitution methods prior to the occurrence of plasticity. Values obtained in this way are included in parentheses in the table below for comparison with the values obtained by contour integration for $EJ/_{10}/(a\sigma_o^2)$, where the subscript $_o$ is used to denote the value at which plasticity is first observed (for σ_o/Y in the range 0.007 to 0.084).

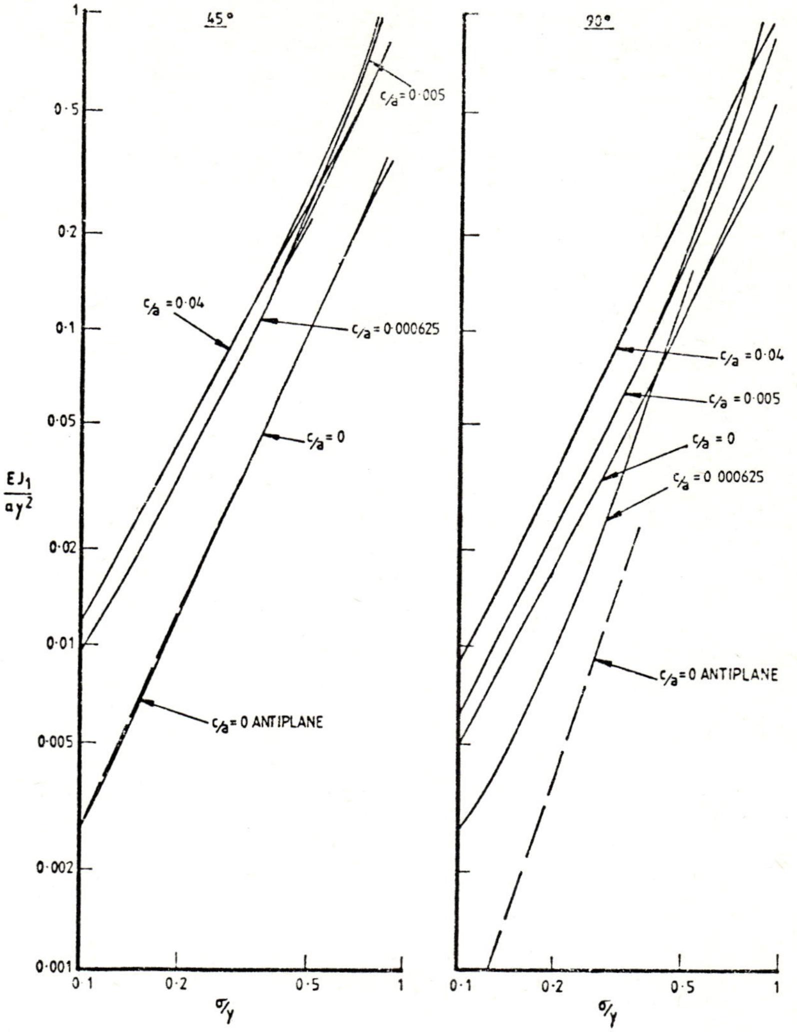

FIGURE 2. Computed values of non-dimensional J_1 and J_1* versus non-dimensional stress. Where the curves bifurcate J_1 is the lower of the two branches.

β	c/a	0	0.000625	0.005	0.04
90°		0.50	0.32(0.60)	0.60(0.61)	0.99
45°		0.34	0.88(0.44)	0.87(0.51)	0.98(0.76)

TABLE 1. LINEAR ELASTIC NON DIMENSIONAL J_1

The values obtained by the contour integral method are thus found to be inaccurate for the smallest cracks. The values for the larger cracks are comparable to the values which may be deduced from the stress intensity factors determined by Hasebe and Iida {5} for the case of cracks at the root of notches with 45° or 90° angles in the surface of a uniformly stressed half plate. As the values of the integral are proportional to σ^2 until σ/Y is approximately ½, this implies that plasticity is unimportant for such stresses on the non zero crack sizes considered.

4. DISCUSSION

When there is no crack and the values of σ/Y are small, J_1 is proportional to plastic zone size. Hence for antiplane strain it would be proportional to $\sigma^{-1/\lambda}$. Analytic results for anti-plane strain are derived in the appendix, and the dashed lines in the above figures represent the asymptotic values for small σ/Y, which are proportional to $\sigma^{-1/\lambda}$ also. If J is assumed to be proportional to $\sigma^{-1/\lambda}$ for small σ in plane strain also, this would be a dependence on σ to a power slightly greater than 2. There will be some error in the corresponding numerical values because of the lack of representation of the singularity at the root of the step.

For a small crack of length c in a linear elastic material $EJ/(a\sigma^2)$ should be proportional to $(c/a)^{1+2\lambda}$ and hence for anti plane strain to $(c/a)^{1/3}$ for a 90° angle and to $(c/a)^{1/7}$ for a 45° angle. The linear elastic finite element results for c/a = 0.04 and 0.005 imply a power dependence of about $(c/a)^{0.24}$ for a 90° angle and $(c/a)^{0.055}$ for a 45° angle, compared with proportionality to $(c/a)^{0.1}$ for a 90° angle and $(c/a)^{0.01}$ for a 45° angle with values of λ as given by Williams {4}.

The asymptotic form of the plasticity correction to these results has not yet been calculated even for anti plane strain, but J will be a function of a $(\sigma/Y)^{-1/\lambda}$ (which applies in the absence of the crack) and $(c/a)^{2\lambda}$ $(\sigma/Y)^2$ c (which applies in the absence of plasticity). For small c/a it follows that the term independent of c dominates when $(c/a)^{1+2\lambda}$ $(\sigma/Y)^{2+1/\lambda}$ is small. For $\lambda = -1/3$, this is when c/a is less than $(\sigma/Y)^3$ while for $\lambda = -3/7$ it is when c/a is less than $(\sigma/Y)^{7/3}$.

CONCLUSIONS

Numerical investigations of contour integrals around the tips of cracks at the roots of steps in plane strain in non work hardening elastic plastic materials have

(1) Found somewhat similar behaviour for in plane and out of plane deformation for small stresses.

(2) Not found plasticity to have much effect for small stresses when there is a crack to step size ratio of 1/16% or more.

(3) Found the stress dependence of the integrals to be approximately the square and the crack size dependence to be about the 0.24 power and the 0.055 power respectively for a 90° and a 45° step.

(4) Not found much effect of crack size for larger values of the ratio of applied stress to yield stress.

REFERENCES

1. BLACKBURN, W S, JACKSON, A D AND HELLEN T K. An integral associated with the state of a crack tip in a non-elastic material. Intnl. Jnl. Fracture Vol.13 No.2 pp 183-199, 1977.

2. ROOKE, D P AND CARTWRIGHT, D J Handbook of stress intensity factors, HMSO, 1976.

3. WILLIAMS, M L. On the stress distribution at the base of a stationary crack. Jnl. Applied Mech.Vol. 24 pp 109-114, 1957.

4. WILLIAMS, M L, Bending stress distribution at the base of a straight crack. Jnl. Applied Mech. Vol. 28, No. 1, pp 78-82, 1961.

5. HASEBE, N and IIDA, J. A crack originating from a triangular notch on a rim of a semi infinite plate. Engng. Fracture Mech. Vol 10 pp 773-782, 1978.

6. HULT, JAH and McCLINTOCK, F A. Elastic plastic stress and strain distributions around sharp notches under repeated shear. Ninth Int. Congr. Appl. Mech. pp 51-58, Brussels, 1956.

7. KOSKININ, M F. Elastic plastic deformation of a single grooved flat plate under longitudinal shear. ASME Jnl. Basic Engng. Vol. 85 pp. 585-594, 1963.

8. RICE, J R. Contained plastic deformation near cracks and notches under longitudinal shear. Intnl. Jnl. Fracture Mech. Vol. 2 pp 427-447, 1966.

9. ARTHUR, P F and BLACKBURN W S. The effect of the shape of plastic zone on zone size and crack opening displacement. Fracture 1969 Ed. R W Nichols, pp 15-24, Chapman and Hall, 1969.

10. SMITH, E. The spread of plasticity from a sharp groove. Int. Jnl. Eng. Sci. Vol.5 pp 791-799, 1967.

11. SMITH, E. Distribution of skew dislocations on two intersecting planes. Intnl. Jnl. Engng. Sci. vol.5, pp 613-620, 1967.

12. BLACKBURN, W S. Calculation of stress intensity factors at crack tips using special finite elements. Mathematics of Finite Elements and Applications. Ed. J R Whiteman, pp 327-336. Academic Press, 1973.

13. HELLEN, T K and PROTHEROE, S J. The BERSAFE finite element system. Computer Aided Design, Vol. 6, No. 1, pp 15-24, 1974.

ACKNOWLEDGEMENT

This paper is published by permission of the Director General, South Eastern Region, Central Electricity Generating Board.

APPENDIX

CONTOUR INTEGRAL AROUND THE ROOT OF A STEP IN ANTI-PLANE STRAIN.

In this appendix the plastic zone size and plastic root opening displacement (analogous to crack tip opening displacement) is calculated at the root of a sharp step of arbitrary angle loaded in anti-plane strain. A semi-infinite region of elastic, perfectly plastic material is assumed containing a step of angle β and projected height a, as shown in fig A.1. The plastic zone is approximated as being confined to the plane of the step and having the component of stress normal to the zone made equal to the yield stress. The contour integral J_1 around the root would equal the yield stress multiplied by this displacement. The displacement at the root of a symmetric notch has previously been calculated by Smith {10} using the same approximation and by Hult and McClintock {6} assuming (as would be appropriate for monotonic loading) a plastic zone consisting of a centred fan of lines along which the stress is constant and equal in magnitude to

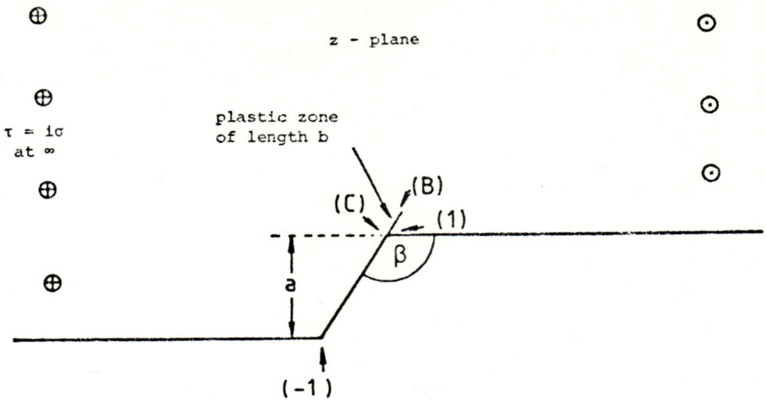

Physical space of semi-infinite body with stepped boundary loaded in antiplane strain. The material is elastic, perfectly plastic with slip confined to the plane of the step. t values for various z boundary values are given in brackets.

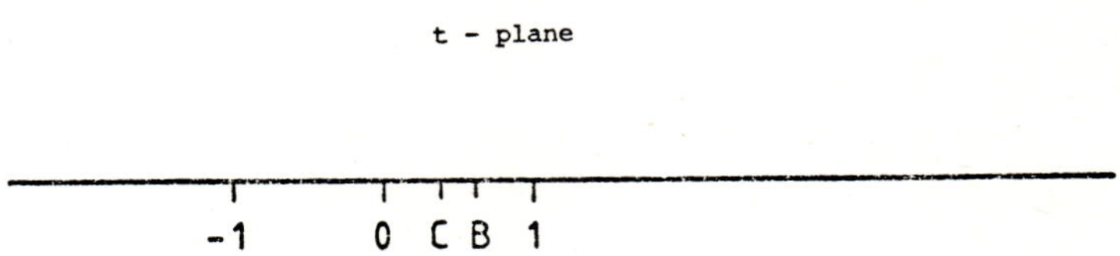

Transformed space

FIGURE A1.

the yield stress.

The physical space containing the stepped material is taken to be the upper half of the complex z plane (see fig A1) and within this space the stresses are represented by the complex variable τ where $\tau = \tau_y + i\tau_x$; $\tau_y = \sigma_{zy} = \mu \frac{\partial w}{\partial y}$ and $\tau_x = \sigma_{zx} = \mu \frac{\partial w}{\partial x}$ where μ is the shear modulus and w is the out of plane displacement. The equations of equilibrium and compatibility for anti-plane strain imply that, within the elastic part of the z plane, τ is an analytic function of $z = x + iy$. Using a Schwarz-Christoffel transformation the stepped boundary in the z plane may be mapped to the real axis in the complex t plane such that the elastic region of the z space is mapped to the upper half of the t plane. Within the t plane the boundary conditions along the real axis easily determine τ as a function of t and thus, by analyticity τ as a function of z.

The Schwarz-Christoffel transformation is defined by

$$\frac{dz}{dt} = A \frac{t-B}{(t+1)^{-\alpha}(t-1)^{\alpha+1}} \tag{A1}$$

where $\pm\infty$ in the z plane maps to $\pm\infty$ in the t plane, the two corners of the step map to ± 1, the end of the plastic zone maps to B, $\alpha = \frac{\beta}{\pi} - 1$ and A is a constant. The position of the root of the step is defined as the origin of the z plane. The value of t corresponding to o_- will be denoted by C and the plastic zone length will be called b. The following five boundary conditions determine the constants A, B, C, b and the arbitrary constant that arises on integration of equation (A1):

I $z = ae^{-i\beta}\operatorname{cosec}\beta$, $t = -1$

II $z = o_-$, $t = C$

III $z = -be^{-i\beta}$, $t = B$

IV $z = o_+$, $t = 1$

V Stress finite at $t = B$

Dealing with I to IV first

I $$A \int_{-1}^{1} \frac{(t-B)\,dt}{(t+1)^{-\alpha}(1-t)^{\alpha+1}} = -a\operatorname{cosec}\beta \tag{A2}$$

II $$\int_{C}^{1} \frac{(t-B)\,dt}{(t+1)^{-\alpha}(1-t)^{\alpha+1}} = 0 \tag{A3}$$

III

$$A\int_B^1 \frac{(t-B)\,dt}{(t+1)^{-\alpha}(1-t)^{\alpha+1}} = b \qquad \text{(A4)}$$

IV defines the origin of the z plane

Boundary condition V refers to stress. The stress has the following boundary conditions on the real axis in the t plane:

$$\tau \to i\sigma \quad , \quad t \to \pm\infty \qquad \text{(A5)}$$

$$\text{Re}\ \tau\frac{dz}{dt} = 0 \quad , \quad \{\ \text{Im}\ t=0,\ \text{Re}\ t \notin (C,1)\ \} \qquad \text{(A6)}$$

$$\text{Re}\ \tau\frac{dz}{dt} = ke^{i(\pi-\beta)}\frac{dz}{dt} \quad , \quad \{\ \text{Im}\ t=0\ ,\ \text{Re}\ t \in (C,1)\ \} \qquad \text{(A7)}$$

where k is the yield stress in shear and σ the applied stress at infinity. Since Re $\tau\frac{dz}{dt}$ is specified almost everywhere on the real axis $\tau\frac{dz}{dt}$ may be determined (up to a constant) as

$$\frac{1}{i\pi}\int_{\infty}^{\infty} \frac{\text{Re}\ \tau\frac{dz}{dt}}{\zeta - t}\,d\zeta \qquad \text{(A8)}$$

On substitution from equations (A1) and (A7) in (A8), it is found that in order to satisfy equation (A5)

$$\tau\frac{dz}{dt} = iA\sigma - \frac{iAk}{\pi}\int_C^1 d\zeta \left\{ \frac{1}{(\zeta+1)^{-\alpha}(1-\zeta)^{\alpha+1}} + \frac{t-B}{(\zeta+1)^{-\alpha}(1-\zeta)^{\alpha+1}(\zeta-t)} \right\} \qquad \text{(A9)}$$

But boundary condition V implies that the LHS of (A9) $\to 0$ as $t \to B$. Thus

$$\int_C^1 \frac{dt}{(t+1)^{-\alpha}\ (1-t)^{\alpha+1}} = \pi\,\bar{\sigma} \text{ where } \bar{\sigma} = \frac{\sigma}{k} \qquad \text{(A10)}$$

Equation (A10) determines C and then equation (A3) may be used to determine B, equation (A2) to determine A and equation (A4) to give the plastic zone size.

Defining

$$I_{1\alpha}(X) = \int_X^1 \frac{t\,dt}{(1+t)^{-\alpha}(1-t)^{\alpha+1}}$$

and $$I_{2\alpha}(X) = \int_X^1 \frac{dt}{(1+t)^{-\alpha}(1-t)^{\alpha+1}}$$

then it follows from equation (A10) that $I_{2\alpha}(C) = \pi\bar{\sigma}$, which may be solved numerically for C.

In terms of C, $$B = \frac{I_{1\alpha}(C)}{\pi\bar{\sigma}}$$

$$A = \frac{\pi\,\bar{\sigma}\,a\,\mathrm{cosec}\,\pi\,(1+\alpha)}{I_{2\alpha}(-1)\,I_{1\alpha}(C) - \pi\bar{\sigma}\,I_{1\alpha}(-1)}$$

But $I_{1\alpha}(-1) = B(-\alpha,2)\ {}_2F_1\,(-\alpha,2;2-\alpha;-1) - B\,(\alpha+1,2)\ {}_2F_1$

$(\alpha+1,2;3+\alpha;-1)$

where B in the Beta function and ${}_2F_1$ the Gauss Hypergeometric function. This may be expanded and reduced to give $I_{1\alpha}(-1) = \pi(1+2\alpha)\ \mathrm{cosec}\ \pi(1+\alpha)$ and similarly $I_{2\alpha}(-1) =$ $B(-\alpha,\ 1+\alpha) = \pi\ \mathrm{cosec}\ \pi\ (1+\alpha)$

Thus $$A = \frac{\bar{\sigma}\ a}{I_{1\alpha}(C) - \pi\bar{\sigma}(1+2\alpha)}$$

The root opening displacement δ may be determined as

$$\mathrm{Im}\ \frac{1}{\mu} \int_C^1 \tau\,\frac{dz}{dt}\,dt$$

Thus in non-dimensional terms the root opening displacement is

$$\frac{\mu\delta}{ka} = \frac{A}{\pi a}\left\{ (C-1)\ \pi\bar{\sigma} - \int_C^1 \frac{(\zeta-B)}{(\zeta+1)^{-\alpha}(1-\zeta)^{\alpha+1}}\ \ell n \left|\frac{\zeta-1}{\zeta-C}\right|\ d\zeta \right\}$$

and the plastic zone size is

$$b = A \int_{B}^{1} \frac{(t-B)\,dt}{(t+1)^{-\alpha}(1-t)^{\alpha+1}}$$

For small values of $\bar{\sigma}$, the root opening displacement may be expanded and approximated to first order in $\bar{\sigma}$ by

$$\frac{\mu\delta}{ka} = \frac{\pi^{-1-\frac{1}{\alpha}}(-\alpha)^{-\frac{1}{\alpha}}\bar{\sigma}^{1-\frac{1}{\alpha}}}{1-\alpha}$$

which for particular vaues of β is as follows

$$\beta = \frac{\pi}{4}, \quad \frac{\mu\delta}{ka} = \frac{3}{7}\left(\frac{3}{4}\pi\right)^{1/3}\bar{\sigma}^{7/3} \simeq 0.39\,\pi^{1/3}\bar{\sigma}^{7/3}$$

$$\beta = \frac{\pi}{2}, \quad \frac{\mu\delta}{ka} = \frac{1}{6}\pi\bar{\sigma}^{3}$$

$$\beta = \frac{3\pi}{4}, \quad \frac{\mu\delta}{ka} = \frac{\pi^{3}}{320}\bar{\sigma}^{5}$$

For small values of $\bar{\sigma}$ the plastic zone size may be expanded to first order in $\bar{\sigma}$ as

$$\frac{b}{a} = (-\alpha)^{-(\alpha+1/\alpha)}(1-\alpha)^{\alpha-2}\pi^{-1/\alpha}\bar{\sigma}^{1-1/\alpha}$$

The contour integral J evaluated on a contour surrounding the plastic zone is equal in this instance to $k\delta$. The non-dimensional value of J referred to in the main text is equal to the non-dimensional root opening displacement defined above (with Poisson's ratio one third). The values of the non-dimensional root opening for small values of non-dimensional stress $\bar{\sigma}$ are plotted in Figure 2 along with graphs of non-dimensional J versus non-dimensional stress generated by the finite element methods as described earlier for in plane strain.

Further calculations are being undertaken to eliminate the assumption that plasticity is confined to the plane of the crack, and to determine the asymptotic form of the plasticity correction to the asymptotic form of the integral for small $\bar{\sigma}$ in the presence of a crack.

Effect of attachment and stiffener damage on plastic strip yielding for a crack in a stiffened sheet.

by

T. P. Rich[1] and D. J. Cartwright[2]

SUMMARY

The complex variable formulation of two dimensional elasticity is used to analyse a crack with symmetrical strip yield zones in a stiffened sheet. The strip yield zone size, the crack opening displacement, the maximum load concentration in the stiffener and the force in the attachment point near the crack have been evaluated. The dependence of these parameters on the stiffness and spacing of the stiffeners is determined for a crack located in the sheet either symmetrically between two stiffeners or under the central stiffener of three stiffeners. When the crack is under a central stiffener this is considered to be unbroken or broken at the crackline. The effect of attachment flexibility and attachment failure is presented for configurations of engineering interest.

INTRODUCTION

Stiffening elements are frequently used to improve the strength and stability of thin sheet structures. Such methods of reinforcement are particularly common in the aircraft industry where it is necessary to utilise high strength, low weight materials in wing and fuselarge structures. The stiffening may be used as a means of limiting the extent of accidental damage, to retard the growth rate of cracks or to improve the level of residual static strength of the damaged structure. The purpose is to lessen the severity of the damage sufficiently so as to avoid complete failure of the structure between routine inspections while keeping the frequency of these inspections as low as possible. It thus becomes necessary to determine the growth rate of the crack and the size of the crack which will cause complete failure of the structure. In fatigue, after initiation of a crack, the growth rate can usually be determined using the stress intensity factor. These have been determined for a variety of stiffened sheet configurations many of which have been reported [1, 2]. In the determination of residual static strength, where extensive yielding may occur around the crack tip, it is necessary to consider the effect of additional loads which will be transferred to the

1. Dept. Mech. Engng.
Texas A and M University
College Station
Texas 77843
U.S.A.

2. Dept. Mech. Engng.
The University
Southampton SO9 5NH
England.

attachment points as this may lead to distortion and possibly failure of the attachments thereby rendering the stiffener ineffective. Fatigue cracks may also initiate in the stiffener itself ahead of a crack in the main sheet. Thus it is necessary to consider the effect of a broken stiffener on the extent of yielding in the sheet so that its effect on the static strength of the structure may be determined; see for example references 7 and 9 to 12 in Ref. [2].

In this paper the complex variable formulation for two dimensional elasticity is used in combination with the method of compatible deformations to analyse a crack with symmetrical strip yield zones [3, 4] in a stiffened sheet. The effect of attachment deflection, attachment failure and stiffener failure on the length of the strip yield zone is examined for a crack in the sheet. The crack is considered to be located centrally between two stiffeners (bay centred crack) or symmetrically across the central one of three equally spaced stiffeners (stiffener centred crack): the central stiffener is considered to be broken or intact at the crackline. The magnitude of the crack tip opening displacement, the stiffener load concentration and the maximum attachment force are determined in a lightly and a heavily stiffened sheet for several strip yield zone lengths as a function of attachment flexibility and number of failed attachments. The results presented extend those previously obtained [2] in which the stiffeners were considered to be both undamaged and connected to the sheet with rigid attachments.

CONFIGURATIONS AND MODEL

The stiffened sheet configurations studied are shown in Fig. 1 (bay centred crack) and Fig. 2 (stiffener centred crack). For both configurations the model consists of an infinite flat sheet of modulus E, Poisson's ratio ν and thickness t which contains a crack of length 2a having a strip yield zone of length s where $(a + s) = a_1$, the distance from the crack centre to the end of the strip yield zone. The stress σ_Y is uniform along the length of the strip yield zone. The sheet, which is subjected to a uniform stress σ perpendicular to and remote from the crack, is stiffened along lines perpendicular to the crackline with stiffeners of area A_s, and modulus E_s, spaced a distance b apart. The stiffeners are assumed to be concentrated at the sheet surface, and to have zero in-plane bending stiffness. They are fixed to the sheet with attachments of diameter d spaced a distance p apart with the distance from the crackline to the first unbroken attachment equal to p. The attachments are assumed to deflect with an elastic compliance q under the action of the attachment forces. Remote from the crack the stiffeners are subjected to a stress σ_s $(= \sigma E_s / E)$ in order to maintain strain compatibility between the sheet and each stiffener. For the stiffener centred crack configuration (Fig. 2) the stiffeners are considered to be either

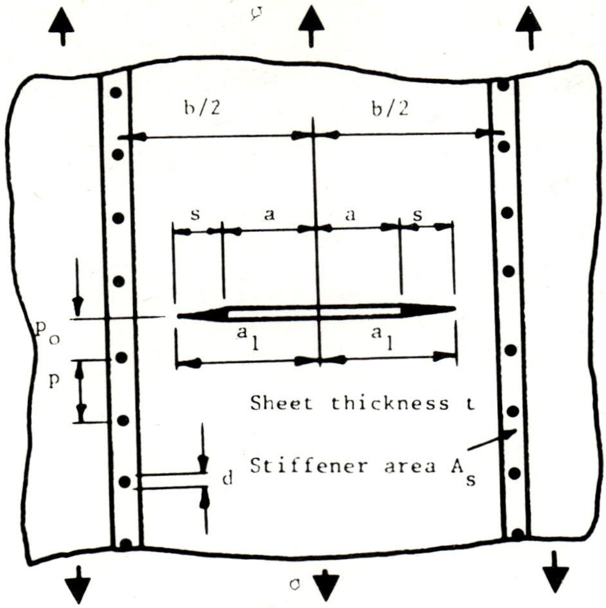

Fig. 1 Strip yield zone crack in a sheet stiffened with two stiffeners.

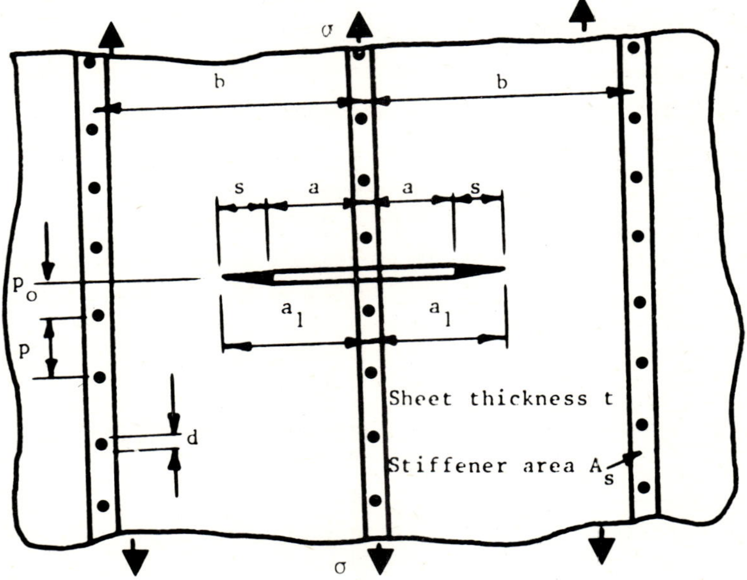

Fig. 2 Strip yield zone crack across the central stiffener in a sheet stiffened with three stiffeners.

entirely intact or the central one is considered to be broken at the crackline.

METHOD OF ANALYSIS

In the analysis the stiffener is represented as a distribution of point forces. The distribution is determined by satisfying compatibility of displacements and equilibrium of forces between the sheet and the stiffener where they are attached. In the case of discretely attached stiffeners the forces occur at the attachment points (rivet locations) and are obtained by solving a series of simultaneous equations which arise from the compatibility conditions and the condition that the stresses are bounded in the strip yield zone. The load concentration in the stiffener is determined by summing the forces and the uniform end load on the stiffener and the crack tip opening displacement is determined by summing the displacements resulting from the attachment forces, the uniform stress and the cohesive stress over the strip yield zone. The compatibility condition between any two adjacent attachment points of the same stiffener can be written, in the notation of Ref.2, as

$$\Delta v^{a}_{m,n} + \Delta^{b}_{m,n} + \Delta v^{c}_{m,n} - \Delta v^{d}_{m,n} = \Delta v^{e}_{m,n} \tag{1}$$

The relative displacement, $\Delta v^{a}_{m,n}$ represents the elongation of the sheet due to the uniform stress σ between the nth and the (n + 1)th attachment points along the line of the mth stiffener. The relative displacement $\Delta v^{b}_{m,n}$ is the elongation of the same interval due to the point forces $P_{i,j}$ in the sheet and $\Delta v^{c}_{m,n}$ is that resulting from the cohesive stresses σ_Y over the strip yield zone. The relative displacement $\Delta v^{d}_{m,n}$ is the elongation of the corresponding interval in the stiffener due to the uniform stress $\sigma\, E_s/E$ and that due to the attachment forces. The term $\Delta v^{e}_{m,n}$ is the difference between the displacement of the nth and the (n + 1)th points resulting from the attachment deflection. The relative displacements can be written in terms of the complex functions defined by Erdogan [5]: this has been done by Cartwright and Rich [2] for a crack of length $2a_1$ lying along the real (x) axis with the centre at $z = 0$ ($z = x + iy$). The relative displacement is defined as

$$\Delta v^{a}_{m,n} = \frac{\sigma a_1(1 + \nu)}{2E} \left[\bar{v}^{a}(z_{m,n+1}) - \bar{v}^{a}(z_{m,n})\right] \tag{2}$$

for the special case of the displacement between the x axis and the first attachment this becomes

$$\Delta v_{m,o}^{a} = \frac{\sigma a_1 (1 + \nu)}{2E} \bar{v}^{a} (z_{m,1}) \tag{2a}$$

where $z_{m,n+1}$ and $z_{m,n}$ are the coordinates of the (n + 1)th and the nth attachment in the mth stiffener and the functions $\bar{v}^{a}(z_{m,n+1})$ and $\bar{v}^{a}(z_{m,n})$ are the dimensionless displacements in the y direction at each of these points respectively. The elongation $\Delta v_{m,n}^{b}$, in the sheet due to the point forces $P_{i,j}$ acting along the line of the stiffener at all the attachment points is given by

$$\Delta v_{m,n}^{b} = \frac{(1+\nu)}{Et} \sum_{i=1}^{M} \sum_{j=1}^{N} P_{i,j} \left[\bar{v}^{b}(z_{i,j};z_{m,n+1}) - \bar{v}^{b}(z_{i,j};z_{m,n}) \right] \tag{3}$$

and

$$\Delta v_{m,o}^{b} = \frac{(1+\nu)}{Et} \sum_{i=1}^{M} \sum_{j=1}^{N} P_{i,j} \bar{v}^{b}(z_{i,j};z_{m,1}) \tag{3a}$$

where $\bar{v}^{b}(z_{i,j};z_{m,n+1})$ and $\bar{v}^{b}(z_{i,j};z_{m,n})$ are respectively the dimensionless displacements at $z_{m,n+1}$ and $z_{m,n}$ resulting from a pair of collinear opposing forces $P_{i,j}$ acting perpendicular to and away from the crackline at $z_{i,j}$ and $\bar{z}_{i,j}$. The summation limit N is the number of attachment points each side of the crackline (the zeroth point n = 0 being on the crackline) and M is the number of stiffeners. The elongation $\Delta v_{m,n}^{c}$, in the sheet due to the cohesive stress σ_Y over the length s ahead of each crack tip is given by

$$\Delta v_{m,n}^{c} = \frac{-\sigma_Y a_1 (1 + \nu)}{2\pi E} \left[\bar{v}^{c}(z_{m,n+1};s) - \bar{v}^{c}(z_{m,n},s) \right] \tag{4}$$

and

$$\Delta v_{m,o}^{c} = \frac{-\sigma_Y a_1 (1 + \nu)}{2\pi E} v^{c}(z_{m,1};s) \tag{4a}$$

where $\bar{v}^{c}(z_{m,n+1};s)$ and $\bar{v}^{c}(z_{m,n};s)$ are the dimensionless displacements at the attachment points $z_{m,n+1}$ and $z_{m,n}$ due to the cohesive stress σ_Y over the length s ahead of each crack tip. The expressions for $\bar{v}^{a}$, $\bar{v}^{b}$ and $\bar{v}^{c}$ have been derived from the work of Erdogan [5] and can be obtained from the work of Cartwright and Rich [2]. The elongation $\Delta v_{m,n}^{d}$ in the stiffener between the nth and the (n + 1)th attachment points is given by

$$\Delta v^{d}_{m,n} = \frac{p\sigma}{E} + \frac{p}{A_s E_s} \sum_{k=n+1}^{N} P^{s}_{m,k} \tag{5}$$

and between the first attachment point and the x axis is given by

$$\Delta v^{d}_{m,o} = \frac{p_o \sigma}{E} + \frac{p_o}{A_s E_s} \sum_{k=1}^{N} P^{s}_{m,k} \tag{6}$$

where $P^{s}_{m,k}$ is the force at the kth attachment point in the mth stiffener i.e. at $z_{m,k}$ and p_o is the distance from the crack line to the first unbroken attachment point in each stiffener. The first term on the right hand side of equations (5) and (6) is the uniform extension of the interval due to the remote stress $\sigma E_s/E$ applied to the stiffener. The second term is the uniform extension of the interval between the nth and the (n + 1)th point, in the mth stiffener, resulting from the attachment forces at and beyond the (n + 1)th point. The final term $\Delta v^{e}_{m,n}$ in equation (1) is the difference in displacement between the attachment point at $z_{m,n+1}$ and that of $z_{m,n}$ due to the elastic deflection of the attachment. It is given by

$$\Delta v^{e}_{m,n} = q(P^{s}_{m,n+1} - P^{s}_{m,n}) \tag{7}$$

and $$\Delta v^{e}_{m,o} = q\, P^{s}_{m,1} \tag{8}$$

where q is the elastic compliance of each attachment.

Equilibrium at each attachment is given by

$$P_{i,j} + P^{s}_{i,j} = 0 \quad (i = 1, M;\ j = 1, N) \tag{9}$$

The substitution of equations (2) to (9) into (1) gives the following series of simultaneous equations in terms of the unknown attachment forces $P_{i,j}$:

$$(1 + \nu) \sum_{i=1}^{M} \sum_{j=1}^{N} \bar{P}_{i,j} \left[\bar{v}^{b}(z_{i,j};z_{m,n+1}) - \bar{v}^{b}(z_{i,j};z_{m,n}) \right] +$$

$$\frac{\lambda_1 \gamma_n}{2\mu_1} \sum_{k=n+1}^{N} \bar{P}_{m,k} - \frac{\lambda(1+\nu)}{4\pi} \frac{\sigma_Y}{\sigma} \left[\bar{v}^{c}(z_{m,n+1};s) - \bar{v}^{c}(z_{m,n};s) \right]$$

$$+ Q\,(\bar{P}_{m,n+1} - \bar{P}_{m,n})$$

$$= \frac{\lambda_1 \gamma_n}{2\mu_1} - \frac{(1+\nu)\lambda}{4}\left[\bar{v}^a(z_{m,n+1}) - \bar{v}^a(z_{m,n})\right] \qquad (10)$$

where

$$\lambda_1 = \frac{2a_1 Et}{A_s E_s},\ \mu_1 = \frac{a_1}{p},\ Q = qEt,\ \bar{P}_{i,j} = \frac{P_{i,j}}{A_s \sigma_s}$$

$$\gamma_o = \frac{p_o}{p} \text{ and } \gamma_n = 1 \text{ for } n \geqslant 1$$

The condition that the stresses be bounded over the strip yield zone is given by ensuring that the singular term at $x = \pm a_1$ is reduced to zero. The condition is satisfied by

$$\sigma\sqrt{\pi a_1} + \frac{1}{t\sqrt{\pi a_1}}\left[\sum_{i=1}^{M}\sum_{j=1}^{N} P_{i,j}G(z_{i,j})\right] - \sigma_Y\sqrt{\pi a_1}\,h(s) = 0 \qquad (11)$$

The first term in equation (11) is the stress intensity factor (magnitude of the singular term) due to the remote stress, the second term is that due to the point loads and the third is the stress intensity factor due to the cohesive stress σ_Y. The Green's function $G(z_{i,j})$ is that for a pair of collinear, opposing unit forces located at $z_{i,j}$ and $\bar{z}_{i,j}$ acting in the sheet in a direction away from, and perpendicular to, the crackline. The dimensionless form of equation (11) is

$$\frac{2}{\pi}\sum_{i=1}^{M}\sum_{j=1}^{N} \bar{P}_{i,j}\, G(z_{i,j}) - \lambda h(s)\frac{\sigma_Y}{\sigma} = -\lambda \qquad (12)$$

The N x M unknown forces $P_{i,j}$ and the stress ratio σ_Y/σ are determined from the M x (N+1) equations given by equations (10) and (12). The crack opening displacement δ is calculated at $x = \pm a$ by summing the displacements due to the force at each attachment point, the uniform stress and the cohesive stress using equation (A1), (A2) and (A3) of Ref. [2] respectively. The load concentration at the crackline in the mth stiffener is given by the ratio $L_m/(\sigma bt)$ where L_m is the load in the stiffener when the sheet is cracked and σbt is the load carried by the sheet across one bay when the sheet is uncracked. The ratio depends on the attachment froces in the following way

$$\frac{L_m}{\sigma bt} = \frac{\mu}{1 - \mu}\left[1 + \sum_{j=1}^{N} \bar{P}_{m,j}\right] \tag{13}$$

where μ, the ratio of the stiffness of the stiffeners to that of the sheet and stiffeners combined, is given by

$$\mu = \frac{A_s E_s}{bt\ E + A_s E_s} \tag{14}$$

When the mth stiffener is broken at the crackline the compatibility condition for the interval adjacent to the crack line in that stiffener is replaced by

$$1 + \sum_{j=1}^{N} \bar{P}_{m,j} = 0 \tag{15}$$

which expresses the condition that the force in the first interval of the broken mth stiffener is zero.

The above formulation is applicable to a configuration in which the stiffeners are spaced an arbitrary distance apart and symmetrically located each side of the y axis where stiffeners are across and perpendicular to the crackline and have an equal number of attachment points symmetrically placed either side of the crackline. However in order to limit somewhat the number of variables, the solutions are obtained for equally spaced stiffeners only.

RESULTS AND CONCLUDING DISCUSSION

Table 1 shows the values of parameters used in the present paper. The value of p/b and d/p were selected as being

	p/b	$^1/_6$
	d/p	0.125
	ν	0.3
N	unbroken stiffeners	10
	central broken stiffener	15

Table 1 Parameter used for stiffened panel configurations.

typical of those used in aircraft structures. The number of

attachment points N either side of the crackline was selected for the case of unbroken stiffeners on the basis of a convergence study [2] and that for a central broken stiffener from the results of the convergence study shown in figure 3.

In figure 4 is shown the effect of a broken central stiffener on the stiffener centred crack for values of s/a = 0.1 and 0.5 in a sheet having $\mu = 0.2$. The crack opening displacement δ/δ_o is given as a function of a/b in figure 4b where

$$\delta_o = \frac{8\sigma_Y a}{E} \ln \left\{ \sec\left(\frac{\pi\sigma}{2\sigma_Y} \right) \right\} \tag{16}$$

is the crack opening displacement for the crack in the unstiffened sheet ($\mu = 0$) subjected to the same applied stress ratio σ/σ_Y as the stiffened sheet: the variation of σ/σ_Y with a/b for the stiffened sheet is given in figure 4a. It can be seen from figure 4b that the broken stiffener initially causes an increase in the crack opening displacement which reduces as the crack tip reaches the outer pair of stiffeners (a/b = 1.0). The results for the first attachment force are shown in figure 4c (intact stiffeners) and figure 4d (central broken stiffener). It can be seen from figure 4d that the initial increase in δ/δ_o for the broken stiffener, seen in figure 4b, is associated with forces in the central stiffener acting so as to open the crack (positive forces) whereas for the intact central stiffener (figure 4c) the opposite occurs. For longer cracks (a/b > 0.8) the central stiffener force becomes less important and the crack opening displacement is dominated by the forces in the outer stiffeners which acts so as to restrict the crack from opening for both the intact and broken stiffener configurations. It can also be seen from figures 4c and 4d that first attachment force in the outer stiffeners increases abruptly as the tip of the strip yield zone reaches the stiffeners.

Figure 5 shows the effect of μ on the applied stress ratio σ/σ_Y, the crack opening displacement δ/δ_o, the load concentration $L/(\sigma bt)$ in the outer stiffeners at the crackline and the force at the first attachment $P_1/(\sigma bt)$ for two values of s/a equal to 0.1 and 0.5. For short cracks increasing μ has the effect of increasing δ/δ_o (fig.4b) for a given value of s/a and a/b: this is reversed for long cracks. Also figures 4c and 4d show that larger values of μ will cause larger loads to be transferred through the attachment points: these increase for a given a/b as s/a increases.

In figure 6 the effect of attachment flexibility qEt on the bay centred crack is shown for values of s/a = 0.3 and 0.9 for $\mu = 0.5$. It can be seen that for both values of s/a the

Key	○	□	★	☆
$a/_b$	0.026		0.500	
$s/_a$	0.30		0.9	
μ	0.2	0.5	0.2	0.5
Value at N = 20: $\frac{\sigma}{\sigma_Y}$	0.273	0.148	0.627	0.578
Value at N = 20: $\frac{\delta}{\delta_o}$	2.781	9.736	1.165	1.638
Value at N = 20: $\frac{P_1}{\sigma bt}$	0.176	0.511	0.031	0.097

(a) Configurations used in convergence tests

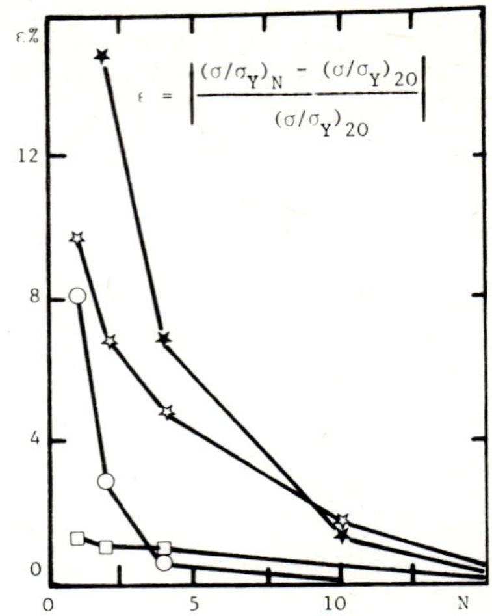

(b) Convergence of applied stress ratio

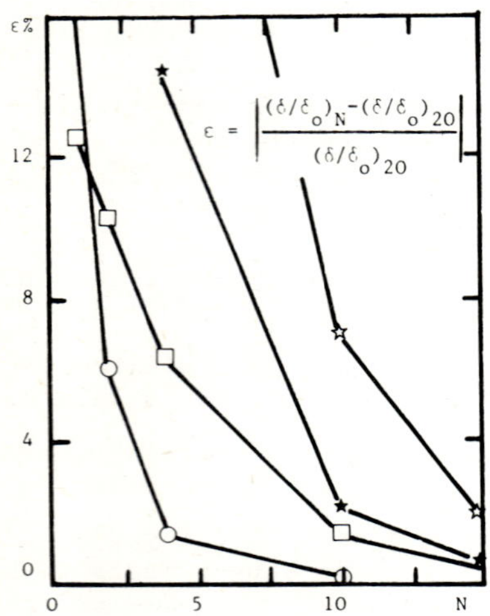

(c) Convergence of crack opening displacement

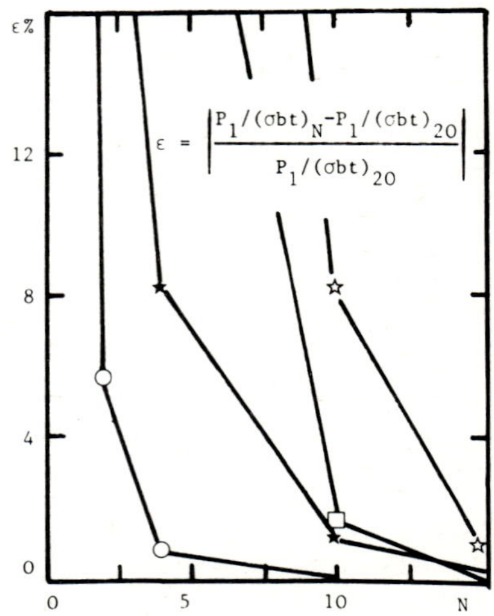

(d) Convergence of force at first attachment from crackline

Fig. 3 Convergence for stiffener centred crack as a function of the number of attachment points either side of the crackline: central stiffener broken at the crackline.

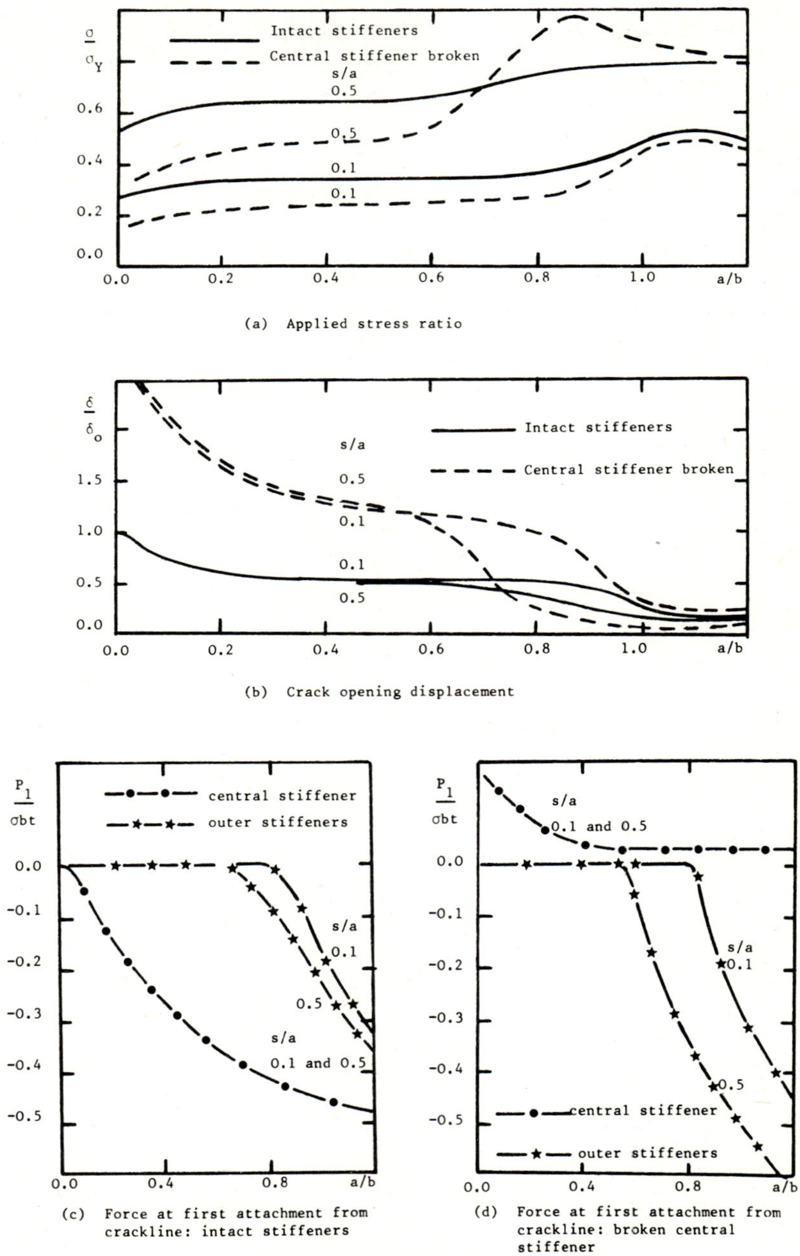

Fig. 4 Effect of the central stiffener broken at the crackline on a stiffener centred crack (μ = 0.2).

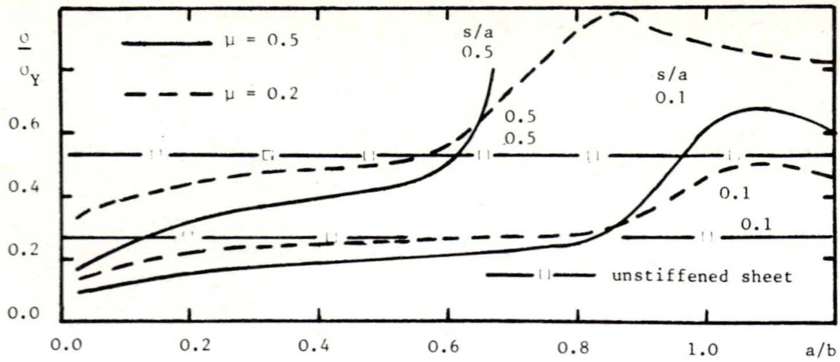

(a) Applied stress ratio

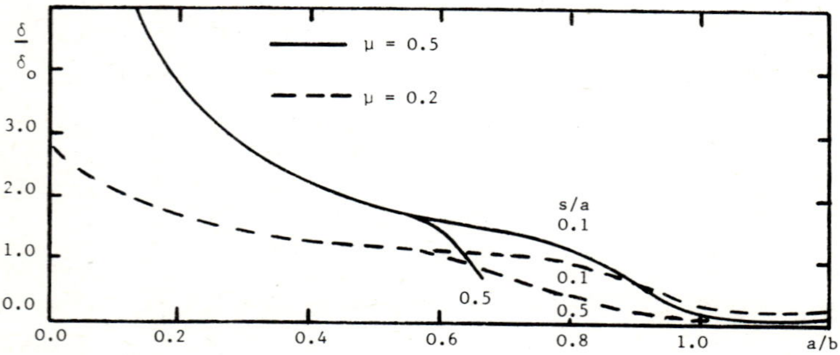

(b) Crack opening displacement

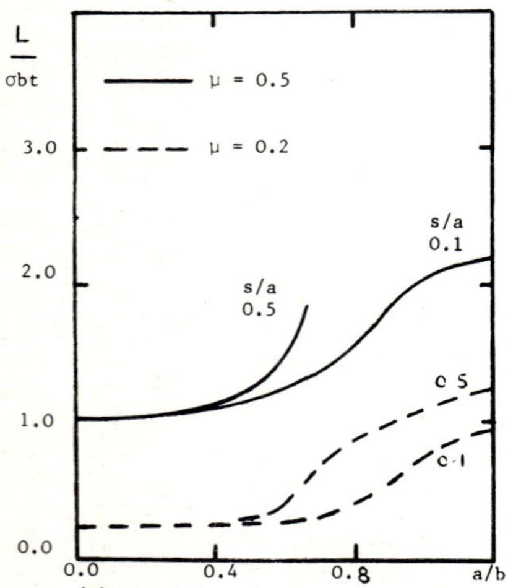

(c) Load in outer stiffener at crackline

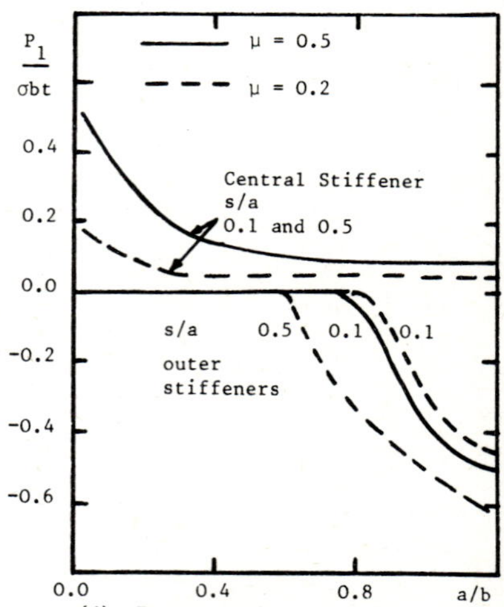

(d) Force at first attachment from crackline

Fig. 5 Effect of stiffness ratio for a stiffener centred crack with central stiffener broken at the crackline.

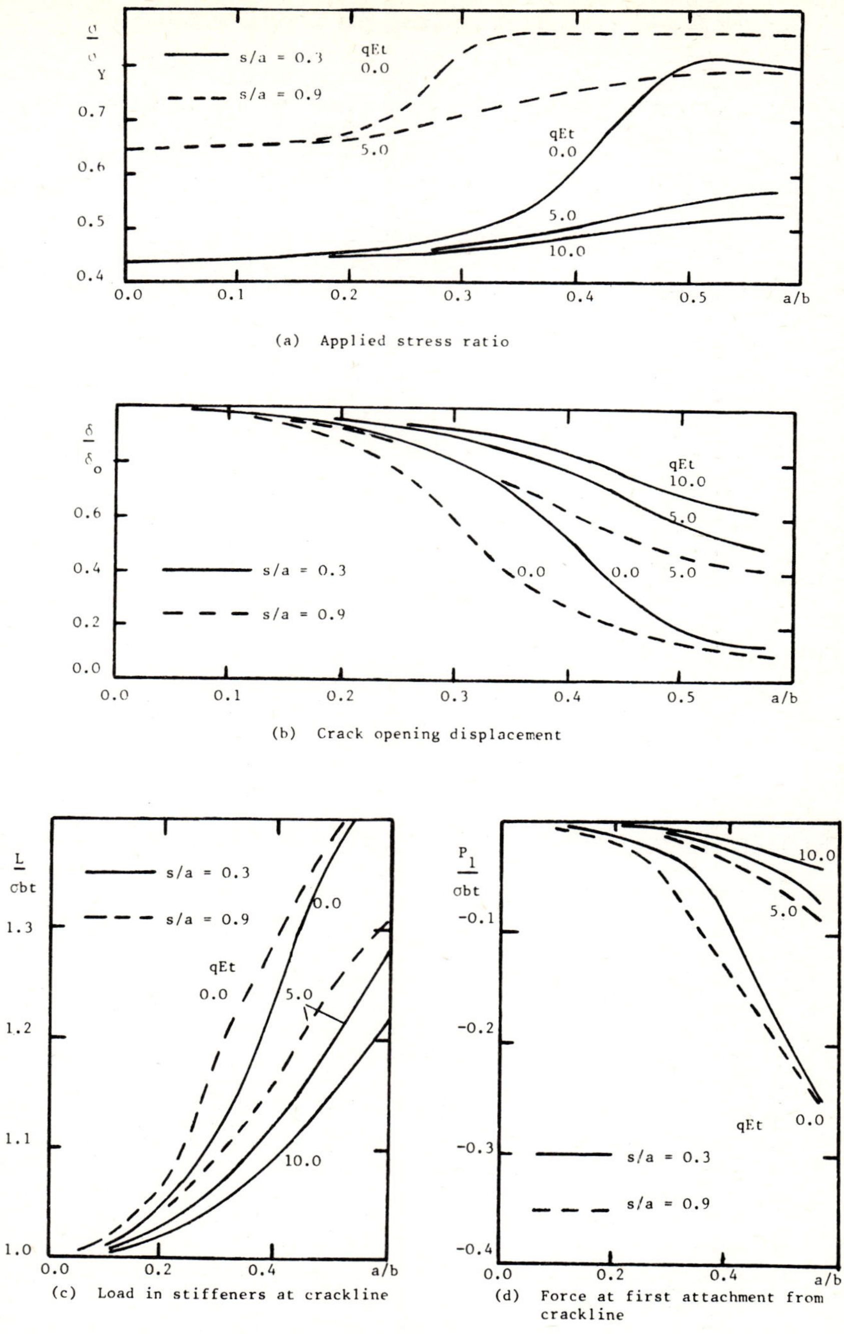

Fig. 6 Effect of attachment flexibility for a bay centred crack (μ = 0.5).

effect of increasing the attachment flexibility from qEt = 0.0 (rigid attachments) to a typical value of qEt = 5.0 causes a marked increase in δ/δ_o (typically a factor of 3 for a/b = 0.5). This increase is accompanied by a corresponding decrease in the magnitude of the stiffener load concentration and force at the first attachment as shown in figures 6c and 6d. Increasing qEt from 5.0 to 10.0 causes only a further 20% increase in δ/δ_o at the same a/b (= 0.5).

The effect of attachment failure is shown in figure 7 for a bay centred crack in which the tips are considered to be just past the stiffeners (a/b = 0.52): the sheet has a value of μ = 0.5. The effect of increasing the number of failed attachments N_f is to allow the crack opening displacement to increase for all values of s/a with a corresponding decrease in the loads carried by the stiffener and at the first attachment point.

For a constant value of applied stress σ/σ_Y figure 8 shows that the strip yield zone length s/a will be reduced as the crack tip approaches an unbroken stiffener. Figure 9 shows the effect of the applied stress on the strip yield zone length s/a for several combinations of stiffener and attachment damage for the bay centred crack (a/b = 0.52).

The effect of yielding in the sheet causes an increase in the load concentration in the stiffener and in the force at the first attachment. These effects are shown in figures 10 and 11 respectively for two values of μ = 0.2 and 0.5. It can be seen that the load concentration L at s = 0.9 compared to that at s = 0 (L_o) is higher for lower stiffening ratios and is typically of order 20%. The increase in the first rivet force P_1 compared to that at s = 0 (P_o) is more marked but for situations of practical significance will be of order 10% (a/b≃0.5).

REFERENCES

[1] CARTWRIGHT, D.J., ROOKE, D.P. - Cracks at or between stiffeners in a reinforced sheet: stiffener attached by a double row of rivets. Technical Report, Royal Aircraft Establishment, Farnborough (in preparation).

[2] CARTWRIGHT, D.J., RICH, T.P. - Plastic strip yielding for a crack in a stiffened sheet, First International Conference on Numerical Methods in Fracture Mechanics, Ed. D.R.J. Owen and A.R. Luxmore, University College Swansea, 1978.

[3] DUGDALE, D.S. - Yielding of steel sheets containing slits, J.Mech. Phys., Solids, Vol. 8, pp. 100-104, 1960.

[4] BILBY, B.A., COTTRELL, A.H., SWINDEN, K.H. - The spread of plastic yield from a notch, Proc. Roy. Soc. (London) A 272, p. 304-314, 1963.

[5] ERDOGAN, F. - On the stress distribution in plates with collinear cuts under arbitrary loads, Fourth U.S. National Congress on Applied Mechanics, pp. 547-553, 1962.

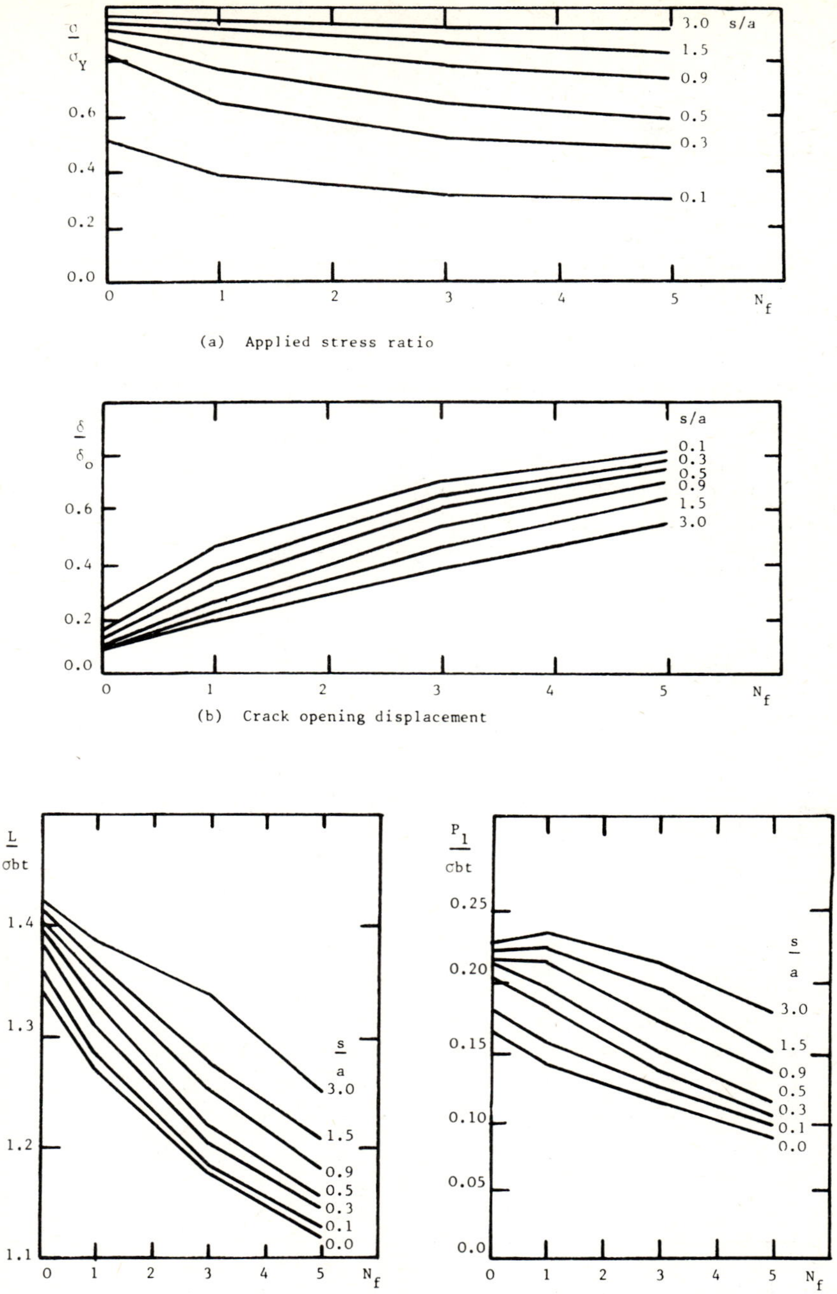

(a) Applied stress ratio

(b) Crack opening displacement

(c) Load in stiffeners at crackline

(d) Force at first attachment from crackline

Fig. 7 Effect of rivet failure for a bay centred crack ($a/b = 0.52$, $\mu = 0.5$).

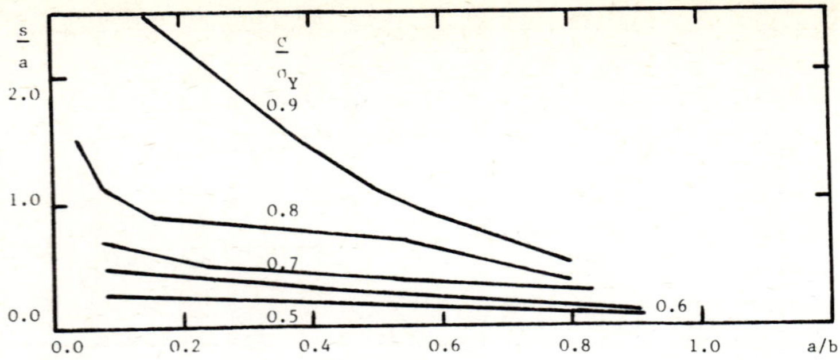

Fig. 8 Variation of strip yield zone length with crack length for a stiffener centred crack: intact stiffeners (μ = 0.5).

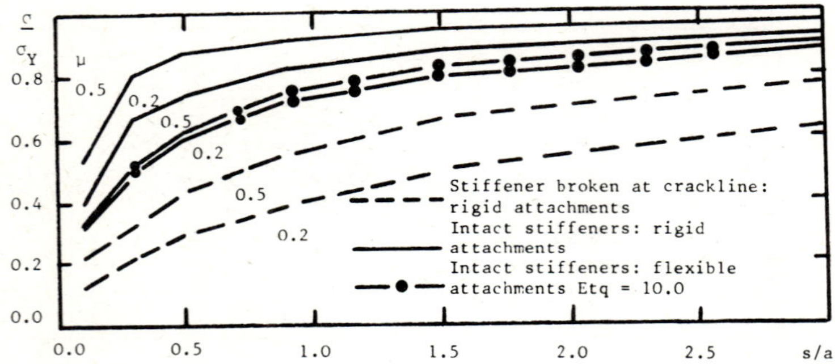

Fig. 9 Effect of applied stress on strip yield zone length for a bay centred crack (a/b = 0.52).

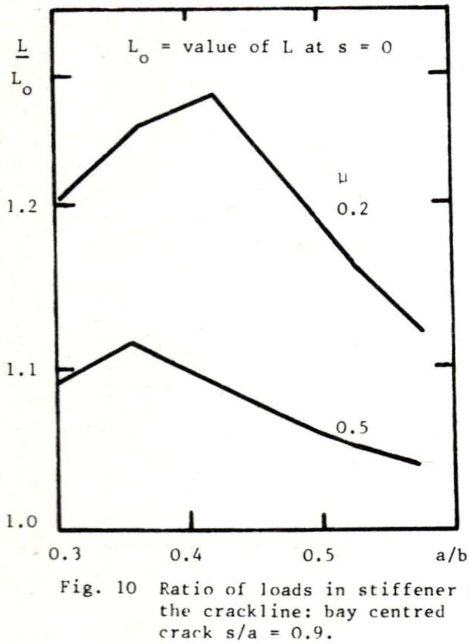

Fig. 10 Ratio of loads in stiffener at the crackline: bay centred crack s/a = 0.9.

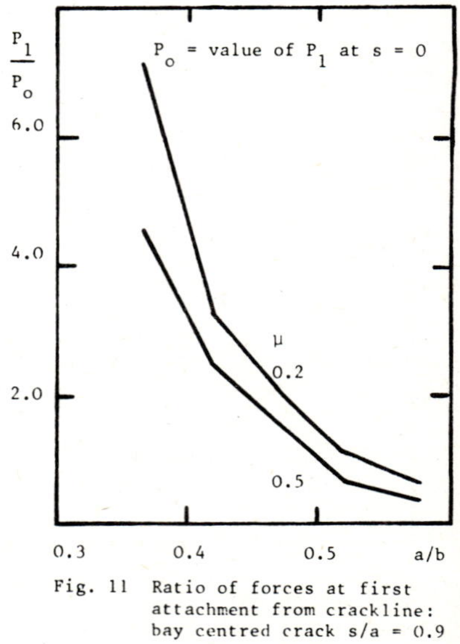

Fig. 11 Ratio of forces at first attachment from crackline: bay centred crack s/a = 0.9

A NUMERICAL COMPARISON OF POST YIELD FRACTURE BEHAVIOUR IN ROTOR STEELS

T. K. Hellen*

SUMMARY

A post-yield fracture mechanics study is described in which experimental and numerical analysis techniques are compared. Experiments have been performed on edge-cracked plates of 1% Cr MoV rotor steel and two particular crack lengths have been analysed using elasto-plastic finite element analysis. The shorter crack gives rise to gross plastic yielding at its failure load, and serves as a good comparison for the numerical methods. Elasto-plastic finite element analyses have been conducted using the BERSAFE system with two dimensional plane stress, plane strain and three-dimensional representations. Fracture parameters have been evaluated and compared with the experimental results, including two types of contour integral, potential and strain energy release rates. Two sets of non-linear stress-strain data are compared.

1. INTRODUCTION

In linear elastic fracture mechanics (LEFM), for a plate of unit thickness, the Griffith [1] criterion states that

$$G = \frac{-dP}{da} = 2\gamma \qquad (1)$$

where G is the potential energy release rate obtained from two crack tip positions a and $a+\delta a$ at constant load, and taken as the limit of $-\delta P/\delta a$ as $\delta a \rightarrow 0$. γ is the fracture surface energy which is constant for a given material and temperature. G also equals the strain energy release rate H and J-integral as proposed by Rice [2] in LEFM.

A basic assumption of LEFM is that the extent of plasticity about the crack tip is very confined. Thus, the classical

*Research Officer, CEGB, Berkeley Nuclear Laboratories, Berkeley, Gloucestershire, U.K.

crack tip equations relating the local displacements and stresses to stress intensity factor K are valid, and K is directly proportional to $\sqrt{G}$. However, modern technology now requires an assessment of the post-yield fracture mechanics (PYFM), the current status of which is described in a text [3]. Its early development is based on the Dugdale strip yield model [4], extended by Bilby, Cottrell and Swinden [5] using dislocation theory, whereby plasticity is confined to the line of the crack extended into the uncracked material. This approach, however uses assumptions which can be circumvented by numerical techniques.

In particular, the finite element method offers a versatile approach to the solution of both LEFM and PYFM problems [3, 6, 7]. In the latter case, the method solves the elasto-plastic equations of equilibrium, constitution and compatibility in an incremental manner, and apart from errors introduced by discretization, can be regarded as an exact solution of the postulated problem. The analysis is usually used to create a datafile of results from which J-integrals may be calculated. Some doubts exist on the validity of J in PYFM, however, since in the elasto-plastic case path independence is not assured and so the value around the theoretically correct contour, situated at the tip, is not equal to the more remote contours which have to be used in practice. Also, J is path dependent when unloading occurs, e.g. during stable crack growth, and is limited to two dimensional plane structures. Another integral, J*, has been proposed by Blackburn [8] which is applicable to these situations, and equals J in LEFM. Several comparisons to date [7, 9] show J* to be a promising candidate as a general fracture criterion.

The present work compares these integrals with another approach to calculating energy release rates. Since the finite element method involves minimisation of potential energy, the stability of the numerical process is such that this energy is very accurately computed as a single scalar quantity, even in non-linear cases (errors will occur due to bad discretizations such as coarse meshes or large load increments). By performing different finite element runs with different crack lengths a and $a+\delta a$, and investigating different values of δa, an assessment of dP/da can be determined. Whatever assumptions are made, such as retaining constant load, or ignoring yield zone shift, should not effect the limit of $-\delta P/\delta a$ as $\delta a \to 0$, so if the calculated value $-\frac{\delta P}{\delta a}$ is constant, it is likely to be very accurate.

This method is an extension of the virtual crack extension method of Hellen [10] in LEFM into non-linear analysis. There, several multi-directional extensions could be made in one computer run, and very accurate analyses have been made. Unfortunately, that technique could not be directly extended into

PYFM without requiring at least two computer runs. Also, unlike LEFM, the use of strain energy release rates in PYFM produces results which are quite different from those using potential energy release rates.

Results are presented comparing these different methods on edge-cracked specimens under tensile loading which have also been tested experimentally. Two different crack length to width ratios are investigated in specimens of a 1% CrMoV steel. The longer crack has limited crack tip yielding and can be analysed reasonably accurately by LEFM, whereas the shorter crack incurs gross crack tip yielding and consequently requires use of PYFM. All finite element analyses use the BERSAFE system [11] with associated post-processor modules for evaluation of the contour integrals.

2. CALCULATION OF STRAIN AND POTENTIAL ENERGIES

A direct evaluation of the potential and strain energies is a straightforward and accurate product of the finite element method. The potential energy is defined as the strain energy less the potential of forces in a deformed linear or non-linear material. Thus, the potential energy is:

$$P = \int_V W(\varepsilon_{ij})dV - \int_V Q_i u_i dV - \int_{S_T} F_i u_i dS \qquad (2)$$

at some deformed state where the total elastic and plastic strain is ε_{ij}, $W(\varepsilon_{ij})$ is the corresponding strain energy so that the stress σ_{ij} is given by $\sigma_{ij} = \frac{\partial W}{\partial \varepsilon_{ij}}$, and Q_i and F_i are body and surface forces, respectively, with associated displacements u_i. The body has volume V and S_T is the loaded part of its surface.

In the finite element method, an equivalent expression to equation (2) is available in elasticity as:

$$P = \tfrac{1}{2}\{u\}^T[K]\{u\} - \{u\}^T\{F\} \qquad (3)$$

where {F} is now the vector of equivalent nodal loads, including body forces, pressures and point loads, {u} is the corresponding vector of displacements, and [K] is the stiffness matrix. For thermal or residual strain analysis, equation (3) is still valid as long as {u} contains total displacements, {F} covers mechanical loads but excludes equivalent "thermal" loads, and an extra thermal term is added.

In elastic finite element analysis, P is efficiently evaluated in the BERSAFE system as:

$$P = -\tfrac{1}{2}\sum \frac{b_i^2}{k_{ii}} \qquad (4)$$

where summation is over all degrees of freedom, and k_{ii} and b_i are the stiffness and load vector terms corresponding to degree of freedom i at the forward elimination stage in the solution of equations [10]. In elasto-plastic analysis, equation (4) is no longer applicable so that a more basic energy formulation is required. Each such analysis has its total applied load broken into suitable load increments, within each of which the program automatically iterates to retain the dual requirements of compatibility and equilibrium. Convergence occurs when the load residuals are less than an input tolerance indicator which is given as an input parameter.

For each iteration, if the stress component i at a point p is initially σ_i and finally $\sigma_i + \Delta\sigma_i$, and similarly for strains, the total contribution to the increase in strain energy is ΔU^p, where

$$\Delta U^p = \sum_i (\sigma_i + \tfrac{1}{2}\Delta\sigma_i)(\Delta\varepsilon_i^T - \Delta\varepsilon_i^o - \Delta\varepsilon_i^\alpha)\delta Vp \qquad (5)$$

where δVp is the effective volume surrounding point p, and superscripts T, O and α denote total, initial and thermal, respectively. Over the whole structure, the strain energy is ΔU, where

$$\Delta U = \sum_p \sum_i (\sigma_{pi} + \tfrac{1}{2}\Delta\sigma_{pi})(\Delta\varepsilon_{pi}^T - \Delta\varepsilon_{pi}^o - \Delta\varepsilon_{pi}^\alpha)\delta Vp \qquad (6)$$

By accumulating this strain energy over iterations and load increments during the elasto-plastic analysis, we can write at load level λ,

$$U(\lambda) = \sum^{\lambda} \Delta U \qquad (7)$$

The corresponding potential energy $P(\lambda)$ is given by:

$$P(\lambda) = U(\lambda) - \{u_\lambda\}^T\{F_\lambda\} \qquad (8)$$

where $\{u_\lambda\}$ and $\{F_\lambda\}$ apply at load level λ. From equations (7) and (8), the energies which are of interest in fracture mechanics are readily obtained at any stage in the loading history.

One important facility in non-linear analysis is the ability to restart the computer run after a pause. In BERSAFE, the results are stored throughout the analysis. In addition to storing stresses, plastic strains and displacements per load increment, energy terms and load vectors are also retained so that the run may be interrupted and restarted at any point, with change in load if required (e.g. unloading), and the energies of equations (7) and (8) are still available. Strain energy can be obtained from current values of stress and strain, whereas potential energy requires extra history - dependent terms.

3. CALCULATION OF ENERGY RELEASE RATES

The potential and strain energy release rates are readily obtained from the calculated energies given by equations (7) and (8). Both are obtained together from two or more finite element elasto-plastic runs with the sole difference being the extra crack length δa. The potential energy release rate is

$$G = \frac{-dP}{da} = -\lim_{\delta a \to 0} \frac{\delta P}{\delta a} \tag{9}$$

so δa is taken as a small extension, typically 10^{-1} or 10^{-2} of the tip element size. Unloading and subsequent reloading are ignored during the crack extension. If for several such δa values, a consistent result $-\delta P/\delta a$ is obtained, then it can be assumed to equal the limiting value as $\delta a \to 0$ as in the elastic virtual crack extension case [10]. Since δa is much smaller than other geometrical parameters, plastic zone movements should have negligible effect on results. In all cases studied so far, no significant variations in $\delta P/\delta a$ have yet been detected. The accuracy of the potential energy calculation is enhanced in BERSAFE since it is stored in high precision of over 16 significent digits, and over very small extensions δa, only the first 3 or 4 digits are the same over the two runs, so an accurate difference δP is obtained. Also, the expression for $\delta P/\delta a$ is valid for many conditions such as general mechanical or thermal loading, two and three dimensions and inhomogenous materials.

The above comments apply also to strain energy release rate H except that the accuracy is inferior to that of G in load controlled situations. In terms of finite element analysis the quantity appears to be more sensitive to the various discretizational assumptions. In strain controlled cases, such as thermal loading, G and H are identical.

When calculating δP or δU for extensions δa_1, δa_2,..., it is important that each run should otherwise be identical, such as the number and size of load increments, first yield load, convergence tolerances, etc., otherwise the variations of energy will not correspond to conditions of constant load (or displacement). A useful guide to the maximum size of extension δa is that, for a given load step, the same number of iterations should exist over the different runs. If this is not so, the extra iteration in one run will give an extra accumulation to the energies which can be significant in the case of strain energy.

4. CONTOUR INTEGRATION

Nearly all finite element analyses conducted hitherto have been concerned with calculating J-integrals, defined as [2]:

$$J = \int_c (W dx_2 - T_i \frac{\partial u_i}{\partial x_i} ds) \qquad (10)$$

integrated around a contour C surrounding the crack tip, where at any point on the contour W is the strain energy density, T_i are traction components corresponding to displacements u_i, and (x_1, x_2) are local axes about the crack tip such that the crack plane lies along negative x. As long as W relates stress and strain in a conservative manner, the integral remains path independent, provided only one singularity exists within the contour. This applies to both elastic and non-linear elastic conditions. In general incremental non-conservative plasticity the integral is path dependent. Since the theoretically correct contour is of zero length at the tip, one would expect the results from contours near the tip to be more accurate than those further away, but inaccuracies occur because in practice the tip region displacements and stresses are insufficiently accurate to produce good contour results. However, many authors have found that J remains sufficiently constant over a range of contours throughout the mesh.

In thermal or initial strain fields, J is path dependent unless an extra term is added [7]:

$$J = \int_c (W dx_2 - T_i \frac{\partial u_i}{\partial x_1} ds) + \iint_S \sigma_{ij} \frac{\partial \varepsilon^{\alpha}_{ij}}{\partial x_1} dS \qquad (11)$$

Other modifications have been proposed to extend J to other situations and retaining the path independence requisite. However, a general integral, J*, has been proposed [8] and applied to a wide range of situations [7, 9], including two and three dimensions, incremental elasto-plasticity, initial and thermal strain, non-homegeneous materials, and creep. It is defined as:

$$J^* = \lim_{\rho \to 0} \int_\rho (\tfrac{1}{2}\sigma_{ij} \frac{\partial u_i}{\partial x_j} dx_2 - T_i \frac{\partial u_i}{\partial x} ds) \qquad (12)$$

around a contour ρ which is suitably close to the crack tip. In practice, it is evaluated on a more convenient contour C away from the tip by use of Green's theorem as:

$$J^* = \int_c (\tfrac{1}{2}\sigma_{ij} \frac{\partial u_i}{\partial x_j} dx_2 - \sigma_{ij} \frac{\partial u_i}{\partial x_1} n_j \, ds) + \lim_{\rho \to 0} \iint_S (\tfrac{1}{2}\sigma_{ij} \frac{\partial^2 u_i}{\partial x_1 \partial x_j} - \tfrac{1}{2} \frac{\partial \sigma_{ij}}{\partial x_1} \frac{\partial u_i}{\partial x_j} - \frac{\partial}{\partial x_3} (\sigma_{i3} \frac{\partial u_i}{\partial x_1})) dS \qquad (13)$$

where n_j are the direction cosine components of contour C and S is the area between contours ρ and C.

The small contour ρ has usually been taken around the outer boundary of the crack tip elements. In elastic cases, J* = J. In incremental elasto-plasticity, J* has shown to be close to J under mode I conditions and exhibits better the path independence property. In the BERSAFE system, both integrals J and J* are evaluated in the post-processor program PLOPPER. For J, the first term in the integrand is calculated as:

$$\tfrac{1}{2}\sigma_{ij} \frac{\partial u_i}{\partial x_j} dx_2$$

instead of Wdx_2. Other investigators have used W as the sum of elastic and plastic energies, which requires a more complicated evaluation. The two formulations are identical for contours lying outside plastic zones, and remain numerically similar for contours within the zones.

In elasticity, the values of J, J*, G and H are theoretically equal. In non-linear elasticity and elasto-plasticity, no theoretical basis for equality of any of these parameters is available, and so the use of numerical techniques provides an interesting method of comparing their relative usefulness. Good comparisons of J* and G (equal to H) have been made in thermo elasto-plastic geometries [12].

5. NUMERICAL AND EXPERIMENTAL COMPARISONS FOR AN EDGE-CRACKED PLATE

5.1 Specimen Geometry and Materials Data

Post-yield fracture mechanics tests have been made on a series of specimens with a single edge crack, with different combinations of thickness, width and crack length [13]. Each specimen was loaded by uniaxial tension applied over the end faces, and was machined from a 1%CrMoV HP steam turbine rotor forging. Precracking was performed to the appropriate British Standard [14].

Two particular specimens have been analysed numerically. They represent the extremes in crack length to width ratios in the series, one extending 0.0012m through the width (of 0.02m), giving an a/w ratio of 0.06, the other extending 0.01m through, so that a/w = 0.5. Both specimens had a thickness of 0.01m.

For both cases, the same finite element mesh was used, with a local refinement including special singularity crack tip elements [15] for each particular tip position, in an elasto-plastic initial stress analysis using BERSAFE. The mesh, shown in Figure 1 (without local crack tip refinements) has been used for other fracture problems, giving stress intensity factors accurate to within 1% for several elastic loading cases and methods of evaluation [6, 7]. It consists of 76 quadratic isoparametric elements, 227 nodes and 454 degrees of freedom. The crack tip element side length along the crack equals a/10.

For the material, Young's modulus E = 210 GN/m^2, Poisson's ratio ν = 0.3 and the initial yield stress σ_y is 570 MN/m^2. Two sets of non-linear stress-strain data were used, one being bilinear with a constant work-hardening slope of 0.1E. The second set of data was made available after completion of most of the present work, and had a variable work-hardening slope of lower average value ($\sim$0.05E) which was more relevant to the current material. Most of the present results refer to the first data.

Under conditions of mathematical plane strain (MPS), plasticity is not so widespread as in plane stress (PS) for the same amount of loading, due to constraints on strains in the out-of-plane direction. The amount of plasticity in a three-dimensional model lies somewhere in between. The plastic zones under MPS are shown in Figure 1 for both specimens studied. For the short crack, both sets of non-linear data were used. The zones were obtained from the finite element output corresponding to the applied load nearest to the experimentally determined critical value. Thus, for the short crack, the load level used was at an applied load of 594 MN/m^2 and 587 MN/m^2 for the two sets of non-linear data, respectively, compared with the experimental value of 580 MN/m^2, both involving a number of load increments. For the long crack, the load level was at an applied load of 156 MN/m^2, compared with the experimental value of 120 MN/m^2, only one load increment being involved above first yield.

The zones indicate that the amount of plasticity around the short crack is several times larger than that from the long crack at the respective failure stresses, with greater plasticity for the second, refined, set of non-linear data. The short cracked specimen obviously requires the use of PYFM.

5.2 Analysis of Specimen with Long Crack

The long crack specimen had a fairly small plastic zone at the applied load level of 156 MN/m^2, and so at the failure load of 120 MN/m^2 little departure from linear elastic fracture mechanics would be expected. For the present material, K_{Ic} has been obtained from valid 50mm thick CT specimens and has a mean value of 54.4 $MN/m^{-3/2}$, which implies a critical J value of J_{Ic} = 0.0128 MN/m in plane strain. From LEFM theory, this would indicate a gross critical stress of 104 MN/m^2. The experimental critical stress is 120 MN/m^2 which gives a non-dimensional value $EJ/a\sigma_y^2$ = 0.81. This result was obtained using J and J* integrals at load levels of 93 MN/m^2 and 102 MN/m^2, respectively. The contour used was round the outer boundary of the mesh, obtained from the elasto-plastic analysis at the load level of 156 MN/m^2. The J and J* results are in fact closer to the LEFM value, predicting less plasticity than actually occurred in the experiment.

5.3 Analysis of Specimen with Short Crack and Bilinear Non-Linear Data

Elasto-plastic analyses of the specimen with the short crack have been made using two-dimensional MPS and PS, and a full three-dimensional model for the first set of work-hardening data. The three-dimensional mesh was identical to the two-dimensional one in the x-y plane, with two layers of elements in the third (z) direction. Elements used were 15 node wedges and 20 node bricks with special elastic singularity elements around the crack profile, and in total there were 152 elements, 833 nodes and 2499 degrees of freedom. The back face ($z = 0$) was fixed in the z-direction to form a plane of symmetry, effectively doubling the through-thickness mesh refinement (Figure 2). Contours of integration were obtained using the J* formulation of equation (13), J being inapplicable in three-dimensions. The contour integration was evaluated at midside nodes along the crack profile, in planes normal to the profile at that point, i.e. nodes A and B in Figure 2. Two contours in each plane were used, the inner one being approximately square of side length 5a, the outer one being around the outside of the mesh. The four contours are labelled (1) to (4) as shown in Figure 2.

The three-dimensional results are shown in Figure 3, which shows the various parameters normalised by $E/a\sigma_y^2$ plotted against increasing applied stress. The J* curve refers to contour (4) and is seen to pass through the experimental point. Six load increments were required above first yield to a load level of 594 MN/m^2. The normalised J* results at this load over the inner and outer contours at node A were 9.69 and 10.11 respectively, whilst the corresponding results for node B were 10.07 and 10.55. Hence good path independence occurs at A and B (within 4%), the latter having a mean value 4% higher than the former. Little variation would be expected in view of the straight propagation observed experimentally.

The potential energy difference value G was obtained for an extension $\delta a = a/1000$ and agrees closely with the J* curve. The strain energy difference value H is seen to increase very rapidly and would attain the critical dimensionless value of $EJ_c/a\sigma_y^2 = 6.89$ at the lower load level of 530 MN/m^2.

The two-dimensional MPS analysis was conducted using 9 load increments up to a load of 690 MN/m^2. The results in Figure 3 are seen to give loads at the critical level $EJ_c/a\sigma_y^2$ =

6.89 in the range 615-625 MN/m^2, some 7% higher than the experimental or three-dimensional results. The J and J* integrals were evaluated using contours round the outside of the mesh, J* being closer to experiment by 2% that J at the 6.89 level. The G result with $\delta a = a/1000$ is virtually identical to J for all but the highest load levels. When using $\delta a = a/100$, results within 1% were obtained at all load levels. The strain energy release rate H again produced much higher results at a given load level, and despite the larger numbers involved, variation within 1% occurred for the two extensions δa.

The two-dimensional plane stress analysis produced more rapid plastic growth and six load increments were required up to an applied load of 595 MN/m^2. The J and J* curves gave higher values at a given load level than the other analyses, being very close to one another except at the highest load where J* was some 20% below J. The G results remained very close to J throughout the range and again variations of less than 1% occurred for the two δa values a/100 and a/1000. The H results were characteristically very high.

5.4 Analysis of Specimen with Short Crack and Refined Non-Linear Data

To compare the previous results, which used constant work-hardening data, with the second set of non-linear data, several computer analyses were repeated. The results for the three-dimensional analysis are shown in Figure 4. Only 4 load increments were used between first yield and applied load of 600 MN/m^2, and at this highest load the four sets of J* results were compared. For node A (Figure 2), the inner and outer contour values were 10.41 and 11.19 respectively, showing a 7% variation, whilst for node B, the corresponding values were 10.67 and 11.35, showing a 6% variation. Thus path independence is not so good as in the previous case, indicating errors due to using larger load increments in areas of gross plasticity. The G curve with $\delta a = a/1000$ is also shown and the two curves shown in Figure 4 pass very close to the experimental point, the G curve being the closer. G was also evaluated using $\delta a = a/100$ and gave a result at load level 600 MN/m^2 of $EJ/a\sigma_y^2 = 14.39$, which is much higher (by 18%) than the $\delta a = a/1000$ result. Since the agreement is much better at lower load levels, the reason is again due to a large final load increment.

The two-dimensional plane stress analysis was repeated and a maximum load of 551 MN/m^2 was achieved in 6 load increments, beyond which convergence was not obtained because of the large amount of plasticity apparent. Since plastic collapse is therefore indicated at a load of about 550 MN/m^2, the measured failure load of 580 MN/m^2 would never be attained and therefore the state of plane stress is not realistic. Even so, the G values with $\delta a = a/100$ and a/1000 agreed within

1% at load level 530 MN/m^2, but were some 30% lower than the J* result and 40% lower than the J result. Since G and J should be similar, a more refined elasto-plastic analysis for plane stress using the second set of non-linear data is indicated.

A similar conclusion holds for MPS. The G values remained close together with $\delta a = a/100$ and $a/1000$, but were considerably lower than J and J*. Using 7 load increments up to a load of 630 MN/m^2, variations in the size of convergence tolerance for load residuals in the initial stress algorithm were made, using values of 8%, 4% and 1%. At the collapse load of 580 MN/m^2, G varied over these three cases by less than 5% whilst J and J* exceeded G by 30%. The number of load increments used was also doubled in one case, and at the load level 580 MN/m^2 variations in G, J and J* never exceeded 2%. These convergence checks indicate that the plasticity part of the analyses is accurate and that therefore the discrepencies are due either to an insufficiently refined mesh, which is unlikely because of the good agreement observed in the use of the first set of non-linear data, or reflect differences in the physical nature of the candidate parameters. In this second set of results, the plasticity extends more than in the first set, as seen in Figure 2 for MPS. For PS and three-dimensions, much more plasticity occurs showing there is gross plasticity at the higher load levels.

6. DISCUSSION OF RESULTS

The experimental tests on the range of specimens studied revealed that the fast fracture progressed retaining a straight profile, with no evidence of slow crack growth. Hence it is reasonable to assume that any relevant fracture parameter would remain fairly constant along that profile, and an energetic parameter such as F or J would apply provided the concept of critical values as used in LEFM extends into PYFM. This is equivalent to the experimental point shown in Figures 3 and 4 lying on the calculated candidate curve.

The specimen with the long crack failed at a relatively low applied load and calculations of a somewhat higher load revealed only a small amount of plasticity. Hence in this case the behaviour should be closely approximated by LEFM theory. The critical load from this theory is 104 MN/m^2, which compares well with the calculated J* result (102 MN/m^2) and J (93 MN/m^2). However, the experimentally obtained failure load of 120 MN/m^2 is nearly 20% higher and is more likely a reflection of experimental scatter rather than real non-linear effects.

The two-dimensional MPS and PS analyses both use assumptions on the z-component of stress or strain, and in view of the thickness of the specimens used the PS model is considered somewhat less realistic than the MPS one. In fact, for the short cracked specimen the experimental point lies very close to the three-dimensional curves, which in turn lie between the PS and MPS curves. The range of load for all these results at the critical value of $EJ/a\sigma_y^2 = 6.89$ is, from Figure 3, 135 MN/m^2 about the experimental load of 580 MN/m^2, a scatter band of 23%. This reduces to 13% if the H curve from plane stress is omitted. This gives an accurate assessment of failure load corresponding to J_c, and is due to the intrinsic shape of the curves obtained - large errors in the dimensionless parameter value producing small errors in load.

For the first set of non-linear stress-strain data, with constant work-hardening slope, the numerical analyses appears to be very accurate and produce J and G values which are very nearly equal. J* differs, increasingly with increased plasticity, because of its different physical interpretation.

The use of the second set of non-linear data causes larger extents of plasticity because the data is closer to elasto-perfect plasticity. In three-dimensions, the main effect was to elevate J*, with G remaining similar to before. The estimated failure load varies only very slightly, again because of the shape of the curves. However, in two-dimensional MPS and PS, the discrepancies in the calculated values arose because of the effect of much more yielding. Little variation occurred when the number of load increments was doubled and the convergence tolerance was reduced, implying that the numerical analysis is accurate well within engineering requirements. It is therefore likely that these analyses reveal physical differences in G, J and J* which are almost non-existent for the first set of non-linear data.

7. CONCLUSIONS

Comparisons have been made of four fracture parameters computed for the elasto-plastic analysis of an edge-cracked plate in tensile loading together with experimental results. The parametric curves, plotted against applied load, showed best agreement with experiment when using a full three-dimensional analysis. Plane strain results gave higher failure loads whilst the opposite was true in plane stress. The parameters increased more rapidly with load as the load increased, implying that large errors in parameter values give only small errors in the predicted failure load. The two sets of non-linear stress-strain data used produced only small changes in predicted failure load in the three-dimensional case, even though the amounts of plastic yielding differed. However, larger differences were apparent in two dimensions. Of the parameters investigated, the J and J* integrals and potential energy release rate generally agreed well with one another and experiment. The strain energy release rate predicted low failure loads and is not considered to be a suitable fracture parameter.

ACKNOWLEDGEMENT

This paper is published by permission of the Central Electricity Generating Board. The author would like to express his appreciation of the co-operation given by Dr. D. Batte of NEI Parsons and Dr. W. S. Blackburn of CEGB in making available and discussing the experimental data.

REFERENCES

1. GRIFFITH, A.A., The Phenomena of Rupture and Flow in Solids, Transactions, Royal Soc. London, Vol.221, 1920.

2. RICE, J.R., A Path Independent Integral and the Approximation Analysis of Strain Concentration by Notches and Cracks, J. Appl. Mechs., Vol.35, pp379-386, 1968.

3. LATZKO, D.G.H.(ed), Post-Yield Fracture Mechanics, Applied Science Publishers, 1979.

4. DUGDALE, D.S., Yielding of Steel Sheets Containing Slits, J. Mech, Phys, of Solids, Vol.8, pp100-104, 1960.

5. BILBY, B.A., COTTRELL, A.H. and SWINDEN, K.H., The Spread of Plastic Yield from a Notch, Proc. Royal Soc., A272, pp304-314, 1963.

6. CHELL, G.G. (ed), Developments in Fracture Mechanics - 1, Applied Science Publishers, 1979.

7. HELLEN, T.K., Finite Element Energy Methods in Fracture Mechanics, Ph. D. Thesis, Univ. London, 1976.

8. BLACKBURN, W.S., Path Independent Integrals to Predict Onset of Crack Instability in an Elastic Plastic Material, Int. J. Frac. Mechs., Vol.8, pp343-346, 1972.

9. BLACKBURN, W.S., HELLEN, T.K., and JACKSON, A.D., An Integral Associated with the State of a Crack Tip in a Non-Elastic Material, Int. J. Fracture, Vol.13, pp183-200, 1977.

10. HELLEN, T.K., On the Method of Virtual Crack Extensions, Int. J. Num. Meths. Eng'g. Vol.9, pp187-207, 1975.

11. HELLEN, T.K., and PROTHEROE, S.J., The BERSAFE Finite Element System, Comp. Aided Design, Vol.6, pp15-24, 1974.

12. MUSCATI, A., and HELLEN, T.K., Elastic-Plastic Finite Element Analysis as Applied to Cracked Structures Subjected to Thermal loading, Num. Meths. in Thermal Problems, ed Lewis, R.W., and Morgan, K, Pineridge Press, Swansea, 1979.

13. BATTE, D., Private Communication, work to be published.

14. Methods for Test for Plane Strain Fracture Toughness (K_{Ic}) of Metallic Materials. British Standard BS 5447, 1977.

15. BLACKBURN, W.S., Calculation of Stress Intensity Factors at Crack Tips Using Special Finite Elements, Maths. of Finite Els. and Appls., ed Whiteman, J.R., Academic Press, 1972.

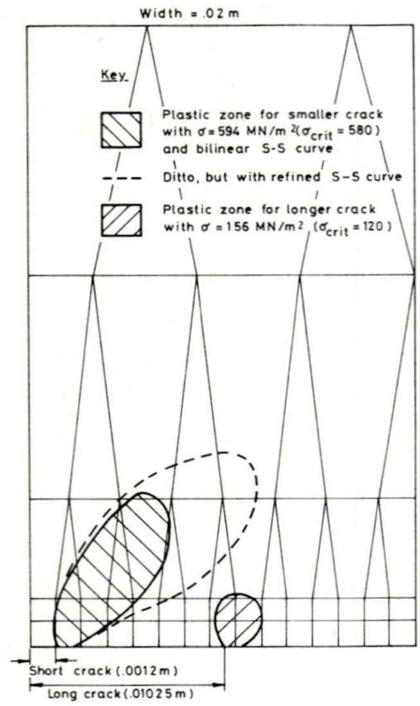

FIG. 1. Mesh and Plastic Zones at Critical Stress for Two Crack Lengths (Plane Strain)

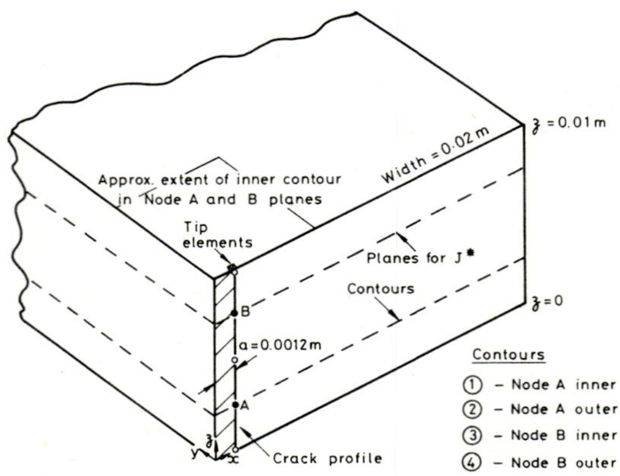

FIG. 2. Definition of Contour Integration Planes in Three Dimensions

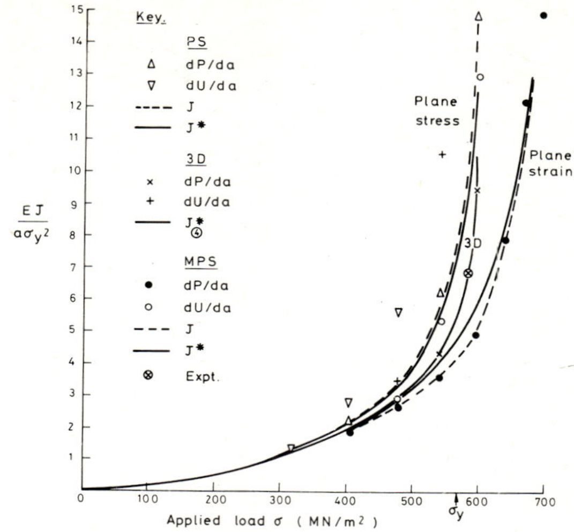

FIG. 3. Results for Short Crack with First Set of Non-Linear Data.

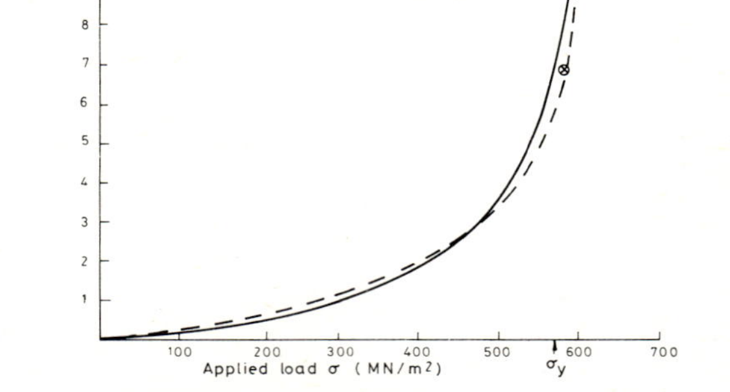

FIG. 4. Results for Short Crack (Three Dimensions) with Second Set of Non-Linear Data.

A COMPARISON OF FINITE ELEMENT SOLUTIONS FOR J-INTEGRAL ANALYSIS OF COMPACT SPECIMEN (A ROUND ROBIN TEST IN JAPAN)

M. Shiratori (I) and T. Miyoshi (II)

SUMMARY

To determine the state of art of the J-integral analysis by finite element method, a round robin test was undertaken in Japan. Seven groups of participants carried out elastic-plastic finite element analyses of a compact specimen for the same geometry, the same loading condition and the same material properties. The deviation of the J-integral was within 5 % in the standard problem where the element breakdown was prescribed. On the other hand, larger deviation was resulted in the non-standard problem where element breakdown was not prescribed. Therefore, it was concluded that the main factor which affects the accuracy of J-integral was element breakdown rather than the finite element program.

1. INTRODUCTION

In the scheme of the elastic-plastic fracture mechanics, J-integral is the most promising parameter for describing the criterion for both of the crack initiation and unstable crack propagation in the tough structural members.

Difficulty comes from the nonlinearity of the material properties in the analysis of J-integral for the cracked specimens or the structural members. It is difficult to obtain closed form solutions for such problems, and therefore, the numerical analysis may often be the only way for obtaining the solutions. It is necessary to get reliable numerical solutions in order that the J-integral might be used as a reliable criterion. Wilson and Osias[1] have reported the results of the

---

(I) Associate Professor of Mechanical Engineering, Yokohama National University, Tokiwadai, Hodogayaku, Yokohama, Japan

(II) Associate Professor of Precision Machinery Engineering, University of Tokyo, Hongo, Bunkyoku, Tokyo, Japan analytical round robin which was carried out as a task under ASTM Committee E24 Task Group E24.01.09. In the round robin test, there were ten numerical solutions for the same specimen geometry, i.e., three point bend bar, the same loading condition and the same material properties. Deviations among these ten solutions were not negligible except for the small scale yielding range, and therefore, it is necessary to clarify what the main factor is that affects the accuracy of the numerical solutions.

A round robin test has been carried out in Japan in order to solve the following problems:

(1) What is the cause of the large deviation observed in the results of the U. S. round robin test?

(2) What is the main factor which affects the accuracy of the numerical solution for J-integral?

(3) How can we expect the accuracy of J-integral for a compact specimen which is also an important specimen in the fracture toughness testing?

A working group for J-integral analysis was organized under JSME Committee PSC21. There were seven groups of participants listed alphabetically in table 1. They worked from December, 1978 till November, 1979, and some important conclusions were obtained.

Table 1 Participants in JSME round robin for J-integral analysis

List is in alphabetical order. There is no relationship between the order of this list and the order of the solutions labeled in table 2 and 4.

(1) H. Okamura, K. Kageyama and Y. Takahata, University of Tokyo
(2) K. Ohji and S. Kubo, Osaka University
(3) H. Ohtsuka, Chiyoda Chemical Engng. & Constr. Co., Ltd.
(4) H. Miyamoto and M. Kikuchi, Science University of Tokyo
(5) M. Sakata*, S. Aoki and K. Kishimoto, Tokyo Institute of Technology
(6) M. Shiratori, Yokohama National University and T. Miyoshi, University of Tokyo
(7) G. Yagawa and Y. Takahashi, University of Tokyo

* Chairman of the working group

2. DESCRIPTION OF THE PROBLEM

The round robin test was conducted under the following condition:

(1) A compact specimen shown in Fig. 1 was selected for the plane strain elastic-plastic analysis. Due to the symmetry, upper half of the specimen was analyzed.

(2) Material for the nuclear reactor vessel, A533B steel, at room temperature was simulated. Fig. 2 shows its stress-strain curve which was approximated by the following equations:

$$\varepsilon = \sigma/E \qquad \text{for} \quad \sigma < \sigma_Y \tag{1}$$

$$\varepsilon = \sigma/E + [(\sigma/\sigma_0)^n - (\sigma_Y/\sigma_0)^n] \qquad \text{for} \quad \sigma > \sigma_Y \tag{2}$$

where E: Young's modulus, ν: Poisson's ratio, σ_Y: tensile yielding stress, σ_0 and n: parameters in eq. (2). The values of these material properties are

$$E = 21000.0 \text{ kgf/mm}^2, \quad \nu = 0.3, \quad \sigma_Y = 56.0 \text{ kgf/mm}^2,$$

$$\sigma_0 = 153.0 \text{ kgf/mm}^2, \quad n = 3.0$$

(3) The problem was classified into two problems, that is, standard and non-standard problems. In the standard problem the solution must be obtained by the finite element method with prescribed element breakdown and prescribed paths for J-integral as shown in Fig. 3. On the other hand the method of calculation was not prescribed in the non-standard problem.

(4) Load vs. load-line-displacement, P-δ, curve, J-δ, curve and J-P curve must be drawn in the prescribed scale. The load-

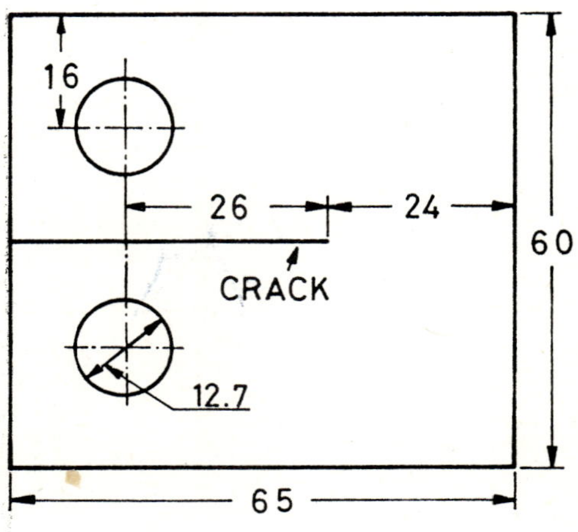

Fig. 1 Geometry of the compact specimen

line-displacement at the crack surface was selected instead of the load-point-displacement because of two reasons; one is that the load-line-displacement corresponds to the clip-gauge-displacement usually obtained in the experiment and the other is that the load-point-displacement includes the plastic deformation caused by the pin-load which was approximated by a point force as shown in Fig. 3.

(5) Summary of the method of the numerical analysis must be attached to the results.

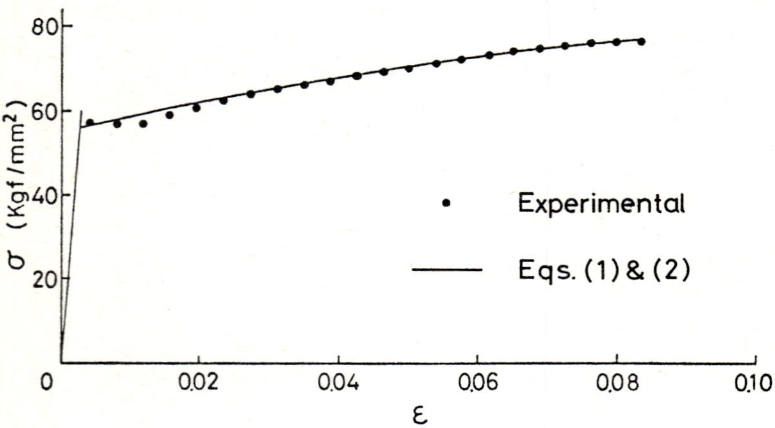

Fiig. 2 Stress-strain curve of A533B steel at room temperature

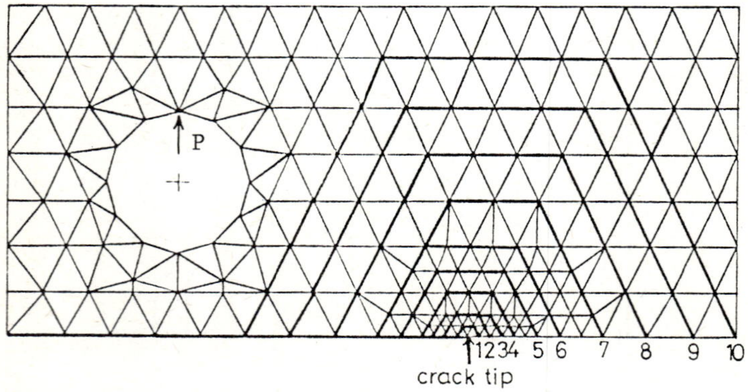

Fig. 3 Element breakdown and J-integral paths for the standard problem

3. RESULTS OF THE STANDARD PROBLEM

Five solutions were contributed for the standard problem. The basic informations for these solutions are listed in table 2. All the finite element programs were based upon the displacement type formulation and the incremental theory of plasticity with Prandtl-Reuss flow rule and von Mises yield criterion. Yamada's and Marcal's methods in the second column of the table mean that the way of choice of the load increment is different to each other, that is, the load increment is adjusted so that each element yields one by one in Yamada's method while it is fixed in Marcal's method. "COEF" in the table means a critical value for the judgement of yielding condition; if the normarized equivalent stress $\bar{\sigma}/\sigma_Y$ of an element exceeds this critical value, the element is regarded as plastic. Forth column of the table describes the means of determining values such as stresses and strains along the J-integral paths.

There are five different methods for evaluating J-integral which are summarized in table 3. The path integral J can be calculated for each path at each stage of the load increment. Two P-δ curves for different crack length by Δa are necessary to obtain J_E but a single P-δ record is enough to calculate J_R, J_M and J_S. The first two methods in the table are applicable to any geometry and Rice's equation to the bend type specimen including the compact specimen, but the application of the rest two methods is confined to the compact specimen only.

In any case the P-δ curve is essential to determine J-integral except for the first method. Fig. 4 shows the solutions of P-δ curves. The symbols A, B, C, D and G correspond to those in table 2. Good agreement has been obtained for $\delta < 0.7$ mm though a little deviations are observed for $\delta > 0.7$ mm. Those deviations may come from the different values of COEF but anyway the deviations are small and negligible (about 5 % for $\delta = 1.0$ mm).

Figs. 5-7 show $\bar{J}$-P, $\bar{J}$-δ and $\bar{J}_M$-δ curves, respectively. These figures show that the agreement among different solutions is quite good and that there are little deviations caused by the difference of the finite element program. The deviations of $\bar{J}$ and J_M in these figures are within 5 % around J_{IC}. Fig. 8 shows the J-δ curves calculated by the methods listed in table 3. $\bar{J}$ is the average of the path integral for all 10 paths and $\bar{J}_*$ for 7 paths; values for the rest three paths along which errors are not negligible are excluded in the evaluation of $\bar{J}_*$. Fig. 8 is the result of the solution C. It shows that $J_S > J_M \simeq \bar{J}_* > \bar{J} > J_E > J_R$. On the other hand, $J_S > \bar{J}_* > \bar{J} \simeq J_E \simeq J_M > J_R$ for the solution A and $J_S > J_M \simeq \bar{J}_* > J_R$ for the solution D. Fig. 9 shows the deviation of the path integral J among 10 paths. Large errors are observed along close path to the crack tip (paths 1-3) though the average value $\bar{J}$ seems to be reasonable. Therefore, care must be taken to avoid the path close to the crack tip.

Table 2 Basic informations for solutions of the standard problem

solution	incremental method	COEF	calculation of J	method for solving linear equations
A	Yamada's method	0.95	$\sigma=(\sigma_1+\sigma_2)/2$	Gaussian elimination
B	〃	0.97	〃	square-root method
C	〃	0.995	$\sigma=\frac{\sigma_1 A_1+\sigma_2 A_2}{A_1+A_2}$	unit partitioning
D	Marcal's method		〃	Choleski's method
G	Yamada's method	0.95	$\sigma=(\sigma_1+\sigma_2)/2$	〃

A_1, A_2 : area of the element

σ_1, σ_2 : stress of the element

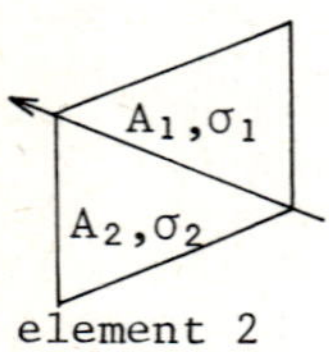

Table 3 Summary of the methods of evaluating J-integral for compact specimen

symbol	methods	algorithm
J	path integral	$J=\int(Wdy-T\frac{\partial u}{\partial x}ds)$
J_E	energy method	$J_E=-\frac{1}{B}\frac{\Delta U}{\Delta a}$
J_R	Rice's equation [2]	$J_R=\frac{2A}{bB}$
J_M	Merkle & Corten's equation[3]	$J_M=\frac{2.25A+0.196C}{bB}$
J_S	Shiratori & Miyoshi's equation[4]	$J_S=\frac{2.36A+0.291C}{bB}$

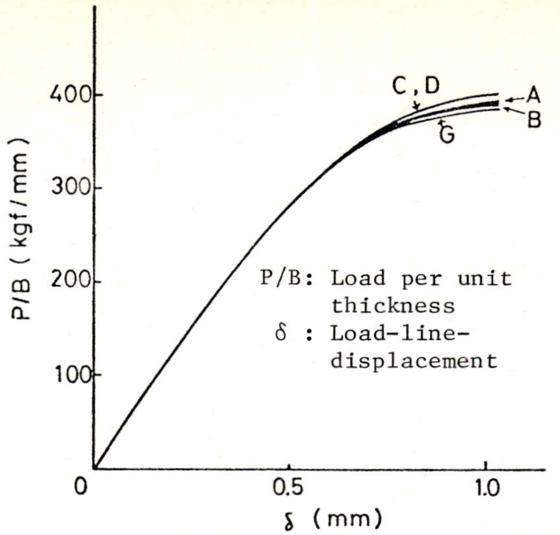

Fig. 4 P/B vs. δ curves for the standard problem

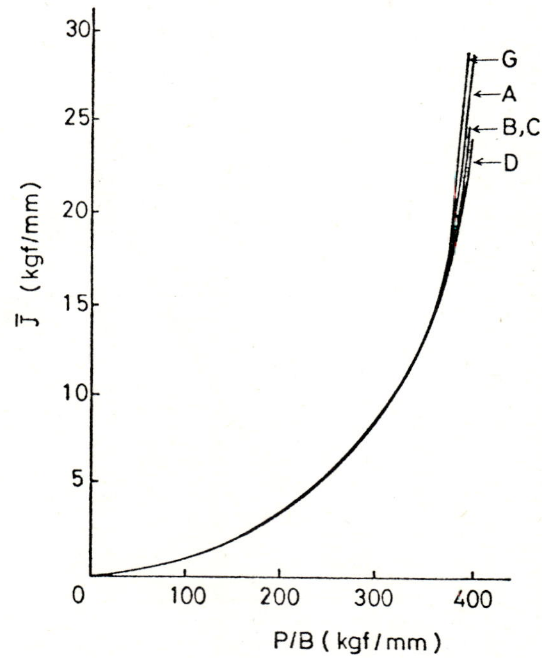

Fig. 5 J vs. P/B curves for the standard problem

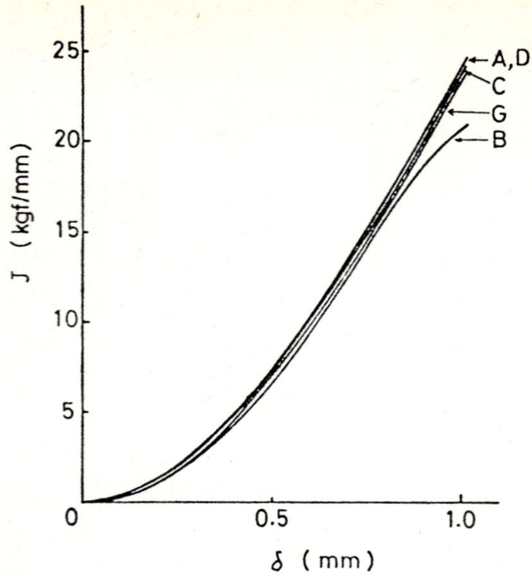

Fig. 6 $\overline{J}$ vs. δ curves for the standard problem

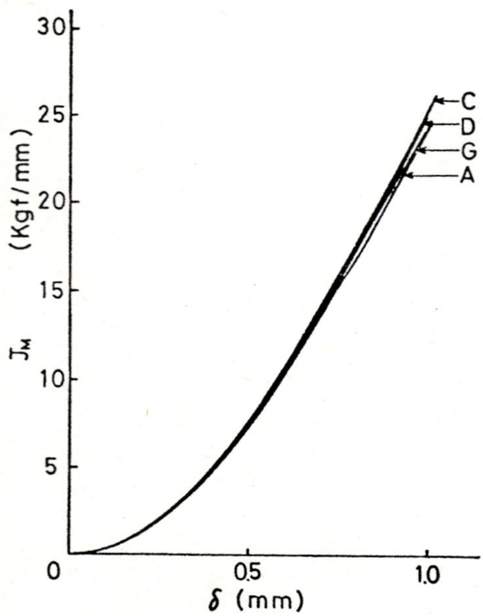

Fig. 7 J_M vs. δ curves for the standard problem

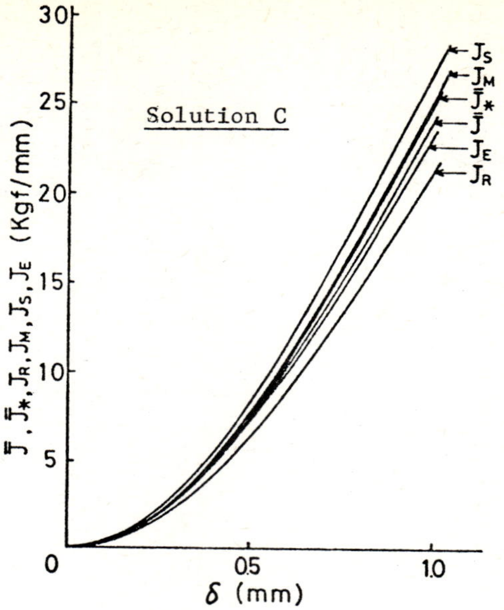

Fig. 8 Comparison of the methods for evaluating J-integral

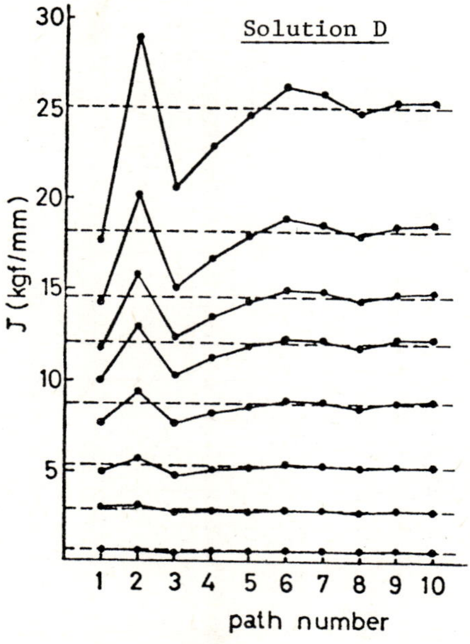

Fig. 9 Path dependence of the J-integral

4. RESULTS OF THE NON-STANDARD PROBLEM

In the non-standard problem, the geometry of the specimen, the loading condition and the material properties were prescribed but all other conditions for analysis were set free. Seven solutions were contributed. The basic informations for these solutions are listed in table 4. Fig. 10 shows the

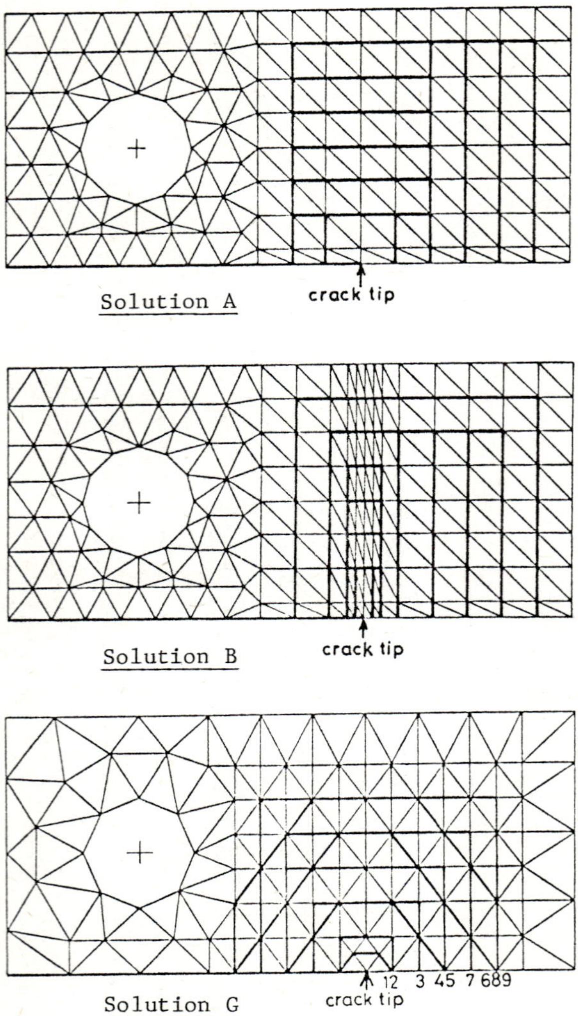

Fig. 10 Element breakdowns and J-integral paths for the non-standard problem

element breakdown for the solutions A_1, A_2, C, F and G. It is the same as that for the standard problem for the solutions E_1 and E_2. Figs. 11-13 show the resulting P-δ, $\bar{J}$-P and $\bar{J}$-δ curves, which obviously have larger deviations than those for the standard problem. The main reason of the large deviations comes from the element breakdown. Coarser element breakdown results higher stiffness in the case of the triangular element. Therefore, it is important to choose the reasonable element breakdown if it is previously known. If a finite element program is based upon the displacement type formulation with compatible shape function, care must be taken to make the element breakdown as fine as possible.

Fig. 14 shows the J_E-δ curves. A_1 and A_2 are the solutions for the non-standard problem, while A and C for the standard problem. A_1 and A_2 are the results for coarser element breakdown than A and C. It is remarkable that all these solutions falled on the same line. It can be concluded that the J_E-δ curve is insensitive to the element breakdown and that the energy method is recommended to evaluate J-integral by the finite element method though it is a little more laborious than other methods.

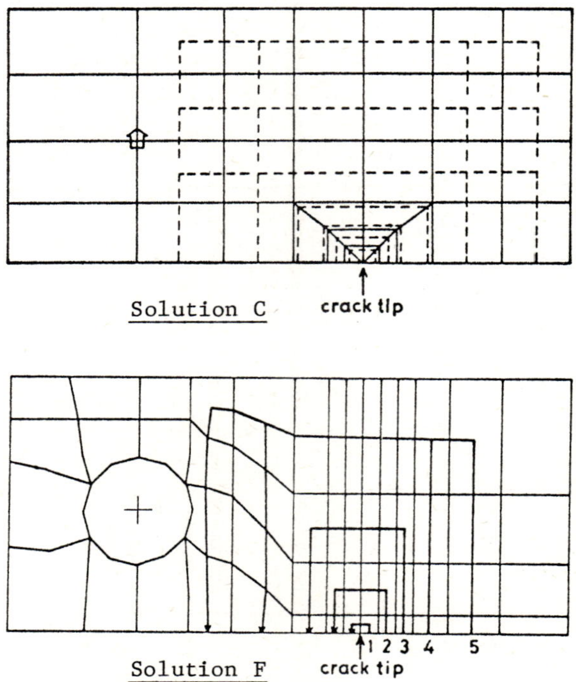

Fig. 10 Element breakdowns and J-integral paths for the non-standard problem (continued)

Table 4 Basic informations for solutions of the non-standard problem

solution	plasticity method	element type	calculation of J	No. of joints No. of elements	(size of element at crack tip)/a*
A_1	Yamada's method	constant strain triangle	$\sigma=(\sigma_1+\sigma_2)/2$	154 248	4/26
A_2	〃	〃	〃	190 312	1/26
C	〃	8 noded isoparametric, singular element at crack tip	integral along Gaussian points	145 38	2/26
E_1	deformation theory	constant strain triangle	$\sigma=(\sigma_1+\sigma_2)/2$	193 317	1/26
E_2	Yamada's method	〃	use of J_{ext}	193 317	1/26
F	〃	8 noded isoparametric	integral along Gaussian points	204 54	2/26
G	〃	constant strain triangle	$\sigma=(\sigma_1+\sigma_2)/2$	126 210	3/26

* a: crack length

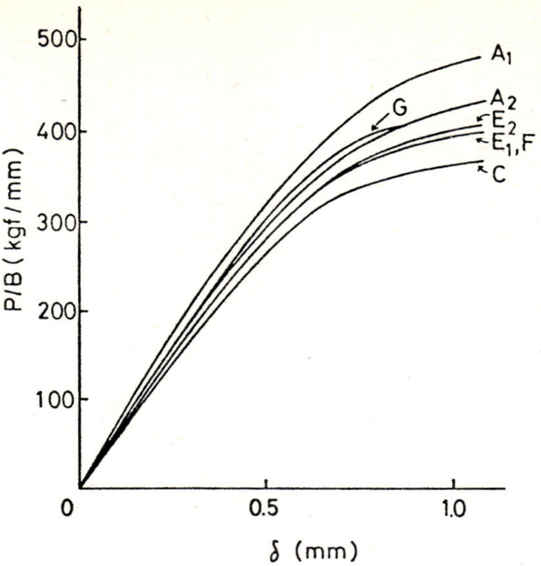

Fig. 11 P/B vs. δ curves for the non-standard problem

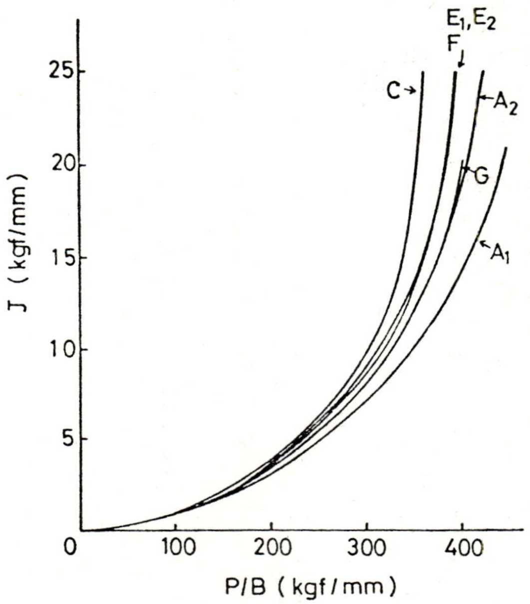

Fig. 12 $\overline{J}$ vs. P/B curves for the non-standard problem

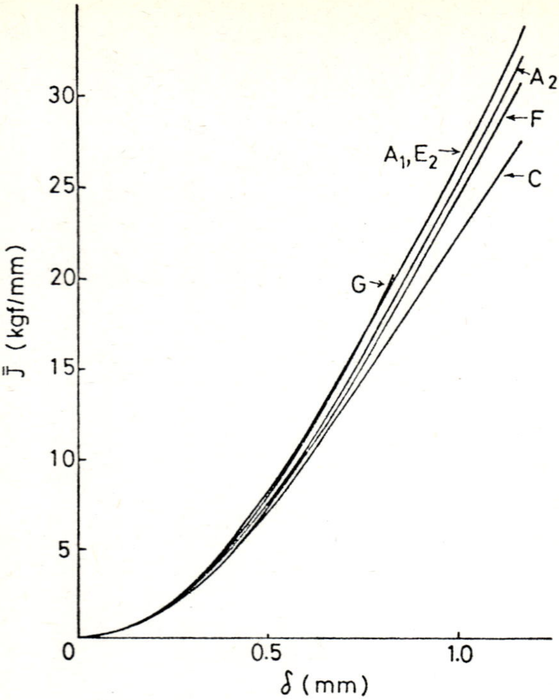

Fig. 13

$\overline{J}$ vs. δ curves for the non-standard problem

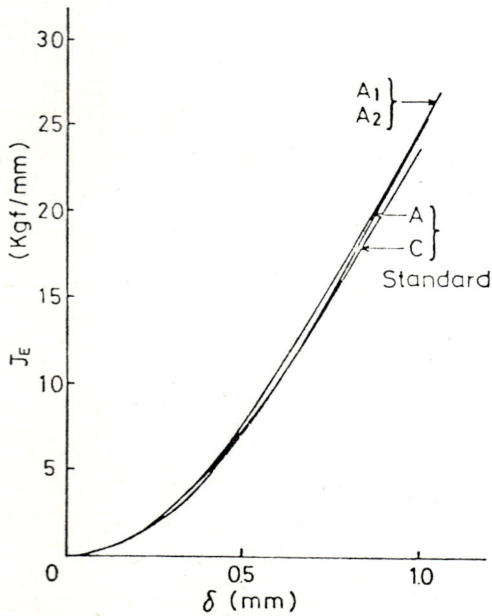

Fig. 14

J_E vs. δ curves for the non-standard problem compared with those for the standard problem

5. CONCLUSIONS

To determine the state of art of the J-integral analysis by finite element method, a round robin test was carried out in Japan. Seven groups of participants carried out elastic-plastic finite element analyses of a compact specimen for the same geometry, the same loading condition and the same material properties. The conclusions are summarized as follows:

(1) The deviation of the J-integral was within 5 % in the standard problem where the element breakdown was prescribed.
(2) Larger deviation was resulted in the non-standard problem where the element breakdown was not prescribed.
(3) Therefore, it was concluded that the main factor which affects the accuracy of the J-integral was element breakdown rather than the finite element program, and that care must be taken to make the element breakdown as fine as possible if the displacement type formulation with compatible shape function was used in the finite element program.
(4) Energy method, by which the J-integral is evaluated by using two P-δ curves, was insensitive to the element breakdown, and therefore, this method seems to be the most reliable.

ACKNOWLEDGEMENT

This description is the summary of the results of the Japanese round robin which was carried out as a task under the JSME Committee PSC21. The authors would like to appreciate the contributions made by the participants listed in table 1, and express their thanks to Professor M. Sakata, chairman of the working group.

REFERENCES

[1] Wilson, W. K., and Osias, J. R., - A Comparison of Finite Element Solutions for an Elastic-Plastic Problem. Int. J. of Fracture, Vol. 14, R95-R109, 1978.

[2] Rice, J. R., Paris, P. C., and Merkle, J. G., - Further Results of J-integral Analysis and Estimates. in Progress in Flaw Growth and Fracture Toughness Testing, Proceedings of 1972 National Symposeum on Fracture Mechanics, ASTM STP 536, pp. 231-245, 1973.

[3] Merkle, J. G. and Corten, H. T., - A J Integral Analysis for the Compact Specimen, Considering Axial Force as Well as Bending Effects. Trans. of the ASME, J. of Pressure Vessel Technology, Vol. 96, No. 4, pp. 286-292, 1974.

[4] Shiratori, M. and Miyoshi, T., - Evaluation of Constraint Factor and J-Integral for Single-Edge Notched Specimen. Proceedings of the Third International Conference on Mechanical Behaviour of Materials. Cambridge, England, Vol. 3, pp. 425-434, 1979.

SECTION 3

DYNAMIC FRACTURE ANALYSIS

WHITHER DYNAMIC FRACTURE MECHANICS?

M. F. Kanninen[(i)]

SUMMARY

The experimental basis for the necessity of a dynamic characterization of crack run/arrest events is reviewed. Current anomalies in the use of linear elastodynamic treatments--apparent geometry and load rate dependence of the dynamic fracture toughness property--are discussed. A review of concurrent work in plastic fracture mechanics is given as a possible basis for circumventing these anomalies.

INTRODUCTION

At the first conference on Numerical Methods in Fracture Mechanics, the author presented an extensive appraisal of the numerical solution techniques used to analyze dynamic fracture mechanics problems [1]. The techniques reviewed were exclusively based on elastodynamic behavior coupled with loading rate dependent (for crack growth initiation) and crack speed dependent (for unstable crack propagation) fracture toughness values. While some new work has appeared in the interim, it is the author's feeling that this does not make a marked departure from the previous trends in the field. Therefore, a reassessment of elastodynamic computational techniques per se is not warranted at this time. The reader interested in this background material can refer to the earlier paper.

An important concern at the time of the first conference was whether a quasi-static or a fully dynamic characteriza-

---

(i) Research Leader and Manager
Fracture Mechanics Projects Office
BATTELLE
Columbus, Ohio U.S.A.

tion of the arrest of a rapidly propagating crack is the more correct. Now, while most workers in the field believe that the dynamic view of crack arrest is more basic, a pragramatic accommodation has been reached with the quasi-static point of view. Hence, this is no longer a critical issue. New issues have emerged to take the place of this controversy, however. These are calling into question basic concepts in dynamic fracture mechanics that were largely taken for granted earlier. A discussion of these with suggestions for possible remedies will form the focal point of this paper. The title chosen for this paper reflects the fact that basic questions about the subject do indeed exist and that further work--possibly in new directions--is called for.

DYNAMIC FRACTURE MECHANICS

Crack Propagation Theories

Until very recently the controversy concerning the proper treatment of the arrest of a rapidly propagating crack dominated work in the field of dynamic fracture mechanics. This controversy centered on whether a dynamic treatment (i.e., one incorporating inertial forces in the equation of motion for the cracked body, stress wave interactions with boundaries, and a crack motion dependent fracture toughness property) or a static post-arrest characterization is basically correct. In a dynamic approach, crack arrest occurs as the terminaton of crack propagation. If this is correct, it follows that, in principle, there can be no direct connection between crack arrest and the quasi-static condition that exists at some long time after arrest. Conversely, if the static condition that corresponds to conditions at the time of arrest (e.g., crack length, applied stresses) uniquely characterizes the arrest process, no consideration of the crack propagation process per se is needed.

Figure 1 shows schematically results obtained by Hahn, et al [2] at Battelle's Columbus Laboratories which revealed clearly the importance of a dynamic-based analysis, at least for the DCB test specimen. As indicated in the figure, crack propagation from an initially blunted crack tip under slowly inserted wedge loading proceeds at an ostensibly constant velocity. This fact, albeit unexpected, made possible a decisive comparison of various possible analysis approaches. The simplest of these possibilities supposes that the crack propagates under quasi-static conditions with a fracture toughness that is always equal to the initiation toughness,

K_{IC}.[(ii)] As shown in the lower part of Figure 1, for quasi-static conditions with $K_{ID} = K_{IC}$, a higher crack speed is predicted. Also, the crack jump length is considerably underestimated. Hence, this approach is clearly invalid.

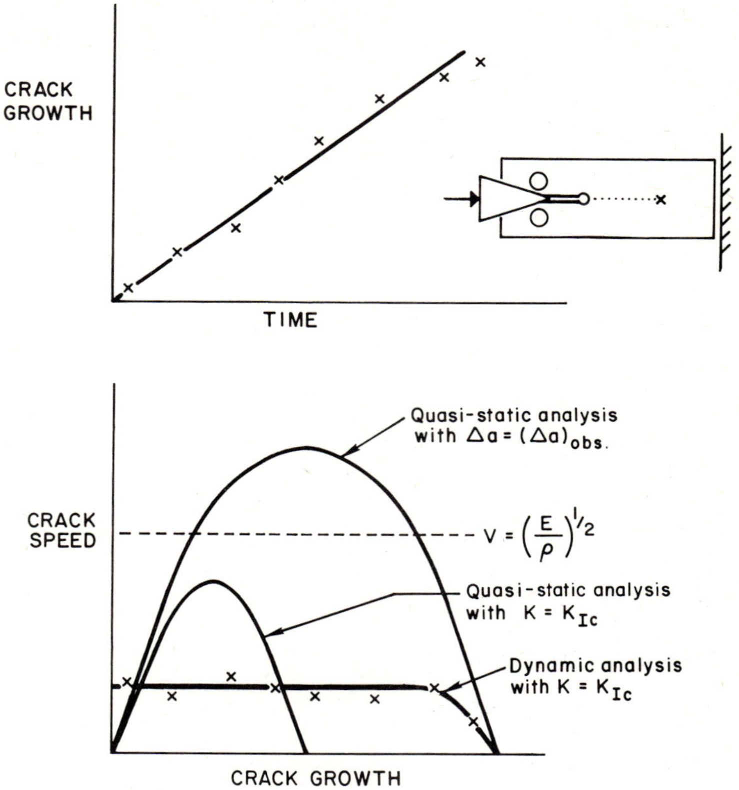

FIGURE 1. QUASI-STATIC VERSUS DYNAMIC ANALYSES OF RAPID CRACK PROPAGATION AND ARREST

(ii) Because of the blunted initial crack tip, the stress intensity factor at the onset of crack growth, K_Q, can be made arbitrarily greater than K_{IC} so that the crack speed and the crack jump length can be systematically varied. Note that with wedge loading the crack propagates into a diminishing stress field and, hence, the arrest of a fast moving crack within a DCB specimen is possible. Moreover, because these can be controlled by the bluntness of the initial crack the DCB specimen is ideally suited to an elucidation of crack arrest principles.

If instead a value of the fracture toughness of the running crack is selected in order to match the observed crack arrest point, much higher crack speeds are obtained. It usually happens that, as indicated in the figure, the predicted crack speeds can exceed the elastic wave speeds for the material. Clearly, therefore, the resolution of this difficulty does not lie in the choice of a fracture toughness property.

Results of the kind shown schematically in Figure 1 provided strong evidence that extended amounts of unstable crack propagation could not be characterized with a quasi-static computational approach. This fact led to the development of a simple dynamic analysis model to study crack propagation in the DCB specimen [3]. A typical result using this approach is also shown in Figure 1 where, to a quite good approximation, the experimental results were reproduced both qualitatively (i.e., a linear crack length-time record virtually from the onset of crack growth to just prior to arrest) and quantitatively.(iii)

The success of the dynamic analysis in predicting crack run/arrest events in DCB specimens exemplified in Figure 1, coupled with the unrealisticness of quasi-static analyses, led to questioning of the then widely accepted static post-arrest characterization of crack arrest. For example, Kanninen [5] performed a series of computations for different initiation conditions in the DCB specimen which showed that the static condition following arrest was a very definite function of the crack jump length in the test. This means that the post-arrest condition--conventionally characterized by the "arrest toughness" K_{Ia}--cannot be related to the material properties controlling the propagation event. Clearly, these two approaches are theoretically incompatible, and on the basis of the foregoing, it appears to be the dynamic approach that is the correct one.

<u>Present Crack Arrest Assessments</u>

Although the work of Hahn, et al [6] accumulated a substantial amount of evidence in support of the dynamic view of

---

(iii) The equations of the one-dimensional model for dynamic crack propagation in the DCB test specimen that were used for the early work in this area were subsequently modified as a result of a more rigorous derivation. This work, together with numerical verifications using a two-dimensional analysis model, can be found in the paper of Gehlen, et al [4]. The correction, it might be noted, is just that anticipated in footnote (vi) of reference [1].

crack propagation and arrest, wide spread acceptance of this view awaited more direct experimental evidence. This was eventually forthcoming in the work of Kalthoff, et al [7]. Their results were obtained using the shadow pattern (or method of caustics) technique which, coupled with flash photography, enables a direct measurement of the stress intensity factor of a fast running crack to be made. If, as assumed in the dynamic point of view, crack propagation occurs only when

$$K_I = K_{ID}(V) \qquad (1)$$

then experimental results such as those of Kalthoff, et al can be used to determine directly the material property K_{ID} as a function of crack speed V. Figure 2 shows their results for DCB specimens using four different K_Q values.

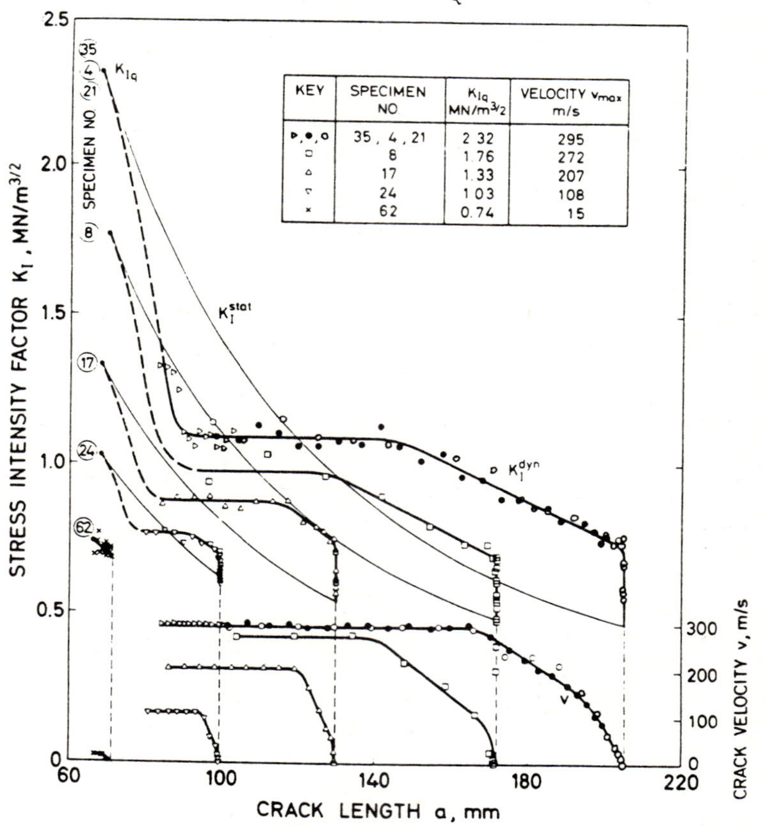

FIGURE 2. STRESS INTENSITY FACTORS FOR CRACK PROPAGATION IN A DCB TEST SPECIMEN FOR VARIOUS DIFFERENT K_Q VALUES - RESULTS OF KALTHOFF, et al [7]

The most important result shown by Figure 2 is that, while the dynamic value of the stress intensity factor at arrest is very nearly the same for all four experiments, the statically computed post-arrest value varies systematically with the crack jump length. This is completely consistent with the dynamic point of view and, of course, at odds with the static arrest characterization. However, it is now generally realized that the DCB specimen is perhaps the most dynamic of all possible structural configurations. Figure 3 illustrates this by comparing a finite difference solution taking into account the finite dimensions of the specimen with Freund's dynamic solution for an infinite medium.(iv)

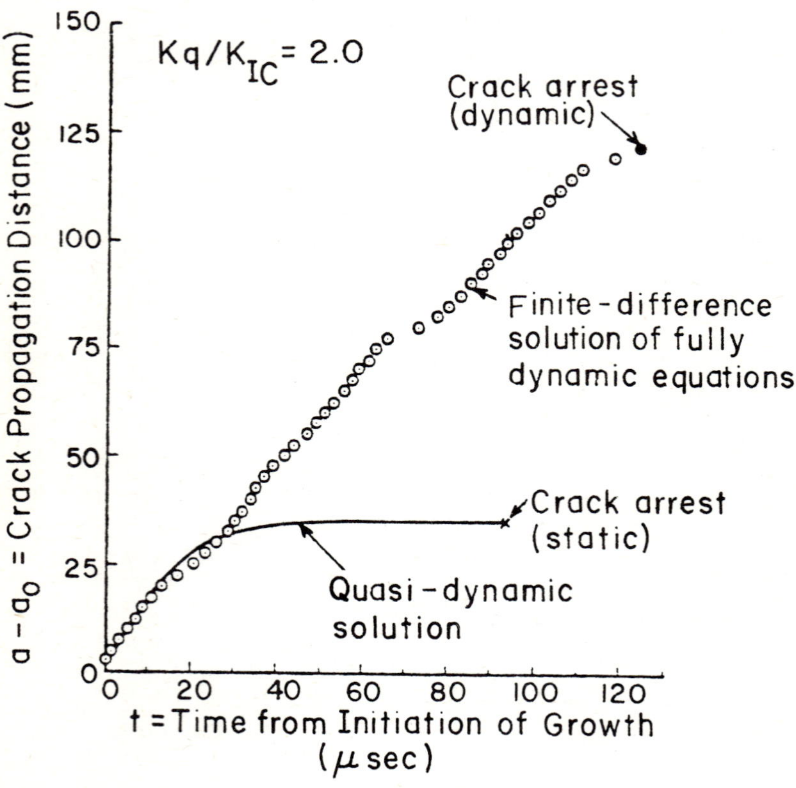

FIGURE 3. COMPARISON OF CALCULATED CRACK LENGTH VERSUS TIME IN A DCB TEST SPECIMEN WITH A DYNAMIC SOLUTION FOR AN INFINITE MEDIUM

(iv) The latter solution was obtained with Equation(11) in reference [1].

The result shown in Figure 3 reveals that the kinetic energy that is reflected back to the crack tip (and is therefore available for use in providing the material's resistance to crack growth) plays a crucial role in crack propagation in a DCB test specimen. That is, the time required for an elastic stress wave to travel from the crack tip to the specimen boundary and return is 26 μsec. It can be seen in Figure 3 that this is just where the infinite medium solution departs and, in fact, predicts arrest. Figure 4 which shows the partitioning of the initial strain energy contained in the specimen during the run-arrest event, further bears this out. It can be seen that the kinetic energy rises to a maximum at about the statically predicted arrest point (i.e., $a - a_o$ = 35 mm). The subsequent decrease indicates the kinetic energy

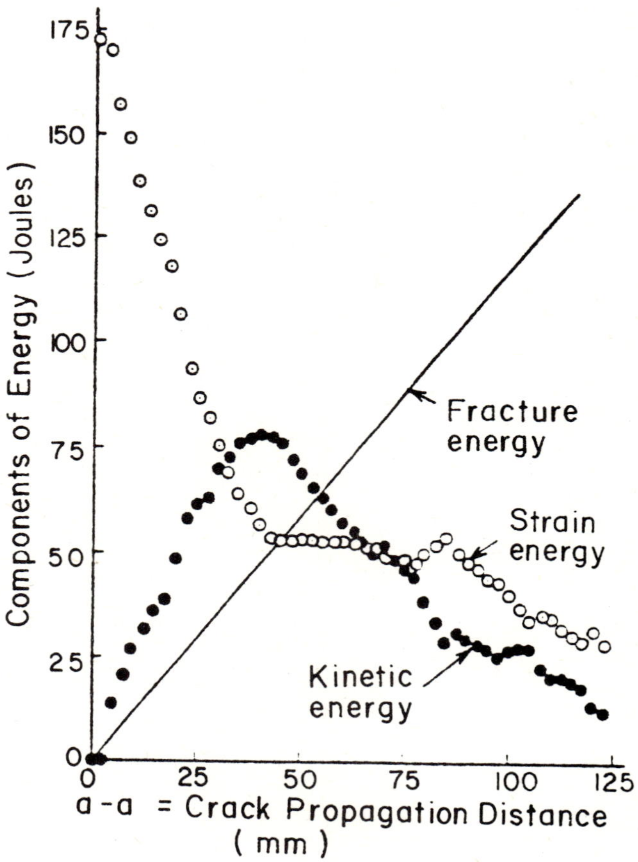

FIGURE 4. DISTRIBUTION OF ENERGY DURING CRACK PROPAGATION IN A DCB TEST SPECIMEN

reflected from the specimen boundaries is being utilized to continue the crack propagation event. Figure 5 shows a result obtained by Kobayashi, et al [8] which indicates that similar behavior occurs in a compact tension specimen.

The accommodation with regard to a static post-arrest characterization mentioned above has been on a pragmatic basis. As examples, Crosley and Ripling [9] and Witt [10] recognize that, because dynamic effects exist in crack arrest, the

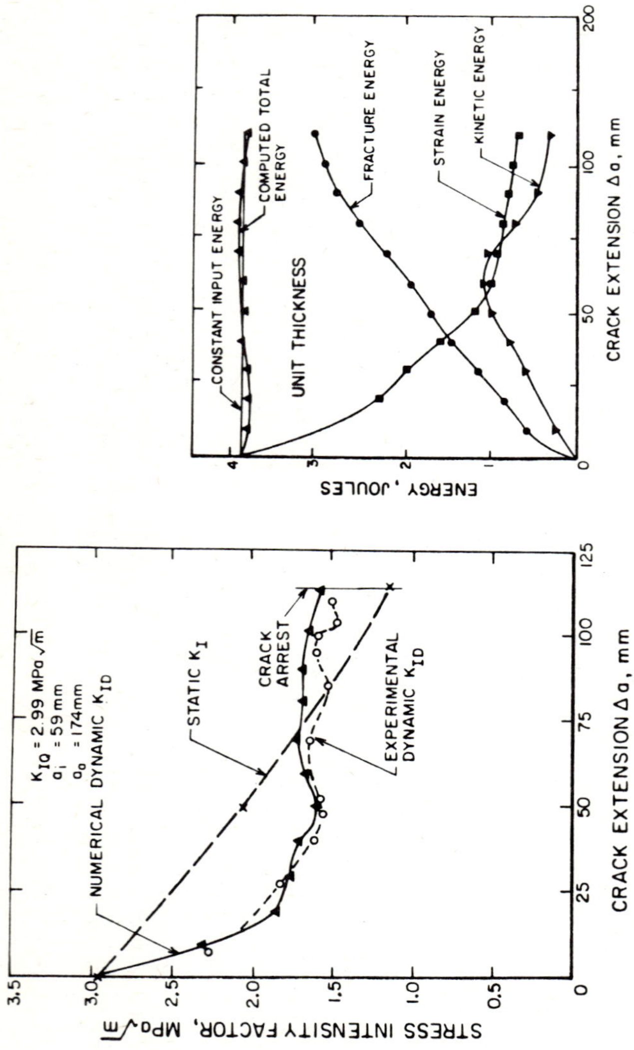

FIGURE 5. COMPARISON OF CALCULATED AND MEASURED RESULTS FOR DYNAMIC CRACK PROPAGATION IN POLYCARBONATE MODIFIED COMPACT TENSION SPECIMENS Results of Kobayashi, et al [8]

quasi-static approach is an over-simplification. Nevertheless, as they assert, reasonably constant statically determined arrest values can be determined experimentally that will suffice for practical purposes if the crack jump length that is allowed is kept small. Moreover, in actual structures, the return of kinetic energy to the crack tip is likely to be small so that the static approach will not be unreasonable.

This same point of view was adopted by Marston, et al [11] in applying a quasi-static approach to assess crack propagation and arrest in a nuclear pressure vessel subjected to thermal stresses in a hypothetical loss-of-coolant accident (LOCA). They concluded that, while dynamic analyses may in general be necessary for crack arrest problems, because of the geometry of the vessel and the anticipated short jump length, a quasi-static analysis should suffice. This assumption is corroborated by the dynamic analysis of the short-jump LOCA event reported by Cheverton, et al. [12] But, as Cheverton, et al also point out, for a hypothetical long crack jump, a dynamic analysis predicts a much deeper penetration than would a quasi-static analysis.

To summarize, while the dynamic approach to crack arrest is clearly the basically correct one, it is also clear that the not inconsiderable computational and experimental complexity required for a fundamentally correct analysis is not always necessary for practical applications. Indeed, for small crack jump lengths, a dynamic fracture mechanics treatment will be indistinguishable from a simpler quasi-static analysis. But, when the two approaches differ, it must be that the dynamic approach is more nearly correct. And, because it generally predicts that the crack will propagate faster and further than will a static analysis, it may be dangerous to assume a priori that quasi-static conditions prevail in any given circumstance.

Adequacy of Elastodynamic Fracture Mechanics

As described in the foregoing, dynamic fracture mechanics has advanced and, in doing so, new critical issues have emerged to replace the crack arrest controversy. Of more prominence is the growing realization that the applicability of even the most rigorous analysis procedures that have been developed may be much more limited than was previously realized. That is, virtually all mathematical solutions and interpretations of experimental results are now made in terms of linear elastic fracture mechanics (LEFM) treatments. However, most work is done on either ductile tough materials like the nuclear pressure vessel steel A533B or on viscoelastic polymeric materials like Homolite 100. While these materials do not satisfy the basic assumptions of LEFM, for lack of an alternative, elastodynamic analyses have been used. Hahn, et al [13] present a crack arrest data base from crack propagation and arrest measurements on various pressure vessel steels. Similar data are given by Francois [14].

A tacit assumption in the collection of a crack arrest data base is that the elastodynamically inferred property $K_{ID} = K_{ID}(V)$ is a material property. As such, it clearly must be independent of the crack/structure geometry and of the manner in which the load is applied. At least two pieces of evidence exist, however, which casts some doubt on this assumption. The first is exemplified by the results of Kalthoff [15] shown in Figure 6. In these experiments cracks were propagated in both rectangular DCB specimens (RDCB) and single edge notch specimens (SEN). It can be seen that the values determined by the method of caustics (see above) were found to be distinctly different in the specimens.

To determine if batch-to-batch material property variations were influencing their results, Kalthoff also used a tee-shaped specimen (RDCB/SEN). In this specimen cracks propagate for a time in each a DCB-like geometry and, later, in a SEN-like geometry (see Figure 6). He found that the results from each portion of the event correlated quite well with the simple specimen results of the corresponding geometry. Hence, material property variations are not important and, he concludes, there is a definite geometry effect.

Other investigators have also reported results which indicate that the K_{ID} property exhibits some geometry dependence; see, for example, Kobayashi, et al [8]. Dahlberg, et al [16] have argued that geometry-dependence and even a dependence on higher order time derivatives must be accepted to avoid the necessity for nonlinear dynamic analyses. They point out that, even if K-dominance (see below) of the inelastic region around the crack tip exists, a dependence of the dynamic fracture toughness on the second and higher order derivatives of the crack length cannot be excluded by any theoretical argument.

While the geometry dependence exhibited in Figure 6 is certainly significant from a conceptual point of view, the practical limitation imposed by these results is probably not debilitating. It can be argued that, in view of the many other uncertainties that are present in any structural analysis problem, this relatively small difference is not significant. One would simply take a lower bound of such results and thereby impose only a modest penalty on the structure.

Of possibility greater significance, therefore, are the results obtained by Kanninen, et al [17] in a study of dynamic crack propagation initiated by impact loading. Specifically, they used AISI 4340 steel three point bend specimens which were instrumented to measure crack length versus time. First,

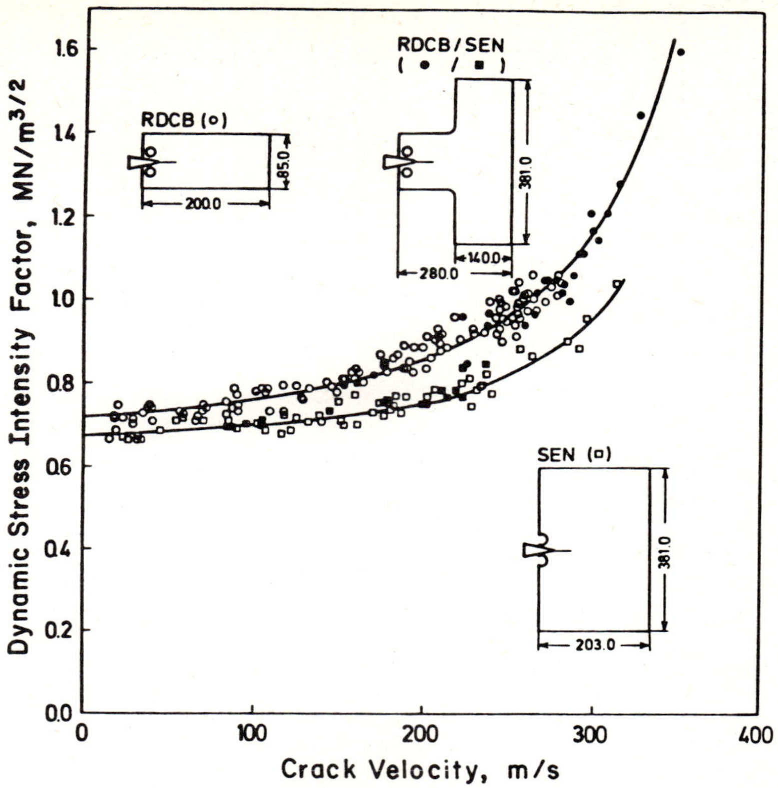

FIGURE 6. DYNAMIC FRACTURE TOUGHNESS VALUES IN ARALDITE B AS DETERMINED BY THE METHOD OF CAUSTICS
Results of Kalthoff [15]

to assess any possible material property variations and to determine the effect of geometry-dependence of the fracture toughness property determined in an earlier program using the DCB test specimen, dynamic crack propagation was initiated under quasi-static loading. A comparison between the experimental results and those predicted with an elastodynamic finite difference calculation using this property is shown in Figure 7. It can be seen that the agreement is excellent.

Because of the agreement shown in Figure 7 and the belief that the material used very well satisfies the basic requirements of a linear elastic fracture theory, it might be logical to expect that the same K_{ID} value would also apply in impact loading. However, as shown in Figure 8, the calculation seriously underestimates the material's resistance to fast

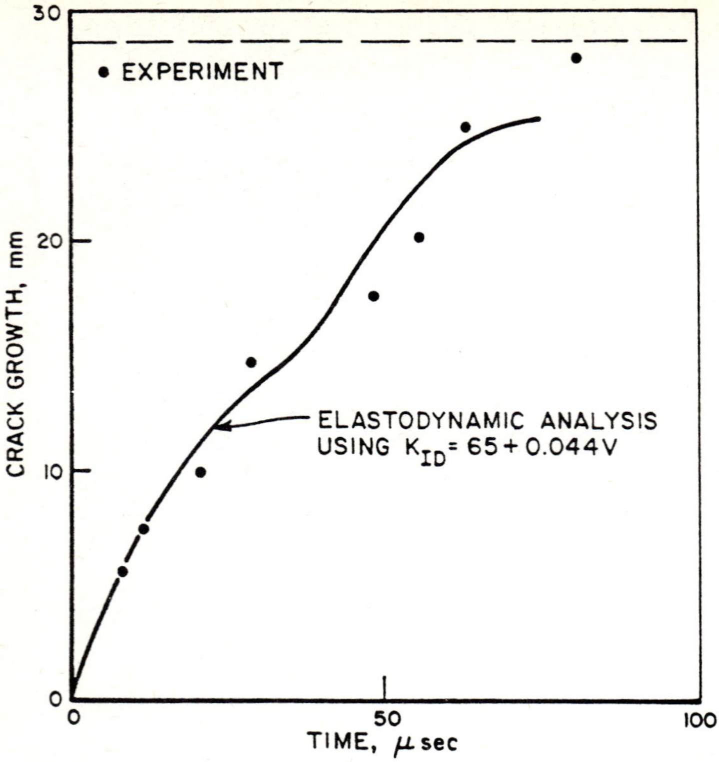

FIGURE 7. COMPARISON OF MEASURED AND CALCULATED CRACK GROWTH VERSUS TIME IN A QUASI-STATICALLY LOADED THREE-POINT BEND SPECIMEN

fracture.[v] A prediction much more in agreement with the experimental results was obtained by back calculating a toughness value from the experimentally determined throw energy (i.e., by deducting the kinetic and strain energy in the

---

(v) In contrast to the quasi-static loading results shown in Figure 7, where zero time corresponds to the initiation of crack growth, zero time in Figure 8 is the time that the striker contacts the specimen. Clearly, in an impact event, a time lapse is required for the crack tip stress intensity to build up to a critical value.

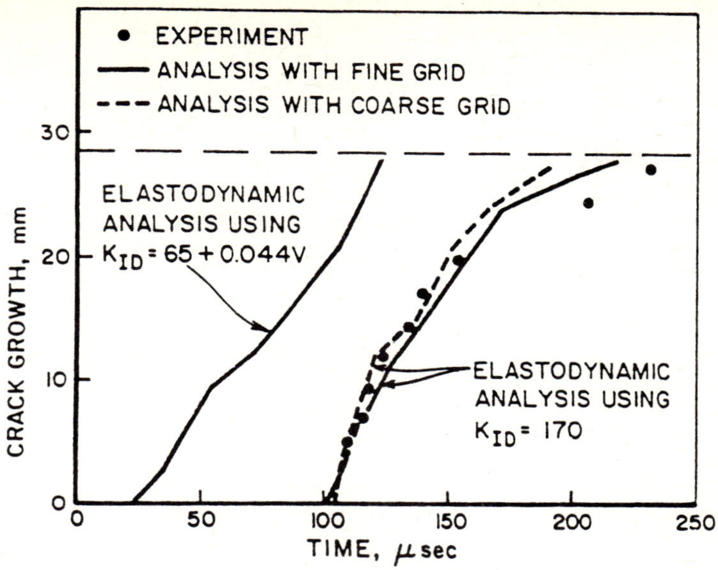

FIGURE 8. COMPARISON OF MEASURED AND CALCULATED CRACK GROWTH VERSUS TIME IN AN IMPACT LOADED THREE-BEND SPECIMEN

broken specimen from the energy supplied by the striker). This gave a value of K_{ID} = 170 $MNm^{-3/2}$, a value roughly double that of the quasi-statistically initiated event. Figure 8 shows that excellent agreement is obtained using this value.

There were several artifacts involved in the impact testing that could possibly cause this anomolous result. As described in reference [17], however, none were found to be significant. In addition, as shown in Figure 8, two different finite difference mesh sizes were used without appreciably altering the result. It was therefore concluded that the difference exhibited between quasi-statically initiated and dynamically-initiated rapid crack propagation does seem to proceed with a markedly different toughness property.(vi)

(vi) It can be seen from the results shown in Figure 7 and 8 that the crack speeds in the two events were similar. Regardless, AISI 4340 steel is not greatly rate-dependent. Thus, there is no simple explanation for the quite different toughness values that seem to be required as a speed-dependence.

The disparity exhibited by the impact testing results, when coupled with the geometry-dependence assessments made by Kalthoff and others, appears to cast serious doubt on the general applicability of linear elasticity-based crack propagation/arrest procedures. At the same time it should be recognized that, in any use of experimental observations to assess the basis of mathematical analysis procedures, no direct measurement of the stress intensity factor is possible. While observations of fringe and shadow patterns associated with a propagating crack can be made, the relation of these measurements to fracture mechanics parameters always requires the use of some mathematical model. And, any such model must be based upon a constitutive relation and other presumptions about the interaction between the propagating crack and the component that contains it.

To assess the possible effects of polymeric materials, Popelar and Kanninen [18] have devised a dynamic visoelastic representation for polymer DCB test specimens. Their results indicated that differences do exist but they can be accounted for by using the correct choice of the modulus--the static or long-term viscoelastic modulus, at least for cracks initiated under quasi-static conditions. However, they were unable to substantiate the finding of Fourney [19] that a substantial portion of the initial strain energy is lost by viscous damping prior to crack arrest in the photoelastic polymer Homolite 100. It is entirely possible that a more appropriate viscoelastic model is needed (Popelar and Kanninen used a three-parameter solid representation) before this can be done.

PLASTIC FRACTURE MECHANICS

Crack Tip Fracture Criteria

While the initial work in fracture mechanics was based upon an energy balance criterion, later work identified more esoteric fracture parameters--principally, the stress intensity factor, the crack opening displacement, and the J-integral parameter. In LEFM, these are all interrelated. Specifically, for plane strain conditions in the "opening" mode

$$G = J = \frac{1-\nu^2}{E} K^2 = Y \cdot \delta \qquad (2)$$

where G is the strain energy release rate, J is the value of the J-integral, K is the stress intensity factor, and δ is the crack tip crack opening displacement while E, ν, and Y are, as usual, the elastic modulus, Poisson's ratio, and yield stress, respectively.

Which of the four basic parameters involved in LEFM is the "most basic" may be thought to be a purely academic question. However, it assumes considerably more importance when it becomes necessary to select a crack tip fracture parameter as the basis of a plastic fracture methodology capable of treating stable crack growth accompanied by extensive crack tip plasticity. Many different choices have been made, all having their origins in one of the LEFM parameters. But, because a set of equalities like (2) for conditions more general than LEFM does not exist, it is important to determine which criterion is on the firmest footing. This, in turn, suggests a more careful study of LEFM.

The modern view of LEFM is contained in Figure 9. It can be shown that, if the body everywhere obeys a linear elastic stress-strain law (see insert in Figure 9), then the stresses at the crack tip can be expressed in terms of a polar coordinate (r,Θ) system with origin at the crack tip as

$$\sigma_{ij} = \frac{K}{\sqrt{2\pi r}} F_{ij}(\Theta) + \ldots. \qquad (3)$$

where the omitted terms are of higher order in r. For small values of r (i.e., very near the crack tip), only the first term is significant. Then, the remote stresses, the crack length, and the external dimensions of the cracked body will affect the stresses at the crack tip only through the parameter K, the stress intensity factor. More definitely, there will be a region--the "K-dominant" region--having the characteristic dimension D in which the first term of the series is a sufficiently good approximation.

To continue this argument, let R denote the size of the inelastic region surrounding the crack tip where the assumption of linear elastic behavior is invalid. It is in this region that the fracture event takes place. While it is not possible to directly characterize the fracture process using a linear elastic formulation, this is not necessary provided the inelastic region is contained in the K-dominant region. That is, if $R<D$, then any event occurring within the inelastic region is controlled by the deformation in the surrounding K-dominant region. Consequently, if crack growth occurs, it must do so at a critical value of the stress intensity factor.

The importance of this result is not for its own sake but rather for the generalization that is suggested for elastic-plastic conditions. In particular, using a power law hardening solution, an analogous argument to that given above can be followed. As illustrated in Figure 10, the crack tip stresses in this situation can be expressed as

$$\sigma_{ij} = J^{\frac{1}{n+1}} r^{-\frac{1}{n+1}} F_{ij}(\Theta,n) + \ldots. \qquad (4)$$

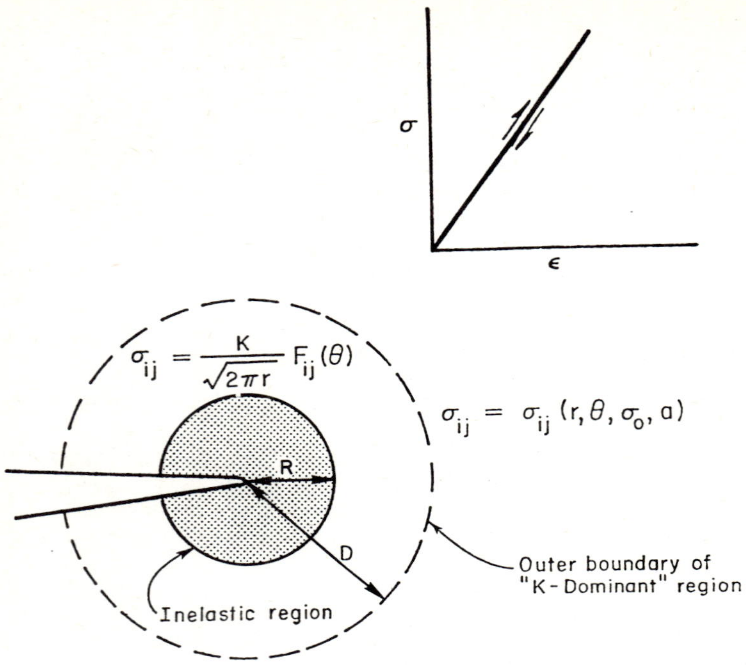

FIGURE 9. BASIS OF LINEAR ELASTIC FRACTURE MECHANICS

where n is a property of the material's stress-strain curve. Now, the effect of the remote stresses, the crack length, and the external dimensions of the body on the stresses within a "J-dominant" region depend only on the parameter J. If this region surrounds the inelastic region, then the conditions governing the fracture event must correspond to a critical value of J. The crack growth criterion can therefore be expressed as

$$J(a,\sigma_o) = J_c \tag{5}$$

where a denotes the crack length and σ_o the applied stresses.

Notice that, in contrast to the LEFM argument, the inelastic region is not the plastic region here. It is instead the much smaller region in which the deformation plasticity approach (see insert in Figure 10) is invalid. That is, the region in which the hole growth and coalescence processes involved in ductile crack extension are occurring--processes that clearly cannot be taken into account directly in a continuum mechanics approach. However, where J-dominance exists,

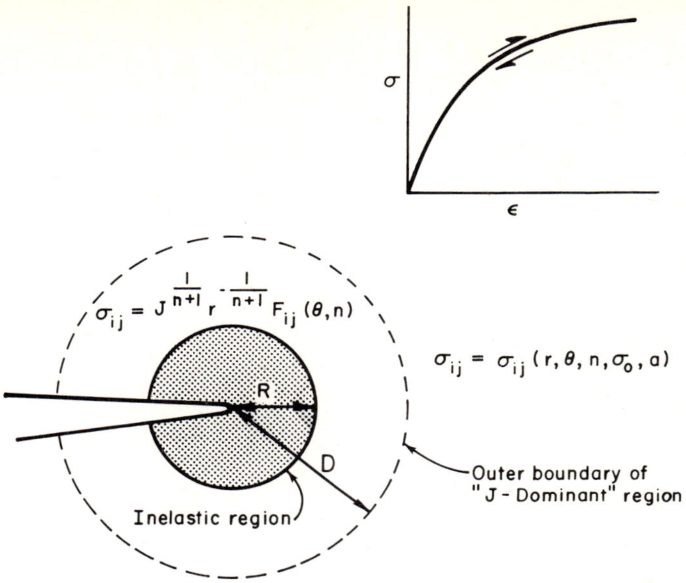

FIGURE 10. BASIS FOR THE USE OF THE J-INTEGRAL IN PLASTIC FRACTURE MECHANICS

the initiation and growth of a crack can be expected to be governed by a material property--the J-resistance curve--which gives critical J values as a function of crack growth.

For a growing crack, elastic unloading takes place in the "wake" plastic region left behind the crack tip. This is also an inelastic process that cannot be addressed within the deformation plasticity-based approach just described. Consequently, it is to be expected that the size of the inelastic region becomes larger and inexorably overtakes the limits of the J-dominant region. At this point the J parameter also becomes invalid. Precise delineations of the amount of crack growth possible before the loss of J-dominance occurs do not now exist. This will depend upon the dimensions of the body and the rate of change of the J-resistance curve. For example, as suggested by Hutchinson and Paris [20], J-dominance will exist provided

$$\omega \equiv \frac{b-a}{J}\frac{dJ}{da} >> 1 \qquad (6)$$

where a denotes the crack length and b is the dimension of the body nearest the crack tip.

The key to developing an analysis procedure for plastic fracture is to identify an appropriate crack tip fracture criterion. Work performed by Kanninen, et al [21] has encompassed three main stages. First, center cracked panels were tested to obtain data on crack growth initiation and stable growth. Second, "generation-phase" analyses were performed in which the experimentally observed applied stress/stable crack growth behavior was reproduced in a finite element model with each of a number of candidate crack initiation and stable growth criteria being evaluated for the material tested. In the third stage, "application-phase" finite element analyses were performed using one of the candidate criteria to determine applied stress/crack growth behavior for a given specimen geometry.

The fracture criteria examined included the J integral, the local and average crack opening angles, the conventional LEFM R curve, and various generalized energy release rates. Each of the candidate criteria is attractive in one way or another. Hence, the task of selecting the best criterion for application to nuclear steels is not an easy one. Clearly, geometry independence is a crucial test of the acceptability of a plastic fracture criterion. Physical relevance is another. Practicality is a third. With these as primary qualifications, some assessments can be drawn from progress made so far.

The advantages of the J integral are its virtual independence of finite element type and element size, the computational ease involved in evaluating it, and, because of its history-independence, its catalogability. However, while the J integral is widely acceptable as a criterion for crack growth initiation, as already noted, it is valid only for a limited amount of stable crack growth. A J integral-based approach is unable to cope with large amounts of stable crack growth attended by large-scale plasticity because it is based upon deformation plasticity. Deformation plasticity (nonlinear elasticity) requires small plastic strains and precludes material unloading. This manifests itself in pronounced specimen dependence after a small amount (e.g., 10% of the remaining ligament size) of stable growth.

The crack opening angle is appealing because of its rapidly grasped physical significance and the opportunity that it offers for direct measurement. However, it should be recognized that there are two different definitions of the crack opening angle: a crack tip value that reflects the actual slope of the crack faces (CTOA), and an average value based on the original crack position (COA). While the critical value of the COA can be measured, it is difficult to see how its value has any direct connection with the fracture process.

Conversely, while the critical value of the CTOA can likely be associated with the fracture process, it presents a formidable measurement task. In addition, there are clearly some difficulties in making either value apply to mixed character shear/flat crack growth.

A proper stable crack growth criterion must differentiate between the energy dissipated in direct fracture-related processes near the crack tip and energy dissipated in geometry-dependent plastic deformation remote from the crack tip. With this in mind, a number of investigators have opted for a generalization of the LEFM energy release rate as the basic plastic fracture methodology. But, there is a basic difficulty inherent in this approach. There is a theoretical basis for expecting a computational step size dependence in an energy release rate parameter that is based on the work of separating the crack faces. It can be argued that this can be handled by appealing to micromechanical considerations. Regardless, it appears that the necessity to arbitrarily circumvent the inherent step size difficulty with any energy release rate parameter makes its use somewhat unattractive.

Dynamic Plastic Crack Propagation

Work in dynamic elastic-plastic crack propagation has been performed by Achenbach, et al [22,23]. The material model used in the work was based on Prandtl-Reuss incremental plasticity and a bi-linear stress-strain relation with irreversible material unloading behind the crack tip. Results have been obtained for crack propagation in plane stress, plane strain and in anti-plane strain conditions. These results show that the order of the crack tip singularity and the position of the plastic unloading interface, while highly dependent on the slope of the stress-strain curve in the plastic regime, are only moderately dependent on the crack speed. Results for plane stress conditions are shown in Figure 11, where $\alpha = E_t/E$.

These findings are important primarily in that they demonstrate that a fundamentally correct elastic-plastic dynamic formulation of a propagating crack can be achieved for use in a finite element program. In addition, because for specified material stress-strain behavior, the crack tip characterizing parameters will be essentially unaffected by modest changes in the crack speed, this work shows that it will be possible to devise an efficient computational model.

Further work will be needed to identify a plastic fracture criterion for dynamic crack propagation. This parameter must be one whose critical values are geometry-independent material property values over a wide range of geometries and crack growth lengths. A strong possibility that has been identified during the course of experiments and

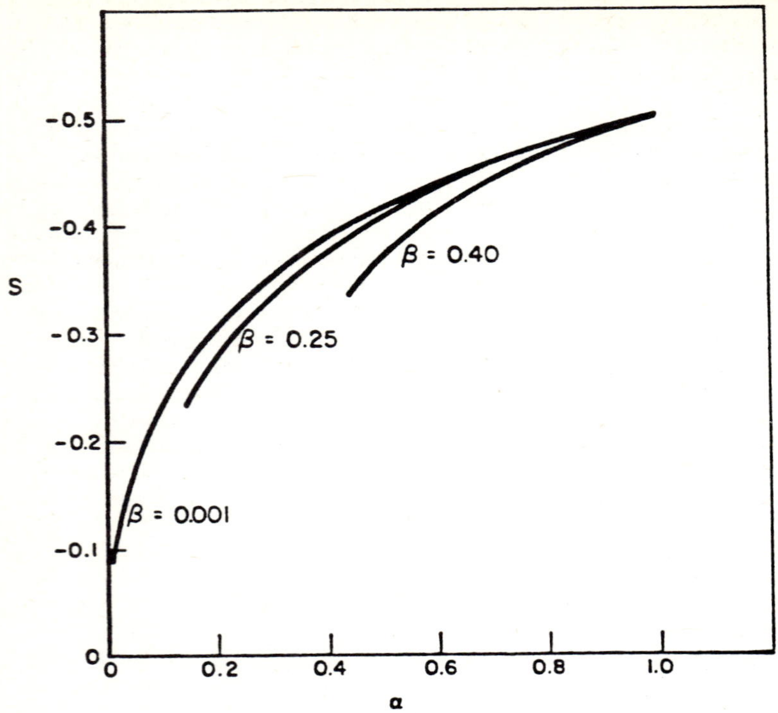

FIGURE 11. ORDER OF THE CRACK TIP SINGULARITY IN PLANE STRESS DYNAMIC CRACK PROPAGATION

generation-phase analyses on compact tension specimens and center-cracked tension panels is the crack tip crack opening angle (CTOA). Kobayashi, et al [8] have shown how effectively this parameter can be used in their analysis of circumferential crack propagation in a pipe.

DISCUSSION

Most potential fracture problems involve cracks emanating from flaws in or near a weld where it is difficult to apply a fracture mechanics assessment for several reasons. First, residual stresses probably exist of unknown magnitude. Second, the toughness of the material in the heat affected zone is uncertain. Third, welding processes generally cause plastic deformation which invalidates the currently available linear elastic fracture mechanics capabilities. These difficulties suggest the use of a crack arrest strategy whereby, even if unstable crack propagation occurs, it will be arrested in the base material. This is the rationale for the dynamic fracture mechanics analysis discussed here.

The schism between the dynamic and the quasi-static characterizations of crack arrest is no longer a critical issue. It is generally agreed that there are conditions where a quasi-static interpretation of a laboratory test result can give an appropriate measure of the arrest toughness property and, under certain types of loading and crack-structure geometries, dynamic effects in crack arrest will indeed be negligible. Specifically, when inertia forces, stress wave reflections, and rate-dependent fracture processes are negligible, then the two approaches will give exactly the same prediction. There appears to be a fairly wide range of conditions in which, because the predictions are not greatly different, for practical purposes, the extra effort required to obtain a dynamic solution is unwarranted. However, where the two predictions are significantly different, it is the dynamic solution which is the more accurate.

It is important to recognize that a quasi-static calculation will generally underestimate the true crack driving force. Consequently, in extracting an arrest toughness value from an experiment, a value that is lower than the actual material property will be obtained. This will be conservative. Applying a quasi-static analysis using a given toughness value to assess the possiblity of crack arrest, on the other hand, will over-estimate the likelihood of crack arrest. While these two errors do tend to offset each other, it would always be prudent in addressing new situations to check the possibility that dynamic effects could be important.

The basic assumption in elastodynamic analyses of crack propagation has been that the dynamic fracture toughness is a unique geometry-independent material property that can, at most, depend upon crack speed, temperature, and plate thickness. However, results presented by several investigators are beginning to seriously question the legitimacy of this assumption. Some investigations indicate that the external dimensions of the component can affect values inferred for the toughness property. Others suggest that unstable crack propagation emanating from impact loading occurs with a toughness which differs markedly from that corresponding to conventional quasi-static loading. Whether or not residual plasticity or other nonlinear effects could play a key role in mollifying this seeming lack of uniqueness cannot presently be determined.

CONCLUSIONS

1. Crack propagation accompanied by significant plastic deformation cannot be rigorously treated at present--only elastodynamic solutions can now be applied. However, these appear to give reasonable predictions even for tough ductile materials (e.g., nuclear vessels, gas pipelines), under certain conditions.

2. Quasi-static predictions of crack arrest can be valid in some circumstances--e.g., when crack jump length is small in comparison to component dimensions--but will give an underestimate when dynamic effects are significant.

3. Controversy now exists on the validity of $K_{ID} = K_{ID}$ (V) as a unique material property. Some geometry dependence has been cited. Also, comparisons of slow versus impact loading have revealed unexplainable differences.

ACKNOWLEDGEMENT

This paper was written in conjunction with a research program on unstable crack propagation in elastic-plastic materials supported by the Office of Naval Research under contract number N00014-77-C-0876. The author would like to thank Dr. Nicholas Perrone, Director of Structural Mechanics Research at ONR for his encouragement of this work. The portion of the paper on plastic fracture mechanics was based in part on research performed at Battelle with support from the Electric Power Research Institute, Palo Alto, California, under RP 601-1, Dr. T. U. Marston, Program Manager.

REFERENCES

[1] KANNINEN, M.F. - A Critical Appraisal of Solution Techniques in Dynamic Fracture Mechanics, Numerical Methods in Fracture Mechanics, Ed. A. Luxmoore and R. Owen, U. Swansea Press, pp. 612-633, 1978.

[2] HAHN, G.T., HOAGLAND, R.G., KANNINEN, M.F. and ROSENFIELD, A.R. - A Preliminary Study of Fast Fracture and Arrest in the DCB Test Specimen, Dynamic Crack Propatation, G. C. Sih, editor, Noordhoff, Leyden, pp. 649-662, 1973.

[3] KANNINEN, M.F. - A Dynamic Analysis of Unstable Crack Propagation and Arrest in the DCB Test Specimen, Int. J. Fracture, Vol. 10, pp. 415-430, 1974.

[4] GEHLEN, P.C., POPELAR, C.H. AND KANNINEN, M.F. - Modeling of Dynamic Crack Propagation: I. Validation of One-Dimensional Analysis, International Journal of Fracture, Vol. 15, pp. 281-294, 1979.

[5] KANNINEN, M.F. - An Analysis of Dynamic Crack Propagation and Arrest for a Material Having a Crack Speed Dependent Fracture Toughness, Prospects of Fracture Mechanics, Ed. G.C. Sih, et al, Noordhoff International Publishing, Leyden, The Netherlands, pp. 251-266, 1974.

[6] HAHN, G.T., et al - Critical Experiments, Measurements and Analyses to Establish a Crack-Arrest Methodology for Nuclear Pressure Vessel Steels, Battelle's Columbus Laboratories Reports to the U.S. Nuclear Regulatory Commission, 1974-1979.

[7] KALTHOFF, J.F., BEINART, J. and WINKLER, S. - Measurements of Dynamic Stress Intensity Factors for Fast Running and Arresting Cracks in Double-Cantilever-Beam Specimens, Fast Fracture and Crack Arrest, Ed. G. T. Hahn and M. F. Kanninen, ASTM STP 627, pp. 161-176, 1977.

[8] KOBAYASHI, A.S., SEO, K., JOU, J.Y., and URABE, Y. - A Dynamic Analysis of Modified Compact-Tension Specimens Using Homolite - 100 and Polycarbonate Plates, Experimental Mechanics, Vol. 20, pp. 73-79, 1980.

[9] CROSLEY, P.B. and RIPLING, E.J. - Comparison of Crack Arrest Methodologies, Crack Arrest Methodology and Applications, Ed. G. T. Hahn and M.F. Kanninen, ASTM STP 711, pp. 211-209, 1980.

[10] WITT, F.J. - Crack Arrest Methodology for Pressurized Water Nuclear Reactor Pressure Vessel Applications, Westinghouse Electric Corporation, Report No. WCAP - 9627, 1979.

[11] MARSTON, T.U., SMITH, E. and STAHLKOPF, K.E. - Crack Arrest in Water-Cooled Reactor Pressure Vessels During Loss-of-Coolant Accident Conditions, Crack Arrest Methodology and Applications, Ed. G. T. Hahn and M. F. Kanninen, ASTM STP 711, pp. 422-431, 1980.

[12] CHEVERTON, R.D., GEHLEN, P.C., HAHN, G.T. and ISKANDER, K. - Application of Crack Arrest Theory to a Thermal Shock Experiment, Crack Arrest Methodology and Applications, Ed. G. T. Hahn and M. F. Kanninen, ASTM STP 711, pp. 392-417, 1980.

[13] HAHN, G.T., ROSENFIELD, A.R., MARSCHALL, C.W., HOAGLAND, R.G., GEHLEN, P.C., KANNINEN, M.F. - Crack Arrest Concepts and Applications, Fracture Mechanics, Ed. N. Perrone, et al, University of Virginia Press, pp. 205-228, 1978.

[14] FRANCOIS, D. - Dynamic Crack Propagation and Arrest, Developments in Pressure Vessel Technology - Flaw Analysis, Ed. R. W. Nichols, Applied Science Pub, pp. 151-167, 1979.

[15] KALTHOFF, J.F. Institut fur Festkorpermechanik, private communication, 1980.

[16] DAHLBERG, L., NILSSON, F. and BRICKSTAD, B. - Influence of Specimen Geometry on Crack Propagation and Arrest Toughness, Crack Arrest Methodology and Applications, Ed. G. T. Hahn and M. F. Kanninen, ASTM STP 711, pp. 89-108,1980.

[17] KANNINEN, M.F., GEHLEN, P.C., BARNES, C.R., HOAGLAND, R.G., HAHN, G.T. and POPELAR, C.H. - Dynamic Crack Propagation Under Impact Loading, Nonlinear and Dynamic Fracture Mechanics, Ed. N. Perrone and S. Atluri, ASME AMD Vol. 35, pp. 185-200, 1979.

[18] POPELAR, C.H. and KANNINEN, M.F. - A Dynamic Viscoelastic Analysis of Crack Propagation and Crack Arrest in a Double Cantilever Beam Test Specimen, Crack Arrest Methodology and Applications, Ed. G. T. Hahn and M. F. Kanninen, ASTM STP 711, pp. 3-21, 1980.

[19] FOURNEY, W.L., University of Maryland, private communication, 1980.

[20] HUTCHINSON, J. and PARIS, P.C. - The Theory of Stability Analysis of J-Controlled Crack Growth, Elastic-Plastic Fracture, Ed. J. D. Landes, et al, ASTM 668, pp. 37-64, 1979.

[21] KANNINEN, M.F., RYBICKI, E.F., STONESIFER, R.B., BROEK, D., ROSENFIELD, A.R., MARSCHALL, C.W. and HAHN, G.T. - Elastic-Plastic Fracture Mechanics for Two-Dimensional Stable Crack Growth and Instability Problems, Elastic-Plastic Fracture, Ed. J. D. Landes, et al, ASTM STP 668, pp. 121-150, 1979.

[22] ACHENBACH, J.D. and KANNINEN, M.F. - Crack Tip Plasticity in Dynamic Fracture Mechanics, Fracture Mechanics, Ed. N. Perrone, et al, University of Virginia Press, pp. 649-670, 1978.

[23] ACHENBACH, J.D., KANNINEN, M.F. and POPELAR, C.H. - An Asymptotic Analysis of Dynamic Elastic-Plastic Crack Propagation, Journal of the Mechanics and Physics of Solids, in press, 1980.

A STUDY OF DYNAMIC CRACK GROWTH IN THE DCB SPECIMEN

K. Hellan(I)

SUMMARY

Crack propagation in a double cantilever beam is studied analytically and numerically. The specimen is modelled as a pair of Timoshenko-beams in direct contact along the prospective crack path. Results are correlated to the simple shear-beam prediction, and are compared with those of previous models assuming some type of elastic layer connecting the beams. The said correlation is suggested as a basis for simple interpolation formulae to include a wider range of geometrical data.

INTRODUCTION

Experiments on dynamic growth are frequently performed using wedge-loaded, double-cantilever beam (DCB) specimens, e.g. Hahn et al. [1-4]. Kanninen's beam model has served to provide independent analytical data and to correlate the experimental findings. In this, the specimen is viewed as a pair of Timoshenko-beams joined along the crack path by a series of closely spaced springs. The model considered herein was introduced by Hellan [5-6], defined simply by two Timoshenko-beams in direct contact along the crack path. Even if this is a limiting case of the spring model (corresponding to a continuous array of springs with infinite extensional, zero rotational stiffness) the case stands alone for analytical and numerical reasons. Firstly, the fracture condition will involve a force singularity at the crack tip, contrary to being kinematically defined in the spring model, and secondly, the stronger discontinuity propagating with the crack tip may have severe numerical consequences. In these respects the present model has properties similar to those of Freund's double shear-beam [7],

(I) Professor, Division of Applied Mechanics, University of Trondheim - NTH, Norway.

which in turn may be viewed as a limiting case when the present bending stiffness is taken to infinity.

Partly suppressing one degree of freedom, Bilek [8] has independently applied the model to evaluate the dynamic fracture resistance from observed crack lengths a versus time t. By contrast , the present paper includes detailed analyses of the crack history until arrest when the fracture resistance is a known function of da/dt. The initial motion from a blunted precrack (a barrier) is also studied, and the proper inclusion of the discontinuities in general is discussed. In the performed finite difference analysis nodal points are placed along the discontinuity lines, which are shear-wave characteristics and the crack-tip trajectory in the a-t plane. The latter being unknown a priori may imply that the grid has to be generated tentatively along with the computation.

A conclusion from the performed calculations is that the predictive capabilities of the present model are essentially the same as those of the spring model. If the present mathematical fundaments may be somewhat easier to articulate, there remains to translate this into any obvious numerical advantage. However, the description of the initial motion becomes unique and definitely simpler. The wave nature of the problem is also enhanced, so that a correlation of the present results to those of the shear-beam prediction becomes natural. On this basis, interpolation formulae to include a wider range of geometrical data may be suggested.

Implicit in the presentation is the assumption of an essentially elastic, linear behaviour, with the dissipative processes confined to a narrow region moving with the crack tip. Griffith energy balance is assumed to govern the separation.

GOVERNING EQUATIONS

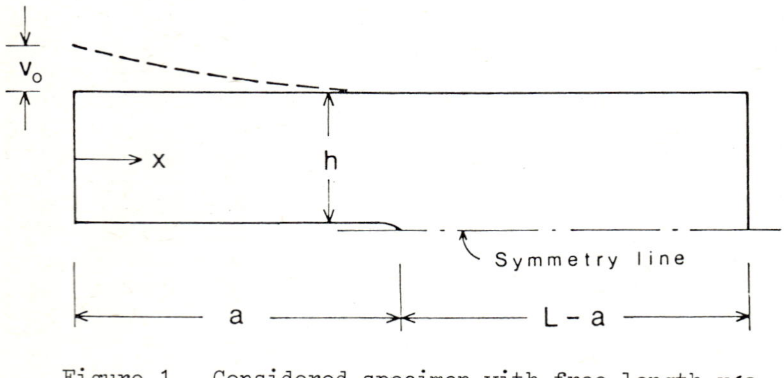

Figure 1. Considered specimen with free length x<a and contact length x>a.

One half of the considered specimen is shown in Fig.1. $v \equiv v(x,t)$ denoting the transverse (upward) displacement and $\phi \equiv \phi(x,t)$ denoting the clock-wise rotation of a cross section, the governing differential equations will read

$$
\begin{aligned}
&0 < x < a: \ \kappa GA\frac{\partial}{\partial x}\left(\frac{\partial v}{\partial x} + \phi\right) = \rho A\frac{\partial^2 v}{\partial t^2}\ ,\\
&0 < x < L: \ EI\frac{\partial^2 \phi}{\partial x^2} - \kappa GA\left(\frac{\partial v}{\partial x} + \phi\right) = \rho I\frac{\partial^2 \phi}{\partial x^2}\ , \qquad (1)\\
&a < x < L: \ v = 0\ ,
\end{aligned}
$$

when the half-specimen is modelled as a Timoshenko-beam supported on a symmetry plane, [5-6]. In (1), ρ is density and t is time, A is the cross-sectional area and I is the second area moment, E is Young's modulus, G is the shear modulus and κ is a shape factor (≈ 0.85 for the rectangular section, e.g. [2]). This set of equations is to be solved subject to initial conditions and end conditions stated later, to the continuity conditions

$$
[\phi] = 0,\ \left[\frac{\partial \phi}{\partial t}\right] = -\dot{a}\left[\frac{\partial \phi}{\partial x}\right] = 0,\ \left[\frac{\partial v}{\partial t}\right] = -\dot{a}\left[\frac{\partial v}{\partial x}\right], \qquad (2)
$$

and to the Griffith energy balance

$$
R = \frac{1}{2}\kappa GA(1 - \dot{a}^2/c_t^2)\left[\frac{\partial v}{\partial x}\right]^2\ ,\ c_t = \sqrt{\kappa G/\rho}. \qquad (3)
$$

As derived in [5], eqs. (2) and (3) include the crack-tip velocity $\dot{a}$ and the shear wave velocity c_t, together with discontinuity terms defined by the notation

$$
[f] \equiv f(a+,\ t) - f(a-,\ t)\ . \qquad (4)
$$

For a given contact width B the resistance or separation energy R will be postulated to depend only upon the velocity of the crack tip in its given environment. To account for a possible overstrain prior to cracking an initiation barrier may be admitted, such that the energy R_c at incipient virtual growth may exceed the limiting value of R as defined by taking $\dot{a}$ to zero from above. For zero Poisson's ratio, R_c would equal $K_c^2B/(2E)$ in terms of a fracture toughness K_c.

Notably, the discontinuity of $\partial v/\partial x$ must correspond to a concentrated holding force acting between the adjacent sides at the crack tip. The maximum crack tip velocity $\dot{a} = c_t$ according to (3) is a close approximation of the Rayleigh wave velocity under plane conditions: $\kappa^{1/2} \approx 0.92$, thus conforming well with the prediction by two-dimensional considerations.

We introduce dimensionless lengths

$$
\{\bar{a},\ \bar{L},\ \bar{x},\ \bar{v}\} = \{a,\ L,\ x,\ v\}/k\ ,\ k = \sqrt{I/A}\ , \qquad (5)
$$

a dimensionless time

$$\bar{t} = tc_t/k \ , \tag{6}$$

the wave velocity ratio

$$\beta = c_t/c_\ell \ , \qquad c_\ell = \sqrt{E/\rho}, \tag{7}$$

and the characteristic coordinates

$$\begin{aligned} \xi_t &= \bar{t} + \bar{x} \ , \quad \eta_t = \bar{t} - \bar{x}, \\ \xi_\ell &= \bar{t} + \beta\bar{x} \ , \quad \eta_\ell = \bar{t} - \beta\bar{x} . \end{aligned} \tag{8}$$

In terms of a dimensionless shear, transverse velocity, bending moment and angular velocity, respectively

$$\begin{aligned} \gamma &= \frac{\partial v}{\partial x} + \phi \equiv \frac{q+p}{2} \ , \quad \frac{\partial \bar{v}}{\partial \bar{t}} \equiv \frac{q-p}{2} \ , \\ m &= \frac{1}{\beta}\frac{\partial \phi}{\partial \bar{x}} \equiv \frac{s+r}{2} \ , \quad \omega = \frac{\partial \phi}{\partial \bar{t}} \equiv \frac{s-r}{2} \ , \end{aligned} \tag{9}$$

the equations of motion (1) may then be broken into the quasi-linear set, [6]

$$0 < x < a: \quad 2\partial p/\partial \xi_t = \omega, \tag{10}$$

$$2\partial q/\partial \eta_t = \omega, \tag{11}$$

$$0 < x < L: \quad 2\partial r/\partial \xi_\ell = \gamma, \tag{12}$$

$$-2\partial s/\partial \eta_\ell = \gamma, \tag{13}$$

$$a < x < L: \quad p = q = \phi \ , \tag{14}$$

$$-2\partial\phi/\partial\eta_\ell = r \text{ or } 2\partial\phi/\partial\xi_\ell = s. \tag{15}$$

The pair of equations (15) is only an alternative way to express the identities (9c, d).

With these new variables the internal conditions (2) - (3) may be restated as

$$[r] = [s] = 0, \ [q]/[p] = (1-\alpha)/(1+\alpha), \tag{16}$$

$$(1-\alpha)/1+\alpha) = \bar{R}/[p]^2 \tag{17}$$

in terms of a dimensionless crack velocity and separation energy

$$\alpha = \dot{a}/c_t \ , \quad \bar{R} = 2R/(\kappa GA). \tag{18}$$

To a given transverse displacement v_o at $x = 0$ and to free lateral surfaces at $x = 0$ and $x = L$ correspond the end conditions

$$x = 0: \quad p = q, \quad r = -s, \tag{19}$$

$$x = L: \quad r = -s. \tag{20}$$

The specimen being originally in a static configuration is stated by the initial conditions

$$t = 0-: \quad p = q = \gamma, \quad r = s = m \tag{21}$$

where by simple analysis

$$\begin{aligned} x < a_o: \quad & \gamma = \gamma_o, \text{ a constant}, \\ & m = \beta\gamma_o\bar{x}, \end{aligned} \tag{22}$$

$$x > a_o: \quad -\gamma = m = \beta\gamma_o\bar{a}_o\exp\{-\beta(\bar{x} - \bar{a}_o)\}. \tag{23}$$

a_o is the original free length, and almost total damping by $x = L$ has been assumed. Since motion starts at the crack tip, and c_ℓ (by realistic assumption) is the largest signal velocity, conditions (21) are valid also along the bounding characteristics

$$\bar{x} = \bar{a}_o - \bar{t}/\beta \qquad \text{and} \qquad \bar{x} = \bar{a}_o + \bar{t}/\beta.$$

INITIAL MOTION

The criterion for initial cracking is taken to be the energy balance for a virtual, quasistatic extension. Then, (17) and (21) - (23) provide

$$[p]^2 = \gamma_o^2(1 + \beta\bar{a}_o)^2 = \bar{R}_c \tag{24}$$

giving the critical shear γ_o. If an initiation barrier is assumed, such as discussed above, a sudden motion $\dot{a} > 0$ is to be expected when (24) is achieved. Such a motion may also be induced by certain continuos forms $R(\dot{a})$ having a negative initial slope. With the local unloading a step-like shear signal will be emitted, propagating down the free length of the beam while exciting bending action. By (11) this is a discontinuity of q on the η_+-line $\bar{x} = \bar{a}_o - \bar{t}$; Fig. 2, whereas p is continuous across the same line, as it is on the contact side $\bar{x} > \bar{a}$ of the crack tip. Hence, (24) must hold even during the first motion, and the introduction of $[p]^2 = \bar{R}_c$ in (17) therefore yields the initial crack velocity $\alpha = \alpha_o$ through

$$(1-\alpha)/(1+\alpha) = R/R_c, \tag{25}$$

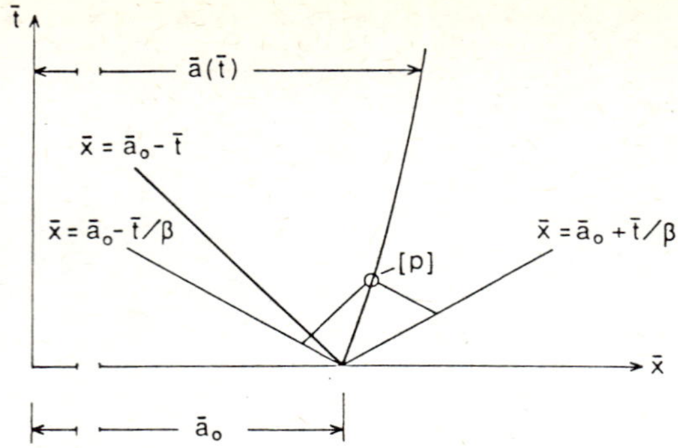

Figure 2. Characteristics and the crack tip trajectory $\overline{a}(\overline{t})$ emanating from the initial crack tip.

R depending on α. This result was first noted in [5], and is equal to the shear-beam velocity obtained in [7].

Using (16c) and (24) we also find the discontinuity of q on the η_t-line $\overline{x} = \overline{a}_o - \overline{t}$, equal to

$$\Delta q = [q]_{t=0-} - [q]_{t=0+} = -\overline{R}_c^{1/2}\left(1 - \frac{1-\alpha_o}{1+\alpha_o}\right) \qquad (26)$$

A further study of the first motion leads to the initial acceleration value, [6],

$$\frac{d\alpha}{d\overline{t}} = \frac{k}{c_t^2}\ddot{a} = -\frac{\beta^2\overline{a}_o}{1+\beta\overline{a}_o}\alpha_o(1-\alpha_o^2). \qquad (27)$$

CONTINUED MOTION: A NUMERICAL BASIS

Discontinuities of p and q are propagated on lines in the $\overline{x}$, $\overline{t}$-diagram representing paths of emitted and reflected wavefronts, and the sought crack tip trajectory $\overline{a}(\overline{t})$. These are the heavy lines in Fig. 4. Such double values may be included in a finite difference scheme only when nodal points are placed along the discontinuity lines. If the crack tip trajectory is sought as a smooth curve, this means that the grid has to be developed along with the computation, which precludes any attempt to construct a regular grid.

In the present tentative numerical study integration along the characteristics has been pursued, the last mentioned disadvantage being somewhat compensated by the rather efficient forward integration which may be performed iteratively in comparatively large time-steps. The linear difference forms of (10) - (13) and (15) are presently

$$
\begin{aligned}
\Delta p/\Delta\bar{t} &= \omega \text{ mean value along } \xi_t . \\
\Delta q/\Delta\bar{t} &= \omega \quad \text{"} \quad \text{"} \quad \text{"} \quad \eta_t . \\
\Delta r/\Delta\bar{t} &= \gamma \quad \text{"} \quad \text{"} \quad \text{"} \quad \xi_\ell . \\
-\Delta s/\Delta\bar{t} &= \gamma \quad \text{"} \quad \text{"} \quad \text{"} \quad \eta_\ell . \\
-\Delta\phi/\Delta\bar{t} &= s \quad \text{"} \quad \text{"} \quad \text{"} \quad \eta_\ell \quad \text{or} \\
\Delta\phi/\Delta\bar{t} &= r \quad \text{"} \quad \text{"} \quad \text{"} \quad \xi_\ell .
\end{aligned} \tag{28}
$$

Essentially, a forward marching procedure is followed in a standard way, sweeping form the original crack tip on the basis of initial data at the bounding characteristics. In the sequel the crack tip conditions and the end conditions are incorporated. The progress may be iterative for three different reasons. First, the governing equations (10) - (15) are quasi-linear, and the desired mean values of the right-hand variables are unknown during the first integration along a grid element. Second, the shear-wave characteristics (ξ_t,η_t) and the bending-wave characteristics (ξ_ℓ,η_ℓ) are not parallel, and the base-point values of the latter have to be estimated by extrapolation and confirmed by interpolation to provide nodal values in the primary shear-grid; Fig. 3. Third, when the crack tip trajectory is unknown a priori, the position of each new point has to be estimated first, as explained in the following. Of course, all sources of iteration may be dealt with simultanously and effectively by hand, but the nature of the computing sequence is not well suited for programming. Views on the latter aspect are presented below.

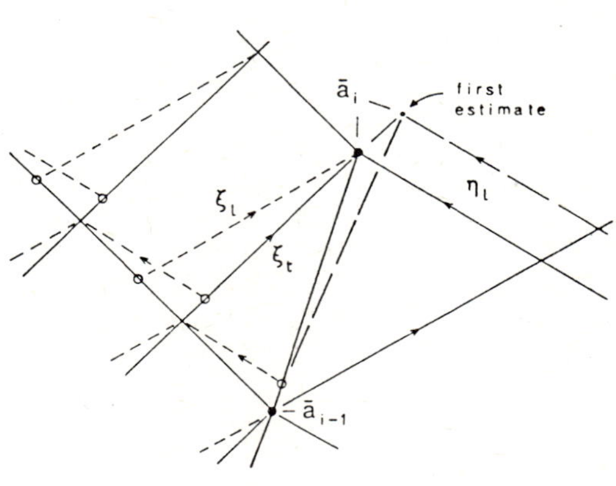

Figure 3. Element of the crack tip trajectory (heavy line) and neighbouring characteristics. Extra/interpolated values of r, s and γ in circled base-point.

Reverting to the crack tip trajectory, a known nodal point $\overline{a}_{i-1}$ and its estimated neighbour $\overline{a}_i$ are indicated in Fig. 3. By the process mentioned, values of p at $\overline{a}_i$ result form the use of (10) along the ξ_t-line and (14) - (15a) along the intersecting η_ℓ-line. Next, α_i follows from (17), and the location of $\overline{a}_i$ may be verified by drawing a secant line through $\overline{a}_{i-1}$ with the inclination average $(\alpha_{i-1} + \alpha_i)/2$. One or two iterations has proved sufficient to fix each location.

We further relate our experience that gradients may be high close to the wave-fronts, which may call for a finer mesh or certain non-linear interpolations in these regions. Also, simple devices may be necessary to accomodate the end conditions to the irregular grid (at x = L when $R_c/R = 4$; Fig. 4).

Finally, the alternative use of a regular finite difference scheme in the $\overline{x},\overline{t}$-directions may be briefly commented on. First, a requirement by the properties of the characteristics is the choice of differences such that $\Delta\overline{t}/\Delta\overline{x} < \beta$. A further requirement due to the discontinuities on certain (ξ_t,η_t)-lines is that nodes should be situated on the directions $d\overline{t}/d\overline{x} = \pm 1$. With a realistic value of β (≈ 0.6) this implies that a regular grid with $2\Delta\overline{t} = \Delta\overline{x}$ is recommendable. Double values on the pertinent characteristics should be implemented in accordance with (26). The crack tip trajectory is now naturally represented by a broken curve (which might be interpreted physically as the result of a fluctuating, rather than continuous resistance). For such a representation to yield reliable results it is necessary that the mesh is considerably finer than implied in the above discussion. However, despite the very special attention which will have to be paid to the wavefront and crack-tip discontinuities, this is probably a strong alternative if a program for large-scale evaluation is to be developed.

PRESENTATION OF RESULTS

Fig. 4 shows the computational grid and the derived crack-tip trajectory corresponding to the input data $a_o/h = 1.45$ (defining $\overline{a}_o = 5$), L/h = 5.65, $R_c/R = 4$ (which implies a velocity-independent separation energy), and $\beta = c_t/c_\ell = 3/5$. This choice would premit a detailed comparison with the results in [9], where 1.45, 5.65, 4 and 0.58 were assumed, respectively. Values of $\overline{R}_c$ or (through (24)) γ_o may be assigned arbitrarily; if $\gamma_o = 1$ then $\overline{R}_c = 4\overline{R} = 16$.

The resulting crack-tip trajectory is drawn smoothly through the (by dots) indicated nodal point values. An initial deceleration period is followed by an interval of nearly constant velocity, whereupon disturbances associated mainly with the shear wave reflection take effect. This is an acceleration before incidence and partial reflection of the wavefront, whereafter the crack tip gradually decelerates to arrest.

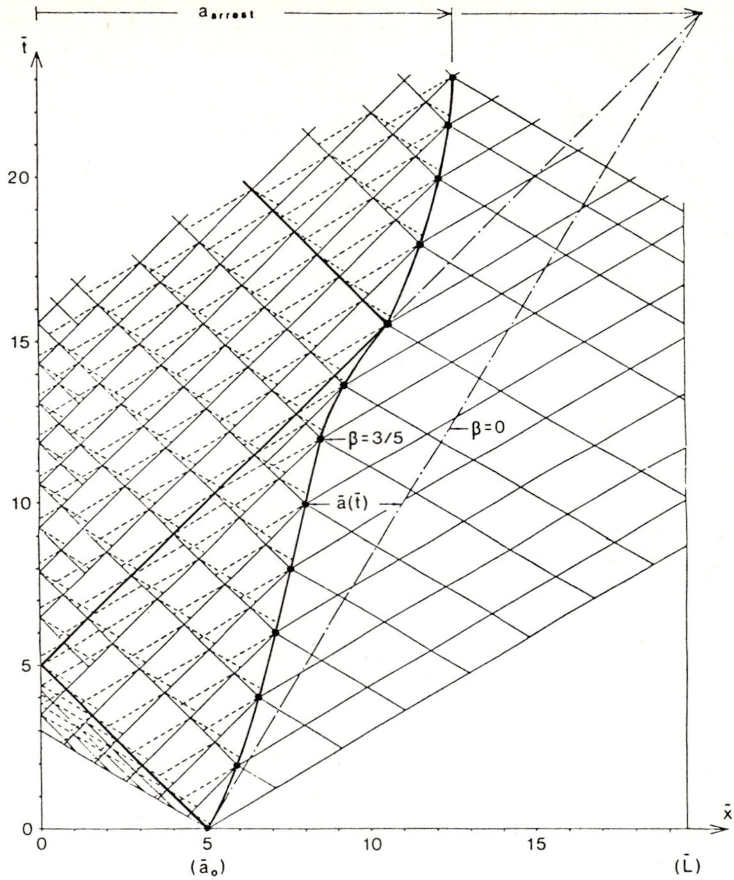

Figure 4. Resulting crack tip trajectory thorugh computed locations (nodes). Other heavy lines denote paths of wave-front discontinuity. $R_c/R = 4$, $a_o/h = 1.45$, $L/h = 5.65$.

Fig.5 compares the present results with the two alternatives discussed in [9]. That pair differs in the way to represent the initiation barrier; such ambiguities being generally associated with the spring models. The present result seems to confirm the earlier trend, reconciling former uncertainties mainly in the initial regime while predicting a crack jump and a time to arrest which are somewhat below the former averages. A discrete effect of wave incidence is visible to a higher extent than in the earlier results. As for reflections from the end x = L, this was found to be a very modest contribution with the data chosen.

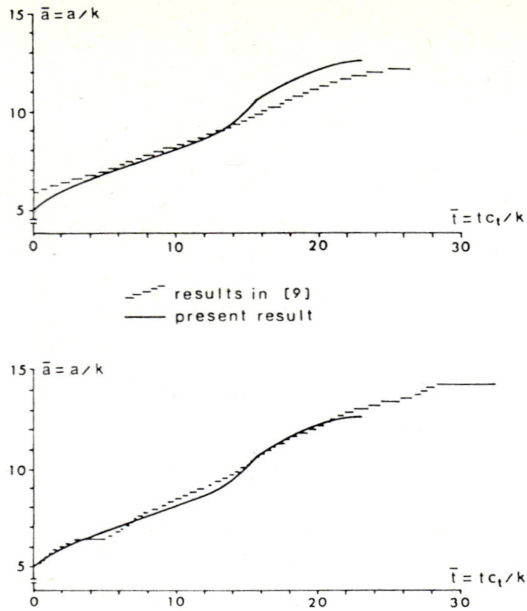

Figure 5. A comparison with two alternative results in ref [9] where β = 0.58. Present and common data as in Fig. 4.

Figs. 6 and 7 are crack motions as obtained with $\beta = 3/5$, $R_c/R = 2$ and, respectively, a_o/h = 1,45 and $a_o/h = 1.2$. In both cases L/h was assumed sufficient to exclude end effects from x = L. A common feature is the shorter life of the crack motion after wavefront incidence, as compared to the result of a higher initiation barrier.

Fig. 8 depicts a numerical study of the limit $a_o/h = \infty$. The motion is obviously too small to be a valid result by our Timoshenko-beam theory anyway, and may only be interpreted as a nearly stationary crack tip. This is a reasonable extension toward the Euler-beam prediction, [10], whereby crack jumps in the present sense cannot operate.

Fig. 9 is a result which complements Fig. 4 in a certain sense, [11]. In Fig. 4, the stationary velocity $\alpha \approx 0.25$ was approached when $R_c/R = 4$ was prescribed, and in Fig. 9 it is seen that about this same resistance ratio is approached when α is prescribed at the above stationary value. There is no exact reason why this should be so, but intuitively the fact might be regarded as an indication of the reproducibility of the method. (Obviously, a result like Fig.9 is not consistent with the physical assumption of a resistance being uniquely

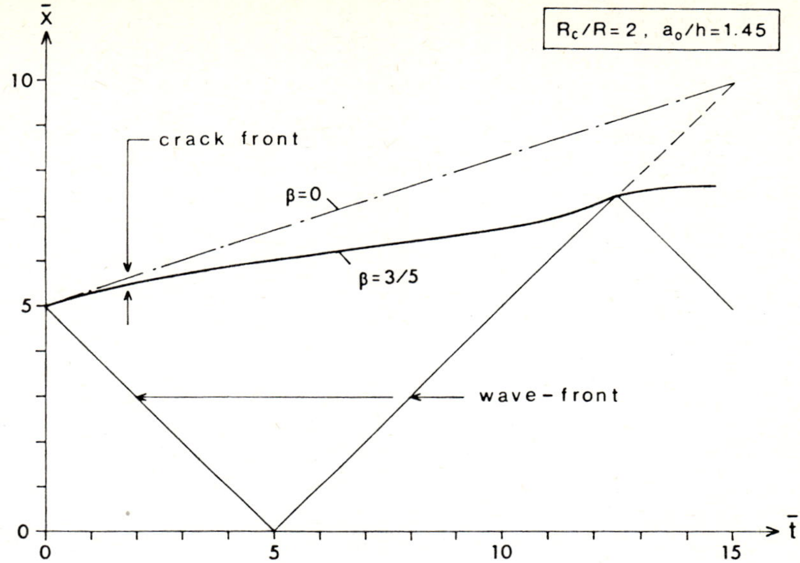

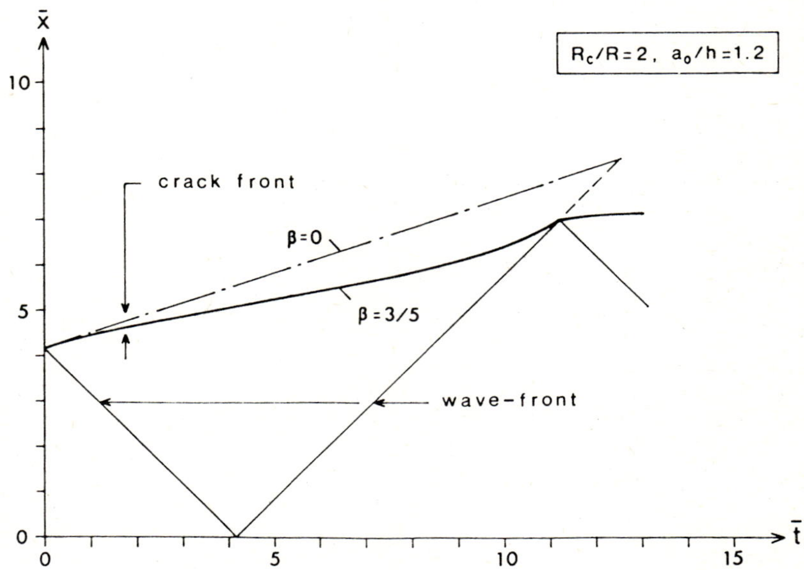

Figures 6-7. Further trajectories of the wavefront and of the crack tip until arrest.

related to the crack velocity.

We finally mention that, for quite hight levels of the initiation barrier, a period of reversed shear may occur at x = 0 if the deflection is prescribed, [6]. Conversely, a small "jump" of the beam end would then be incurred by wedging conditions. Effects of such behaviour are considered to be negligible in the following discussion (as are the effects of some extension into x<0 of real test specimens).

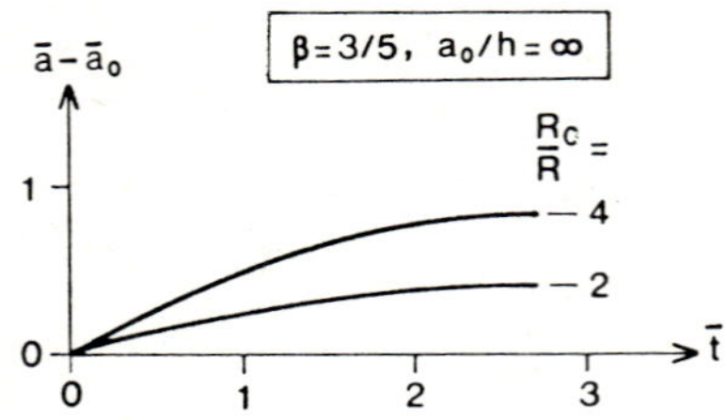

Figure 8. The computed small crack-tip motion of a very slender beam specimen.

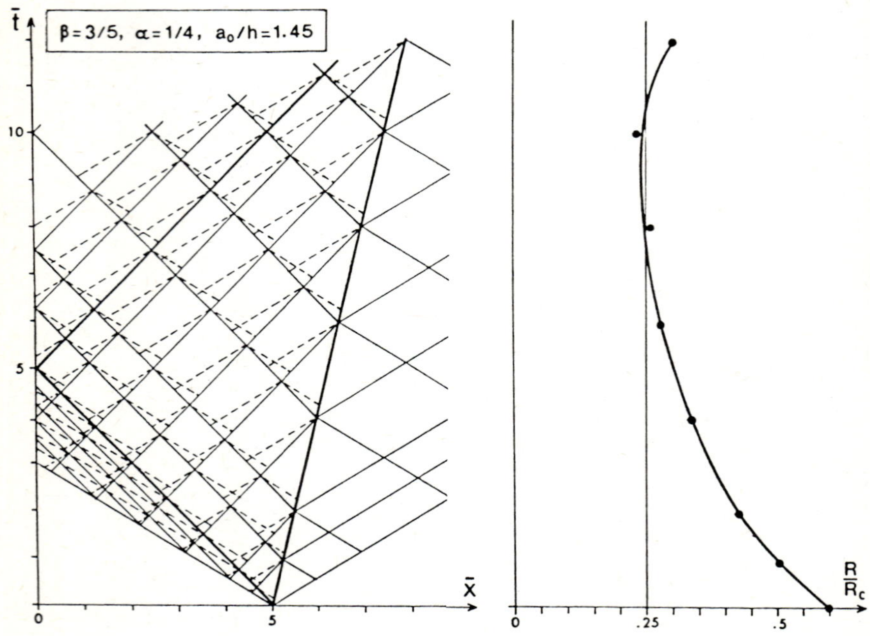

Figure 9. Result from [11], giving a variation of the separation energy to comply with a constant crack-tip velocity.

A SUGGESTED INTERPOLATION. FURTHER DISCUSSION

The lines $\beta = 0$ in Figs. 4, 6, 7 and their intersections with the extended wavefront lines are the shear-beam predictions of crack trajectories and arrest points, [7]. According to this assumption, there is a constant crack velocity and a crack jump which are simply expressed in terms of the resistance ratio as

$$\dot{a} = c_t(R_c/R - 1)/(R_c/R + 1),$$
$$\Delta a = a_o(R_c/R - 1), \tag{29}$$

$R \equiv R(\dot{a})$ being the current resistance value. As is seen in the figures, neither of these predictions is very successful under the present circumstances. However, it might be hoped that the effect of the resistance ratio is tolerably well accounted for by the formulae, so that scaling factors depending only on the bluntness h/a_o may provide the most necessary modification. Thus

$$V = \eta_1 c_t(R_c/R - 1)/(R_c/R + 1),$$
$$\Delta a = \eta_2 a_o(R_c/R - 1), \tag{30}$$

where V denotes the average velocity of the crack tip and where η_1 and η_2 are functions og h/a_o. A preliminary support of this assumption may be drawn from Figs. 4 and 6, whereby

$$\text{for } R_c/R = 4: \quad \eta_1 = 0.55, \quad \eta_2 = 0.51,$$

$$\text{for } R_c/R = 2: \quad \eta_1 = 0.55, \quad \eta_2 = 0.54,$$

or approximately, for $a_o/h = \underline{1.45}$:

$$\eta_1 = \eta_2 = 0.55.$$

To adapt the assumption to the form of reference curves used to extract V and R from observed values of Δa and R_c, [3], [12], we rearrange equations (30) to read

$$\frac{V}{c_t} = \frac{V}{\beta c_\ell} = \frac{\eta_1 \Delta a/a_o}{2\eta_2 + \Delta a/a_o},$$
$$\frac{R}{R_c} = \frac{\eta_2}{\eta_2 + \Delta a/a_o}. \tag{31}$$

In [12], reference curves have been derived by two-dimensional considerations for a case when the bluntness far exceeds the range of a beam theory, i.e. $a_o/h = \underline{0.52}$. However, taking

$$\eta_1 = 0.60, \quad \eta_2 = 1.18$$

together with $\beta = 1/\sqrt{3} = 0.58$ as in [2], we obtain the dotted alternatives to the original, solid curves in Fig. 10(I). At the extreme slenderness, on the other hand, recall the earlier observation for $a_o/h = \underline{\infty}$,

$$\eta_1 \approx \eta_2 = 0$$

(with some care in the interpretation of η_1). A parabolic fit to the three data point of each η gives

$$\eta_1 = 0.4\frac{h}{a_o}(2.7 - \frac{h}{a_o}),$$
$$\eta_2 = 0.15\frac{h}{a_o}(6 - \frac{h}{a_o}). \qquad (32)$$

The results by (31) - (32) with $\beta = 0.58$ are finally compared in Fig. 11 with the results from a version of Kanninen's theory, [3]. Data points from Fig. 7 are included as circles.

It has been tacitly assumed that the ligament length $L - a_o$ is sufficient to exclude strong end effects from $x = L$; however, this may not definitely be the case for the largest crack jumps in Fig. 10. Even within this assumption it is quite possible that the formulae (32) will have to be revised with the arrival of more data. Also, the recognized practical need of a starter section to permit large velocities within a linear frame of behavior may call for further analyses of two-material specimens, [3]. In any case, the possible use of the shear-beam to model interpolation formulae seems to have some support in the present results.

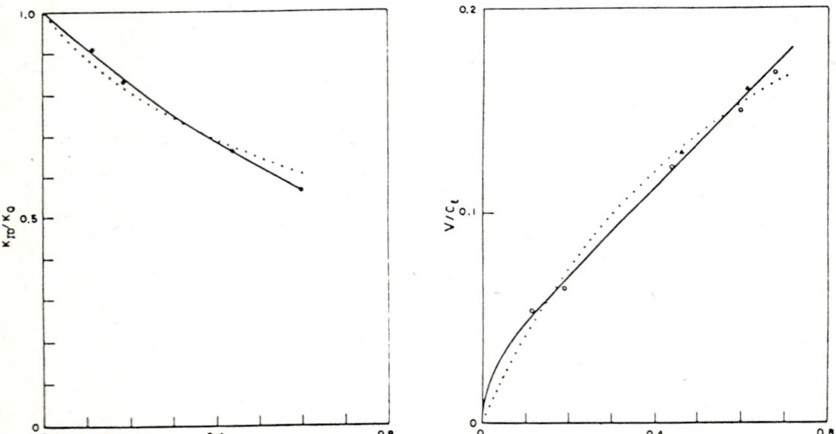

Figure 10. Reference curves from [12], the solid lines, with present approximations included by dotted lines. $a_o/h = 0.52$, $a_o/L = 0.312$.

(I) $K_{ID}/K_Q = \sqrt{(R/R_c)}/\sqrt{A}$, where A is a velocity dependent function ≈ 1 under the present conditions, e.g. [12].

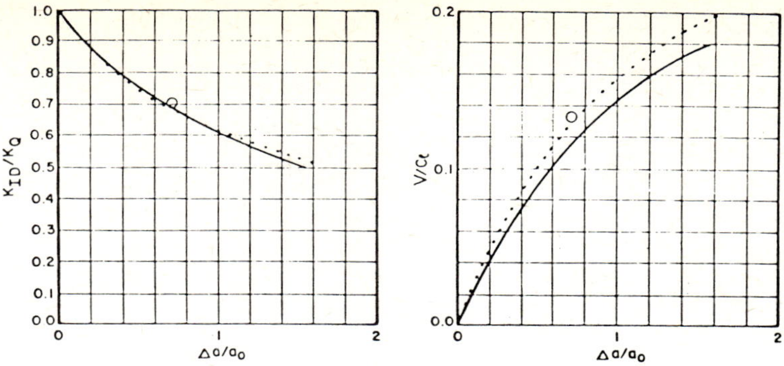

Figure 11. Reference curves from [3], the solid lines, with present approximations included by dotted lines. $a_o/h = 1.2$.

Each point of the reference curves in Figs. 10-11 has been evaluated on the assumption that R is a constant, and the use of the curves in cases when $\dot{a}$ and R do vary has been recognized as a source of some error, e.g. [13]. More satsifactory, in principle, is the use of a measured trajectory a(t) as an input for evaluating the resistance, e.g. [8] and Fig. 9. However, the initial transient (α_o being theoretically even higher than by (25); [5, p.183]) is hardly observed experimentally, which suggests that the computed transient of R should be disregarded accordingly. If possible, then, a correlation of R and $\dot{a}$ should apply to stationary values. (This is a velocity somewhat smaller than the average V in the present results). From such a point of view the correspondence between Figs. 4 and 9 is encouraging.

REFERENCES

1. HAHN, G.T., HOAGLAND, R.G., KANNINEN, M.F. and ROSENFIELD, A.R.- Engng. Fracture Mechanics, Vol.7, pp.583-591, 1975.
2. KANNINEN, M.F.- Int. J. Fracture, Vol.10, pp.415-430, 1974.
3. HAHN, G.T. et al.- Report NUREG/CR-0057, BMI-1995, Appendix B, Battelle-Columbus Laboratories, 1977.
4. GEHLEN, P.C., POPELAR, C.H. and KANNINEN, M.F.- Int. J. Fracture, Vol.15, pp.281-294, 1979.
5. HELLAN, K.- Int. J. Fracture, Vol.14, pp.91-100,173-184, 1978.
6. HELLAN, K.- Int. J. Fracture, Vol.16, 1980 (in press).
7. FREUND, L.B.- J. Mech. Phys. Solids, Vol.25, pp.69-79, 1977.
8. BILEK, Z.- Scripta Metallurgica, Vol.12, pp.1101-1106, 1978.
9. MALLUCK, J.F. and KING, W.W.- Int. J. Fracture, Vol.13, pp.655-665, 1977.
10. BILEK, Z.J. and BURNS, S.J.- J. Mech. Phys. Solids, Vol.22, pp.85-95, 1974.
11. HERBERG, J.- Diploma Thesis, NTH 1979.
12. GEHLEN, P.C., HOAGLAND, R.G. and POPELAR, C.H.- Int. J. Fracture, Vol.15, pp.69-84, 1979.
13. HAHN, G.T. et al.- Report NUREG/CR-0825, BMI-2025, pp.2.49-54, Battelle-Columbus Laboratories, 1978.

EXPLICIT TIME INTEGRATION IN FEM-ANALYSIS OF DYNAMIC CRACK PROPAGATION

Björn Brickstad[1] and Fred Nilsson[2]

SUMMARY

Stress-intensity factors for a given dynamic crack growth history are determined by FEM with aid of gradual nodal relaxation technique. The time integration of the equations of motion is accomplished by an explicit scheme based on the central difference. The inverse problem, i.e. determination of the crack growth history for given loading and material behaviour is also considered. Here, an iterative procedure based on the Illinois algorithm is used. A number of examples obtained from previous experimental investigations are analysed and the results are compared with results from calculations based on implicit time integration. It is found that the accuracy of the explicit scheme is comparable to that of the implicit method. The explicit method is, however, much more efficient from the point of computing costs. It is also found that the inverse problem can be effectively treated with the proposed iterative procedure.

INTRODUCTION

A few years ago dynamic crack propagation problems could be analysed only for very idealized geometries involving such unrealistic assumptions as infinite bodies. Since then it has become quite generally accepted that more realistic problems can only be treated by aid of numerical methods. The development of such methods has been very rapid, and today elastic problems can be dealt with successfully. Both finite element (FEM) and finite difference (FDM) methods have been used, although in most cases the former seems to have been preferred.

The main problem in this context is to model a continous crack growth through a discrete grid. Several methods to overcome this difficulty have been suggested. One of these is the gradual

1) Research Assistant 2) Researcher
Department of Strength of Materials and Solid Mechanics,
The Royal Institute of Technology, S-100 44 Stockholm, Sweden

nodal relaxation technique (cf. Keegstra [1], Rydholm et al [2], Malluck and King [3]). This technique has given good results when compared with analytical solutions. No theoretical analysis of convergence properties has yet been made, but in our opinion the results obtained so far justify the use and further development of this method.

While thus it is possible to perform numerical analyses with fairly good accuracy, it still remains quite costly with respect to computer time and storage capacity. It is therefore of importance to increase the efficiency of the methods if this type of calculations is to become a practical tool in design work.

It is well known that the choice of time integration method in a FEM model can have significant effects on the computer cost. Implicit methods, which have mostly been used, are unconditionally stable and the time step can be chosen large without affecting the stability of the solutions. Explicit methods on the other hand, put a stability limit on the size of the time step. The calculation time in each step is, however, much shorter than for implicit methods. When accuracy requirements limit the time step size, explicit methods can give a considerable cost reduction. These questions have for example been investigated for ordinary wave-propagation problems by Nilsson [4].

The object of the present paper is to compare an explicit method with previous calculations by Newmark's implicit method (Brickstad and Nilsson [5]). A method for solving the problem of calculating the crack growth history is also presented, and the results obtained by this method are compared with the real crack growth histories obtained in the experiments reported in [5].

STATEMENT OF THE PROBLEM

We shall here consider crack growth under mode I conditions in a sheet under plane stress. It is assumed that the material is isotropic, linear elastic and that the crack growth process can be described by eqn. (1).

$$\begin{aligned} G(t) &= G_{pc}(\dot{a}) \qquad && \dot{a} > 0 \\ G(t) &< G_{pc}(\dot{a}=0) \qquad && \dot{a} = 0 \end{aligned} \tag{1}$$

G is the energy-release rate, t is time, $\dot{a}$ is the crack-tip velocity and G_{pc} is a material function dependent only on $\dot{a}$. Between G and the stress-intensity factor K the relation (2) exists

$$G = \frac{K^2}{E}\, g(\dot{a}/C_2,\ \nu) \tag{2}$$

Here E is the Young's modulus, ν is Poisson's ratio and C_2 is the velocity of equivoluminal waves. g is a universal function and is for example given by Nilsson [6]. Because of (2), eqn. (1) can equally well be formulated in terms of the stress-intensity factor, i.e.

STATEMENT OF THE PROBLEM

$$K(t) = K_{pc}(\dot{a}) \qquad \dot{a} > 0$$
$$K(t) < K_{pc}(\dot{a}=0) \qquad \dot{a} = 0 \tag{3}$$

In order to verify the relevance of eqs. (1) and (3) experiments like those reported in [5] have to be performed. A notched specimen is subjected to some kind of well-defined loading. When fracture occurs, the crack growth history is recorded by some suitable method. Together with knowledge of the loads on the specimen during the process, K(t) can be calculated and thus the validity of (3) can be checked. We will here term the problem of calculating K(t) when the crack growth history is known for the direct propagation problem.

Another problem occurs when $K_{pc}(\dot{a})$ together with the boundary conditions are known and one wants to calculate the crack growth history. This is the case in design situations and we will term it the inverse problem.

The numerical treatment of both these problems with aid of an explicit FEM technique will be examined in the present paper.

THE DIRECT PROBLEM AND EXPLICIT TIME INTEGRATION

The formulation of an elastodynamic FEM problem results in an equation system of the form (4).

$$[M][\ddot{u}] + [K][u] = [F] \tag{4}$$

[u] is the vector of nodal displacements, [ü] is the vector of nodal accelerations, [M] is the global mass-matrix, [K] is the global stiffness matrix and [F] is the vector of applied forces. Together with appropriate boundary conditions the solution of this system gives an approximation of the continuum solution. Of particular interest here is the conditions in the crack-tip vicinity. We will here follow the method described in [2], i.e. a regular element grid is employed. The crack growth is modelled by introduction of a closing force F_c perpendicular to the crack surface at the node immediately behind the crack tip. The magnitude of this force is given by

$$F_c = F_o\,(1 - \delta/d)^{1/2} \tag{5}$$

where F_o is the nodal force acting at this node just before the crack tip reaches it, δ is the distance travelled by the tip along the next element side and d is the length of this side. The energy release rate is calculated from eqn. (6)

$$G = \frac{2}{d} \int_{t_c}^{t_c+d/\dot{a}} F_c(t)\,\dot{u}_c\,dt \tag{6}$$

where t_c is the time when the tip passes the node c and $\dot{u}_c$ is the velocity of this node. K is then given by eqn. (2).

When using the explicit central difference time integration method, it is advantageous to have a diagonalized mass matrix (cf. Nilsson [4]). At the time instant t_K the equation of motion

(4) for the j:th degree of freedom can then be written as eqn.(7).

$$^{t_k}\ddot{u}_j = \frac{1}{M_j}\left({}^{t_k}F_j - \left[K\right]\left[{}^{t_k}u_j\right]\right) \tag{7}$$

Assuming the quantities in the right member to be known at time t and the velocities known at time $t_k - \Delta t_{k-1}/2$ (Δt_{k-1} is the time-step), we obtain the following relations by the central difference approximation

$$^{t_k+\Delta t_k/2}\dot{u}_j = {}^{t_k-\Delta t_{k-1}/2}\dot{u}_j + \frac{1}{2}(\Delta t_{k-1} + \Delta t_k)\ {}^{t_k}\ddot{u}_j \tag{8}$$

$$^{t_k+\Delta t_k}u_j = \Delta t_k\ {}^{t_k+\Delta t_k/2}\dot{u}_j + {}^{t_k}u_j \tag{9}$$

For t = 0 eqn. (8) is replaced by eqn. (10).

$$^{\Delta t_1/2}\dot{u}_j = {}^{0}\dot{u}_j + \frac{1}{2}\Delta t_1\ {}^{0}\ddot{u}_j \tag{10}$$

By these relations the state at time t_{k+1} is obtained.

No inversions of the matrices are needed thus reducing the required computer time. For this reason it is not necessary to perform the calculations in double precision which also reduces the need for computer storage capacity. When evaluating the accelerations in eqn. (7) only those elements of the stiffness matrix which are coupled to the particular degree of freedom u_j appear in the corresponding equation. It is not necessary to assemble the global stiffness matrix and for the uniform grids used in this investigation only a small part of the stiffness matrix needs to be stored.

The main drawback of this integration method is the stability requirement. The theoretical stability condition is

$$\Delta t_k < \frac{d_{min}}{C_1} \tag{11}$$

where d_{min} is the smallest element dimension in the grid and C_1 is the velocity of irrotational waves. In practice it is found that Δt_k must be chosen somewhat smaller than this theoretical limit. However, for the present applications this stability requirement is not a severe restriction since in wave propagation analysis it appears that a relatively small time step is necessary in order to avoid large errors in the solution, that are caused by smoothing effects on the wave fronts, see [4].

A FORTRAN program was developed to handle the presented explicit method together with the special nodal relaxation technique mentioned above. The initial conditions ^0u_j were determined by statical calculations with the FEM-program NONSAP.

THE INVERSE PROBLEM

In this section the numerical technique of the inverse problem is considered, i.e. finding the root α of the non-linear equation (12).

$$f(\dot{a}) = K(\dot{a}) - K_{pc}(\dot{a}) = 0 \tag{12}$$

The root then represents the momentaneous crack velocity at the considered time instant. It is here assumed that the function $K_{pc}(\dot{a})$ is known for a certain material. $K(\dot{a})$ is, however, in general an unknown function of $\dot{a}$, and the root α cannot be expressed in closed form. For this reason one is obliged to use iterative methods which should meet the following demands.

i) Convergence even for poor initial approximations
ii) Good convergence properties in all parts of the considered interval
iii) Only one function evaluation in each iteration step.

The last requirement is due to an ambition of reducing the cost, since for every approximation $\dot{a}_n$ one has to evaluate $K(\dot{a}_n)$ over a finite element along the crack plane.

We are now going to discuss briefly some aspects of a few iterative methods which mainly follow a work by Dahlquist and Björk [7]. Consider the secant method which is defined by the following scheme denoting the n:th evaluation $f(\dot{a}_n)$ by f_n

$$\dot{a}_{n+1} = \dot{a}_n + h_n \; ; \quad h_n = -f_n \cdot \frac{\dot{a}_n - \dot{a}_{n-1}}{f_n - f_{n-1}} \qquad (13)$$

The geometrical meaning of this method is that $\dot{a}_{n+1}$ is determined as the abscissa of the point of intersection between the secant through $(\dot{a}_{n-1}, f_{n-1})$ and $(\dot{a}_n, f_n)$, and the $\dot{a}$-axis. The secant method has good convergence properties but requires good initial approximations. If instead one always chooses the secants through points on each side of the root so that one point is retained, the regula falsi method is obtained. It has the advantage of always being convergent for continuous functions $f(\dot{a})$, but has poor convergence properties near a root.

It would be desirable to improve the rate of convergence for regula falsi and yet guarantee its global convergence for all initial approximations $\dot{a}_0$ and $\dot{a}_1$ with $f_0 \cdot f_1 < 0$. For this purpose we again consider the secant method. Assuming that $f'(\dot{a})$ and $f''(\dot{a})$ have constant sign in a studied interval and denoting the error in the estimate $\dot{a}_n$ by ε_n, i.e. $\varepsilon_n = \dot{a}_n - \alpha$, one can easily

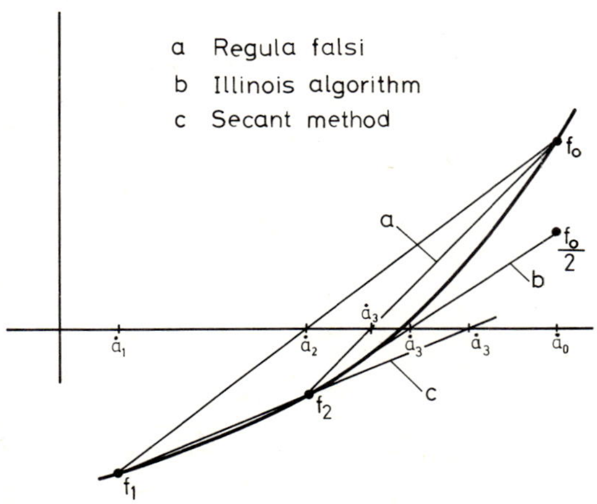

Fig. 1 - Comparison between iterative methods.

show that the quotient

$$\frac{\varepsilon_{n+1}}{\varepsilon_n \cdot \varepsilon_{n-1}}$$

has constant sign for all n. Now if $f_o \cdot f_1 < 0$, i.e. $\varepsilon_o \cdot \varepsilon_1 < 0$, it follows that the signs for the series ε_o, ε_1, ε_2, ... must vary according to one of the following sequences:

i) ...+−++−++−++...
ii) ...+−−+−−+−−+...

This means that for the secant method all steps except every third are regula falsi steps, i.e. secants through points on each side of the root. Assuming that for a certain n we have $f_n \cdot f_{n-1} < 0$ then we use eqn. (13) to calculate $\mathring{a}_{n+1}$. If now $f_{n+1} \cdot f_n < 0$ we can repeat this procedure. On the other hand, if $f_{n+1} \cdot f_n > 0$, then obviously $f_{n+1} \cdot f_{n-1} < 0$. We then make a modified step and take $\mathring{a}_{n+2}$ as the root of the linear function through $(\mathring{a}_{n+1}, f_{n+1})$ and $(\mathring{a}_{n-1}, \beta f_{n-1})$, where β is a parameter, $0 < \beta < 1$. $\beta = 1$ would give a pure regula falsi step and then usually cause $f_{n+2} \cdot f_{n+1} > 0$. $\beta = 0$ gives $\mathring{a}_{n+2} = \mathring{a}_{n-1}$ and hence $f_{n+2} \cdot f_{n+1} < 0$. By choosing β in a suitable way we can obviously get an $\mathring{a}_{n+2} \neq \mathring{a}_{n-1}$ so that $f_{n+2} \cdot f_{n+1} < 0$, and then proceed to the next step with the secant method. The choice of $\beta = 1/2$, sometimes called the Illinois algorithm, has proved successful for many applications (Fig. 1). If we obtain the desired result with every modified step then from the sign variation in the secant method it follows that we have to use a modified step every third iteration. If still after a modified step with $\beta = 1/2$, we obtain $f_{n+2} \cdot f_{n+1} > 0$ we can for the next step use the secant through $(\mathring{a}_{n+2}, f_{n+2})$ and $(\mathring{a}_{n-1}, (1/2)^2 \cdot f_{n-1})$ and so on until we obtain two points on each side of the root α.

The Illinois algorithm satisfies all the requirements we set up for an iterative method and is always convergent if one chooses $\mathring{a}_o$ and $\mathring{a}_1$ so that $f_o \cdot f_1 < 0$. These initial approximations can soon be found, when applying this method to our actual problem, by guidance partly of the expected physical behaviour of $K(\mathring{a})$ and partly of the sign and the magnitude of f for a sequence of initial guesses.

SPECIMEN SHAPE AND FEM-MODELLING TECHNIQUE

Two different specimen geometries were investigated. Both were rectangular with a width w of 406 mm and a thickness of one half millimeter. The material used was a hardened carbon steel with yield strength σ_{ys} = 1500 MPa. The static fracture toughness K_c, evaluated at initiation of slow stable crack growth was 125 $MN/m^{3/2}$.

1. Specimen type A: Effective length 2h between grips was 400 mm. The specimen ends were given a uniform displacement in a special testing machine where fixed grip conditions could be realized. In [8] more detailed information about the experimental procedure is given. The FEM-mesh used is shown in Fig. 2 and consists of 200 plane elements of uniform size resulting in 441 degrees of

freedom. This uniformity of the mesh is of importance in the numerical calculations for reducing spurious reflections at element interfaces which otherwise would influence the solution, cf. Bazant [9].

2. Specimen type B: Effective length 2h between grips was 1600 mm. For these specimens fracture was initiated by slowly increasing the load in a hydraulic testing machine, giving force controlled boundary conditions. The FEM-mesh used together with the explicit time integration method is shown in Fig. 3 and contains 820 uniform elements and a total of 1763 degrees of freedom. The upper row of long elements represents the mounting plate of the testing machine and was assigned a very high stiffness together with a low mass aiming to simulate the kinematical constraint by the grips at the specimen ends. When using the implicit integration method for this specimen a somewhat coarser FEM-grid, see [5], was implemented some distance from the crack plane in order to reduce the cost, resulting in 429 elements and 841 degrees of freedom. This change in element size was found to be of small significance as regards the accuracy in the calculations if only the close uniform mesh size was kept near the crack plane.

Fig. 3

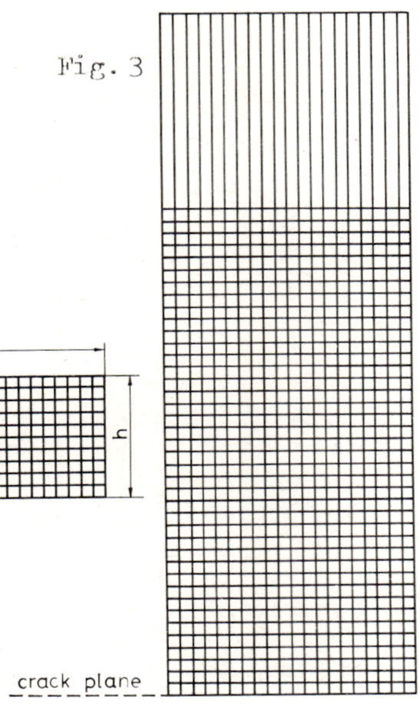

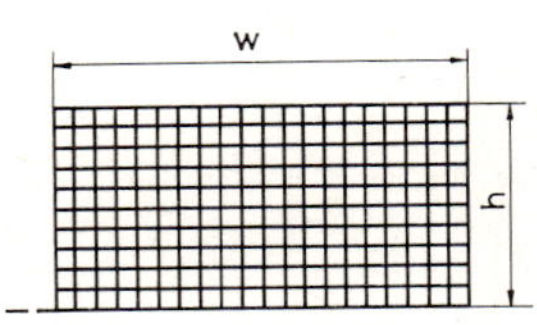

Fig. 2

Figs. 2-3 - FEM-mesh for specimen type A and B, respectively.

COMPARISON OF COMPUTING PARAMETERS

In Table 1 some important computer parameters are shown for typical computations of the dynamic stress intensity factor using the implicit and the explicit time integration method. The figures should be considered in the light of the following specifications.

In the implicit scheme, using Newmark's time integration method [10], the nondiagonal structural stiffness matrix appears and hence the use of a nondiagonal mass matrix was justified. In the current version of the FEM-program all the non-zero elements of these matrices need to be available and stored during the solution phase. Also, the majority of the calculations, i.e. the formation of the structure stiffness matrix and the main steps in the solution of the incremental equations of motion, needed to be carried out in double precision on the IBM 370/165 computer used. Otherwise the truncation errors when too few digits are used may cause large errors in the solution as well as numerical instabilities. When using the explicit integration method there is no need for assembled structural matrices and thus the needed core space is reduced to a minimum.

It is clear from the table that the explicit method is superior as far as storage requirement and computer time is concerned. Furthermore, the typical running time for this method was in these applications estimated to

$$T \approx 1.45 \cdot 10^{-4} \cdot F \cdot N$$

where T = number of CPU-seconds, F = number of degrees of freedom and N = number of used timesteps.

Time integration method	Number of degrees of freedom	Core space requirements [K-bytes]	CPU-time per time-step [s]
Implicit	441	546	1.40
Implicit	841	582	1.92
Explicit	441	76	0.064
Explicit	1763	115	0.25

Table 1. Computing parameters for calculations of K.

RESULTS AND DISCUSSION

From a great number of experiments, reported in [5, 8], six experiments were evaluated and of these, three were chosen to be presented in this paper from space considerations. This section is divided in three parts. In the first part a comparison between the explicit and the implicit time integration method is made for the direct problem. The two latter parts consider the inverse problem. Among others, this problem has been studied by Keegstra et al. [11]. They used a set of holding back forces near the crack tip to simulate the crack growth, related to an introduced fracture model. Predicted values of the crack velocity was then compared with experimental results.

1. Figs. 4-5 (specimen type B) and Fig. 6 (specimen type A) show the stress intensity factor K as function of the relative crack length. The open and filled circles represent K evaluated with the explicit and the implicit time integration method, respectively. Also plotted are the experimental values of the momentaneous crack velocity $\dot{a}$ (open squares). Every value derives from an evaluation within a distance corresponding to a finite element along the crack plane. The evaluations were not performed for large values of the relative crack length since earlier static investigations with these FEM-grids, see [5, 8], have shown discrepancies

compared to exact solutions for small remaining ligaments of the specimen. In the figures the ratio σ_o/σ_{ys} is shown, ranging from 0.18 to 0.47, where σ_o is the applied mean stress at the initiation of crack growth and σ_{ys} is the yield strength.

In Figs. 4-5 the velocity $\dot{a}$ principally follows the same pattern. Initially $\dot{a}$ increases rapidly after which it, for larger crack lengths, tends to stabilize or at least fluctuate around a more or less constant value. Partly, these fluctuations are rather a consequence of the evaluation procedure from the oscilloscope recordings, than reflecting the real physical behaviour. In the evaluated crack length interval K computed with the two methods are in remarkably good agreement with each other. In fact, the relative mean deviation between the curves was estimated to be less than two percent which in the light of the different qualities of the integration methods must be considered as a very good result. It is therefore difficult to make qualitative comparisons but it seems as if the agreement is better for the lower velocity case (Fig. 4) and in the low velocity region in Fig. 5. This is consistent with expectations since if the process is slow then the errors of the FEM-calculations ought to be small. Furthermore, the explicit method tends to give somewhat smaller values of K than the implicit method. In the authors' opinion, this can possibly be attributed to a phase difference between the methods. In the explicit method the state at the next time step is taken from the previous step, whereas the implicit method belongs to a predictor-corrector class of schemes, i.e. the state at the next time step is determined by an implicit manner in the integration scheme. This may influence the calculation of the displacements as well as the determination of the stress intensity factor.

Fig. 6 shows the same variables for specimen type A where displacement controlled boundary conditions were imposed. Here, the velocity $\dot{a}$ is essentially constant during the propagation. The two integration methods result in K-values that mainly run together around a constant value. Dahlberg et al. [8] performed a quasi-static calculation of K with satisfactory result indicating that inertia effects play a minor role in this kind of experiments, due to the fixed displacement controlled boundary conditions. Among others, this influence of the boundary conditions in the calculation of K has been demonstrated by Urabe et al. [12].

In summary one can conclude that for this kind of applications the explicit time integration method compares well with the implicit method. It seems that neither different geometries and boundary conditions, nor fluctuations of the crack velocity has any essential influence on the difference between the K-curves evaluated with the two methods. Therefore, the explicit method was used for the remainder of this work in calculating K.

2. We now turn to the inverse problem, that is to predict the crack velocity and the dynamic stress intensity factor from known information of the propagation toughness. This information was obtained in the following way: from an experiment, where the conditions at the boundaries and the crack velocity $\dot{a}$ were measured during the propagation event, the dynamic stress intensity factor

Figs. 4-6 - Comparison of the stress-intensity factor calculated with explicit and implicit time integration method.

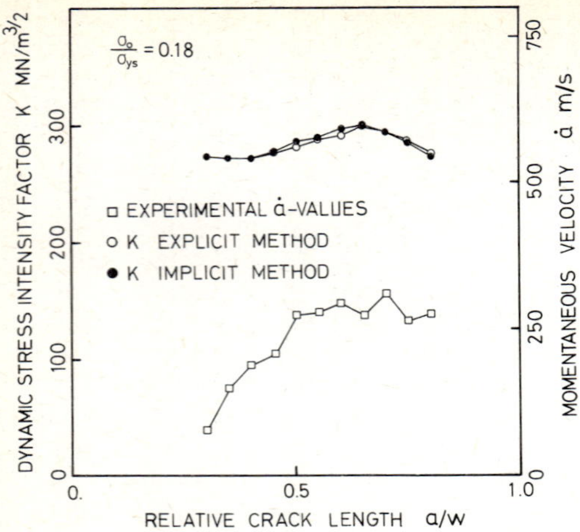

Fig. 4

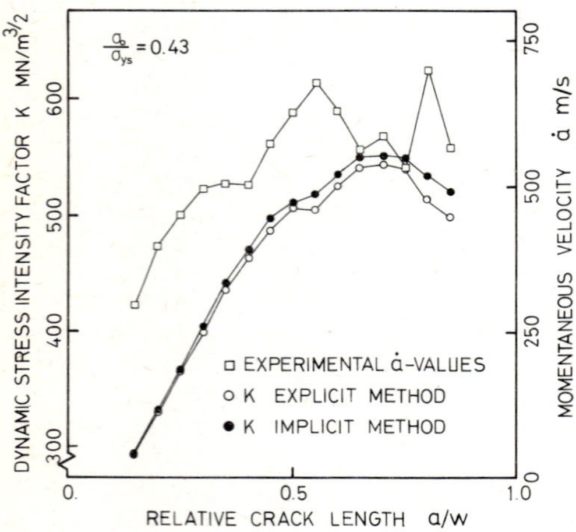

Fig. 5

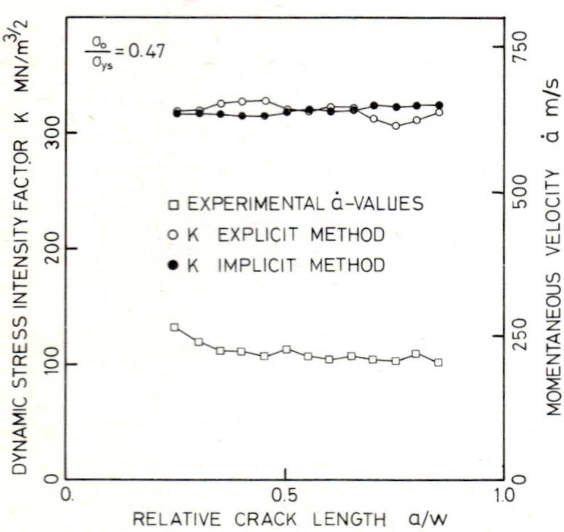

Fig. 6

Figs. 7-9 - Comparison between experimental and predicted results when individual K_{pc}-curves are prescribed.

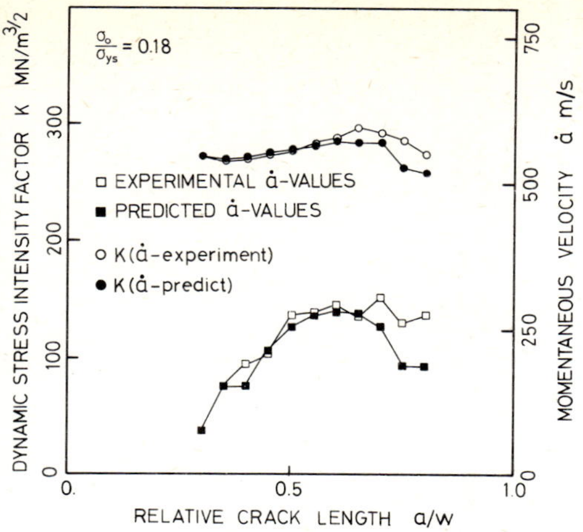

Fig. 7

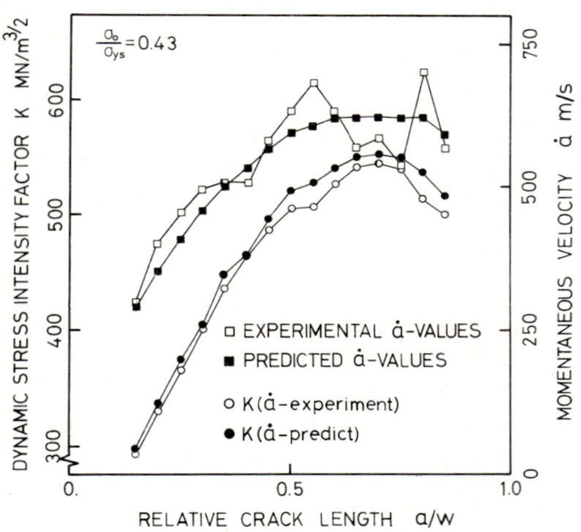

Fig. 8

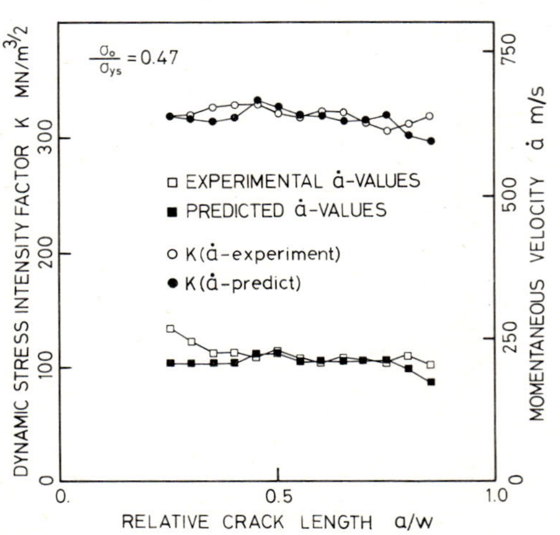

Fig. 9

Figs. 10-12 - Comparison between experimental and predicted results when mean curves of K_{pc} are prescribed.

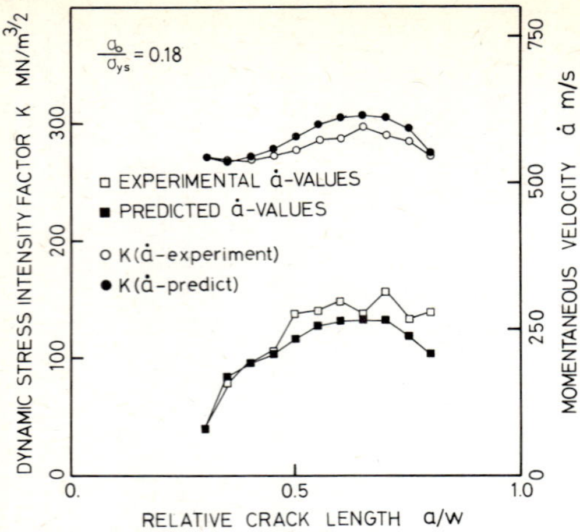

Fig. 10

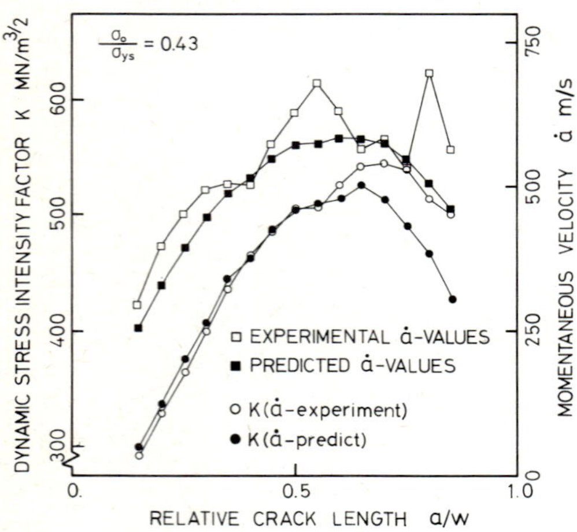

Fig. 11

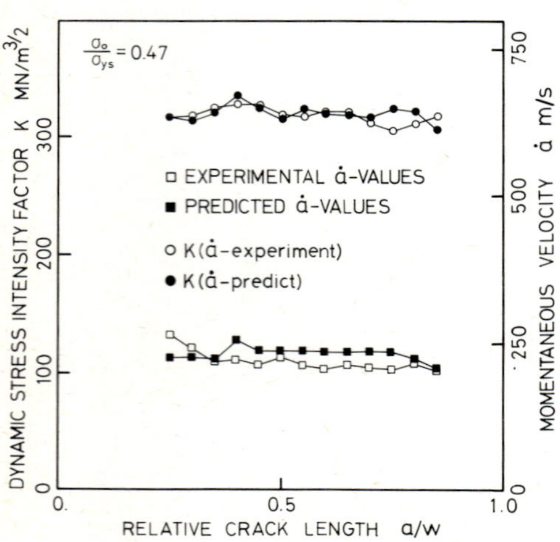

Fig. 12

RESULTS AND DISCUSSION

K was determined. Assuming a linear theory K(t) then equals the crack propagation toughness $K_{pc}(\dot{a})$ for a certain value of $\dot{a}$ during the process. To this discrete information of K_{pc} as function of $\dot{a}$, a second degree polynomial was fitted by the method of least squares. This $K_{pc} - \dot{a}$ relation was then used as input in a subsequent numerical prediction analysis of the particular experiment. Figs. 7-8 and Fig. 9 show these predictions for specimen type B and A, respectively. The predicted values of $\dot{a}$ (filled squares), determined iteratively by means of the earlier discussed Illinois algorithm, are here compared with the experimental values of $\dot{a}$ (open squares). Since only the K_{pc}-values from each experiment in question has been used for each figure, the curves are a measure of how well the experiments are reproduced with this iterative prediction technique. Also indicated in Figs. 7-9 are the dynamic stress intensity factor evaluated from experimental (open circles) and predicted (filled circles) values of the crack velocity. In general the predicted values of $\dot{a}$ compare well with the experimental results. It is worth noting that the predicted curve tends to smoothe out the crack velocity which is consistent with expectations since a smoothed K_{pc}-curve was prescribed as input. Hence there is a certain discrepancy in the region of large fluctuations of the measured crack velocity (see Fig. 8). This is also reflected in the amount of iterations in the Illinois algorithm, which tends to decrease when the crack velocity is stabilized. Consequently, very few iterations were necessary for specimen type A (Fig. 9) where the velocity is approximately constant during the process. Furthermore, it is reasonable to adjust the used criteria of terminating the iterations to the accuracy in determining the $K_{pc} - \dot{a}$ relation, thus avoiding a costly, unnecessarily high accuracy in the iteration process when using a crude K_{pc}-curve. In general the rate of convergence of the process was very high and normally only a few iterations were needed to reach the desired accuracy, most likely a consequence of the root α, of the iterative equation, being well conditioned.

As to the comparison between the K-curves extracted from the two sets of $\dot{a}$-values, it is difficult to make any significant conclusions since they only reflect the values of the velocities. Hence a close agreement between the curves could here likewise be seen.

3. Finally we consider the same problem as in the previous section with the distinction that now the material data was obtained from a whole set of experiments performed on the particular material. This is the case of a possible future physical application as to predict the growth history for a crack in a structure when information of the material is given. In Fig. 13 (specimen type A) and Fig. 14 (specimen type B) the regression curves for the crack propagation toughness K_{pc} are shown as functions of $\dot{a}$. The discrete values were obtained from a great number of experiments reported in [5, 8]. The results for the two geometries are displayed in Figs. 10-12, where now the material relations given in Figs. 13-14 are prescribed. As expected, somewhat larger discrepancies occur in this case between the predicted and measured crack speed. The deviation seems to increase in the high velocity

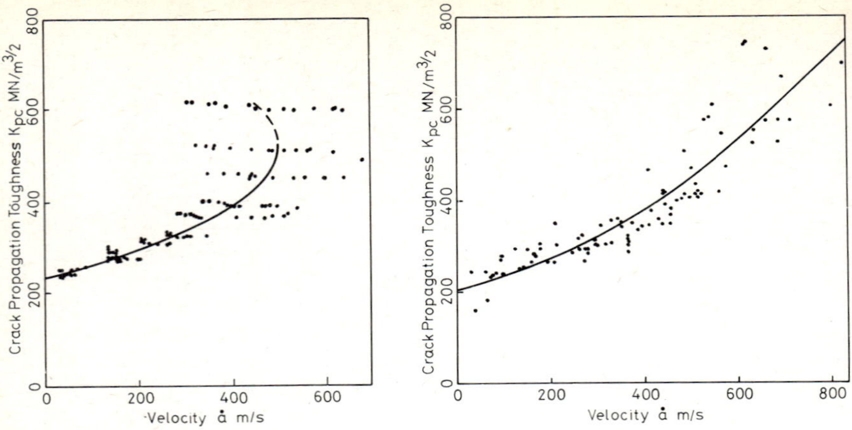

Figs. 13-14 - Crack propagation toughness as function of crack velocity for specimen type A and B, respectively.

region at high load levels which could be attributed to larger scatter in the K_{pc}-curves indicating that non-linear effects are influencing the process. The relative mean deviation in $\dot{a}(t)$ between the predicted and experimental values was estimated to be less than ten percent. This relatively good accuracy together with fairly low computer costs open possibilities of using this method for industrial purposes, for example checking the safety of a structural component. In connection with this the problem of crack arrest is of great importance and the authors are currently working with this subject.

CONCLUSIONS

a) The accuracy of the central difference scheme is comparable to implicit time integration in the present examples.

b) For the type of problems considered here, explicit time integration is significantly more efficient than implicit methods.

c) With use of the proposed iterative procedure, the inverse problem can be solved efficiently which is important for design applications.

ACKNOWLEDGEMENT

This work was financed through a grant from the Swedish Board of Technical Development (STU). The authors want to express their gratitude for this support.

REFERENCES

1. KEEGSTRA, P.N.R. - A Transient Finite Element Crack Propagation Model for Nuclear Pressure Vessel Steels. J. Inst. Nuc. Eng., Vol. 17, No. 4, pp. 89-96, 1970.

2. RYDHOLM, G., FREDRIKSSON, B., and NILSSON, F. - Numerical Investigations of Rapid Crack Propagation. Numerical Methods in Fracture Mechanics, pp. 660-672, Swansea, 1978.
3. MALLUCK, J.F. and KING, W.W. - Fast Fracture Simulated by a Finite Element Analysis which Accounts for Crack-Tip Energy Dissipation. Numerical Methods in Fracture Mechanics, pp. 648-659, Swansea, 1978.
4. NILSSON, L. - Impact Loading on Concrete Structures, Publication 79:1, Department of Structural Mechanics, Chalmers University of Technology, 1979.
5. BRICKSTAD, B. and NILSSON, F. - Numerical Evaluation by FEM of Crack Propagation Experiments. Int. J. of Fracture, Vol. 16, No. 1, pp. 71-84, 1980.
6. NILSSON, F. - A Note on the Stress-Singularity at a Non-Uniformly Moving Crack Tip. J. of Elasticity, Vol. 4, pp. 73-75, 1975.
7. DAHLQUIST, G. and BJÖRK, Å. - Numerical Methods, Prentice-Hall, N.J., 1974.
8. DAHLBERG, L., NILSSON, F. and BRICKSTAD, B. - Influence of Specimen Geometry on Crack Propagation and Arrest Toughness. Crack Arrest Methodology and Applications, ASTM-STP 711, Ed. Hahn, G.T., USA, 1980.
9. BAZANT, Z.P. - Spurious Reflections of Elastic Waves in Non-uniform Finite Element Grids. North-Western University, Evanston, 1977, to be published in Computer Methods in Applied Mechanics and Engineering.
10. NEWMARK, N.M. - A Method of Computation for Structural Dynamics. J. of Eng. Mech. Div., Vol. 85, No. EM3, pp. 67-94, 1959.
11. KEEGSTRA, P.N.R., HEAD, J.L. and TURNER, C.E. - A Two-Dimensional Dynamic Linear-Elastic Finite Element Program for the Analysis of Unstable Crack Propagation and Arrest. Numerical Methods in Fracture Mechanics, pp. 634-647, Swansea, 1978.
12. URABE, Y., KOBAYASHI, A.S., EMERY, A.F. and LOVE, W.J. - Further Dynamic Finite Element Analysis of the Tapered DCB Specimen. J. of Engineering Materials and Technology, Trans. ASME, pp. 324-328, Oct. 1977.

MOVING SINGULARITY-FINITE-ELEMENT MODELING OF FAST FRACTURE IN FINITE BODIES: "GENERATION" & "PROPAGATION" STUDIES.

T. Nishioka*, R.B. Stonesifer*, and S.N. Atluri*

Georgia Institute of Technology, Atlanta, Georgia

SUMMARY:

Results of a "generation" study (ie., determining the dynamic stress intensity factor for a crack propagating with a given velocity history), using moving singularity-elements, wherein, alternatively, propagating-crack eigen-functions, static-crack eigen-functions, or distorted-isoparametric approximations are employed, are reported. The relative accuracies of the three types of moving singular-elements are assessed. Also, results of a "propagation study" (ie., determining the crack-tip propagation history based on a given dynamic-fracture-toughness versus crack-velocity relation), involving a rectangular double cantilever beam specimen, and their comparison with the previously reported experimental results of Kalthoff et al, and numerical results of Kobayashi and others, are included.

INTRODUCTION:

In Refs [1,2,3] the authors have presented a "moving singular-element" procedure for the dynamic analysis of fast crack-propagation in finite bodies. In this procedure, a singular-element, within which a large number of analytical eigen-functions corresponding to a steadily propagating crack are used as basis functions for displacements, may move by an arbitrary amount $\Delta\Sigma$ in each time-increment Δt of the numerical time-integration procedure. The moving singular element, within which the crack-tip always has a fixed location, retains its shape at all times, but the mesh of "regular" (isoparametric) finite elements, surrounding the moving singular element, deforms accordingly. An energy-consistent

*Center for the Advancement of Computational Mechanics, School of Civil Engineering, Georgia Institute of Technology, Atlanta, Georgia 32332, USA.

variational statement was developed in [1,2] as a basis for the above "moving singularity-element" method of fast fracture analysis. It has been demonstrated in [1,2] that the above procedure leads to a direct evaluation of the dynamic K-factor (s), in as much as they are unknown parameters in the assumed basis functions for the singular-element.

Solutions to a variety of problems, in the category of "generation" studies in the sense defined earlier, were obtained by using the above procedure and were discussed in detail in [1,3]. These problems included, among others: (*i*) constant-velocity, self-similar propagation of a finite central crack in a finite panel [analogous to the well-known Broberg's problem],(*ii*) stress-wave loading of a stationary central crack in a finite panel [analogous to the well-known problems of Baker; Sih et al; and Thau et al],(*iii*) constant-velocity propagation of a central crack in a panel, wherein, the propagation starts at a finite time after stress-waves from the loaded edge reach the crack, and (*iv*) constant velocity propagation of an edge-crack in a finite panel, whose edges parallel to the crack are subjected to prescribed, time-independent, displacements in a direction normal to the crack-axis [analogous to the well-known problems of Nilsson [4] (*v*) etc. In Ref [5], the results of numerical simulation of experimentally measured crack-tip versus time history in a rectangular-double-cantilever-beam (RDCB), as reported by Kalthoff et al [6], were reported. Also, in Ref [5], the authors' results for the computed dynamic K-factor for RDCB were compared with the experimental (caustics) results of [6], and the independent numerical results of Kobayashi, [7] who uses a node-release technique in fast fracture simulation.

In the present paper both the "generation" and "propagation" aspects of dynamic crack-propagation (in the sense defined earlier) are treated. In the "generation" study, attention is focused on: (*i*) the effect of using only the static eigen-functions (or the well-known Williams'solution) in the moving singular-element for dynamic crack propagation, and (*ii*) the use of a distorted isoparametric element (or the quarter-point triangular element) in the analysis of dynamic crack propagation. In both the cases (*i*) and (*ii*), results are obtained, for several values of propagation-velocity, for the above described Nilsson's problem [4]. These results are compared with the authors' previous results [3], for the same problem, obtained by using the moving singularity-element, wherein dynamic eigen-functions for a steadily propagating crack were embedded. In the"propagation"study, the problem of determining the crack-tip time history in a RDCB, with a given dynamic-fracture-toughness (K_{ID}) versus crack-velocity ($\dot{\Sigma}$) curve, is treated. The assumed (K_{ID}) vs ($\dot{\Sigma}$) curve is identical to that in [7], and the presently computed results are compared with the experimental results [6] and numerical results of [7].

Only a brief description of the above studies, and pertinent results, are given in the following pages. For the omitted mathematical details, the authors' earlier work [1,2] may be consulted.

A. GENERATION STUDY

I Variational Formulation

The principal deviation of the specially developed variational formulation from the normal statement of the principle of the virtual work, is that the deformations and stresses both before and after an increment in crack length are included. The thus modified principle of virtual work can be written as:

$$\int_{V_2}\{(\sigma^2_{ij} + \sigma^1_{ij})\delta\varepsilon^2_{ij} + \rho(\ddot{u}^2_i + \ddot{u}^1_i)\,\delta u^2_i\}dv$$
$$=\int_{S_{\sigma 2}} (\bar{T}^1_i + \bar{T}^2_i)\delta u^2_i ds + \int_{\Sigma^+_1} (\bar{T}^2_i + \bar{T}^1_i)^+\,(\delta u^2_i)^+ ds$$
$$+\int_{\Sigma^-_1} (\bar{T}^2_i + \bar{T}^1_i)^-(\delta u^2_i)^- ds + \int_{\Delta\Sigma^+} (\bar{T}^2_i + \sigma^1_{ij}\nu^1_j)^+(\delta u^2_i)^+ ds$$
$$+\int_{\Delta\Sigma^-} (\bar{T}^2_i + \sigma^1_{ij}\nu^1_j)^-\,(\delta u^2_i)^- ds \qquad \text{(I.1)}$$

Where t_1 corresponds to the time just prior to creation of the new crack surface $\Delta\Sigma$ and t_2 is the time for which the solution is desired (ie., after the crack has extended by $\Delta\Sigma$). The above variational principle results in two rather attractive features. First, the modification of boundary conditions in the crack tip region come as a result of the formulation without any of the arbitrariness associated with the usual procedures of removing tractions on the new crack surfaces. Secondly, the formulation results in terms which can be interpreted as corrections for disequilibrium induced in the fininte element model through mesh translation (or remeshing). This latter feature is more readily seen in terms of the finite element equations given later.

It should be noted that a similar variational formulation for quasi-static crack growth in an elastic-plastic body has also been presented in reference [1]. The work presented here will, however, be limited to the dynamic linear elastic case.

II Singular Element for Dynamic Crack Propagation

A singular crack tip element was also developed in [1,2] and used in conjunction with the new formulation for the analysis of mode I dynamic crack propagation in linear elastic two dimensional bodies. This singular element uses an arbitrary number of the displacement eigen-functions, which come from the solution of a crack in an infinite body. For dynamic crack propagation, these eigen-functions are taken as those for the corresponding steady-state dynamically propagating crack in an infinite body. That is, the eigen-functions which are used in the singular element are dependent on the crack propagation speed and therefore, in the general case of non-uniform crack speed, will differ from one time step to the next. There are two inherent complications associated with the implementation of this singular element for routine analysis of fast fracture problems. First, there is a need for special techniques in assuring compatibility of displacements, velocities, and accelerations at the interface between the singular element and the standard eight noded isoparametric elements. The procedure of least squares matching at the boundary of the singular element was used [1,2] to overcome this difficulty. The second drawback with this singular element is that the crack speed dependence results in non-symmetric matrices [1,2]. Since the unsymmetry is limited to degrees of freedom associated with the singular element, it was found in [1,2] that the usual symmetric equation solver could be used with a simple iteration procedure and it was observed that the first iteration gave satisfactory convergence in all cases.

It should be noted that this element has several attractive characteristics which tend to offset the two drawbacks mentioned above. First, the crack surface traction boundary conditions as well as (steady-state) stress equilibrium are satisfied inherently by the eigen-functions. Secondly and perhaps more importantly, the stress intensity factor is obtained directly from the solution as the coefficient of the first $(1/\sqrt{r})$ eigen-function (ie., no interpolation, extrapolation, curve fitting or energy balancing is required).

III Implementation of the Variational Formulation and Singular Element

The modeling of crack propagation with this singular element requires that the element be translated along the path of crack propagation. To accomplish this translation, a procedure was developed [1,2] in which nodal points surrounding the singular elements are shifted. It is important to note that the amount of crack growth per time step $(\Delta\Sigma)$ is not in any way related to the distance between finite element nodes. Figure 1 shows how the singular element is shifted and the surrounding mesh adjusted and also shows how the mesh is periodically regenerated so as to limit the distortion of the element surrounding the singular element.

The finite element equations resulting from the special variational formulation and the above singular element can be written as:

$$\underset{\approx}{K}\, \underset{\sim}{q}_2 + \underset{\approx}{m}\, \ddot{\underset{\sim}{q}}_2 = \bar{\underset{\sim}{Q}}_2 + \bar{\underset{\sim}{Q}}_1 - \underset{\approx}{K}\, \underset{\sim}{q}_1 - \underset{\approx}{m}\, \ddot{\underset{\sim}{q}}_1 \quad \text{for } V_2 - V_s$$

$$\underset{\approx}{K}^*_s\, \underset{\sim}{q}_{s2} + \underset{\approx}{D}^*_s\, \dot{\underset{\sim}{q}}_{s2} + \underset{\approx}{m}^*_s\, \ddot{\underset{\sim}{q}}_{s2} = \underset{\sim}{Q}^*_s \quad \text{for } V_s \qquad \text{(III.1)}$$

where V_s refers to the volume encompassed by the singular element and subscripts 1 and 2 refer to quantities at t_1 and t_2 respectively. The terms on the right hand side of the first equation which are from time t_1 constitute the correction terms mentioned previously. If the interpolation involved in translating the mesh were perfect, these terms would be identically zero. The matrices K^*_s and D^*_s are unsymmetric while all others are symmetric. The Newmark-Beta time scheme was used in [1,2] for solving (III.1).

IV Consideration of Efficiency and Practicality

Having developed the singular element with crack speed dependent basis functions, and having become aware of the inherent complications with a routine use of this procedure, it is perhaps natural to ask if all the sophistication is warranted in the solution of practical crack propagation problems. For instance, one might question the use of dynamic† eigen-functions (and the resulting nonsymmetric stiffness) based on the moderate speed dependence of these functions for crack speeds of practical interest (ie., less than half the shear stress wave speed of the material). It was also pointed out that obtaining compatibility at the interface between the singular elements and the regular elements was done using a least-squares procedure. This procedure generally involves the calculation and storage of several additional matrices of dimensions on the order of the number of eigen-functions* and also involves the inversion of one of these. Since this extra effort results from using two types of elements, one might ask what decrease in accuracy would re-

†The term "dynamic eigen-functions" will be used to denote that the eigen-function is dependent on the speed of crack propagation.

*The calculations here and in [1-3] and [5] used 19 eigen-functions plus one rigid body mode. Note that for the problem of a dynamically loaded crack which does not propagate, the "dynamic eigen-functions" reduce to the usual static (Williams') eigen-functions.

sult from using no singular element or perhaps using eight noded isoparametric elements around the crack tip with their midside nodes shifted to the quarter point so as to give the desired $1/\sqrt{r}$ stress singularity.

an effort to answer some of these questions, this pa... considers two variations on the originally proposed [1,2] formulation and solution procedure. In the first, the singular element is used with static eigen-functions. This results in symmetric matrices. In the second, the special singular element is replaced by standard eight noded isoparametric elements with midside nodes shifted appropriately so as to give the desired singularity. Of course, this latter procedure eliminates the problem of compatibility that is inherent with the special element.

V Description of the Problem and the Finite Element Meshes

The primary purpose of these computations is to compare results using several levels of modeling sophistication. To keep the problem from obscuring the basic differences in modeling techniques, a rather simple combination of geometry and loading was selected. The problem is that of the constant velocity propagation of an edge crack in a square sheet whose edges parallel to the direction of crack propagation are subjected to uniform displacements $\bar{u}_2$ in the direction normal to that of crack propagation. The dimensions and mesh configuration are depicted in Figure 2. This problem is analogous to that treated by Nilsson [4] who obtained an analytical solution for the steady-state stress-intensity factor for the constant velocity propagation of a semi-infinite crack in a finite-height, infinite-width strip. In the following, the material properties are $\nu=0.286$, $\mu=2.94 \times 10^4$ N/mm^2 and $\rho=2.45 \times 10^{-6}$ kg/mm^3. Three levels of constant crack velocity propagation, $(v/C_s)=0.2, 0.4, 0.6$ are considered. In each case, the initial crack length, (Σ_o/W) is 0.2. For each case, the crack growth increment size is maintained constant (as opposed to time step size) at a value of $\Delta\Sigma/W=0.005$ which corresponds to 2.5 percent of the singular element's dimension in the crack propagation direction. The singular element is represented by the shaded region of the finite element mesh shown in Figure 2.

The above explanation of the details of the analysis is also applicable to the analyses using quarter-point isoparametric elements. The region occupied by the singular element is replaced as shown in Figure 2, by four isoparametric elements, each of which has been degenerated to a triangular element so that the degenerate side corresponds to the node at the crack tip. Each midside node which is adjacent to the crack tip is then shifted toward the crack tip node so as to occupy the quarter point of the element side, thereby giving the desired $1/\sqrt{r}$ singularity in stress at the crack tip. In all of the analyses presented here, the uniform prescribed displacement is applied statically at t=0 sec. While maintaining this prescribed displacement, the crack is then made to propagate at a uniform speed. This procedure therefore results in crack acceleration over the first time step. Since the time step size is varied so as to maintain the crack growth increment size constant, this acceleration will also differ for each value of the uniform crack speed.

VI Procedures for Computing Stress Intensity Factors

Before proceeding with the results of the analyses, a description of the various procedures which are used for computing the dynamic stress-intensity factor is in order. As already pointed out, when the singular element is used with the dynamic eigen-functions, the stress intensity factor is computed directly as the coefficient of the $1/\sqrt{r}$ dependent eigen-function (ie., β_1).

If one chooses to use the static eigen-functions for the case of a propagating crack, one is left with either accepting the error of using the first coefficient (β_1) of the static functions as the dynamic stress intensity factor or using an alternate (indirect) procedure for its determination. One candidate procedure is the global energy balance method. In this method one computes the energy dissipated (through the fracture process) between two time steps t_1 and t_2 for which the corresponding crack lengths are Σ_1 and Σ_2. If the work done on the system at t_i is W_i, the total strain energy U_i, and the total kinetic energy T_i, then the energy release rate (per unit of new crack area) is given by:

$$G = \frac{(W_2 - W_1) - (T_2 - T_1) - (U_2 - U_1)}{(\Sigma_2 - \Sigma_1)b} \qquad \text{(VI.1)}$$

where b is the length of the propagating crack front. If the fracture is pure mode I, then the stress intensity factor can be obtained from v the crack speed, μ the shear modulus, ν poissons ratio, ρ the mass density, and the formula:

$$K_I = \sqrt{\frac{2\mu G}{f(\alpha_d, \alpha_s)}} \qquad \text{(VI.2)}$$

where $f(\alpha_d, \alpha_s) = \alpha_d(1 - \alpha_s^2)/[4\alpha_d\alpha_s - (1 + \alpha_s^2)^2]$

$\alpha_d^2 = 1 - (v/C_d)^2$; $\alpha_s^2 = 1 - (v/C_s)^2$

$C_s^2 = \mu/\rho$; $C_d^2 = 2\mu\ (1-\nu)/\rho(1-2\nu)$ (plane strain)

Another approach to computing the dynamic stress intensity factor involves the direct calculation of the fracture energy through evaluation of the work done by the system in releasing the tractions on the newly created crack surface $\Delta\Sigma[\Delta\Sigma=\Sigma_2(t_2)-\Sigma_1(t_1)]$. This calculation involves the tractions on $\Delta\Sigma$ existing at t_1 and the displacements at $\Delta\Sigma$ existing at t_2. For pure mode I fracture behavior one has only (u_y and T_y present at $\Delta\Sigma$ and the energy release rate is given by:

$$G = \frac{1}{2\Delta\Sigma b} \int_{\Delta\Sigma} T_y\ (x,t_1)\ u_y\ (x,t_2)dx \tag{VI.3}$$

Having obtained G, one is then able to compute the stress intensity factor using (VI.2). In this paper, only mode I crack behavior is considered. In the following this method of obtaining K_I will be referred to as the crack tip integral method.

The last procedure which may be used for evaluating the dynamic stress intensity factor involves "fitting" the near crack tip field with the appropriate crack speed dependent eigen-functions. As in the case of the energy methods, this method does not depend on using any particular procedure for obtaining the solution. Rather, once a solution is obtained, one selects any arbitrary number of points around the crack tip regions and determines the coefficients of the eigen-functions which give the best fit to the stresses and/or displacements at those points through a general least-squares procedure.

VII Results of Computations With Singular Elements Using Dynamic and Static Eigen-Functions

The first comparisons to be made are between the results from the singular element with dynamic eigen-functions and the singular element with static eigen-functions. The first aspect of the analysis to be compared is the stress intensity factor as a function of crack length (or equivalently time) for three crack speeds. In Figure 3, the comparison is made for crack speed v equal to $0.2C_s$ (C_s is shear wave speed). Also in this figure is Nilsson's steady-state value of K_I for the infinite strip problem. All of the stress intensity factors have been normalized with respect to the static value for the infinite strip problem given by $K_s^\infty=(\bar{u}_2E)/(1-\nu^2)\sqrt{h}$. It is important to note that the value of K_I plotted in Figures 3 and 4, for both the static eigen-function case and dynamic eigen-function case, are just the coefficients of the respective $1/\sqrt{r}$ dependent terms (ie., β_1). Therefore, strictly speaking, the K_I from the static eigen-function case is <u>not</u> the actual dynamic stress intensity factor. One would need to resort to some alternate procedure to get the true value of K_I. However, noting the small differences in Figure 4, this extra effort is probably not warranted in this particular instance ($v=0.2C_s$). Figure 4 shows a similar comparison for crack propagation speeds of $0.4C_s$ and $0.6C_s$. It is clear that as the crack propagation speed increases, the difference between the static eigen-function results and the dynamic eigen-function results increases.

Figure 5 contains crack profiles for several different crack lengths and crack propagation speeds. The top three plots are for a crack length of $\Sigma/W=0.3$, and the crack propagation speeds of $v/C_s=0.2, 0.4$ and 0.6. The plot for $v/C_s=0.2$ confirms the observation made before with regard to the two solutions being nearly identical. For the case of $v/C_s=0.4$ there is also little evidence of the solution being any different in terms of displacement, even though the stress intensity factors of Figure 4 are appreciably different. By the speed of $0.6C_s$, however, there is an easily discernible difference between the two solutions. In fact, in the vicinity of the crack tip, the displacements from the static eigen-function solution are roughly 10 percent less than those for the dynamic eigen-function solution. Though a plot of stresses is not included, a comparison of σ_y at 2mm in front of the crack indicates that the stresses for the static eigen-function case are greater than those for the dynamic eigen-function solution by about 6 to 11 percent. The stress intensity factor difference can be seen from Figure 4 to vary from approximately 16 to 22 percent and therefore seems to be roughly twice the difference in stress (or displacement). Similar comparison of stress for $v/C_s=0.4$ showed that σ_y at 2mm ahead of the crack for the static eigen-function solution exceeded that for the dynamic eigen-function solution by approximately 1 to 3 percent. From Figure 5, the stress intensity factors for $v/C_s=0.4$ can be seen to differ by about 3 to 8 percent which again is roughly twice the difference seen in stress.

The bottom two plots in Figure 5 show the crack profiles at $\Sigma/W=0.5$ and 0.7 for cases $v/C_s=0.4$ and 0.6, respectively. These show that the near crack tip displacements for both cases are smaller when static eigen-functions are used with the difference being greater at the higher rate of propagation. Therefore, in summary, it was found that the use of static eigen-functions for this crack propagation problem resulted in an increase in near field stresses and a decrease in the near field displacements with the deviation from the dynamic eigen-function solution going from essentially zero at $v/C_s=0.2$ and finally to roughly 10 percent at $v/C_s=0.6$.

VIII Results of Computations with Quarter Point Isoparametric Elements

Having considered the effect of using the static eigen-functions in place of the dynamic eigen-functions, we will now proceed by comparing these results with those using the quarter-point isoparametric elements. It will be noticed by looking at Figures 3 and 4, that two procedures were used for computing K_I from the quarter-point isoparametric element results. The first used the global energy balance of equation VI.1 and the second used the crack tip integral given by equation VI.3. Then both used equation VI.2 to obtain K_I from G.

Before considering the dynamic crack propagation problem, a solution was obtained for the quasi-static crack propagation problem. The results of the first fifteen growth steps are shown in Figure 6. It can be seen that rather significant oscillations were encountered, with the oscillation of the crack-tip-integral evaluated K_I being somewhat less than that based on the global energy. It has been determined that this oscillation stems from the variational formulation I.1. As mentioned in the introduction, this formulation results in terms which are,in effect, corrections for the disequilibrium which is introduced by the interpolation procedures used in translating the mesh. It has been found in this study the Newmark-Beta time integration scheme is not capable of using these correction terms as they were intended. Rather, the correction from one time step ends up being included as part of the error in the subsequent time step. A similar but much less severe oscillation of this type was shown in Figure 1 of reference [2] which is also for a quasi-static problem. This solution in reference [2] was obtained through use of the singular element with static eigen-functions. It is thought that the oscillations using this special element were much less severe than those of Figure 6 because there is an inherently smaller interpolation error with the more sophisticated element. The reason this oscillation is not visible for dynamic problems is that the nodal forces due to the correction terms are insignificant when compared to the nodal forces due to inertial effects. The effect of removing these correction from have the formulation for the quasi-static propagation problem is shown in Figure 7. One characteristic of the solution of Figure 7 which shows up in all those using the quarter-point elements, is that the crack tip integral procedure leads to a smaller K_I than the global energy. It is thought that the crack tip integral value is the less accurate of the two and that this is due to the unavoidable inaccuracy of evaluating tractions at the boundary of the singular isoparametric element.

For all subsequent calculations using the quarter-point element, the correction terms of III.1 are left out. It was found that the omission of these terms had two effects on the results of the dynamic crack propagation analyses. First, the effect of increasing the crack propagation speed from zero to the uniform value in one time step seemed to result in less oscillation of K_I in the initial time steps. Second, the value of K_I from the crack tip integral was smaller by approximately 8 percent. The effect on the global energy, however was, found to be insignificant.

The results of the calculations using the quarter-point isoparametric elements are included in Figures 3 through 5. In Figure 3, it can be seen that for v/C_s=0.2, the quarter-point elements gave results which agree as well with the dynamic eigen-function singular element as did the static eigen-function singular element. It is interesting to note however, that the isoparametric element results and the static eigen-function results fall on opposite sides of the dynamic eigen-function results. Here and in the following, the global energy results will be discussed exclusively due to the presumed inferiority of the crack tip integral procedure when used with the singular isoparametric elements.

In Figure 4, the results for v/C_s=0.4 and 0.6 are presented. Here again, the isoparametric elements fall on the opposite side of the dynamic eigen-function results from those using the static eigen-functions. It is interesting to note that the difference between the isoparametric element results and the dynamic eigen-function results is very nearly equal in magnitude to the difference between the static eigen-function results and the dynamic eigen-function results. It is also interesting to note that the effect of regenerating the mesh (denoted by arrows in the figures) is more noticeable for the isoparametric elements and increases with an increase in crack propagation speed. It is perhaps surprising that the crack tip integral procedure was less sensitive to remeshing than the global energy procedure.

Figure 5 allows us to determine how the observed differences in stress intensity factor reflect actual differences in the solution in terms of displacement. As before, the difference in displacements for v/C_s=0.2 are very small. For v/C_s=0.4, it can be seen that the differences in displacements increase as the crack increases in length. At v/C_s=0.4 and Σ/W=0.3 and 0.7, it is interesting to note that the isoparametric elements and static eigen-function singular element are in close agreement near the crack tip and both give a smaller crack opening than the dynamic eigen-function element. As the crack elongates, the isoparametric element tends to give larger displacements at locations more distant from the crack tip.

B. PROPAGATION STUDY

It is clear from the results presented in the previous section that the best results [in comparison with the analytical results as in Figs 3-6] are obtained by using a moving singularity-element with embedded propagating-crack-eigen-functions. Using such a dynamic-eigen-function singularity-element, "generation" calculations were performed earlier by the authors [5], using the experimentally measured [6] crack-tip velocity history in test specimens. It was shown [5] that the dynamic stress-intensity factor computed in the "generation" calculation was sensitive to the computer-input data for $\dot{\Sigma}$ vs t curve. Further, in the simulation of experimental data for the wedge-loaded rectangular double cantilever specimen [WL-RDCB], the computed K-factor agreed to within 1-2% of the experimentally determined K-factor during the first (2/3) of the propagation history. However, the computed results for the K-factor exhibit a maximum during the latter third of the propagation history, prior to arrest, while the experimentally determined ones [6] do not.

Here we present the results of a "propagation" calculation (in the sense defined earlier) using the procedure of the moving-singularity-element with embedded propagating-crack-eigen-functions. Similar results using moving singular isoparametric element, and singular element with eigen-functions, will be presented elsewhere.

The present propagation calculation for the WL-RDCB Specimen, identical to Specimen No. 4 of Kalthoff, et al [6], is carried out using the K_{ID} versus $\dot{\Sigma}$ relation as shown in Fig. 8 with a crack initiation fracture toughness of K_{IQ}=2.32 $MNm^{-3/2}$. It is noted that the curve in Fig. 8, as well as the value of K_{IQ} are identical to the ones used by Kobayashi [7]. The results of this propagation calculation, vis., the curves for K_I vs t, Σ vs t, and $\dot{\Sigma}$ (or v) vs t, are shown in Fig. 9 along with the appropriate comparison experimental results of Kalthoff et al [6] and numerical results of Kobayashi [7].

It is seen from Fig. 9, that the agreement between the present numerical results and the experimental ones [6] for K_I is excellent during the first (2/3) of the propagation history. Even though both the present results and those of Kobayashi [7] for the dynamic K-factor exhibit a peak as compared to the experimental results, it is seen from Fig. 9 that this peak in the results of [7] is much larger and occurs much sooner in the propagation history, as compared to the present results. It should be noted that while the Araldite B used in the experiments [6] may have rate-dependent viscoelastic properties, the present calculations as well as those in [7] were based on the assumption of linear elastic material behavior.

From Fig. 9, and the authors' earlier "generation phase" results presented in [5], it is seen that the computed K_{ID} results in the "propagation" and "generation" calculations agree excellently; thus indicating the consistency of the present procedures of fast fracture simulation using a moving singularity-element with propagating-crack eigen-functions.

Further "propagation-calculation" results for other fracture specimens, the effect of changes in K_{ID} vs. Σ relations, as well as the mathematical (and algorithmic) details of the "propagation" calculation, will be presented elsewhere shortly.

C. CONCLUDING REMARKS:

It seems clear from the results obtained in the present generation study, that the best accuracy, especially at higher speeds of crack propagation, is by using the propagating-crack eigen-functions in conjunction with the moving singularity-element procedure. While at lower speeds of propagation ($v \simeq 0.3C_s$), a comparable accuracy is obtained by using either a singular-element with static-crack eigen-functions or a quarter-point isoparametric element, the use of these two simplified elements results in errors the order of 10% at higher speeds of crack propagation [$v \simeq 0.6C_s$]. It is to be noted that the above statements are relevent in the present context when roughly equal numbers of degrees of freedom are employed in all the 3 cases of formentioned singular elements. For the computations made here, it was found that the cost of using the singular-element with static or dynamic eigen-functions was roughly 1.5 or 2.0 times as expensive as the singular-isoparametric element. The trade-offs however are: (*i*) for any crack-speed, accuracy on the order of ± 1% (as compared to analytical results) can be obtained with far fewer degrees of freedom when dynamic eigen-functions are used in the moving singular-element, and (*ii*) when dynamic eigen-functions are used, the dynamic K-factor can be computed directly. The case of the quarter-point isoparametric element is perhaps a very attractive alternative to the special crack-tip element for many engineering applications. It appears from the present analyses [Fig 3] that this is a viable procedure, but that as the crack velocity increases ($>0.3C_s$) one must increase the number of isoparametric elements at the crack-tip in order to achieve accuracy comparable to that for a singular-element with dynamic eigen-functions. (Note that the present results with a dynamic eigen-function element were obtained with the same number of degrees of freedom for all crack speeds)

The results from the propagation study indicate, in addition to the excellent correlation of the present results with the cited experimental ones, that the present moving-singularity-element procedure of simulating fast fracture is inherently consistent.

ACKNOWLEDGEMENTS:

The authors gratefully acknowledge support for this work from the U.S.A.F.OSR under contract No. F49620-78-C-0085. The authors thank Colonel J.D. Morgan, III for his encouragement and Peggy Eiteman for her careful typing of this manuscript.

REFERENCES:

[1] Atluri, S.N., Nishioka, T., and Nakagaki, M., "Numerical Modeling of Dynamic and Nonlinear Crack Propagation in Finite Bodies, By Moving Singular Elements" in Nonlinear and Dynamic Fracture Mechanics ASME-AMD vol 35, Eds. N. Perrone and S.N. Atluri, ASME, NY, 1979, pp 37-67.

[2] Nishioka, T., and Atluri, S.N., "Numerical Modeling of Dynamic Crack Propagation in Finite Bodies, By Moving Singular Elements - Part I. Formulation", ASME Journal of Applied Mechanics 1980 (In Press).

[3] Nishioka, T., and Atluri, S.N., "Numerical Modeling of Dynamic Crack Propagation in Finite Bodies, By Moving Singular Elements - Part II. Results" ASME Journal of Applied Mechanics, 1980 (In Press).

[4] Nilsson, F., "Dynamic Stress-Intensity Factors for Finite Strip Problems" Int. Jnl of Fracture, vol 8, No. 4, 1972, pp 403-411.

[5] Nishioka, T., and Atluri, S.N., "Efficient Computational Techniques for the Analysis of Some Problems of Fracture in Pressure Vessels and Piping" To be presented the Pressure Vessels and Piping Conference, ASME, 1 Aug. 1980, San Fransisco.

[6] Kalthoff, J.F., Beinert, J., and Wingkler, S., "Measurements of Dynamic Stress Intensity Factors for Fast Running and Arresting Cracks in Double-Cantilever-Beam Specimens" Fast Fracture and Crack Arrest ASTM STP 627, Eds. T.T. Hahn, and M.F. Kanninen, ASTM 1977, pp 161-176.

[7] Kobayashi, A.S., "Dynamic Fracture Analysis by Dynamic Finite Element Method-Generation and Propagation Analyses" in Nonlinear and Dynamic Fracture Mechanics ASME-AMD vol 35, Eds. N. Perrone and S.N. Atluri, ASME, NY, 1979, pp 19-37.

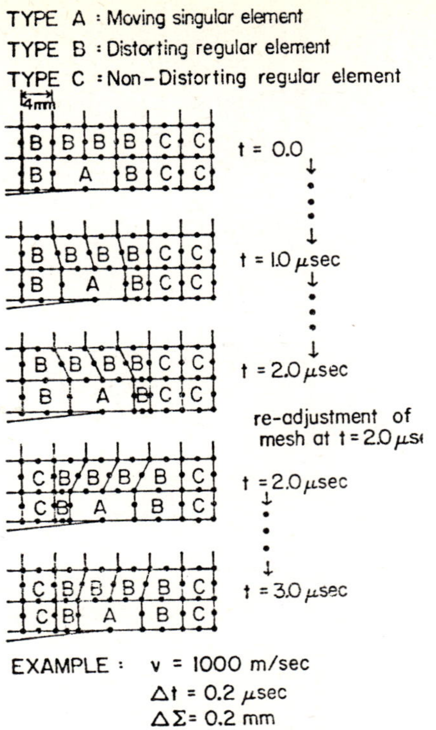

Figure 1. Crack Tip Mesh Translation and Re-adjusting Procedure.

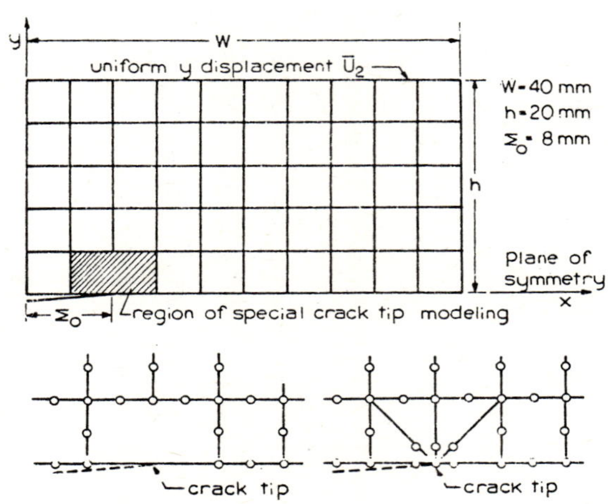

Figure 2. Geometry and Finite Element Meshes

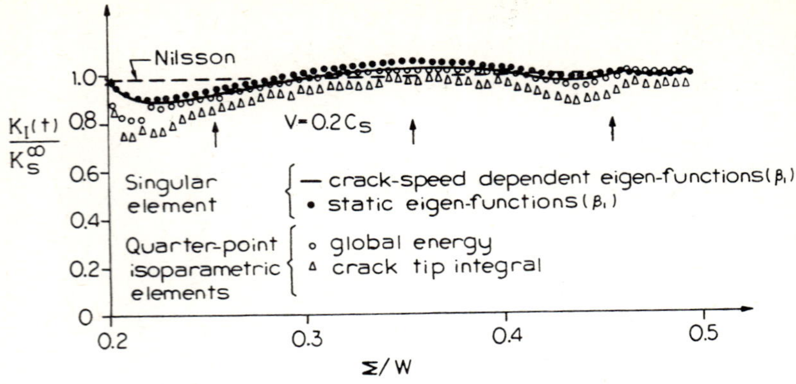

Figure 3. Comparison of stress intensity factors for crack propagation speed of 0.2 C_s.

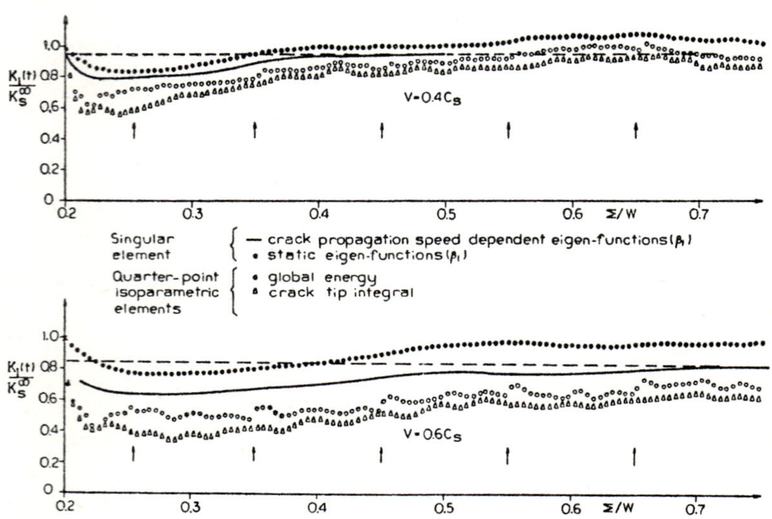

Figure 4. Comparison of stress intensity factors for crack propagation speeds of 0.4 C_s and 0.6 C_s.

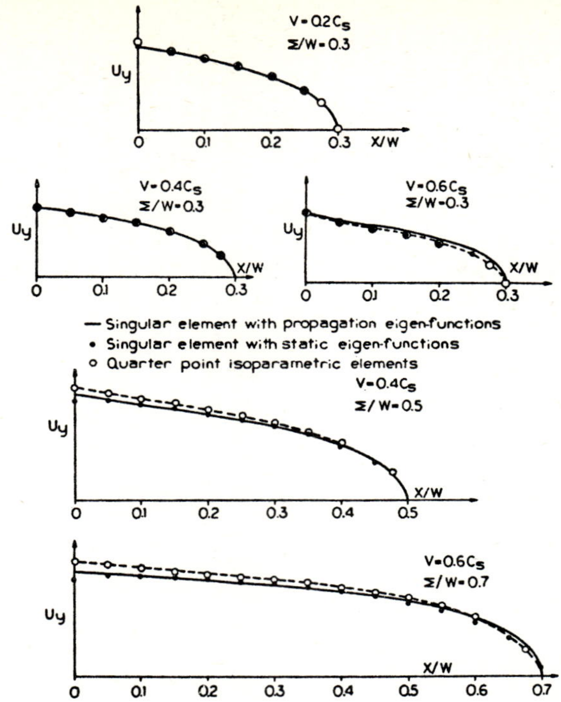

Figure 5. Comparison of crack opening profiles for several crack lengths and crack propagation speeds

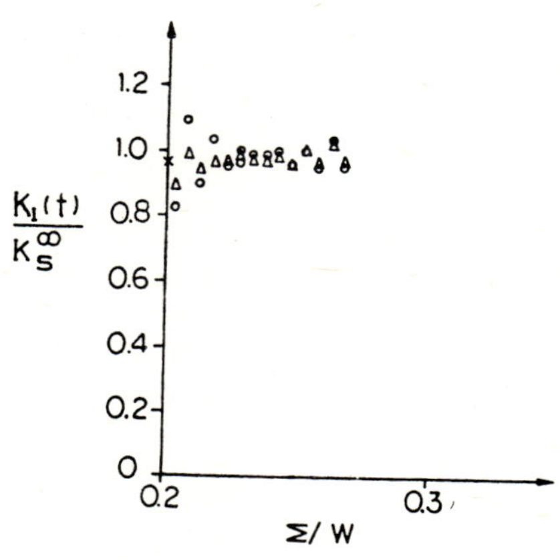

Figure 6. Quasi-static crack growth with quarter point elements and the formulation of reference [1].

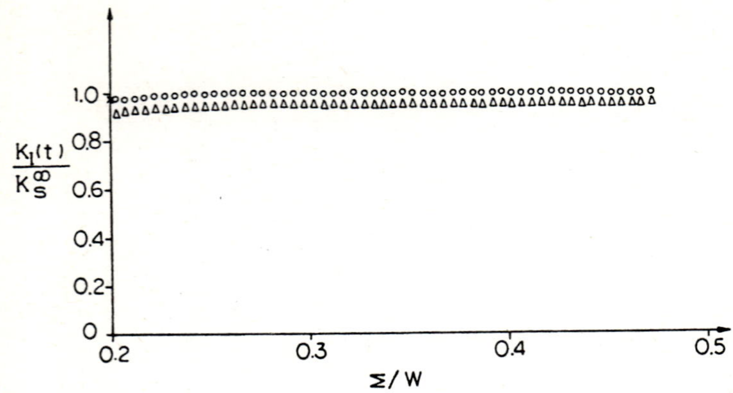

Figure 7. Quasi-static crack growth using quarter point isoparametric elements with the correction terms omitted.

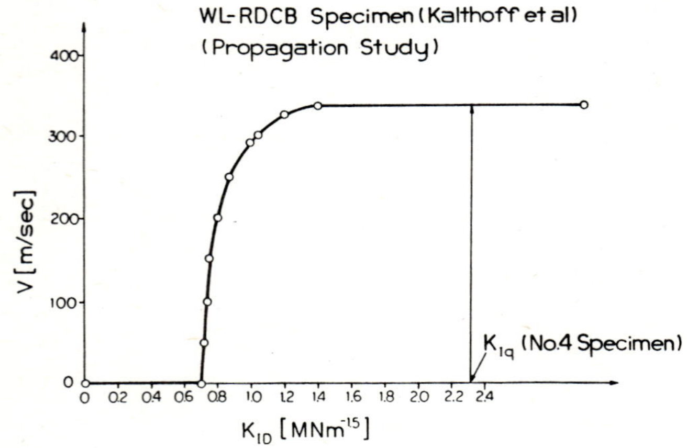

Figure 8. Dynamic Fracture Toughness Versus Crack Velocity Relation (Kalthoff et.al.)

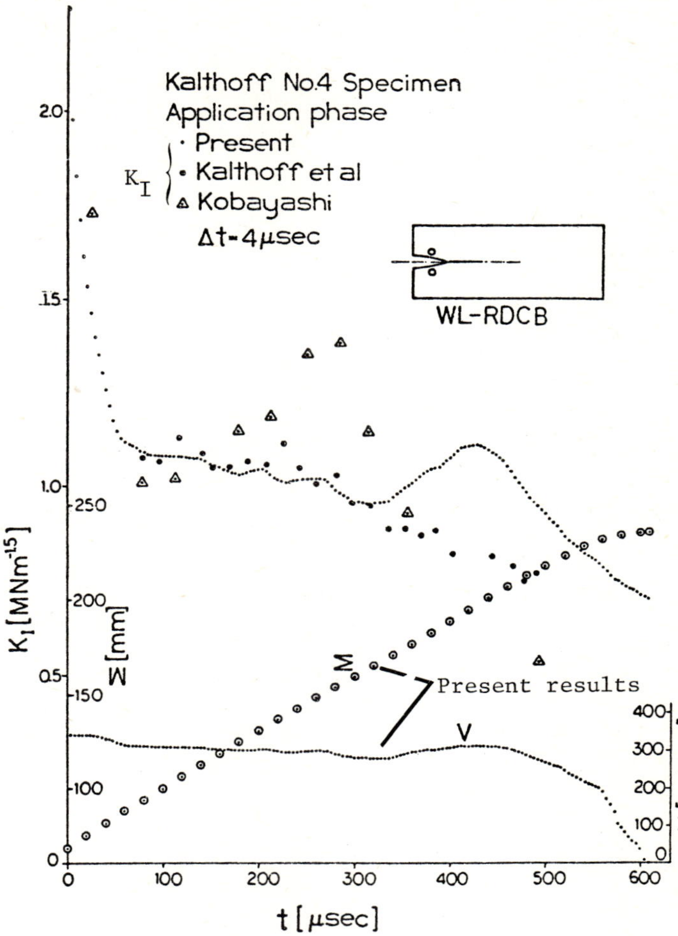

Figure 9. Present Predictions (and Comparisons) for $K_I(t)$; $\Sigma(t)$; and v.

AN IMPROVED DYNAMIC ANALYSIS OF AN INITIATING AND FAST-RUNNING GRIFFITH CRACK IN A FINITE RECTANGULAR PLATE

M. Perl*

ABSTRACT

Crack Initiation and dynamic crack propagation in a centrally cracked plate subjected to constant tension is investigated, employing two-dimensional linear elastic fracture mechanics. The analysis is performed using the SMF2D code which is based upon the simultaneous employment of both a stationary and a moving coordinate system. The different parameters e.g.: the stress intensity factor, the energy release rate as well as the energies distributions were evaluated throughout the various simulations. Special attention was given to the influence of the energy supplied to the body by the work carried out by the external tractions during crack propagation.

INTRODUCTION

In a previous paper by Perl et al. [1] a preliminary analysis of dynamic crack propagation in the Griffith specimen was reported. The present paper further elaborates on this problem yielding a much better understanding of the role of the different dynamic parameters: the dynamic stress intensity factor, the energy release rate and the various energies distributions. Furthermore, a comprehensive comparison between the Griffith specimen and the SEN (Single Edge Notch) one is made and discussed.

* Research Associate, Department of Mechanical Engineering, University of Washington, Seattle, Washington 98195.

Following a brief presentation of the field equations and the numerical scheme, both static and dynamic results for the Griffith specimen are reported. The various results are compared to experimental findings as well as to the solution for the SEN specimen.

THE FIELD EQUATIONS

Only an outline of the field equations and the numerical procedure is revised here. A detailed description of the method employed, the numerical approximations as well as the SMF2D code is given in [1-3].

The SMF2D code is based on the simultaneous employment of two coordinate systems (see Fig. 1). The stationary coordinate system which is attached to the treated body, defining the stationary domain Ds, and the moving coordinate system which originates at the moving crack tip, defining the moving domain Dm.

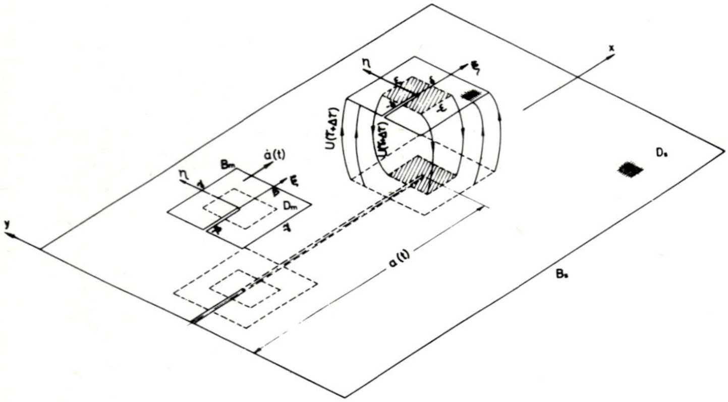

Fig. 1 The stationary and the moving grids

Assuming an elastic medium under plane strain conditions, the dimensionless displacement field $\underset{\sim}{U}$ in the stationary domain Ds is governed by the following equations of motion:

$$\frac{1}{2(1-\nu)}\left[U_{k,ki} - (1-2\nu)U_{i,kk}\right] = U_{i,\tau\tau} + \psi U_{i,\tau} \qquad (1)$$

$$i=1,2$$

Where: $\underset{\sim}{U}$ was normalized to an arbitrary unit of length, H, say for convenience the height of the specimen investigated. The dimensionless time τ was normalized to H/C_1 where C_1 is the dilatational or the plate wave velocity depending upon whether plane strain or plane stress conditions prevail. ν is the Poisson Ratio of the material investigated, but it is replaced by the apparent Poisson Ratio $\nu^* = \nu/1 + \nu$ in plane stress conditions. The last term in Eqs. (1) has been added for solving the static case using dynamic relaxation [4]. ψ is a parameter by which the code is switched from the static case ($\psi>0$) to the dynamic one ($\psi=0$). Eqs. (1) together with the appropriate boundary conditions for the particular problem being solved (e.g. [2]) completely define the problem in the stationary domain.

The moving coordinate system being attached to the moving crack tip satisfies the relation:

$$\begin{aligned} \xi &= x - \alpha(\tau) \\ \eta &= y \end{aligned} \tag{2}$$

Where α (τ) is the non-dimensional time dependent crack length. Therefore, in the moving domain Dm the equations of motion (1) being expressed in terms of ξ and η become

$$\begin{aligned} \frac{1}{2(1-\nu)}\left[U_{k,ki} + (1-2\nu)U_{i,kk}\right] &= U_{i,\tau\tau} - 2\,\dot{\alpha}(\tau)U_{i,1\tau} + [\dot{\alpha}(\tau)]^2 U_{i,11} \\ &\quad - \ddot{\alpha}(\tau)U_{i,1} + \psi U_{i,\tau} \qquad i=1,2 \end{aligned} \tag{3}$$

Where the dot denotes derivation with respect to time. The only boundary condition which applies to the moving domain is the traction free surface of the crack:

$$\sigma_{\eta\eta} = \sigma_{\xi\eta} = 0 \text{ on } \eta = 0 \text{ for } -\beta \leq \xi < 0 \tag{4}$$

where:

σ_{ij} = the stress tensor components.

The field equations for both the stationary and the moving domains (1,3) subjected to the appropriate boundary conditions are solved by the finite difference method. The equations and the boundary conditions are approximated by finite differences to a second order accuracy thus yielding an explicit three level time-step algorithm for solving the static and the dynamic displacement fields. As mentioned before, a detailed description of the exact discretization of the equations and the integration procedure is given in [1-3].

THE MATHEMATICAL MODEL

The Griffith specimen is described in Fig. 2. It is a centrally cracked ($2a_o$) rectangular specimen (2H x 2W) subjected to a constant, uniform, remote tension stress σ_∞ applied, perpendicularly to the crack plane, to the ends ED and FC. Since the specimen has two axis of symmetry only the upper right quarter is considered ($X \geq 0$, $y \geq 0$), provided the boundary conditions for that part, i.e.:

$$\sigma_{yy} = \sigma_{xy} = 0 \quad \text{on} \quad y = 0 \quad \text{for} \quad 0 \leq x < a_o \tag{5a}$$

$$\sigma_{xx} = \sigma_{xy} = 0 \quad \text{on} \quad x = W \quad \text{for} \quad 0 \leq y \leq H \tag{5b}$$

$$\sigma_{yy} = \sigma_\infty \; \sigma_{xy} = 0 \text{ on} \quad y = H \quad \text{for} \quad 0 \leq x \leq W \tag{5c}$$

are supplemented by the symmetry conditions:

$$U_y = \tau_{xy} = 0 \quad \text{on} \quad y = 0 \quad \text{for} \quad a_o \leq x \leq W \tag{5d}$$

$$U_x = \tau_{xy} = 0 \quad \text{on} \quad x = 0 \quad \text{for} \quad 0 \leq y \leq H \tag{5e}$$

Equations (1) together with their boundary conditions (5) and equations (3) together with condition (4) define the problem completely. Bearing in mind that the available experimental data does not supply enough information so as to be compared to numerical findings, the various parameters of the specimen were chosen to enable at least a qualitative comparison with the SEN specimen [5]. Thus, the specimen was assumed to be under plane stress conditions, to be made of PMMA with a Poisson Ratio of $\nu = 0.395$, and to have a height to width ration of W/H = 1.

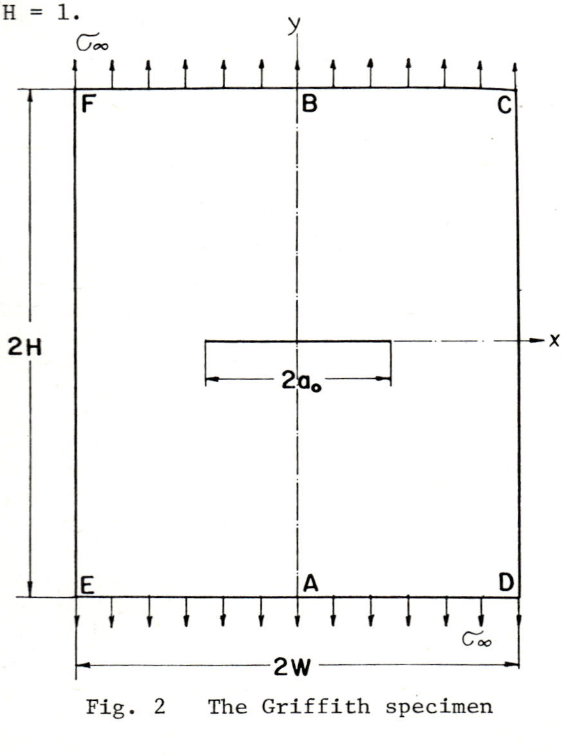

Fig. 2 The Griffith specimen

In order to maintain the same accuracy as in [5] the same mesh size of h=H/75 was employed. Hence, the model consists of 5929 mesh points. The dynamic domain contains 140 mesh points originating from a 10 x 14 net.

THE STATIC CASE

The static problem was solved in order to provide initial conditions for the dynamic case as well as to reconfirm the validity and accuracy of the numerical scheme. The problem was solved for six crack lengths:

$$a_o/W = 0.107,\ 0.2,\ 0.3,\ 0.4,\ 0.6,\ 0.8$$

The static stress intensity factor K_I^{stat} was calculated by means of two independent methods. First, the cleavage stress, prevailing at the nearest mesh point ahead of the crack tip in the moving domain, was used:

$$K_I^{stat} = \sigma_{\eta\eta}\ (h/2,0) \cdot \sqrt{2\pi \cdot (h/2)} \qquad (6)$$

This approximation is based on the assumption that the singular term is dominant in the vicinity of the crack tip. The second method employs the Jin integral and the relation:

$$K_I^{stat} = \sqrt{JinE} \qquad (7)$$

where E is Young modulus. Equation (7) referes to plane stress conditions.

The numerical results, using both methods, are given in Table 1. A comparison showing a very good agreement between the numerical results and the analytical solution by Isida [6] is also presented in Table 1. The static curve is given by a dashed line in Fig. 4.

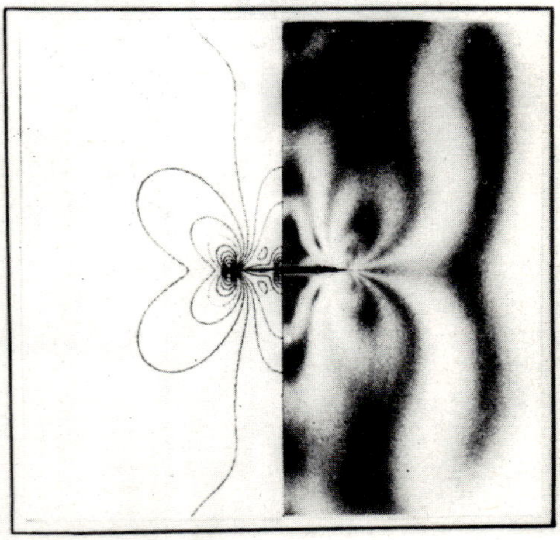

Fig. 3 The numerical principal stress difference field compared with the fringe pattern observed in a circular polariscope

TABLE 1

The Static Results for the Griffith Specimen

	a_o/W	$\sigma_{\eta\eta}(h/2,0)$	Jin	$K_I^{(1)} = \sigma_{\eta\eta}\sqrt{\pi h}$	$K_I^{(2)} = \sqrt{JinE}$	$K_I^{(1)}/K_I^{(2)}$	$K_I^{(2)}/K_I$(Ref.[6])
1	0.107	2.97192	0.32521	0.60825	0.50252	1.21	0.87
2	0.2	4.29907	0.73307	0.87987	0.75423	1.17	0.90
3	0.3	5.64492	1.30144	1.15532	1.00495	1.15	0.92
4	0.4	7.12901	2.10737	1.45906	1.27880	1.14	0.95
5	0.6	11.21192	5.31095	2.29469	2.03010	1.13	1.00
6	0.8	13.60780	7.80716	2.78504	2.46137	1.13	--

For the completeness of the presentation Fig. 3 of ref. [1] is given, showing a good matching between the numerical principal stress difference field and the observed isochromatic fringe pattern.

THE DYNAMIC CASE

In order to enable a comparison between the Griffith and the SEN specimen under dynamic cracking the same procedure used for the SEN specimen as specified in ref. [5] was adopted. Three groups of simulations were run to evaluate the influence of the initial crack length a_o/W and the initial loading, and to study the terminal velocity phenomenon. Only one of the moving crack tips in the Griffith specimen will be hereafter considered.

The Influence of the Initial Crack Length a_o/W

The first three simulations were run with different initial crack lengths of $a_o/W = 0.107, 0.2, 0.3$. The crack velocity functions were identical to those used for the SEN [5] namely, the crack was accelerated to its terminal velocity of $0.5C_R$ while the crack extended to $3.2\ a_o/W$. This approximation is experimentally justified. (See Kobayashi et al. [6]). K_I^{dyn} and G_I^{dyn} as a function of crack length are given in Figs. 4-5, and as a function of crack velocity in Figs. 6-7. The distribution of the strain energy U, the kinetic energy T, the fracture energy F, the external work W and the total energy E are given in Fig. 8.

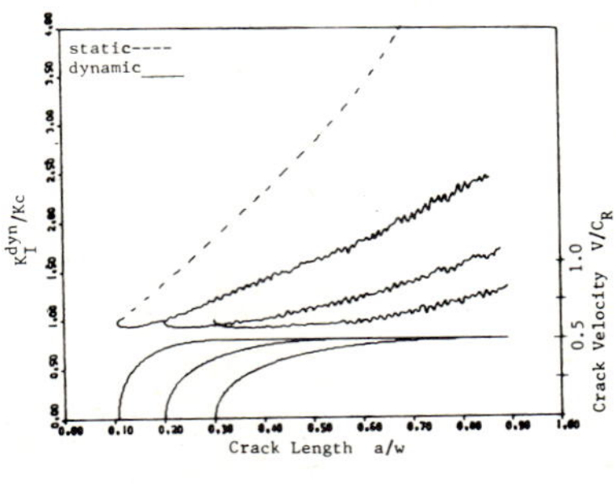

Fig. 4 The normalized dynamic and static stress intensity factors

In view of the above results the following conclusions can be drawn:

a) The smaller the initial crack length a_o/W, the higher K_I^{dyn} and G_I^{dyn} values are encountered. A similar result was reached for the SEN.

b) K_I^{dyn} and G_I^{dyn} are crack-velocity independent up to a velocity of about $0.45C_R$, after which they become highly dependent on this parameter. This is consistent with a similar finding for the SEN specimen as well as with the experimental measurements by Green et al. [7].

c) During most of the crack propagation, apart from the initial zone within which the infinite body effect prevails [8], K_I^{dyn} and G_I^{dyn} are proportional to the remote stress σ_∞ and its square respectively.

d) The total fracture energy F, sunk at the crack tip, is proportional to $\sigma_\infty{}^2$.

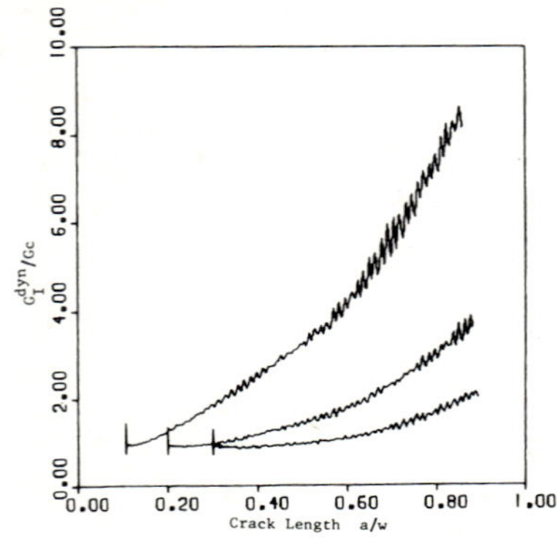

Fig. 5 The normalized dynamic energy release rate

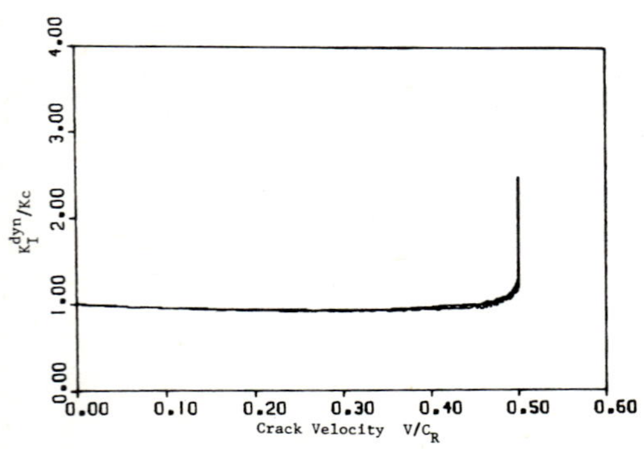

Fig. 6 The dynamic stress intensity factors

e) As in the case of the SEN specimen, K_I^{dyn} and G_I^{dyn} reach a minimum value of 0.9 K_c and 0.86 G_c respectively, at the end of the infinite-body-effect zone.

f) During the initial stage of cracking the energies distributions are similar to those found in the SEN specimen. Once, the first dilatational wave emitted at the crack tip at initiation reaches the loaded surfaces, the external tractions start carrying out the work W on the body. This work, being accumulated by the body changes the other energies distribution. While in the SEN specimen the conversion of strain energy into kinetic and fracture energy continues, in the Griffith specimen the strain energy remains almost constant, and the further demand for the fracture energy as well as the built up of the kinetic energy is supplied by the external work W.

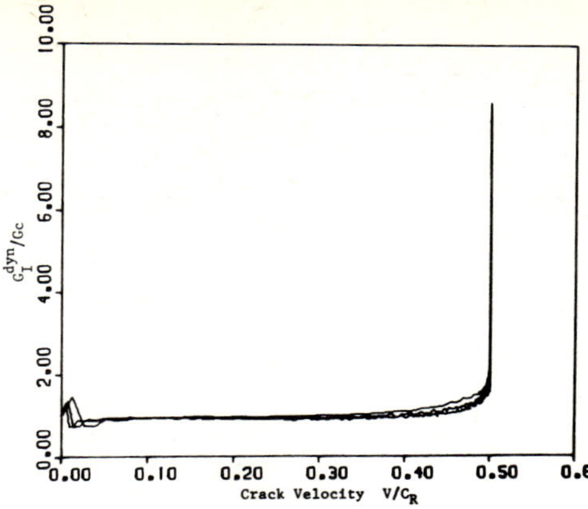

Fig. 7 The energy release rates

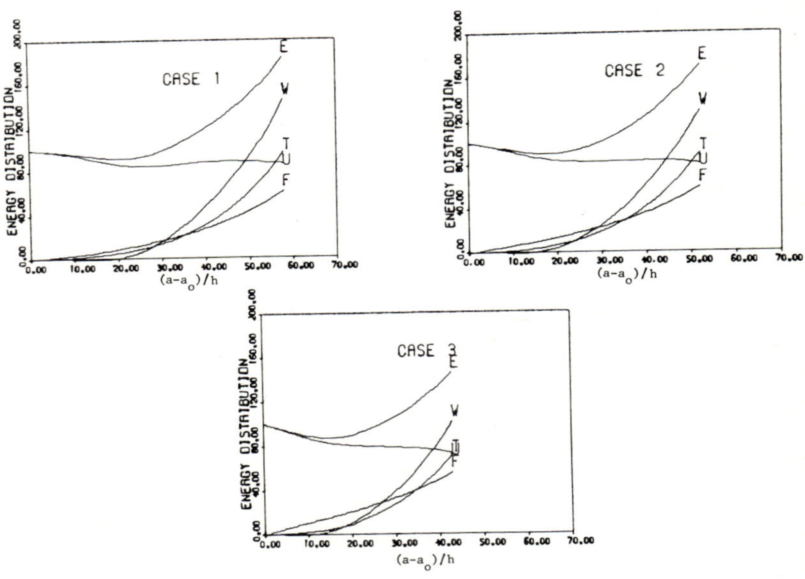

Fig. 8 The energies distribution

The Influence of the Initial Loading K_{Iq}

As for the SEN specimen three simulations were run in order to evaluate the influence of the initial loading K_{Iq}. The different velocity profiles are given in Fig. 9. K_I^{dyn} and G_I^{dyn} as functions of the crack length are given in Figs. 9-10, and as functions of the crack velocities in Figs. 11-12. The various energies distributions for this case are given in Fig. 13. In view of the above results, which exhibit a similar pattern to those for the SEN specimen, the following conclusions can be made:

a) For identical specimens with different bluntnesses, K_I^{dyn} and G_I^{dyn} are proportional to the initial loading K_{Iq} and its square respectively, throughout most of the crack propagation period, apart from the initial zone which is dominated by the infinite-body-effect.

b) The total fracture energy is proportional to K_{Iq}^2, Fig.13.

c) The energies distributions are substantially influenced by the work done by the external tractions-W. (see also conclusion f of the previous section).

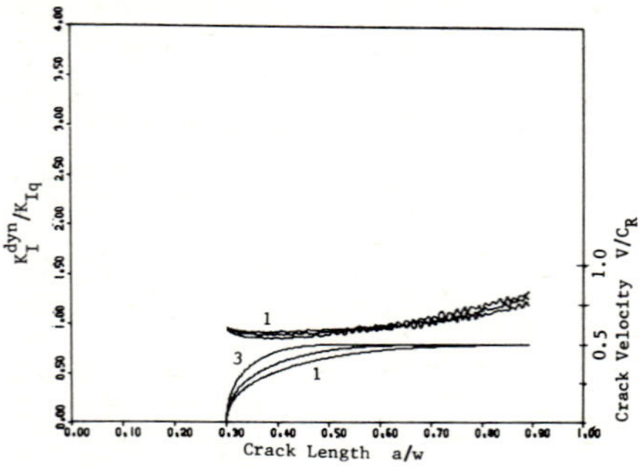

Fig. 9 The dynamic stress intensity factors

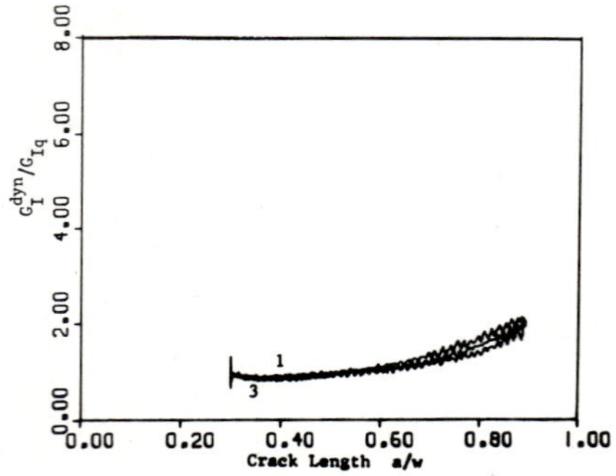

Fig. 10 The dynamic energy release rates

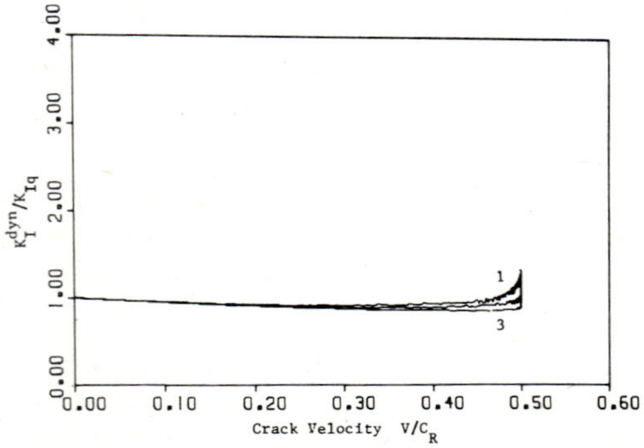

Fig. 11 The dynamic stress intensity factors

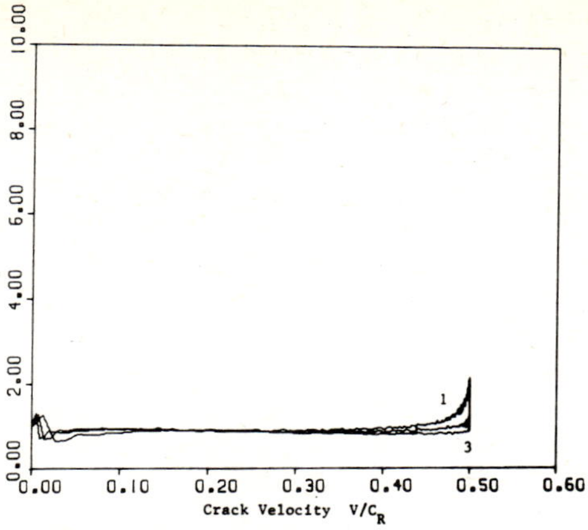

Fig. 12 The dynamic energy release rates

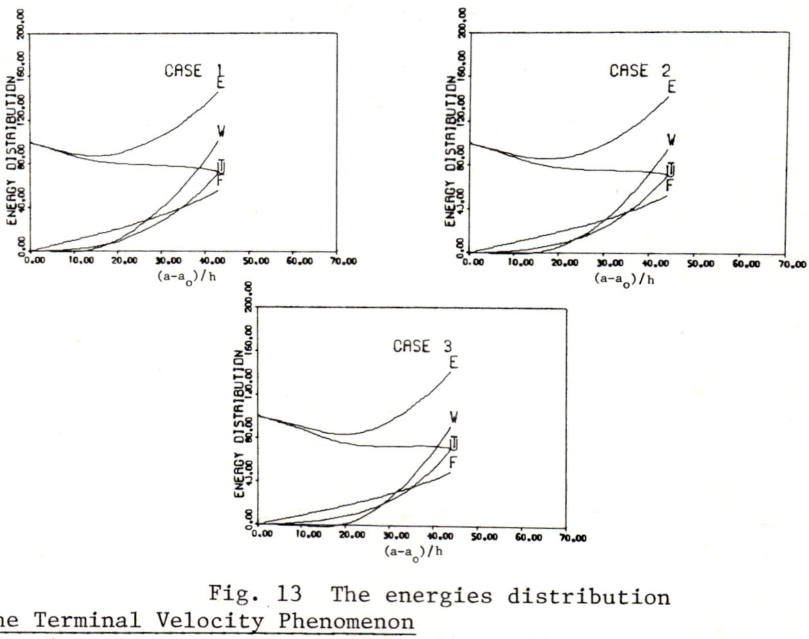

Fig. 13 The energies distribution

<u>The Terminal Velocity Phenomenon</u>

An attempt to improve the understanding of the terminal velocity phenomenon in the SEN specimen has been reported on recently. This analysis [5] led to the conclusion that the terminal velocity is not a universal phenomenon, and that it merely results from the loading conditions which prevailed during the different experiments.

In order to either cooraborate or to reject this conclusion the same attempt was repeated with the Griffith specimen. Four simulations utilizing the Griffith specimen with an initial crack length of a_o/W =0.107, were run. All cracks were assumed to have similar acceleration profiles, but different hypothetical terminal velocities of 0.3 C_R, 0.5 C_R, 0.75 C_R and 1.0 C_R. The velocity functions are given in Fig. 14. K_I^{dyn} and G_I^{dyn} are plotted as function of the crack length in Figs. 14-15, and as a function of crack velocity in Figs. 16-17.

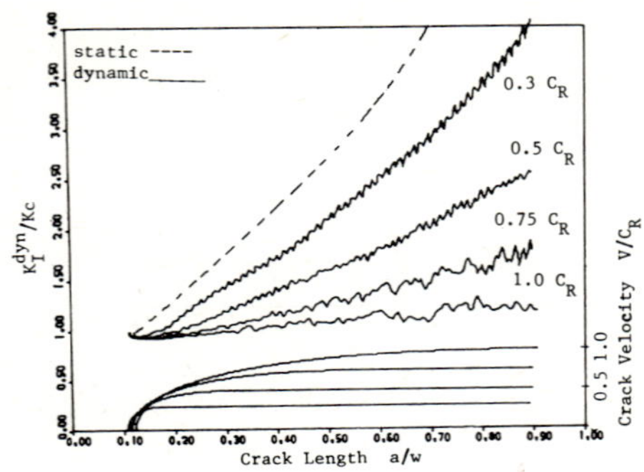

Fig. 14 The dynamic stress intensity factors

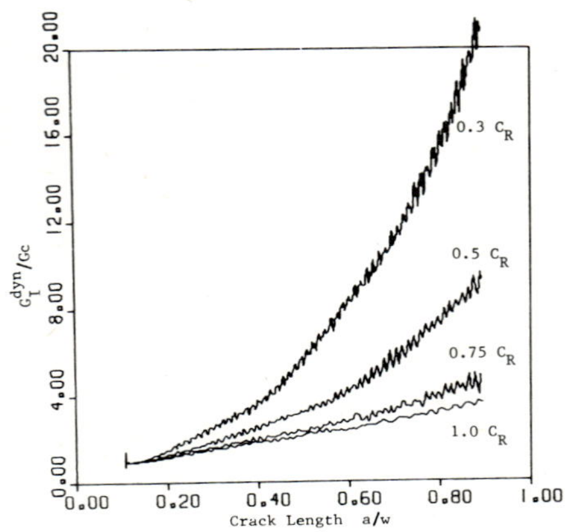

Fig. 15 The dynamic energy release rates

The above results show that in respect to the terminal velocity phenomenon both the Griffith and the SEN specimens exhibit similar behavior bringing to the follwing conclusions:

a) Higher terminal velocity results in lower dynamic stress intensity factor and dynamic energy release rate.

b) In a finite body, contrary to an infinite one [9], energy is supplied to the crack tip even when the crack travels with the Rayleigh wave velocity speed.

c) Since the energy release rate decreases with the terminal velocity but increases with the inital loading K_{Iq}, and since it is known from various experimental works [10] that the material dynamic toughness increases sharply at about 0.5 C_R, the following can be concluded: It is always possible to drive a crack at any velocity up to the Rayleigh wave one provided K_{Iq} is sufficiently increased thus supplying enough energy to meet any demand of the material toughness whatsoever.

d) The above conclusion accounts for the high crack velocities: 0.5-1.0 C_R measured by Cotterell [11] and Green [7] under high impact loading as well as for the fact that most of the other experimentalists, who have used conventional loading means, report a terminal velocity of about 0.4-0.6 C_R.

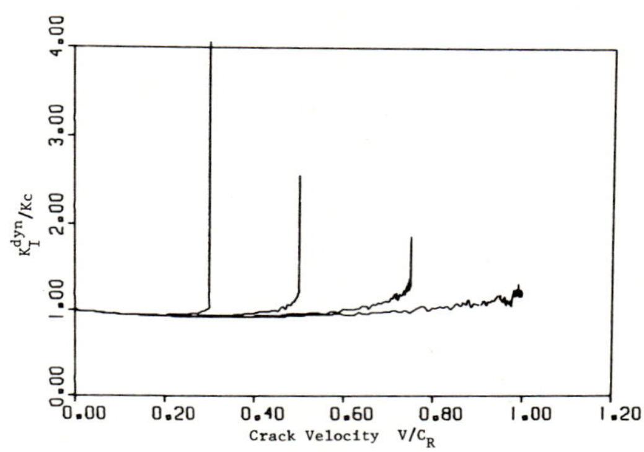

Fig. 16 The dynamic stress intensity factors

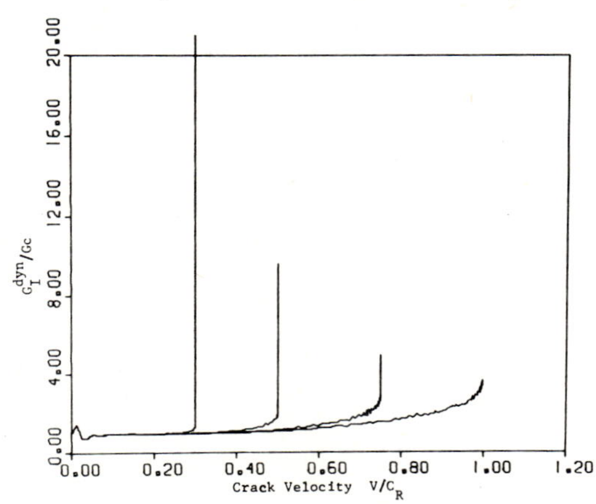

Fig. 17 The dynamic energy release rates

CONCLUDING REMARKS

The results, regarding dynamic crack propagation in the Griffith specimen, were found to be qualitatively similar to the ones previously reported for the SEN specimen:

a) The smaller the initial crack length is the following dynamic parameters are amplified: the crack acceleration, the dynamic stress intensity factor, the energy release rate and the absolute and relative kinetic energy.

b) The dynamic stress intensity factor and the dynamic energy release rate are directly proportional to the inital stress intensity factor K_{Iq}, and its square respectively.

c) The dynamic stress intensity factor and the energy release rate are considerably influenced by both the geometry and the loading conditions, and are not a unique function of the crack velocity. Hence, it seems that these two parameters, in themselves, do not constitute the fracture criterion.

d) It is possible that the terminal velocity phenomenon results from the loading conditions that prevailed in the experiments in which it was measured. It seems that in finite bodies, under high loading conditions this "velocity barrier" might be passed.

Apart from these similarities some differences between the two configurations can be noticed:

e) At initiation the energy stored in half a Griffith specimen is higher than that stored in an identical SEN specimen. This ratio varies from 1.03 - 1.30 for initial crack lengths of a_o/W = 0.3-0.107 respectively.

f) The fracture energy F and the kinetic energy T in the SEN specimen are built due to the initial strain energy, while in the Griffith specimen, their main source is W-the work carried out by the external tractions.

g) The amount of kinetic energy built up in the Griffith specimen is five times higher than the one in the SEN case.

h) The total energy consumed on fracture-F in both specimens is proportional to the strain energy stored in the body at initiation. This reconfirms the fact that F depends on initial loading.

REFERENCES

1. PERL, M., SHMUELY, M. and BETSER, A.A. - On the Dynamic Propagation of a Griffith Crack in a Finite Rectangular Plate. Submitted to Eng. Frac. Mech.

2. SHMUELY, M. and PERL, M. - The SMF2D Code for Proper Simulation of Crack Propagation. Crack Arrest Methodology and Applicatons, ASTM 1978.

3. PERL, M. - Dynamic Crack Propagation Analysis by a Numerical Method. Doctoral Dissertation, Technion-Israel Institute of Technology, 1979.

4. DAY, A.S. - An Introduction to Dynamic Relaxation. The Engineer, 1965, pp. 218-221.

5. PERL, M. - Dynamic Crack Propagations Analysis of the SEN Specimen by a Numerical Method, in preparation.

6. KOBAYASHI, A.S. and ENGSTROM,W.L. - Transient Analysis in a Fracturing Aluminum Plate. Office of Naval Research, NR 064-478-(3A), 1967.

7. GREEN, A.K. and PRATT, P.L. - Measurement of Dynamic Fracture Toughness of PMMA by High Speed Photography. Eng. Frac. Mech., Vol. 6, p. 71, 1974.

8. PERL, M. - The Infinite Body Effect in Dynamic Fracture Mechanics. Submitted to J. Appl. Mech.

9. FREUND, L.B. - Crack Propagation in an Elastic Solid Subjected to General Loading - II. Non Uniform Rate of Extension. J. Mech. Phys. Sol., Vol. 20, pp. 141-152, 1972.

10. FREUND, L.B. - Dynamic Crack Propagation, The Mechanics of Fracture, ed. F. Erdogan, ASME AMD - Vol. 19, pp. 105-139, 1976.

11. COTTERELL, B. - Velocity Effect in Fracture Propagation, Apl. Mech. Revs., Vol. 4, p. 227, 1965.

INFLUENCE OF SMALL-SCALE YIELDING ON DYNAMIC FRACTURE

A.S. Kobayashi, M. Ramulu and B.S-J Kang
Professor Graduate Student Graduate Student
University of Washington
Department of Mechanical Engineering
Seattle, Washington 98195, USA

SUMMARY

Dynamic fracture and crack-arrest responses of the well-discussed wedge-loaded rectangular, double cantilever beam (WL-RDCB) specimens machined from a 3.2 mm thick annealed polycarbonate sheet were studied by dynamic photoelasticity and dynamic finite element analysis. A small, but finite plastic yield zone surrounded the propagating crack which resulted in crack velocities, dynamic fracture toughness and crack arrest toughness lower than those observed previously in 6.4 mm thick polycarbonate fracture specimens.

INTRODUCTION

Current experimental and numerical investigations in dynamic fracture involve extensive use of dynamic photoelasticity [1,2], dynamic caustics [3], dynamic finite difference analysis [4,5], and dynamic finite element analysis [6,7]. The results published in the above are based on the theory of elasto-dynamics and have evoked a controversy regarding the existence of a material property, i.e., a unique dynamic fracture toughness, K_{ID}, versus crack velocity relation, $\dot{a}$, as a material property which governs dynamic crack propagation and arrest [1,2,6,8]. Despite this unresolved controversy on elastodynamic fracture, the increasing research efforts in this field should eventually lead to a phenomenological dynamic fracture criterion which can be used in either finite difference or finite element codes to predict crack propagation histories in actual structures. For example, recent numerical simulation of dynamic fracture in a thermally shocked structure [9] correlated well with experimental results [10] and undoubtedly such practical applications will increase as the controversies in elastodynamic fracture are resolved.

Unlike the efforts in elastodynamic fracture, little basic research in dynamic ductile fracture, particularly in the presence of large scale yielding has been conducted to date. Earlier investigations on dynamic ductile fracture, including those of Kanninen et al [11], Hahn et al [12] and Ogasawara et al [13] as well as the recent papers by Koshiga et al [14] and Nora et al [15] involved phenomenological modelings of specific experimental observations. While these results are useful in predicting dynamic ductile fracture responses in specific structures, they fail to address the basic laws governing dynamic ductile fracture.

One major difficulty in dynamic ductile fracture research is the lack of suitable experimental technique with which the dynamic states of stress, strains, plastic yield zones, etc. can be measured directly for detailed scrutiny. In contrast, dynamic photoelasticity and dynamic caustics provide optical imaging from which the dynamic stress intensity, K_I^{dyn}, among others can be extracted for elastodynamic fracture studies. Lacking a pure experimental approach, one method which can be used is the combined experimental-numerical procedure for computing dynamic fracture parameters. In this procedure, experimental results such as crack position versus time in a fracturing specimen, is used to drive a crack in a finite element model of the fracture specimen. The numerical results, such as the equivalent dynamic stress intensity factor for elastic analysis, dynamic J-integral and crack opening displacement, COD, among others can then be studied as candidate dynamic ductile fracture criteria. The soundness of this procedure when used in elastic analysis has been verified by one of the authors [16, 2]. It was then used in a series of sensitivity studies on the influence of various dynamic fracture toughness, K_{ID}, versus crack velocity, $\dot{a}$, relations [17,18]. An obvious extension of this elastic procedure is its use in elasto-plastic analysis.

This paper reports some preliminary results obtained in the course of developing an experimental-numerical procedure for determining dynamic ductile fracture parameters of a propagating crack tip. As an exploratory study, it is limited to small scale yielding where the dynamic stress intensity factor with a plasticity correction factor results in an apparent dynamic fracture toughness which could be used to characterize the dynamic ductile fracture.

EXPERIMENTAL ANALYSIS

The concurrent experimental analysis in this investigation consisted of dynamic photoelasticity using thin, i.e., 3.2 mm thick, polycarbonate wedge-loaded, rectangular double cantilever beam (WL-RCB) specimens which is shown by Figure 1. The annealed polycarbonate, from which the specimens were machined,

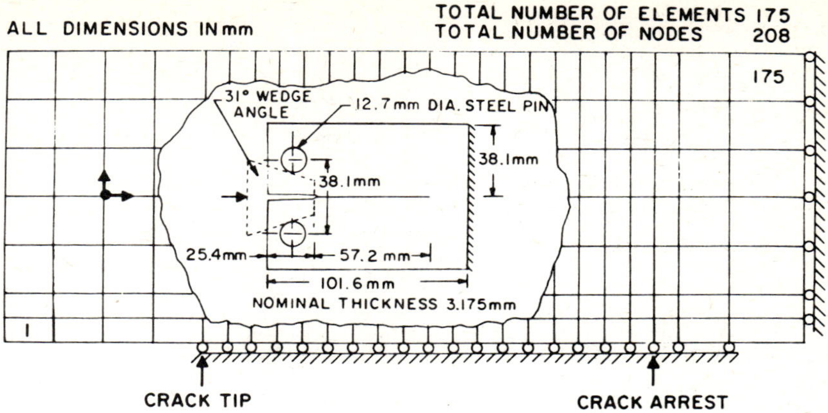

Figure 1. Finite Element Breakdown of WL-DCB Specimen

has a well-defined yield point of 44 MPa, flows under high stresses, and exhibits tensile instability with accompanying Lueder's bands. Under high strain rate loading, such as in the vicinity of a propagating crack tip, the strain-rate sensitive polycarbonate fractures in a cleavage mode without the large plastic zone ahead of the propagating crack tip. Such ductile to brittle transition in dynamic loading of polycarbonate is discussed in Reference [19]. Despite this cleavage fracture appearance, the dynamic photoelastic patterns in Figure 2(a) and 2(b) show that the propagating crack tip is preceded by a small yield zone of 1.2 mm and 0.4 mm diameters, respectively,

(a) 1ST FRAME 7 μ SECONDS (b) 7TH FRAME 71 μ SECONDS

Figure 2. Typical Dynamic Photoelastic Patterns in Polycarbonate WL-DCB Specimen No. K031380-P.

and thus, represent dynamic fracture in the presence of small scale yielding. The elastic region surrounding this small yield zone also provides the elasto-dynamic state which can be used to extract apparent elastic fracture parameter for characterizing such dynamic ductile fracture. For example, the isochromatics form the standard loops from which apparent dynamic stress intensity factor can be extracted through the use of well-known elasto-dynamic near-field solutions [20]. Possible errors involved in characterizing the elasto-plastic dynamic field with such straight forward application of elastic analysis will be discussed later. The 16-spark gap Cranz-Shardin camera and the associated dynamic photoelastic systems used to record 16 discrete dynamic photoelastic patterns in the fracturing polycarbonate WL-RDCB specimen has been described in many previous papers and thus will not be repeated here.

In the as-received condition, the polycarbonate sheet exhibits considerable residual stress distribution and thus the sheets were annealed after rough cutting. This annealing which caused some distortion and shrinkage consisted of overnight heating at 160°C, followed by gradual cooling at the rate of 5°C per hour. The starter crack consisted of 0.4 mm wide edge crack approximately 25 mm length which was machined and then chiseled to simulate a somewhat blunt crack tip. Static and dynamic stress-fringe constant and Poisson's ratio at various strain rates, all of which were determined previously [21], were used in data reductions. Static modulus of elasticity of E = 2.00 GPa and the yield strength of 44 MPa were determined separately for this particular polycarbonate sheet.

CRACK TIP ISOCHROMATICS IN THE PRESENCE OF SMALL SCALE YIELDING

The influence of small scale yielding on the dynamic isochromatics around the crack tip was studied by a static Dugdale strip yield model (DM) as shown in Figure 3. Since the Westergaard's stress function for a partially pressurized semi-infinite crack tip is known [22], the crack tip isochromatics, τ_m, for a stationary crack with a Dugdale strip yield of length r_y, can be expressed as

$$\tau_m = \frac{1}{2}[(\sigma_{xx} - \sigma_{yy})^2 + 4\tau_{xy}^2]^{\frac{1}{2}} \tag{1a}$$

where

$$\sigma_{xx} - \sigma_{yy} = K_I \frac{\sin\theta_{DM}}{\sqrt{2\pi r_{DM}}} \frac{\frac{r_y}{r_{DM}} \sin\frac{\theta_{DM}}{2} + \sin\frac{3\theta_{DM}}{2}}{(\frac{b}{r_{DM}})^2 + 1 + 2\frac{r_y}{r_{DM}}\cos\theta_{DM}} \tag{1b}$$

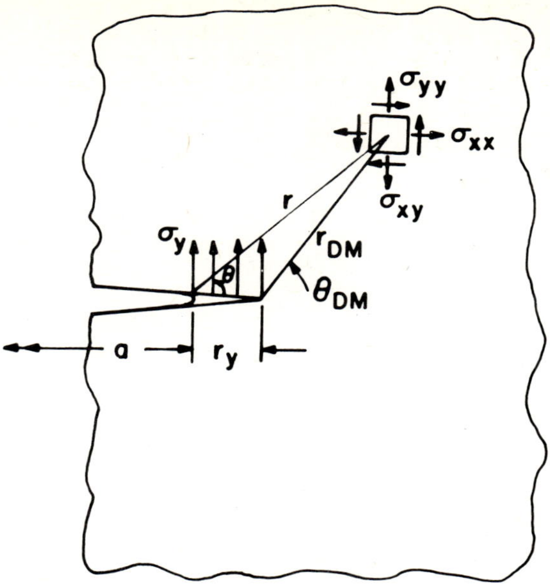

Figure 3. Dugdale Strip Yield Zone at the Crack Tip of a Semi-Infinite Crack.

$$2 \; {}_{xy} = -K_I \frac{\sin\Theta_{DM}}{\sqrt{2\pi r_{DM}}} \frac{\frac{r_y}{r_{DM}} \cos\frac{\Theta_{DM}}{2} + \cos\frac{3\Theta_{DM}}{2}}{\left(\frac{r_y}{r_{DM}}\right)^2 + 1 + 2\frac{r_y}{r_{DM}}\cos\Theta_{DM}} \qquad (1c)$$

where r_{DM} and Θ_{DM} are the polar coordinates with origin at the DM crack tip in Figure 3. The stress intensity factor, K_I, corresponds to the negative of the stress intensity factor at the DM crack tip without the Dugdale strip yield zone. From this equality, the length of the DM zone becomes

$$r_y = \frac{\pi}{8}\left(\frac{K_I}{\sigma_y}\right)^2 \qquad (2)$$

where σ_y is the yield strength of the material.

The resultant crack tip isochromatics in the presence of a DM zone can be obtained numerically for the highly loaded crack tip at the onset of crack propagation. For estimated stress intensity factor of K_I = 2.44 MPa$\sqrt{m}$, yield strength of σ_y = 44 MPa and an isochromatic fringe order of 7-1/2 or τ_{max}=84.8 MPa.

in the 3.2 mm WL-RDCB specimen. The DM strip yield length is r_y = 1.3 mm. Figure 4 shows the estimated crack tip isochromatics with the DM zone and the experimentally observed isochromatics of Figure 2(a). The significant difference between the estimated and experimental isochromatics indicate the amount of plasticity correction necessary to increase the elastic K_I in the presence of small scale yielding.

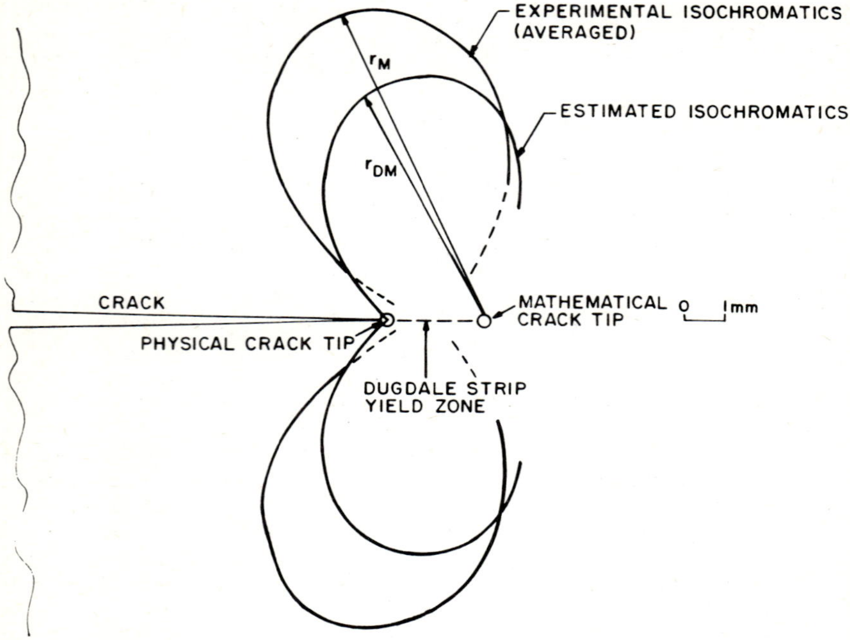

Figure 4. Experimental and Estimated Isochromatics.

The procedure then for estimating the apparent stress intensity factor from the isochromatics in the presence of small scale yielding is to first estimate the location of the far end of the crack tip yield zone or the DM crack tip location in the photoelastic pattern. An overdeterministic least square method [23] is then used to fit at the DM crack tip a two parameter, elastic crack tip stress field to the experimentally recorded isochromatics. Using the elastic stress intensity factor thus computed, the DM strip yield length r_y is determined by equation (2) and the estimated isochromatics is reconstructed numerically by the use of equation (1). Should the computed r_y be substantially different from the measured r_y, then r_y must be remeasured and the above numerical procedure is repeated. Fortunately, such repeated measurement was not necessary in the test analyzed for this paper. The ratio of the maximum radial distances, r_M and r_{DM} in Figure 4, is the plasticity correction factor. This procedure must be repeated until the succeeding plasticity correction factor approaches unity. However, in view of the approximate nature of the entire procedure, this iteration procedure was not programmed into this data reduction scheme.

The above procedure is based on static crack tip yielding and obviously must be modified for dynamic analysis. It, however, provides a first order estimate of the plasticity correction involved in deducing the apparent dynamic stress intensity factor from the dynamic isochromatics in the presence of small-scale yielding.

DYNAMIC FINITE ELEMENT ANALYSIS

The dynamic finite element algorithm used in this study differs with that presented two years ago [24]. The new fracture package [2] includes a mechanism for gradual release in crack tip nodal force as well as an iteration routine to assure agreement between the prescribed and computed crack tip nodal forces in this explicit code. Figure 1 shows the finite element breakdown used in this plane stress analysis which is considered more appropriate for these thin fracture specimens, despite the cleavage fracture surfaces. The dynamic finite element code was executed in its elastic-plastic mode. The coarseness of finite element breakdown, i.e., 3.2 mm square in comparison to the observed maximum plastic yield zone of 1 mm, however, forced the actual computation into its elastic mode except for the initial 1~2 microseconds in crack propagation time, As a result, the dynamic stress intensity factor, K_I^{dyn}, was computed elastically using the dissipated energy at the released node with Freund's equation [25].

RESULTS OF DYNAMIC PHOTOELASTICITY

A total of seven dynamic photoelastic tests were conducted. Figure 2 shows two typical dynamic photoelastic patterns of a fracturing polycarbonate WL-RDCB specimen. These dynamic photoelastic patterns show crack tip shadow patterns which are believed to be caustics [3] generated by the large plastic strains in the confined plastic region ahead of the propagating crack tip. The size of such a plastic zone ranges from r_y = 1.3 mm to 0.04 mm. Minor distortions of the otherwise elastic isodhromatics in the vicinity of this plastic region is also noted.

Apparent dynamic stress intensity factor, K_I^{dyn}, was computed by using the data reduction proecdure described previously. Figure 5 shows the apparent dynamic fracture toughness, which were obtained by this procedure, during dynamic fracture and crack arrest in a polycarbonate WL-DCB specimen. The general variation of these experimental results differs slightly with those obtained by others [3,7]. The apparent dynamic

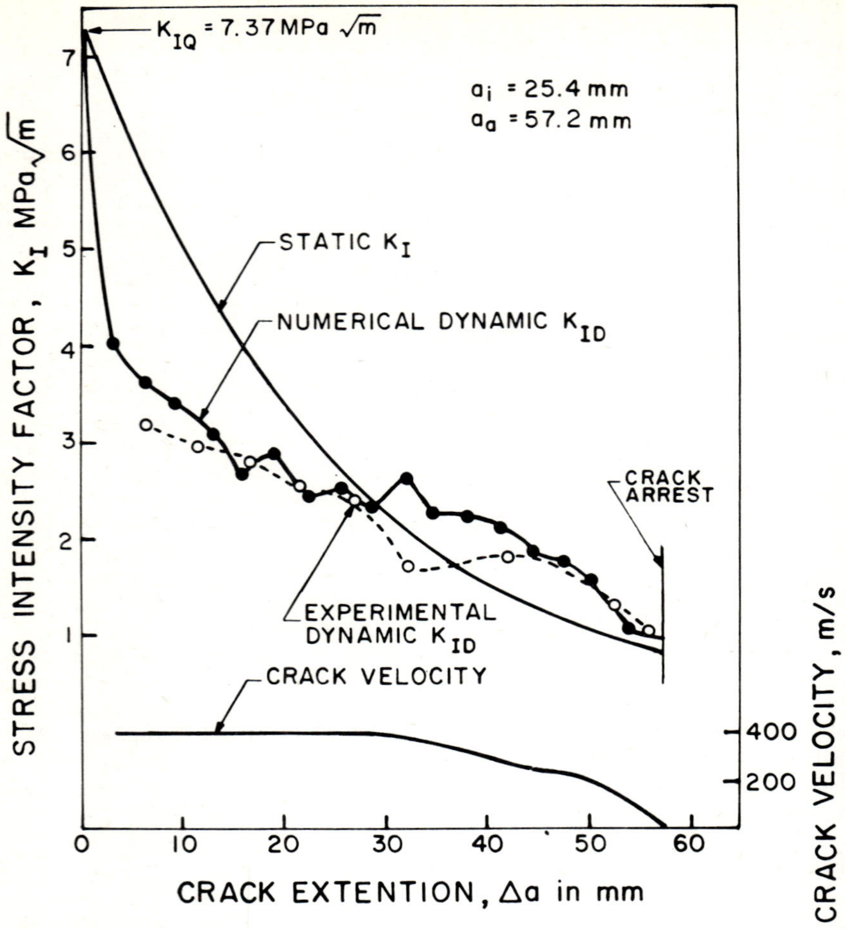

Figure 5. Stress Intensity Factors of a Fracturing Poly-Carbonate WL-DCB Specimen No. K031380-P.

fracture toughness continues to decrease in the former while distinct regions of constant K_{ID} after crack propagation were observed in the latter. The crack initiation stress intensity factor of K_{IQ} = 7.37 MPa$\sqrt{m}$ is more than twice of the pop-in fracture toughness of K_{IC} = 3.4 MPa$\sqrt{m}$ measured previously [21]. The measured apparent crack arrest fracture toughness of K_{Ia} = 1.0 MPa$\sqrt{m}$, however, is considerably lower than this pop-in fracture toughness and lower than the K_{Ia} = 1.4 MPa$\sqrt{m}$ reported for the 6.4 mm polycarbonate fracture specimens [2].

Also shown in Figure 5 is the crack velocity change with crack extension in this WL-RDCB specimen. This crack velocity history was duplicated in the other six specimens and are similar in shape to those reported by others [3,26] but the terminal velocity is lower than that in Reference [2].

RESULTS OF DYNAMIC FINITE ELEMENT ANALYSIS

Figure 5 shows the calculated dynamic stress intensity factor obtained through elasto-dynamic finite element analysis. As mentioned previously, the finite element analysis was executed in its elastic-plastic mode, but elastic analysis prevailed except for the initial 1~2 microseconds of crack propagation. Agreements between calculated dynamic stress intensity factors and the measured dynamic fracture toughness during the crack propagation history are apparent. The precipitous drop in calculated dynamic stress intensity factor at the initial phase of crack propagation can be in part attributed to the infinite plate syndrome discussed by Perl [27]. It was also found that this precipitous drop in dynamic stress intensity factor could be moderated by assigning a gradually increasing crack velocity during the first 30-50 microseconds of crack propagation following Reference [27]. Recent experimental results [20,28,29], however, show that the crack velocity of a blunt starter crack immediately reaches the terminal velocity upon propagation and does not gradually approach the terminal velocity as postulated in Reference [27] during this crucial initial period of crack propagation. These experimental evidences thus precluded adjustments of the crack velocity to reduce the precipitous drop in K_I^{dyn}.

Figure 6 shows the energy partitions during crack propagation in the WL-RDCB specimens discussed previously. Although the dissipated plastic energy at the plastic yield zone was not computed in this coarse finite element model, the decreasing total energy in Figure 6 indicated qualitatively, the small plastic energy dissipated in the fracture process of polycarbonate WL-RDCB specimen.

CONCLUSIONS

1. Small-scale yielding at the propagating crack tip does change the shape of the crack-tip dynamic isochromatics.

2. Crack velocity, dynamic fracture toughness and crack arrest toughness are lower in the presence of small-scale yielding at the propagating crack tip.

3. Dynamic stress intensity factor obtained by elasto-dynamic finite element analysis are in good agreement with the apparent dynamic fracture toughness obtained through an elasto-plastic data reduction of the dynamic isochromatics in the presence of small scale yielding. Elasto-plastic dynamic finite element analysis, does not appear necessary for such dynamic fracture problems.

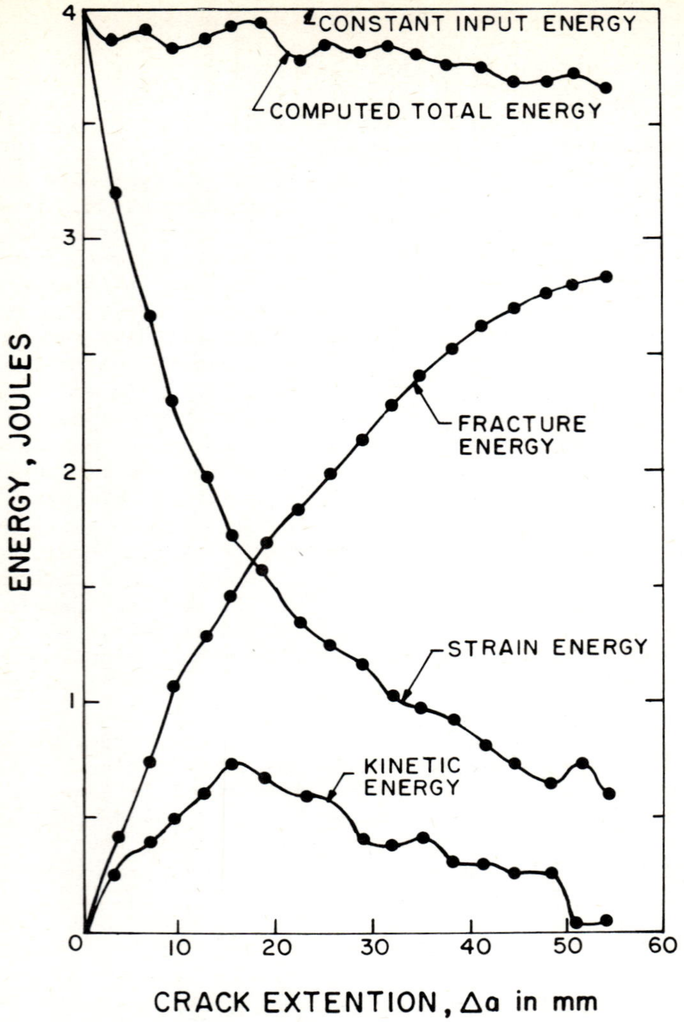

Figure 6. Energy of a Fracturing Polycarbonate WL-DCB Specimen No. K031380-P.

4. The decreasing total energy indicate the influence of dissipated plastic energy in the presence of a small plastic zone.

ACKNOWLEDGEMENTS

The results of this investigation were obtained in a research contract funded by the Office of Naval Research under Contract No. N0014-76-C-0600 NR 064-478. The authors wish to acknowledge the support and encouragement of Dr. N. Perrone of ONR during the course of this investigation.

1. DALLY, J.W., "Dynamic Photoelastic Studies of Fracture," Experimental Mechanics, No. 1, 19, No. 10, pp. 349-367, October 1979.

2. KOBAYASHI, A.S., SEO, K., JOU, J.Y. and URABE, U., "A Dynamic Analyses of Modified Compact-Tension Specimens Using Homalite-100 and Polycarbonate Plates," Experimental Mechanics, Vol. 20, No. 3, pp. 73-79, March 1980.

3. KALTHOFF, J.F., BEINERT, J. and WINKLER, S., "Measurements of Dynamic Stress Intensity Factors for Fast Running and Arresting Cracks in Double-Cantilever-Beam Specimen," Fast Fracture and Crack Arrest, ASTM STP 627, pp. 161-176, July 1977.

4. SCHMUELY, M., and PERL, M., "The SMF2D Code for Proper Simulation of Crack Propagation," to be published in Crack Arrest Methodology and Applications, ASTM STP.

5. POPELAR, C.H. and GEHLEN, P.C., "Modeling of Dynamic Crack Propagation II Validation of Two-Dimensional Analysis," International Journal of Fracture, Vol. 15, pp. 159-178, 1979.

6. KOBAYASHI, A.S., "Dynamic Fracture Analysis by Dynamic Finite Element Method-Generation and Propagation Analysis," Nonlinear and Dynamic Fracture Mechanics, edited by N. Perrone and S.N. Atluri, ASME, pp. 19-36, 1979.

7. ATLURI, S.N., NISHIOKA, T. and NAKAGAKI, M., "Numerical Modeling of Dynamic and Nonlinear Crack Propagation in Finite Bodies, by Moving Singularities," ibid loc cit, pp. 37-66.

8. KANNINEN, M.F., GEHLEN, P.C., BARNES, C.P., HOAGLAND, R.G., HAHN, G.T. and POPELAR, C.H., "Dynamic Crack Propgation Under Impact Loading," ibid loc cit, pp. 195-200.

9. HAHN, G.T., HOAGLAND, R.G., GEHLEN, R.C. and ROSENFIELD, A.R., "Analysis of Crack Arrest in Reactor Pressure Vessels," ASME Paper No. 78-MAT-16.

10. CHEVERTON, R.D. and ISKANDER, S.K., "Application of Static and Dynamic Crack Arrest Theory to Thermal Shock Experiment TSE-4," NUREG/CR-0767, ORNL/NUREG-57, June 1979.

11. KANNINEN, M.F., MUKHERJEE, A.K., ROSENFIELD, A.R. and HAHN, G.T., "The Speed of Ductile-Crack Propagation and the Dynamics of Flaw in Metals," Mechanical Behavior of Materials Under Dynamic Loads (ed. by U.S. Lindholm), Springer-Verlag, pp. 96, 1968.

12. HAHN, G.T., SARATE, M., KANNINEN, M.F. and ROSENFIELD, A.R. "A Model for Unstable Shear Crack Propagation in Pipes Containing Gase Pressures," International Journal of Fracture, Vol. 9, pp. 209-222, June 1973.

13. OGASAWARA, M., MIMURA, M., and ISHIGURO, K., "Laboratory Test for Unstable Ductile Fracture," Proceedings of International Symposium on Crack Propagation in Pipelines, Newcastel-Upon-Tyne, March 1974.

14. KOSHIGA, F., KURITA, Y., AKIYAMA, T. and KAWAHARA, M., "Shear Fracture Propagation by DCB Testing," ASME Journal of Pressure Vessel Technology, Vol. 100, pp. 18-23, February 1980.

15. NORA, Y., FUKUDA, M., NOGAKI, N., KOGA, T., and TAKEUCHI, I., "Study on Resisting of Various Types of Steels Against Propagating Shear Fracture by Modified West Jefferson Type Burst Test," ASME Preprint 78-PVP-71, 1980.

16. KOBAYASHI, A.S., "Experimental-Numerical Analysis of a Running Crack," Recent Research on Mechanical Behavior of Solids, University of Tokyo Press, pp. 291-313, 1979.

17. HODULAK, L., KOBAYASHI, A.S., EMERY, A.F., "Influence of Dynamic Fracture Toughness on Dynamic Crack Propagation," to be published in Engineering Fracture Mechanics.

18. HODULAK, L., KOBAYASHI, A.S., EMERY, A.F., "A Critical Examination of a Numerical Fracture Dynamic Code," to be published in Fracture Mechanics (12th) ASTM STP.

19. YEE, A.F., "The Yield and Deformation Behavior of Some Polycarbonate Blend," General Electric Technical Information Series, 76 CRD 130, August 1976.

20. KOBAYASHI, A.S., and MALL, S., "Dynamic Fracture Toughness of Homalite-100, Experimental Mechanics, Vol. 18, No.1, pp. 11-18, January 1978.

21. MALL, S., and KOBAYASHI, A.S., "Dynamic Photoelastic and Dynamic Finite Element Analyses of Polycarbonate Dynamic Tear Specimens," Fracture Mechanics edited by C.W. Smith, ASTM STP 677, pp. 498-510, 1979.

23. TADA, H., PARIS, P.C. and IRWIN, G.R., The Stress Analysis of Cracks Handbook, Del Research Corporation, Hellertown, PA, p. 37. 1973.

24. KOBAYASHI, A.S. and RAMULU, M., "Dynamic Stress Intensity Factors for Unsymmetric Dynamic Isochromatics," to be published in Experimental Mechanics.

24. KOBAYASHI, A.S., MALL, S., URABE, Y. and EMERY, A.F., "A Numerical Dynamic Fracture Analysis of Three-Wedge-Loaded DCB Specimens," Proceeding of the International Conference on Numerical Analysis in Fracture Mechanics, University College of Swansea, pp. 673-684, January 9-13, 1978.

25. FREUND, L.B., "Crack Propagation in an Elastic Solid Subjected to General Loading I, Constant Rate of Extension," Journal of Mechanics and Physics of Solids, Vol. 20, pp. 129-140, 1972.

26. KOBAYASHI, T., DALLY, J.W. and FOURNEY, W.L., "Influence of Specimen Geometry on Crack Propagation and Arrest Behavior," Proceedings of the 6th International Conference on Experimental Stress Analysis, Munchen, 1978.

27. PERL, M., "An Improved Dynamic Analysis of an Initiating and Fast-running Griffith Crack in a Finite Rectangular Plate," this conference proceeding.

28. MALL, S., KOBAYASHI, A.S. and URABE, Y., "Dynamic Photoelastic and Dynamic Finite-Element Analyses of Dynamic-Tear-Test Specimens," Experimental Mechanics, Vol. 18, No. 12, pp. 449-456, December 1978.

29. DALLY, J.W. and SHUKLA, A., "Dynamic Crack Behavior at Initiation," Mechanics Research Communication, Vol. 6 (4), pp. 239-244, 1979.

FINITE ELEMENT ANALYSIS OF STATIONARY CRACKS IN TIME DEPENDENT STRESS FIELDS

S. MALL*

SUMMARY

The effect of inertia force is taken into account to derive a modified J-integral for computing the transient stress intensity factor for elastodynamic problems with stationary cracks. The application of the modified J-integral is demonstrated with two problems. The first problem is a theoretical problem which involves a centrally cracked rectangular strip subjected to suddenly applied and maintained tension. The second problem is a practical problem which involves a 3-point bend specimen subjected to the drop weight impact. The results obtained with the modified J-integral are in excellent agreement with those obtained with other procedures (numerical and experimental).

INTRODUCTION

The problem of a stationary crack in a finite elastic body, subjected to transient loading, has been analyzed by several investigators with the finite element and finite difference methods [1-8]. Aberson et al [5], Glazik [6] and, Kelly and Sun [7] have solved various stationary crack elastodynamic problems with the help of special singular crack tip elements. The time dependent stress intensity factors were determined directly from these singular element formulations. Also, an alternate and simple singular element, i.e. quarter-point triangular element at the crack tip, was utilized to analyze this class of problems [1]. On the other hand, conventional elements at the crack tip have been also successfully employed for dynamic fracture analyses of various

* Assistant Professor, Department of Mechanical Engineering, University of Maine, Orono, ME 04469, USA.

notched bend specimens in the drop weight impact testing mode [2-4]. The dynamic stress intensity factor was determined with an indirect procedure in these finite element analyses with conventional elements as well as with quarter-point singular element [1-4]. This procedure involved the comparison between the dynamic crack opening displacement (COD) of the node close to the crack tip to its static counterpart obtained from the same finite element mesh. The ratio of dynamic and static COD is also the ratio of dynamic stress intensity factor to the static value. This procedure yielded results (with comparatively coarse mesh) which were in excellent agreement with their counterparts obtained either experimentally or from previous numerical analyses [1-4]. This COD method is, however, an indirect procedure which requires the exact static solution of the problem (which is commonly available for many problems), and further applicable only for cases where crack is subjected to a single mode loading.

As an alternative, transient stress intensity factors for dynamic crack problems can be directly evaluated through the near field solution (e.g. COD) which obviously would require fine mesh near the crack tip and thus high computation cost. While, in case of finite element analysis of static crack problems, stress intensity factor is commonly evaluated either from J-integral or strain energy release rate with a comparatively coarse mesh. The energy release rate procedure cannot be employed in elastodynamic problems with stationary crack, since there is no dissipation of energy unlike the rapidly propagating crack [9]. The conventional J-integral can be, however, modified to take into the account of inertia force present in these problems. The objective of the present paper is to show this modification of the conventional J-integral to evaluate the dynamic stress intensity factor of a stationary crack subjected to dynamic loading. The application of the modified J-integral is demonstrated with two problems.

J-INTEGRAL FOR ELASTODYNAMIC STATIONARY CRACK PROBLEMS

The modification in the conventional J-integral to take into the account of inertia force, which is the case in elastodynamic problems with stationary crack, is presented here. This modification will follow the same methodology as employed by Wilson and Yu [10] for the thermal stress crack problems. The conventional J-integral will be, at first, evaluated over any closed path Γ in the presence of inertia force. The following analysis is restricted to a plane, homogeneous and elastic body subjected to plane loading. The conventional J-integral having a line crack along the x_1-axis is defined as:

$$J = \int_{\Gamma} [W\, dx_2 - \sigma_{ij}\, n_j \frac{\partial u_i}{\partial x_1} ds] \quad (i, j = 1, 2) \quad \text{-----(1)}$$

where W is the strain energy density, u_i and σ_{ij} are displacement and stress components, respectively. Using Gauss' theorem, Eq. (1) can be transformed into an area integral:

$$J = \int_A \left[\frac{\partial W}{\partial x_1} - \frac{\partial}{\partial x_j}\left(\sigma_{ij}\frac{\partial u_i}{\partial x_1}\right)\right] dA \qquad \text{---------(2)}$$

where A is the area enclosed by the path Γ. Also,

$$\frac{\partial W}{\partial x_1} = \frac{\partial W}{\partial \varepsilon_{ij}}\frac{\partial \varepsilon_{ij}}{\partial x_1} = \sigma_{ij}\left[\frac{\partial}{\partial x_j}\left(\frac{\partial u_i}{\partial x_1}\right)\right] \qquad \text{---------(3)}$$

Combining Eqs. (2) and (3),

$$J = -\int_A \left(\frac{\partial u_i}{\partial x_1}\right)\left(\frac{\partial \sigma_{ij}}{\partial x_j}\right) dA \qquad \text{---------(4)}$$

Substituting the equilibrium equations, i.e. $\sigma_{ij,j} = \rho\ddot{u}_i$, in the presence of inertia force; the conventional J-integral, Eq. (4), can be expressed as the following:

$$J = -\int_A \rho\ddot{u}_i \left(\frac{\partial u_i}{\partial x_1}\right) dA \qquad \text{---------(5)}$$

When the inertia force is zero, i.e. static case, the J-integral over any closed path is equal to zero (since integrad of Eq. (4) or (5) will be equal to zero). Here ρ is the mass density.

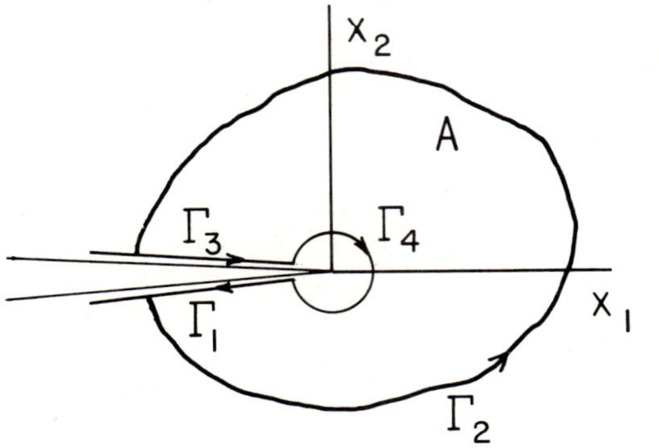

Figure 1. A Path for J-integral at the Crack Tip

Eq. (5) can be utilized to compute the stress intensity factor in case of elastodynamic stationary crack problems in the following manner. Consider a closed path ($\Gamma = \Gamma_1 + \Gamma_2 + \Gamma_3 + \Gamma_4$), as shown in Figure 1, which encloses an area A. The conventional J-integral over this contour Γ can be written as:

$$\int_{\Gamma}[\ ] = \int_{\Gamma_1}[\ ] + \int_{\Gamma_2}[\ ] + \int_{\Gamma_3}[\ ] + \int_{\Gamma_4}[\ ] \qquad \text{---------(6)}$$

where,

$$[\ ] = Wdx_2 - \sigma_{ij}\, n_j \frac{\partial u_i}{\partial x_1} ds \qquad \text{---------(7)}$$

Here, the path Γ_2 be an arbitrary, and let path Γ_4 be a circular with radius r about the crack tip where $r \to 0$. Using the near field solution, i.e. considering only singular term, we have for the stationary crack,

$$\int_{\Gamma_4}[\ ] = -\kappa \frac{K^2{}_{Id}}{E} \qquad \text{---------(8)}$$

where $\kappa = (1-\nu^2)$ and 1 for plane strain and plane stress conditions, respectively. Further, if the crack surface is free of traction,

$$\int_{\Gamma_1}[\ ] = \int_{\Gamma_3}[\ ] = 0 \qquad \text{----------(9)}$$

Thus, combining Eqs. (5), (6), (7), (8) and (9), we can compute the time dependent stress intensity factor, in the elastodynamic stationary crack problems, from the following expression

$$\kappa \frac{K^2{}_{Id}}{E} = J_{\Gamma_2} + \int_A \rho \ddot{u}_i \left(\frac{\partial u_i}{\partial x_1}\right) dA \qquad \text{----------(10)}$$

where J_{Γ_2} is the conventional J-integral over any arbitrary path surrounding the crack tip, and the second term of Eq. (10) is the area integral over the area (A) enclosed by this arbitrary path. The application of this modified J-integral, i.e. Eq. (10), will be presented in the following two examples.

A CENTRALLY CRACKED RECTANGULAR STRIP

This problem was originally analyzed by Chen [8] with finite difference method employing a very fine grid (i.e. 50 zones/cm), and subsequently by various investigators with finite element method [1, 2, 5, 6]. This problem consists of a rectangular strip (20 x 40 mms) with a central crack (2a = 4.8 mm), loaded dynamically in plane strain condition with an uniform tension $\sigma H(t)$, where H(t) is Heaviside step function. Figure 2 shows the finite element breakdown of one quadrant of

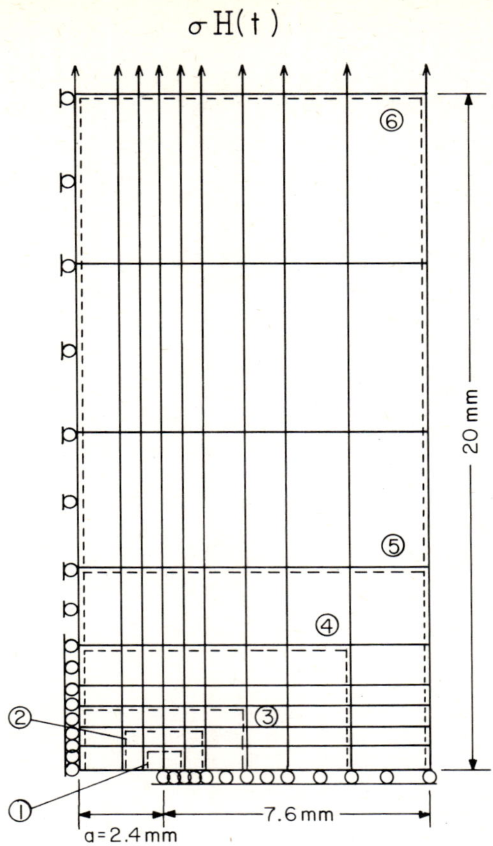

Figure 2. Finite element mesh and integration paths for J-integral of a centrally cracked strip.

this strip which consists of 81 quadratic elements and 280 nodes. The material is a linear elastic having Young's modulus of 200 GPa, Poisson's ratio of 0.3 and density of 5 g/cm^3. The finite element NONSAP Code [11] was employed in the present analysis. The time integration scheme was Newmark's method (with parameter α and δ equal to 0.25 and 0.5, respectively), and the time step was equal to the transit time of longitudinal wave across the smallest size of element. Further, only lumped mass formulation was employed in the present investigation.

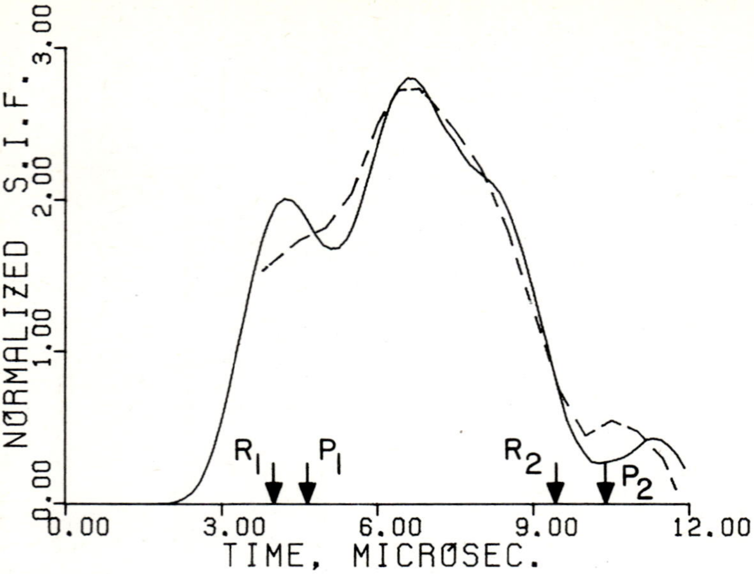

Figure 3. Dynamic stress intensity factor for a centrally cracked strip from the present finite element method (solid line) and from finite difference method (dashed line - Reference 8).

Figure 3 shows the time dependence of the normalized* dynamic stress intensity factor obtained from the indirect crack opening displacement method (i.e. by comparison of dynamic and static COD of the node close to the crack tip) from the present analysis as well as the same obtained by Chen [8] with finite difference method. Both are in excellent agreement with each other except for deviations over two time periods which correspond to the arrival times of the Rayleigh surface wave (from the other tip) and pressure wave (from nearest boundary reflection) at the crack tip and their subsequent reflection, as marked R_1, P_1, R_2 and P_2 in Figure 3. These local differences in the results obtained from finite element and finite difference analyses in the vicinity of R_1, P_1, R_2 and P_2 were also observed in previous investigations

---

* In this section, stress intensity factors are normalized with a factor $\sigma\sqrt{a}\sqrt{\pi}$, where σ is applied stress and a is the half crack length, i.e. 2.4 mm.

TABLE I

Rectangular Strip with a Central Crack - Normalized Stress Intensity Factor $\underset{\sim}{K} = K/\sigma\sqrt{a}\sqrt{\pi}$.

Path Number	Static		Peak Dynamic Value			
			Including area integral of Eq. (10)		Excluding area integral of Eq. (10)	
	$\underset{\sim}{K}$	$\underset{\sim}{K}/1.028$ *	$\underset{\sim}{K}$	$\underset{\sim}{K}/2.804$ **	$\underset{\sim}{K}$	$\underset{\sim}{K}/2.804$ **
1	0.947	0.921	2.540	0.906	2.510	0.895
2	1.025	0.997	2.817	1.005	2.700	0.963
3	1.025	0.997	2.889	1.030	2.613	0.932
4	1.023	0.995	2.920	1.041	2.530	0.902
5	1.022	0.994	2.842	1.014	2.504	0.893
6	1.023	0.995	3.142	1.121	2.316	0.826

[1, 2, 5, 6]; and may be, thus, considered as a point of departure between finite element and finite difference analyses for this problem. The dynamic stress intensity factor obtained from the indirect COD method will, therefore, be assumed as a standard to compare the following J-integral analysis.

The six contours, over which the J-integrals were calculated, are shown in Figure 2. The normalized static stress intensity factors ($\underset{\sim}{K}_I = K_I/\sigma\sqrt{a}\sqrt{\pi}$) obtained from the conventional J-integral are listed in Table I, which is reported to be equal to 1.028 in Reference 8. The normalized dynamic stress intensity factors ($K_{Id}/\sigma\sqrt{a}\sqrt{\pi}$) computed from the modified J-integral (i.e. Eq. (10)) over these six contours are compared with the counterpart from indirect COD method in Figure 4. These results are in excellent agreement, with the most striking differences occuring around sharp local maxima and minima of the dynamic stress intensity factor. The modified J-integral analysis with integration path 1 underestimates the peak dynamic stress intensity factor by 10 percent than that obtained from the COD method, path 6 overestimates the same by 12 percent, while paths 2, 3, 4, and 5 give peak dynamic intensity factor within 4 percent of the counterpart obtained from COD method. The contribution of the area integral in computing

* The normalizing factor 1.028 is the static solution of this problem as reported in Reference 8.

** The normalizing factor 2.804 is the value of peak dynamic normalized S.I.F. obtained from indirect COD method in the present analysis.

dynamic stress intensity factor (i.e. second term in Eq. (10)) was found to be directly related to area enclosed by the contour. As a demonstration, Figure 5 compares the normalized dynamic stress intensity factor computed from J-integral with path 3, including and excluding area integral term of Eq. (10) with that obtained from the indirect COD method. Here, the omission of area integral (i.e. in other way, the conventional J-integral) estimates the peak dynamic stress intensity factor 10 percent lower than the same computed with the inclusion of area integral. This difference is negligible with path 1, and maximum (i.e. 36 percent) with path 6 which covers the external boundary of the strip. Table 1 illustrates these comparisons for all integration paths. The present analysis, thus, clearly demonstrates that an integration path, neither very close to nor very far from the crack tip, should be preferred with the modified J-integral to evaluate the dynamic stress intensity factor. A further application of this analysis is shown in the following example.

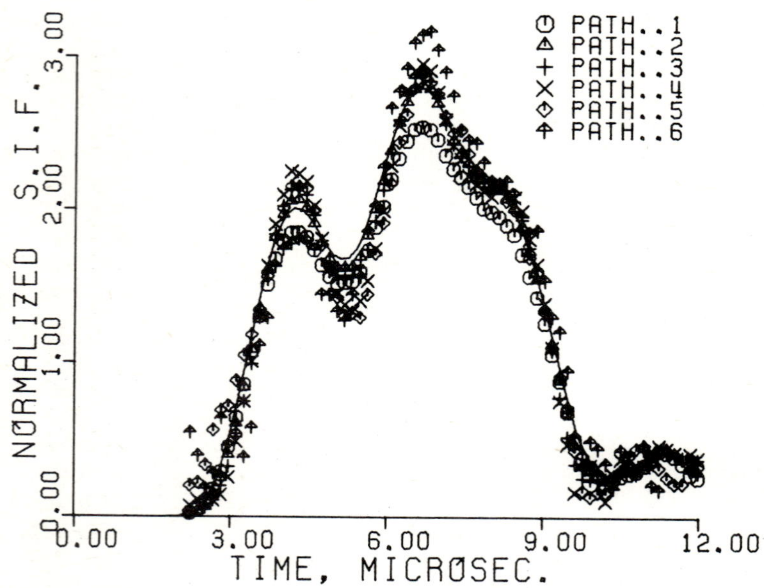

Figure 4. Dynamic Stress intensity factor for a centrally cracked strip from J-integral for six paths (data points) and from indirect COD method (solid line).

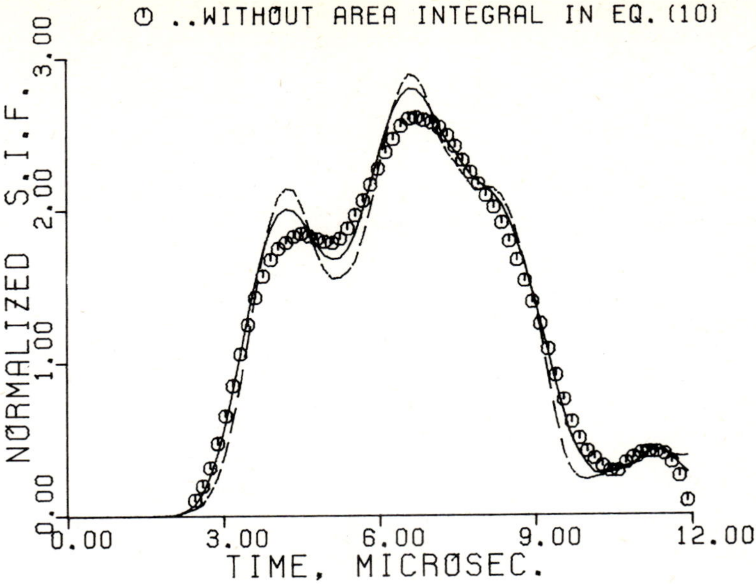

Figure 5. Dynamic stress intensity factor for a centrally cracked strip from J-integral with path 3 of Figure 2 (dashed line), without area integral term in Eq. (10) (data points) and from indirect COD method (solid line).

KALTHOFF'S EXPERIMENT

A 3-point bend specimen of an epoxy resin, Araldite B (under drop weight impact testing) was analyzed to compare the dynamic stress intensity factor from the present J-integral analysis with that obtained experimentally by Kalthoff et al [12]. They conducted a series of drop weight impact tests of notched bend specimens with different sharpness of the starter notch. The impact load history and the transient stress intensity factor history of three reported specimens closely correspond to each other except for the time of crack initiation (see Figure 2 of Reference 12). All three specimens can be, thus, treated as a single specimen. Due to symmetry, half of the specimen was analyzed with the finite element breakdown, shown in Figure 6. The material properties of Araldite B were: Young's modulus equal to 3660 MPa, Poisson's ratio equal to 0.392 and density equal to 1.15 g/cm^3 [12]. The smoothed impact (tup) load-time history, used in the present finite element analysis as shown in Figure 7, is the average of three experimental curves in Figure 2 of Reference 12. The node corresponding to the impact

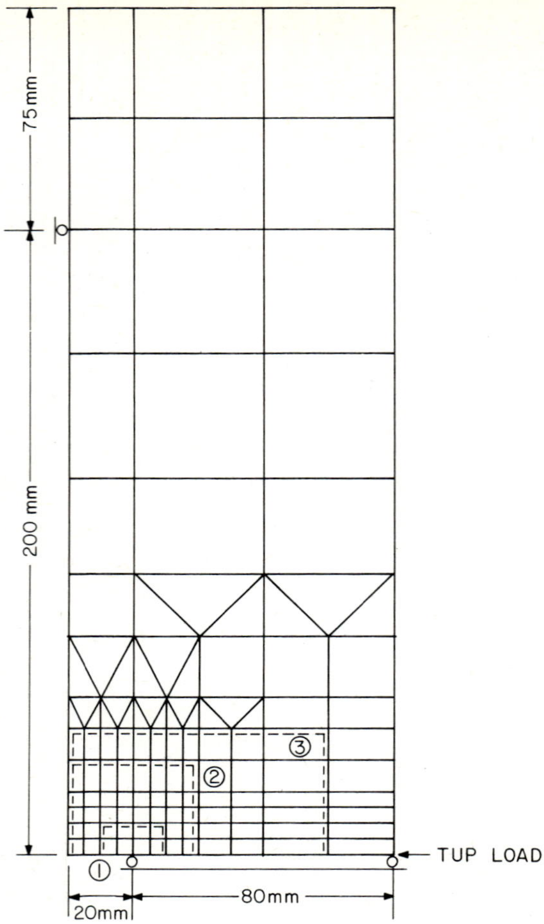

Figure 6. Finite element mesh for Kalthoff's specimen.

point was given an initial velocity of 5 m/s, corresponding to the free fall drop of the impact tup.

The dynamic stress intensity factor-time history obtained from the indirect crack opening displacement method is in excellent agreement with the experimental counterpart as shown in Figure 8. The experimental stress intensity factor-time history, shown by dashed line in Figure 8, is the average of experimental data for three specimens given in Figure 2 of Reference 12. Figure 8, also, shows the dynamic stress intensity factor-time history obtained from J-integral (i.e. Eq. (10)) for three paths (shown in Figure 6) which are in excellent agreement with the counterpart obtained either experimentally or from the COD method. This analysis further demonstrates that the present J-integral analysis is a viable alternative for computing transient stress intensity factor in the dynamic finite element analysis of stationary crack.

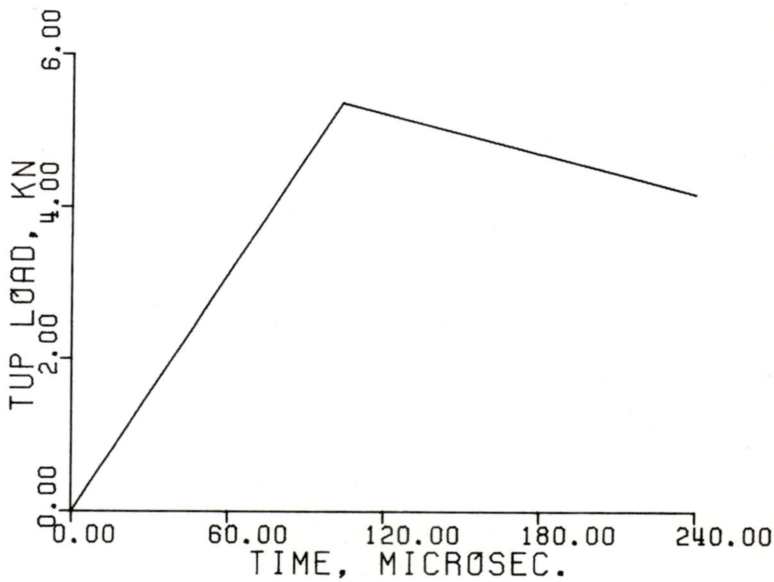

Figure 7. Smoothed impact load-time history for Kalthoff's specimen.

CONCLUSIONS

The conventional J-integral was extended to compute the transient stress intensity factor for elastodynamic problems with stationary crack. The indirect procedure to determine the dynamic stress intensity factor by comparing static and dynamic crack opening displacement is, no doubt, a simple, straightforward and accurate method for this class of problems. However, this requires the exact solution of static stress intensity factor and is applicable to a single mode situation only. On the other hand, modified J-integral is a direct procedure and can be easily extended to the mixed mode situation. Further, the modified J-integral is also a simple procedure which can be easily incorporated in the existing subroutine of conventional

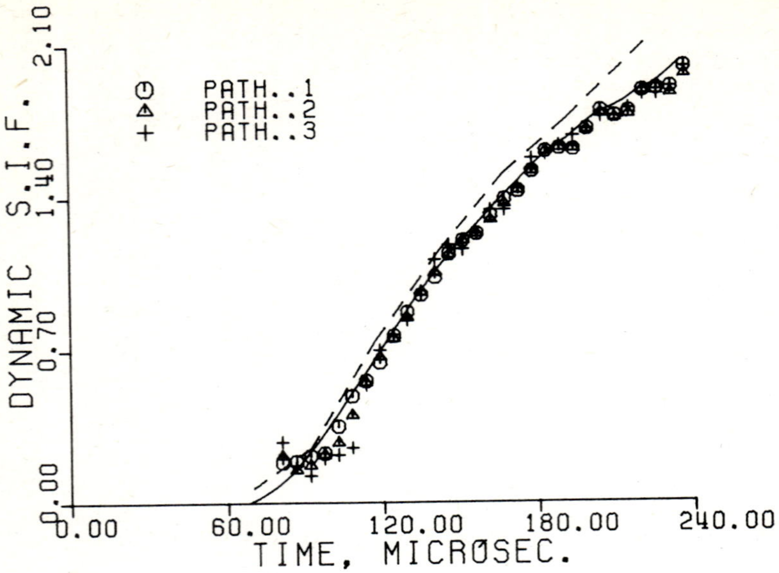

Figure 8. Dynamic stress intensity factor (MPa$\sqrt{m}$) for Kalthoff's specimen from COD method (solid line), J-integral analysis (data point) and obtained experimentally (dashed line).

J-integral by introducing the area integral term of Eq. (10). Limited experience, with the modified J-integral analysis as presented here, demonstrated that an integration path neither very close nor far from the crack tip yielded the most accurate dynamic stress intensity factor which was in excellent agreement with its counterpart obtained either from the indirect crack opening displacement procedure or experimentally.

REFERENCES

1. MALL, S. - Dynamic Finite Element Analysis of Cracked Bodies with Stationary Crack, Fracture Mechanics: Twelfth Conference, ASTM STP 700, American Society for Testing and Materials (in press).

2. MALL, S., KOBAYASHI, A. S. and LOSS, F. J. - Dynamic Fracture Analysis of Notched Bend Specimens, Crack Arrest Methodology and Applications, ASTM STP 711, American Society for Testing and Materials (in press).

3. MALL, S., KOBAYASHI, A. S. and URABE, Y. - Dynamic Photoelastic and Dynamic Finite Element Analyses of Polycarbonate Dynamic Tear Test Specimens, Fracture Mechanics, ASTM STP 677, Ed. Smith, C. W., American Society for Testing and Materials, 1979.

4. MALL, S., KOBAYASHI, A. S. and URABE, Y. - Dynamic Photoelastic and Dynamic Finite Element Analyses of Dynamic Tear Test Specimens. Experimental Mechanics, Vol. 18, No. 12, pp. 449-456, 1978.

5. ABERSON, J. A., ANDERSON, J. M. and KING, W. W. - Dynamic Analysis of Cracked Structures Using Singularity Finite Elements, Elastodynamic Crack Problems, Ed. Sih, G. C., Noordhoff International, 1977.

6. GLAZIK, J. L. JR. - Dynamic Finite Element Analysis of Cracked Bodies. ASME Preprint 78 - PVP - 94, American Society of Mechanical Engineers.

7. KELLEY, J. W. and SUN, C. T. - A Singular Finite Element for Computing Time Dependent Stress Intensity Factors. Engineering Fracture Mechanics, Vol. 12, pp. 13-22, 1979.

8. CHEN, Y. M. - Numerical Computation of Dynamic Stress Intensity Factors by a Lagrangian Finite - Difference Method (The Hemp Code). Engineering Fracture Mechanics, Vol. 7, pp. 653-660, 1975.

9. KOBAYASHI, A. S., MALL, S., URABE, Y. and EMERY, A. F. - A Numerical Dynamic Fracture Analysis of Three Wedge-Loaded DCB Specimens, Numerical Methods in Fracture Mechanics, Eds. Luxmoore, A. R. and Owen, D. R. J., University College Swansea, 1978.

10. WILSON, W. K. and YU, I. W. - The Use of the J-integral in Thermal Stress Crack Problems. Int. J. of Fracture, Vol. 15, No. 4, pp. 377-387, 1979.

11. BATHE, K. J., WILSON, E. L. and IDING, R. H. - NONSAP, A Structural Analysis Program for Static and Dynamic Response of Nonlinear Systems. Report No. UCSEM 74-3, Structural Engineering Laboratory, Univ. of California, Berkeley, Cal., U.S.A.

12. KALTHOFF, J. F., WINKLER, S. and BEINERT, J. - The Influence of Dynamic Effects in Impact Testing. Int. J. of Fracture, Vol. 13, No. 4, pp. 528-531, 1977.

GAS DEPRESSURISATION MODELLING FOR CRACK PROPAGATION STUDIES IN PRESSURISED PIPELINES

E. S. Caldis, Research Assistant, Department of Civil Engineering, University of Wales, Swansea, U.K.

D. R. J. Owen, Reader, Department of Civil Engineering, University of Wales, Swansea, U.K.

SUMMARY

Transient dynamic crack propagation problems in pressurised pipelines are considered. The pipeline is modelled by use of the Semiloof thin shell element in which a large deformation viscoplastic response is permitted. A stress criterion for propagation is utilised and the effect of incremental crack tip nodal release is investigated.

The gas depressurisation characteristics are modelled by semi-empirical relationships and also by means of unidimensional compressible flow equations. Both approaches are compared by application of the computational model to the solution of a practical pipeline problem.

1. INTRODUCTION

Although the use of dynamic transient solution techniques is widespread in structural mechanics, their use in the study of dynamic fracture problems is quite recent. To date, most of the work reported has been restricted to linear elastic behaviour[1-4] with the energy sink required for crack extension by an energy criterion being introduced by suitable manipulation of the crack tip nodal force. This approach may well be adequate for the double cantilever beam (DCB) specimen but for the case of a dynamically propagating crack in a pipeline, it can be shown that the inclusion of elasto-plastic material behaviour can have a profound effect on the pipe-opening profile [5,6]. Although Emery et al [7] conclude that for the particular case examined (thick shell formulation), the inclusion of large deformation effects does not significantly alter the crack characteristics, it is difficult to ascertain to what extent this conclusion holds for the general case.

This is particularly true when the gas depressurisation characteristics are themselves of prime interest. Since the problem is essentially a coupled fluid-structure problem with the gas depressurisation characteristics being a direct function of the crack-opening area, both large deformation effects and plastic material behaviour are included in analysis.

The gas depressurisation characteristics can be modelled in two ways. Firstly empirical expressions based on the crack-opening area can be employed to predict the pressure distribution. However, parameters used in this approximation are problem dependent and require a priori knowledge of the depressurisation characteristics [16]. Alternatively uni-dimensional finite elements can be used to provide a discretised solution to the compressible flow equations. In this approach the gas flow inside the pipe is assumed to be parallel to the longitudinal axis and lateral flow is based on the pipe opening area and assumed to correspond to that of a "choked" nozzle condition. Note should be taken however of the limitations of this approach as pointed out in Ref. [9]. A coupled fluid/structure finite element analysis is then undertaken.

The main disadvantage of employing conventional plasticity expressions in dynamic transient calculations is that the classical theory does not permit stress levels to instantaneously rise above the static yield value. The strain-rate dependence of the plastic response can be included numerically by the use of viscoplastic theory.

In this paper a brief summary is given of the essential details relating to the structural aspects of the problem. Solution techniques for the gas depressurisation are then presented. An updated version of the node release mechanism described in Ref. [23] is utilised for crack propagation. Finally results are presented and some conclusions drawn on the relative advantages of the two gas models as well as the node release mechanism.

2. TRANSIENT DYNAMIC FINITE ELEMENT ANALYSIS

The dynamic problem invariably gives rise to a set of ordinary differential equations of the form [10].

$$\underline{M}\,\ddot{\underline{a}} + \underline{C}\,\dot{\underline{a}} + \underline{P}\,(\underline{a}) + \underline{F}(t) = 0 \tag{1}$$

in which dots denote differentiation in time, $\underline{a}$ stands for a set of parameters describing the displacements, $\underline{M}$ is the mass matrix (generally independent of time and displacement), $\underline{C}$ is the damping matrix which may well be a function of $\dot{\underline{a}}$ and $\underline{F}$ are the applied or activating forces. The term $\underline{P}$ is the internal set of forces opposing the displacement of the structure, which for nonlinear situations involving large deformation and/or plasticity behaviour must be calculated from the stress distribution satisfying all nonlinear conditions. Therefore

$$\underline{P} = \int_V \underline{B}^T \underline{\sigma} \, dv \tag{2}$$

in which $\underline{\sigma}$ are the nonlinear stresses and $\underline{B}$ is the appropriate matrix expressing the strains in terms of the nodal displacements. For large deformation problems $\underline{B}$ itself is a function of the displacements $\underline{a}$.

If the solution at time t_{n+1} is obtained by considering equilibrium conditions at time t_n the method is explicit and only conditionally stable. The most common solution process of this type is the central finite difference algorithm which when applied to (1) results in

$$\underline{a}_{n+1} = (\underline{M} + \underline{C}\Delta t/2)^{-1}[\Delta t^2(-\underline{P}_n - \underline{F}_n) + 2\underline{M}\,\underline{a}_n - (\underline{M} - \underline{C}\Delta t/2)\,\underline{a}_{n-1}] \tag{3}$$

The computation becomes trivial if matrices $\underline{M}$ amd $\underline{C}$ are made diagonal by use of an appropriate lumping procedure. With this approach there are however severe restrictions on the time step length.

In the present formulation, large strains with small rotations are admitted. For the total Lagrangian approach adopted here [11,12,21] the original coordinate system is employed throughout the analysis together with the full nonlinear Green-Lagrange strain expressions.

In order to introduce a material response which is strain rate dependent into the dynamic model the use of viscoplastic theory is advantageous. The total strain rate $\dot{\underline{\varepsilon}}$ is assumed to be the sum of the elastic, $\dot{\underline{\varepsilon}}_e$ and viscoplastic $\dot{\underline{\varepsilon}}_{vp}$ components. The viscoplastic strain rate is defined according to Perzyna's associated form of the viscoplastic flow rule [13] as

$$\dot{\underline{\varepsilon}}_{vp} = \gamma < \phi(F) > \frac{\partial F}{\partial \underline{\sigma}} \tag{4}$$

where F is a scalar yield function and γ is a function of the material viscosity coefficient. The notation < > implies a zero value for all non-positive values of $\phi(F)$.

The stress change occurring in a time interval $\Delta t_n = t_{n+1} - t_n$ is given, by use of the Euler forward difference time stepping algorithm, to be

$$\Delta\underline{\sigma}^n = \underline{D}[\underline{B}_n - \dot{\underline{\varepsilon}}^n_{vp} \, \Delta t_n] \tag{5}$$

For the explicit scheme summarised by (3) it is only necessary to accumulate the total stresses from the incremental values

(5), at each time step, in order to evaluate $\underline{P}_n$ according to (2). Considerable computational economies are offered by solving pipeline problems by means of shell elements and in this work the Semiloof thin shell element is employed which is described in detail in Ref. [18].

For shell analysis a yield function governing the onset of viscoplastic deformation has been proposed by Crisfield [19] which is expressed in terms of the membrane stresses (N_{xx},N_{yy}, N_{xy}) and bending moments (M_{xx},M_{yy},M_{xy}) as

$$F = Q_t + \frac{Q_m}{\alpha^2} + \frac{1}{\sqrt{3}\alpha} |Q_{tm}| - 1 \tag{6}$$

in which

$$Q_t = \frac{1}{N_o^2} (N_{xx}^2 + N_{yy}^2 - N_{xx}N_{yy} + 3N_{xy}^2)$$

$$Q_m = \frac{1}{M_o^2} (M_{xx}^2 + M_{yy}^2 - M_{xx}M_{yy} + 3M_{xy}^2)$$

$$Q_{tm} = \frac{1}{M_oN_o} (M_{xx}N_{xx} + M_{yy}N_{yy} - \tfrac{1}{2}M_{xx}N_{yy} - \tfrac{1}{2}M_{yy}N_{xx} + 3M_{xy}N_{xy})$$

where $N_o = \sigma_o h$ is the uniaxial yield force, $M_o = \sigma_o h^2/4$ is the uniaxial yield moment and σ_o is the uniaxial yield stress of the material. The parameter α is a function of the visco-plastic curvature and the limiting value, $\alpha = 1$, results in the Ilyushin criterion [20].

3. CRACK-TIP NODAL RELEASE MECHANISM

In the stress criterion of crack propagation employed here the steady state normal stress component at the Gaussian integrating point nearest the crack tip is designated σ_g. To initiate crack propagation the nodal restraint at the crack tip is removed, either instantaneously or gradually. The crack tip is assumed to have now progressed to the next node and further propagation is governed by the normal stress value at the integrating point immediately adjacent to the new crack-tip position. If this stress exceeds $\sigma_c = \sigma_g/S$, then the crack is assumed to propagate to the next node. The factor S can be interpreted as a measure of the initial bluntness of the crack tip and σ_c is a measure of the material toughness. Crack arrest is deemed to have occurred when either no propagation takes place for a considerable time or even if propagation is continuing it does so at a markedly reduced velocity. A knowledge of the crack velocity during propagation between any two adjacent nodes can be used to automatically preset and adjust the rate of nodal force release for crack advance.

The justification for this node release mechanism is discussed in Ref. [23] where some results for the problem shown in Fig. 1 are presented. Fig. 2 shows the corresponding crack advance with time for different nodal release rates.

4. DUCTILE CRACK PROPAGATION IN PRESSURED PIPELINES

The situation leading to dynamic crack propagation in a pressurised pipeline can be summarised as follows. A short axial crack is initially developed along a generator of the pipe. Under the action of the internal pressure, P_I, the crack begins to open out, as shown in Fig. 3 resulting in a pressure drop to a value, P_o, at the crack tip. A pressure drop also occurs along the cracked zone, the value of which is a function of the axial distance z and the angle of crack opening 2θ. This pressure variation has been quantified [14-16] in previous investigations but only under constant velocity steady-state assumptions.

One approach is to base the drop in crack tip pressure on the plan area of the crack configuration according to the following semi-empirical relationship [8]

$$P_o/P_I = 1 - 0.7\ A/A_p \tag{7}$$

in which A is the crack opening area and A_p is the cross-sectional area of the pipe. The variation in pressure in the cracked zone is taken to be

$$P = P_o\, e^{-\alpha\theta\left(\frac{z_c}{R_i}\right)} \tag{8}$$

where α is a constant (assumed as 2.28 in the present study[8]), θ is the half angle of crack opening measured in radians, R_i the internal pipe radius and z_c the axial distance measured from the current position of the crack tip. The pressure at and ahead of the crack tip is assumed to be P_o.

The solution sequence followed is summarised below:

(i) The static solution for the pressurised pipeline in the absence of any crack is first obtained. This is achieved by means of a transient dynamic analysis with a damping factor approaching the critical damping value.

(ii) The damping factor is then set to zero and an initial (or starter) crack of predetermined length is introduced along the axis of the pipe. Numerically this is achieved by the sudden release of selected nodal points. Physically this situation corresponds to accidental pipeline damage and is the method by which a propagating crack is usually introduced in experimental studies. At this stage depressurisation as well as axial pressure variation are accounted for either according to expressions (7) and (8) or by a compressible flow finite element analysis.

5. FINITE ELEMENT MODELLING OF GAS DEPRESSURISATION

The unidimensional continuity equation for gas flow in a pipeline can be written in the form [17]

$$\frac{\partial \bar{s}}{\partial t} + u\frac{\partial \bar{s}}{\partial x} + (1+\bar{s})\frac{\partial u}{\partial x} + (1+\bar{s})\frac{u}{A}\frac{\partial A}{\partial x} + \frac{1}{\rho_o}\frac{q}{A} = 0 \tag{9}$$

in which

$$\bar{s} = \frac{1}{\rho_o}(\rho - \rho_o) \tag{10}$$

and ρ_o is a datum density
A is the local cross-sectional area of the pipe
u is the local gas velocity
q is the pointwise lateral mass flow rate per unit crack length over the cracked pipe section.

The momentum equation, again unidimensional, can be written

$$\frac{\partial u}{\partial t} + u\frac{\partial u}{\partial x} + \frac{1}{\rho}\frac{\partial p}{\partial x} - \frac{uq}{\rho A} = 0 \tag{11}$$

where p is the local gas pressure. A simultaneous solution of (9) and (11) is required throughout the crack opening history to evaluate the instantaneous values of the above variables - particularly the gas pressure distribution.

In the approach described here, equations (9) and (11) are uncoupled and solved iteratively. Also the convective term in momentum equation (11) is linearised. One dimensional finite elements are employed in solution and a Galerkin weighted residual formulation adopted. Denoting approximated quantities by ($\hat{}$), we have the standard finite element interpolation expressions, typified by

$$u(x) = \hat{u}(\zeta) = \Sigma\, N_i(\zeta)\, \hat{u}_i \tag{12}$$

in which

$\hat{u}(\zeta)$ is the interpolated value at any position within the element defined by the natural coordinate ζ where $-1 \le \zeta \le 1$. Once again $N_i(\zeta)$, i=1,2 are the element shape functions. Applying the standard Galerkin process to (9) and (11) results in

$$\left[\int_\Omega N_i N_j \, d\Omega\right]\frac{d\hat{s}_i}{dt} + \left[\int_\Omega \hat{u}\frac{dN_i}{dx} N_j \, d\Omega + \int_\Omega \left(\frac{d\hat{u}}{dx} + \frac{\hat{u}}{A}\frac{d\bar{A}}{dx}\right) N_i N_j \, d\Omega\right]\hat{s}_i$$

$$+ \left[\int_{\Omega} \left(\frac{d\hat{u}}{dx} + \frac{\hat{u}}{\hat{A}} \frac{d\hat{A}}{dx} + \frac{1}{\rho_o} \frac{\hat{q}}{\hat{A}} \right) N_j \, d\Omega \right] = 0 \qquad (13)$$

$$\left[\int_{\Omega} N_i N_j \, d\Omega \right] \frac{d\hat{u}_i}{dt} + \left[\int_{\Omega} \left(\hat{u}^* \frac{dN_i}{dx} N_j - \frac{\hat{q}}{\hat{\rho}\hat{A}} N_i N_j \right) d\Omega \right] \hat{u}_i$$

$$+ \left[\int_{\Omega} \frac{1}{\hat{\rho}} \frac{d\hat{\rho}}{dx} N_j \, d\Omega \right] = 0 \qquad (14)$$

For the iterative solution process, equations (13) and (14) are uncoupled so that in (13) $\hat{u}$ and $d\hat{u}/dx$ are known from the previous solution and input explicitly. Similarly in (14) $\hat{\rho}$ and $d\hat{\rho}/dx$ are also input explicitly. The variables $\hat{A}$, $d\hat{A}/dx$, $\hat{q}$ are evaluated independently. In equation (14) u* is the linearised form of the convective term and is the value of u obtained from the previous iteration so that

$$[\hat{u}^*]^n = [\hat{u}]^{n-1} \qquad (15)$$

in which n is the iteration counter.

Equations (13) and (14) are a special case of the transient dynamic equilibrium equations (1) and (2) where now $\underline{M} = \underline{0}$. Consequently various finite difference time integration schemes can be employed in solution. In particular the backward difference expression for the velocity is simply

$$\dot{u}_{t+\Delta t} = \frac{1}{\Delta t} \left[u_{t+\Delta t} - u_t \right] \qquad (16)$$

Using this temporal approximation, equations (13) and (14) reduce to the standard implicit form

$$\left[\frac{\underline{C}}{\Delta t} + \underline{K}_{t+\Delta t} \right] \underline{u}_{t+\Delta t} = \frac{\underline{C}}{\Delta t} \underline{u}_t - \underline{F}_{t+\Delta t} \qquad (17)$$

where matrix $\underline{C}$ does not vary with time in this application.

6. APPLICATION TO PIPELINE FRACTURE PROBLEMS

The numerical solution of a pipeline crack propagation model is now described. Details of the geometry and material properties are included in Table 1. Both viscoplastic material behaviour and large deformation effects are considered. A symmetric quadrant of the model was idealised by a total of 44 Semiloof shell elements and the developed mesh is shown in Fig. 4.

The static solution is first obtained by carrying out a damped transient dynamic analysis. The starter crack of prescribed length is then introduced by the release of selected nodal points and at this stage the damping parameter is set to zero.

A stress criterion of crack propagation is considered with a bluntness parameter S = 1.5 and a critical fracture stress σ_c = 414000 kN/m^2 (60,000 lb/in^2) being assumed.

Pipe internal radius, R_i (mm)	294.64
Pipe wall thickness, h(mm)	10.16
Starter crack length, $2a_o$ (mm)	609.6
Internal pressure (kN/m^2)	2090.0
Elastic modulus, E(GN/m^2)	200.0
Poisson's ratio, ν	0.3
Yield stress (kN/m^2)	276000
Ultimate tensile stress (kN/mm^2)	414000
No strain hardening	0.0
Mass density (kg.sec^2/m^4)	800000
Bluntness parameter, S	1.5
Original density, ρ_o (kg/m^2)	1.23
Coefficient of specific heat γ	1.4
Original pressure wave celerity C_o (m/s)	571.6
Discharge coefficient, C_d	0.65 and 0.90

TABLE 1

An attempt is first made to assess the effect that different gas depressurisation models have on crack propagation. The results obtained by using a finite element model for gas flow and performing a fully coupled fluid/structure analysis are compared with those given by use of the semi-empirical relations (7), (8). An updated version of the node release mechanism reported in Ref. [23] is subsequently included in both analyses. Limited comparison, at this stage, is also made with experimental results reported in Ref. [22]. The four numerical cases considered are summarised below:

(i) Gas depressurisation by semi-empirical formulation
(ii) Gas depressurisation by finite element discretisation
(iii) As Case (i) but with incremental node release mechanism
(iv) As Case (ii) but with incremental node release mechanism

The crack velocity and arrest lengths for Cases (i)-(iv) are presented in Fig.5. Crack propagation occurred 0.001 seconds after the starter crack was initiated. The results indicate that the gas depressurisation response during the propagation phase does not significantly alter the crack velocity and the arrest length. It is however difficult at this stage to estimate to what degree this observation is problem dependent. Inclusion of the node release model has a significant effect on the crack velocity; predicting an average velocity more consistent with experimental observation. The short pipe length modelled in the analysis (for reasons of economy) does not allow any conclusions on arrest length to be drawn, as in each case the crack propagated as far as element 37 (Fig. 4) with mesh instability then becoming evident. Fig. 6 indicates the extent of viscoplastic flow for Case (iv) for two different time intervals illustrating the problems faced in basing a crack propagation criterion on energy balance.

Depressurisation curves for Cases (iii) and (iv) are given in Figs. 7 and 8 respectively. The pressure curves are drawn for the symmetric half of the one dimensional mesh at 0.0003 second intervals from the initialisation of the starter crack at t=0 secs up to t=0.0045 secs. The different depressurisation rates in Cases (iii) and (iv) do not significantly alter the structural response. The results for Case (iv) are consistent with the experimental results reported in Ref. [22] in that no significant depressurisation is evident up to the time of crack arrest. The empirical gas depressurisation curves plotted in Fig. 8 alongside the despressurisation response of the finite element gas model are for comparison purposes only and are the axial pressure gradients given by expressions (7), (8) but based on the geometric crack opening of Case (iv).

Crack opening displacements for the analyses considered are of the order of 2% of the pipe diameter thus satisfying the choked nozzle assumption which forms the basis of the uniaxial finite element discretisation of gas flow. For larger response times when crack opening displacements are of the order of the pipe diameter this assumption will not be valid [9]. The crack profile at t=0.002 secs is shown in Fig. 9 for Case (iv).

7. DISCUSSION

The numerical examples presented in this paper for dynamically propagating cracks in pipelines indicate that the modelling of the gas depressurisation characteristics has no significant influence on the structural response during the crack advance phase. Therefore the finite element discretisation of the compressible flow equations, as compared to the simpler semi-empirical pressure model, is only justified for situations where the gas depressurisation characteristics are of major interest such as in nuclear applications.

The results also show the marked influence of different crack tip nodal release mechanisms and it is appreciated that considerable work is still required in order to establish an unequivocal thermodynamically based crack extension criterion.

In the present gas dynamic solutions the coefficient of discharge has been assumed constant along the crack length. This quantity is in fact a function of the local crack opening geometry and such a dependence could be readily included in the finite element gas flow model.

8. REFERENCES

1. KEEGSTRA, P. N. R., HEAD, J. L. and TURNER, C. E., - A transient finite element analysis of unstable crack propagation in some two-dimensional geometries. Proc. 4th Int. Conf. on Fracture, Waterloo, 1977.

2. MALLUCK, J. G. and KING, W. W. - Fast fracture simulated by a finite element analysis which accounts for crack-tip energy dissipation. Proc. 1st Int. Conf. on Num. Meth. in Fracture Mechanics, Swansea, Jan. 9-13th 1978.

3. KOBAYASHI, A. S., MALL, S., URABE, Y. and EMERY, A. F. - A numerical dynamic fracture analysis of three wedge-loaded DCB specimens. Proc. 1st Int. Conf. on Num. Meth. in Fracture Mechanics, Swansea, Jan. 9-13th, 1978.

4. ERDOGAN, F., DELALE, F., OWCZAREK, J. A. - Crack propagation and arrest in pressurised containers. J. Press. Vessel Technology, Vol. 99, 90-99, 1977.

5. OWEN, D. R. J. and SHANTARAM, D. - Numerical study of dynamic crack growth by the finite element method. Int. J. Fracture, Vol. 13, No.6, 821-837, 1977.

6. AYRES, D. J. - Elastic plastic dynamic analysis of opening area of an axial crack in a pipe under pressure. Paper L7/6 3rd Int. Conf. on Structural Mechanics in Reactor Technology London, Sept. 1975.

7. EMERY, A. F., LOVE, W. J., KOBAYASHI, A. S. - Fracture in straight pipes under large deflection conditions - Part I and II. J. Press. Vessel Technology, Vol. 99,122-136, 1977.

8. OWEN, D. R. J. and SHANTARAM, D. - Dynamic crack propagation in two and three dimensional solids by use of finite element methods. Euromech Colloquium No. 77, Paris, Sept. 1976.

9. PARKS, D. M. and FREUND, L. B. - On the gas dynamics of running ductile fracture in a pressurised line pipe. J. Press. Vessel Technology, Vol. 100, 13-17, 1978.

10. SHANTARAM, D., OWEN, D. R. J. and ZIENKIEWICZ, O. C. - Dynamic transient behaviour of two and three dimensional structures including plasticity, large deformation effects and fluid interaction. Int. J. Earthquake Engg. & Struct. Dynamics, 4, 561-578, 1976.

11. CALDIS, E. S. - A preliminary study of large deformation transient dynamic finite element analysis. M.Sc. Thesis Univ. College Swansea, Ref. C/M/137/78.

12. DINIS, L. M. S. - Quasi-static and dynamic transient non-linear analysis of plates and arbitrary thin shells by use of the Semiloof element. Ph.D. Thesis, University College Swansea, C/Ph/49/78, 1978.

13. PERZYNA, P. - Fundamental problems in viscoplasticity. Recent Advances in Applied Mechanics, Vol. 9, Academic Press, New York, 1966.

14. ALDER, G. M. - The flow through a propagation shear fracture in a high pressure gas main. Research report, Dept. of Mech. Engg. University of Edinburgh, Aug. 1972.

15. HAHN, G. T., SARRATE, M., KANNINEN, M. F. and ROSENFELD, A.R. International Journal of Fracture, 9, No.2, 209-222, 1973.

16. KANNINEN, M. F., SAMPATH, S. G. and POPELAR, C. - Steady state crack propagation in pressurised pipelines without back-fill. Research report, Battelle Columbus Labs., Ohio, Jan. 1975.

17. SHAPIRO, A. H. - The dynamics and hydrodynamics of compressible fluid flow. The Ronald Press Company, New York, 1953.

18. IRONS, B. M. - The Semiloof shell element. Chapter 11 of Finite Elements for thin shells and curved members. (Eds. R. H. Gallagher and D. G. Ashwell) John Wiley & Sons, London, 1976.

19. CRISFIELD, M. A. - On an approximate yield criterion for thin steel shells. Dept. of Environment, TRRL Report 658 Crowthorne, 1974.

20. ROBINSON, M. - A comparison of yield surfaces for thin shells Int. J. Mech. Sci. 13, 345-354, 1971.

21. NAYAK, G. C. - Plasticity and large deformation problems by the finite element method. Ph.D. Thesis, University College Swansea, C/Ph/15/71.

22. BAUM, M. R. and BUTTERFIELD, J. M. - Initial depressurisation rates of pneumatically ruptured pipes Pt.I - Axial defects in ductile pipes. C.E.G.B. Berkeley, Ref.RD/B/N3742.

23. CALDIS, E. S., OWEN, D. R. J. and ZIENKIEWICZ, O. C. - Nonlinear dynamic transient methods in crack propagation studies. Symposium Nonlinear/Dynamic Fracture, ASME Winter Annual Meeting, New York, Dec. 1979.

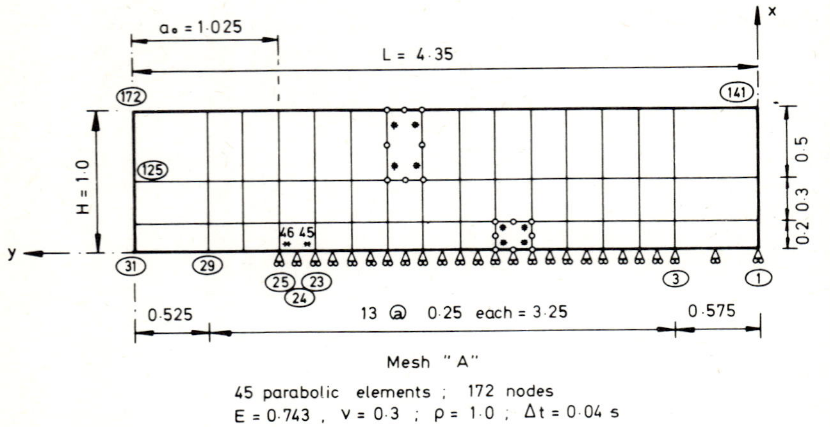

Fig. 1 Details of mesh for DCB problem (Plane Strain)

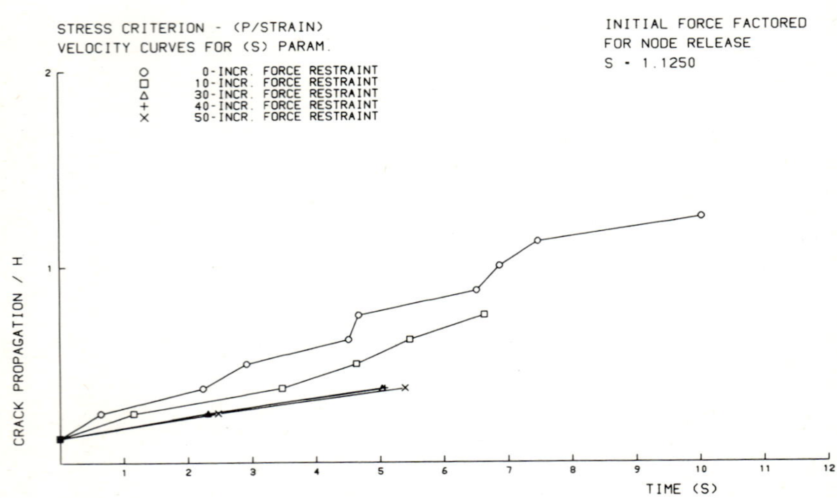

Fig. 2 Propagation - time curves (DCB specimen) for different nodal release rates

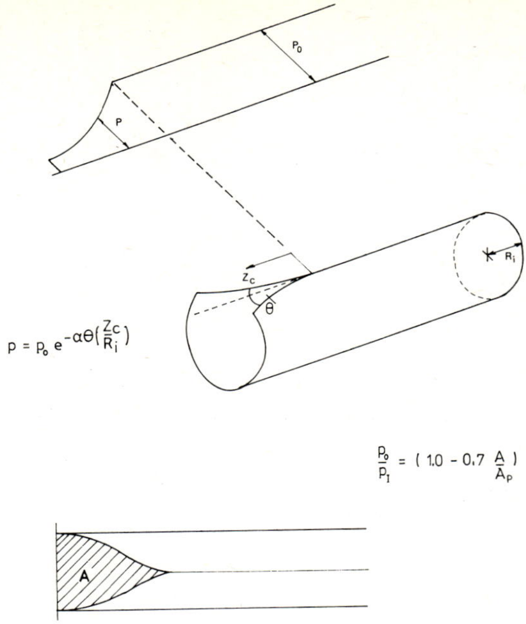

Fig. 3 Opening Profile and Empirical Pressure Distribution in a Cracked Pipe Section

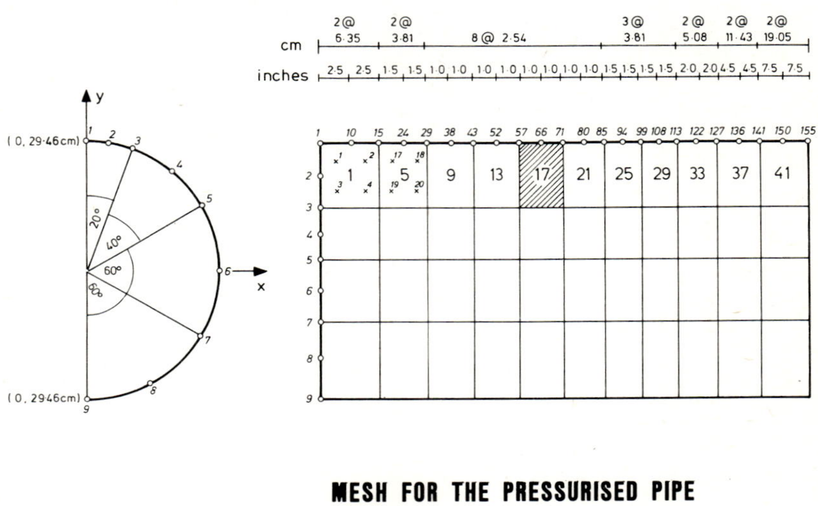

Fig. 4 The Developed Pipe Mesh

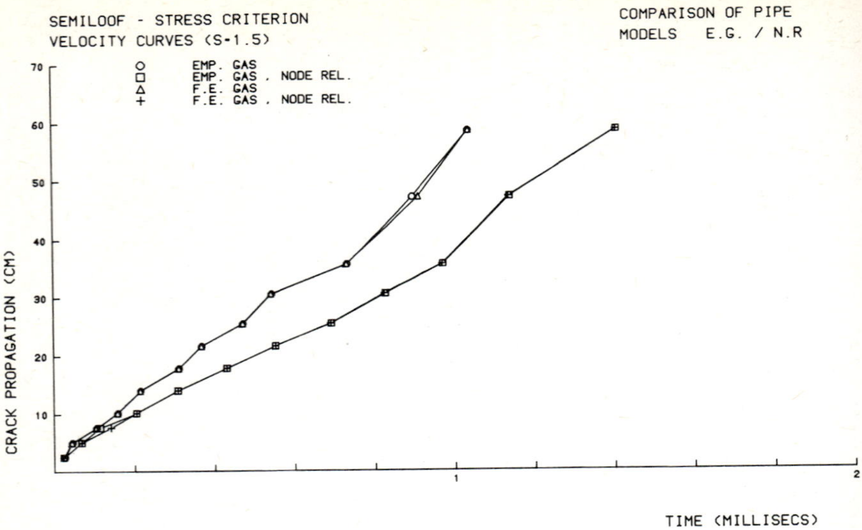

Fig. 5 Propagation-time curves for different gas depressurisation and nodal release models (pipe problem)

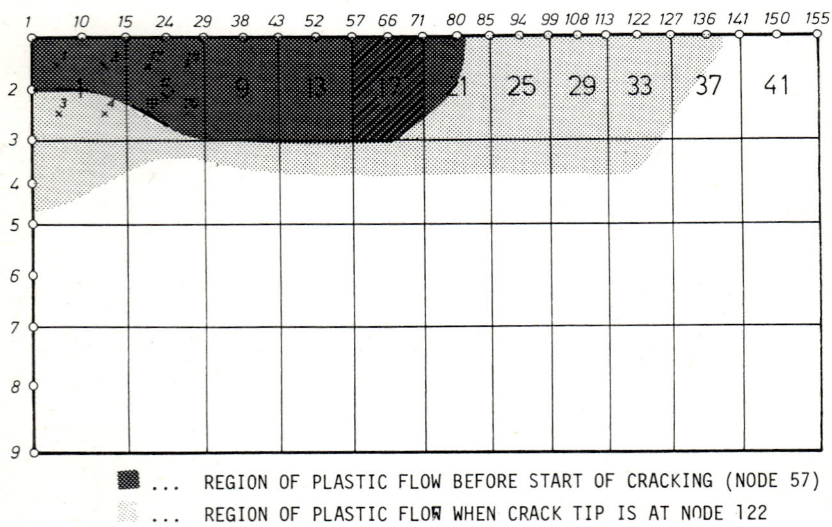

Fig. 6 Development of viscoplastic zones at various time intervals

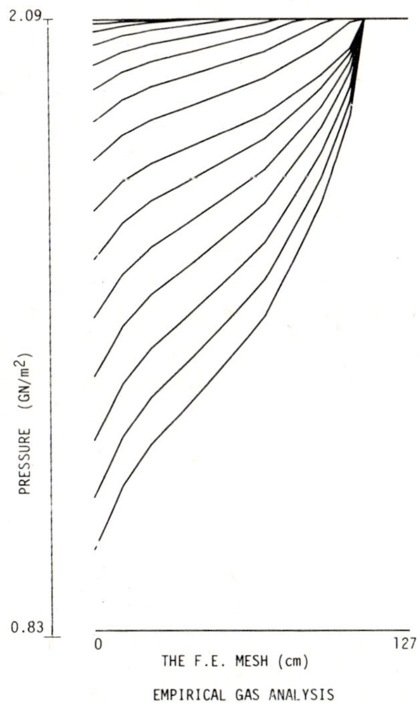

Fig. 7 Pressure-time curves for empirical gas model

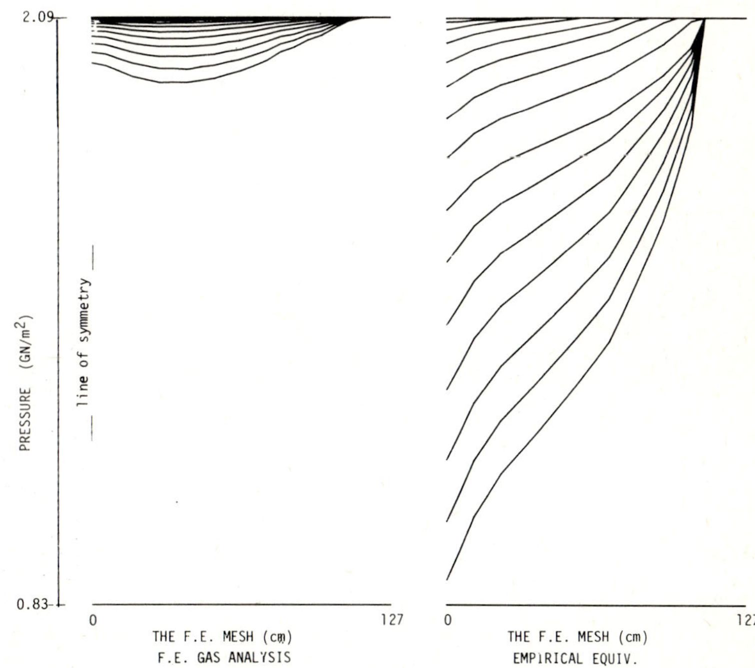

Fig. 8 Pressure-time curves for finite element gas model

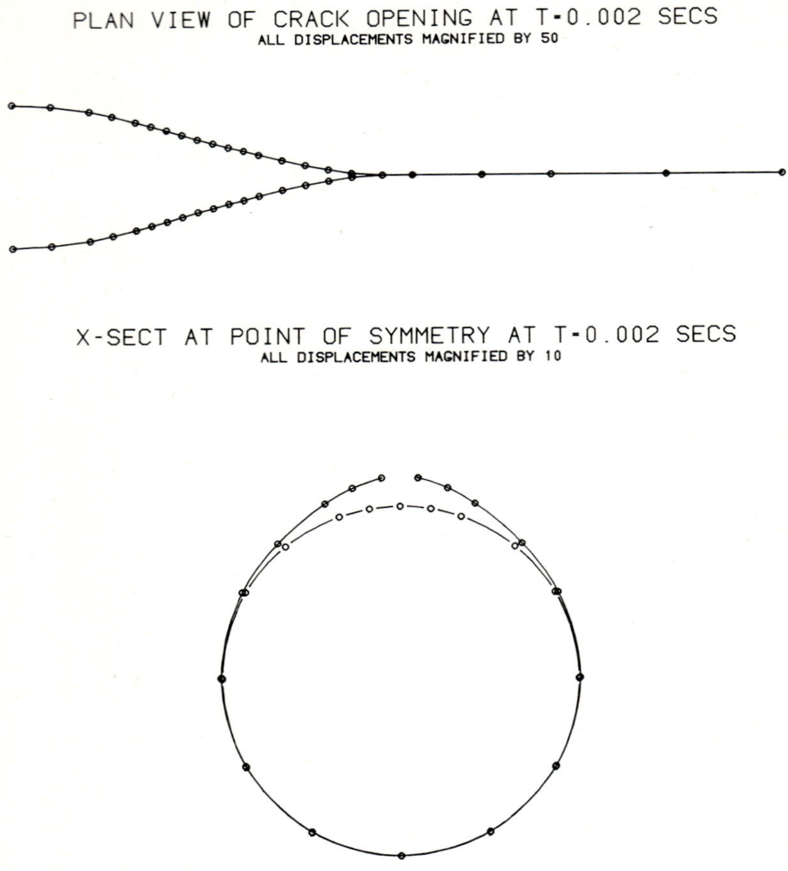

Fig. 9 Magnified view of pipe deformations

SECTION 4

FATIGUE LIFE PREDICTION

GROWTH OF FATIGUE CRACKS AT HOLES SUBJECTED TO PIN OR REMOTE LOADING

A.P. PARKER* and D.J. CARTWRIGHT**

SUMMARY

The fatigue growth rate of two equal length, diametrically opposed cracks emanating from a region of stress concentration is measured. The region of stress concentration is caused by either a central stress-free hole in a uniformly stressed strip, or a central pin-loaded hole near one end of the strip, the other end being subjected to uniform stress. Analysis of the crack growth rates in terms of the range of the opening mode stress intensity factor, calculated by the use of a point matching technique, indicates that the results for unloaded holes conform with existing data, and with data from a calibration specimen. A similar agreement occurs for pin-loaded specimens providing that the pin-loaded hole is sufficiently far from the stress-free end of the specimen. When the stress free end is close to the pin-loaded hole, crack growth is no longer normal to the loading axis, as assumed in the theoretical model.

1. INTRODUCTION

The failure of structural components by the growth of cracks is a problem of increasing importance in various fields of engineering. Failures of this type may occur as a result of both monotonic and cyclic loading. The use of a fail-safe approach to the design of such components creates the need for in-service inspection, which implies an accurate

---

* Department of Civil Engineering, Royal Military College of Science, Shrivenham, Swindon, England.

** Department of Mechanical Engineering, University of Southampton, England.

knowledge of crack growth rate and critical flaw size. These are required in order to estimate residual strength and optimum inspection intervals of damaged components. The stress intensity factor, which controls crack growth, is thus required for these structural components.

In many structures load is transferred through a pin or rivet joint. Several workers have considered these configurations over a period of many years. Despite this activity, failures originating at pin and rivet loaded joints are still commonplace. The lack of knowledge of stress fields around cracks in such configurations has inhibited their safe and efficient design.

In this paper results for stress intensity factors are presented for configurations which are intended to model through-the-thickness cracks emanating from a single hole in a rectangular sheet subjected to the following loading conditions:

a. A uniform tensile stress applied to each end of the sheet.

and b. A traction on the hole boundary in one direction, balanced by a uniform stress applied to the end of the sheet in the opposite direction.

The uniformly stressed configuration (a) is used to model a stress free cracked hole in a specimen subjected to end loads remote from the hole (remotely loaded type 'A' specimen). Configuration (b) is used to model a pin-loaded cracked hole (pin-loaded type 'B' specimen).

Fatigue crack growth rates are determined experimentally on both types of specimen as a function of the stress intensity factor range, ΔK, calculated from the numerical stress intensity solutions. These results are compared with available crack growth rate data, and with additional measurements on a calibration specimen.

For the remotely loaded specimens the crack growth rates are shown to agree with available data and calibration specimen results. The same general agreement is noted in the case of pin-loaded specimens, although crack growth rates are influenced by the presence of sliding mode crack deformation, and a realignment of the crack growth direction.

2. STRESS INTENSITY FACTORS FOR FATIGUE CRACK GROWTH RATE SPECIMENS

The overdetermined (least squares) point matching technique has been used previously {1}, {2} in the determination of opening mode stress intensity factors, K_I for configurations involving two, equal length, diametrically opposed radial cracks emanating from a central circular hole of diameter D in a rectangular sheet of height H and width W. The distance between cracks tips is 2a. These results included the case of remote, uniform uniaxial stress σ, Figure 1a, and of uniform stress applied to one end, the sheet being held in equilibrium by a normal pressure applied to half of the hole boundary, Figure 1b.

The details of the numerical technique and solutions are available {2}. The stress intensity factors for the model specimens necessary for the analysis of the experimentally measured crack growth rates, are also illustrated in Figures 1a and 1b. The results for D/W = 0.15, and H/W = 0.5 and 0.75 are additional to those available in reference {2}.

3. EXPERIMENTAL MEASUREMENT OF FATIGUE CRACK GROWTH RATES

Measurements have been made of the number of cycles N, versus crack length a for the specimens shown in Figure 2. These results have been analysed using the stress intensity factors presented in section 2. The objectives of the experimental programme were:

a. To check the stress intensity factors for remotely-loaded and pin-loaded specimens predicted on the basis of the theoretical model (see section 2).

b. To compare the rates of crack growth in geometrically similar remotely-loaded and pin-loaded specimens (see Figures 2a and 2b respectively).

c. To determine the crack path in both remotely-loaded and pin-loaded specimens.

Fatigue crack growth rates of a pre-existing crack at a hole (inset on Figure 2) were measured on both types of specimen. In the remotely loaded, type 'A' specimens (Figure 2a) the pre-cracked hole was located centrally with the loads applied through pins remote from the hole. For the pin-loaded, type 'B' specimens the load was applied at one end through the pre-cracked hole, this being reacted remotely at the other end of the specimen. The dimensions for the type 'A' specimens are also given in Figure 2a. The ratios of hole diameter, D to the width, W listed in Figure 2

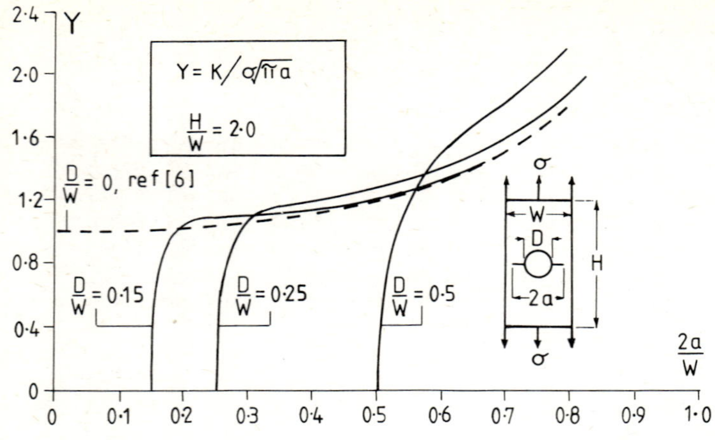

(a) Remote-Loading

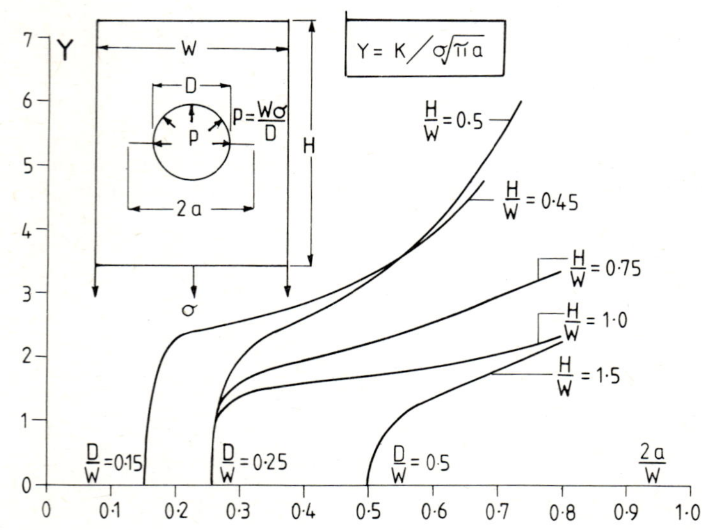

(b) Pin-Loading

Figure 1: Stress Intensity Factors for Model Specimens

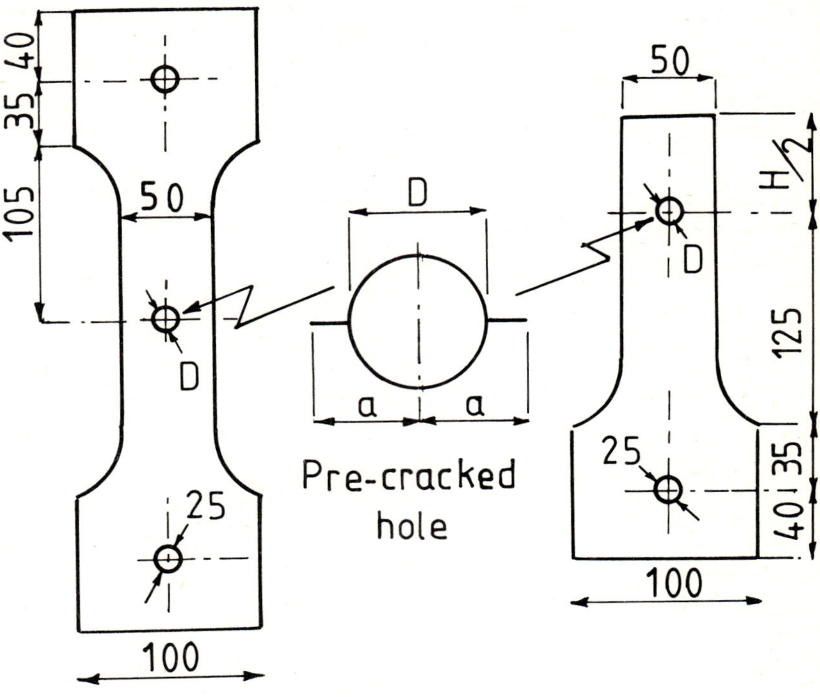

Diameter 'D' (mm)	D/W	Number Tested
7.5	0.15	3
12.5	0.25	3
25.0	0.5	3

(a) Remote Loading (Type 'A') Specimens

Diameter 'D' (mm)	D/W	H (mm)	H/W	Number Tested
7.5	0.15	11.25	0.45	3
12.5	0.25	12.5	0.5	3
12.5	0.25	18.75	0.75	3
12.5	0.25	25.0	1.0	3
25.0	0.5	37.5	1.5	3

(b) Pin Loading (Type 'B') Specimens

Figure 2: Fatigue Crack Growth Rate Specimens and Numbers

were selected so as to include a wide range of values for D/W covering typical applications.

The type 'B' specimens are of width W and contain a hole of diameter D with a distance from the pin centre to the stress-free end of H/2. Fatigue tests were carried out on a total of 15 pin-loaded specimens. Specimen dimensions for the pin-loaded configuration are given in Figure 2b.

The specimens were made from 3.25mm thick L72, an unclad, solution-treated aluminium-copper alloy. This material is a commonly used aircraft alloy and its properties are well known. Crack growth data is available for this material {3}.

The stress levels for all tests were selected to be typical of those used in aircraft structures. The ratio of minimum to maximum stress in the fatigue cycle, R, was held constant at 0.1, which is also typical of that used in aircraft structures. A Schenck fatigue testing machine was used for the experimental programme. This has an electronic oscillating and static force regulating system, and operates at approximately 45Hz. The crack length was monitored using a travelling microscope mounted on a base plate fitted to the fatigue machine beneath the specimen. The accuracy and ease of crack length measurement was improved by using a variable frequency strobe light set to the same frequency as the fatigue loading. The microscope/strobe combination was enclosed in a reflective foil box, which eliminated external light, but reflected the light from the strobe. Using this arrangement, measurement could be made of the increments in crack length down to 0.05mm.

After each fatigue crack growth test an acetate replica of the fatigue crack was made so as to provide a permanent, accurate record of the crack path. The same replica technique was employed {4} to monitor growth of small cracks. Since surplus acetone could influence crack growth rate this method was only used at the conclusion of each test.

The crack was initiated by drilling an initial under-sized (4mm diameter) hole and cutting equal length slots from each side of this hole. The ends of the slots were then sharpened in order to increase the stress concentration effect. Initial crack growth was at a relatively high stress level, the net stress at the minimum cross section being approximately one third of the yield stress, with cracks growing out of small plastic zones. This procedure gave an even nucleation through the thickness of the specimen. The stress level was then reduced and cracks grown until they were 0.5 - 0.75mm beyond the proposed hole boundary. Any tendency towards the development of unequal crack lengths was eliminated by subjecting the specimen to in-plane bending until equal length cracks were obtained. Finally, the initial 4mm hole was enlarged to produce an accurate circular hole with short, equal length fatigue cracks.

The specimen, prepared as above, was re-mounted in the fatigue machine. The hole diameter was measured with the travelling microscope and after applying the mean load, the initial crack tip coordinates were measured relative to the centre of the hole. The appropriate cyclic loading was applied and each crack tip coordinate was measured alternately, every 10000 cycles initially, this interval being reduced as crack growth rates increased during the test.

The main problem with the remotely loaded (type 'A') specimens was their sensitivity to very small differences in initial crack length, and the subsequent tendency for these cracks to grow to unequal lengths. In all the specimens reported the crack tip coordinates did not differ by more than 5%: in fact unequal crack growth rarely occurred in the pin loaded specimens, and differences in initial crack length tended to be eliminated.

4. RESULTS AND DISCUSSION

4.1 REMOTELY LOADED SPECIMENS

The cyclic crack growth rate (da/dN) was obtained numerically from the a versus N data for each specimen. This was done using a computer programme {5} in which the slope (da/dN) is determined, for a crack of length a, as the weighted average of the slope of the straight line joining the previous data point to that at a, and the slope of the straight line joining that at a to the next data point. The da/dN data so obtained was in close agreement with that obtained by constructing tangents to the curves of a versus N. The crack tip stress intensity factor range, ΔK was calculated at each value of a from the relationship:

$$\Delta K = Y(\sigma_{max} - \sigma_{min})\ (\pi a)^{\frac{1}{2}} = Y(\Delta\sigma)(\pi a)^{\frac{1}{2}} \qquad (1)$$

where σ_{max} and σ_{min} are the maximum and minimum values of the average stress in the loaded section width W, and Y is the appropriate geometrical factor determined from Figure 1. The values of Y for a D/W ratio of zero (central crack) were obtained from reference {6}, all others were obtained from the theoretical results given in section 2. The results in Figure 1a, for H/W = 2, were used for H/W ⩾ 2 since the numerical results were insensitive to any further increase in H/W {1}.

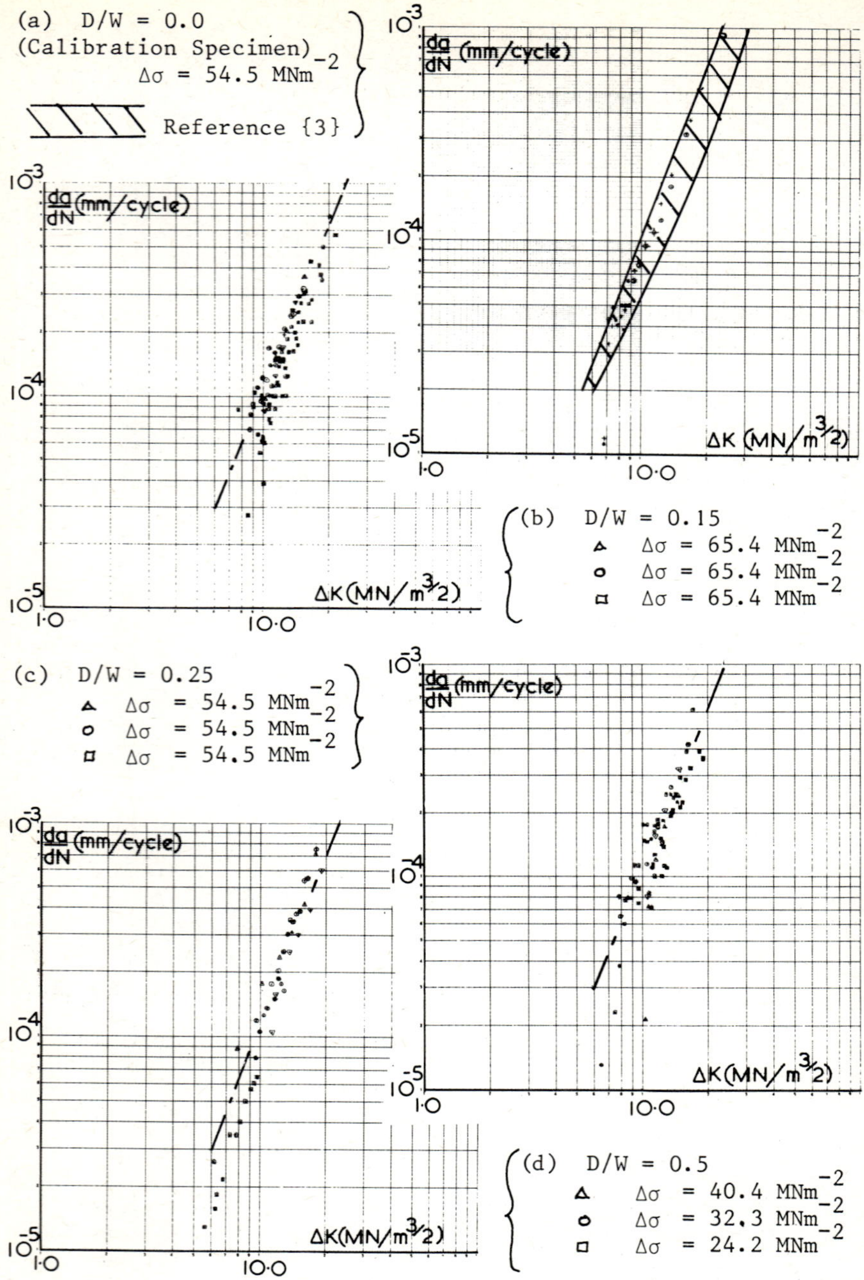

Figure 3: Crack Growth Rate Results for Remotely-Loaded (Type 'A') Specimens

Figure 3a shows a plot of da/dN versus ΔK for the centre cracked sheet (D/W = 0) calibration specimen. These results which exhibit a characteristic straight line behaviour {7}, are compared with published data {3} for L72 alloy. All results for the calibration specimen fall within the 90% confidence limit given for the data in reference {3}, and for further comparison this curve is repeated on all plots for the type 'A' and 'B' specimens as a single dashed line. Figures 3b-3d show da/dN versus ΔK for all remotely loaded, type 'A' specimens.

4.2 PIN LOADED SPECIMENS

Crack growth measurements were made on type 'B' (pin-loaded) specimens having D/W ratios of 0.15, 0.25, 0.5, the details of which are given in Figure 2. In these specimens the crack may grow at an angle (δ) to the line of the minimum section. For these situations the length a in equation (1) was taken to be the perpendicular distance from the crack tip to the loading axis.

The results for H/W = 1.5 and D/W = 0.5 are shown in Figure 4a. Only a limited number of results were obtained for this specimen because of the difficulty of keeping the crack lengths on either side of the hole within 5% of each other. However the results show agreement with the calibration curve, although they are generally below the curve. As with all the type 'A' specimens the angle of the crack to the line of minimum section (δ) was approximately zero.

Results for H/W = 1.0 and D/W = 0.25 are shown in Figures 4b and 4c. For these specimens no problems were experienced in maintaining equal length cracks. All the growth rates show agreement with the claibration curve.

The results for H/W = 0.75, 2R/W = 0.25 are shown in Figures 4d, 5a and 5b in which the average crack growth angle (δ) was 11.5°. For H/W = 0.5, 2R/W = 0.25 results appear in Figure 5c, the average crack growth angle (δ) being 13.7°, and those for H/W = 0.45, 2R/W = 0.15 in Figure 5d, with average crack growth angle (δ) of 20.6°.

5. DISCUSSION

For the type 'A' specimens it can be seen from Figures 3b-3d that the crack growth rates are in close agreement with results for the calibration specimen. This implies that the model is a reasonable representation of the specimen behaviour for remote loading.

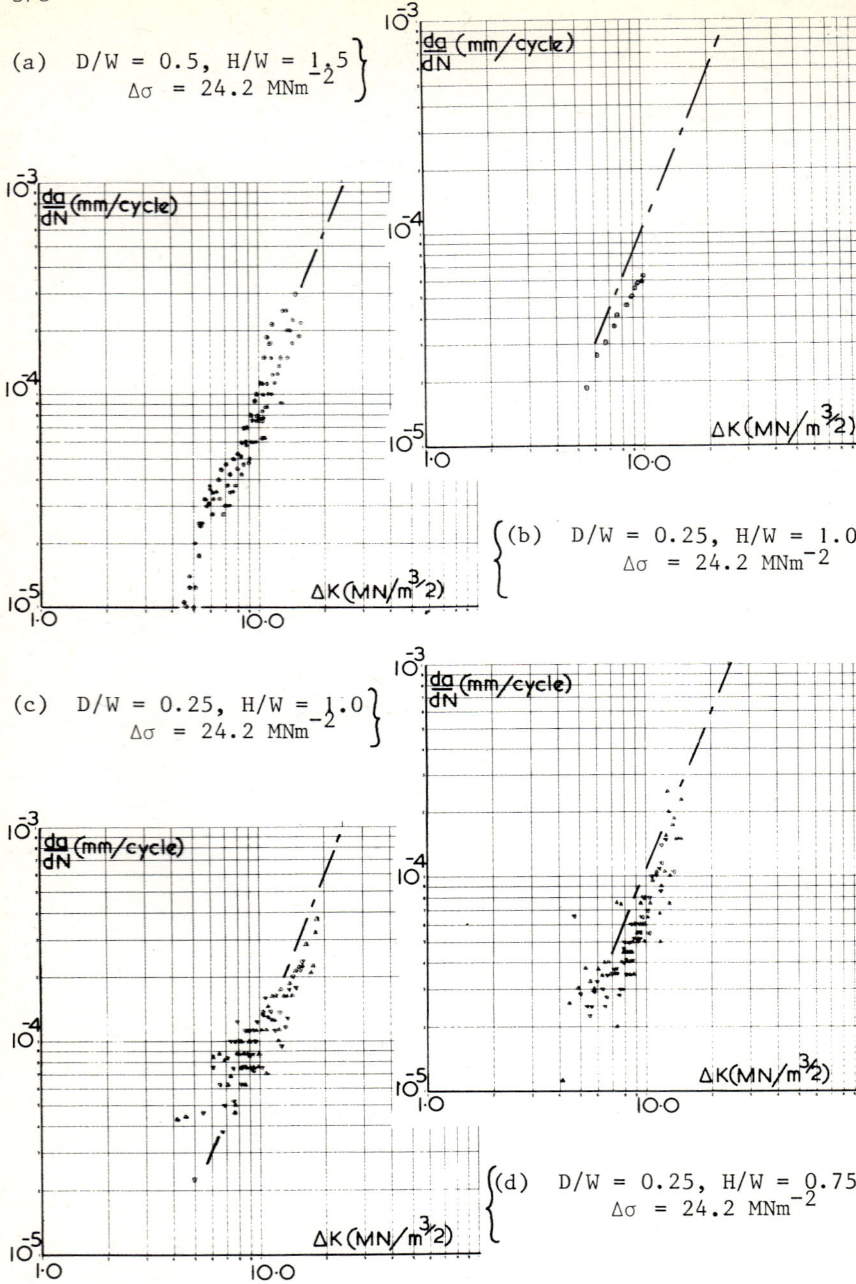

Figure 4: Crack Growth Rate Results for Pin-Loaded (Type 'B') Specimens

(a) D/W = 0.25, H/W = 0.75
$\Delta\sigma$ = 24.2 MNm^{-2}

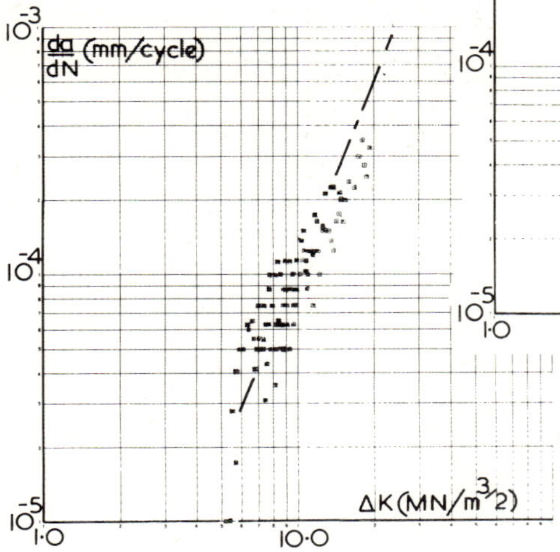

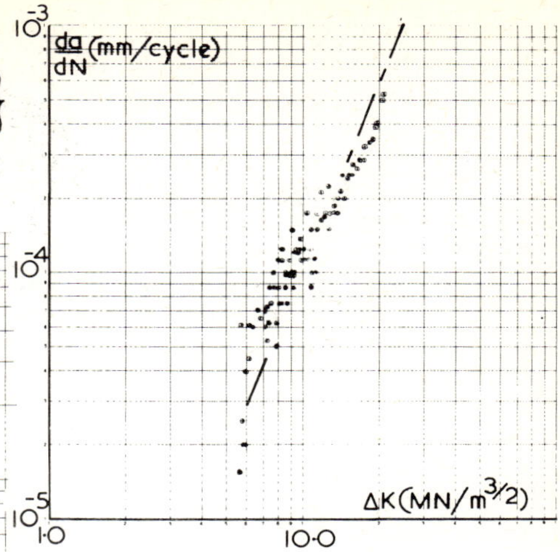

(b) D/W = 0.25, H/W = 0.75
$\Delta\sigma$ = 24.2 MNm^{-2}

(c) D/W = 0.25, H/W = 0.5
△ $\Delta\sigma$ = 24.2 MNm^{-2}
○ $\Delta\sigma$ = 24.2 MNm^{-2}
□ $\Delta\sigma$ = 24.2 MNm^{-2}

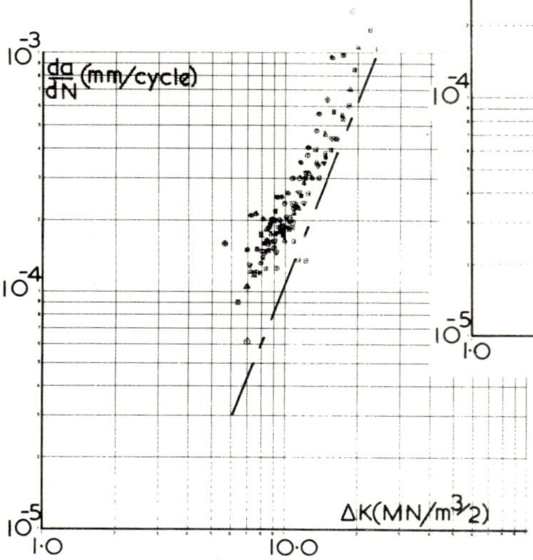

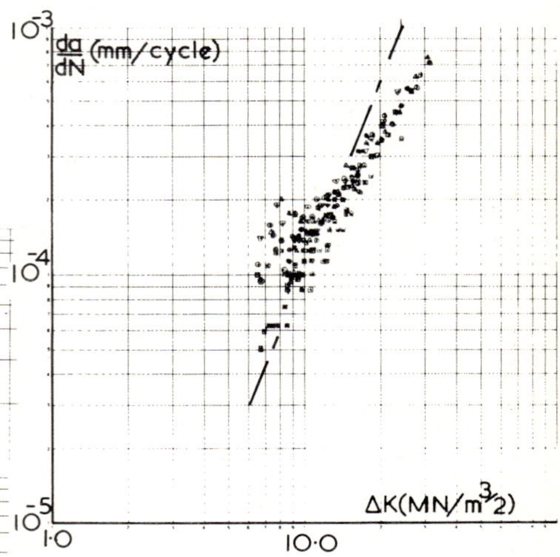

(d) D/W = 0.15, H/W = 0.45
△ $\Delta\sigma$ = 24.2 MNm^{-2}
○ $\Delta\sigma$ = 24.2 MNm^{-2}
□ $\Delta\sigma$ = 24.2 MNm^{-2}

Figure 5: Crack Growth Rate Results for Pin-Loaded (Type 'B') Specimens

For the pin loaded, type 'B' specimens there are several differences between the assumed model and the actual specimen. These may be summarised as:

a. The hole boundary loading will not be a simple uniform pressure distribution, and may include shear tractions.

b. The remotely applied loading will not produce a uniform direct stress at a distance H/2 below the hole centre. (This effect may be relatively insignificant, no evidence of such an effect is apparent in the case of remotely loaded specimens in which this is the only discrepancy between the model and the physical situation).

c. The cracks may grow at an angle (δ) to the direction assumed in the model.

d. The cracks were initiated under symmetric, mode I loading. However, because of the asymmetry caused by the pin loading, there will be a sliding mode contribution to crack growth. This will be more significant initially, and will tend to reduce as the crack realigns and grows.

All the pin loaded specimens with $D/W < 0.5$ exhibited a clear initial increase in da/dN with the number of cycles, followed by a decrease and then a gradual increase. This effect, which is depicted schematically in Figure 6, is seen in Figures 4c, 4d, 5a-5d: it tends to be more pronounced with increasing crack angle, δ. Initially the crack growth rate is higher than that for the calibration specimen and following a decrease establishes a rate lower than that for the calibration specimen.

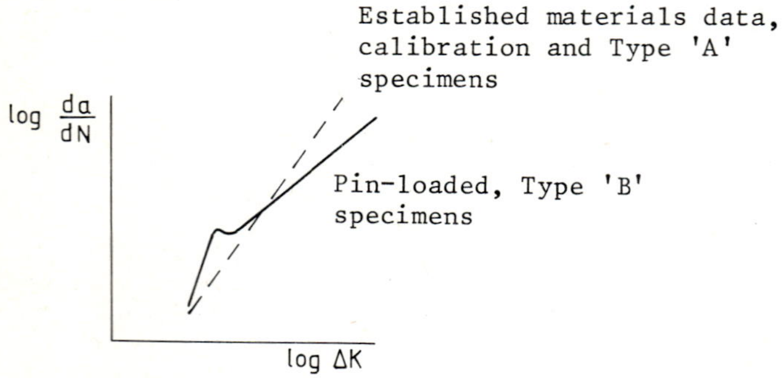

Figure 6: Schematic Form of Crack Growth Rates for Pin-Loaded type 'B' Specimens.

An interpretation of the initial rise in crack growth rate may be associated with crack growth under mixed mode conditions (see (d) above), whilst the subsequent decrease occurs as the crack realigns itself perpendicular to the direction of maximum principal stress near the hole boundary.

The tendency to unequal length crack growth in the case of all remotely loaded specimens, and pin loaded specimens with D/W = 0.5, may be contrasted with equal length crack growth for smaller pin loaded holes. (This phenomenon is explained on the basis of asymptotic K values in reference {1}). This suggests that in typical lug geometries the tendency will be for single cracks, whilst in riveted configurations, with relatively small D/W ratios, the tendency will be for the initiation of numerous, equal length cracks. Such crack growth is observed in practice {8}.

6. CONCLUSIONS

It has been shown that the growth rate of cracks emanating from a traction-free hole in a uniformly-stressed sheet is consistently in agreement with those predicted by use of stress intensity factors calculated by a point matching technique.

In the case of cracks at pin-loaded holes the growth rates are in general agreement with those predicted from the numerical results, provided the distance, on the loading axis between pin centre and free boundary is not less than half the strip width. This ratio includes most typical lug geometries. For values of this ratio less than 0.5 the crack growth deviates substantially from the line of minimum section through the pin centre, and hence the model in which the crack was assumed to lie along the line of minimum section is inappropriate.

For all remotely-loaded cracked holes and for pin-loaded holes with D/W = 0.5 there is a tendency to develop unequal length cracks at either side of the hole. However, for the pin loaded holes having D/W = 0.25 there is a clear tendency to develop equal length cracks. Typical rivet diameter/ spacing ratios are close to 0.25, so there will be a tendency towards numerous, equal length cracks in such structures.

7. REFERENCES

{1} PARKER, A.P. - The Mechanics of Fracture and Fatigue in Common Structural Configurations. PhD Thesis, Dept. of Mechanical Engineering, University of Southampton, 1979.

{2} CARTWRIGHT, D.J. and PARKER, A.P. - Opening Mode Stress Intensity Factors for Cracks in Pin Loaded Joints. Int. J. Frac., (To be submitted).

{3} Crack Propogation in Sheet. Engineering Sciences Data Unit, Item Number 73021, 1973.

{4} CARDRICK, A.W. and GARDINER, R.W. - A Replica Method for Recording the Surface Growth of Fatigue Microcracks, Technical Report TR 72210, Royal Aircraft Establishment, Farnborough, 1973.

{5} MILLER, M. - Stress Intensity Factors and Crack Growth Rates in Integrally Stiffened Panels. PhD Thesis, Dept. of Mechanical Engineering, University of Southampton, 1975.

{6} NEWMAN, J.C. - An Improved Method of Collocation for the Stress Analysis of Cracked Plates with Various Shaped Boundaries, NASA Technical Note TN D-6376, 1971.

{7} PARIS, P.C. and ERDOGAN, F, - A Critical Analysis of Crack Propogation Laws, Trans ASME, J Basic Engng, Vol 85, pp. 528-534, 1963.

{8} KIRKBY, W.T. - Examples of Aircraft Failure, Fracture Mechanics of Aircraft Structures, Ed Liebowitz, H., AGARD-AG-176, pp. 8-13, 1974.

8. ACKNOWLEDGEMENT

The authors wish to acknowledge support from the Procurement Executive of the Ministry of Defence to the University of Southampton during the course of the research.

HIGH ORDER ELEMENTS AND FATIGUE CRACK GROWTH

I.J. Smith†

SUMMARY

Methods to calculate first and second mode stress intensity factors for cracks in welds were examined and a finite element displacement formulation using high order isoparametric elements was chosen. Cubic quadrilateral elements were used and the crack singularity modelled by node shifting and side collapse. This method allows calculations of K_I and K_{II} to within a few per cent for well known test pieces using only the displacement of the first pair of nodes on the free crack surface.

A program was produced, especially orientated towards modern small computers, using this approach for the 2 dimensional study of crack growth and direction during the fatigue of a welded joint.

The reasons for choosing this approach and its advantages over other singularity elements are discussed as is its extension to 3 dimensions and elasto-plasticity. The results for known geometries are presented and its use in the fatigue studies of weldments shown.

INTRODUCTION

Metal components and structures, when subjected to repeated loadings and unloadings, can fail at stresses well below the stresses they can withstand in a single, static loading. This phenomenon is known as fatigue and has been the subject of investigations for over a century now [1,2].

The failure of components usually occurs at some stress concentrator, for example at a re-entrant corner where a change of section is made. Structures can show fatigue problems too, and these occur likewise at stress concentrators. With modern fabrication methods this usually means at the welds used to join the individual parts of the structure together [3]. The reasons for this are twofold. There are initial crack-like defects in the weld toe, which is a region of severe stress concentration due to the weld geometry. There are also high residual stresses present. These residual stresses may be of yield stress magnitude and are caused by the shrinkage of the weld metal as it cools down to ambient temperature.

---

† Senior Research Engineer, Structures and Computation Dept, The Welding Institute, Abington, Cambridge

When such a welded joint is subject to cyclic loading, the defect can grow. It has been established that fracture mechanics provides a powerful way of describing such growth via a single parameter, the stress intensity factor (K). The fatigue crack growth laws, the best known of which is due to Paris [4], relate the rate of crack growth ($\frac{da}{dN}$) to the range of the stress intensity factor (ΔK). For example, Paris' law is usually written

$$\frac{da}{dN} = C\ \Delta K^m \tag{1}$$

where C and m are material constants. If the logarithms of each side of equation (1) are plotted against each other, a straight line of slope m and intercept C is produced. In the new British fatigue design rules for welded joints (BS 5400 Pt 10) a value of m = 3.0 is chosen to be representative of most of the common welded joints. (For steel, C has a value around 10^{-13} - 10^{-12} when crack growth is measured in mm/cycle and ΔK in $Nmm^{-3/2}$) [5].

Integration of the crack growth law between the initial crack size a_i and the final crack size a_f results in an equation relating fatigue life to range of stress intensity factor thus:

$$N = \frac{1}{C}\int_{a_i}^{a_f} \frac{da}{\Delta K^m} \tag{2}$$

From equation (2), once the relationship between ΔK and a is known, the fatigue life can be estimated.

The path the crack takes, and the rate at which it grows, in biaxial loading is more complicated [6,7] and requires a knowledge of both K_I, the mode 1 stress intensity factor, and K_{II}, the mode 2 stress intensity factor.

This paper treats the problems in obtaining values for these stress intensity factors, given a particular weld joint geometry, loading and crack size. To allow any joint geometry and loading to be considered a numerical method of analysis is obviously required. The two main methods which have shown sucess in dealing with this type of problem have been the finite element method (FEM) [8,9,10] and the boundary element (or boundary integral) method (BEM) [11,12].

The BEM approach only requires discretisation of the surface (at least in the elastic case) of the body being considered, unlike the FEM which requires discretisation of the entire body. Therefore, the BEM approach shows most promise in 3-dimensional work where the reduction in problem size becomes marked. In 2-dimensional work the difference is not so clear, due to the extra complexity of the BEM's formulation. Also, the handling of elasto-plasticity and initial (residual) stresses and strains by the BEM are still the subject of much research.

The finite element method is, of course, a well known numerical method in stress analysis and has also been applied to problems in a great number of different disciplines. As a result of both the work in establishing elements with good modelling properties for crack tips, and the emergence of literature devoted to the efficient programming of the FEM, it was thought that this approach would prove advantageous in dealing with the estimation of stress intensity factors in fatigue crack growth studies. It can deal with non-linearities in both material properties and geometry and also with initial (residual) stresses and strains. This paper reports the formulation and use of a FEM computer program in the 2-dimensional linear elastic fracture mechanics (LEFM) studies of cracks, and shows how such a program can be extended to 3-dimensions and to elasto-plastic crack growth analysis.

ANALYSIS REQUIREMENTS

The analysis method chosen for this work had to satisfy a number of different criteria. It had to evaluate both the first and second mode stress intensity factors (K_I and K_{II}) for an arbitrary crack in an arbitrary 2-dimensional body, under any combination of traction or displacement loadings. Also, the method should be efficient in computer resources and be easy to use, especially with respect to the preparation of the data. It should also have the capability to be extended to 3-dimensions and to non-linear material properties.

The accuracy required of the numerical method is quite severe due to the form of the fatigue crack growth laws. In these the stress intensity factor appears raised to a power, usually around 3, and we therefore need to be able to evaluate ΔK to better than 1-2% if possible. This contrasts with studies in fracture where the material toughness properties are usually subject to large scatter, and stress intensity factors of 5-10% accuracy are sufficient.

Fortunately, an accurate crack tip singularity element recently proposed is ideally suited for incorporation in one of the modern and efficient FEM programming approaches. Specifically, the collapsed cubic quadrilateral isoparametric element suggested by Pu, Hussain and Lorensen [13] was used for crack tips, and a program based on the displacement formulation of the FEM using isoparametric elements and a frontal-based solver [14] was written incorporating this crack tip element.

CRACK TIP ELEMENTS

The representation of the singularity in the stress field at a crack tip has been the subject of much effort and a variety of approaches have been tried. These approaches have been reviewed extensively [15] and will not be discussed further except to say that most of them use some form of special element around the crack tip. Henshell and Shaw [16], and independently Barsoum [17], first pointed out that by moving the mid-side nodes of the two adjacent sides of a quadratic isoparametric quadrilateral element to the quarter points, the required singularity was produced along these sides. Subsequently, it was shown that the singularity could be produced along any ray from a corner of the element if the quadrilateral had one side collapsed to form a triangle [17]. This degenerate element has a much superior performance as compared with the uncollapsed one, especially for mixed mode cracks [13].

A comparison has been made recently between various crack tip singularity elements by Fawkes, Owen and Luxmoore [18]. They came to the conclusion that an analytical crack tip element was superior to elements based on the shifted node quadrilateral (among others). Their main criticisms of the shifted node approach were that it does not adequately calculate K_{II} in mixed mode situations, and that additional effort in extrapolating the stress intensity factors from the displacement solutions was required. They suspected that the cause of the mixed mode problem was due to the singularity modelling only being applicable along the sides of the element, but they did not test a degenerate element. Their suspicions are probably correct for, as is shown in this paper and elsewhere [13], when degenerate isoparametric elements are used, K_I and K_{II} can be evaluated accurately in mixed-mode cases.

Most of the work with shifted node elements has involved the use of the quadratic quadrilateral (8 nodes) or the quadratic triangle (6 nodes), mainly because these are popular, widely used elements in general finite element applications. The quadratic element is, of course, only one of a series of elements known as the 'Serendipity' family [10], each having a different order of shape function used in their formulation. This element, of order 2, is the first of the family having mid-side nodes moved to form a singularity. The parabolic variation in displacement, and linear variation in stress, that can be represented by the quadratic element, is probably adequate for general stress analysis, but it falls short of that needed in the presence of a singularity. Even though the immediate crack tip elements take account of the tip singularity, there is a high stress gradient in the region of the crack. Higher order elements would allow a coarser mesh in this near crack region as well as reducing the number of elements required throughout the body. The next higher order element in the 'Serendipity' family, of order 3, is the cubic, i.e. it can represent a cubic variation in displacement or a parabolic variation in stress. It has 2 mid-side nodes per side and therefore a total of 12 nodes per quadrilateral. It also can have its mid-side nodes shifted, this time to the one-ninth and four-ninth positions, to produce the crack tip singularity. The use of cubic elements, with either the degenerate isoparametric element [13] or a special analytical element for the crack [19], has been shown to require very few elements to model cracked bodies. It was therefore decided to use the degenerate, shifted-node cubic quadrilateral element as a crack tip element in this work. Figure 1 shows the degenerate element and the arrangement of these elements used around a crack tip.

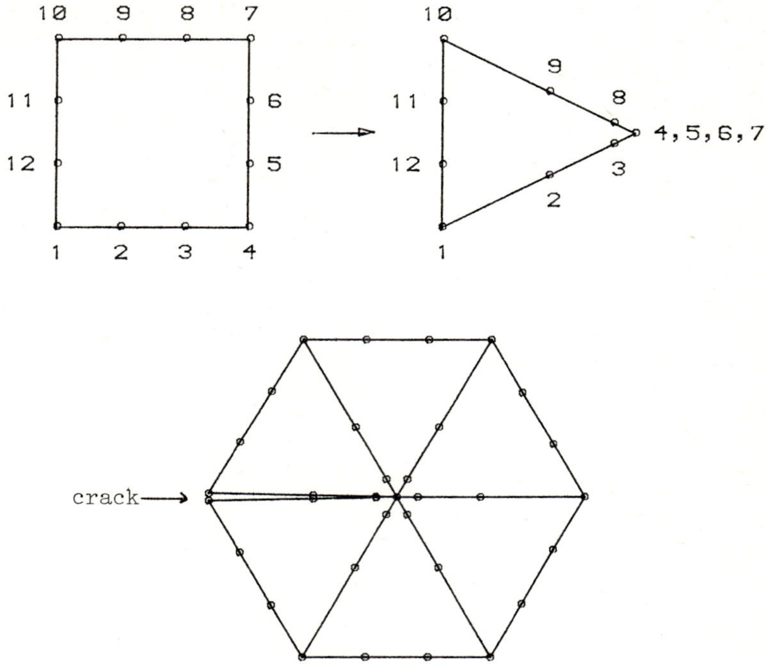

Fig. 1 The Crack Tip Element

COMPUTER PROGRAM

To make use of the degenerate cubic crack tip element, a FEM computer program called PLIPE (Plane Isoparametric Problem solver, Elastic version) was written. This program follows the approach of Hinton and Owen's text on finite element programming [14], but is generalised for a range of different element orders. It also has a number of changes in the file arrangements and certain routines were totally rewritten to improve efficiency, and allow the calculation of stress intensity factors.

The program used the displacement formulation of the finite element method, i.e. it assembles the contribution of each element to produce an overall equation for the body relating displacements with applied loads thus

$$[K]\ \{\delta\} = \{P\} \tag{3}$$

where [K] is the global stiffness matrix, $\{\delta\}$ is the vector of displacements and {P} is the vector of applied loads. The program then solves equation (3) to find displacements and thus stresses, strains, stress intensity factors, etc.

The actual assembly and solution of the global stiffness matrix is never carried out, as the frontal solution method [14] assembles, and reduces as far as possible, all the variables from each element in turn. The major advantage with a frontal solver is that node numbering can be totally arbitrary, as the frontwidth is dependent upon the element numbering, which is easier to control. The stiffness matrix for each element is formed and stored before all the matrices are assembled and reduced and if a series of problems is being tacked (e.g. different crack lengths) only those elements that change need be regenerated. This also reduces the input data required.

Substructuring is also easy with a frontal solver because when the front has advanced through the mesh by a certain amount, the reduced equations are effectively those of a substructure. This facility will be implemented in PLIPE in due course.

Because of the high order of element formulation, stress intensity factors are evaluated either directly from the displacements of the nearest pair of crack tip nodes or by linear regression (LR) through the first three pairs of nodes. The relationship used to relate the displacements u and v to K_I and K_{II} are given by a truncated form of the general expression for the displacement field around the crack tip singularity [13]. Thus K_I and K_{II} are given, for any node on the crack face, by

$$K_I = \frac{v}{\sqrt{r}} \cdot \text{const} \tag{4}$$

$$K_{II} = \frac{u}{\sqrt{r}} \cdot \text{const} \tag{5}$$

where $\text{const} = \dfrac{E\sqrt{2\pi}}{4}$ for plane stress

or $= \dfrac{E\sqrt{2\pi}}{4(1-\nu^2)}$ for plane strain

and
v = displacement perpendicular to the crack
u = displacement parallel to the crack
r = distance from the crack tip
ν = Poisson's Ratio
E = Young's Modulus.

Both the methods (nearest nodes or LR) give good results if the length of the crack tip elements are around 13-14% of the crack size, the LR results being slightly more accurate.

In Hinton and Owen's text the use of sequential disc files, for the storage of the element stiffness matrices and the reduced equations, is shown. PLIPE however does not use sequential disc files. It exists in two versions, one of which uses random access disc files and was produced for use on fixed memory machines or where the programmer wants total control of the storage of these variables. The other version takes advantage of the virtual memory arrangements on machines such as the Welding Institute's DEC 2020, and does not use storage files at all. This effectively allows the computer system to control the disc storage itself and results in slightly faster run times.

RESULTS FOR TEST GEOMETRIES

The following cracks in finite plates were chosen as test problems for analysis by PLIPE

(i) Centre crack in tension
(ii) Single edge crack in tension
(iii) Single edge crack in bending
(iv) Double edge crack in tension
(v) 45° slant edge crack in tension
(vi) 45° slant edge crack in bending.

The results obtained from PLIPE were compared with curves presented in Rooke and Cartwright's collection of stress intensity factors (R & C) [20]. Problems (i) to (iv) are identical except for different boundary conditions and/or loads and (v) and (vi) only differ in applied loads. As each problem was analysed twice, four sets of meshes in total were used. A mesh set consisted of a number of nearly identical meshes, each differing only in crack length. The mesh sets are summarised in Table 1 and typical members of each mesh set are shown in Figures 2 to 5.

Mesh set	Problems	No. of elements	Crack sizes	No. of degrees of freedom
1	(i) - (iv)	12	0.7 - 0.05	176
2		14	0.35 - 0.05	202
3	(v) & (vi)	16	0.505-0.345	216
4		22	0.6 - 0.3	282

Table 1 The Mesh Set Details

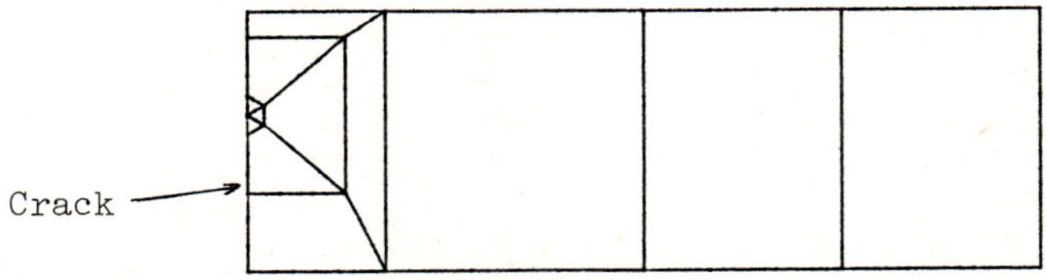

Fig. 2 Mesh Set 1

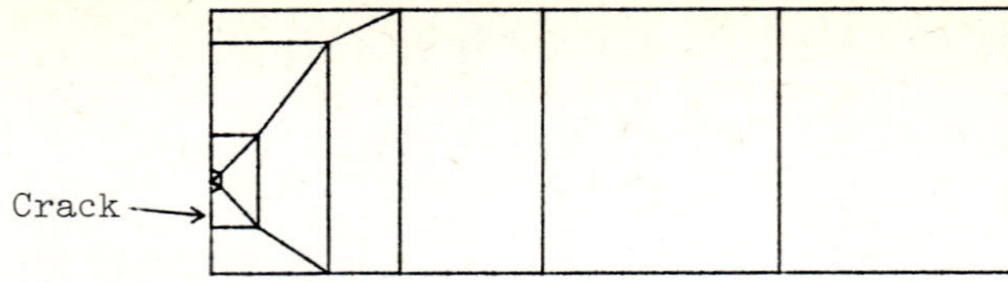

Fig. 3 Mesh Set 2

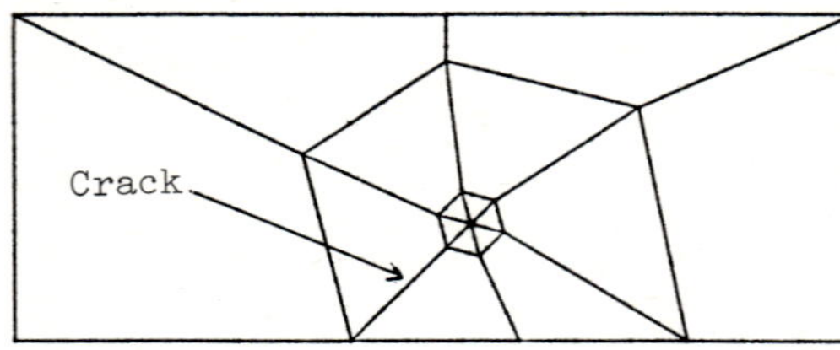

Fig. 4 Mesh Set 3

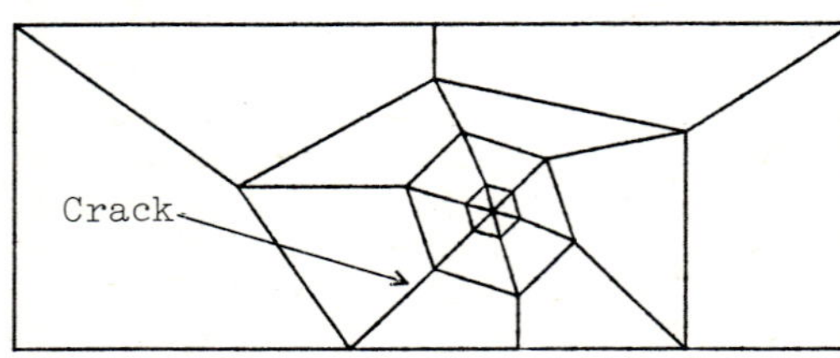

Fig. 5 Mesh Set 4

In mesh sets 1 and 2 (Figures 2 and 3) the loading was applied to the right-hand side edge of the mesh, and the crack ran from the bottom left-hand corner up to the centre of the crack tip zone. The application of the appropriate boundary conditions allows these meshes to represent the test problems (i) to (iv). The results obtained are shown on Figures 6 to 9 as K_I divided by the stress intensity factor for an infinite plate under the same stress conditions, plotted against the crack size as a ratio of the width of the plate. The length to width ratio (h/b) is 3 in all cases.

The solid lines are from R & C, the inverted triangles are from runs using mesh set 1 and the crosses from mesh set 2. The main difference between the meshes is that set 2 has an additional layer of elements near to the crack region, and these elements allow realistic modelling of the cracked plates down to the smallest crack sizes studied.

Another way of viewing the difference between mesh set 1 and 2 is that set 1 has 3 element layers before reaching the full width of the plate and set 2 has 4 layers. The two mesh sets for the 45° slant crack problems were designed similarly.

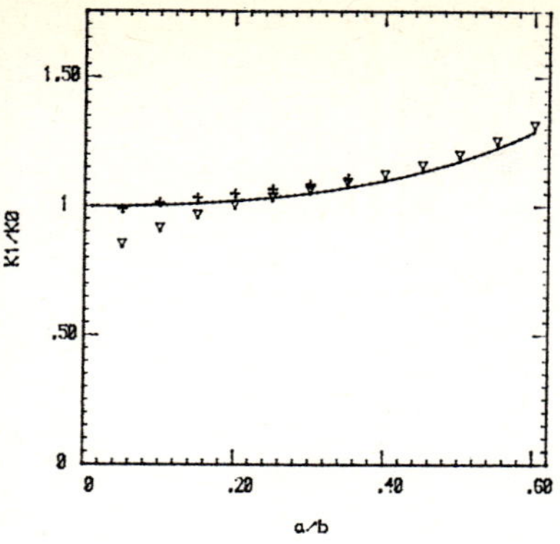

Fig. 6 Centre Crack in Tension

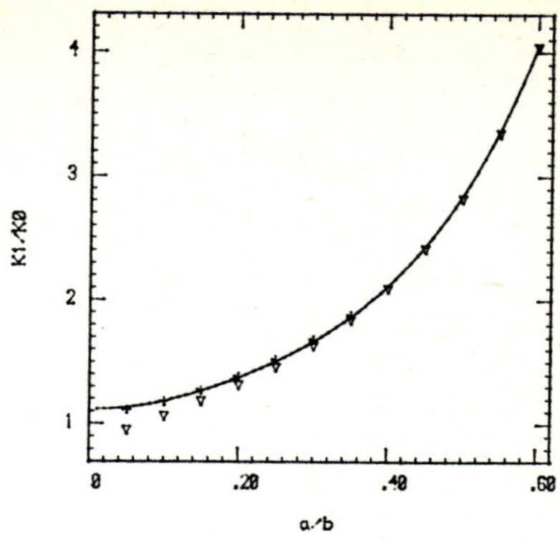

Fig. 7 Single Edge Crack in Tension

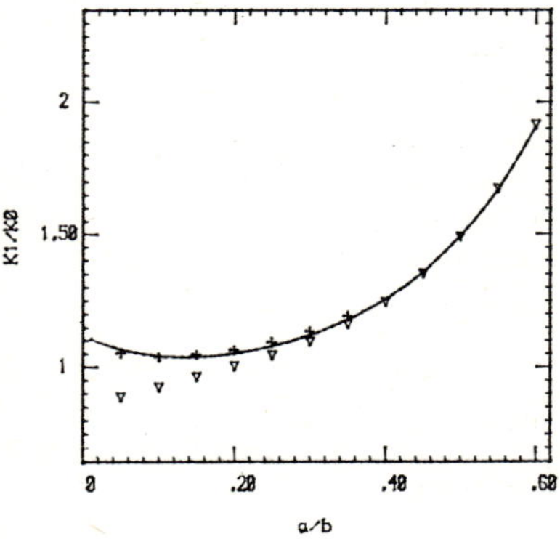

Fig. 8 Single Edge Crack in Bending

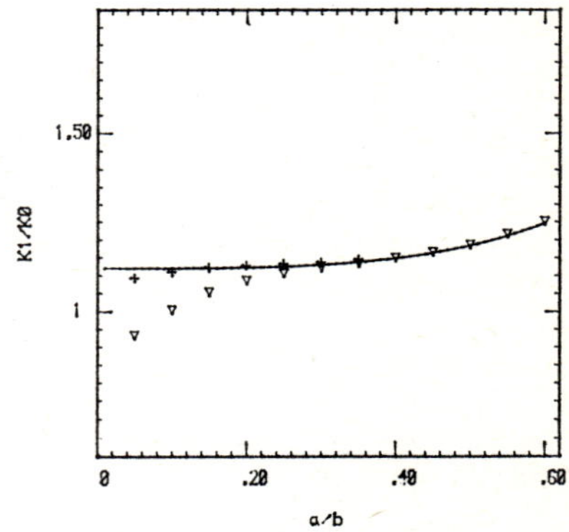

Fig. 9 Double Edge Crack in Tension

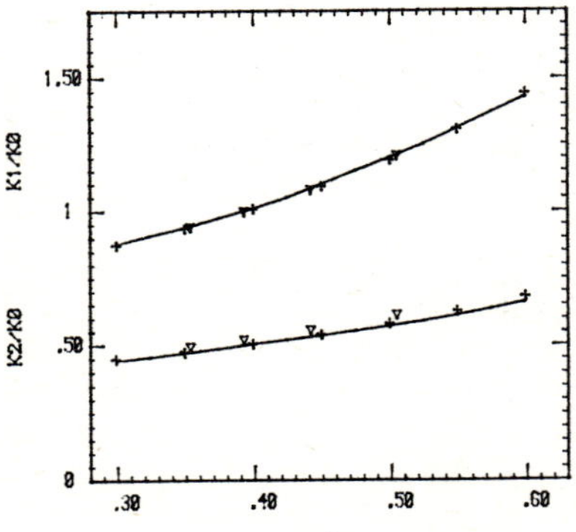

Fig. 10 Slant Crack in Tension, 45°

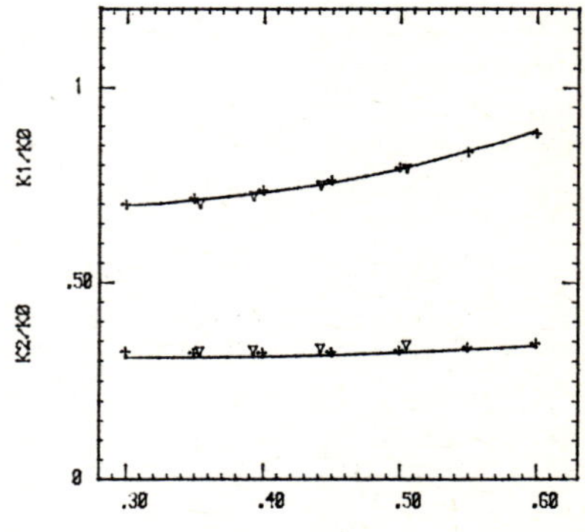

Fig. 11 Slant Crack in Bending, 45°

Mesh set 3 has 3 layers of elements between the crack and the boundary and mesh set 4 has 4 layers. The results obtained for K_I and K_{II} are shown in Figures 10 and 11.

The inverted triangles are the mesh set 3 results and the crosses are from mesh set 4 runs. The continuous line is once again from R & C.

DISCUSSION OF TEST PROBLEM RESULTS

Problems (i) to (iv)

Mesh set 1, the simplest, gives good agreement with R & C down to a/b values about 0.3 and mesh set 2 continues the agreement down to a/b = 0.05. In virtually all cases the agreement is within 2% and a considerable number of the points are less than 1% different from the R & C values.

Problems (v) and (vi)

For the 45° slant crack problems, only a/b values from 0.3 to 0.6 were considered as that was the extent of the results available in R & C. The results using mesh set 4 are all within 3% and most are within 1.5% of the values in R & C. The mesh set 3 results are not so good for the K_{II} results but are still within about 6% of the values in R & C.

Implications for mesh design

For more complicated geometries, these test problem results suggest the following guide to mesh design.

(a) Surround the crack tip by six degenerate elements whose length is around 14% of the crack length.

(b) Fit 3 or 4 element layers around this tip region, each layer's element sizes increasing by roughly 3 times the previous ones. The edges of the final layer should reach the boundaries of the body, or at least be sufficiently far from the crack to be possible boundaries.

(c) Complete the discretisation of the body according to the usual mesh considerations, bearing in mind the high order of the elements.

(d) Cycle through steps (b) and (c) until the mesh appears 'reasonable' for the problem under consideration.

Users are, of course, expected to use their experience and engineering judgement to decide when a mesh is 'reasonable' but, as can be seen from the test problem results, answers only slightly (i.e. ~10%) in error can be obtained from even very 'unreasonable' looking meshes.

Computation costs

Mesh design, data input and mesh verification takes a few hours at most for problems of this size. These problems have been run on a DEC 2020 in single precision (effective word length = 36 bits). In all cases computer run times were short, averaging 1.2 - 1.6 CPU secs per element per problem, i.e. the points on Figure 6 took a total of approximately 300 CPU secs to produce. This, depending upon computer resource allocations and accounting procedure, represents at most only a few tens of pounds and is trivial compared with the staff costs in preparing the data and examining the results.

ANALYSIS OF A WELDED JOINT

The application of the program PLIPE to the fatigue analysis of a particular non-load-carrying fillet weld (NLCFW) joint will now be discussed. Figure 12 shows outline of a quarter model of the NLCFW joint with a mesh used for the stress analysis of an uncracked joint.

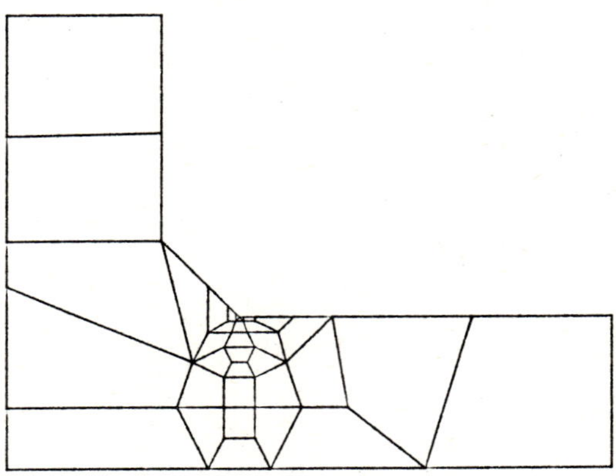

Fig. 12 Fillet Welded Joint

The weld is assumed to have the same elastic properties as the parent metal and a leg length of 0.25 the plate thickness. The load-carrying plate (horizontal) and the non-load-carrying attachment (vertical) are assumed to be the same thickness, 2B. The joint is loaded in tension on the right-hand side vertical edge, and is expected to crack from the lower weld toe vertically downwards. This joint is one of a series of joints studied by Gurney and Johnston [21] using a constant strain triangle (CST) finite element program.

To concentrate attention upon the weld it is possible to normalise the K results for a welded joint with respect to the results for a plate with no weld present. In the case of the NLCFW joint this would be a double edge cracked plate of width 2B, the weld being absent and the attachment therefore immaterial. The stress intensity factor is given by

$$K = Mk \times K' \qquad (6)$$

where K' would be the K value for a double edge cracked plate

and Mk is the correction factor to account for the presence of the weld.

In this analysis the Mk based results are presented. In the work by Gurney and Johnston, the Mk values are obtained first by evaluating the stress along the path of the crack in an uncracked joint, and then by applying Albrech's method [22] to obtain the Mk results. This was necessary as their program could not model cracks directly.

Program PLIPE was used to carry out a stress analysis of the uncracked joint, using the mesh shown in Figure 12, to allow direct comparison with the earlier work. Excellent agreement was obtained even though the mesh by PLIPE consisted only of 35 elements and 430 degrees of freedom as compared with the 763 elements and 876 degrees of freedom in the other program.

PLIPE was then used with mesh sets W1 and W2 to obtain K_I values for a range of crack sizes from a/B = 0.7 to 0.05 (N.B. B is half the plate thickness). Mesh set W1 covered the range 0.7 to 0.3 and W2 the range 0.3 to 0.05. These mesh sets are shown in Figures 13 and 14.

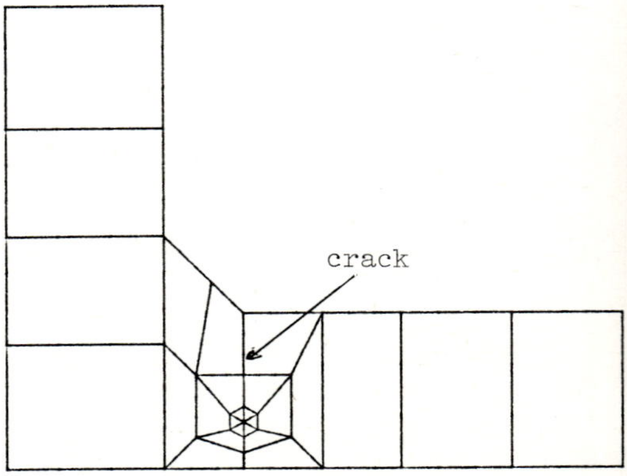

Fig. 13 Mesh Set W1

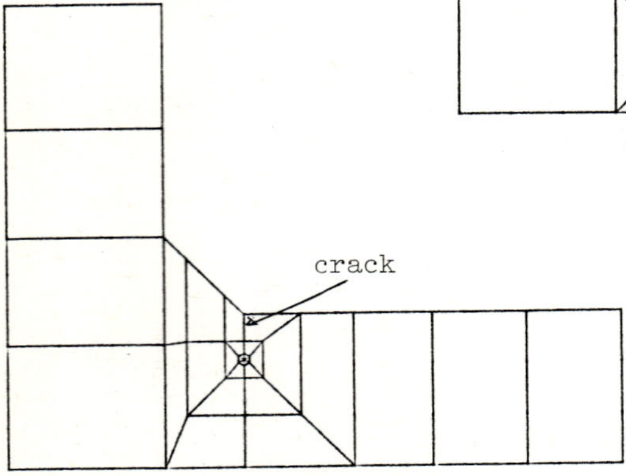

Fig. 14 Mesh Set W2

Mesh set W1 consisted of 26 elements and 358 degress of freedom and mesh set W2 was slightly more refined with 31 elements and 414 degrees of freedom. The total CPU run time to produce these 14 data points was 767 secs. This, once again, represents only a few tens of pounds computing costs and is small compared with the normal staff costs in performing such investigations.

The K_I results were converted to Mk results using an analytical expression for K' [23] and are shown in Figure 15 along with the results using Albrecht's method.

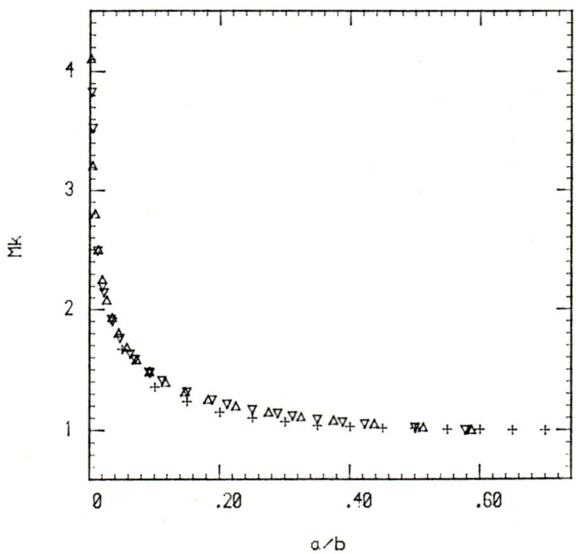

Fig. 15 Results for fillet welded joint

The triangles are the results of the CST program, the inverted triangles the results produced by PLIPE using Albrecht's method and the crosses the results for the cracked joint using PLIPE.

The difference between the cracked and uncracked results is about 6% for a/B values around 0.20 - 0.25, with the earlier results being conservative.

PROGRAM EXTENSION

3-dimensions

A 2-dimensional elastic program like PLIPE can be extended to 3 dimensions fairly easily, the necessary changes being obvious. This will, of course, increase the storage requirements considerably and will result in unacceptably high run times for normal problems unless care is taken in how the changes are made.

In the element assembly phase, there are alternative algorithms [24] which can be used to reduce the arithmetic. PLIPE already uses explicit expressions for some of the matrix multiplications instead of DO loops and such approaches are well worth investigating at every stage. In addition, element stiffness matrices are invariant under translation or rotation and if a body can be discretised using mainly identical elements then the stiffness of these elements are identical and need only be formed once.

The frontal solution technique used is very efficient, especially in 3 dimensions where there is a high degree of element connectivity. Improvements can be made in the inner-most DO loops by moving one division outside the loop [25] and by using machine code routines for these loops. Also, sub-structuring will save time when a number of similar problems are being considered.

It can be well worth examining the operating system of the computer, especially with the modern virtual memory machines in use today. The program can then be carefully designed to make the best use of the available facilities. Such techniques are highly machine-dependent and an example has recently been described by Kamel, McCabe and DeShazo [26].

2-dimensional elasto-plasticity

Many of the points made in the preceding section are also applicable here. PLIPE is in the process of extension to a 2-dimensional elasto-plastic version called PLIPP. The main use envisaged for this version will be in cyclic elasto-plastic behaviour in the region of a crack, i.e. elasto-plastic effects near a growing fatigue crack [27,28]. The 'initial stress' or 'residual force' technique is being used with the original tangent stiffness matrix [29].

In such work, where the material non-linearity is localized, the frontal solution method is once again extremely useful. Sub-structuring can be used to separate the elastic and elasto-plastic regions [30] with a corresponding saving in time.

CONCLUSIONS

It is concluded that crack tip elements, based on a degenerate, shifted node isoparametric cubic quadilateral, allow the calculation of both K_I and K_{II} for cracks accurately and that modern programming techniques have resulted in programs, an example of which has been discussed, that are quick to write and easy and cheap to operate.

REFERENCES

1. KOCANDA, S. - Fatigue Failure of Metals, Sijthoff & Noordhoff, 1978.
2. MADAYAG, A.F. - Metal Fatigue: Theory and Design, John Wiley & Sons, 1969.
3. GURNEY, T.R. - Fatigue of Welded Structures, 2nd edn, Cambridge Univ. Press, 1979.
4. PARIS, P.C. - The Fracture Mechanics Approach to Fatigue, 10th Sagamore Army Materials Research Conference, Syracuse Univ. Press, 1964.
5. GURNEY, T.R. - The Basis of the new Fatigue Design Rules for Welded Joints, Proc. Conf. on The New Code for the Design of Steel Bridges, Cardiff, 1980.
6. SIH, G.C. - Mechanics of Fracture: Linear Response, Proc. 1st Int. Conf. on Num. Meth. in Frac. Mech., Ed. Luxmoore, A.R. and Owen, D.R.J., Swansea, 1978.
7. BARR, B.I.G., EVANS, W.T., WATKINS, J. and ROUHI, M.R. - Fracture Path Prediction using Numerical and Experimental Techniques, Proc. 1st Int. Conf. on Num. Meth. in Frac. Mech., Ed. Luxmoore, A.R. and Owen, D.R.J., Swansea, 1978.
8. FENNER, R.T. - Finite Element Methods for Engineers, Macmillan, 1975.
9. HUEBNER, K.H. - The Finite Element Method for Engineers, John Wiley & Sons, 1975.
10. ZIENKIEWICZ, O.C. - The Finite Element Method, 3rd edn, McGraw-Hill, 1977.
11. CRUSE, T.A. and MEYERS, G.J. - Three-dimensional Fracture Mechanics Analysis, J. of the Struc. Div., Proc. ASCE, Vol. 103 (ST2), pp 309-320, 1977.
12. TAN, C.L. and FENNER, R.T. - Three dimensional Stress Analysis by the Boundary Integral Equation Method, J. of Strain Analysis, Vol. 13, pp 213-219, 1978.
13. PU, S.L., HUSSIAN, M.A. and LORENSEN, W.E. - The Collapsed Cubic Isoparametric Element as a singular Element for Crack Problems, Int. J. Num. Meth. in Engrg, Vol. 12, pp 1727-1742, 1978.
14. HINTON, E. and OWEN, D.R.J. - Finite Element Programming, Academic Press, 1977.
15. GALLAGHER, R.H. - A Review of Finite Element Techniques in Fracture Mechanics, Proc. 1st Int. Conf. on Num. Meth. in Frac. Mech., Ed. Luxmoore, A.R. and Owen, D.R.J., Swansea, 1978.
16. HENSHELL, R.D. and SHAW, K.G. - Crack Tip Elements are Unnecessary, Int. J. Num. Meth. in Engrg, Vol. 9, pp 495-509, 1975.
17. BARSOUM, R.D. - On the Use of Isoparametric Finite Elements in Linear Fracture Mechanics, Int. J. Num. Meth. in Engrg, Vol. 10, pp 25-37, 1976.
18. FAWKES, A.J., OWEN, D.R.J. and LUXMOORE, A.R. - An Assessment of Crack Tip Singularity Models for Use with Isoparametric Elements, Engrg Frac. Mech., Vol. 11, pp 143-159, 1979.

19. GIFFORD, L.N. and HILTON, P.D. - Stress Intensity Factors by Enriched Finite Elements, Engrg Frac. Mech, Vol. 10, pp 485-496, 1978.
20. ROOKE, D.P. and CARTWRIGHT, D.J. - Compendium of Stress Intensity Factors, HMSO, 1976.
21. GURNEY, T.R. and JOHNSTON, G.O. - A Revised Analysis of the Influence of Toe Defects on the Fatigue Strength of Transverse non-load-carrying Fillet Welds - To be published in Welding Research Int.
22. ALBRECHT, P. and YAMADA, K. - Rapid Calculation of Stress Intensity Factors, J. of Struc. Div. Proc. ASCE, Vol. 103 (ST2), pp 377-389, 1977.
23. TADA, K., PARIS, P.C. and IRWIN, G.R. - The Stress Analysis of Cracks Handbook, Del Research Corp., 1973.
24. IRONS, B.M. - Economical Computer Techniques for Numerically Integrated Finite Elements, Int. J. Num. Meth. in Engrg, Vol. 1, pp 201-203, 1969.
25. TAYLOR, R.L. - Computer Procedures for Finite Element Analysis, Chapter 24 of The Finite Element Method, 3rd edn, by Zienkiewicz, O.C., McGraw-Hill, 1977.
26. KAMEL, H.A., McCABE, M.W. and DeSHAZO, P.G. - Optimum Design of Finite Element Software subject to Core Restrictions, Comp. & Struc., Vol. 10, pp 63-80, 1979.
27. OHJI, K., OGURA, K. and OHKUBO, Y. - Cyclic Analysis of a Propagating Crack and its Correlation with Fatigue Crack Growth, Engrg Frac. Mech., Vol. 7, pp 457-464, 1975.
28. NAKAGAKI, M. and ATLURI, S.N. - Fatigue Crack Closure and Delay Effects under Mode 1 Spectrum Loading: An Efficient Elasto-Plastic Analysis Procedure, Fatigue of Engrg Materials and Struc., Vol. 1, pp 421-429, 1979.
29. NAYAK, G.C. and ZIENKIEWICZ, O.C. - Elasto-Plastic Stress Analysis: A Generalization for Various Constitutive Relations including Strain Softening, Int. J. Num. Meth. in Engrg, Vol. 5, pp 113-135, 1972.
30. ALIZADEH, A. and WILL, G.T. - A Substructural Frontal Solver and its Application to Localized Material Non-linearity, Comp. & Struc., Vol. 10, pp 225-231, 1979.

STRESS INTENSITY FACTORS FOR CRACKS WITH CIRCULAR AND ELLIPTIC CRACK FRONTS, DETERMINED BY 3D FINITE ELEMENT METHODS.

A.C. PICKARD*

SUMMARY

Stress intensity factors have been derived for corner, surface and embedded cracks, with circular and elliptic crack fronts, in cuboid blocks, for a variety of crack length to block width ratios. 3D finite element methods have been used to model the problems, using the MARC finite element stress analysis program. The solutions obtained are compared with those in the literature, and are found to be substantially in agreement with those derived by finite element methods. Simple approximation formulae are derived to predict stress intensities for arbitrary sized elliptic cracks in cuboid blocks, up to crack length to block width ratios of 0.5. These are used, with crack propagation information generated from compact tension specimens, to predict the propagation lives of some part-elliptic bucket groove cracks in an industrial gas turbine power turbine disc.

INTRODUCTION

In many engineering components, fatigue failure occurs by the initiation and propagation of corner, surface or sub-surface cracks with elliptic or near-elliptic crack fronts. If the crack propagation life of a component is to be evaluated, for growth from a given initial crack (usually determined by the minimum detectable defect size), it is essential that accurate stress intensities are available for arbitrary sized elliptic or part-elliptic cracks. Also, if macroscopic crack propagation information has been generated using standard specimens such as compact tension specimens, it must be shown that this information is applicable to the growth of small elliptic cracks. The latter may be checked by initiating and growing elliptic-fronted cracks in specially designed laboratory specimens and measuring crack

*Research Engineer, Rolls-Royce Limited, Aero Engine Division, Moor Lane, Derby, England.

propagation rates directly, or by investigating the propagation of this form of crack during the testing of full-scale components.

This paper describes the evaluation of stress intensity factors for corner, surface and subsurface cracks with elliptical crack fronts, and the application of these factors to the prediction of the crack propagation behaviour of some part-elliptical bucket groove cracks in an industrial gas turbine power turbine disc.

METHOD

3D finite element methods were used to model various cracked block configurations, using the MARC finite element stress analysis program [1]. Figures 1 and 2 show typical element configurations used to model cracks with circular and elliptic crack fronts respectively. 20-node isoparametric brick elements were used throughout. These were degenerated to 15-node wedge-shaped elements around the crack tip. The midside nodes of the crack tip elements were moved to the quarter-point positions on the sides adjacent to the crack tip, to model the stress-strain singularity at the crack tip [2]. The differential stiffness method [3] was used to evaluate energy release rates for small crack extensions at various positions on the crack front, as shown in figure 3. Surface and subsurface cracks were modelled by imposing boundary conditions to enforce the required symmetry. Solutions were obtained for remote tensile and bending stress fields, in the latter cases only for corner and surface cracks, with maximum tensile stress at the cracked surface.

Stress intensities were derived from energy release rates in all cases using the relationships:

$$G = 2\frac{\Delta U}{\Delta A} \qquad (1)$$

where ΔU is the energy released in the half-body modelled, and ΔA is the increase in crack area, and

$$K = \left[\frac{EG}{1-v^2}\right]^{\frac{1}{2}} \qquad (2)$$

where E is Young's modulus, and v is Poisson's ratio. In all cases, E was taken as 26.9E6 psi, and v as 0.3. The incremental area ΔA was evaluated analytically for extensions to circular cracks, and numerically for extensions to elliptical cracks.

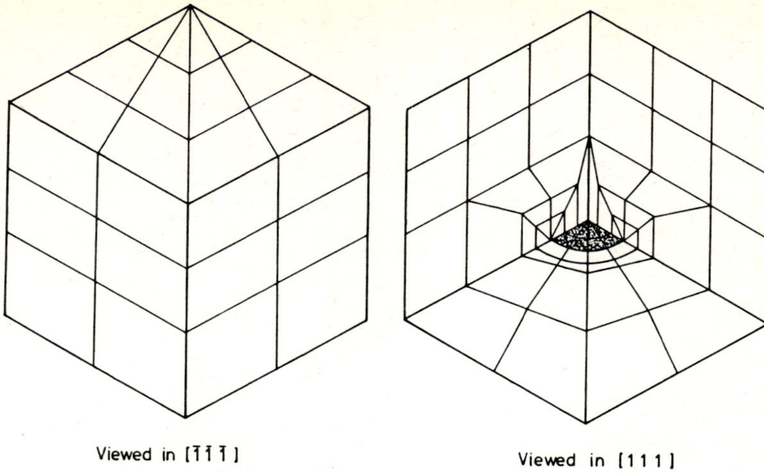

Figure 1. Finite element mesh, H/W = 1, a/W = b/W = 0.2 block.

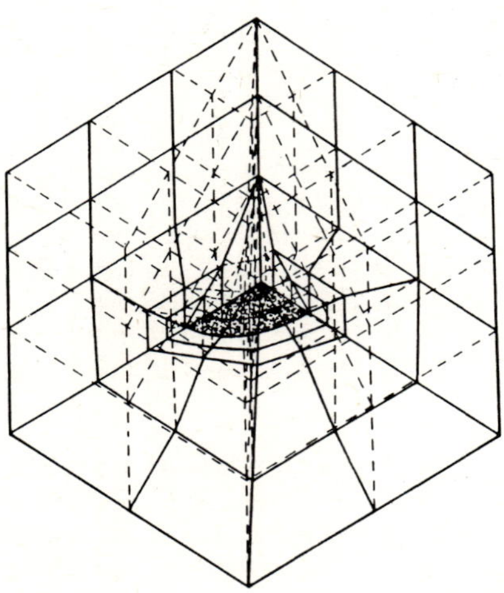

Figure 2. Finite element mesh, H/W = 1, a/W = 0.30, b/W = 0.20 block.

RESULTS

Stress intensities were obtained for circular and elliptic-fronted cracks in various cuboid-shaped blocks,

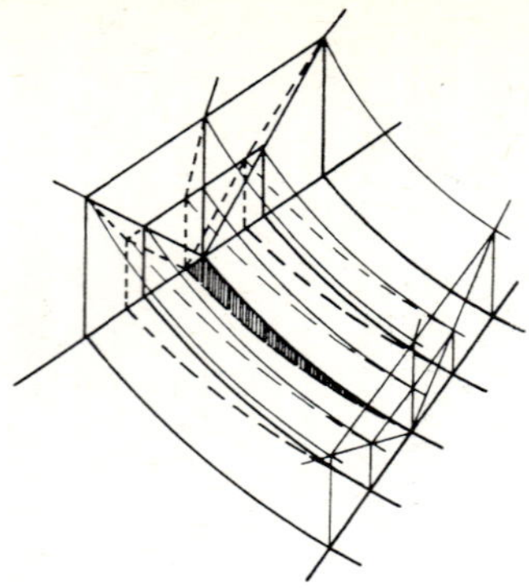

Figure 3. Mesh distortion (exaggerated) around the crack front associated with the differential stiffness K evaluation. Change in crack area is shown shaded.

with block height equal to block width, W. Figure 4 shows the variation of normalised stress intensity with angle around the crack front for a quarter-circular corner crack in a square section block, with a/W = 0.2,0.3,0.4 and 0.5, under remote tensile loading. Results obtained by Hellen and Blackburn [4] using finite element methods, and Tracey [5] (for a quarter-circular crack in a square section block with a/W = 0.2) are also shown. Figure 5 shows the variation of normalised stress intensity with angle around the crack front for a semicircular surface crack in a rectangular block, with a/W = 0.2,0.3,0.4 and 0.5, for remote tensile loading. Results obtained by Hellen and Blackburn [4], Hilton, Kiefer and Sih [6], Pian and Moriya [7], Tracey [8] (for a semicircular crack in a half-cylinder) and Smith [9] (for a semicircular crack in a simi-infinite solid) are also shown.

Figure 6 shows the variation of normalised stress intensity with normalised crack length, for remote tensile loading of H/W = 1 blocks, for quarter-circular, semicircular and circular embedded flaws. Predicted curves (see below) are also shown. Figures 7 to 9 show similar results for H/W = 2 and 4 blocks, for quarter-circular and two forms of semicircular cracks respectively. For comparison, figure 10 shows the variation of normalised stress intensity with normalised crack length for quarter- and semicircular cracks in H/W = 1 blocks, for tensile loading, predicted by Liu [10], Smith, Emery and Kobayashi [11], Kobayashi and Moss [12], Shah and Kobayashi [13], and Maddox [14]. Paris and Sih [15] predict similar results, for surface cracks, to reference [11], but with a tangent-formula back face

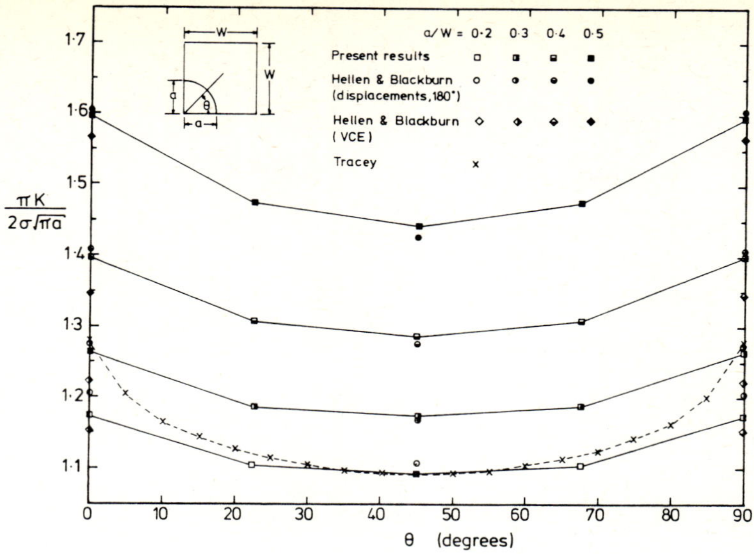

Figure 4. Angular variation of K, quarter-circular crack.

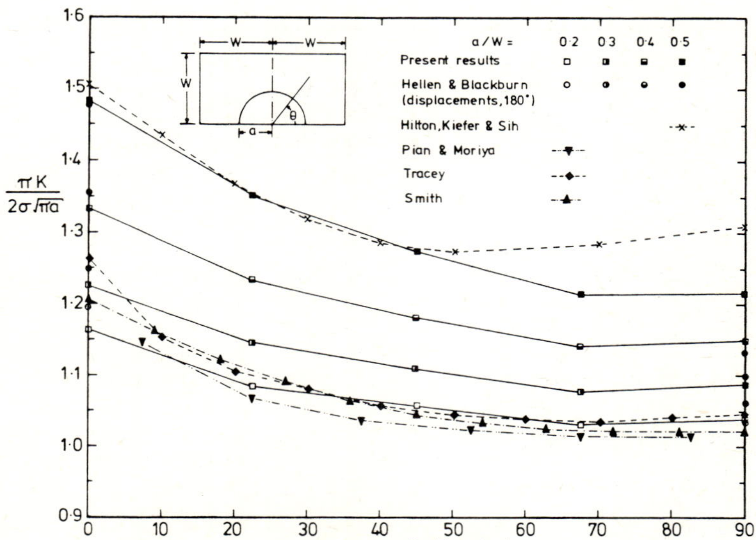

Figure 5. Angular variation of K, semicircular crack.

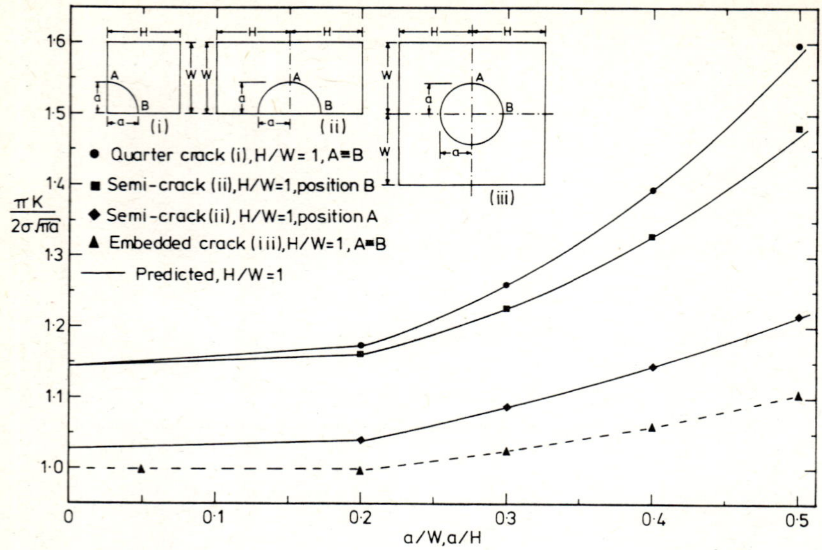

Figure 6. Normalised stress intensities, H/W = 1 blocks.

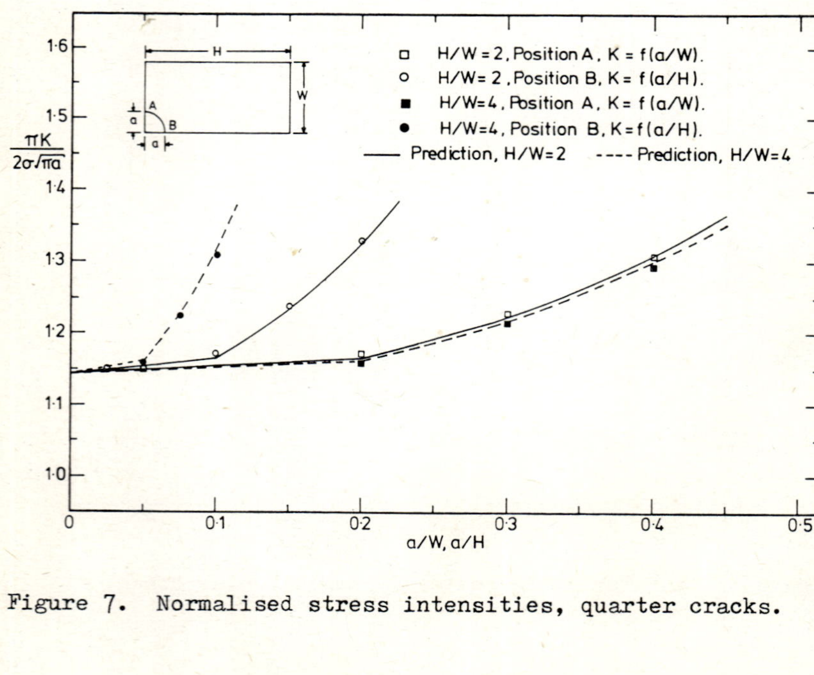

Figure 7. Normalised stress intensities, quarter cracks.

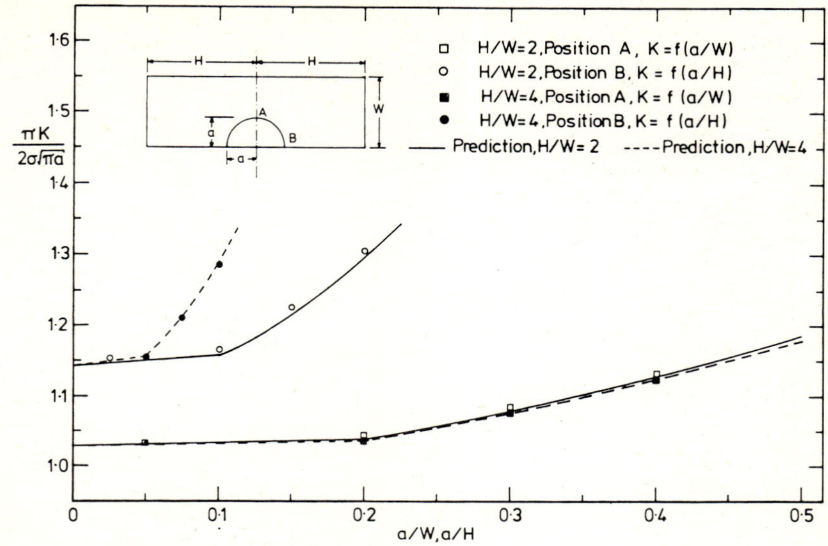

Figure 8. Normalised stress intensities, surface cracks.

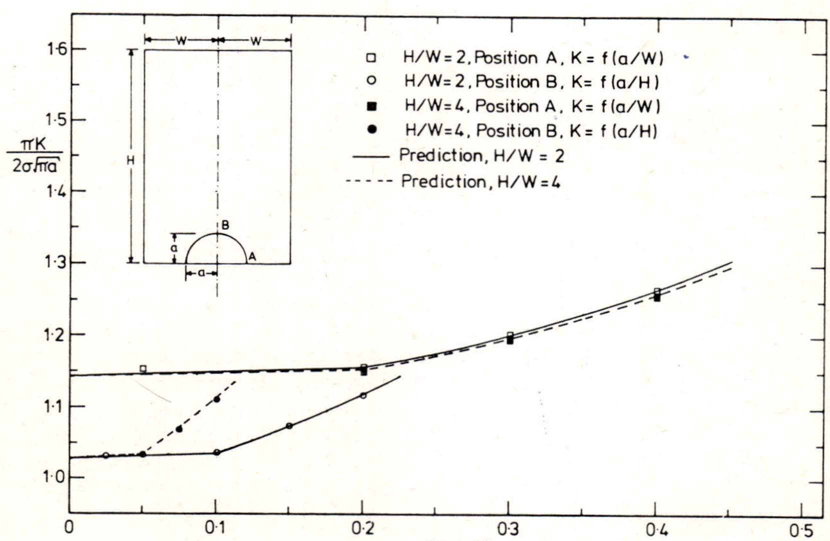

Figure 9. Normalised stress intensities, surface cracks.

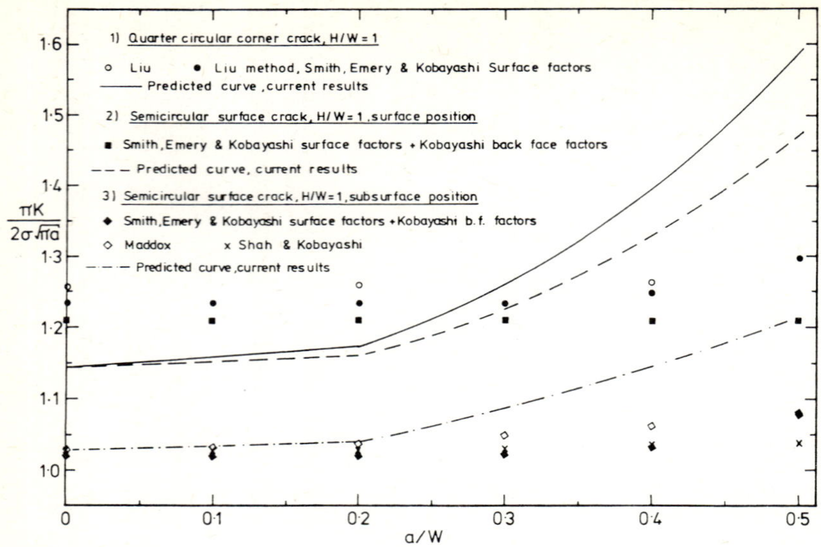

Figure 10. Comparison of results, corner and surface cracks, H/W = 1.

correction factor. Randall [16] predicts a constant normalised stress intensity for surface cracks. All of these show a much smaller variation of normalised stress intensity with normalised crack length than that predicted by the present results, but it is significant that most [10,11,12,13] are based on the Kobayashi back face correction relationship.

Analyses of various elliptic cracks were performed, and the results of these are shown in figure 11. The ratio of predicted normalised stress intensity (see below) to finite element calculated stress intensity is shown as a function of minor/major axis ratio. Within the range $0.6 \leqslant a/b \leqslant 1.65$, the stress intensity ratio is equal to $1.0 \pm 1\%$.

Figure 12 shows the variation of stress function (see below) with normalised crack length for remote linear bending stress distributions, with maximum tensile stress on the crack side of the neutral axis, for both quarter - circular and semicircular cracks. The polynomial stress function method due to Chell [17] predicts a similar linear relationship between stress function and normalised crack length, but with values of $F(\sigma)/\sigma_{max}$ of 0.284 and 0.815 respectively at $a/L = 1$.

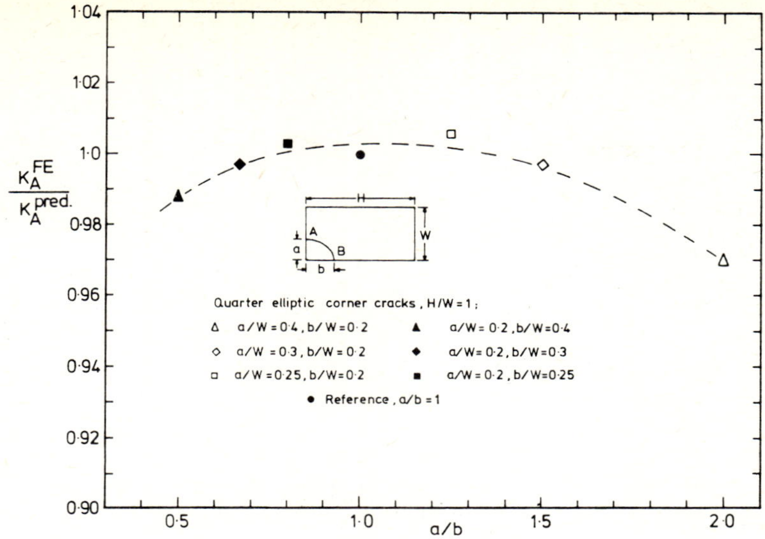

Figure 11. Correction factors for elliptic cracks.

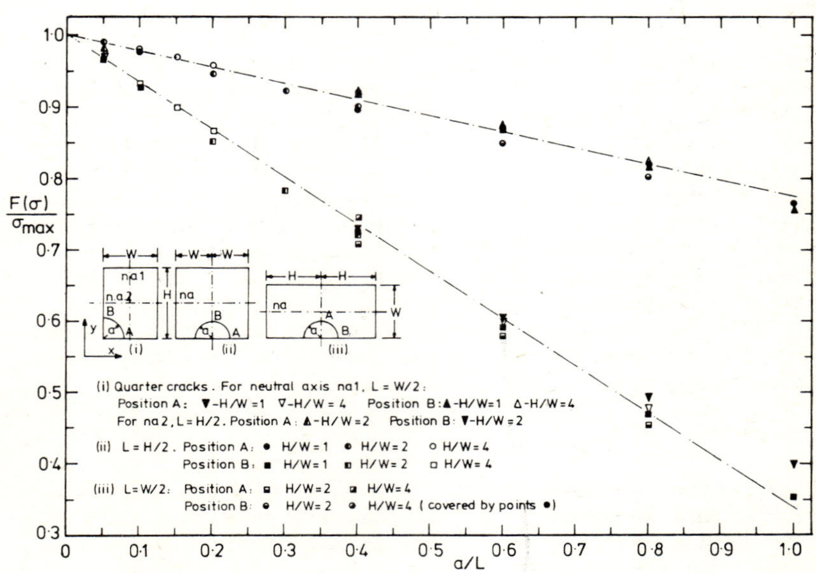

Figure 12. Stress functions for linear bending fields.

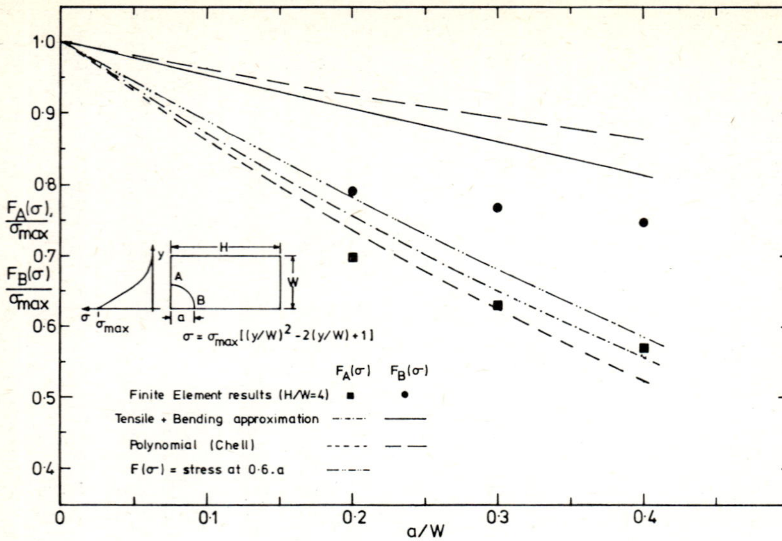

Figure 13. Stress functions for parabolic stress fields.

Figure 13 shows the variation of stress function with normalised crack length for a parabolic bending stress distribution of the form:

$$\sigma = \sigma_{max}((y/W)^2 - 2(y/W) + 1) \qquad (3)$$

Stress functions were derived by finite element methods, for a/W = 0.2,0.3 and 0.4 blocks, by applying pressure loading on the face remote from the crack to produce the required stress distribution, and by three approximation methods. These are the 'tensile + bending' approximation, where the stress distribution over the crack is replaced by a tensile stress, equal to the crack tip stress, and a linear bending stress, equal to zero at the crack tip, such that the 'tensile + bending' field carries the same load as the real field; the polynomial approximation, based on the work of Chell [17]; and an approximation based on assuming a driving stress for the crack equal to the stress at 0.6 x the crack length (for maximum depth position only).

DISCUSSION

From the results presented above, simple approximation formulae were derived to predict stress intensities for cracks with elliptic crack fronts in arbitrarily - sized blocks. The stress intensity is expressed in the form:

$$K = M_G . M_B . M_S . F(\sigma) . \Phi . (\pi c)^{\frac{1}{2}} \qquad (4)$$

where M_G is a general correction factor, M_B is a back-face correction factor, M_S is a side face correction factor, $F(\sigma)$ is a stress function related to the remote stress distribution, Φ is the standard elliptic correction factor, and c is the crack length.

For a crack of the form shown in figure 11, with $b \geqslant a$, at position A,

$$\Phi_A = \left(\int_0^{\pi/2} (1-(1-(a/b)^2)\sin^2\phi)^{\frac{1}{2}} d\phi\right)^{-1} \tag{5}$$

and at position B,

$$\Phi_B = \Phi_A.(a/b) \tag{6}$$

At position A, crack length c = a, whereas at position B, c = b.

The correction factors M_G, M_B and M_S depend on the form of crack being analysed (i.e. corner, surface or subsurface). At position A, for $H/W \geqslant 1$;

$$M_G = f(a/W),\ M_B = f(a/W),\ M_S = f(b/H) \tag{7}$$

At position B, for $H/W \geqslant 1$;

$$M_G = f(b/W),\ M_B = f(b/H),\ M_S = f(a/W) \tag{8}$$

For a corner crack, surface position (a/W, b/W, b/H=x);

$$M_G = 1.143 \quad (0<x\leqslant 0.2) \tag{9}$$

$$M_G = 0.1x^2 + 0.29x + 1.081 \quad (0.2<x\leqslant 0.5) \tag{10}$$

$$M_B = 1 + 0.06x \quad (0\leqslant x\leqslant 0.2) \tag{11}$$

$$M_B = 0.75x^2 - 0.185x + 1.019 \quad (0.2<x\leqslant 0.5) \tag{12}$$

$$M_S = 1 + 0.07x \quad (0\leqslant x\leqslant 0.2) \tag{13}$$

$$M_S = 0.9x^2 - 0.21x + 1.019 \quad (0.2<x\leqslant 0.5) \tag{14}$$

For a surface crack, surface position;

$$M_G = 1.143 \quad (0\leqslant x\leqslant 0.2) \tag{15}$$

$$M_G = 0.1x^2 + 0.29x + 1.081 \quad (0.2<x\leqslant 0.5) \tag{16}$$

$$M_B = 1 + 0.03x \quad (0\leqslant x\leqslant 0.2) \tag{17}$$

$$M_B = 0.5x^2 - 0.16x + 1.018 \quad (0.2<x\leqslant 0.5) \tag{18}$$

$$M_S = 1 + 0.05x \quad (0 \leqslant x \leqslant 0.2) \tag{19}$$

$$M_S = 0.65x^2 - 0.135x + 1.011 \quad (0.2 < x \leqslant 0.5) \tag{20}$$

For a surface crack, subsurface (maximum depth) position;

$$M_G = 1.028 \quad (0 \leqslant x \leqslant 0.2) \tag{21}$$

$$M_G = 0.3x + 0.968 \quad (0.2 < x \leqslant 0.5) \tag{22}$$

$$M_B = 1 + 0.04x \quad (0 \leqslant x < 0.2) \tag{23}$$

$$M_B = 0.25x^2 - 0.025x + 1.003 \quad (0.2 < x \leqslant 0.5) \tag{24}$$

$$M_S = 1 + 0.02x \quad (0 \leqslant x < 0.2) \tag{25}$$

$$M_S = 0.2x^2 - 0.04x + 1.004 \quad (0.2 < x \leqslant 0.5) \tag{26}$$

The stress function, $F(\sigma)$, has been derived for tensile and bending fields. For tensile fields, $F(\sigma)$ is a constant equal to the remote tensile stress; Figures 12 and 13 show the variation of stress function with crack length for linear and parabolic bending fields respectively. It is noted that in the latter case, both the polynomial and tension + bending approximations are reasonable for position A, but both seriously overestimate stress functions at position B.

The above expressions have been used to produce the 'prediction' curves in figures 6 to 10, and are valid in the ranges $0 \leqslant a/W, b/W, b/H \leqslant 0.5$ and $0.6 \leqslant a/b \leqslant 1.65$. Figures 4, 5 and 10 show that the present results generally show good agreement with other finite element solutions, but disagree with approximation methods based on the Kobayashi back face correction factors. The embedded crack solution, for a/W = 0.05 (see figure 6) is within 0.2% of the analytical solution for an embedded crack in an infinite body.

BUCKET GROOVE CRACK ANALYSIS

During the cyclic spin testing of a ferritic (12% Chromium) steel industrial gas turbine power turbine disc at 190°C, part - circular corner cracks were formed at the acute corners of all the bucket grooves. Figure 14 shows the location of the cracks relative to the blade fixing features. Subsequently, 2046 cycles were performed at room temperature. These were followed by an overspeed test in which all of the bucket groove cracks propagated by a ductile extension mechanism. Figure 15 shows the appearance of the fracture surface of one of the cracks. The dark area of cracking emanating from the corner of the bucket groove corresponds to crack propagation at 190°C, and the light band to the 2046 cycles at room temperature.

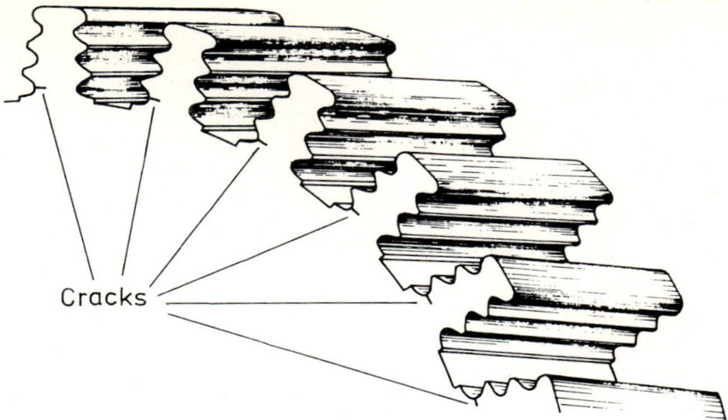

Figure 14. Position of bucket groove cracks in disc rim.

Figure 15. Appearance of fracture surface and measurement definitions.

The elastic stress distribution in the bucket groove was evaluated by photoelastic analysis. Crack propagation information was obtained from tests on compact tension specimens at room temperature, using material taken from the rim region of another disc.

For the purposes of the crack propagation analysis, the cracks were modelled as corner cracks in a rectangular section

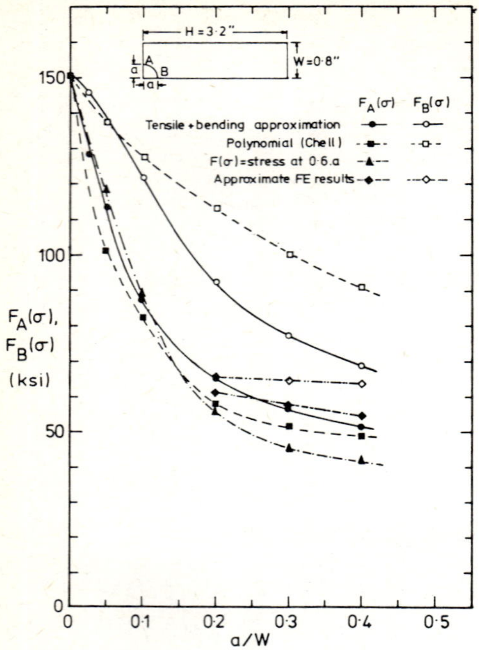

block, and the stress distribution normal to the crack plane was considered. Stress functions $F_A(\sigma)$ and $F_B(\sigma)$ were derived, and are shown in figure 16; the approximation methods used are described above. Finite element analyses were performed, for three crack lengths, of an H/W = 4 block with a quarter - circular corner crack, subjected to the stress field at the root of the bucket groove. As noted for the results shown in figure 13, the estimates of $F_A(\sigma)$ show reasonable agreement, but those of $F_B(\sigma)$ do not.

Figure 16. Stress functions for bucket groove cracks.

Based on the above stress functions, the correction factors for quarter - elliptical cracks evaluated above, and the crack sizes before room temperature cyclic testing, the number of cycles taken to grow the crack to its final length at position A, and the final crack length at position B were estimated. The results, for several different cracks, are shown in table 1. Both the final crack lengths and number of cycles predicted using the polynomial stress function appear to be too large, while those predicted using both the finite element and 'tension + bending' stress functions are in good agreement with the experimental results.

CONCLUSIONS

Correction factors have been derived to estimate stress intensity factors for cracks with elliptical crack fronts in arbitrary blocks, in the range $0 \leq a/W, b/W, b/H \leq 0.5$ and $0.6 \leq a/b \leq 1.65$, for remote tensile loading. Stress functions have been investigated for linear and quadratic bending fields.

Measured Dimensions				Predicted Lives and Final Crack Lengths					
				1		2		3	
a_i (in)	a_f (in)	b_i (in)	b_f (in)	N_f	b_f (in)	N_f	b_f (in)	N_f	b_f (in)
0·132	0·160	0·237	0·271	1765 ×	0·256	1467	0·257	1829	0·280
0·176	0·218	0·295	0·338	2128	0·323	2022	0·328	2489	0·361
0·211	0·247	0·282	0·317	1804	0·312	1838	0·321	2175	0·357
+ 0·240	0·296	0·571	0·648	1605	0·594	1740	0·637	2116	0·600
0·213	0·246	0·277	0·313	1683	0·305	1716	0·314	2029	0·349
0·150	0·185	0·250	0·285	2122 ×	0·273	1847	0·279	2300	0·310
0·225	0·260	0·275	0·310	1769	0·307	1843	0·318	2134	0·356
0·200	0·235	0·275	0·310	1817	0·303	1807	0·311	2162	0·347
0·180	0·215	0·230	0·265	2152	0·261	2016	0·272	2402	0·313
0·190	0·225	0·275	0·310	1843	0·302	1792	0·309	2170	0·343
0·175	0·205	0·270	0·300	1654	0·291	1542	0·296	1907	0·325
0·100	0·125	0·110	0·145	- ×	-	1899	0·149	2135	0·172
0·200	0·240	0·390	0·440	1610	0·412	1624	0·416	2015	0·433

Case 1: Stress functions derived by finite element methods.

Case 2: Stress functions derived by 'tensile + bending' approximation

Case 3: Polynomial stress function.

▪ Crack illustrated in figure 15

+ Double origin crack

× Outside the range of validity of the stress functions

Table 1. Estimates of final crack length and number of cycles of propagation for bucket groove cracks.

The correction factors derived have been used, with various stress function estimates, to predict the crack propagation behaviour of some bucket groove cracks in an industrial gas turbine power turbine disc.

ACKNOWLEDGEMENTS

The author would like to thank Mr. J.D. Gibbons for provision of experimental results and for many helpful discussions, and Rolls-Royce Limited for the provision of research facilities.

REFERENCES

1. MARC - CDC, Developed by MARC Analysis Research Corporation, Providence, RI.
2. BARSOUM, R.S. - On the Use of Isoparametric Finite Elements in Linear Fracture Mechanics. Int. J. Num. Meth. Engng., Vol. 10, pp. 25 - 37, 1976.
3. PARKS, D.M. - A Stiffness Derivative Finite Element Technique for Determination of Crack Tip Stress Intensity Factors. Int. J. of Fracture, Vol.10, pp. 487 - 502, 1974.
4. HELLEN, T.K. and BLACKBURN, W.S. - The Calculation of Stress Intensity Factors in Two and Three Dimensions Using Finite Elements. CEGB report RD/B/N3378, 1975.

5. TRACEY, D.M. - from Conference on Finite Element Techniques in Fracture Mechanics, Stuttgart, Germany, April 16 - 17, 1973.
6. HILTON, P.D., KIEFER, B.V. and SIH, G.C. - Specialised Finite Element Procedures for Three Dimensional Crack Problems, Numerical Methods in Fracture Mechanics, Ed. Luxmoore, A.R. and Owen, D.R.J., Quadrant, 1978.
7. PIAN, T.H.H. and MORIYA, K. - Three Dimensional Fracture Analysis by Assumed Stress Hybrid Elements, Numerical Methods in Fracture Mechanics, Ed. Luxmoore, A.R. and Owen, D.R.J., Quadrant, 1978.
8. TRACEY, D.M. - 3D Elastic Singularity Element for Evaluation of K Along an Arbitrary Crack Front. Int. J. of Fracture, Vol. 9, pp. 340 - 343, 1973.
9. SMITH, F.W. and ALAVI, M.J. - Stress Intensity Factors for a Penny-Shaped Crack in Halfspace. J. Eng. Fract. Mech., Vol. 3, pp.241 - 254, 1971.
10. LIU, A.F. - Stress Intensity Factor for a Corner Flaw. Eng. Fract. Mech., Vol. 4, pp.175 - 179, 1972.
11. SMITH, F.W., EMERY, A.F. and KOBAYASHI, A.S. - Stress Intensity Factors for Semicircular Cracks, part 2 - Semi-Infinite Solid. Trans. ASME J. Appl. Mech., 953, 1967.
12. KOBAYASHI, A.S. and MOSS, W.L. - Stress Intensity Magnification Factors for Surface-Flawed Tension Plate and Notched Round Tension Bar, Fracture 1969, Chapman and Hall, 1969.
13. SHAH, R.C. and KOBAYASHI, A.S. - Stress Intensity Factor for an Elliptical Crack Approaching the Surface of a Plate in Bending, Stress Analysis and Growth of Cracks, ASTM STP513, 1972.
14. MADDOX, S.J. - Analysis of Cracks in Fillet Welded Joints, Int. J. Fracture, Vol. 11, pp.221 - 243, 1975.
15. PARIS, P.C. and SIH, G.C. - Stress Analysis of Cracks, Fracture Toughness Testing and its Applications, ASTM STP 381, pp. 30 - 81.
16. RANDALL, P.N. - in Plane Strain Crack Toughness Testing of High Strength Metallic Materials, ASTM STP 410, Eds. Brown, W.F. Jr. and Srawley, J. E., pp.88 - 129.
17. CHELL, G.C. - The Stress Intensity Factors for Part Through Thickness Embedded and Surface Flaws Subject to a Stress Gradient. CEGB Report RD/L/N 31/74, 1974.

A NUMERICAL STUDY OF THE EFFECT OF WELDING RESIDUAL STRESS UPON FATIGUE CRACK PROPAGATION WITH ATTENTION PAID TO THE PHENOMENON OF CRACK OPENING AND CLOSURE

Shuichi Fukuda* and Michio Sugino**

SUMMARY

As there seem to be few systematic investigations on the effect of welding residual stress on fatigue crack propagation, this paper attempts to analyze its effect numerically. Since similar phenomenon of crack opening and closing as Elber's effect which is caused by remaining cyclic plastic strain on crack surfaces is duly expected to occur by plastic strain caused by welding heat input as well, the effect of welding residual stress is studied in terms of crack opening and closing. As it is quite difficult to evaluate plastic strain caused by heat input quantitatively by experimental methods and that general applicability of the method is desired for design, finite element method is used. The numerical results agree fundamentally well with the experimental results, hence the validity of crack opening and closing approach is confirmed.

1. INTRODUCTION

Up to now, fatigue strength of welds has been discussed mainly on the basis of fatigue crack initiation life. But due to such reasons as (1) the operating conditions of welded structures have become severer and severer, and cases are increasing in which cracks are found in the operating conditions, (2) with the remarkable progress of nondestructive inspection, the determination of allowable defect size becomes much more important, and (3) in some cases of weldments, enough weld penetration is hard to be secured, and in some cases partial penetration is employed for the purpose of reducing the amount of work. The estimation of the fatigue strength of these partially penetrated welded joints is quite important for practical

---

* Associate Professor, ** Graduate Student
Welding Research Institute, Osaka University, JAPAN

applications, it has become widely accepted that it is more reasonable for some type of weldments to assess the fatigue life on the basis of crack propagation life rather than crack initiation life. Therefore, much efforts are being paid to clarify the behavior of fatigue crack propagation. But as to the effect of welding residual stress, which is considered to be the most important effect to be clarified first at least from the standpoint of mechanics, few , if any, systematic investigations have been carried out to study its effect upon fatigue crack propagation, although enormous amount of work can be found with respect to its effect on fatigue life.

Recently Elber [1] pointed out that the crack tip plastic strain caused by cyclic loading moves onto the crack surface with the extension of a fatigue crack, and that the residual deformation due to this plastic strain affects the crack opening and closing behavior. Since then, many researchers [2] - [5] have confirmed the importance of crack opening and closure in the analysis of fatigue crack propagation in base metal specimens. Although Elber's effect is caused by the remaining cyclic plastic strain on the crack surface, it is duly expected in the case of welded specimens that similar phenomenon can be observed due to the remaining plastic strain caused by welding heat input. Therefore, this paper analyzes the effect of welding residual stress upon fatigue crack propagation numerically with attention paid to the crack opening and closing behavior.

2. FATIGUE CRACK PROPAGATION IN WELDED SPECIMENS AND BASE METAL SPECIMENS: EXPERIMENTAL RESULTS

First, we describe briefly the results of our experiment we carried out [6] which provide the basis for this numerical analysis. In this series of experiments, fatigue crack propagation behavior was observed using high tensile strength steel of 80 kg/mm^2 grade and JIS SS41 steel center-notched plate specimen of 6 mm thickness, 105 mm width and 500 mm length. Crack propagation rate was measured by crack gages, and crack opening and closing behavior was observed by wire strain gages. The welded specimen is a specimen corresponding to a butt-welded joint which is made by preparing a groove of 2 mm depth and 5 mm width in the center surface of the specimen, and welding using a microwire submerged arc welding process. The center notch was introduced after welding. Fig.1 shows the da/dN - Δ K relation for base metal specimens and Fig.2 shows the da/dN - Δ K relation for welded specimens. Although the da/dN - Δ K relation differs for different stress ratio R in the case of base metal specimens, the

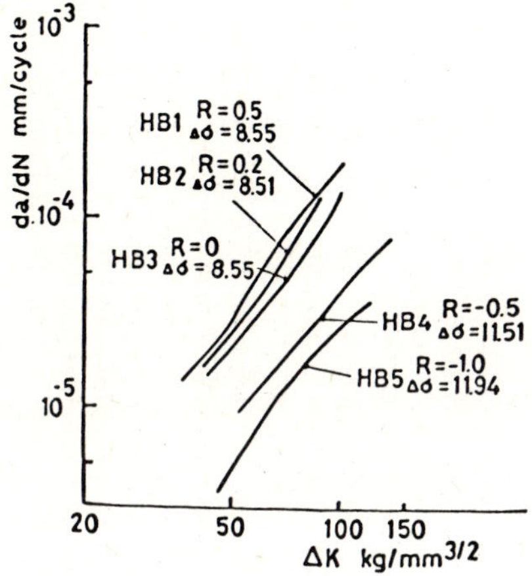

Fig. 1 da/dN -Δ K relation of base metal specimens (high strength steel)

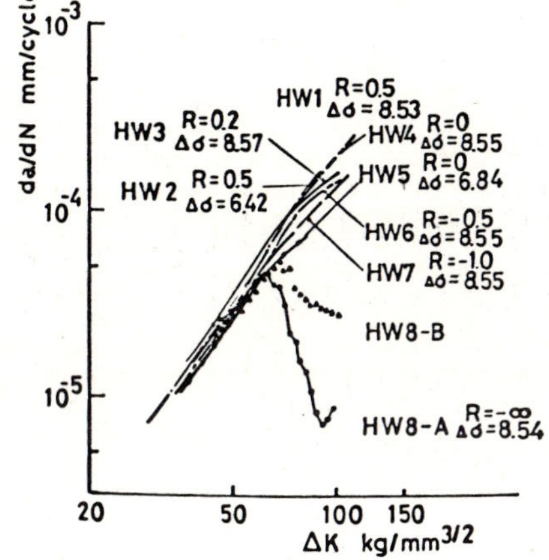

Fig. 2 da/dN -Δ K relation of welded specimens (high strength steel)

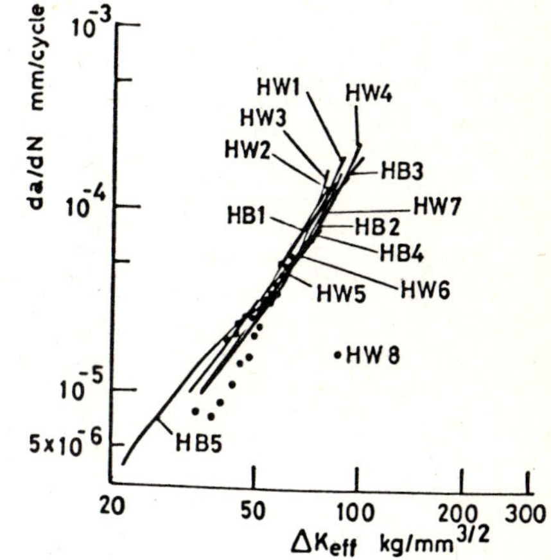

Fig. 3 da/dN -Δ Keff relation of base metal specimens and welded specimens (high strength steel)

results of welded specimens corresponding to the region of small crack length fall on a small scatter band in spite of the differences in stress ratio R and stress range $\Delta\sigma$. Fig.3 shows the replot of the results of Fig.1 and Fig.2 using experimentally evaluated effective stress intensity factor range Δ Keff. It can be observed from Fig.3 that the results of Fig.1 and Fig.2 fall almost on a straight line if Δ Keff is used. i.e., The effect of stress ratio R in base metal specimens and the effect of welding residual stress in welded specimens can be integrated into a small scatter band, if Δ Keff is used. Therefore, with these results in mind, this paper analyzed fatigue crack propagation with chief attention paid to the crack opening and closing behavior.

3. ADVANTAGES OF NUMERICAL ANALYSIS APPROACH

In this paper, fatigue crack propagation was studied numerically using thermo-elasto-plastic FEM. Although this is a basic work, one of the objectives of this analysis is to provide the basis for welding design in practical engineering applications. By adopting thermo-elasto-plastic FEM, it becomes possible (i) to extend the analysis in a rather straightforward manner to practical cases and (ii) to discuss quantitatively. The more concrete advantages of adopting numerical analysis approach in this case are considered to be as follows;

(1) As the dimensions of a specimen must be larger than a certain critical size in order to study the effect of welding residual stress, some of the experimental techniques already developed for observing crack opening and closure are quite difficult to apply, so that it is considered necessary to develop or to improve the technique in order to experimentally clarify the effect of welding residual stress in terms of crack opening and closure. But in the case of fillet welded joints, more fundamental problem arises, because the technique of measuring welding residual stress in this type of joints has not been established yet. Therefore, it seems quite difficult to obtain enough information useful for welding design by experimental approach alone.

(2) Plastic strain caused by welding thermal cycle is quite difficult to evaluate experimentally. Therefore, even if crack opening and closure might be evaluated experimentally with high accuracy, it seems difficult to correlate it quantitatively to thermally induced plastic strain. But by using FEM, both plastic strain caused by welding thermal cycle and plastic strain caused by cyclic loading can be evaluated quantitatively, so that they can be correlated to crack opening and closure quantitatively.

(3) If we consider an actual welded structure, it is expected that cracks in 3 dimensional welding residual

stress field are more important. But at least for the present, no other technique than numerical technique can analyze 3 dimensional welding residual stress field and the behavior of a crack in such stress field quantitatively and without loss of general applicability.

4. NUMERICAL ANALYSIS PROCEDURE

In this paper, the base metal specimen and butt-welded specimen with the dimensions and configurations shown in Fig.4 are analyzed with the aim of comparing numerical results with experimental results decribed in section 2. (a) and (b) in Fig.4 are analyzed in order to study the effect of the presence of welding residual stress under the condition of the length of the initial crack being the same. (b) and (c) in Fig.4 are analyzed in order to examine the effect of the difference of the length of the initial crack (half crack length a=14 mm and a=24 mm respectively) under the condition of the welding residual stress field being the same. As a material, mild steel is selected for analysis, because in the case of welding high tensile strength steel, the problem of matching between deposit metal and base metal arises. The procedure of numerical analysis is as follows; Heat conduction analysis is first carried out using finite difference method by assuming that the part corresponding to a weld bead is heated instantaneously to 1500°C and that the other part is at room temperature of 15°C. Then the distribution of welding residual stress before the initial crack is introduced is evaluated using thermo-elasto-plastic FEM based on thus obtained temperature distribution [7]. In this paper, the computation time is greatly reduced by introducing Marcal's method [8]. (Masubuchi [9] calculated welding residual stress by the similar method.) The increments of temperature differ from element to element, but each computation stage is selected so that the greatest increment of element temperature be less than 30°C. The mesh division used for heat conduction analysis and thermo-elasto-plastic analysis is shown in Fig.5 (a) and (b). The material properties used in heat conduction analysis is shown in Table 1, and the mechanical properties used in thermo-elasto-plastic analysis is shown in Fig.6. With respect to Young's modulus E, yield stress σ_y, and coefficient of linear thermal expansion α , such temperature dependency is assumed as shown in the figure. But Poisson's ratio ν and strain hardening rate H' are assumed to be independent of temperature. After obtaining the initial residual stress distribution, the meshes corresponding to the crack tip regions shown in Fig.5 (a) and (b) are further divided into finer meshes as shown in Fig.5 (c) and (d), and the nodal forces are redistributed

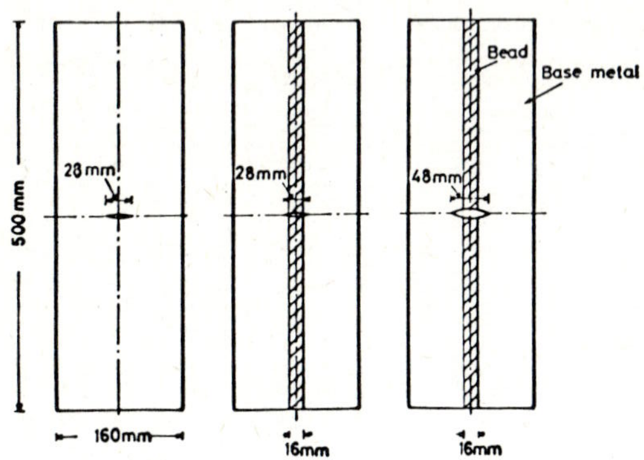

Fig. 4 Configurations and sizes of specimens analyzed

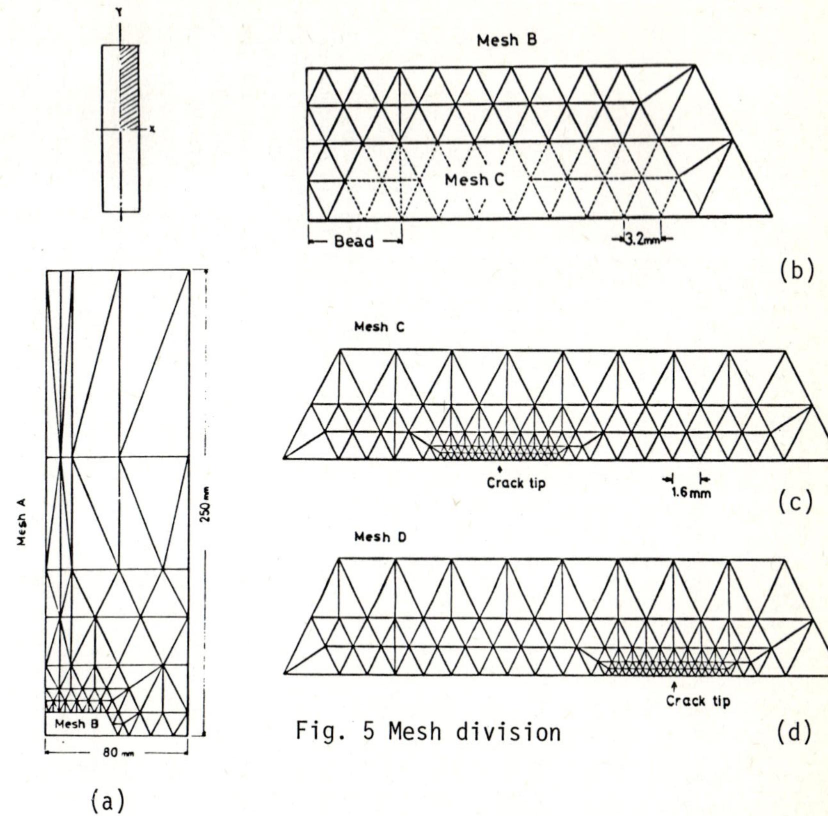

Fig. 5 Mesh division

Thermal conductivity	Specific heat	Density	Coefficient of heat transfer
0.1 (cal/sec cm °C)	0.16 (cal/g °C)	8.0 (g/cm^3)	0.0001 (cal/sec cm^2°C)

Table 1 Material properties used in heat conduction analysis

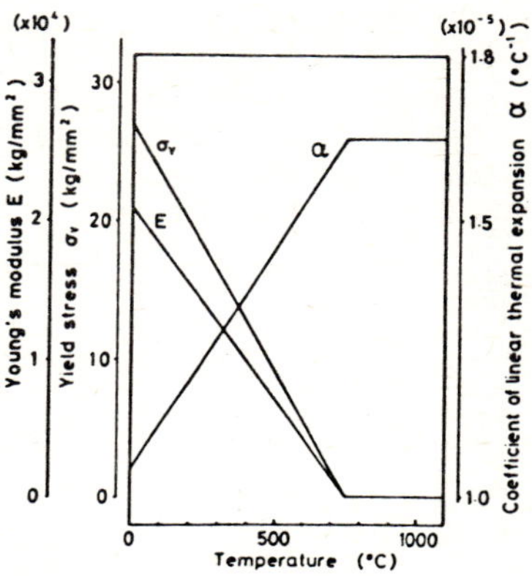

Fig. 6 Mechanical properties
(ν =0.3, H'=1200 kg/mm2)

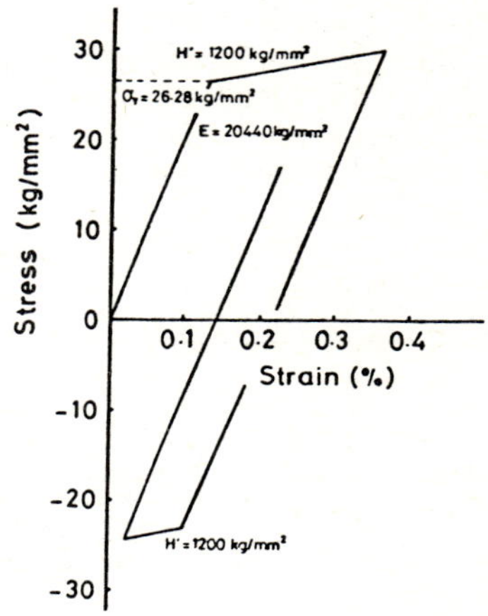

Fig. 7 Cyclic stress-strain curve

	INITIAL CRACK LENGTH a (mm)	STRESS RATIO R
BASE METAL SPECIMEN	14	0 -1
WELDED SPECIMEN	14	0 -1
	24	0 -1

$\sigma_{max} = 7$ kg/mm^2

Table 2 Cases analyzed

correspondingly. Then the initial crack is introduced by letting the crack surface be free surface, and fatigue crack propagation is analyzed by means of cyclic elasto-plastic FEM using Yamada's [10] incremental theory of plasticity and Ziegler's [11] kinematic hardening rule. A fatigue crack is extended one element for each loading cycle when the nodal reaction force at the crack tip changes from negative to zero [12]. The stress-strain relation shown in Fig.7 is used in the analysis for that corresponding to room temperature of Fig.6. Table 2 shows the cases analyzed.

5. RESULTS AND DISCUSSION

5.1 WELDING RESIDUAL STRESS DISTRIBUTION

Fig.8 shows the change of temperature distribution calculated by heat conduction analysis. Based on these results, the welding residual stress distribution is obtained using thermo-elasto-plastic FEM as shown by the solid marks in Fig.9. The open marks in Fig.9 (a) and (b) show the welding residual stress distribution after the initial crack of a=14 mm and a=24 mm are introduced respectively. In Fig.9, the displacement of the initial crack surface is also shown. By comparing the open marks in (a) and (b), it is observed that the case of longer initial crack length shows higher stress value at the crack tip.

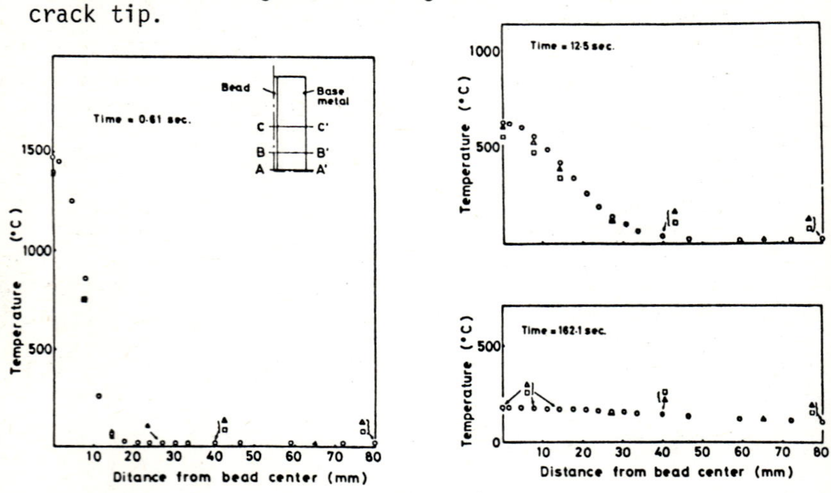

Fig. 8 Change of temperature distribution

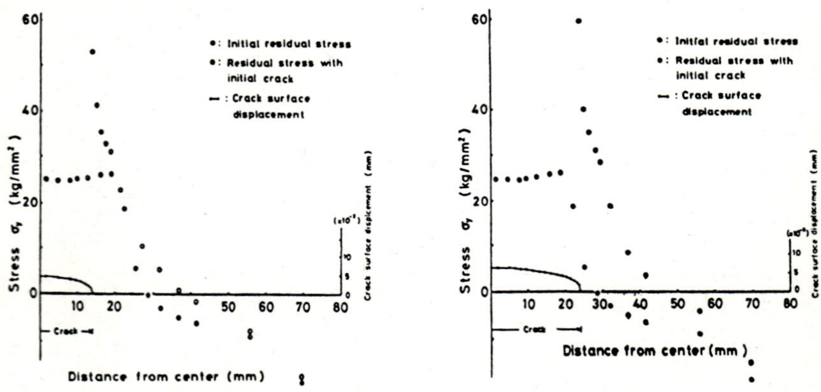

Fig. 9 Welding residual stress distribution before and after crack introduction

5.2 CHANGE OF WELDING RESIDUAL STRESS DISTRIBUTION DUE TO CYCLIC LOADING

Residual stress not only changes due to cyclic loading, but it also is redistributed due to crack extension. Therefore, the change of residual stress in the cracked welded specimen with the initial welding residual stress shown in Fig.9 is evaluated using cyclic elasto-plastic FEM. Fig.10 shows the residual stress distribution at the minimum load of each cycle for the condition shown in the figure. It is observed that the distribution of residual stress changes from the initial shape to a flatter and more gently changing distribution, and the stress decreases sharply in the neighbourhood of the crack tip. In the subsequent cycles, the shape of the distribution as a whole does not change appreciably, but the stress at the crack tip decreases cycle after cycle. By comparing (a) and (b), it is observed that although the residual stress value at the crack tip is higher in the case of a=24 mm than in the case of a=14 mm, it becomes lower after one cycle of loading. In the case of (c) where R=-1, the compressive residual stress is produced at the crack tip. These numerical results seem to suggest that in reality residual stress decreases by a large amount within the first few cycles and that the shape of the distribution as a whole does not change appreciably in the subsequent cycles, and that the value of residual stress near the crack tip approaches zero with the repetition of loading cycles.

5.3 CRACK OPENING AND CLOSING BEHAVIOR

Fig.11 (a), (b), and (c) shows the crack surface displacement in the welding residual stress field. The dotted curve shows the crack surface displacement when the initial crack is introduced, and at this stage, the crack surface displacement is 5 - 6 times greater than that of the base metal specimen at the minimum load. This result indicates that the effect of welding residual stress is remarkable. But the difference of crack opening and closure between a base metal specimen and a welded specimen does not become very clear by comparing the crack surface displacements alone. Therefore, in order to understand the difference more clearly, the crack opening stress and crack closing stress at each cycle is shown in Fig.12. Here, the effective stress range ratio U is defined as $U = \Delta\sigma_{eff}/\Delta\sigma$, where $\Delta\sigma_{eff}$ is the effective stress range which denotes the available portion of the stress range $\Delta\sigma$ to propagate a fatigue crack, i.e., the portion of the stress range above an open mark (crack opening stress level) or the portion of the stress range above

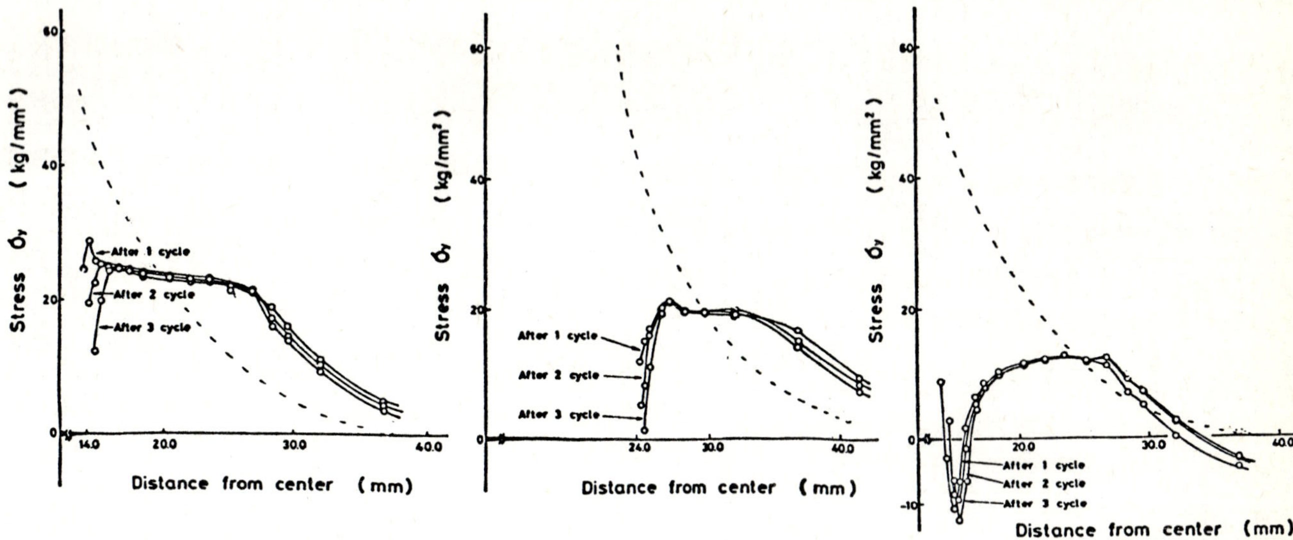

dotted curve = welding residual stress after initial crack is introduced

(a) R=0, a=14 mm (b) R=0, a=24 mm (c) R=-1, a=14 mm

Fig. 10 Change of welding residual stress with fatigue crack extension

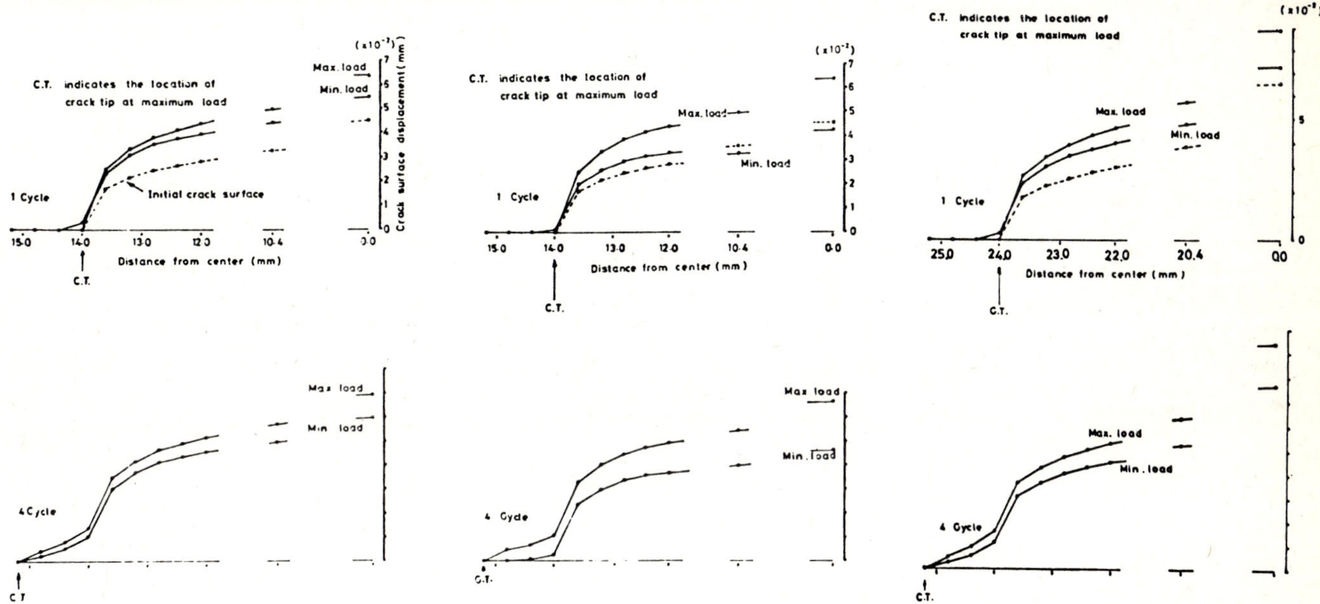

(a) R=0, a=14 mm (b) R=-1, a=14 mm (c) R=0, a=24 mm

Fig. 11 Fatigue crack surface displacement in welding residual stress field an solid mark (crack closing stress level). In the case of a base metal specimen of R=0, U is obtained approximately as 0.6 if based on an open mark (crack opening stress level) and approximately as 0.85 if based on a solid mark (crack closing stress level). In the case of R=-1, U is approximately 0.4 if based on crack opening stress level and U is approximately 0.45 if based on crack closing stress level. In the case of a welded specimen, U≑1 for both a=14 mm and a=24 mm, whether it is based on

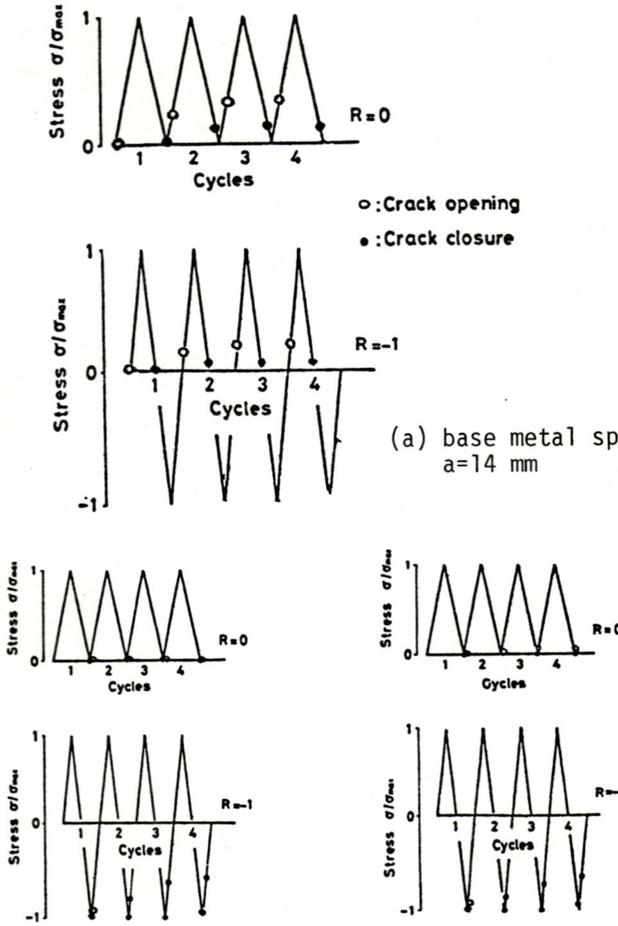

(a) base metal specimen
a=14 mm

(b) welded specimen
a=14 mm

(c) welded specimen
a=24 mm

Fig. 12 Change of crack opening and closing stress

crack opening stress level or on crack closing stress level. In the case of R=-1, U is also nearly 1 for both a=14 mm and a=24 mm, if based on crack closing stress level. But in spite of the case being the same, i.e., R=-1, it is observed that if it is based on crack opening stress level, U has a tendency to decrease with the repetition of loading cycle for both a=14 mm and a=24 mm. The similar tendency can also be seen in the case of R=0 and a=24 mm. For reference, the changes of U in the case of the experiments on high tensile strength steel described in section 2 are shown in Fig.13. Although the materials are different, since in Fig.13 the material is high tensile strength steel and in Fig.12 the material is mild steel (as the results of mild steel in the experiments of section 2 are widely scattered, less scattered results of high tensile strength steel is cited here, because the qualitative and quantitative trends are nearly the same.), the results of numerical analysis agree quite well both qualitatively and quantitatively with experimental results. It is seen from the experimental results that the U of a welded specimen approaches that of a base metal specimen with the extension of a fatigue crack. In the numerical analysis, the similar result is also obtained in the case of the change of crack opening stress level for R=-1 of a

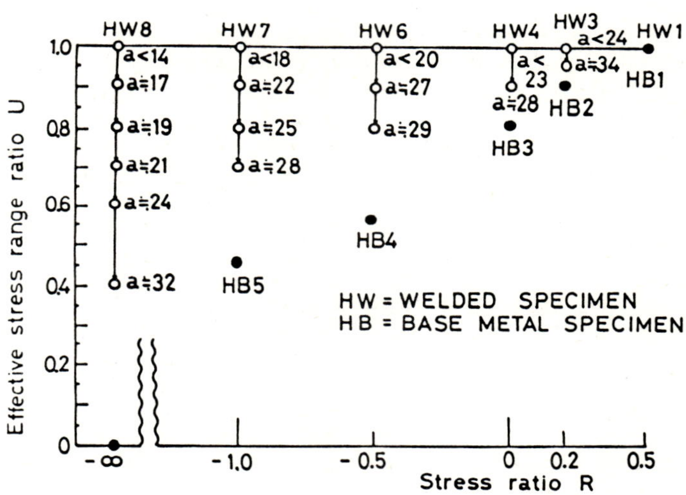

Fig. 13 Change of U of welded specimens with crack extension and U of base metal specimens (experimental results of high tensile strength steel)

welded specimen. But as the computation time increases, calculation is not carried out in this paper to such an extent that the U of a welded specimen decreases to the level corresponding to that of a base metal specimen. In order really to serve for the determination of welding condition for fabrication or to serve for the determination of allowable defect size in nondestructive inspection, more convenient method is desirable, including the reduction of computation time. But to accurately estimate the value of U in welding residual stress field, the accurate estimation of plastic strain caused by welding heat input and plastic strain caused by cyclic loading is necessary, because the value of U is determined by the quite complicated combination of these factors. Therefore, it seems quite difficult, but is is absolutely necessary to develop a practical and convenient method in the future to serve the above mentioned purposes.

6. CONCLUSIONS

(1) When an initial crack is introduced in the welding residual stress field, its crack surface displacement is 5 - 6 times greater than that of a base metal specimen with the same crack length at the maximum load. Thus, remarkable effect of welding residual stress is observed from the initial stage.
(2) If the effective stress range ratio U is evaluated based on crack closing stress, U is obtained as $U \doteqdot 0.85$ for $R=0$ and $U \doteqdot 0.45$ for $R=-1$ in the case of a base metal specimen, but in the case of a welded specimen $U \doteqdot 1$ for all cases so that the U of a welded specimen is greater than that of a base metal specimen. Thus, remarkable effect of welding residual stress is observed.
(3) If U is evaluated based on crack opening stress, $U \doteqdot 0.6$ for $R=0$ and $U \doteqdot 0.4$ for $R=-1$ in the case of a base metal specimen, but in the case of a welded specimen $U \doteqdot 1$ for $R=0$. In the case of $R=-1$ of a welded specimen, it is observed that U decreases with the extension of a fatigue crack or with the repetition of loading cycles. These U values, however, are greater than those of base metal specimens, so that the effect of welding residual stress is on the unsafe side within the limits of this analysis.
(4) Welding residual stress in a cracked specimen decreases sharply in the neighbourhood of the crack tip and the distribution changes from the initial shape to the flatter one, but its shape does not change appreciably in the subsequent loading cycles. Welding residual stress at the cracktip decreases with the extension of a crack.

REFERENCES

1. ELBER, W. - The Significance of Fatigue Crack Closure. ASTM STP 486, pp. 230-242, 1971

2. KIKUKAWA, M. et al. - Measurement of Fatigue Crack Propagation and Crack Closure at Low Stress Intensity Level by Unloading Elastic Compliance Method. J. SMSJ, Vol. 25, No. 276, pp. 899-903, 1976 (in Japanese)

3. OGURA, K. et al. - Analytical Study of Fatigue Crack Propagation under Varying Load. Trans. JSME, Vol. 42, No. 358, pp. 1615-1624, 1976 (in Japanese)

4. TAIRA, S. et al. - Opening and Closure of Fatigue Crack in Quenched-and-Tempered Carbon Steel. J. SMSJ, Vol. 26, No. 286, pp. 92-98, 1976 (in Japanese)

5. SHIMADA, H. and FURUYA, Y. - Relation between the Phenomenon of Fatigue Crack Opening and Closure and the Strain History at the Crack Tip. Preprint JSME, No. 780-4, pp. 157-159, 1978 (in Japanese)

6. FUKUDA, S. et al. - An Experimental Study of Effect of Welding Residual Stress upon Fatigue Crack Propagation Based on Observation of Crack Opening and Closure. TRANS. JWRI, Vol. 8, No. 2, 1979

7. UEDA, Y. and YAMAKAWA, T. - Analysis of Thermal Elastic-Plastic Stress and Strain during Welding by Finite Element Method. TRANS. JWS, Vol.2, No. 2, 1971

8. MARCAL, P.V. and KING, I.P. - Elastic-Plastic Analysis of Two-Dimensional Stress System by the Finite Element Method. Int. J. Mech. Sci., Vol. 9, pp. 143-155, 1967

9. MASUBUCHI, K. - Analysis of Welded Structures, Pergamon Press, New York, 1980

10. YAMADA, Y. - Plasticity and Viscoelasticity; Computer Approach to Structural Engineering II-2, Baifukan, 1972 (in Japanese)

11. ZIEGLER, H. - A Modification of Prager's Hardening Rule, Quart. Appl. Math., Vol. 17, pp. 55-65, 1959

12. OHJI, K. et al. - Study of Nonpropagating Fatigue Crack from the Standpoint of Mechanics (Finite Element Approach). TRANS. JSME, Vol.40, No. 337, pp. 2505-2512, 1974

Numerical Techniques in Computer-Aided Fatigue Crack-Growth Analysis

By

J. B. Chang(I), R. M. Engle(II), and M. Szamossi(I)

SUMMARY

A review and evaluation of the state-of-the-art numerical techniques employed in the commonly used fatigue crack-growth computer programs have been performed recently in order to select a cost-effective method to incorporate into an improved fatigue crack growth analysis computer code which is being developed at Rockwell. The numerical techniques evaluated included the close-form solution, mean-value and Runge-Kutta methods, Taylor series expansion technique, and linear approximation method.

The approach taken for the evaluation was to determine the predictive accuracy of a method by correlating the predicted crack growth lives to the test data and to estimate the computation costs through the charged central processing units (CPU). Two most commonly used techniques, Runge-Kutta and linear approximation, were evaluated extensively. Results showed that the linear approximation method is most cost-effective. This paper summarizes the review and presents the results of the evaluation.

I. Program Manager and Technical Staff, respectively, Rockwell International Corporation, North American Aircraft Division, Los Angeles, California, United States.

II. Aerospace engineer, Air Force Wright Aeronautical Laboratories, Flight Dynamics Laboratory, Wright-Patterson Air Force Base, Ohio, United States

INTRODUCTION

In recent years, the computer-aided fatigue crack-growth analysis has gained popularity as a tool for evaluating the behavior of cracks or crack-like flaws contained in primary structures under cyclic service loadings. The majority of the commonly used crack growth analysis computer codes, such as CRACKS[1] and EFFGRO[2], are based on linear elastic fracture mechanics (LEFM) concepts and employ a damage accumulation scheme which interrelates the following elements:

1. Fatigue crack-growth-rate equation and the corresponding growth-rate constants, fracture properties, and other related material properties

2. Cyclic service loading descriptions, including the maximum and minimum stresses of each load cycle and the number of cycles

3. Crack configurations, sizes, locations, and geometries of the cracked structure

4. Crack-tip stress-intensity-factor solutions

5. Numerical integration procedures

Fatigue crack-growth-rate relationships used in most of the computer codes are in general expressed as $da/dN = F(\Delta K, R, \ldots)$, where a is the crack size, ΔK is the stress-intensity-factor range, and R is the cyclic stress ratio. However, the numerical integration techniques employed in each code for solving this first-order differential equation vary somewhat from one to another, depending on the usages and applications.

Currently, a research and development program is being undertaken by Rockwell for the U.S. Air Force. This program is aimed at upgrading the crack-growth prediction technology required for the implementation of the damage tolerance control procedures throughout the life cycle of aircraft systems, as specified in the currently issued U.S. military standards and specifications[3,4]. To accomplish this objective, a review and evaluation of the state-of-the-art numerical techniques was deemed necessary in order to establish the most reliable crack-growth analysis procedure that is within the state-of-the-art and yet cost-effective. A task conducted in this research program was to perform such a review and evaluation[5,6]. The evaluation approach was to determine the predictive accuracy of a specific method by correlating the analytical results to the test data generated from the aforementioned research program[7] and to estimate the computational cost through the CPU's charged by each computer run.

STATE-OF-THE-ART TECHNIQUES

In general, two categories of crack-growth analyses are commonly performed. Category I is the calculation of the elapsed time (number of cycles) which is required to "grow" a crack from an initial size (a_o) to a final size (a_f) when subjected to cyclic loadings. Category II is the calculation of the incremental growth of a crack within a period of time (number of cycles). In mathematical form, these two categories can be expressed as:

Category I:

$$N = \sum_{i=1}^{N-1} \Delta N_i = \sum_{i=1}^{N-1} \left(\int_{a_o}^{a_f} \frac{da}{F(\Delta K, R \ldots)} \right)_i \tag{1}$$

Category II:

$$\Delta a = a_f - a_o = \sum_{i=1}^{N} da/dN \Big|_i \tag{2}$$

Numerous techniques have been used to solve the preceding mathematical problems. The following methods represent the commonly used ones in the area of fatigue crack-growth analysis:

CLOSED-FORM INTEGRATION

In category I, for a simple case such as a center-through crack contained in a wide panel subjected to a constant amplitude loading, it can be shown that the required number of cycles to grow the crack from a_o to a_f is:

$$N = \frac{2}{C(2-n)\left[(1-R)^{m-1}(\sigma_{max} - \sigma_{min})\sqrt{\pi}\right]^n} \left(a_f^{\frac{2-n}{2}} - a_o^{\frac{2-n}{2}} \right) \tag{3}$$

Equation 3 resulted by employing the following fatigue crack-growth-rate equation and the crack-tip stress intensity factor:

$$da/dN = C\left[\Delta K/(1 - R)^{1-m}\right]^n \tag{4}$$

$$K = \sigma\sqrt{\pi a} \tag{5}$$

where C and n are the crack-growth-rate constants, m is the Walker stress ratio collapsing exponent[8], ΔK is the stress-intensity-factor range, σ is the applied cyclic stress, and a is the crack size.

NUMERICAL INTEGRATION

The stress-intensity-factor solutions for cracks contained in common structural components are much more complicated than is shown in equation 5. In general, it can be expressed as:

$$K = \sigma\beta(a)\sqrt{\pi a} \tag{6}$$

where $\beta(a)$ is a function of crack configuration and structural geometry. It can be either a simple mathematical form or a complex function. For a center-through crack contained in a finite-width plate case, $\beta(a)$ can be expressed as:

$$\beta(a) = \sqrt{\sec\frac{\pi a}{W}},$$

where W is the width of the plate. Geometrical correction functions for other crack configurations contained in various geometries are available in several handbook-type documents[9,10,11]. Some of them are presented in polynomial forms and some are written in compound functions. In any case, the closed-form integration solution as expressed in equation 3 cannot be directly applied since $\beta(a)$ is also a function of the crack dimension, a.

The numerical integration technique employed in the TRKGRO computer program[12] is a typical method to solve this type of problem. The procedure is briefly described as follows:

1. Assume an increment of crack growth $\Delta a = a_{i+1} - a_i$

2. Calculate $\bar{a} = (a_{i+1} + a_i)/2$

3. Determine expression for ΔK in terms of the average crack size, $\overline{a}$

$$\Delta K = \beta(\bar{a})\sqrt{\pi}(\sigma_{max} - \sigma_{min})\sqrt{\bar{a}} \tag{7}$$

4. Solve for ΔN for each increment of crack growth

$$\Delta N = \frac{\Delta a}{C\left[\beta(\bar{a})\sqrt{\pi}(\sigma_{max} - \sigma_{min})(1 - R)^{m-1}\sqrt{\bar{a}}\right]^n} \tag{8}$$

5. Repeat the procedure for $a_j = a_{j-1} + \Delta a$ and then sum up the incremental crack-growth cycle to obtain the total number of cycles.

$$N = \sum_{j=1}^{N-1} \Delta N_j \tag{9}$$

Runge-Kutta Integration Method

To calculate the increase in crack size, Δa, resulting from a known period of time, ΔN, the Runge-Kutta integration method was the original technique adopted in CRACKS[1]. The Runge-Kutta method has been used very successfully for solving ordinary differential equations with initial conditions on a digital computer. Detailed derivations are given in mathematical textbooks[13,14]. The Runge-Kutta integration technique calculates the increment of crack size, Δa, from the current size, a_i, within ΔN cycles as follows:

$$\Delta a = a_{n+1} - a_n = \frac{1}{6}(k_o + 2k_1 + 2k_2 + k_3) \tag{10}$$

where

$$k_o = \Delta N \cdot \left.\frac{da}{dN}\right|_{a_n}$$

$$k_1 = \Delta N \cdot \left.\frac{da}{dN}\right|_{a_n + k_o/2}$$

$$k_2 = \Delta N \cdot \left.\frac{da}{dN}\right|_{a_n + k_1/2}$$

$$k_3 = \Delta N \cdot \left.\frac{da}{dN}\right|_{a_n + k_2}$$

Figure 1 illustrates the geometrical interpretation of the values of k_o through k_3. All four k-values represent the slopes at various points. k_o is the slope of the starting point; k_3 is the slope at the right-hand point whose ordinate is $a_n + k_2\ (\Delta N)$; k2 is one of the two slopes considered at the midpoint with ordinate $(a_n + 1/2\ k_1\ \Delta N)$; and, finally, k_1 is the second slope at the midpoint whose ordinate is $(a_n + 1/2 k_o \Delta N)$.

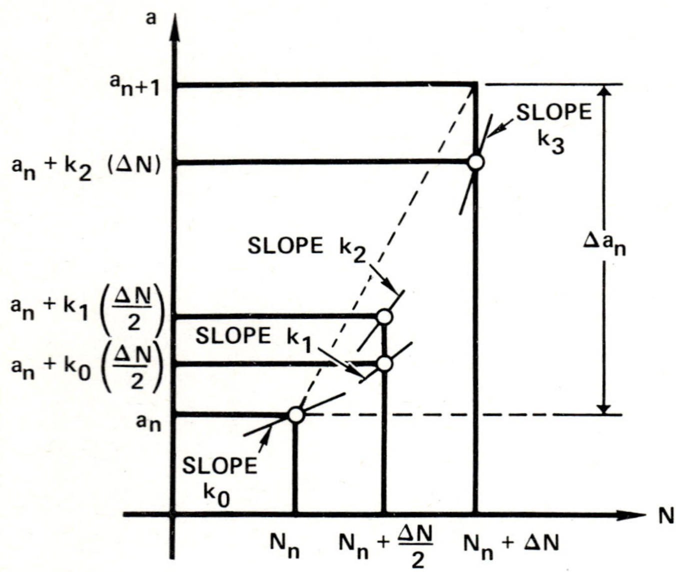

Figure 1. Four Slopes Used in Runge-Kutta Method

Taylor Series Approximation Method

A damage accumulation technique identified as the Taylor Series Approximation Method[15] is being employed in a crack-growth program CGR-GD[16]. A brief description of this method is presented as follows:

For a given crack-growth-rate equation (i.e., Paris equation)

$$\frac{da}{dN} = C(\Delta K)^n \tag{11}$$

Separating terms yields

$$C \Delta N = \int_{a_i}^{a_f} \frac{da}{[\Delta K(a)]^n} \tag{12}$$

Define

$$\mu = \int_{a_i}^{a_f} G(a)da \tag{13}$$

$$\mu = C\Delta N \quad \text{and} \quad G(a) = \frac{1}{[\Delta K(a)]^n}$$

Approximate equation 13 with

$$\mu \cong G(a_i) \int_{a_i}^{a_f} da = G(a_i)(a_f - a_i) \tag{14}$$

Expand G(a) into the Taylor series (about $a = a_i$) and integrate the Taylor series yield

$$\mu = \sum_{m=0,1,2} P_m(\Delta a)^{m+1} \tag{15}$$

where

$$P_m = \frac{a_i^{m+1}}{(m+1)!} \left. \frac{d^m G(a)}{da^m} \right|_{a=a_i} \quad \text{and}$$

$$\Delta a = \frac{a_f}{a_i} - 1$$

Assume a power series expansion in the "small parameter" as:

$$\Delta a = \sum_{P=1,2,\ldots} \mu^P a_P \tag{16}$$

where each a_p is the to-be-determined parameter

Substitute equation 16 into 15 and equate to zero the coefficients of every integral power of μ in order to get

$$P_o a_1 = 1$$

$$P_o a_2 + P_1 a_1^2 = 0$$

$$P_o a_3 + 2P_1 a_1 a_2 + P_2 a_1^3 = 0 \ , \ \ldots \tag{17}$$

a_1, a_2, a_3, etc, can thus be determined as:

$$a_1 = 1/P_o,\ a_2 = -P_1/P_o^3,\ a_3 = (-P_oP_2 + 2P_1^2)/P_o^5,\ \ldots \tag{18}$$

The crack size a_f is then calculated as

$$a_f = a_i\,(1 + \Delta a) = a_i\left(1 + \sum_{P=1,2,3,\ldots} \mu^P a_P\right) \tag{19}$$

Linear Approximation Method

The linear approximation method was formulated by Vroman[2] in the early 1970's. The Rockwell in-house-developed computer program, EFFGRO, is the first to adapt this method for damage accumulations when an input stress spectrum is formulated in load steps as shown in Figure 2. The Vroman linear approximation method assumes that the growth rate is constant throughout a load step in such a stress spectrum as shown in table I, so that the crack size is in a linear relationship with the number of load cycles.

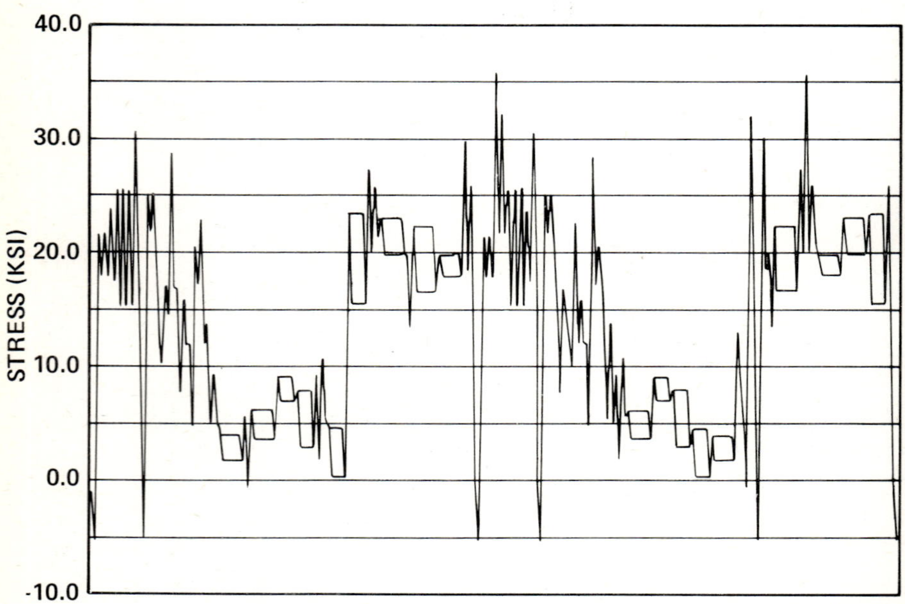

Figure 2. A Flight-by-Flight Spectrum

Table I

A TYPICAL STRESS SPECTRUM IN TABULAR FORM

Step	Max Stress	Min Stress	No. of Cyc/ Block (Flight)
1	σ_{max_1}	σ_{min_1}	N_1
2	σ_{max_2}	σ_{min_2}	N_2
3	σ_{max_3}	σ_{min_3}	N_3
.	.	.	.
.	.	.	.
.	.	.	.
i	σ_{max_i}	σ_{min_i}	N_i

The damage-accumulation scheme proceeds by considering a load step (i) and taking the values of σ_{max_i} and σ_{min_i} to calculate the cyclic growth rate, da/dN. In EFFGRO, the rate equation used is the Paris equation, with Walker's stress ratio collapsing exponent, m, as expressed in equation 4.

The value of $(\delta a)/(da/dN)$ is then compared to N_i, where "a" is the crack size. If $(\delta a)/(da/dN)$ is greater than N_i, then the crack growth for that particular load step is $\Delta a = N_i \times (da/dN)_i$; "a" is increased by Δa, and the program proceeds to the next load step (i + 1).

If $(\delta a)/(da/dN)$ is less than or equal to N_i, then the number of cycles to grow (δa) is $(\delta a)/(da/dN)$. This value is subtracted from N_i, the crack size "a" is increased by (δa), and this load step is reconsidered. This process continues with $(\delta a)/(da/dN)$ being compared to the remaining cycles in this load step. When all the load steps in a block (or flight) are exhausted, the program proceeds to the first load step of the next block (or flight).

EVALUATIONS

Among the techniques described previously, it was decided to select two commonly used schemes for extensive evaluation. Based on this rationale, the Runge-Kutta and the linear approximation methods were chosen. Crack-growth analyses of a center-through crack contained in the 2219-T851 aluminum plate subjected to various types of loading cases were used for the evaluation. Figure 3 illustrates the test specimen. The loading cases included constant amplitude, single overload, block loading, and a random cycle-by-cycle flight spectrum. Crack-growth predictions were obtained by using two computer programs, CRACKS and EFFGRO. Since these two programs use identical rate equation and load interaction models, they provide very close results. The predicted fatigue crack-growth lives were summarized and presented in Table II. The CRACKS predictions represented results using the Runge-Kutta technique, while the EFFGRO predictions represented linear approximation method results.

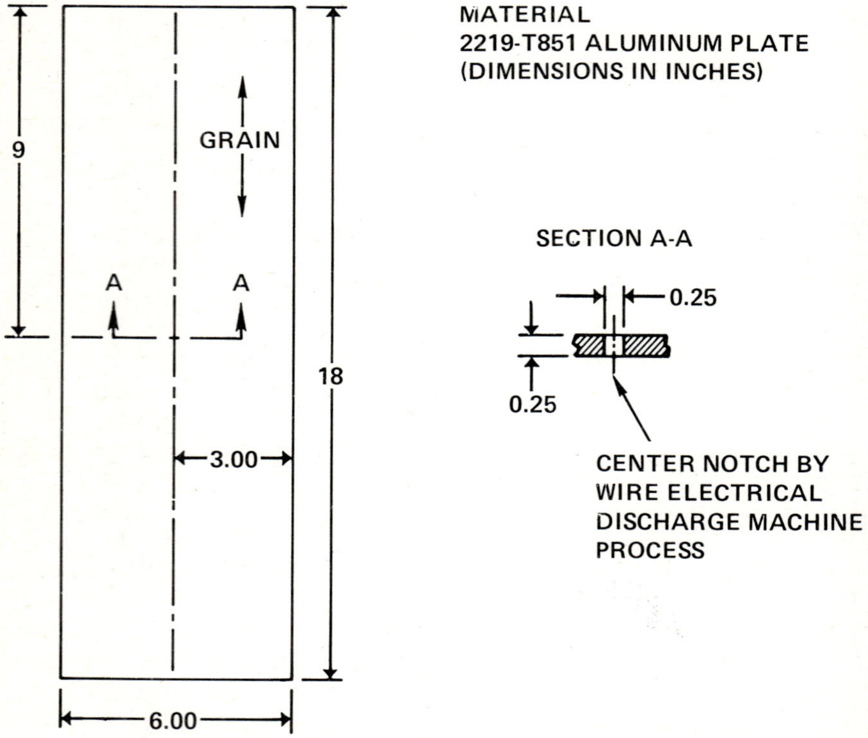

Figure 3. Test Specimen Configuration

Table II

SUMMARY OF COMPUTER COST COMPARISON - RUNGE-KUTTA METHOD VERSUS LINEAR APPROXIMATION METHOD

Loading Type	CRACKS		EFFGRO		CPU(CRACKS)/CPU(EFFGRO)
	Predicted Life Cycles	CPU	Predicted Life Cycles	CPU	
Constant Amplitude	1,009,500 2,199,750	0.1352	1,004,270 2,184,830	0.0260	5.2
	1,010,325 150,000	1.0023	1,001,420 150,000	0.0683	14.7
	11,325 24,675	0.0177	11,310 24,630	0.0098	1.8
Single Overload	14,105 14,105	0.0390	14,105 12,860	0.0370	1.05
	40,015 12,505	0.0092	12,515 12,505	0.0067	1.37
	35,010 25,010	0.0235	35,010 25,010	0.0068	3.45
Block Loading	11,500 9,250 17,620 25,500	0.0218	10,885 8,560 17,705 23,080	0.0260	0.84
	10,485 8,000 17,620 23,000	0.0237	10,495 6,850 17,470 22,135	0.0250	0.51
	74,750 8,800 102,000 12,050	0.0308	74,625 8,210 101,760 10,970	0.0177	1.74
	65,750 6,550 95,750 9,900	0.0128	65,625 6,000 95,760 9,850	0.0177	0.72
Random cycle-by-cycle flight spectrum	156,650 90,070	2.6667 0.4998	155,290 90,500	0.4598 0.2267	5.8 2.2

The approach taken in rating the efficiency of these two schemes was to compare their computer cost through the recorded CPU charged for each comparable run. This approach, of course, can only provide relative comparisons. Yet, from the results shown in table II, the linear approximation scheme is more efficient than the Runge-Kutta method in general.

CONCLUDING REMARKS

The numerical techniques used for performing fatigue crack-growth analyses were reviewed. These include the closed-form solution, mean-value and Runge-Kutta methods, Taylor series expansion technique, and linear approximation scheme. An extensive evaluation on two most commonly used methods, Runge-Kutta and linear approximation were conducted. Results showed that by using these two methods, fatigue crack-growth life predictions are very close. The linear approximation method is apparently more efficient for constant-amplitude-type loadings. For block-loading types, computer costs for these two methods are comparable. For flight-by-flight spectrum loads, the linear approximation method is still more efficient.

ACKNOWLEDGEMENT

This report is based on part of the results of a study performed on Contract F33615-77-C-3121, sponsored by the United States Air Force, Air Force Wright Aeronautical Laboratories, Flight Dynamics Laboratory, Wright-Patterson Air Force Base, Ohio, United States.

REFERENCES

1. ENGLE, R. M. - CRACKS, A Fortran IV Digital Computer Program for Crack Propagation Analysis. AFFDL-TR-70-107, Air Force Flight Dynamics Laboratory, WPAFB, Ohio, 1970.

2. SZAMOSSI, M. - Crack Propagation Analysis by Vroman's Model, Program EFFGRO. NA-72-94, Rockwell International, North American Aircraft Division, Los Angeles, 1970.

3. ANON - Military Standard, Aircraft Structural Integrity Program, Airplane Requirements, MIL-STD-1530A, 1975.

4. ANON - Military Specification, Airplane Damage Tolerance Requirements. MIL-A-83444A, 1974.

5. CHANG, J. B., STOLPESTAD, J. H., SHINOZUKA, M., and VAICAITIS, R. - Improved Methods for Predicting Spectrum Loading Effects - Phase I Report, Volume I - Results and Discussions. AFFDL-TR-79-3036, Vol. I, United States Air Force Flight Dynamics Laboratory, Wright-Patterson Air Force Base, Ohio, 1979.

6. CHANG, J. B. - Improved Methods for Predicting Spectrum Loading Effects, Fourth Quarterly Report. NA-78-491-4, Rockwell International, North American Aircraft Division, Los Angeles, 1979.

7. CHANG, J. B., and STOLPESTAD, J. H. - Improved Method for Predicting Spectrum Loading Effects - Phase I Report, Vol. II - Test Data. AFFDL-TR-79-3036, Vol. II, United States Air Force Flight Dynamics Laboratory, Wright-Patterson Air Force Base, Ohio, 1979.

8. WALKER, K. - The Effect of Stress Ratio During Crack Propagation and Fatigue for 2024-T3 and 7075-76 Aluminum. ASTM STP 462, American Society for Testing and Materials, 1970.

9. TADA, H., PARIS, P. C., and IRWIN, G. R. - The Stress Analysis of Cracks Handbook. Del Research Corp., Hellertown, Pennsylvania, 1973.

10. SIH, G. C. - Handbook of Stress Intensity Factors for Researchers and Engineers. Institute of Fracture and Solid Mechanics, Lehigh University, Bethlehem, Pennsylvania, 1973.

11. ROOKE, D. P., and CARTWRIGHT, D. J. - Compendium of Stress Intensity Factors. Her Majesty's Stationery Office, London, 1976.

12. ANON - Technical Proposal for Improved Methods for Predicting Spectrum Loading Effects. NA-77-648, Rockwell International, Los Angeles, 1977.

13. ROMANELLI, M. J. - Runge-Kutta Method for the Solution of Ordinary Differential Equation. Mathematical Methods for Digital Computers, John Wiley and Sons, New York, 1960.

14. KUO, S. S. - Numerical Methods and Computers. Addison-Wesley, 1965.

15. JOHNSON, W. S. - Prediction of Constant Amplitude Fatigue Crack Propagation in Surface Flaws. ASTM STP 687, J. B. Chang Ed., American Society for Testing and Materials, 1979.

16. JOHNSON, W. S. - CGR, An Improved Computerized Model to Predict Fatigue Crack Growth Under Spectrum Loading. Report 4577, Naval Ship Resarch and Development Center, 1975.

Fatigue Crack Growth Analysis for Random Loading Based on a Generalised Memory Criterion

H. Führing

Fraunhofer-Institut für Betriebsfestigkeit (LBF)
Darmstadt, F.R. Germany

SUMMARY

A new model for predicting non-linear crack growth due to load sequence effects in random loading is presented which considers the acceleration as well as the retardation effect. The retardation effect (factor Q_r) is being quantified by a ratio of plastic zone sizes. This procedure represents a generalisation of the memory criterion published elsewhere. The acceleration factor, Q_a, is formulated as a function of the affected crack length assumed to equal the contact length which is due to crack closure. Though the load sequence effect has been correlated to the stress intensity range, $\Delta K_{LS} = Q_{LS} \cdot \Delta K$ with $Q_{LS} = Q_a \cdot Q_r$, the new model is implicitly based on the crack closure mechanism. Comparisons of predicted crack growth lives with experimental data reveal deviations of less or equal to 100 % (in cycles).

INTRODUCTION

Crack growth in a structure subjected to a given load-time function of arbitrary shape is a topical field of fracture mechanics application in advanced reliability and safety studies. In the past, the major macro-mechanisms of fatigue crack growth (crack closure, acceleration, retardation) have been explored and described by means of continuum mechanics and simple failure models, e.g. [1] [2] [3]. Nontheless, because of the complex nature of those "exact" analyses and high costs of manpower and computer time there exists a need for rapid, inexpensive analytical models to be useful for practical application. This paper describes a new model for cycle-by-cycle crack growth prediction which meets the mentioned requirements. Comparisons between predicted and experimentally observed crack growth are presented.

CONCEPT FOR A NEW CRACK GROWTH MODEL

For a given cracked structure under repeated loading the crack length at an arbitrary load cycle k=j is obtained by summing up all crack growth increments due to preceding load cycles

$$a_j = a_0 + \sum_{k=1}^{j-1} \Delta a_k = \underbrace{\left(a_0 + \sum_{k=1}^{j-2} \Delta a_k\right)}_{a_{j-1}} + \Delta a_{j-1} \qquad (1)$$

with a_0 being the initial crack length. Fig. 1 shows the load sequence (L.S.) serving as an example. The fat lines indicate that the crack is assumed to grow only on the increasing part of the load cycle. Each

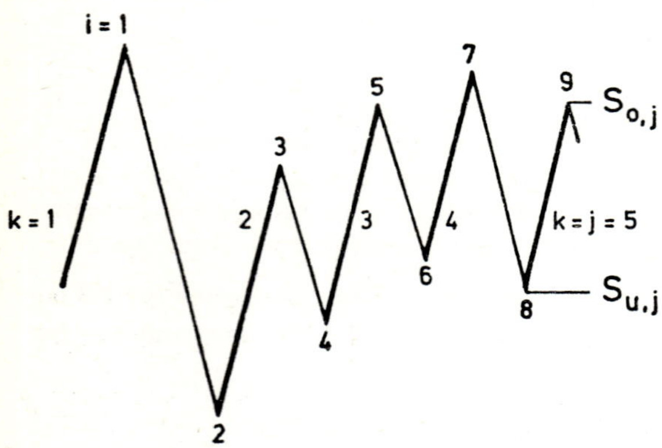

Δa_1	Δa_2	Δa_3	Δa_4	Δa_5	Ranges for Δa_k Calculation
a_1 $= a_0$	a_2 $= a_1 + \Delta a_1$	a_3 $= a_2 + \Delta a_2$	a_4 $= a_3 + \Delta a_3$	a_5 $= a_4 + \Delta a_4$	Correlation of Crack Lengths to Load Reversal Points

Fig. 1: Load Sequence Notations

cycle is described by a pair of load reversal points: $S_{u,j}$ and $S_{o,j}$, the respective stress ratio being $R_j = S_{u,j}/S_{o,j}$. The crack growth increment Δa_j generally is unknown. Regarding constant amplitude (C.A.) loading the increment can be expressed by a Forman material law

$$\Delta a_j = \frac{C \cdot \Delta K_j^{\,n}}{(1-R_j)\cdot K_c - \Delta K_j} \qquad (2)$$

or any other fatigue crack growth law based on linear fracture mechanics. As, factually, crack growth under nonC.A. loading does not respond to such a linear approach one has to establish a new or modified law applicable to more general purposes. As to the new model the following modified Forman law is used

$$\Delta a_j = \frac{C \cdot \Delta K_{LS,j}^{n}}{(1-R_j)\cdot K_c - \Delta K_{LS,j}} \tag{3}$$

Any other C.A. law with ΔK replaced by ΔK_{LS} can be used, too. $\Delta K_{LS,j}$ is the modified stress intensity range of the load cycle j including all effects of load history:

$$\Delta K_{LS,j} = Q_{LS,j} \cdot \Delta K_j = Q_{a,j} \cdot Q_{r,j} \cdot \Delta K_j \tag{4}$$

This definition is the key to the new model: the crack growth increment is a product of a load sequence factor Q_{LS} and the stress intensity range. The Q_{LS} factor itself is split up into two separate factors: the acceleration factor Q_a and the retardation factor Q_r. Eqns. (3) and (4) show that the crack growth prediction for arbitrary load sequences needs the linear fracture mechanics prediction as a supporting function. The Q_{LS} factor depends on the actual sequence of load reversals. In the special case of C.A. loading it becomes $Q_a = Q_r = Q_{LS} = 1$. The retardation factor Q_r is a function of a plastic zone size ratio: $Q_r = f(\omega_{LS}/\omega_{CA}) \leq 1$ with ω_{LS} being the plastic zone size due to the load history (ω_{CA}: without load history). Only Q_r-values less or equal to one are considered. The acceleration factor Q_a is a function of a crack growth increment over contact length ratio $Q_a = f(\Delta a/\lambda) \geq 1$ with λ being the length of the contact zone due to crack closure. Only Q_a-values greater or equal to one are considered. The formulae describing Q_r and Q_a are established in the following sections. As will be seen, the model is implicitly based on the crack closure mechanism.

RETARDATION FACTOR Q_r

From experiments it is a well-known phenomenon that crack growth is retarded after singlepeak overload (SPOL) application. The crack grows slower than it does without overload. Continuum mechanics studies [2] revealed that the retardation phenomenon can be described only if the crack closure mechanism is taken into account. Fig. 2. schematically demonstrates the analytical results. According to the local approach which correlates fatigue damage to the straining of the material element just in front of the crack, fatigue crack growth can be described by the following law

$$da/dN = \tilde{C} \cdot \Delta v_{a,LS}^{\tilde{n}} \tag{5}$$

with displacement variation of the crack tip element, Δv_a, obtained through a Dugdale analysis. Subscript "LS" denotes that the Δv_a-values are affected by the sequence of the loading. As a result of an analysis neglecting crack closure, Fig. 2a, we obtain a Δv_a greater than the respective C.A. value. Looking at Eqn. (5) this means an acceleration of crack growth in contrast with the experimental observations. If crack

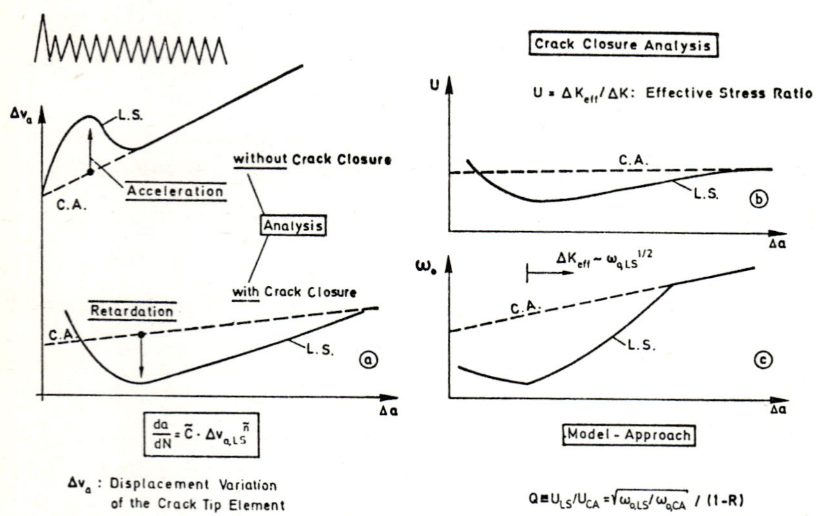

Fig. 2: Retardation Effect Analyzed with the DUGDALE-Method [2]

closure is taken into account the Δv_a-values obtained are smaller than the C.A. value. This, indeed, results in retardation as experimentally observed. The development of the effective stress intensity ratio [4] $U = \Delta K_{eff} / \Delta K = (S_o - S_{op})/(S_o - S_u)$ is shown in Fig. 2 b. Since analytical formulations for the Δv_a function as well as the U-ratio function are missing the author once suggested [5] that the retardation effect might be expressed by the alteration of the plastic zone size. Fig. 2 c shows the ω-curve of the SPOL case. The mentioned suggestion is based on the fact that $\omega_{o,LS}$ is proportional to ΔK_{eff}^2 [6]. This relation, strictly speaking, applies to the rising part of the curve only. Observing the definition $Q \equiv U_{LS}/U_{CA}$ the acceleration factor is derived by means of the respective plastic zone size functions. The model approach is:

$$Q_r = \sqrt{\omega_{o,LS}/\omega_{o,CA}}/(1-R) \quad \text{if } R \geq 0$$
$$Q_r = \sqrt{\omega_{o,LS}/\omega_{o,CA}} \quad \text{if } R \leq 0 \qquad (6)$$

The Dugdale formula of the plastic zone size at the upper (lower) reversal points of C.A. loading reads

$$\omega_{o,CA} = \frac{\pi}{8}\left(\frac{K_o}{\bar{\sigma}_y}\right)^2 \qquad \omega_{u,CA} = \frac{\pi}{8}\left(\frac{\Delta K}{2\bar{\sigma}_y}\right)^2 \qquad (7)$$

where K_o is the stress intensity factor at maximum load and $\bar{\sigma}_y$ is the yield strength of the material

$$\bar{\sigma}_y = \sigma_y \quad \text{for plane stress}$$
$$\bar{\sigma}_y = 2.5\ \sigma_y \quad \text{for plane strain} \qquad (8)$$

The plastic zone size affected by the load sequence is

$$\omega_{LS} = \left(\frac{\Delta a_{i,m}}{\omega_{i,m}} - 1\right)^2 \cdot \omega_{i,m} \qquad (9)$$

where

$$\Delta a_{i,m} = a_i - a_m \qquad (10)$$

$$\omega_{i,m} = \frac{\pi}{8}\left(\frac{K_i - K_m}{2\bar{\sigma}_y}\right)^2 \tag{11}$$

Subscript i refers to the i-th reversal point lying in the j-th cycle. For the sake of simplicity subscript j was left out in all terms of Eqns. (6), (9) to (11). Subcript m refers to the memory element m* valid for the stress state to be treated. Eqns. (9) and (11) can be used for upper load reversal states ($\omega_{o,LS}$: tensile plastic zone sizes, uneven numbered) and lower load reversal states ($\omega_{u,LS}$: compressive plastic zone sizes, even numbered) as well. Eqn. (9) is the generalised writing of the expressions (15), (19), (46) and (47) of Ref. [5]. I).

Some comments on the generalised memory criterion which resulted in Eqns. (9) - (11) are required. Fig. 3 shall serve as an illustrative example. Between each pair of load reversals S_i and S_{i+1}, the crack is extended by a small amount. The stress states at the reversal points are depicted at the bottom of the figure. The stress states at the lower reversal points, $S_u > o$, are assumed contact-free. The plastic zone size for lower reversal points with compressive loads, $S_u < o$, is corrected by defining a fictitious compressive load according to

$$\tilde{S}_u = (1 - \sqrt{0.2(1-R)^2 + 0.8})\ S_o \tag{12}$$

Eqn. (12) takes into account that even in case that the fatigue crack closure effect is neglected cracks close in compression ("elastic contact problem" [2]).

The extent ω of the plastic zone sizes is depicted in the middle of the figure. For clarity reasons the one-dimensional extent of the plastic zone is depicted two-dimensionally by means of fictitiously designed elliptical contours.

Regarding point i=5 the table indicates that four active memory elements m=1,2,3,4 exist. The characteristic quantities a_m, x_m, K_m of each of them must be stored. The value m=m* which is valid for the ω-state under consideration is found by checking all

---

I) ω^t in [5] corresponds to $\omega_{o,LS}$ of this paper and ω^c to $\omega_{u,LS}$, respectively.

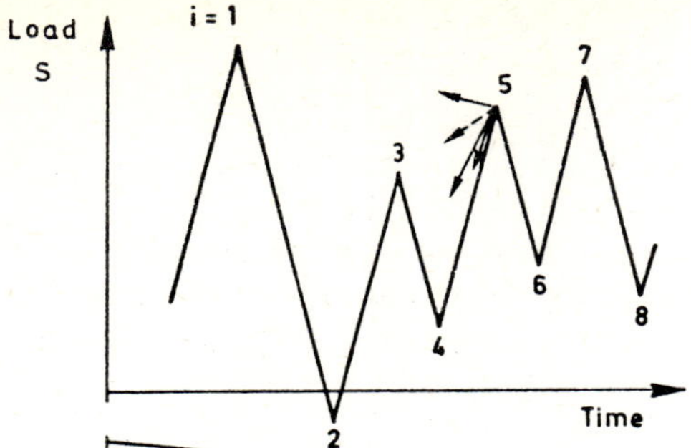

Load - Time - Function

i	1	2	3	4	5	6	7	8
M	0	1	2	3	4	3	4	1
i_m	–	1	1, 2	1,2,3	1, 2, 3, 4	1, 2, 5	1, 2, 5, 6	7
m^*	-	1	2	3	2	5	-	7

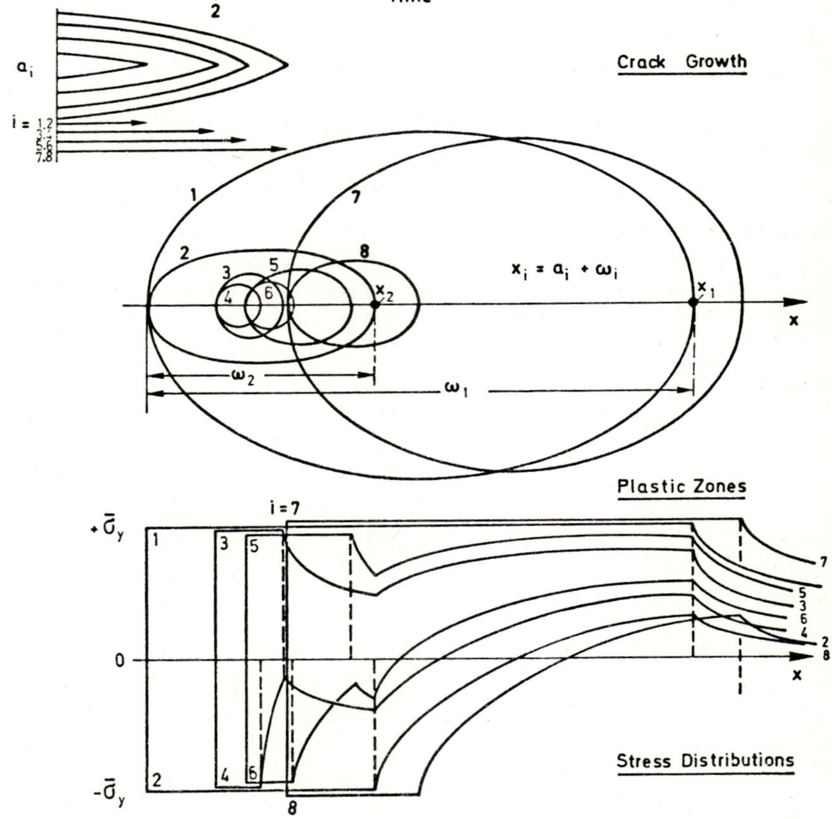

Fig. 3: Illustrative Example Describing the Memory Behaviour

m-values of the active memory in a descending order. The valid memory number is the one which first fulfills the unequaltiy

$$a_i + \omega_i < x_m \tag{13}$$

with x_m being the location of the memory induced elastoplastic boundary

$$x_m \equiv a_m + \omega_m \tag{14}$$

Fig. 4 shows four examples of a single peak overload followed by a single peak underload. The computations were carried out by extending the crack fictitiously. The four cases are characterized by identical maximum loads but different R-ratios as well as peak cycle ratios [6]

$$P \equiv \frac{S_1 - S_2}{S_3 - S_2} \tag{15}$$

Experiments with increasing P-ratio at same R-ratio revealed that the crack growth retardation increases. The limiting case is reached with $P = P^*$ where the crack arrests. P^* is assumed to be a material constant. It must be noted that the P^* effect is not accurately modelled by the retardation formalism given by Eqn. (9). The P^* effect has not been takten into account here because it is unlikely to occur with service load-time functions. This sets a restriction to the applicability of the prediction model in case of extreme SPOL sequences.

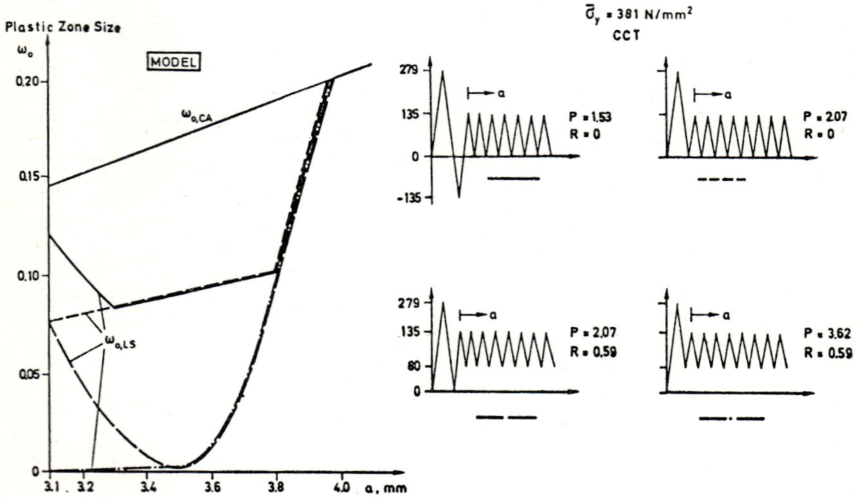

Fig. 4: Plastic Zone Size Development of Four Examples with Varying Peak Overloads

ACCELERATION FACTOR Q_a

The effect that sometimes fatigue cracks are growing faster than expected after linear theory is called crack growth acceleration. This effect, typical with so-called low-high block sequences, can be explained only by the crack closure mechanism. Analyses [2] showed that each low-high step is accompanied by an "initial contact problem": the residual displacements of the material elements at the wake of the crack are initially smaller than they ought to be according to C.A. loading, thus lowering the contact stresses which act on the crack surfaces during crack closure. A certain amount of crack extension is needed to stabilize the residual displacements. During this period the crack opening load increases to the level which applies to C.A. loading, thus the effective ratio decreases. The left hand side of Fig. 5 illustrates the initial contact problem by means of the effective ratio $U = \Delta K_{eff} / \Delta K$. The U(Δa)-curve approaches the stable U_{CA}-curve from above. Relationships of the affected crack length where acceleration holds because of $U_{LS} > U_{CA}$, have not been investigated satisfactorily up to now. Regarding the model, the affected crack length was approximated by the length of the C.A. contact zone, λ. This is the region behind the crack tip where the crack closes at minimum loads.

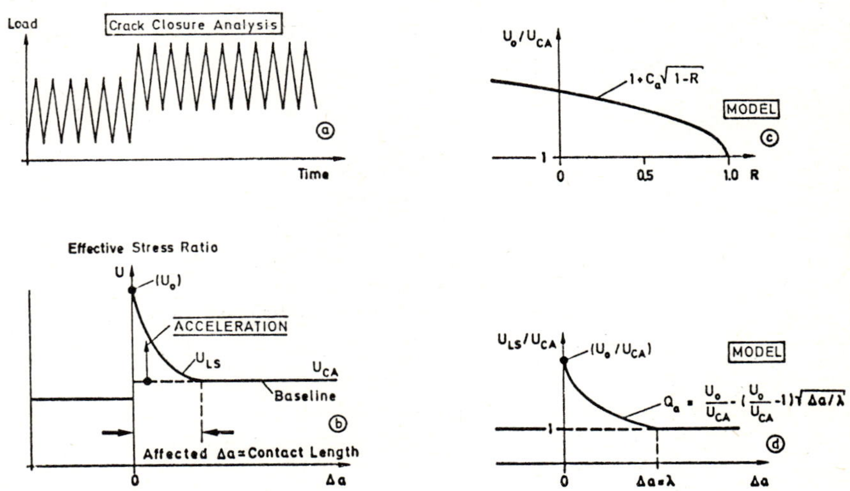

Fig. 5: Acceleration Effect Analyzed with the DUGDALE-Method [2] and Model Approach

The sketches on the right hand side of Fig. 5 illustrate two assumptions of the model approach concerning the instantaneous U-value following the load jump and the U_{LS}/U_{CA}-function. Combining both we obtain the acceleration factor

$$Q_a=1+C_a\sqrt{1-\bar{R}}\cdot(1-\sqrt{\Delta\bar{a}/\bar{\lambda}}) \tag{16}$$

C_a is the only empirical value which is introduced into the model. As will be shown in the next section it is considered to be a material constant. The bars in Eqn. (16) are used in connection with random load sequences. Then, a stress ratio of an "equivalent" C.A. loading $\bar{R}=\bar{S}_u/\bar{S}_o$ is defined with $\bar{S}_o$ ($\bar{S}_u$) being the average maximum (minimum) load within the acceleration range. The acceleration starts at each load cycle the maximum load of which forms a plastic zone being more than 10 % larger than the average plastic zone holding before. Thus, the acceleration effect always starts when the unequality

$$\omega_{o,CA} > 1.10\ \bar{\omega}_{o,CA} \tag{17}$$

is being fulfilled. The plastic zone sizes were calculated according to Eqn. (7) where in case of $\omega_{o,CA}$, S_o is replaced by $\bar{S}_o$ within the stress intensity formula. The extent of the contact zone of the equivalent C.A. loading is approximated by

$$\bar{\lambda} = \left(1.25\ (1-a/x_L)^2 + 1.5(1-\bar{R})(S_L/\bar{S}_o)^2\right)\bar{\omega}_{u\,CA}/(1-\bar{R})^2 \tag{18}$$

with S_L being the plastic limit load of the structure with respect to the current crack length and x_L being the coordinate of the elasto-plastic boundary at limit load. If the λ-value calculated according to Eqn.(18) exceeds the crack length, then $\lambda = a$ is used.

Eqns. (6) to (11) and (16) to (18) represent the new model for fatigue crack growth prediction. In the following, typical load sequences that can be found in literature are treated with the aim to judge over the model's prediction accuracy.

CRACK GROWTH PREDICTION (EXAMPLES)

Results of about one hundred computations are presented in Fig. 6 and 7 in comparison with results from literature. The comparison is made on the basis of failure lives N_f, i.e. the number of cycles reached at critical crack length. If graphical symbols lie on the dashed-dotted diagonale it means that the prediction is accurate. If symbols lie within the indicated band it means that the deviation is less than 100 %. The test represented by open symbols in Fig. 6 are SPOL-tests on CCT speciment with a fixed overload but variable load ranges [7]. The P-ratio varies from 1.0 to 2.1. Only two points lie outside the ± 100 band, both are cases of extreme overloads (P≥2). Regarding these tests the load sequence effect on crack growth furnishes life factors $N_{f,th}/N_{f,\Delta K}$ of 2 (R= -1) and 2 to 4 (R=0), respectively. The tests represented by the full symbols are tests on CCT-specimen subjected to programme loading at three R-ratios including two different cumulative frequency shape functions [8]. Within each test series the maximum programme load was the variable parameter. Each symbol represents the mean value of three repetitive tests, at least.

Neglecting the very short lives (.5 to 1.5 blocks) only three points lie outside the ± 100 % band. The respective tests were carried out at

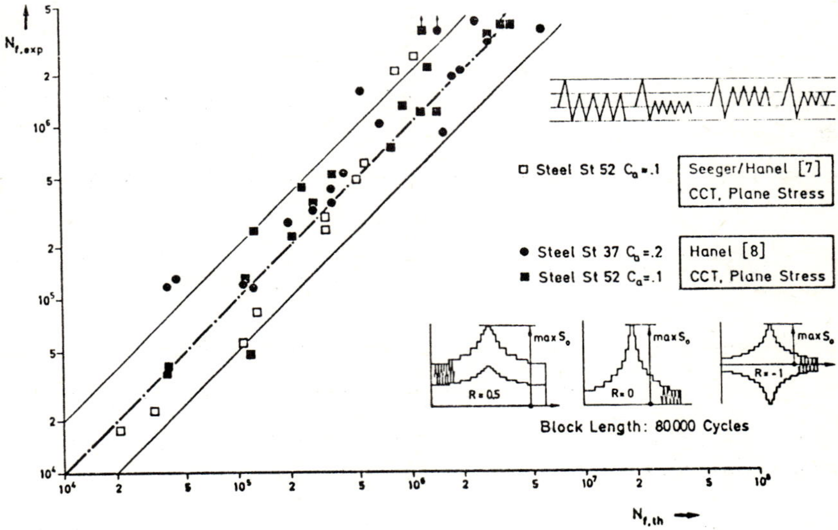

Fig. 6: Comparison of Experimental and Theoretical Results, Single Peak Overload and Programme Loading

very low loads with R= -1. This is a range where the C.A. crack growth law becomes inaccurate. The two steels tested showed a slightly different acceleration value C_a. The tendency that the lower the yield stress (St 37 : σ_y = 299 N/mm², St 52: σ_y = 416 N/mm²) the higher the C_a value was confirmed for aluminium alloys, too.

Fig. 7 is based on experimental results by Schijve [9] on 2024-T3 and 7075-T6 aluminium alloy thin CCT sheets and on results by Alzos and Hillberry [10] on 2024-T3. Both the papers concentrate on the effect of a single peak underload following an overload. In those cases the linear theory reveals a drastical underestimation of crack growth life, $N_{f,\Delta K} \ll N_{f,exp}$ see the dashed lines. With the new model a comparatively good accordance between theory and experiment is obtained. As a restriction it must be stated that it is difficult to judge over the accuracy of the prediction model if original C.A. crack growth data are not available [9] [10]. In the preliminary computations the acceleration effect was not taken into account ($C_a = 0$). However, this effect is rather unimportant for SPOL cases since retardation extremely prevails.

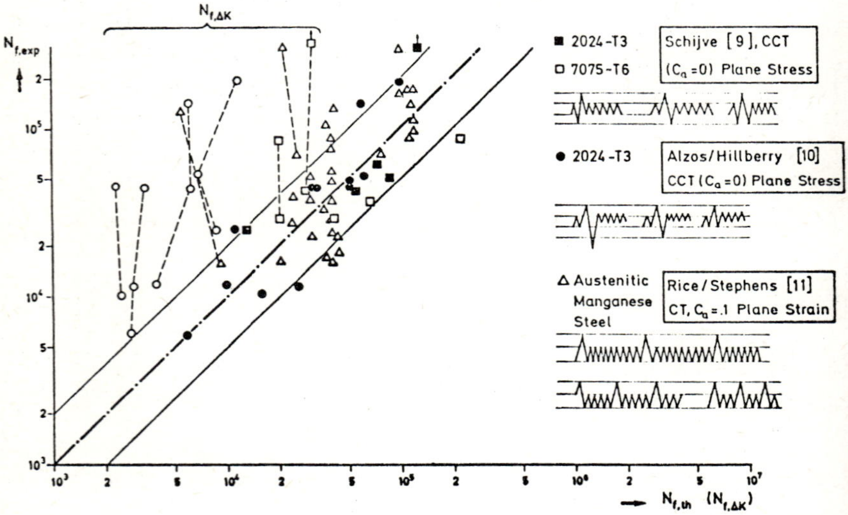

Fig. 7: Comparison of Experimental and Theoretical Results, Single Peak Overload/Underload and Muliple Intermittend Overloads

Additionally, in Fig. 7 you can find comparisons of another test group carried out by Rice/Stephens [11] on CT-specimen of an austenitic manganese steel. The special aspect of the experiments quoted is that the block length given by one single overload and subsequent C.A. loading is varied. With decreasing block length the number of cycles to failure decreases, too. Again, the computational results are within the ± 100 % band except the SPOL cases (of infinite block length) with high peak cycle ratios.

CONCLUSION

The presented model proved to be a suitable tool for crack growth prediction of general load-time functions. Although comparisons are made here only on the basis of final lives it is assured that the accuracy of the analytical results is approximately the same if the comparison is made at any state of crack length. However, under all circumstances the quality of the prediction strongly depends on the quality of the C.A. crack growth law available.

Moreover, several random load sequences (e.g. flight-by-flight loading TWIST [12]) have been checked by means of the model showing as good results as presented here. Those results will be published separately.

The most outstanding distinctions of the new model to others may be formulated as follows: the model takes into account both acceleration as well as retardation effects. It considers not only the maximum peak of a load sequence to effect crack growth but also all other loads of the history according to a generalised memory criterion. The model uses only basic material data and no adjustment factors. The only empirical quantity introduced is kept at a fixed value for all investigations into the same material.

REFERENCES

1. NEWMAN Jr., J.C. - A Finite Element Analysis of Fatigue Crack Closure, Mechanics of Crack Growth, American Society for Testing and Materials, STP 590, 1974.

2. FÜHRING, H. - Elastic - Plastic Crack Closure Analysis of Dugdale Crack Plates Based on Fatigue Fracture Mechanics (in German), Institut für Statik und Stahlbau, TH Darmstadt, 1977.

3. OHJI, K., OGURA, K. and OHKUBO, Y. - Cyclic Analysis of a Propagating Crack and its Correlation with Fatigue Crack Growth, Engng. Fracture Mechanics, Vol. 7, pp. 457 - 464, 1975.

4. ELBER, W. - The Significance of Fatigue Crack Closure, Damage Tolerance of Aircraft Structures, American Society for Testing and Materials, STP 486, 1971.

5. FÜHRING, H. and SEEGER, T. - Structural Memory of Cracked Components Under Irregular Loading, Fracture Mechanics, American Society for Testing and Materials, STP 677, 1979.

6. FÜHRING, H. and SEEGER, T. - Sequence Effects in Fatigue Crack Growth Described on a Continuum Mechanics Basis, Second European Colloquium on Fracture, VDI - Fortschrittsberichte Reihe 18, 1979.

7. SEEGER, T. and HANEL, J.J. - Der Einfluss von plastischen Verformungen sowie von last- und temperaturinduzierten Eigenspannungen auf das Dauerfestigkeitsverhalten von Kerb- und Risstäben, Werkstoff- und Bauteilverhalten unter Schwingbeanspruchung, VDI-Berichte 268, 1976.

8. HANEL, J.J. - A Crack Closure Model to Predict Crack Propagation in Cracked Plates Under Cyclic Loading, (in German), Institut für Statik und Stahlbau, TH Darmstadt, 1975.

9. SCHIJVE, J. - Fatigue Damage Accumulation and Incompatible Crack Front Orientation, Engng. Fracture Mechanics, Vol. 6, pp. 245 - 252, 1974.

10. ALZOS, W. X. and HILLBERRY, B. M. - Effect of Single Overload/ Underload Cycles on Fatigue Crack Propagation, Fatigue Crack Growth Under Spectrum Loads, American Society for Testing and Materials, STP 595, 1976.

11. RICE, R. C. and STEPHENS, R. I. - Overload Effects on Subcritical Crack Growth in Austenitic Manganese Steel, Progress in Flaw Growth and Fracture Toughness Testing, American Society for Testing and Materials, STP 536, 1973.

12. LOWAK, H., de JONGE, J.B., FRANZ, J. and SCHÜTZ, D. - MINITWIST - A Shortened Version of TWIST, Fraunhofer Institut für Betriebsfestigkeit (LBF) Darmstadt/National Aerospace Laboratory (NLR), The Netherlands, 1979.

AN APPLICATION OF LINEAR FINITE ELEMENT FRACTURE ANALYSIS TO FATIGUE CRACK GROWTH IN THE DESIGN OF DIESEL ENGINE PISTONS

C. PATTERSON[I], M. C. OLDALE[II] AND N. I. BAGDATLIOGLU[II]

SUMMARY

Fatigue failure of a diesel engine piston is investigated. A finite element analysis is performed, aimed at improving the design by means of a parameter survey of the effect of ellipticity of the cooling gallery and crack length on the stress intensity factor. Use of Paris' expression for fatigue crack growth is made to allow the fatigue life of the modified designs to be estimated and improvement assessed.

INTRODUCTION

The investigation reported here was undertaken as a result of piston failure in a new diesel engine undergoing commissioning trials. This medium speed (750 r.p.m.) engine features an aluminium piston of 10 inch bore x 12 inch stroke which is oil cooled by a cocktail-shaker action with oil splashing around an elliptical cooling gallery located near the piston top ring. The oil is fed into the gallery by a slipper arrangement on the piston undercrown, being taken from the small end of the connecting rod. The oil is then fed through the centre of the piston and out to the gallery via a spoke type arrangement.

Investigations showed that the mode of failure was fatigue crack growth at the junction of the oil cooling gallery to the drain holes, which recirculate the coolant, leading to brittle crack propagation. Failure was occurring at approximately one half of the specified design life.

I Senior Lecturer, Department of Mechanical Engineering, University of Sheffield, UK

II Graduate Student, Department of Mechanical Engineering, University of Sheffield, UK

The piston is subjected to a complex loading pattern duri the engine cycle involving tensile forces produced by inertial effects during the induction stroke in addition to compressive and thermal stresses. However, the contributive factor to fai ure must be the tensile loads as any compressive stresses will have the effect of closing the crack. The temperature field around the trouble area is low (approx. 150°C) and uniform and an analysis was carried out with thermal corrections showing that they did not significantly affect life predict- ions. Section 2 deals at greater length with details of pisto loadings during the engine cycle.

A finite element analysis was performed, aimed at improv ing the life. Although the piston is inherently of a complex three dimensional nature, resort to such an expensive view- point will be shown to be unnecessary by use of a much simplifield two dimensional model which still retains the salient features. The problem was modelled as an elliptical hole in a rectangular plate under uniformly distributed load with a crack emanating from the end of the major axis. General dimensions arranged to agree with the real situation.

The intention of the finite element analysis was to perform a parameter survey of the effect of ellipticity of the oil cooling gallery and crack length on the stress intensity factor at the crack tip. Then, by application of principles of linear elastic fracture mechanics, to achieve a quant- itative appreciation of the expected improvement in fatigue life on increasing the radius of curvature of the gallery.

All the necessary computer software was in existence for the analysis and had been applied to planar-cracked geometries [1]

PISTON LOADING

Discussion will be made of the actual loading applied to pistons during engine operation. This will provide a basic understanding of the philosophy behind piston design, the reasons for oil cooling and the causes of piston failure.

The loading applied to the piston can be considered to consist of two parts,

(i) Mechanical loading due to the gas pressure in the cylinder during the combustion process and inertia loads set up as a result of the reciprocative nature of the engine motion. A simple engine cycle of starting-up, constant load running and shutting down is shown in Fig. 1. A schematic representation of the corresponding loading conditions is given in

Fig. 2.

(ii) Thermal loading due to the temperatures and heat transfer conditions in the cylinder and at other boundaries to the piston.

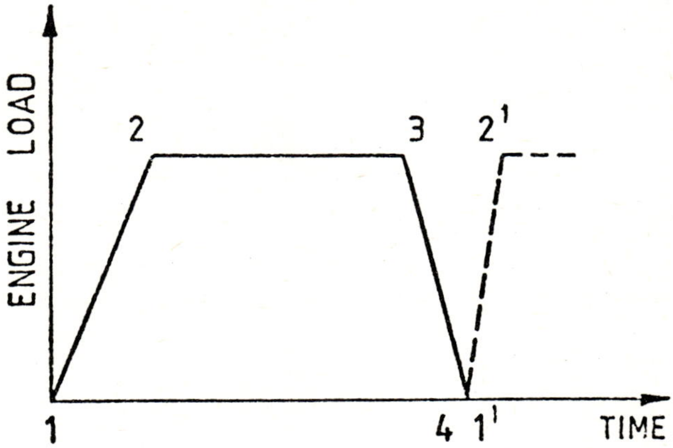

FIG. 1 Simple Engine Loading Cycle

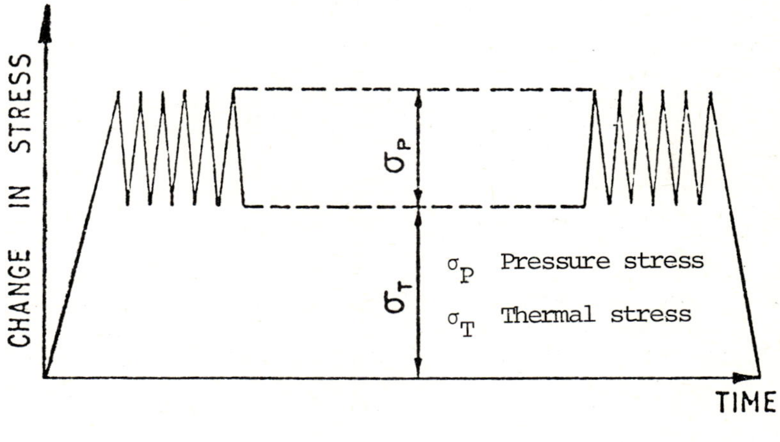

FIG. 2 Cyclic Nature of the Stresses

There is an alternating or cyclic nature to both the thermal and the mechanical loads on the piston. The thermal loading will vary due to the starting and stopping of the engine load and speed changes. However, these thermal stresses become steady for full load operation of the engine and only a small error is introduced by assuming them constant. How the two types of loading are separated into high and low cycle components is shown in Figs. 3, 4.

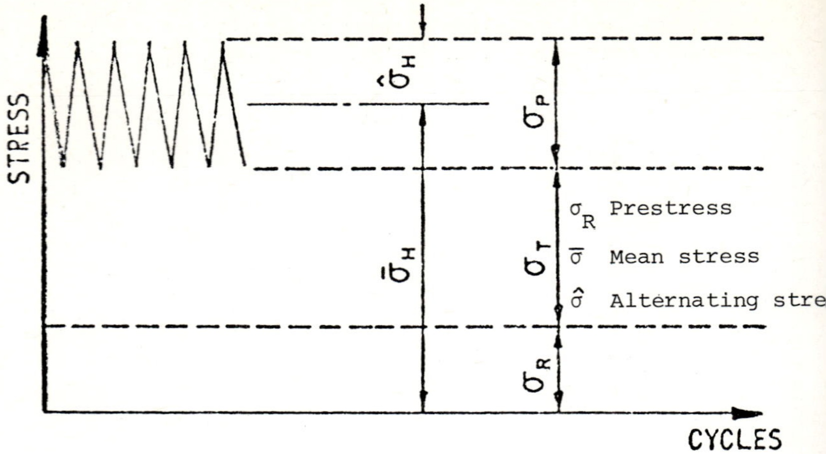

FIG. 3 High Cycle

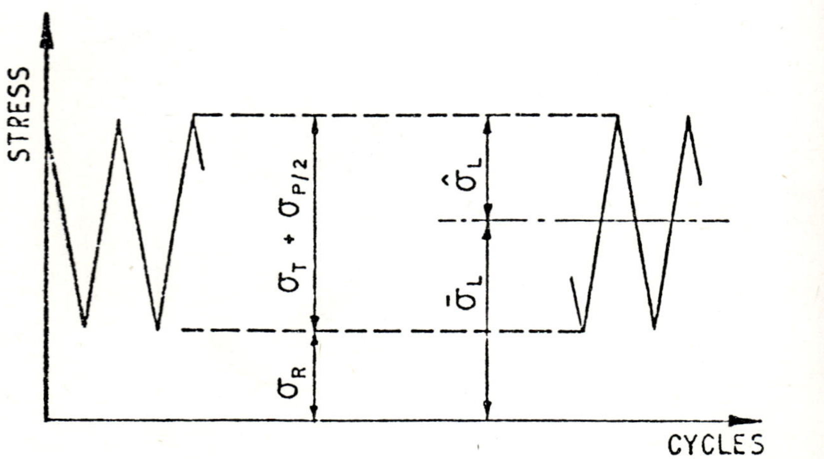

FIG. 4 Low Cycle

These two fatigue conditions form the basis of the analysis for any engine component and must be treated separate A related analysis has been carried out [2], [3], for the situation where cracks were occurring in the oil cooling ring of the piston. The problem was dominated by thermal gradients causing high thermal stresses. A different approach was used to evaluate the safety of the piston under the combined thermal and mechanical loads by means of a Goodman diagram.

The main emphasis in this analysis is on the inertia forces which are the contributive factor in the failure of the piston. To show the inertia effects of the piston on the net force, a typical p-V diagram for this type of engine is shown in Fig. 5, together with the resultant inertia loading indicated by the dotted line.

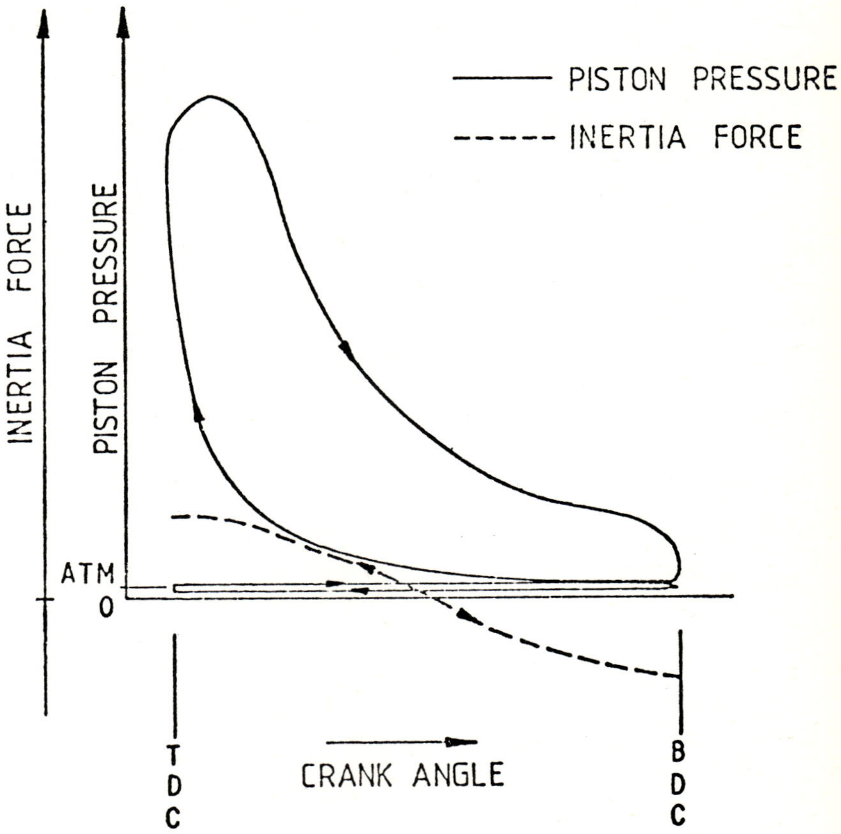

FIG. 5 ENGINE DIAGRAM

MODELLING THE PISTON

The piston, shown in cross-ection in Fig. 6, is a complex 3-dimensional design. To try to reproduce this component in a fully 3-dimensional finite element analysis would prove to be an extremely costly procedure both in terms of time and money. On consideration of the actual mechanics of the problem, however, it is possible to reduce the analysis to a much simplified two-dimensional model which still retains the salient features of the original. The problem area was represented by an elliptical hole in a rectangular plate under uniformly distributed tensile load with cracks emanating from the ends of the major axis of the ellipse. The model is shown in Fig. 7, of which, due to symmetry, only a quarter was analysed.

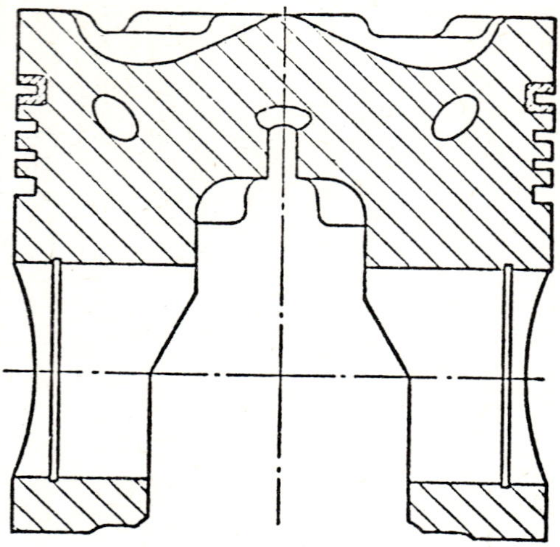

FIG. 6 CROSS-SECTION OF THE PISTON

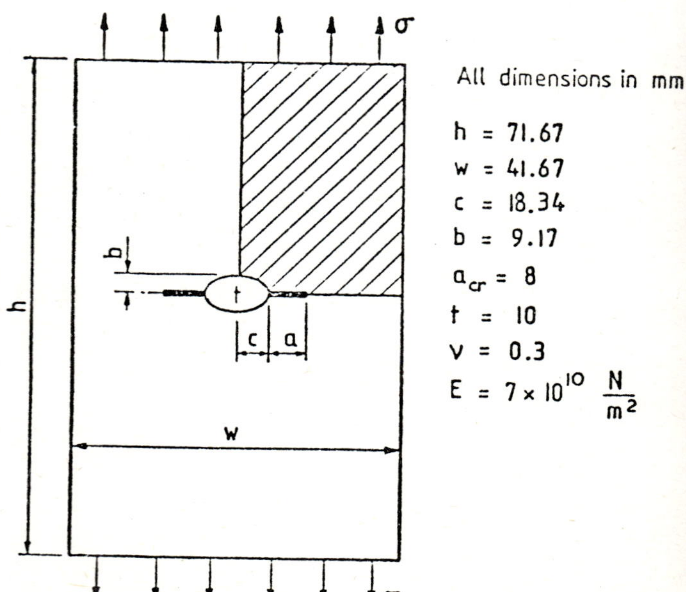

FIG. 7 MODEL OF THE PISTON FOR F. E. ANALYSIS

The finite element mesh consists of 105 eight node isoparametric quadrilateral elements with 9 embedded crack tip elements, these special elements including asymptotic displacement functions in their formulation. The length of the crack was varied in the analysis between 0.5 and 8mm, these being the assumed initial defect in the material, i.e. the grain size at the manufacturing stage and the observed critical crack size at the onset of brittle crack propagation.

The only remaining unknown in the analysis is the value of the load. Due to the complex loading conditions on the piston, an analogical approach is adopted in order to calculate the magnitude of the tensile stresses which cause crack propogation. A test load of magnitude 150N was applied to the model to obtain the full solution. Having obtained the test stress, the true stress is found by a linear relation

$$\sigma_t = \lambda . \sigma_o \qquad \ldots \qquad \ldots \qquad (1)$$

where

σ_t = true value of the tensile stress acting on the piston.

σ_o = test stress on the model

λ = scaling factor

The displacement fields for a series of crack lengths wer obtained and the stress intensity factors at the crack tip for the opening mode were calculated from these using linear elastic fracture mechanics [4]. Similarly to eqn. (1), the stress intensity factor under test load is related to the critical stress intensity factor by

$$K_{It} = \lambda . K_I \qquad \ldots \qquad \ldots \qquad (2)$$

where

K_{It} = true value of the mode I stress intensity factor at the crack tip

K_I = test value of SIF.

A similar finite element analysis was carried out by replacing the elliptical hole in the plate by a circle, keeping the hole area constant. The relationship between stress intensity factor and crack length for the two cases is shown in Fig. 8.

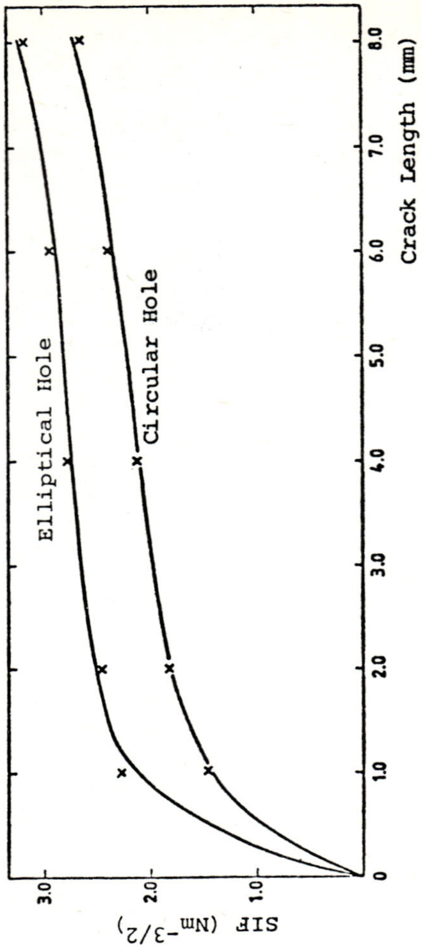

FIG. 8. Stress Intensity Factor v. Crack Length

LIFE ESTIMATE OF PISTON

The results of the static finite element analyses are use[d] to perform a study of the effect of fatigue crack growth on the piston by use of Paris' Power Law [5],

$$\frac{da}{dN} = C\ (\Delta K)^n \quad \ldots \quad \ldots \quad (3$$

where

a is the crack length
N the number of load cycles
ΔK the difference between stress intensity factors at minimum and maximum load during the cycle.

The material constants C, n may be determined using the manufacturers specification, see Fig. 9.

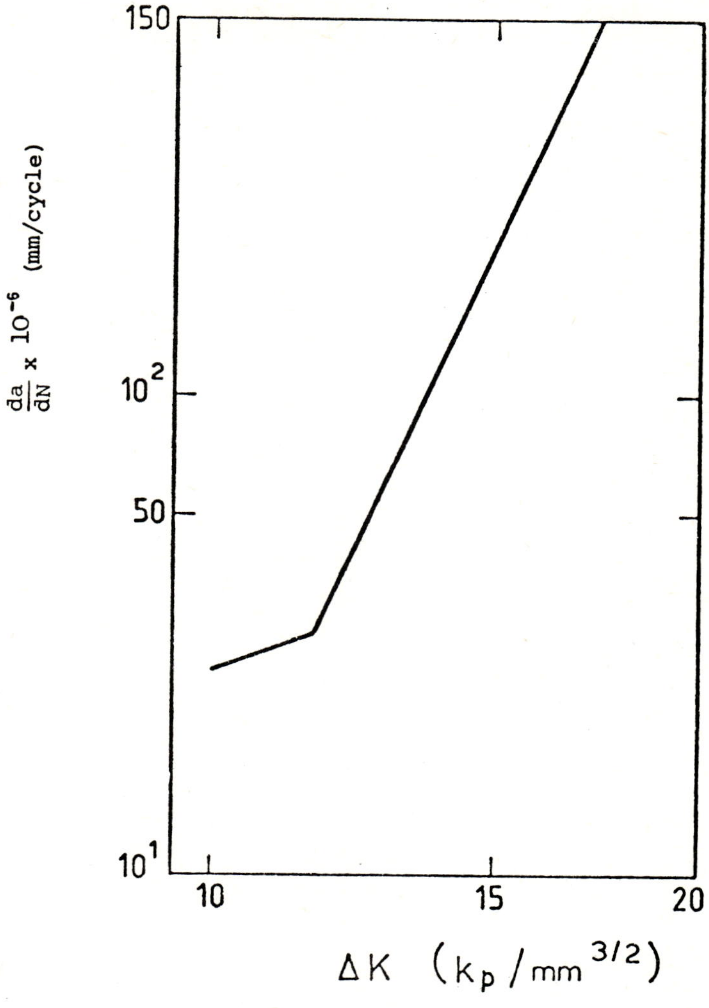

FIG. 9. Crack Propagation Rate Curve

For the particular problem of the tensile load on the piston, we have

$$K_{min} = 0$$

$$\Delta K = K_{max} = K_I \quad \ldots \quad \ldots \quad (4)$$

Using equns. (2), (3), and (4) and integrating,

$$\int_{a_1}^{a_2} \frac{da}{C(\lambda K_I)^n} = \int_{N_1}^{N_2} dN \quad \ldots \quad (5)$$

The constant of proportionality, λ, may now be determined by performing the integration. A step by step method is applied, using an average value of the stress intensity factor over each step. Equating the number of cycles to the average service life of the piston will yield λ. The integration is performed in Table 1.

a_1 (mm)	a_2 (mm)	$\bar{K}_I$ ($Nm^{-3/2}$)	$N.\lambda^n$ (x 10^{-27})
0.5	1.0	2.31	0.66
1.0	2.0	2.40	1.17
2.0	4.0	2.70	1.60
4.0	6.0	2.92	1.25
6.0	8.0	3.19	0.94
		$\Sigma N.\lambda^n$	5.62

Table 1 Numerical Integration

where $C = 7.25 \times 10^{-10}$

$n = 3.19$

Given a total average life for the piston of 20000 hours, this gives the constant λ as,

$$\lambda = 9.86 \times 10^5 \quad \ldots \quad (6)$$

The maximum stress intensity factor at the critical crack length was found to be 3.19 $Nm^{-3/2}$ under the test load. Using equns. (4), (6), the true stress intensity factor is given by 3.15 $MNm^{-3/2}$. This is to be compared with the fracture toughness of the material used in the piston which, after temperature corrections, is 9.56 $MNm^{-3/2}$.

The analysis was repeated for a piston with circular cross-section cooling gallery and using eqn. 6 the life of the modified piston was estimated, see Table 2.

a_1 (mm)	a_2 (mm)	K_I ($Nm^{-3/2}$)	N($x10^{-8}$)
0.5	1.0	1.45	2.20
1.0	2.0	1.80	2.21
2.0	4.0	2.15	2.51
4.0	6.0	2.39	1.79
6.0	8.0	2.61	1.35
		ΣN	10.06

Table 2 Life Prediction

The number of load cycles to reach the maximum permissible crack length for the Circular gallery is 10^9 cycles, i.e. 44000 hours, an improved life of 125%.

The above calculations were made at room temperature. An analysis was carried out using thermal corrections to the material properties but this proved to have no significant effect on life predictions.

DISCUSSION

The analysis described in the previous sections is typical of an engineering design approach where a complex component is represented by a relatively crude model in the hope of obtaining reasonably accurate results without having to resort to long and expensive analyses. In this case the aluminium alloy piston was replaced by a straightforward two dimensional model.

This type of approach is validated here by the agreement of results to theory. The maximum stress intensity factor at the critical crack length for the elliptical hole was found to be well within a factor 3 of the fracture touchness, K_{Ic}, of the material used in the piston. In view of the uncertainties in both the defining of the problem, e.g. specifying the material properties, and the inherent assumptions involved with the model, this agreement is surprisingly good.

This work follows on from reports already published, [2], [3], but whereas those analyses were based on the identification of the weakest area in a piston design where high thermal gradients were present, here the accent is on a quantitative approach to life prediction with regard taken to the fact that in this analysis those thermal gradients are not involved.

The analysis has concluded that by increasing the radii of curvature of the critical areas, the stress concentration will be reduced, thereby inhibiting crack initiation and subsequent fatigue growth. A significant life improvement of the pistons was indicated by replacement of elliptical cooling galleries with circular galleries.

A series of modified pistons have been built according to the proposed design and are currently under test. The tests are continuing, but further validation of the method is made by the fact that they have already significantly outperformed the original design.

REFERENCES

1. PATTERSON, C. - On the use of the Finite Element Energy Method in Calculating Stress Intensity Factors, Fracture Mechanics in Engineering Practice, SAG Conf. Proc., Ed. Stanley, P., Appl. Sci. Publ., 1977.

2. MUNRO, R. GRIFFITHS, W.J. and INGHAM, A.P. - Open Gallery Pistons for Highly Rated High Speed Diesel Engines, 10th Int. Congress on Combustion Engines, Washington, 1973.

3. MUNRO, R. and GRIFFITHS, W.J. - Diesel Piston Design and Performance Prediction, 11th Int. Congress on Combustion Engines, Barcelona, 1975.

4. LIEBOWITZ, H. - Fracture, vol. 2, Academic Press, New York, 1968.

5. POOK, L.P. - Analysis and Application of Fatigue Crack Growth Data. J. of Strain Analysis, Vol.10, No. 4, pp. 242-250, 1975.

NUMERICAL AND EXPERIMENTAL EVALUATION OF STRESS INTENSITY FACTORS FOR CURVED CRACK FRONTS

Elangovan[x], C.J. Lof[xx], G. Reibaldi[xxx] and S.B. Welt[x]

ABSTRACT

A detailed study was made of stress intensity distributions along curved crack fronts in actual structural elements. The growth of corner flaws in pin-loaded lugs and in bolted lap-joints and of surface flaws in a round bar loaded in bending was carefully monitored during constant amplitude fatigue tests. In addition to visual observation of surface crack length, crack front marking was applied using a smaller stress amplitude to produce contrasting fracture surface marks.

Next, a detailed finite element analysis was performed in which the method of virtual crack extension was applied to the curved crack front shapes recorded during the fatigue tests. Modified elements of the ASKA-library were used to model the strain singularity in the crack front zone.

The crack growth rates observed in the fatigue tests were also reduced to stress intensity values by means of a calibrated relationship between the crack growth and the stress intensity variation per cycle.
In general a good correlation between the two sets of stress intensity data was obtained but the results suggest that certain computational results are very sensitive to variations in the assumed load or displacement boundary conditions.

---

x Fokker-VFW, Schiphol - Amsterdam

xx National Aerospace Laboratory NLR - Amsterdam

xxx European Space Technology Centre - Noordwijk

1. INTRODUCTION

The damage tollerance concept has been used since many years in the design practice of aerospace components. To apply such a concept it is very important to determine the stress intensity factor (SIF) in the most accurate way for the actual configurations to be analyzed. In the present work three dimensional crack situations with curved crack front are considered. Curved crack front is common to the majority of flaws encountered in structural components and for such cracks the analytical solutions are very scarce. To evaluate SIF the Finite Element method is considered a suitable tool.
The aim of the present paper is to obtain SIF solutions using a standard finite element package (ASKA) in order to verify the numerical results with experimental test data. The method used to derive the SIF from the finite element result is the "Virtual crack extension", [1] and [5]. When it has been possible, a comparison with existing data available in literature has been performed. The particular crack configurations chosen were considered relevant for the Spacelab-Shuttle fracture mechanics analysis.

2. FINITE ELEMENT ANALYSIS

The crack configurations chosen for the analysis were parts of the Spacelab hardware.

They are:

i) Cylindrical pin in bending, surface flaw in a fillet radius; material: Forged Ti-6Al-4V-STA, see Fig. 1.
ii) Hole in a flat plate in tension and bending loaded by a steel fastener, corner flaw in the hole - Material: 2219-T851, see Fig. 2.
iii) Lug loaded by a pin, corner flaw at one side of a hole in the lug - Lug material: Al 7075-T7351. Pin material custom 455, see Fig. 3.

All these three configurations have surface flaw with curved crack front.

The finite element program used to model all these configurations was the ASKA package. Only few modifications (mid-side nodal shifts, suppression of the singularity check, special APC) were performed on the standard package in order to get the stress field singularity at the crack tip.
In Fig. 4, the virtual crack extension method is shown the range of the parameters employed to obtain SIF solutions.

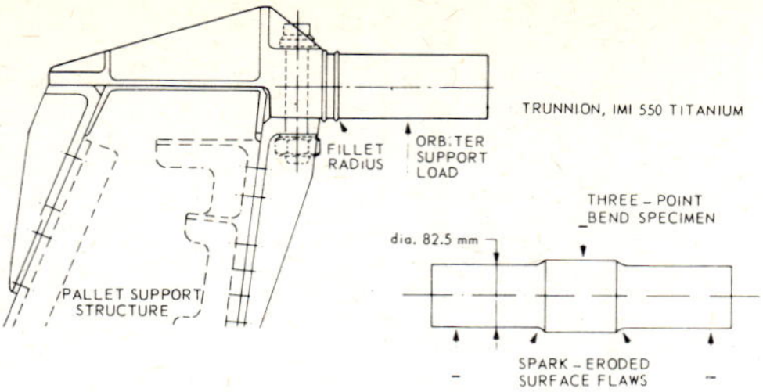

Fig. 1 Spacelab pallet support trunnion and corresponding three-point bend specimen

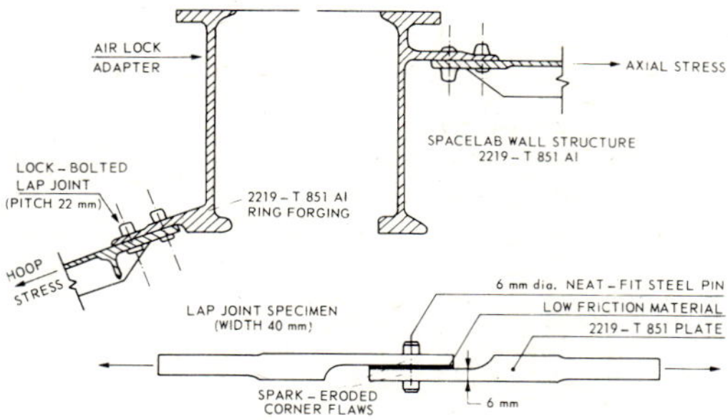

Fig. 2 Spacelab air-lock adapter structure and corresponding lap joint specimen

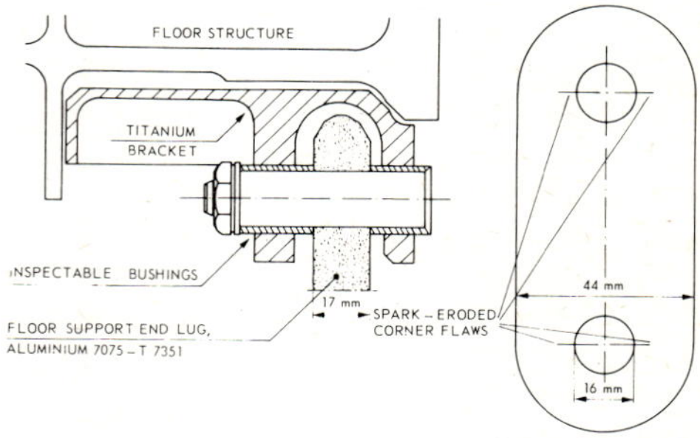

Fig. 3 Spacelab floor support end structure and corresponding lug specimen

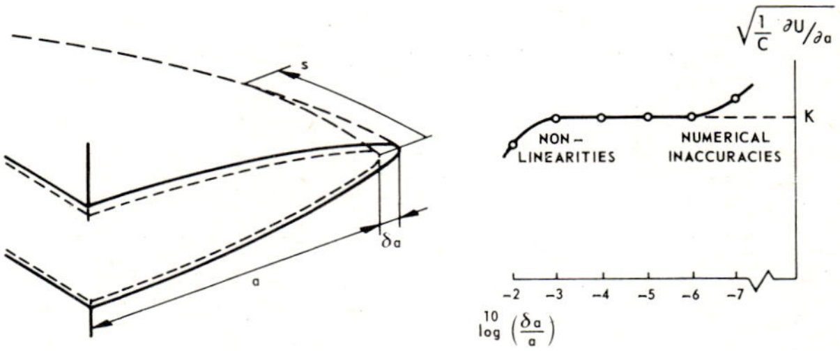

Fig. 4 Virtual crack extension method and range of parameter used

The equation used to drive the SIF is the following:

$$K\ (a) = \sqrt{\frac{E}{1-\nu^2}\ \frac{\Delta U}{\Delta A}} \qquad (1)$$

for the plane strain condition
where:

- K - is the combination of the three modes K_I, K_{II}, K_{III}
- ΔU - is the strain emergy variation due to ΔA
- ΔA - is the total variation of areas introduced along the crack front shown in Fig. 4.
- E - is the Young's module
- ν - is the Poisson ratio

In the eq. (1) the term $\Delta U/\Delta A$ is computed from the Finite Element results. In order to improve the accuracy of the results, special modified ASKA element surrounding the crack tip was used, as in [2] and [3]
To reach engineering accuracy, four elements were considered along each crack front as it shown in [4].
The substructing feature available in ASKA has been advantageously employed for such computations grouping the substructures away from the crack plane end the crack front substructures.
For each of the three configurations chosen, three different marked crack sizes were studied, in order to have an extensive comparison with the test program.
Each of the three different crack sizes have been modelled using three actual crack fronts obtained from the specimens.
For comparison the crack Opening Displacement Method (COD), it has been used to obtain SIF solutions using the displacements between the free surface of the crack. The stress intensity factor (SIF) has been computed for five different nodes, for each size of the three crack front. To derive the SIF, the equation (1) has been used for all the nodes along the crack front, even at the surface. The material properties used for the Finite Element Models are shown in Table 1.
In Fig. 5 the details of the pin-lug specimen model are shown with data used for such model in Table II. In Fig. 6 the results of such calculations are showed. In Fig. 7 the finite element model of the lap-joint specimen is shown with the data in Table III and the results in Fig. 8. In Figs. 9 and 10 the finite element model of the trunnion specimen is shown. In Table V the data for the same model are presented. In Fig. 11 the finite element results are presented.

3. EXPERIMENTAL K-VERIFICATION

On three specimen types, introduced before, stress intensity values were checked by means of crack growth. For that

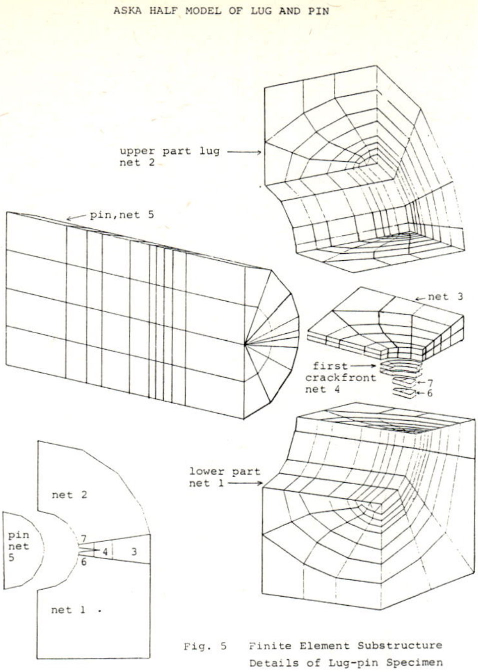

Fig. 5 Finite Element Substructure
Details of Lug-pin Specimen

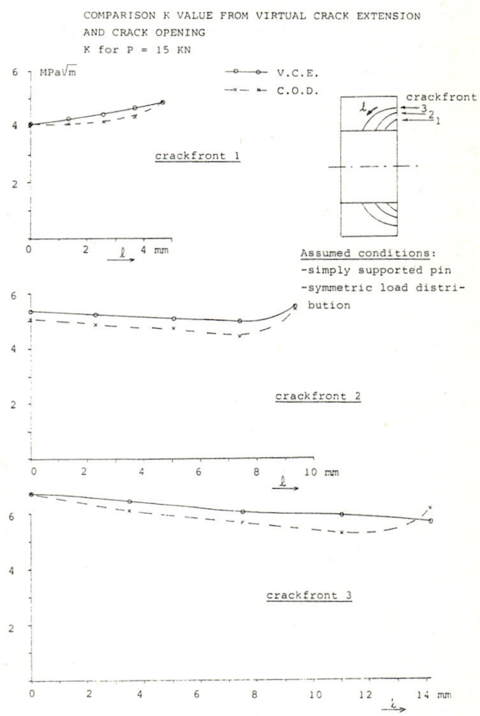

Fig. 6 'K' values from f.e. analyses: pin-lug specimen.

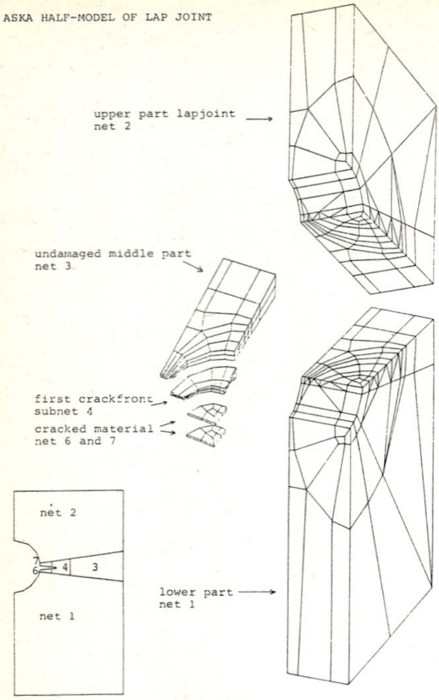

Fig. 7 Finite Element Substructure Details of Lap Joint Specimen

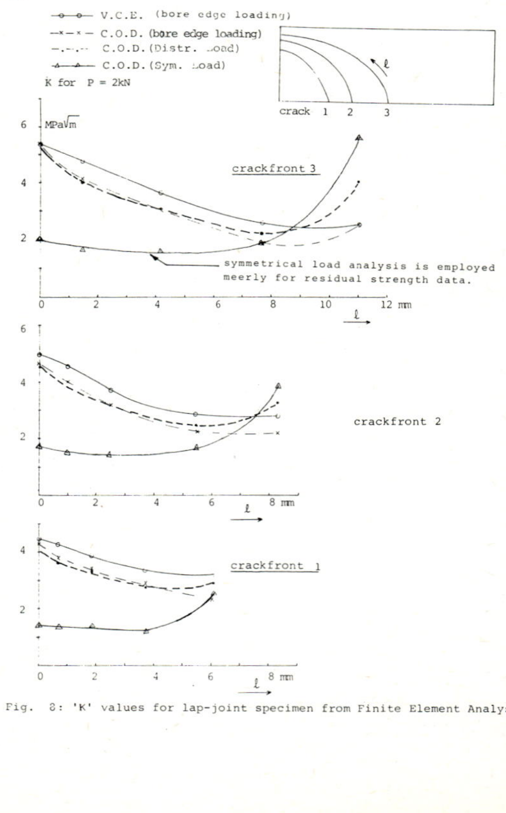

Fig. 8: 'K' values for lap-joint specimen from Finite Element Analyses

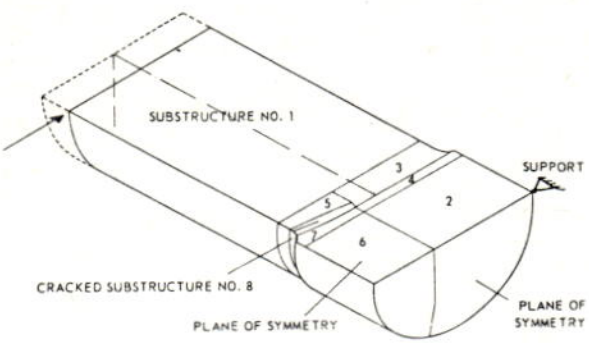

Fig. 9 . Finite Element Model of Trunnion.

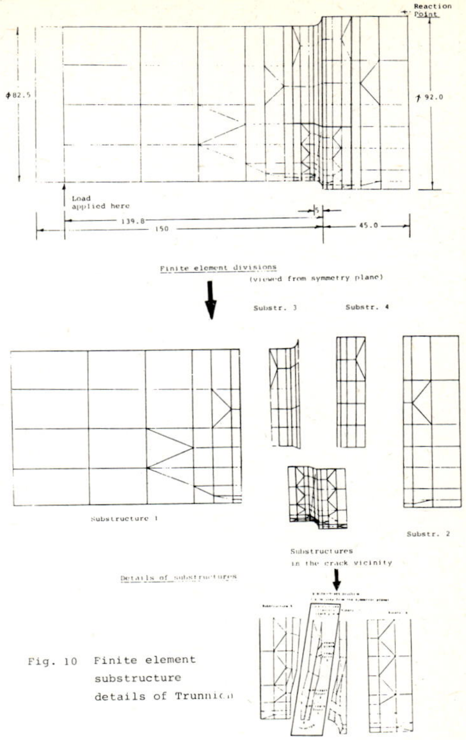

Fig. 10 Finite element substructure details of Trunnion

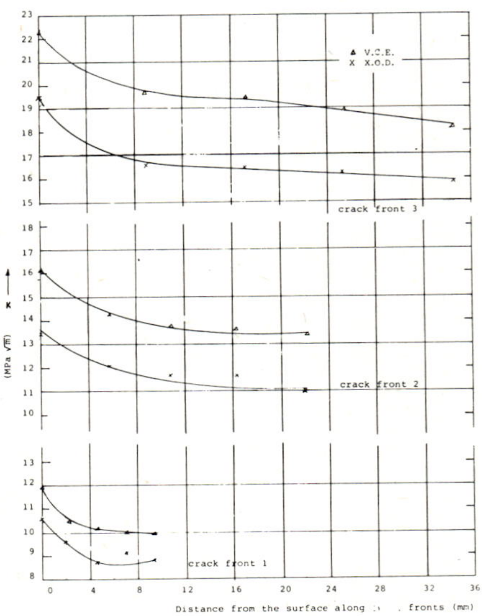

Fig 11 Finite Element Results of Trunnion .

EXPERIMENTAL K–VERIFICATION TEST PROGRAM SUMMARY			
SPECIMEN IDENTIFICATION		TYPE OF TEST	MAXIMUM TEST LOAD OR STRESS
	JR1, LR1, TR1 THROUGH–CRACK	CRACK GROWTH RATE	28 MPa 70 MPa 90 MPa
	JM1, LM1, TR2 THROUGH – CRACK JR2, JM2, LR2, LM2, TM1 SURFACE – NOTCH	CRACK GROWTH RATE + MARKING	70 MPa 100 MPa 90 MPa 130 MPa
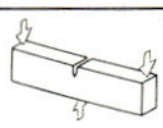	TK1, TK2, LK1, LK2 TK3, TK4 (AT 175 K) ALL SPECIMENS	FRACTURE TOUGHNESS PRECRACKING	– – 10 kN
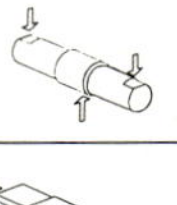	TA1 TA2, TA3 TA2 TA3 (AT 175 K)	CRACK GROWTH RESIDUAL STRENGTH	190 kN 160 kN 236 kN 302 kN
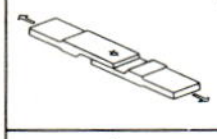	JA5 – JA10, JA13 – JA16 JA4 – 10 JA5 – 11 JA6 – 12 JA13 – 14 JA15 – 16	CRACK GROWTH RESIDUAL STRENGTH	4 kN 18.5 kN 36 3 – 37.8 kN 28. 5 – 33.9 kN 39 6 – 40.– kN 42.2 kN
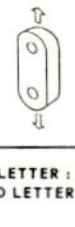	LA1 – LA3 LA1 LA2 LA3	CRACK GROWTH RESIDUAL STRENGTH	30.– kN 55 8 kN 99.7 – 107.5 kN 121.1 – 130 1 kN

FIRST LETTER : T = TRUNNION, J = LAP – JOINT, L = LUG
SECOND LETTER : K = FRACTURE TOUGHNESS, R,M = CRACK GROWTH RATE (AND MARKING)
A = ACTUAL SPECIMEN

Fig. 12 Summary of the test program

purpose, for each material a calibrated relation between the crack growth rate and the stress intensity factor must be available as well as fracture toughness data. A test program has been developed including both calibration of the materials used and testing of the specimens. A summary of this test program is shown in Fig. 12.
Fatigue crack growth calibration curves were derived from constant amplitude load tests on centre cracked (thick) plate specimens (Fig. 13). Center through cracks as well as center surface cracks were initiated at a spark eroded starter notch. A load cycle, with a ratio of .1 was applied for the initiation of the fatigue cracks. Then a crack-growth fatigue load was applied with the same maximum however with ratio .5 (Fig. 14). In order to obtain crack growth data for the surface cracks also in thickness direction, a number of crack front markings were applied. These markings were obtained by interrumptions of the fatigue load cycle by a load cycle with reduced amplitude. (Fig. 14)
From the obtained crack growth data the da/dN-a function was derived by a simple graphical differentiation. By using the well-known K-crack size relationship of the calibration specimen, the da/dN - ΔK function was found for that specific material and specific load cycle ratio. This is shown in Fig. 15. An example of graphical calibration test data presentation is shown in Fig. 16. A "best-fit-calibration-curve" can be drawn through the data points of this logarithmic plot. The data suggest a linear relationship for the central part of the stress intensity domain sloping downward and upward at the lower and upper ends, respectively. Approximations have been proposed by Forman, Colliprist and Feddersen, however, the familiar relationship $da/dN = C_1 . \Delta K^{C_2}$; known as Paris' law provided the best approximation . The constants C_1 and C_2 are listed in Table V.

3.1 Testing of the Actual Specimen

All fatigue testing was done under normal laboratory conditions (295°K and 40-60 percent relative humidity). The load-time history applied to the trunnion and lug specimens was the same as used in the calibration tests (see Fig. 14). For the lap-joint specimens the marker loads were omitted as crack front marking was not successful during calibration. Subsequent crack front shapes were found after residual strength tests of different specimens with different fatigue lifetimes. Detailed description of the tests are given in [2].

3.2 Crack Growth Data Processing

Crack growth along the surface was followed during testing by visual observation. However, after failure of the specimens, either a fatigue or during subsequent residual strength

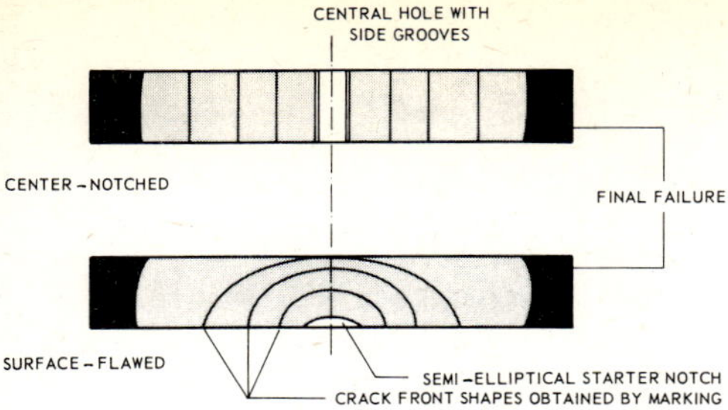

Fig. 13 Starter notch and crack front shapes in fatigue calibration testing

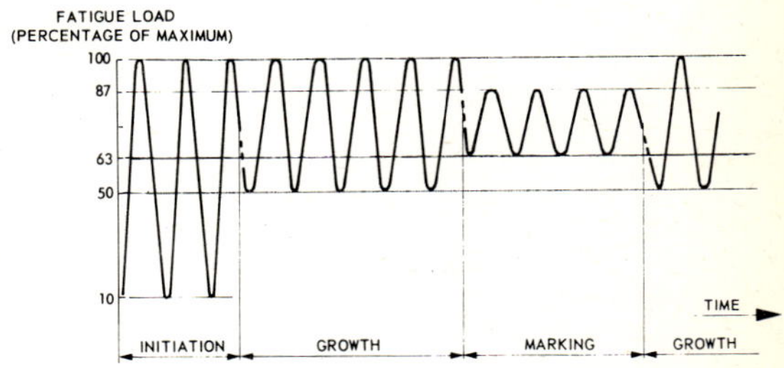

Fig. 14 Fatigue load history applied in calibration tests

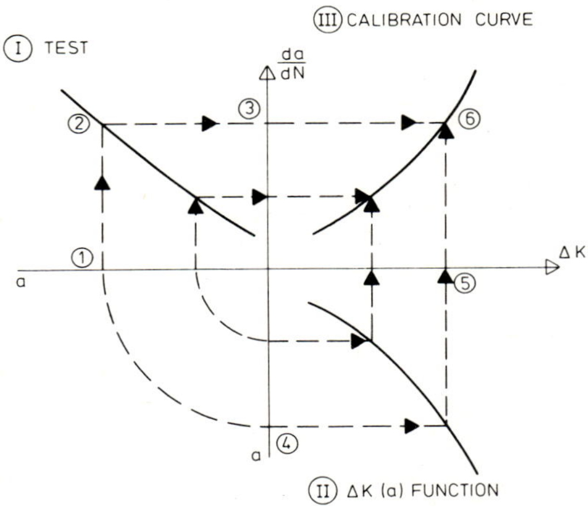

FIG. 15: CALIBRATION PROCEDURE

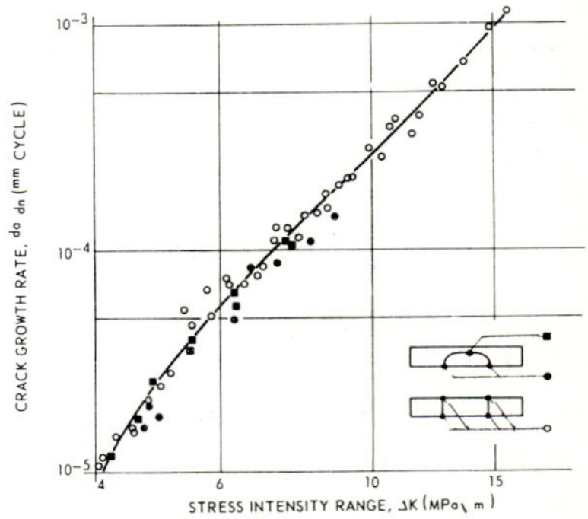

Fig. 16 Crack growth calibration data for 7075-T7351 aluminium

testing, the crack front markings on the fracture surface were projected on paper using a magnification factor of 10. Examples of such markings are shown in Figures 17, 18 and 19.

For a correlation with stress intensity values in planes normal to the crack front the crack growth rate in directions normal to the crack front must be determined. This is accomplished by constructing a set of curves normal to the crack front markings and by measuring the curve length. A plot of this length versus the number of fatigue load cycles, both corrected by eliminating the crack front marking intervals, provides normal crack growth rates for any crack size.

4. COMPARISON

A comparison between the stress intensity distribution derived by the analysis using the ASKA package and the test data is presented in the Figs. 21, 22, 23 for the three configurations. The results for the crack opening displacement method are shown in the same Figures. Numerical data are shown in Tables VI to VIII.

a) Pin-lug specimen

As it is possible to see in Fig. 21 and Table VI that the results for the lug specimen show a good correlation with respect to the average value but the variation along the crack front is different from the ASKA and the test results. Except for the smaller crack size the finite element values for the stress intensity at the surface are higher than the experimental values and the opposite applies to the situation of the box. When evaluating this trend, it was recognized that the stiffners of the cracked side of the lug is steadily decreasing as the crack grows.
This results in a gradual shift of pin load from the cracked to the uncracked side of the lug. In the ASKA model however, a fifty-fifty distribution was presented for all three crack sizes. A trial run with a forty-sixty distribution completely reversed the situation and it turned at that a shift of only three per cent of the total load from one side of the lug to the other, provided parallel curves for the ASKA results and the test results.
These findings underline the sensitivity of the stress intensity solution to variations in the boundary conditions. This is not surprising, of course, as the loading occurs in the immediate vicinity of the crack.

b) Lap-joint specimen

In Fig. 22 and Table VII.

A similar effect might be expected in the lap joint speci-

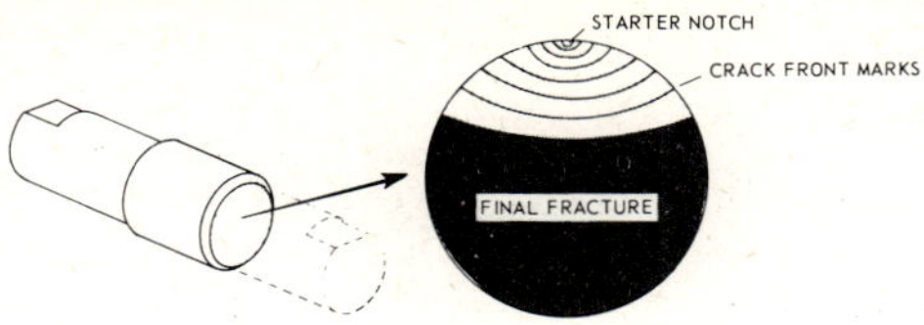

Fig. 17 Crack front markings on fracture surface of trunnion specimen

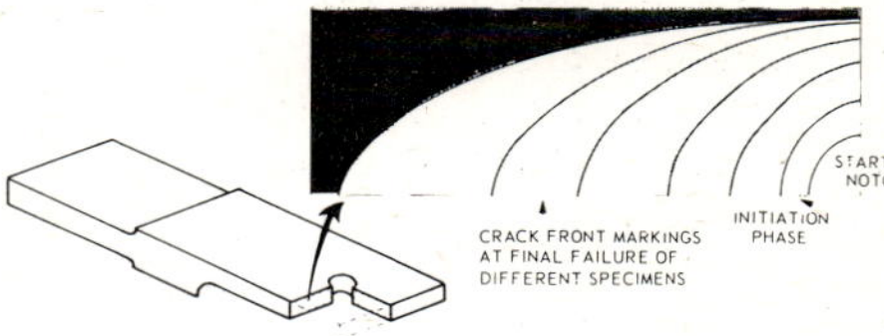

Fig. 18 Crack front markings at failure of different lap joint specimens

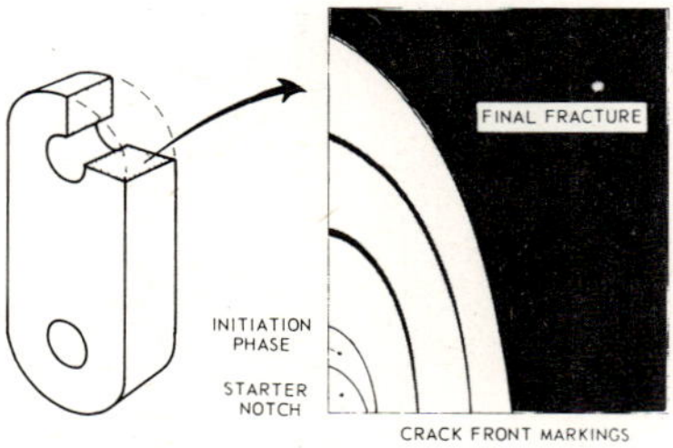

Fig. 19 Crack front markings on fracture surface of lug specimen

provides normal crack growth rates for any crack size

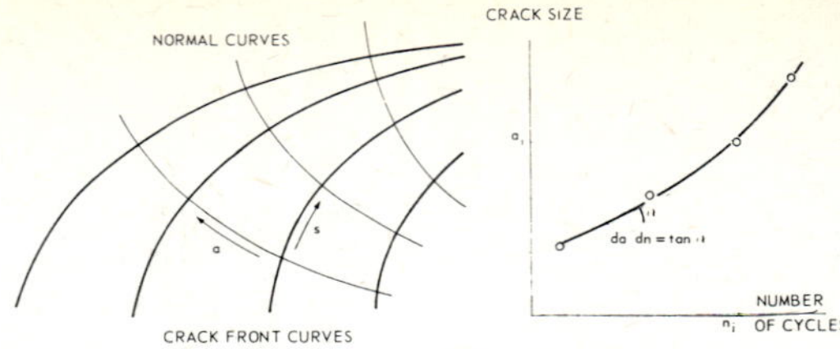

Fig. 20 Graphical derivation of the crack growth rate normal to a crack front

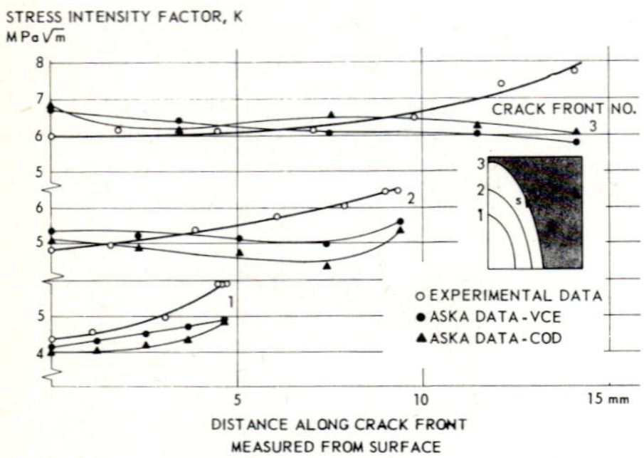

Fig. 21 Stress intensity distributions along three different crack fronts in the lug specimen at a load of 15 kN

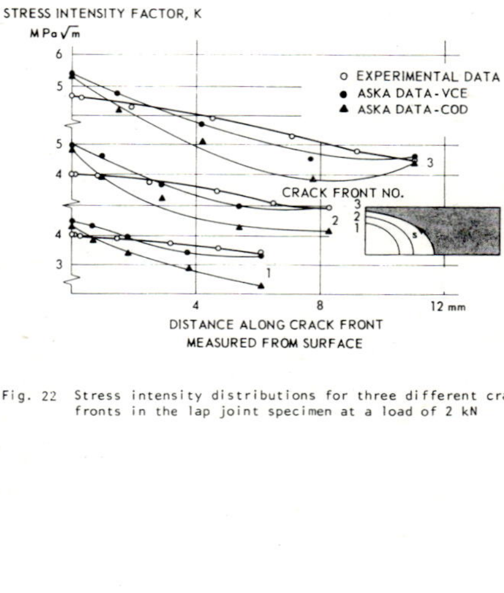

Fig. 22 Stress intensity distributions for three different crack fronts in the lap joint specimen at a load of 2 kN

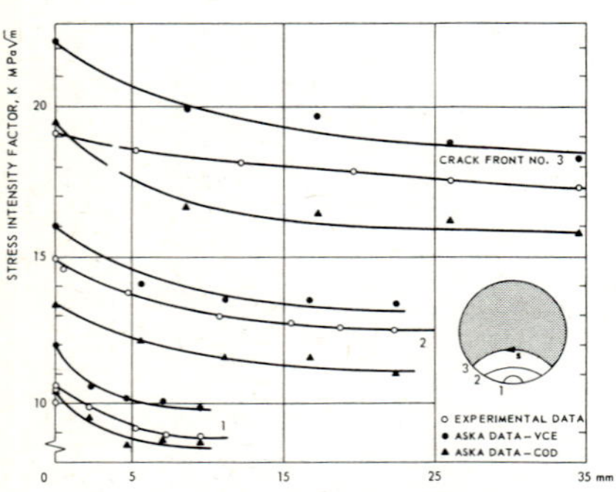

Fig. 23 Stress intensity distributions along three different crack fronts in the trunnion specimen at a bending moment of 5.8 kNm

men and Fig. 22 seems to confirm this conjecture. In this case however, a redistribution of the pin loads, maintaining the excentricity, did little to reduce the differences. There are two additional effects, that were not included in the analysis. Although they could be considered to be of minor significance they might explain part of the discrepancies found.
A possible load transfer between the lap joint holes by contract pressure normal to the free surface has been ignored in the analysis as well as friction between the hole and the pin. The stress intensity factors were derived under the assumption of plane strain and in particular in this thin specimen and plane stress situation may have prevailed for a considerable part of the thickness. Although both effects cannot be quantified they offer room for reconciliation.

c) Trunnion specimen

The trunnion specimen results are shown in Fig. 23 and Table VIII. There is no good correlation between test and analysis. There is a possibility that due to dissimilar cracking of the two notched sections in the symmetric specimen centrally supported on a rigid block, a redistribution of the two end loads occurred. On the basis of the specimen compliance derived from the ASKA analysis a nine per cent reduction of the bearing moment is presented assuming specimen rotation at the central support to be completely suppressed. For the intermediate crack size a two per cent reduction is predicted and for the smaller crack the effect is negligible.

5. CONCLUSIONS

An extensive comparison has been performed between analysis and test data for three different configurations. For three configurations, difficulties were encountered in the formulation of load boundary conditions.
It is advisable in these situations to consider one or more sets of load boundary conditions. They can easily be carried along simulatenously as different load vectors and they allow linear combinations to be formed to account for example, for compliance variations due to crack growth.
If the crack growth is considered dependent on the stress intensity factor along the entire length of the crack front, then the integrated result of stress intensity value are considered, as in Table IX. If the fracture process is thought as an energy conversion phenomenon, the K-squared integral must be taken as in Table IX.
The differences are in the order between 4 and 8% for the linear and between 4 and 15% for the squared distribution.

A good agreement has been found between analysis and data, showing the possibility to be confident in the results derived by a Finite Element Analysis within an engineering range. Standard Finite Element package has been used in all the calculations.

ACKNOWLEDGEMENT

This paper is published with permission of European Space Agency. The work has been carried out under ESA Contract No. 3411/77/NL/PP (SC).

Table I

Material Properties of Structural Specimens

Specimen Description	Material used	Young's Modulus	Poissons' ratio
Lap joint	Al 2219-T85	73500N/mm^2	0.33
Lug-pin	Lug: Al-alloy 7035	71000N/mm^2	0.33
	Pin: custom 455	21 0000N/mm^2	0.32
Trunnion	Titanium alloy IMI 550	112815 N/mm^2	0.32

Table II

DATA PIN-LUG MODEL,FIRST CRACKFRONT

net	nodes	degr. of freedom			elements		
		unknown	extern.	suppr.	PENTA18	PENTAX18	HEXE27
1	1260	3185	420	175	12		112
2	1260	3083	556	141	12		112
3	475	774	651				38
4	243	270	459			32	
5	1428	3941	153	190	90		90
6	54	54	108		4		1
7	54	54	108		4		1
mainnet		1165					
total:		12526			122	32	354

Table III

DATA LAP JOINT MODEL,FIRST CRACKFRONT

net	nodes	degr. of freedom			elements		
		unknown	extern.	suppr.	PENTA18	PENTAC18	HEXE27
1	854	1898	567	97	36		50
2	854	1859	642	61	36		50
3	575	948	777				46
4	243	270	459			32	
6	111	165	168				7
7	111	165	168				7
mainnet		1410					
total:		6715			72	32	160

Table IV

DETAILS OF F.E. CALCULATIONS OF TRUNNION

DESCRIPTIONS	NO. OF NODAL POINTS	NO. OF ELEMENTS	DEGREES OF FREEDOM		
			LOCAL	SUP-PRESSED	EXTER-NAL
SUBSTRUCTURE 1	761	90	1766	172	345
SUBSTRUCTURE 2	568	64	1177	183	344
SUBSTRUCTURE 3	838	95	1777	118	619
SUBSTRUCTURE 4	678	69	1371	96	567
SUBSTRUCTURE 5	964	125	1953	124	815
SUBSTRUCTURE 6	829	102	1615	109	763
SUBSTRUCTURE 7	823	94	1468	107	894
CRACK SUBSTRUCTURES	640	236	1320	117	1362
FIRST LEVEL MAINTEST	(COMBINING SUBSTR. 1 TO 7)	–	1654	—	895
FINAL LEVEL MAINNET	FIRTS LEVEL MAINNET CRACK STRUCTURES	–	1077	--	—

TABLE V

Fatigue crack growth calibration parameters according to equation 11 for the three different materials used in the demonstration configurations

Material type	Paris - parameters	
titanium, IMI 550	$C_1 = 2.377 \times 10^{-8}$	$C_2 = 3.34$
aluminium, 7075-T7351	$C_1 = 1.573 \times 10^{-7}$	$C_2 = 3.22$
aluminium, 2219-T851	$C_1 = 6.537 \times 10^{-8}$	$C_2 = 3.61$

When the stress intensity factor is given in MPa$\sqrt{m}$ - units the crack growth rate obtained is in mm per cycle

Table VI

Lug-pin: Comparison of Test and Analyses results

Crack Front 1				Crack Front 2				Crack Front 3			
s	'K' values			s	'K' values			s	'K' values		
	Test	vce	cod		Test	vce	cod		Test	vce	cod
0	4.38	4.10	4.06	0	4.82	5.33	5.04	0	6.01	6.71	6.76
1.2	4.52	4.29	4.08	2.4	5.08	5.22	4.84	3.5	6.05	6.45	6.11
2.5	4.85	4.45	4.17	5.1	5.48	5.05	4.73	7.6	6.32	6.03	5.65
3.7	5.35	4.69	4.32	7.4	6.0	4.94	4.38	11.6	6.98	5.91	5.26
4.7	5.95	4.85	4.93	9.5	6.50	5.49	5.39	14.2	7.90	5.65	6.09

s - distance along crack front (in mm) from free surface
vec - virtual crack extension method
cod - crack opening displacement
K - in Pa$\sqrt{m}$

Table VII

Lap joint Comparison of Test and Analysed Results

Crack Front 1				Crack Front 2				Crack Front 3			
s	'K' values			S	'K' values			S	'K' values		
	Test	vce	cod		Test	vce	cod		Test	vce	cod
0.0	4.02	4.47	4.31	0	4.01	4.99	4.87	0	4.68	5.38	5.41
0.7	4.0	4.29	3.78	1.0	3.98	4.58	3.94	1.5	4.50	4.75	4.11
1.8	3.87	3.87	3.42	2.8	3.80	3.77	3.74	4.2	4.0	3.64	3.03
3.7	3.70	3.35	2.90	5.4	3.35	2.93	2.25	7.7	3.20	2.51	1.84
6.1	3.42	3.22	2.27	8.3	2.98	2.81	2.15	11.1	2.50	2.57	2.38

s - distance along crack front (in mm) from free surface
vec - virtual crack extension method
cod - crack opening displacement method
K - in Pa$\sqrt{m}$ for 2000 N

Table VIII

Trunnion: Comparison of Test and Analyses Results.

Distance from surface (along crack front)	Crack front 1			Crack front 2			Crack front 3		
	Test	vce	cod	Test	vce	cod	Test	vce	cod
0.0L	10.6	12.0	10.7	14.9	16.1	13.4	19.2	22.3	19.5
0.25L	9.9	10.6	9.6	13.6	14.1	12.1	18.3	19.9	16.6
0.50L	9.3	10.2	8.6	13.0	13.7	11.6	17.9	19.7	16.5
0.75 L	9.0	10.1	9.2	12.6	13.5	11.6	17.6	18.9	16.2
1.0L	8.9	9.9	8.8	12.5	13.4	11.0	19.3	18.3	15.0
Existing comparative analytical result	At 1.0L c = 5.25mm K_1=10.01 a = 7.5			At 1.0L c = 12.3 mm K_1=13.32 a = 19.5 mm			At 1.0L c = 23.5 mm K_1=15.32 a = 19.5 mm		

L : total length of any crack front; vce - virtual crack extension method
cod : crack opening displacement method; K in Pa$\sqrt{m}$

Table IX

Comparison between test and analyses.

Specimen type and crack front no.		$\int K(s)ds$ ASKA	test	relative diff.	$\int K^2(s)ds$ ASKA	test	relative diff.
lug	I	21.-	22.7	.080	78.7	91.2	.147
	II	48.5	51.3	.057	267.7	303.-	.124
	III	87.5	91.4	.045	536.5	583.4	.084
lap joint	I	22.2	22.3	.005	81.7	81.4	.004
	II	30.3	30.1	.007	100.4	103.0	.026
	III	38.-	39.4	.037	146.1	140.2	.041
trunnion	I	195.2	181.-	.073	2031.6	1710	.158
	II	613.8	591.5	.038	8466.-	7863	.071
	III	1348.3	1241.2	.079	25021.	22355	.119

REFERENCES

1. PARKS, D.M. - "A Stiffness Derivative Finite Element Technique for Determination of Crack Tip Stress Intensity Factors", Int. Journal of Fracture, Vol. 10, 1974, pp. 487-502.

2. BARTELDS, L. and DE KONING, A.V. - "Application of Finite Element Methods to the Analysis of Cracks", Phase I, NLR-TR-78138 U.

3. HIBBITT, H.D. - "Some Properties of Singular Isoparametric Element", Int. Journal for Num.Meth. in Eng., Vol. 11, pp. 180-184, 1977.

4. BARTELDS L. - "Application of Finite Element Methods to the Analysis of Cracks", Final Report NLR-TR-79118U.

5. HELLEN, T.K. and BLACKBURN W.S. - "The Calculation of Stress Intensity Factors at a Crack Tip using Special Finite Elements", The Mathematics of Finite Elements and Applications, Academic Press, ed. Whitenon.

6. BARSOM, R.S. - "On the Use of Isoparametric Finite Elements in Linear Fracture Mechanics", Int. Journal Num.Meth. in Eng.,Vol 10, pp. 25-37, 1976.

7. ROOKE, D.P. and CARTWRIGHT, D.J. - "Compendium of Stress Intensity Factors", Section 4.2.4,pp. 297, 1976.

SECTION 5

FRACTURE OF BRITTLE MATERIALS AND COMPOSITES

ON STABILITY OF THE GROWTH OF INTERACTING CRACKS, AND CRACK KINKING AND CURVING IN BRITTLE SOLIDS

S. Nemat-Nasser

Professor of Civil Engineering and Applied Mathematics, Northwestern University, Evanston, Ill. 60201, USA.

SUMMARY

The theory of stability of the growth regimes of interacting tension cracks in brittle solids is reviewed, and the essential features not common in already established fracture mechanics considerations, are emphasized. Then recent progress in calculation of crack branching and crack curving is summarized. In particular, it is pointed out that the assessment of crack configuration, whether it is that of a system of tension cracks or a kinked-curved extension of a preexisting one, requires the consideration of the second law of thermodynamics. This will involve calculation of the second derivatives of the total potential energy with respect to crack lengths and other parameters characterizing the crack paths, which are not commonly used in fracture mechanics, and which require different, more accurate computational schemes. Several solution methods recently worked out by the author and his associates are briefly discussed.

1. INTRODUCTION

In recent years considerable effort has been devoted at Northwestern University by the author and his associates to the application of the theory of stability to various problems in fracture mechanics. This includes stability of configuration of a highly interacting system of tension cracks, crack kinking and crack curving. Whereas the commonly used fracture theories essentially rest on an energy balance concept, i.e. the first law of thermodynamics, examination of the stability of the growth regimes requires the consideration of the second law of thermodynamics. In a recent work [1] I have briefly mentioned the use of the second law of thermodynamics in the form of Gibbs' stability statement for establishing the crack paths and, in general, the configuration of a system of cracks in brittle solids.

According to Gibbs' statement of the second law for an isolated system, among all neighboring equilibrium states that the system can attain with the same entropy, the actual state (which is hence the most stable one) will be the one with minimum total energy. It is this statement that has direct application to the calculation of kinked or curved crack paths, and in general, to the establishment of the configuration of a system of cracks.

In this paper I shall illustrate this approach which I believe is fundamental, in terms of two classes of problems: (1) stability of two interacting tension cracks; and (2) crack kinking or curving (both in plane strain). Some basic results are presented, but for more detailed accounts the cited references should be consulted.

2. INTERACTING TENSION CRACKS

Because of cooling, externally applied loads, residual-stress build-up due to creep, loss of moisture and consequent shrinkage, or other natural or imposed processes, cracks often form in solids. Problems of this kind include the formation of surface cracks in aging wood, thermal cracks in flanks of spreading oceanic ridges, thermal cracks in nuclear reactor fuel elements, crevasses in glaciers, shrinkage cracks in drying concrete, shrinkage cracks in adhesively bonded materials such as paints, polymer films, etc., thermal cracks in thermal shields, and dessication cracks in deserts and at the bottom of dried up lakes.

Mechanisms which induce cracks of this kind are "strain-controlled" in the sense that at each stage of crack growth, the total elastic strain energy available to each crack is finite. Upon crack extension, the elastic energy is released, and therefore the crack growth is self-arresting.

However, in a suitable setting, two or several tension cracks may interact which may lead to abrupt changes in their growth pattern. For example, as cracks grow with a continuous supply of elastic energy (e.g. because of continuous loss of moisture or heat), a critical state may be reached, where some cracks stop growing, as others grow at a faster rate. Another example of a change in the growth pattern is when some cracks actually snap closed while others extend by a finite amount. Unstable crack growth of this kind was first examined by Nemat-Nasser [2, 3], and with detailed calculations by Nemat-Nasser et al. [4], and Keer et al. [5]; the basic theory has been first published in [6]. More recently, a number of interesting related phenomena such as imperfection-sensitivity and general variational formulations have been examined and illustrated by this writer and his associates [7-9].

In the present work the theory is presented for a system of parallel edge cracks in plane strain, induced in an elastic half-plane by the removal of heat at the free surface. This problem arises in consideration of heat extraction from hot dry rock masses, a project which recently has been initiated and to a large extent, successfully implemented at the Los Alamos Scientific Laboratory in New Mexico; for discussion and references, see Nemat-Nasser et al. [4], and Nemat-Nasser [10].

2.1 Statement of Problem

Consider an homogeneous isotropic elastic half-plane which is initially at a uniform temperature, T_0, and is then continuously cooled at its free surface. Because of this cooling, a thermal boundary layer of thickness Δ, measured from the free surface along its normal, develops in the solid. This may then lead to the formation of tension cracks emanating perpendicularly from the free surface of the half-plane. We confine attention to a state of plane strain, assume a set of initially equally spaced edge cracks, and regard the temperature profile prescribed.

Initially, the cracks grow in a stable manner, as the thickness of the thermal layer, Δ, is increased; Δ is used as a measure of the load parameter. As long as the crack spacing is large compared with the common crack length h, there is very weak interaction between adjacent cracks, and such a crack growth pattern in inherently stable. In this case, each crack may be regarded as an isolated one, and since for a fixed Δ, there is a fixed amount of elastic energy available, an extension of a crack at constant Δ would release a certain amount of elastic energy, which results in a reduction of the corresponding stress intensity factor at the crack tip, and therefore crack growth will be arrested. If K is the stress intensity factor, then we have, in this case, $\partial K/\partial h < 0$; see Fig. 3 of [4].

As the common crack length increases with increasing Δ, the interaction between adjacent cracks becomes more important. In Fig. 1 two interacting cracks are shown in a unit cell, with h_1 and h_2 as their corresponding lengths, and K_1 and K_2 the stress intensity factors; the critical values of the stress intensity factors are denoted by K_i^c, i = 1,2, which may depend on crack lengths, i.e. $K_i^c = K_i^c(h_i)$, no sum on i. The corresponding energy release rates are G_i, having critical values G_i^c, which may depend on h_i.

Since we deal with plane strain, we consider a slice of unit thickness in all subsequent statements. Let the corresponding strain energy be denoted by E, and set E = -U. With S_1 and S_2 denoting the surface energies for Crack 1 and Crack 2, respectively, we write for the total potential energy (elastic energy plus the surface energy),

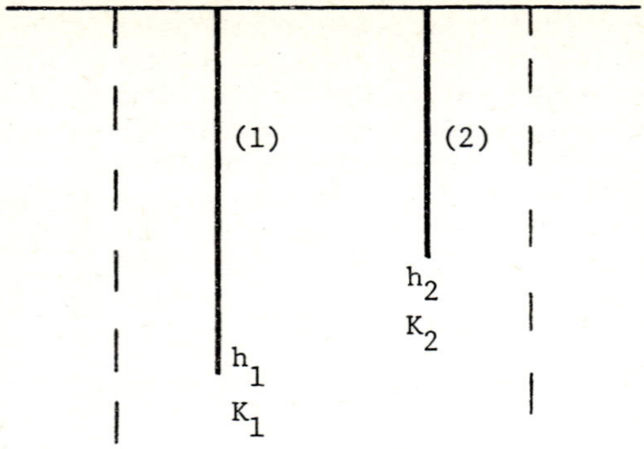

Figure 1: A typical unit cell with two interacting cracks.

$$\Pi = -U + S_1 + S_2 \tag{2.1}$$

where we have

$$U = U(h_1,\ h_2;\ \Delta),\ S_i = S_i(h_i),\ i = 1,2 \text{ (no sum on i).} \tag{2.2}$$

For a constant value of Δ, the equilibrium state is defined by the vanishing of the first variation of the total potential energy Π; the corresponding equilibrium equations are given by

$$\frac{\partial U}{\partial h_1} = \frac{\partial S_1}{\partial h_1} \equiv G_1^c(h_1) > 0\ ,\ \frac{\partial U}{\partial h_2} = \frac{\partial S_2}{\partial h_1} \equiv G_2^c(h_2) > 0. \tag{2.3}$$

In this equation we set

$$\frac{\partial U}{\partial h_i} \equiv G_i(h_1,\ h_2;\ \Delta) \tag{2.4}$$

which is the energy release rate. For a given $\Delta = \Delta_0$, Eqs. (2.3) then define the crack length $h_i = h_i^0$, $i = 1,2$, which correspond to the state at which each crack has attained the value of its critical energy release rate at its tip. The locus of points in the h_1, h_2, Δ-space corresponding to the equilibrium states (2.3) defines equilibrium paths. When two paths intersect, a bifurcation point is established, at which two crack configurations become possible. The objectives of the stability analysis are: (1) to identify all equilibrium paths, i.e. all possible crack configurations; (2) to calculate all bifurcation points and establish whether they are stable, neutral or unstable states. In [8, 11] a complete and general theory encompassing these aspects is presented, where the influence of imperfections is also studied. In particular, it is shown that small imperfections in material properties (or geometry) can lead to large changes in equilibrium paths in the h_1,h_2,Δ-space.

Since in the present example of two interacting cracks only Mode I is involved (because of symmetry), we can formulate the problem in terms of stress intensity factors or the energy release rates, which are related by $G_i = K_i^2(1-\nu^2)/E$, where ν is the Poisson ratio and E is Young's modulus.

Consider now an equilibrium state at which (2.3) is satisfied, and let a_i be the (small) increase in the crack length associated with a given small change λ in the load parameter. The corresponding change in the total potential energy then is

$$\pi = \Pi - \Pi^0 = \Pi_{,i}\, a_i + \Pi'\lambda + \frac{1}{2}\Pi_{,ij}\, a_i\, a_j + \Pi'_{,i}\, a_i\lambda + \frac{1}{2}\Pi''\lambda^2 + \dots, \tag{2.5}$$

where primes denote differentiation with respect to the load parameter Δ, a comma followed by an index denotes differentiation with respect to the corresponding crack length, and all partial derivatives are evaluated at the considered equilibrium state prior to the increase in the load parameter, at which the value of the potential energy is denoted by Π^0.

A complete stability analysis is made by the examination of (2.5), subject to the constraints that the crack ceases to grow if its stress intensity factor falls below critical, and that, if it continues to grow, then its stress intensity factor must remain at critical. Moreover, a crack begins to close if its stress intensity factor tends to become negative.

To outline the basic results emerging from (2.5), we first note that, in view of equilibrium, $\Pi_{,i} = 0$; this is the same as Eqs. (2.3). Then, setting $\delta\pi = 0$, we obtain a system of equations, the first of which is

$$\Pi_{,ij}\, a_j + \Pi'_{,i}\lambda = 0; \tag{2.6}$$

repeated indices summed. If λ is prescribed, and if $\det|\Pi_{,ij}| \neq 0$, then (2.6) can be solved to yield a unique a_i. Subject to the above-mentioned constraints, the first critical state is obtained when

$$\det|\Pi_{,ij}| = 0. \tag{2.7}$$

It is easily shown, [8, 11], that this defines a stable bifurcation point from which two stable equilibrium paths in the h_1, h_2, Δ-space emanate. After this point there are two possible equilibrium paths, one when both cracks continue to extend, and the other when one crack remains stationary and the other crack grows at a faster rate.

To be more specific we consider an example.

2.2 Example

Let the material be homogeneous and the critical values of the stress intensity factors at both cracks be equal. Then, because of complete symmetry in Eqs. (2.3), a fundamental equilibrium path characterized by

$$h_1 = h_2, \; K_1 = K_2 = K_c \tag{2.8}$$

results; path AB in Fig. 2. This fundamental equilibrium path is followed as long as

$$\frac{\partial K_1}{\partial h_1} = \frac{\partial K_2}{\partial h_2} < \frac{\partial K_1}{\partial h_2} = \frac{\partial K_1}{\partial h_2} < 0; \tag{2.9}$$

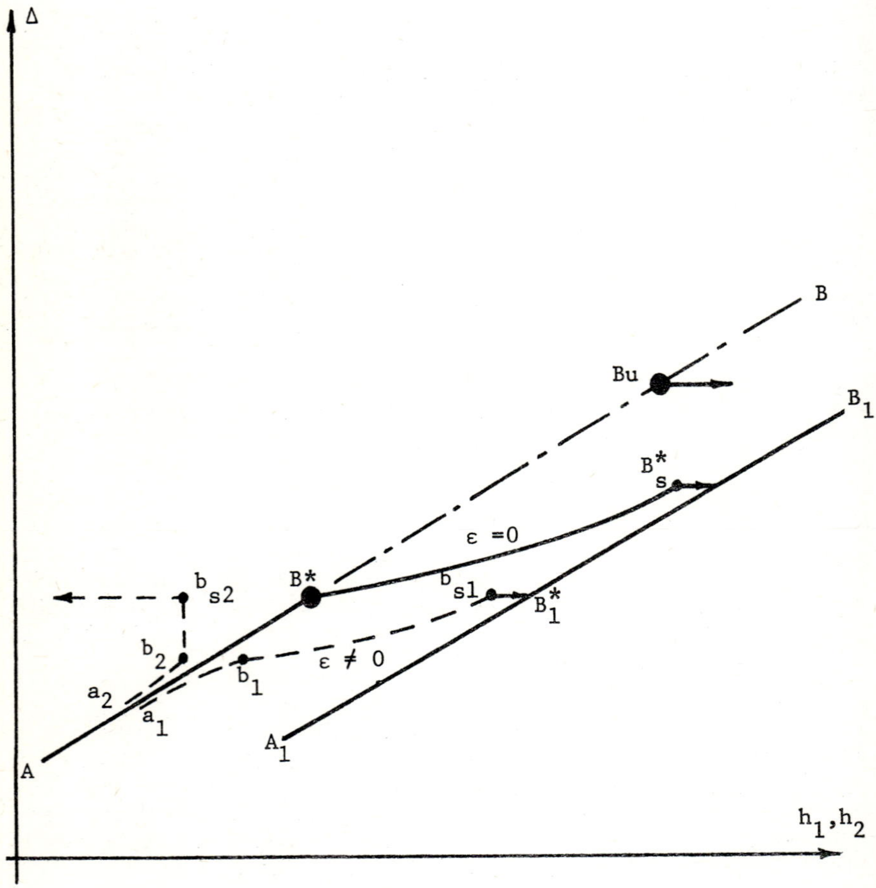

Figure 2: Various equilibrium paths in the h_1, h_2, Δ-space.

segment AB* in Fig. 2. The first critical point is a stable bifurcation point (point B* in Fig. 2) attained when

$$\frac{\partial K_1}{\partial h_1} = \frac{\partial K_2}{\partial h_2} = \frac{\partial K_1}{\partial h_2} = \frac{\partial K_2}{\partial h_1} < 0 \; . \tag{2.10}$$

All points on the fundamental equilibrium path (B* to Bu) for which

$$0 > \frac{\partial K_1}{\partial h_1} = \frac{\partial K_2}{\partial h_2} > \frac{\partial K_1}{\partial h_2} = \frac{\partial K_2}{\partial h_1} \tag{2.11}$$

holds, are stable bifurcation points. The unstable bifurcation point (point Bu) is first reached when

$$\frac{\partial K_1}{\partial h_1} = \frac{\partial K_2}{\partial h_2} = 0 \; . \tag{2.12}$$

All points on the fundamental equilibrium path (Bu to B) for which

$$\frac{\partial K_1}{\partial h_1} = \frac{\partial K_2}{\partial h_2} > 0 \tag{2.13}$$

holds, are unstable bifurcation points. In all cases we always have on the fundamental equilibrium path,

$$\frac{\partial K_1}{\partial h_2} = \frac{\partial K_2}{\partial h_1} < 0 \; ; \tag{2.14}$$

this has been proved in [4]. On the bifurcated path ($B^*B_s^*$), on the other hand, $K_1 \neq K_2$ and, instead of (2.14), we have

$$K_1 \frac{\partial K_1}{\partial h_2} = K_2 \frac{\partial K_2}{\partial h_1} \leq 0, \tag{2.15}$$

the equality sign corresponding to the snap-through instability point (point B_s^*) at which, say, $K_2 = 0$ and hence Crack 2 snaps closed as Crack 1 extends by a finite amount (to new path A_1B_1 in Fig. 2).

The introduction of a very small material imperfection results in deviation of the equilibrium path from the fundamental equilibrium path for all finite values of Δ, and this deviation increases with increasing Δ. Along the new equilibrium paths (a_1b_1 and a_2b_2 in Fig. 2), the crack with the larger value of the critical stress intensity factor, say Crack 2, stops at the critical state (point b_2) defined by

$$\frac{\partial K_2}{\partial h_1} \frac{\partial K_1}{\partial \Delta} - \frac{\partial K_1}{\partial h_1} \frac{\partial K_2}{\partial \Delta} = 0 \; . \tag{2.16}$$

After this, Crack 1 continues to grow (along $b_1 b_{s1}$) with increasing Δ, until the snap-through critical state (point b_{s1}) defined by

$$\frac{\partial K_1}{\partial h_2} = 0 \tag{2.17}$$

is reached. An infinitesimal increase in Δ then results in a snap closure of Crack 2 (at b_{s2}), as Crack 1 extends by a finite amount (from b_{s1} to B_1^* in Fig. 2).

In references [11] it is proved that if $K_2^c = (1 + \varepsilon)K_c$ and $K_1^c = K_c$, where ε is a small parameter, then the value of the load parameter Δ, at which Crack 2 ceases to grow, changes from its value associated with $\varepsilon = 0$, by the amount that is proportional to $\varepsilon^{1/2}$. In [8] it has been illustrated numerically that, for example, with an ε of 3%, the change in the load parameter exceeds 20%.

2.3 Methods of Analysis

As is well known, the stress field at the tip of a crack in a brittle solid has a singularity of the order of $1/\sqrt{r}$ where r measures distance from the tip of the crack. In a numerical calculation, this singularity must be removed by analytical procedures, in order to obtain a reasonable estimate of the value of the stress intensity factor which is actually the strength of the singularity. Finite-element procedures with special elements at the tip of the crack have been developed for this purpose.

The stability analysis on the other hand, requires estimates of the derivatives with respect to the crack lengths of the stress intensity factors at crack tips. These quantities correspond to the strength of singularities which are of the order of $1/r^{3/2}$. For this reason, usual finite-element methods with special elements which are designed to take care of $1/\sqrt{r}$ singularity, are not adequate, and cannot yield accurate enough estimates.

In an effort to overcome the above-mentioned difficulty, we have developed and perfected two methods which directly take care of the corresponding singularities, and therefore, yield very accurate results [4, 7]. These methods are briefly discussed below.

2.4 Analytical Method

For two interacting cracks, Fig. 3 shows the basic elasticity boundary value problem. For a given temperature profile the tractions on the faces of the cracks are prescribed.

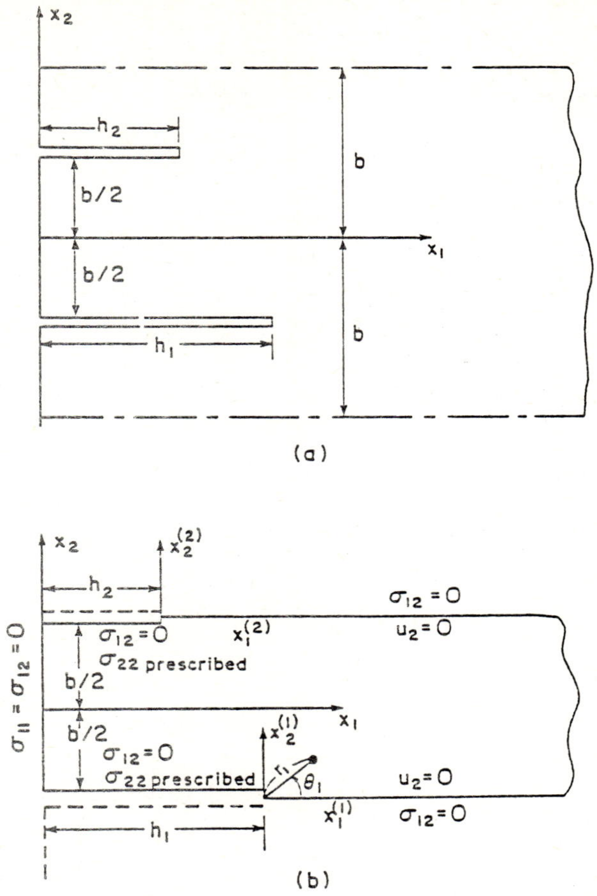

Fig. 3 (a) Typical unit cell consisting of two interacting edge cracks;
(b) The basic elasticity boundary-value problem.

Therefore, the basic elasticity problem can easily be reduced to a pair of coupled singular integral equations with the distribution of the dislocation densities along Cracks 1 and 2 being two unknown functions. These integral equations can then be solved by a collocation scheme. The method of solution is outlined in [4], where the accuracy of the results is also illustrated.

2.5 Combined Analytical and Numerical Method

Let V denote the region occupied by the elastic body. In the absence of body forces, the field equations are,

$$\sigma_{ij,j} = 0 \text{ in } V,$$
$$\sigma_{ij}n_j = t_i \text{ on } S_t, \qquad (2.18)$$
$$u_i = \overline{u}_i \text{ on } S_u,$$

where comma followed by an index denotes partial differentiation with respect to the corresponding (fixed rectangular Cartesian) coordinate, the summation convention on repeated indices is used, and n_j denotes the components of the exterior unit normal to the boundary of the solids; S_t is the part of the boundary on which some components of tractions, t_i, are prescribed, and S_u is the part on which some components of the displacement, $\overline{u}_i$, are given. The unit cell and the corresponding elasticity boundary value problem is shown in Fig. 3. The equivalent thermal tractions are

$$t_1(x_1, -b/2) = 0,$$
$$t_2(x_1, -b/2) = -\hat{\beta}(T-T_0), \quad 0 \leq x_1 \leq h_1; \qquad (2.19)$$
$$t_1(x_1, b/2) = 0,$$
$$t_2(x_1, b/2) = \hat{\beta}(T-T_0), \quad 0 \leq x_1 \leq h_2;$$

where $\hat{\beta} = \hat{\alpha}\overline{E}/(1-\nu)$, $\hat{\alpha}$ being the coefficient of thermal expansion.

According to [4], we first determine the relation between crack lengths (h_1,h_2) and the penetration depth Δ in such a manner that the stress intensity factors of each growing crack maintain its critical value. Then we calculate the derivatives of the stress intensity factors, $\partial K_m/\partial h_n$, for the purpose of stability analysis.

The dominant terms of stresses around the mth crack tip can be represented in the following form:

$$\sigma_{ij}^{(m)} = a_1^{(m)}\sigma_{ij}^{1}(r_m,\theta_m) + a_2^{(m)}\sigma_{ij}^{2}(r_m,\theta_m) + b_1^{(m)}\overline{\sigma}_{ij}^{1}(r_m,\theta_m)$$
$$+ b_2^{(m)}\overline{\sigma}_{ij}^{2}(r_m,\theta_m) + 0(r_m^{3/2}), \qquad (2.21)$$

where (r_m,θ_m) is a local polar coordinate whose origin is located at the mth crack tip, and $\sigma_{ij}^{k}(r_m,\theta_m)$ and $\overline{\sigma}_{ij}^{k}(r_m,\theta_m)$, $k = 1,2$, are given in [7]. The coefficients $b_1^{(m)}$ and $b_2^{(m)}$ can easily be determined as the Taylor expansion coefficients of applied tractions at the mth crack tip, while $a_1^{(m)}$ and $a_2^{(m)}$ are determined by the method of superposition of analytical and finite-element solutions. The stress intensity factor K_m is expressed as

$$K_m = \frac{3\sqrt{2\pi}}{4} a_1^{(m)} , \tag{2.22}$$

and the load parameter Δ is so determined that K_m equals the critical value at the growing crack tip.

If we differentiate Eq. (2.18) with respect to the crack length h_n, the solution of the problem, $\partial\sigma_{ij}/\partial h_n$, is obtained as

$$\frac{\partial\sigma_{ij}^{(m)}}{\partial h_n} = \begin{cases} -\frac{3}{2} a_1^{(n)}\sigma_{ij}^0 + (\frac{\partial a_1^{(n)}}{\partial h_n} - \frac{5}{2} a_2^{(n)})\sigma_{ij}^1 + (\frac{\partial b_1^{(n)}}{\partial h_n} \\ \qquad - 3b_2^{(n)})\bar{\sigma}_{ij}^1 + 0(r_m^{1/2}) \text{ for } m = n, \\ \frac{\partial a_1^{(m)}}{\partial h_n}\sigma_{ij}^1 + \frac{\partial b_1^{(m)}}{\partial h_n}\bar{\sigma}_{ij}^1 + 0(r_m^{1/2}) \text{ for } m \neq n, \end{cases} \tag{2.23}$$

near the mth crack tip, where σ_{ij}^0 has a singularity of $0(r_m^{-3/2})$. Here we define new variables.

$$\sigma_{nij} = \frac{\partial\sigma_{ij}}{\partial h_n} + \frac{3}{2} a_1^{(n)}\sigma_{ij}^0 ,$$

$$u_{ni} = \frac{\partial u_i}{\partial h_n} + \frac{3}{2} a_1^{(n)} u_i^0, \; n = 1,2 , \tag{2.24}$$

where u_i^0 represents the displacement field corresponding to the stress field σ_{ij}^0. These variables are chosen in such a manner that the singularity of theorder $r_n^{-3/2}$ is removed from the derivative problem. With the aid of (2.24) and upon differentiation of (2.18) with respect to h_n, we write

$$\sigma_{nij,j} - \frac{3}{2} a_1^{(n)}\sigma_{ij,j}^0 = 0 \qquad \text{in } V,$$

$$\sigma_{nij}\, n_j = \frac{\partial t_i}{\partial h_n} + \frac{3}{2}a_1^{(n)}\sigma_{ij}^0 n_j \qquad \text{on } S_t,$$

$$u_{ni} = \frac{\partial \bar{u}_i}{\partial h_n} + \frac{3}{2} a_1^{(n)} u_i^0 \qquad \text{on } S_u, \; n = 1,2 . \tag{2.25}$$

The stress distribution around the mth crack tip can be expressed in the same manner as (2.21); here, this becomes

$$\sigma_{nij}^{(m)} = \alpha_{n1}^{(m)}\sigma_{ij}^{1}(r_m,\theta_m) + \alpha_{n2}^{(m)}\sigma_{ij}^{2}(r_m,\theta_m) = \left(\frac{\partial b_1^{(m)}}{\partial h_n} - 3b_2^{(m)}\right) \times$$

$$\bar{\sigma}_{ij}^{1} + O(r_m). \tag{2.26}$$

The unknown coefficients $\alpha_{nk}^{(m)}$, k, m, n = 1, 2, can be determined by the same method used for the analysis of stress intensity factors. Comparing (2.23) and (2.26), the derivatives of the stress intensity factors are obtained as

$$\frac{\partial K_m}{\partial h_n} = \begin{cases} \frac{3\sqrt{2\pi}}{4}\left(\alpha_{n1}^{(n)} + \frac{5}{2} a_2^{(n)}\right) & \text{for } m = n, \\ \frac{3\sqrt{2\pi}}{4}\alpha_{n1}^{(m)} & \text{for } m \neq n. \end{cases} \tag{2.27}$$

As has been exemplified in [7] this method is efficient and yields very accurate results.

3. CRACK KINKING AND CURVING

Crack kinking and curving has received some attention in recent years [12-22]. There are two essential problems involved: (1) what is the proper physical fracture criterion that should be used; and (2) analytical procedures to calculate relevant quantities. I shall address these questions from a fundamental point of view in some detail in this section, and then summarize some recent results I have obtained in collaboration with my associates. For greater detail, the reader should consult [12-24], where additional relevant references are also given.

The crack kinking problem was first examined by Erdogan and Sih [12] who used the stress criterion, and subsequently Sih [13] introduced the concept of "minimum strain energy density". The criterion of local symmetry has been considered by Soviet researchers, e.g. Banichuk [21] and others [25, 26], and has recently been revived by Cotterell and Rice [20]. These criteria do not seem to have as solid a foundation as the so-called maximum energy release rate criterion, that obviously stems from Gibbs' stability statement mentioned in the introduction. The maximum energy release rate criterion was first used by Hussain *et al*. [14], and subsequently employed and extensively discussed by a number of researchers [15-19].

From a mathematical point of view, crack kinking and curving represent a rather complicated problem. In the literature various approaches have been used, which are essentially based on small kinked or curved crack lengths. For the kinked crack, the numerical solution by Lo [27] has considerable advantage, as it uses Green's function which automatically satisfies the boundary conditions on the primary crack. We have used a similar method [22-24] for both a complete numerical calculation [22, 24] and a second-order analytical perturbation approach [23], obtaining explicit results in the latter case. Banichuk [21] and Cotterell and Rice [20] use a perturbation method based on Muskhelishvili's complex potentials, giving first-order perturbation results. We have also used [28] a similar technique, giving both the first- and second-order perturbation results for crack curving.

3.1 A Fundamental Approach

Figure 4 shows a preexisting crack, L_0, and its possible kinked and curved extension, ℓ. We assume that ℓ/L_0 is small (not necessarily infinitesimal), and denote in the x,y-plane its equation by $y = y(x)$.

Let kinked-curved extension, ℓ, be caused by a given change in loads applied far away from the crack, so that the extension ceases when a stable equilibrium condition at the tip of the extended crack is attained; note that for strain-controlled loading, this happens spontaneously, whereas for certain stress-controlled ones it may never happen. We wish to be able to calculate the kinked-curved extension ℓ.

To this end we first observe that since the displacement field existing prior to kinking in the neighborhood of the tip of the primary crack starts with a term proportional to $\sqrt{x}$, it is natural to expect that the function y should be representable for sufficiently small x as

$$y = \omega x + a x^{3/2} + b x^{2} + \ldots, \qquad (3.1)$$

where ω is the initial kink angle, which together with coefficients a, b, ..., are to be calculated in terms of the values of the stress intensity factors, k_1 and k_2, existing at the tip of the primary crack prior to kinking, and their derivatives with respect to crack extension in the x-direction, i.e. $\partial k_i/\partial L_0$, $i = 1,2$; note, since plane strain is considered, only nodes 1 and 2 denoted by subscripts 1, 2, are involved.

To be specific, consider the strain-controlled crack extension regime. A problem of this kind is shown in Fig. 4 for plane strain, where a linearly elastic brittle solid is clamped along a part of its boundary and is subjected to a prescribed infinitesimal displacement, $\tilde{u}$, at point P. The body contains a preexisting crack L_0, and we wish to find for a prescribed displacement $\tilde{u} = \rho\tilde{n}$ what would be the resulting crack path; here ρ is the load parameter and $\tilde{n}$ is a fixed unit vector.

For convenience we shall assume that the body is homogeneous and that the critical value of the energy release rate is constant, denoted by G_c.

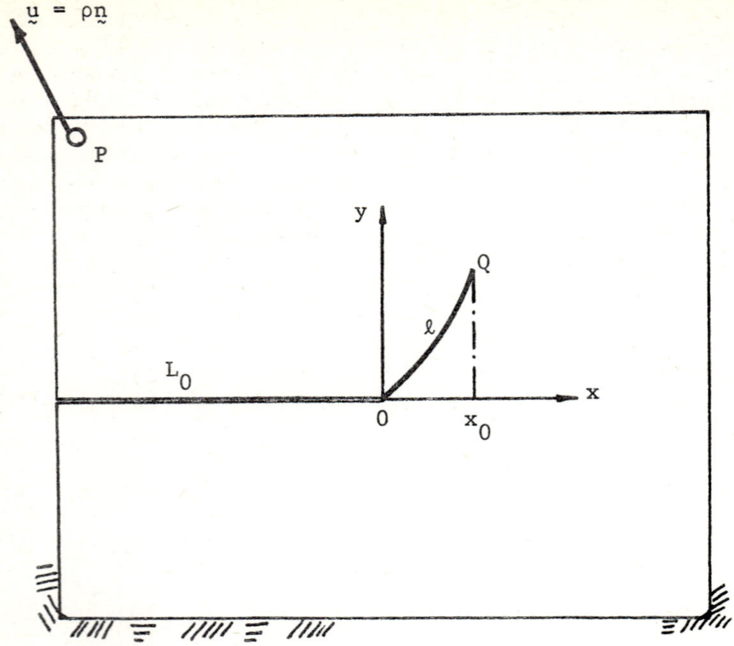

Figure 4: Preexisting crack L_0 and its kinked-curved extension ℓ.

Assume for a moment that the function y(x) is known. Then we can calculate the total elastic energy stored in the body,

$$E = E(y,\rho) = -U(y,\rho) \tag{3.2}$$

which is a functional of y and a function of ρ (note that $\underset{\sim}{n}$ is fixed). Let ℓ measure length along the crack path, starting from the origin 0. Then <u>for equilibrium</u> we must have

$$\frac{\partial U}{\partial \ell} = (1 + y'^2)^{-1/2} \frac{\partial U}{\partial x} = G_c \quad \text{at} \quad x = x_0 \, , \tag{3.3}$$

which states that the energy release rate at the tip of the crack must be equal to the critical value. We note that (3.3) can be obtained by setting equal to zero the first variation of the total potential energy,

$$\Pi = E + \int_0^{\ell} G_c \, d\ell' \, . \tag{3.4}$$

We note further that to each assumed crack path y(x), a different x_0 will be assigned by the equilibrium condition (3.3).

Let $Y \equiv \{y(x), 0 \leq x \leq x_0\}$ define all continuous crack paths which satisfy the equilibrium condition (3.3). The question is, among the functions in this class, which is the actual crack path.

To answer this question, we invoke Gibbs' stability statement mentioned in the introduction. For fixed ρ, each crack path defines a configuration. Among all possible crack paths (i.e. among all possible configurations) which satisfy the equilibrium condition (3.3), the actual one renders the total stored elastic energy a minimum.

It is more convenient to pose the problem in the following equivalent form. Suppose, instead of fixing the load parameter ρ, we fix the length, ℓ, of the crack path (crack extension along path 0Q in Fig. 5), i.e. we set $\ell = \ell_0$, and then seek among all admissible functions y, the one which renders the total stored elastic energy an absolute minimum; note that since $\ell = \ell_0$, for each y, the corresponding load parameter is obtained from (3.3). The total surface energy associated with each crack path y in this case has the following common value:

$$S = \int_0^{\ell_0} G_c \, d\ell = G_c \, \ell_0, \tag{3.5}$$

the last equality following from the assumption that the material is homogeneous. Hence the crack path which makes the total potential energy Π in Eq. (3.4) an absolute minimum, will also render the stored elastic energy E an absolute minimum. We therefore require

$$\delta_y E = -\delta_y U = 0 \text{ for } \delta y \neq 0 ,$$
$$\delta_y^2 E = -\delta_y^2 U > 0 \text{ and } \int_0^{x_0} (1 + y'^2)^{1/2} \, dx = \ell_0. \tag{3.6}$$

In this formulation the load parameter ρ is calculated from condition (3.3) for each crack path.

In practice the theory lends itself to an easy numerical implementation: one first chooses a suitably small crack length, ℓ_0, then one characterizes the crack path by a finite number of parameters, and obtains these parameters in such a manner that (3.6) is satisfied. In a finite-element implementation of this procedure crack path (3.1) must be used. We are exploring this at the present time.

3.2 Perturbation Solutions

Let K_I and K_{II} be the stress intensity factors at the tip of the extended crack of length ℓ, where ℓ is very small compared with L_0. For simplicity, we choose L_0 as the unit of length, so that ℓ actually represents the dimensionless crack extension measured in terms of L_0. Then it is natural to expect that the stress intensity factors at the tip of the extended crack should have the following functional form:

$$K_I = A_0 + \ell^{1/2}A_1 + \ell A_2 + \ell^{3/2}A_3 + \dots,$$

$$K_{II} = B_0 + \ell^{1/2}B_1 + \ell B_2 + \ell^{3/2}B_3 + \dots, \tag{3.7}$$

where the coefficients A and B depend on k_1 and k_2, their derivatives with respect to L_0, the coefficients in the expression for the crack path (3.1), and the load parameter. These A's and B's can be calculated by perturbation methods [20-23, 28].

If the condition of local symmetry is used [21], then one requires that crack extension occurs in only Mode I, putting $K_{II} = 0$. Since this must hold for all suitably small ℓ, we obtain a system of equations,

$$B_a = 0, \; a = 0, 1, 2, \dots \tag{3.8}$$

which gives the coefficients of the crack path, $y = y(x)$, defined in Eq. (3.1), in terms of the stress field existing prior to crack extension, i.e. in terms of k_i, $\partial k_i/\partial L_0$, etc., $i = 1, 2$. Then the crack extension length ℓ is obtained from (3.7), by requiring that $K_I = K_c$, for a given load parameter.

It is more fundamental to use the maximum energy release criterion. To this end let G be the energy release rate at the tip of the extended crack. Then, in view of what has been said, we must have

$$G = C_0 + C_1\ell^{1/2} + C_2\ell + C_3\ell^{3/2} + \dots . \tag{3.9}$$

Again, coefficients C depend on the stress field prior to crack extension as well as on ω, a, b, ..., which define the crack path. In accordance to the Gibbs stability statement, we obtain parameters ω, a, b, ... in such a manner that $\int_0^\ell G \, d\ell$ is maximized. Then the crack extension length ℓ is obtained so that $G = G_c$ at its tip.

3.3 Crack Kinking and Irwin's Formula

Irwin's formula gives the energy release rate at the tip of a vanishingly small kink in terms of the corresponding stress intensity factors by

$$G = \frac{1-\nu^2}{\overline{E}} (K_I^2 + K_{II}^2) . \tag{3.10}$$

The validity of this equation has been questioned in the literature [19]. However, in a recent paper, Hayashi and this writer [24] have shown that this equation holds for all kink angles, provided that the stress intensity factors are those that exist at the tip of a vanishingly small kink. To this end we formulated the kinked extension in terms of a continuous distribution of dislocations, and then solved the corresponding integral equations by the collocation method of Erdogan and Gupta. Our formulation differs from that of Lo [27], in the sense that we use the square-root singularity at the tip of the extended crack, as does Lo, but an appropriate weaker singularity at the root of the extended crack, for which Lo uses, again, a square-root singularity as an approximation. Therefore we are able to consider very small crack extensions.

In an earlier paper, Hayashi and I [23] obtained a second-order perturbation solution, using exactly the same formulation. This perturbation solution yields the following explicit results for the stress intensity factors:

$$K_I = k_1(1 - \frac{3}{8}\omega^2) + k_2(-\frac{3}{2}\omega) + k_2(-\frac{8\sqrt{2}}{\pi}\ell^{1/2}\omega^2)$$

$$+ O(\omega^3),$$

$$K_{II} = k_1(\frac{1}{2}\omega) + k_2(1 - \frac{7}{8}\omega^2) + k_2(\frac{8\sqrt{2}}{\pi}\ell^{1/2}\omega) + O(\omega^3), \tag{3.11}$$

which are in complete agreement with those obtained by Wu [29] using a different approach. For a vanishingly small kink, Irwin's formula gives

$$G = \frac{1-\nu^2}{E}(k_1^2 + k_2^2 - 2k_1k_2\omega - \frac{1}{2}k_1^2\omega^2) . \tag{3.12}$$

This equation yields relatively accurate results for the energy release rate, up to a fairly large value for the kink angle. Note that from the maximum energy release rate criterion the critical kink is obtained from (3.12) to be

$$\omega_c = -2\frac{k_2}{k_1} . \tag{3.13}$$

ACKNOWLEDGMENT

This work has been supported by National Science Foundation Grant No. ENG 77-22155 to Northwestern University.

REFERENCES

1. NEMAT-NASSER, S. - Variational Methods for Analysis of Stability of Interacting Cracks, Proceedings of IUTAM Symposium on Variational Methods in the Mechanics of Solids, Evanston, Ill., Sept. 11-13, 1978, Ed. Nemat-Nasser, S., Pergamon Press, to appear.

2. NEMAT-NASSER, S. - Overview of the Basic Progress in Ductile Fracture, (Invited Lecture), Transactions 4th International Conference on Structural Mechanics in Reactor Technology, Eds. Jaeger, T. A., and Boley, B.A., Vol. L: Inelastic Analysis of Metal Structures, paper L2/1, (see Sec. 5) (1977).

3. NEMAT-NASSER, S. - Stability of System of Interacting Cracks, Letters in Applied Engineering and Science, 16 (1978) 277-285.

4. NEMAT-NASSER, S., KEER, L. M., and PARIHAR, K.S. - Unstable Growth of Thermally Induced Interacting Cracks in Brittle Solids, Int. J. of Solids and Struct. 14 (1978) 409-430.

5. KEER, L.M., NEMAT-NASSER, S., and ORANRATNACHAI, A. - Unstable Growth of Thermally Induced Interacting Cracks in Brittle Solids: Further Results, Int. J. Solids and Struct. 15 (1979) 111-126.

6. KEER, L. M., NEMAT-NASSER, S., and PARIHAR, K.S. - Growth and Stability of Thermally Induced Interacting Cracks in Brittle Solids, Geothermal Energy Research Report, Northwestern University, Sept. (1976).

7. SUMI, Y., NEMAT-NASSER, S., and KEER, L.M. - A New Combined Analytical and Finite-Element Solution Method for Stability Analysis of the Growth of Interacting Tension Cracks in Brittle Solids, Int. J. Engrg. Sci. 18 (1980) 211-224.

8. NEMAT-NASSER, S., SUMI, Y., and KEER, L.M. - Unstable Growth of Tension Cracks in Brittle Solids: Stable and Unstable Bifurcations, Snap-Through and Imperfection Sensitivity, Int. J. Solids and Struct. to appear.

9. SUMI, Y., KEER, L.M., and NEMAT-NASSER, S. - Thermally Induced Radial Cracking in Fuel Element Pellets, submitted for publication.

10. NEMAT-NASSER, S. - Geothermal Energy: Extracting Heat from Hot Dry Rock Masses, Proceedings of the 16th Midwestern Mechanics Conference, Kansas State University, Sept. 19-21, (1979): Developments in Mechanics, Eds. Sinha, S.C. Kirmser, P.G. and Huang, C.L.D., Vol. 10, Kansas State

University, (1979) 285-297.

11. SUMI, Y., NEMAT-NASSER, S., and KEER, L.M. - On Stability and Postcritical Behavior of Interactive Tension Cracks in Brittle Solids, Earthquake Research and Engineering Laboratory Tech. Report. No. 80-4-27, Dept. Civil Engrg., Northwestern University, Evanston, Ill., April 1980.

12. ERDOGAN, F., and SIH, G.C. - On the Crack Extension in Plates under Plane Loading and Transverse Shear, J. Basic Engrg. 85 (1963) 519-527.

13. SIH, G.C. - A Special Theory of Crack Propagation, Mechanics of Fracture 1, Noordhoff (1972).

14. HUSSAIN, M.A., PU, S.L., and UNDERWOOD, J. - Strain Energy Release Rate for a Crack under Combined Mode I and Mode II, ASTM-STP-560 (1974) 55-63.

15. PALANISWAMY, K. and KNAUSS, W.G. - On the Problem of Crack Extension in Brittle Solids under General Loading, Mechanics Today 4, Ed. Nemat-Nasser, S., Pergamon Press (1978) 87-148.

16. GUPTA, G.D. - Strain Energy Release Rate for Mixed Mode Crack Problem, ASME Paper No. 76-WA/PVP-7.

17. WU, C.H. - Elasticity Problems of a Slender Z-Crack, J. of Elasticity 8 (1978) 183-205.

18. WU, C.H. - Maximum-Energy-Release-Rate Criterion Applied to a Tension-Compression Specimen with Crack, J. of Elasticity 8 (1978) 235-257.

19. WU. C.H. - Fracture under Combined Loads by Maximum-Energy-Release-Rate Criterion, J. Appl. Mech. 45 (1978) 553-558.

20. COTTERELL, B. and RICE, J.R. - Slightly Curved or Kinked Cracks, Int. J. of Fracture, to appear.

21. BANICHUK, N.V. - Determination of the Form of a Curvilinear Crack by Small Parameter Technique, Izv. AN SSSR, M.T.T. (Mechanics of Solids) 7 (1970) 130-137 (In Russian).

22. KARIHALOO, B.L., KEER, L.M., and NEMAT-NASSER, S. - Crack Kinking Under Nonsymmetric Loading, Engrg. Frac. Mech., to appear.

23. HAYASHI, K. and NEMAT-NASSER, S. - Energy Release Rate and Crack Kinking, Earthquake Research and Engineering Laboratory Tech. Rept. No. 80-1-23, Dept. Civil Engrg. Northwestern University, Evanston, Ill., January 1980; submitted for publication.

24. HAYASHI, K. and NEMAT-NASSER, S. - Energy Release Rate and Crack Kinking under Combined Loading, in preparation.

25. GOLDSTEIN, R.V., and SALGANIK, R.L. - Plane Problem of Curvilinear Cracks in an Elastic Solid, Izv. AN SSSR, M.T.T (Mechanics of Solids) 7 (1970), 69-82 (In Russian).

26. GOLDSTEIN, R.V., and SALGANIK, R.L. - Brittle Fracture of Solids with Arbitrary Cracks, Int. J. Fracture 10 (1974) 507-523.

27. LO, K.K. - Analysis of Branched Cracks, J. Appl. Mech. 45 (1978) 797-802.

28. KARIHALOO, B.L., KEER, L.M., NEMAT-NASSER, S., and ORAN-RATNACHAI, A. - Approximate Description of Crack Kinking and Curving, in preparation.

29. WU. C.H. - Explicit Asymptomatic Solution for the Maximum-Energy-Release-Rate Problem, Int'l. J. Solids & Struct. 15 (1979) 561-566.

A MODEL FOR CALCULATION OF CRACK GROWTH IN CONCRETE-LIKE MATERIALS

P E Petersson (I) P J Gustavsson (I)

SUMMARY

A calculation model is presented which is suitable for concrete and similar materials. By the model, which is based on fracture mechanics and the finite element method (FEM), it is possible to analyse the fracture process when a microcracked fracture zone develops in front of a crack. The fracture zone can be treated as a fictitious crack that can transfer stress and therefore the model is called the Fictitious Crack Model (FCM). The stress-transferring capability of the fictitious crack decreases when the crack width increases.

Calculation results are presented which indicate that the method of determining the critical stress intensity factor (K_c) from the fracture load at a three-point bend test on a notched specimen is strongly affected by specimen dimensions and notch depth. Also the influence of beam depth on the flexural tensile strength is analysed.

Finally the application of FCM to the shear fracture of reinforced concrete is discussed.

1 CRACK INITIATION AND CRACK GROWTH IN NON-YIELDING MATERIALS

When a notched specimen of a linear elastic material is subjected to load, the stress in front of the notch will, theoretically, approach infinity, see line a in Fig 1, but of course this is impossible for a real material. For a yielding material, a plastic flow due to shear deformations will take place in front of the crack tip and a probable stress distribution for this material is presented by line b in Fig 1 (Dugdale). For a non-yielding material, for example concrete, a zone of microcracks

---

(I) - Res.Ass. at Div of Building Materials, Lund Institute of Technology, Lund, Sweden.

will develop in front of the notch. This reduces the stress concentration and a stress-distribution according to line c in Fig 1 is probable for such a material.

For real materials the stress distribution in front of a crack-tip is affected by a fracture zone. However, if the length of the zone is small compared with the specimen dimensions and the crack depth then linear elastic fracture mechanics may be useful but otherwise other calculation methods have to be used.

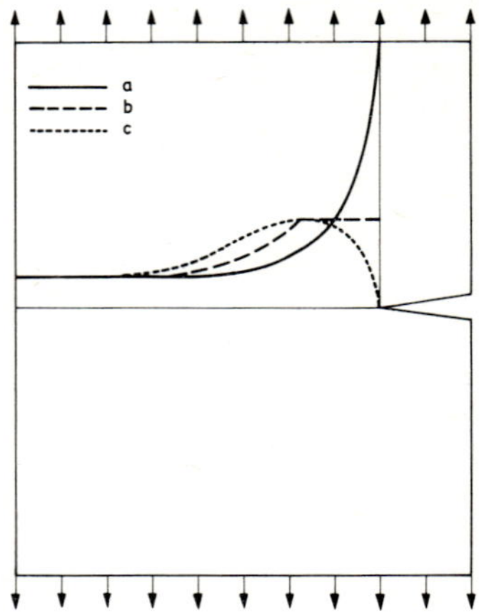

FIG 1. Probable stress distribution in front of a notch for a
a) linear elastic material
b) yielding material
c) non-yielding material.

In /1, 2/ strain measurements are carried out in order to study strains in front of notches in large concrete specimens subjected to load. The results of these tests indicate that microcracks develop hundreds of millimeters in front of the notch before the maximum load is reached. This implies that the fracture zone is never small compared with normal specimen dimensions and also that linear elastic fracture mechanics cannot be used for such a material. In order to carry out relevant calculations one has to consider the properties of the fracture zone.

The discussion in this paper mainly deals with the opening mode, mode I. In this mode the fracture zone develops in a tensile stress field perpendicular to the crack and consequently it is possible to obtain a great deal of information about the fracture zone from the stress-strain curve derived from a stable, direct tensile test, see Fig 2.

In Fig 2 the stress is continuously increasing up to the maximum load and therefore the material must be uniformly strained on this part of the curve. The non-linearity of the increasing part of the curve is due to microcracks. When the stress reaches its maximum value, one cross-sectional area cannot carry more load and it is fair to assume that the development of microcracks will be concentrated to a small material volume close to this cross sectional area when the specimen becomes further deformed. The strain of the material outside this small microcracked material volume, or fracture zone, decreases elastically when the stress diminishes.

For a yielding material a local reduction of the cross sectional area takes place when the tensile strength is reached. This fracture zone, or necking, is due to shear deformations, which are greatly affected by the state of stress; plane stress or plane strain. This means that the properties of the fracture zone of a yielding material are affected by specimen dimensions. For a non-yielding material, the difference between plane stress and plane strain is very small and therefore it is fair to assume that the properties of the fracture zone of this type of material are almost unaffected by the cross sectional dimensions of the specimen.

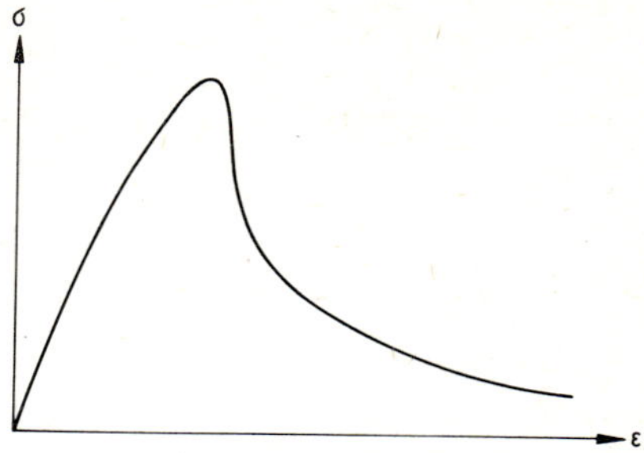

FIG 2. A stable tensile stress-strain curve for concrete. σ= stress, ε= relative strain.

When a fracture zone starts developing the stress decreases and consequently no more fracture zones will be created. As the properties of the fracture zone are unaffected by specimen dimensions, this means that the decreasing part of the σ-ε-curve in Fig 2 is not only related to material properties but also to specimen length. A better way to describe the deformation properties of a non-yielding material therefore is to use two relations; one relation between stress and relative strain up to the maximum stress, see Fig 3a, and one relation between stress and the absolute deformation of the fracture zone after the maximum stress is reached, see Fig 3b. These relations are unaffected by specimen dimensions and therefore they can be considered as material properties.

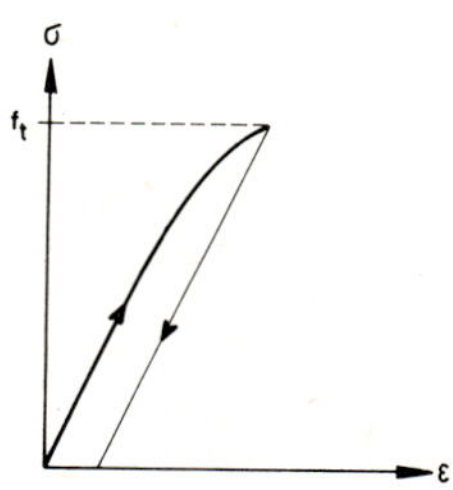

FIG 3a. A relation between stress and realtive strain (ε) up to the maximum load (σ-ε-curve).

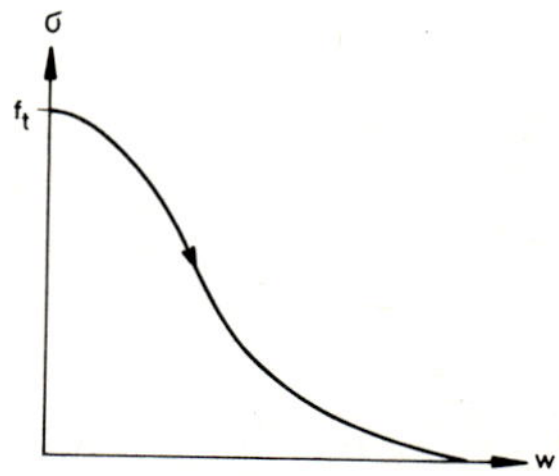

FIG 3b. A relation between stress and absolute deformation (w) of the fracture zone (σ-w-curve).

The curve in Fig 3b decribes the deformation properties of the fracture zone in direct tension but the same relation ought also to be relevant for the fracture zone in front of a mode I crack /10/. The properties of the fracture zone probably are slightly different in a two- or three-dimensional stress field but often it is sufficient to know the one-dimensional properties. A fracture mechanical model by which it is possible to take the relations in Fig 3 into consideration, is described below.

2 THE FICTITIOUS CRACK MODEL

To describe complicated fracture processes one has to use approximative models. In the model described below the following assumptions are made.

* The fracture zone starts developing at one point when the first principle stress reaches the tensile strength. Of course other more complex fracture criterions can be used but often a simple criterion is satisfactory.
* The fracture zone develops perpendicular to the first principle stress.
* The width of the fracture zone in the stressed direction equals the widening of the zone, i.e. the width of the zone is zero when it starts developing. For non-yielding materials this is a fair assumption.
* The stress-transferring capability of the fracture zone always depends on its widening according to the σ-w-curve in Fig 3b. This means that the amount of energy necessary to create one unit of area of a crack is always constant and equals the fracture energy, G_c.
* The properties of the materials outside the fracture zone are given in the σ-ε-curve in Fig 3a.

These assumptions make it possible to describe the real fracture process, Fig 4a, in a simplified way according to Fig 4b. In the model the fracture zone is treated as a tied crack, i.e. a crack that has a stress-transforning capability. As a tied crack is not a real crack but a fictitious crack, the model is called the Fictitious Crack Model (FCM).

Tied crack models have been used before by Dugdale /3/ and Barenblatt /4/. In the Dugdale model, which is suitable for yielding materials, the attractive stress across the fictitious crack is always constant and equals the yield stress. Consequently this model is a special case of the fictitious crack model. In the Barenblatt model the attractive stress is caused by cohesive forces and therefore the fracture zone in this model is of another magnitude than it normally is in Fictitious Crack Model.

The Fictitious Crack Model normally cannot be treated analytically but numerical methods have to be used. The most suitable method seems to be the Finite Element Method (FEM). When FEM is used the stresses acting across the fictitious crack are replaced by nodal forces, see Fig 4c. These forces decrease continuously according to the σ-w-curve when the width of the fictitious crack increases. When the tensile strength is reached in the top node, see Fig 4c, this node is "opened" and forces start acting across the crack in this point. In this way it is possible to follow the crack growth through the material. By using this model it is possible to analyse the formation of cracks in an uncracked material and also the crack growth. The model is discussed in detail in /5,6/.

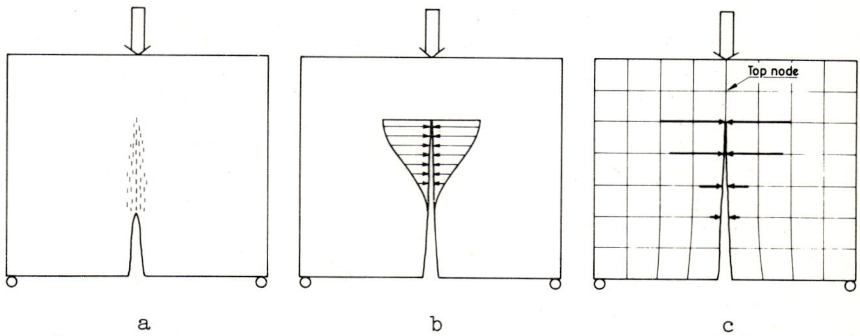

FIG 4a. The fracture zone in a real material.
4b. In the fictitious crack model the fracture zone is represented by a tied crack, i.e. a crack that can transfer stress.
4c. FEM-representation of the fictitious crack model

3 MATERIAL PROPERTIES

When FEM is used it is convenient to approximate the σ-ε-curve and the σ-w-curve in Fig 3. Normally the σ-ε-curve for non-yielding materials is almost linear up to the maximum stress and therefore a good method of approximation is to replace the curve with one straight line, see Fig 5a. The position of this line is defined by the tensile strength (f_t) and the Young's modulus (E).

To obtain the true, descending σ-w-curve one has to carry out stable, direct tensile tests. This is difficult for non-yielding materials and to be successful one has to use very stiff testing machines and small specimens. Only a few results are given in the literature /7,8,9/. This means that one normally has to assume a shape of the σ-w-curve. For plain concrete a good approximation seems to be one straight line, see line 1 in Fig 5b. The position of the line is defined by the tensile strength and the area under the curve. This area represents the amount of energy necessary to create one unit of area of a crack and consequently equals the fracture energy (G_c). For fibre-reinforced concrete an approximation according to Dugdale see line 2 in Fig 5b, is often better. The position of this line is also defined by f_t and G_c.

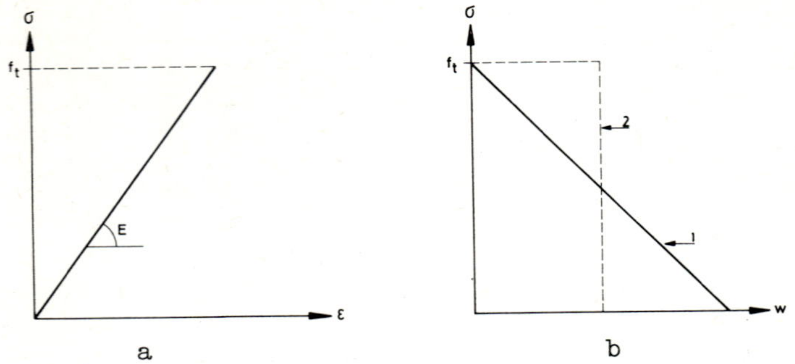

FIG 5a. A straight-line approximation of the σ-ε-curve for non-yielding materials.
5b. Straight-line approximations of the σ-w-curve for plane concrete (line 1) and fibre-reinforced concrete (line 2).

The only parameters necessary to carry out calculations by using the Fictitious Crack Model are f_t, E and G_c. In /10, 11/ suitable methods are presented for the determination of these material parameters.

Another material parameter used in this paper is the characteristic length (l_{ch}) defined by:

$$l_{ch} = G_c E / f_t^2 \qquad (1)$$

l_{ch} can be considered as a measure of the "brittleness" of a material; the lower the value l_{ch} is, the more "brittle" the material is. A normal l_{ch}-value for concrete is 200 mm.

4 CONVERGENCE OF FICTITIOUS CRACK MODEL CALCULATIONS

In fracture mechanical FEM analysis in which the softening of the material is not taken into account, the calculated stress in front of the crack tip will grow unlimited as the element mesh is refined. If a stress criterion is used for crack propagation the calculations will thus not converge, i.e. computational results will be strongly dependent on the element size chosen. As pointed out in /12/ this has been neglected in many calculations of the load carrying capacity of concrete structures.

When using the Fictitious Crack Model convergence is attained, which is demonstrated by an example in Table 1. In this calculation the fracture zone was modeled by short bar elements with negative stiffness and the continuum by rectangular plate elements according to Turner et al /13/.

Elements in fracture zone	Tot number d o f	σ/f_t	Deviation
1	132	1.401	3.2%
2	286	1.375	1.3%
4	546	1.365	0.6%
20	2652	1.357	-

Table 1. Calculated load (F) necessary to create a 0.2 l_{ch} long fictitious crack in a beam in three point bending. $d/l_{ch}=1.0$, $d/l=0.3$, $\sigma=3F/2bd^2$. (d=beam depth, b=beam width, l=beam length)

As the material model relates stress not only to strain, a dimensionsless quantity, but also to the absolute widening of the fracture zone, a large number of elements have to be used when analysing large structures. The number of elements necessary to achive a certain degree of accuracy is dependent on the quotient d/l_{ch}, in which d is some characteristic dimension of the structure.

5 CALCULATION RESULTS

5.1 Critical Stress Intensity Factor (K_c)

Since 1960, when Kaplan /14/ carried out fracture mechanical tests on concrete, many researchers have tried to determine the critical stress intensity factor (K_c) for concrete. Normally one has used linear elastic fracture mechanical relations to calculate K_c from the fracture load at a bend test carried out on a notched beam. As concrete is not a linear elastic material this value of K_c, here denoted as K_c', is an approximation. By using the Fictitious Crack Model it is possible to calculate the difference between K_c' and the real K_c ($K_c=\sqrt{G_cE}$) and in Fig 6 the ratio of K_c'/K_c is shown as a function of the beam depth (d). The ratio of G_c'/G_c is also shown, where $G_c' = (K_c')^2/E$, in the Figure. The real value of G_c corresponds to the area under the σ-w-curve. The curves in Fig 6 are relevant for a three-point bend test on a notched beam with a notch-depth/beam-depth ratio of 0.5. The σ-w-curve is approximated to one straight line.

A normal value of l_{ch} for concrete is 200 mm. Normally one has used specimens with beam depths between 50 and 300 mm when determining K_c'. These values correspond to d/l_{ch}-values between 0.25 and 1.5 and consequently K_c' then becomes only 35-65% of the the real K_c and G_c' only 15-40% of the real G_c. In order to obtain relevant results one has to use specimens with beam depths greater than 10 l_{ch}, i.e. for concrete about 2 m. The results presented in Fig 6 are confirmed by the results in /1,2,11/.

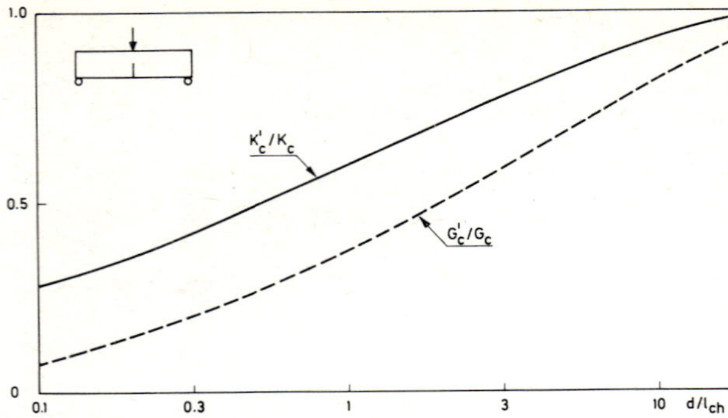

FIG 6. K_c'/K_c and G_c'/G_c as functions of d/l_{ch} for a three-point bend test on a notched beam.

The notch-depth/beam depth -ratio (a/d) also affects the K_c'-value. In Fig 7 K_c'/K_c as a function of a/d is shown for one quality of particle board. The l_{ch}-value of this material is about 100 mm and is of the same magnitude as the concrete value. For this material it is easy to determine the shape of the σ-w-curve and at the calculations this curve is used. Laboratory test results (dots) are also presented in the Figure. As can be seen the test results are in good agreement with the calculated results.

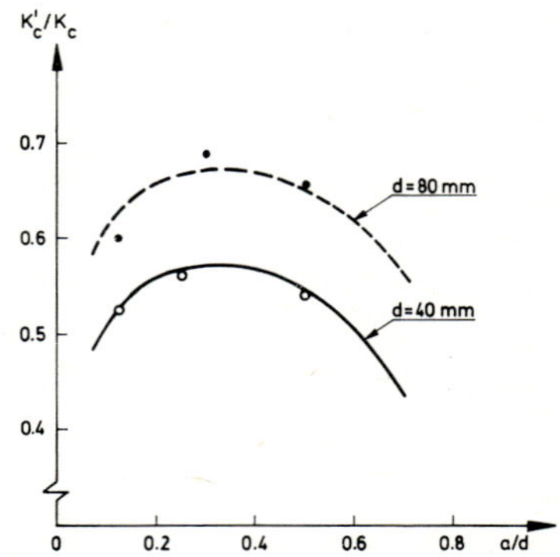

FIG 7. K_c'/K_c as a function of a/d for one quality of particle board. The dots represent test results.

5.2 Flexural Tensile Strength

The failure of non-yielding materials is always caused by cracks. This means that fracture mechanics must often also be used for analysing unnotched specimens.

One frequently used test that is carried out on unnotched specimens is the determination of the flexural tensile strength in a three-point bend test. Results from many investigations, for example /15,16/, indicate that the flexural tensile strength is strongly affected by specimen depth, but no one has been able to give a satisfying explanation for this. By the Fictitious Crack Model it is possible to analyse the bend test and in Fig 8 it is shown how the ratio flexural tensile strength/tensile strength (f_f/f_t) depends on beam depth. In the calculations the σ-w-curve is approximated to one straight line. The upper curve in the Figure represents beams subjected to no shrinkage stresses, i.e. wet specimens, while the lower curve represents beams with a certain degree of shrinkage stresses. In the Figure test results are also presented and the agreement between the calculated curves and the test results is good, in spite of the simplified assumptions. The curves in the Figure confirm earlier results and they also indicate that the shrinkage stresses increase the effect beam depth has on the flexural tensile strength.

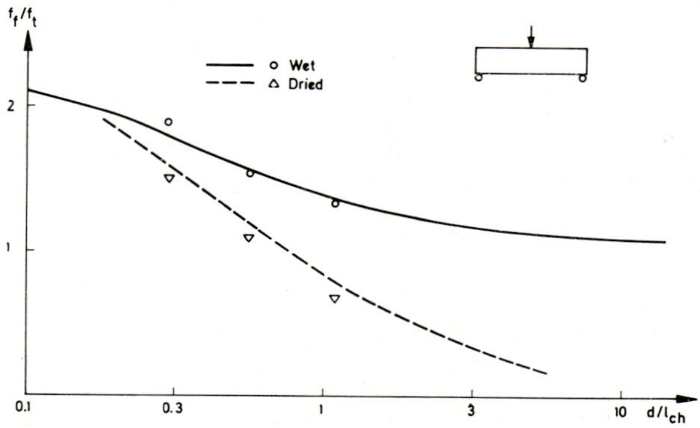

FIG 8. f_f/f_t as a function of d/l_{ch}.

5.3 Notch Sensitivity

Notch sensitivity is a parameter often used in the discussion about fracture mechanics and concrete. The parameter can be defined as f_{net}/f_f, see Fig 9a and b.

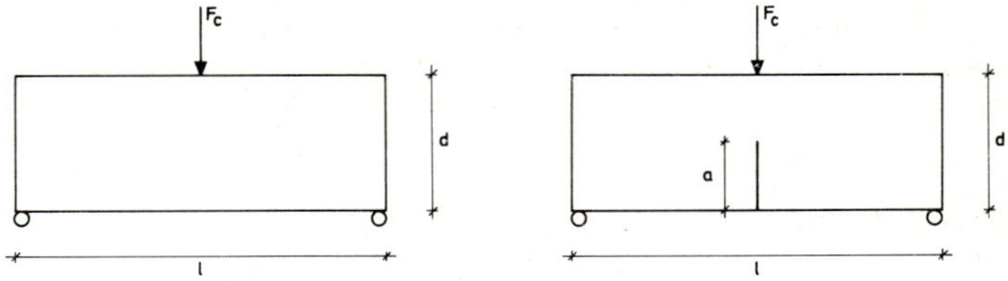

FIG 9a. The flexural tensile strength (f_f)=

$$= \frac{3\ F_c\ l}{2b\ d^2}$$

(b=beam width)

FIG 9b. The net flexural tensile strength (f_{net})=

$$= \frac{3\ F_c\ l}{2b(d-a)^2}$$

It is possible to study the notch sensitivity of different materials by using the Fictitious Crack Model. In Fig 10 results are presented that show how the notch sensitivity depends on notch depth/beam depth and the dimensionsless parameter parameter d/l_{ch}. In the calculations the σ-w-curve is approximated to one straight line.

Many researchers have stated that concrete is a notch-insensitive material. However, according to Fig 10, notch sensitivity is not a real material property, but it is strongly affected by specimen dimensions. For example a concrete beam with a depth of 50 mm ($d/l_{ch} \approx 0.25$) is almost notch insensitive while a concrete beam with a depth of 800 mm ($d/l_{ch} \approx 4$) is significantly affected by a notch. The l_{ch}-value of cement paste is about 10 mm and consequently this material is always notch sensitive for normally used dimensions. The results in Fig 10 also imply that for notched beams with d/l_{ch}-values less than about 1 a usable fracture criterion is $f_{net} = f_f$. When using this simple criterion it should be remembered that f_f depends on specimen dimensions and fracture mechanical properties of the material, see Fig 8.

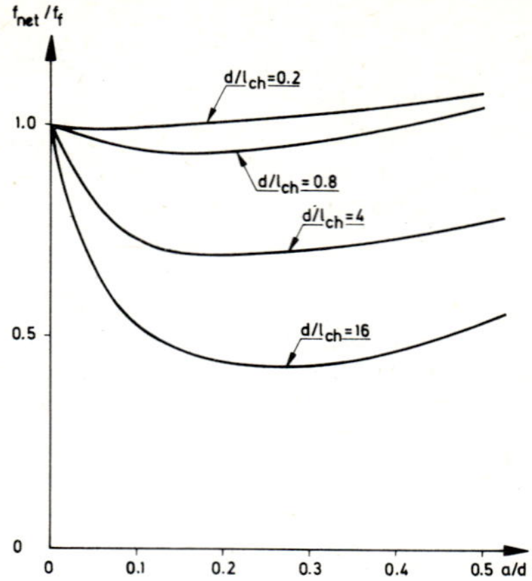

FIG 10. Notch sensitivity (f_{net}/f_f) as a function of a/d for different values of d/l_{ch}.

6 THE SHEAR FRACTURE OF REINFORCED CONCRETE BEAMS

Considering a beam without shear reinforcement, the fracture mechanical properties of the concrete is decisive to the fracture load as soon as any crack or fracture zone has begun to develop. And as such zones always can be expected in the concrete due to inhomogeneity, shrinkage or bending, a fracture mechanics approach to the analysis of the shear fracture of beams with only longitudinal bending reinforcement seems natural.

The fracture mechanic FEM calculation presented below is with some modifications based on the material model previously described. The concrete is assumed to behave in a linear elastic manner until the fracture criterion of Coulomb-Mohr (with tension cut off) is reached in any of the nodal points along an à priori chosen crackpath. When reaching this criterion a fictitious crack is introduced along the finite element edge in front of the nodal point perpendicular to the first principle stress. The reinforcement is assumed to be linear elastic, and a layer modelling the interaction between steel and concrete (the bond-slip-stress relation) is assumed to be ideal elastic-plastic. The crack path is chosen from the principle stress directions of the unfractured beam, and shear force carried by dowel action and aggregate interlock is not taken into account.

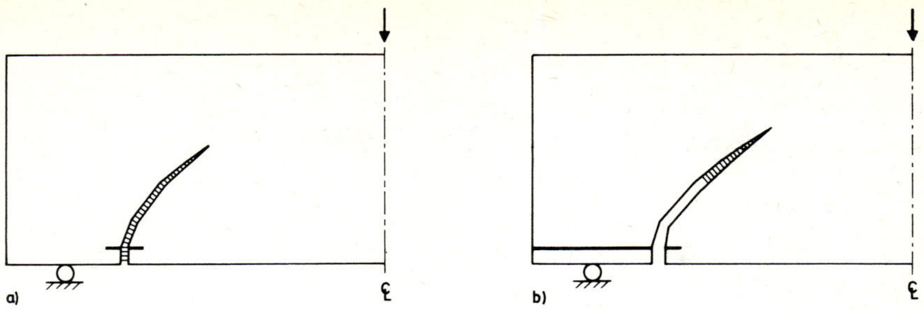

FIG 9. Tied crack, open crack and slip between reinforcement and concrete at loads corresponding to points a) and b) in Fig 10.

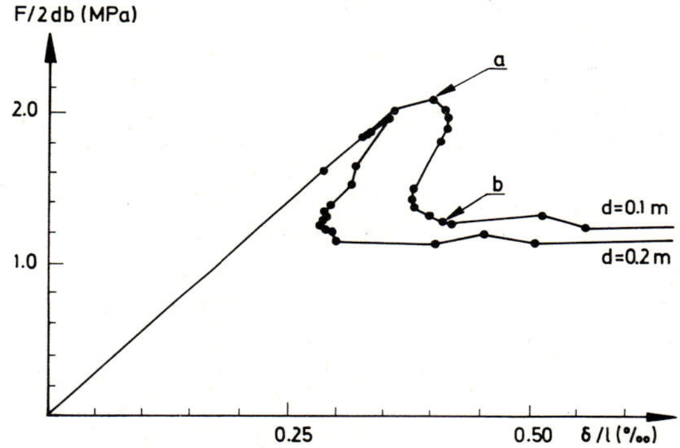

FIG 10. Load vs deflection for two beams of identical shape but different size.

Fig 9 and Fig 10 show results from an analysis of a beam loaded in three point bending and with the following characteristics: $d \times l = 0.1 \times 0.3\ m^2$, $f_c = 10f_t = 30$ MPa, $G_c = 80$ N/m, $E_c = E_s/7 = 30000$ MPa, 1% reinforcement ϕ 6 mm, and plasticity between steel and concrete at bond stress 3.0 MPa and slip 0.008 mm. Fig 10 also shows load vs deflection for a beam with all geometric characteristics doubled, i.e. d = 0.2 m etc. The load vs deflection curves would coincide if the calculations were based on a material model which related stress only to strain.

At the point of maximum load no open, i.e. not force transferring, crack had opened in neither beam. Such cracks developed at increased deflection and final collaps occured due to slip between reinforcement and concrete. The calculated simultaneous decrease in load and deflection is due to neglection of dynamic forces at instable crack propagation.

REFERENCES

1. ENTOV, V. M. and YAGUST, V. T. - Experimental Investigation of Laws Governing Quasi-static Development of Macrocracks in Concrete. Mechanics of Solids, Vol. 10, No. 4, pp. 87-95, 1975.

2. SOK, C. and BARON, J. - Mecanique de la Rupture Appliquee au Beton Hydraulique. Cem. and Concr. Res., Vol. 9, No. 5, pp. 641-648, 1979.

3. DUGDALE, D. S. - Yielding of Steel Sheets Containing Slits. J. Mech. Phys. Solids, Vol. 8, pp. 100-104, 1960.

4. BARENBLATT, G. J. - The Mathematical Theory of Equilibrium Cracks in Brittle Fracture. Advances in Appl. Mech., Vol. 7, pp. 55-125, 1962.

5. HILLERBORG, A., MODÉER, M. and PETERSSON, P. E. - Analysis of Crack Formation and Crack Growth in Concrete by Means of Fracture Mechanics and Finite Elements. Cem. and Concr. Res., Vol. 6, No. 6, pp. 773-781, 1976.

6. MODÉER, M. - A Fracture Mechanics Approach to Failure Analyses of Concrete Materials, Report TVBM-1001, Division of Building Materials, University of Lund, Sweden, 1979.

7. HUGHES, B. P. and CHAPMAN, G. P. - The Complete Stress-Strain Curve for Concrete in Direct Tension. Bulletin RILEM, No. 30, pp. 95-97, 1966.

8. EVANS, R. H. and MARATHE, M. S. - Microcracking and Stress-Strain Curves for Concrete in Tension. Materiaux et Construction, No. 1, pp. 61-64, 1968.

9. HEILMAN, H. G., HILSDORF, H. and FINSTERWALDER, K. - Festigkeit und Verformung von Beton unter Zugspannungen. Deutscher Ausschuss für Stahlbeton, Heft 203, 1969.

10. PETERSSON, P. E. - Fracture Energy of Concrete. Method of Determination. Cem. and Concr. Res., Vol. 10, No. 1, pp. 78-89, 1980.

11. PETERSSON, P. E. - Fracture Energy of Concrete. Practical Performance and Experimental Results. Cem. and Concr. Res., Vol. 10, No. 1, pp. 91-101, 1980.

12. BAŽANT, Z. and CEDOLIN, L. - Fracture Mechanics of Reinforced Concrete. Structural Engineering Report No. 79-9/64 Dep. of Civ. Eng., Northwestern University, Evanston, Illinois, USA, 1979.

13. TURNER, M. J. et al. - Stiffness and Deflection Analysis of Complex Structures. J. of Aeronautical Sci., Vol. 23, pp. 805-823, 1956.

14. KAPLAN, M. F. - Crack Propagation and the Fracture of Concrete. J. of Am. Concr. Inst., Vol. 58, pp. 591-609, 1961.

15. MAYER, H. - Die Berechnung von Durchbiegung von Stahlbetonbauteilen. Deutscher Ausschuss für Stahlbeton, Heft 194, 1967.

16. SABNIS, G. and MIRZA, S. M. - Size Effects in Model Concretes. ASCE, J. of the Struc. Div, pp. 1007-1020, June 1979.

NUMERICAL STUDIES OF ROCK FRAGMENTATION(i)

Marlin E. Kipp(ii) and Dennis E. Grady(ii)

Sandia National Laboratories,(iii) Albuquerque, NM 87185

SUMMARY

The dynamic fracture process in rock has been modelled as a continuous accrual of damage, where the damage is defined to be the volume fraction of material that has been stress relieved by multiple crack growth. A rate equation, governing the growth of the damage, has been formulated based upon a microstructural analysis of crack growth rates and Weibull crack distribution concepts. For a prescribed tensile loading history, damage levels and fragment sizes can be obtained. In most geometries of explosive applications, there are large spatial variations in loading histories which lead to a variation in fragment sizes. Using a two-dimensional finite-difference wave-propagation code and fracture parameters for oil shale, calculations of several configurations of explosives illustrate the resulting damage to surrounding rock.

INTRODUCTION

A requirement common to a wide range of resource recovery applications is the efficient fragmentation of geologic materials. Current methods of explosive rock breakage rest heavily upon traditions of mining experience. As resources become less accessible, the means to evaluate the effectiveness of conventional techniques is mandatory. The potential for computational evaluation of new explosive methods is practically and

---

(i) This article sponsored by the U. S. Department of Energy under Contract DE-AC04-DP00789.

(ii) Member of Technical Staff, Sandia National Laboratories, Albuquerque, New Mexico, USA.

(iii) A U. S. Department of Energy Facility.

economically attractive, but such studies demand a thorough understanding of the rock fracture processes and physically sound computational models.

In an explosive event, the early-time blast-induced stress wave in the rock is the primary factor that determines the extent of fracture [1]. Typical explosive charges induce divergent tensile strain rates ranging from 10-1000/s, whereas free-face wave interaction results in tensile strain rates up to 10^4/s. Rock fracture strength may vary by an order of magnitude over these strain rates [2]. The existence of an inherent distribution of fracture-producing flaws [3] has been shown to be the basis for the observed variation of tensile strength with strain rate [4,5,6].

Instead of explicitly treating individual crack response to wave impingement [7], the dynamic fracture process is more conveniently modelled as a time-dependent accrual of damage, where the level of damage characterizes the current load-carrying capability of the rock. The concept of continuum measures of damage to model dynamic fracture, particularly metal spallation, has been addressed by many authors, including Tuler and Butcher [8], Davison and Stevens [9], Seaman, Curran, and Shockey [10], Cochran and Banner [11], and Davison, Stevens, and Kipp [12].

The concept of continuum damage has also been used as a measure of fracture in rocks [10,13]. The present work is based on a model of dynamic fracture which explicitly incorporates concepts of inherent fracture-producing flaws in rock and observed strain-rate effects in rock fracture.

The complexity of the waves that are generated by explosive charges, and the appearance of relief waves from regions that have fractured, necessitate the use of wave-propagation codes to address realistic geometries. The codes numerically integrate the conservation equations of mass, momentum, and energy, along with the constitutive equations for the materials. Using model parameters for the fracture of oil shale, several calculations are described which illustrate the fracture resulting from explosive loading.

FRACTURE AND FRAGMENTATION MODEL

When a rock mass responds to an applied load by fracturing, the size distribution of the resulting fragments is closely linked to the nature of the loading conditions. For example, in a medium grade of oil shale, the average fragment size is observed to vary with strain rate as illustrated in Figure 1. At low strain rates (quasi-static loading) large fragments form, because only a few cracks dominate the fracture process. At the other extreme developed under plate impact conditions, tensile strain rates of 10^4-10^5/s develop

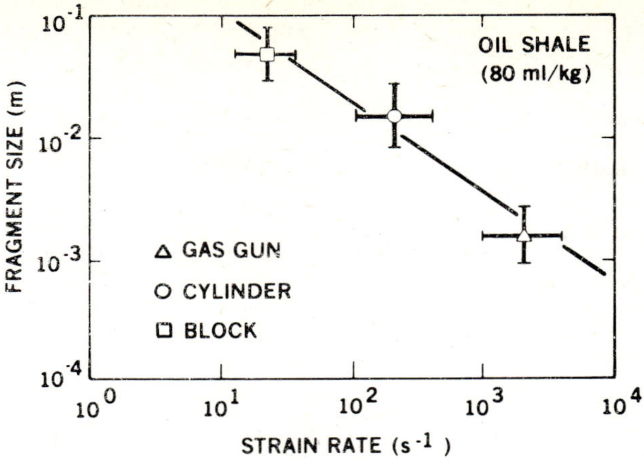

Figure 1. Fragment size dependence on strain rate.

for a very short time duration. As a result, tiny fragments are formed by the interaction of many small cracks growing simultaneously. The overall trend of fragment size as a function of strain rate portrayed in Figure 1 is of particular importance in blasting situations, since the rock surrounding an explosive charge may experience the entire range of these strain rates. The tensile strain rate due to the divergence of the explosive-induced shock wave may be 10^2-10^3/s near the source, and decay to 10^0-10^1/s at large distances. The presence of a nearby free surface may lead to strain rates of 10^3-10^4/s. Consequently, one expects to observe fine fragments near the explosive charge, and fragment size increasing with distance from the charge. Fine fragments are also expected near a free surface.

Although the prediction of the spatial distribution of fragments is the major goal of the numerical simulation, the competent rock must be modelled so that the change in load-carrying ability is taken into account as the fracture state progresses. To model dynamic fracture and the subsequent time-dependent strength of the rock, it is necessary to consider the tensile fracture stress behavior. The fracture stress as a function of the loading strain rate determined by several experimental methods for oil shale is shown in Figure 2. The grade of oil shale is the same as that in Figure 1. At low strain rates, the fracture stress is small, because only a few large cracks are participating. At large strain rates, the fracture stress is larger by an order of magnitude over the quasi-static values. This variation in strain rate has been shown [6,14] to directly reflect the presence of cracks in the rock, based on linear elastic dynamic fracture mechanics considerations.

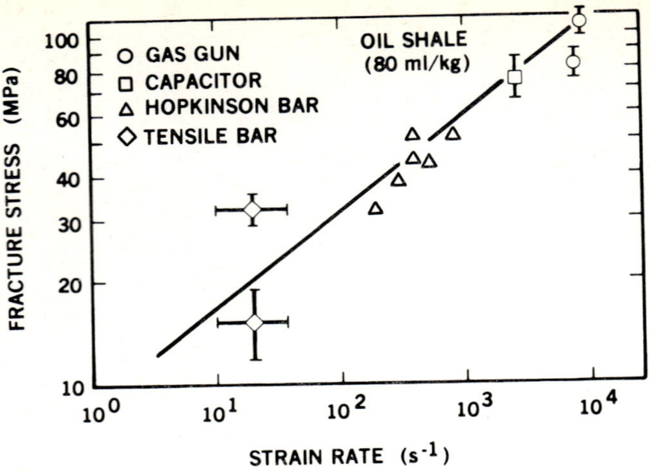

Figure 2. Fracture stress dependence on strain rate.

The model that will be described here explicitly utilizes the experimental fracture results summarized in Figures 1 and 2. As a measure of the current state of fracture, the damage, D, is defined as the volume fraction of material that has lost its load-carrying capability. The damage varies from zero, initially, to a maximum of 1. In order to degrade the material strength, both the bulk modulus, K, and the shear modulus, G, are isotropically reduced by a factor (1-D) [15]. In the linear elastic relations for tension, the stress, σ_{ij}, is related to the strain, ϵ_{ij}, by

$$\sigma_{ij} = K(1 - D)\epsilon\ \delta_{ij} + 2G(1 - D)(\epsilon_{ij} - \frac{1}{3}\ \epsilon\ \delta_{ij}) \ , \qquad (1)$$

where

$$\epsilon = \epsilon_{ii} \ . \qquad (2)$$

The number of flaws per unit volume, n, that are active at a given strain level is assumed to be described by a Weibull distribution,

$$n(\epsilon) = k\ \epsilon^m \ , \qquad (3)$$

where k and m are constants. The actual rate of flaw activation, $\dot{N}$, is the rate obtained by time-differencing Equation (3), multiplied by a factor (1 - D), to account statistically for those flaws that will be shadowed by other flaws:

$$\dot{N} = n'(\epsilon)\dot{\epsilon}(1 - D) \ . \qquad (4)$$

A spherical region of influence, v(t), is assumed to exist around flaws which are growing at a constant velocity, C_g:

$$v(t) = \frac{4}{3}\pi(C_g t)^3 . \quad (5)$$

Hence, the total damage at time t is the convolution of the number of active cracks with the volume influenced by those cracks:

$$D(t) = \int_0^t \dot{N}(\tau)v(t - \tau)\, d\tau . \quad (6)$$

Rewriting to include Equations (3), (4), and (5) results in,

$$D(t) = \frac{4}{3}\pi C_g^3 km \int_0^t \varepsilon^{m-1}\, \dot{\varepsilon}(1 - D)(t - \tau)^3\, d\tau . \quad (7)$$

This integral defines the damage that will be used in Equation (1) to degrade the stresses as a strain history progresses.

In order to describe fragmentation, a measure of the average fragment size will be derived. Assuming the surface area of a fracture is a(t), where

$$a(t) = 2\pi(C_g t)^2 , \quad (8)$$

a total fracture area, A_f, may be written as

$$A_f(t) = \int_0^t \dot{N}(\tau)a(t - \tau)\, d\tau , \quad (9)$$

or

$$A_f(t) = 2\pi C_g^2 km \int_0^t \varepsilon^{m-1}\, \dot{\varepsilon}(1 - D)(t - \tau)^2\, d\tau . \quad (10)$$

In Equation (9), a mean value $\overline{a}(t)$ may be chosen so that

$$A_f(t) = \int_0^t \dot{N}(\tau)\overline{a}(t)\, d\tau = \overline{a}(t)N(t) . \quad (11)$$

Similarly, in Equation (6), a mean value $\overline{v}(t)$ may be chosen, resulting in

$$D(t) = \int_0^t \dot{N}(\tau)\overline{v}(t)\, d\tau = \overline{v}(t)N(t) . \quad (12)$$

If we approximate $\overline{a}(t)$ and $\overline{v}(t)$ using an average radius, $\overline{r}(t)$,

$$\bar{a}(t) = 2\pi\bar{r}^2 , \tag{13}$$

and

$$\bar{v}(t) = \frac{4\pi}{3}\bar{r}^3 . \tag{14}$$

When Equations (13) and (14) are included in Equations (11) and (12) respectively, the average radius, $\bar{r}$, may be eliminated to obtain

$$\frac{1}{N^{1/3}} = \frac{D^{2/3}}{A_f}(9\pi/2)^{1/3} . \tag{15}$$

Equation (15) represents a measure of the average crack spacing, and we will use this expression to approximate the mean fragment size, L_m,

$$L_m \doteq 1/N^{1/3} . \tag{16}$$

When Equations (1), (7), and (10) are specialized to the case of constant strain rate, the fracture stress and fragment size can each be expressed in terms of strain rate, so the data in Figures 1 and 2 can be used to evaluate the constants k, m, and C_g [5]. The constants are determined to be:

$$\begin{aligned} k &= 1.7(10^{27})/m^3 \\ m &= 8 , \\ C_g &= 1.3(10^3)m/s . \end{aligned} \tag{17}$$

Although k and m are not easy to justify physically, the crack growth velocity, C_g, is close to the shear wave velocity for this oil shale, and, hence, within the range of reasonable propagation velocities.

Simulating one-dimensional loading at a constant tensile strain rate of 10^2/s, Figure 3 shows representative stress and damage histories. The stress increases linearly until the rate of damage increase is sufficiently large to cause unloading. The maximum stress occurs at a level of damage of about 0.1, and that stress level reflects the fracture stress amplitude in the data in Figure 2, as it should.

This tensile fracture and fragmentation model was incorporated into a two-dimensional Lagrangian wave-propagation code, TOODY [16]. The compressive behavior of the oil shale was modelled as elastic-perfectly plastic, with initial density of 2.27 Mg/m^3, bulk sound speed of 3000 m/s, Poisson ratio of 0.4, and yield strength of 200 MPa. The damage and fracture surface area integral equations were reduced to

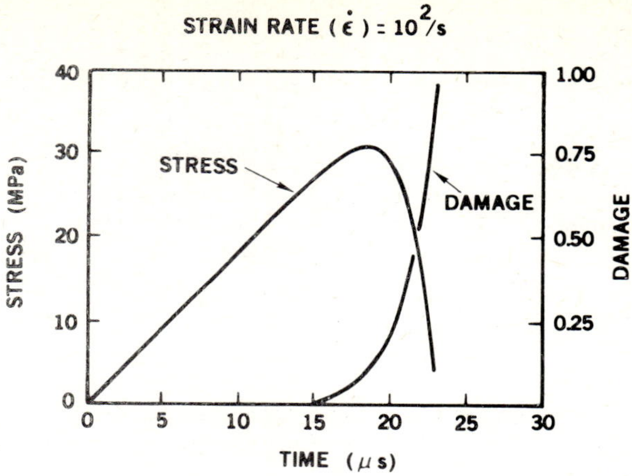

Figure 3. One-dimensional constant strain-rate stress and damage histories.

approximate ordinary differential rate equations to be integrated in time in the wavecode. The explosive energy source is treated as an isentropic release of energy in a gas from the Chapman-Jouguet point of the explosive [16].

NUMERICAL SIMULATIONS

One of the purposes of using wave-propagation codes in blasting simulations is to obtain insight into the early stress wave behavior that leads to the fracture of the rock. Such studies may lead to more efficient use of the available explosive energy. Four examples will be considered in this section: (a) Simulation of the detonation of a small explosive charge in a meter-sized block, (b) Damage variation with explosive burial depth (burden), (c) Damage variation with charge shape and detonation scheme, and (d) Simulation of an experiment in the floor of a mine.

(a) BLOCK EXPERIMENT. A block of oil shale, of nominally 1 m size, was prepared for an 80 gm charge of C-4 explosive to be placed 0.4 m from the nearest free face [17], as shown in Figure 4. The block was surrounded by an impedance-matching grout to minimize internal wave reflections. In the post-event analysis, the block was slabbed to determine fragment distributions.

A calculation made of this experiment permitted comparisons to be made of the fragment sizes. Figure 5 summarizes the average fragment size variation with distance from the explosive charge. The dotted bars are experimental ranges parallel to the free face, and the dotted line is the corresponding calculated size. The trends agree, in that the

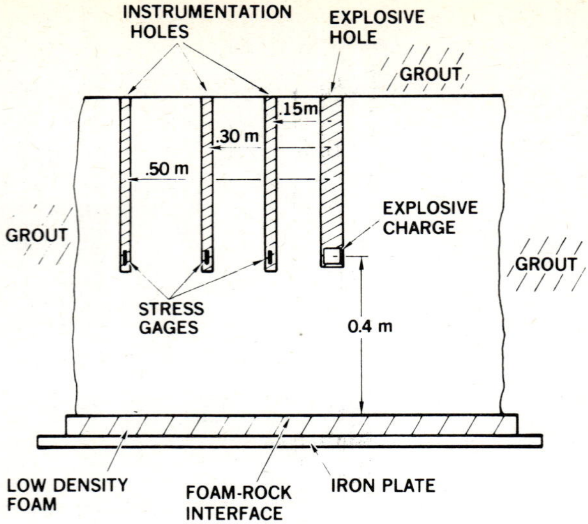

Figure 4. Experimental configuration of oil shale block.

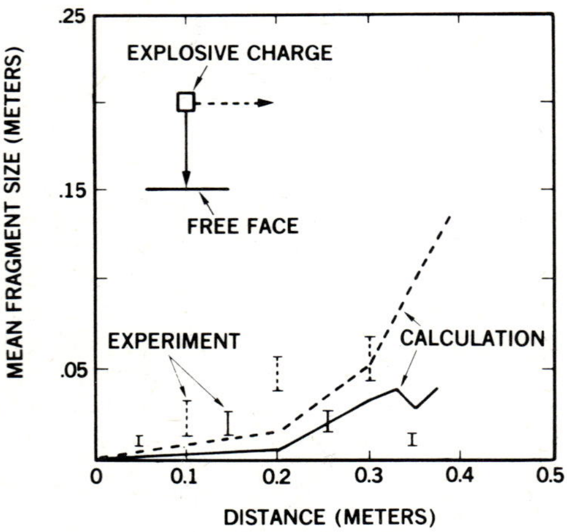

Figure 5. Calculated and experimental fragment sizes.

fragment size tends to increase with distance from the charge, as mentioned earlier, and the magnitude of the calculated sizes is about twice the measured values. The size variation between the explosive and the free face (solid bars and line) distinctly shows that the average size is smallest near the explosive and the free face--both regions in which the strain rates are large. At intermediate locations, the size increases, because of the lower tensile strain rate in this region.

(b) SPHERICAL CHARGES. The configuration in the previous example was taken as the starting point for a series of calculations to examine the effect of varying the burial depth of the explosive (burden, B). Figure 6 illustrates the region of oil shale that experienced damage levels greater than 0.2 from the 80 gm spherical charge. The volume encompassed by this damage contour was chosen to characterize the fractured

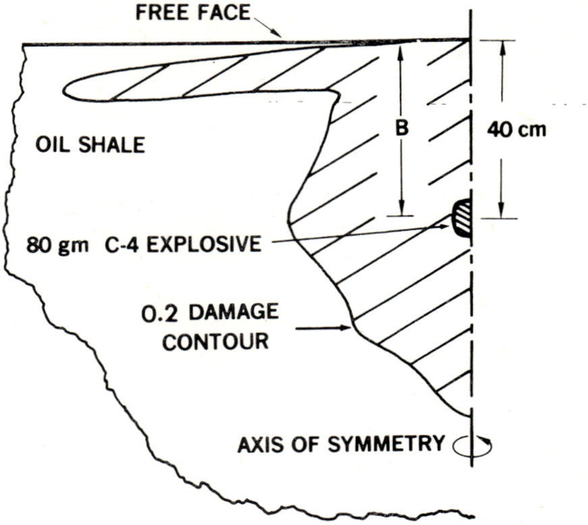

Figure 6. Damage configuration with charge 40 cm from free face.

region, which when normalized with the mass of the explosive is called the powder factor. As the burden is changed, the volume of material that is damaged is expected to change, since the relative size of the regions subjected to divergent and spall tensions will change. In particular, there should be some burial depth which optimizes the amount of fractured rock.

Calculations were performed for burdens ranging from 10 cm to 90 cm. The results are summarized in Figure 7. As the burden is made larger, the volume of damaged material approaches a constant level, since the presence of a free surface no longer has any influence. There is clearly a broad minimum, 20-40 cm burden, in which the energy of the explosive provides the optimum fracture damage. When the charge is near the free surface, there is a significant decrease in the volume of fractured oil shale. The behavior reflected in Figure 7 has been long known in blasting tradition, and these calculations are consistent with that knowledge.

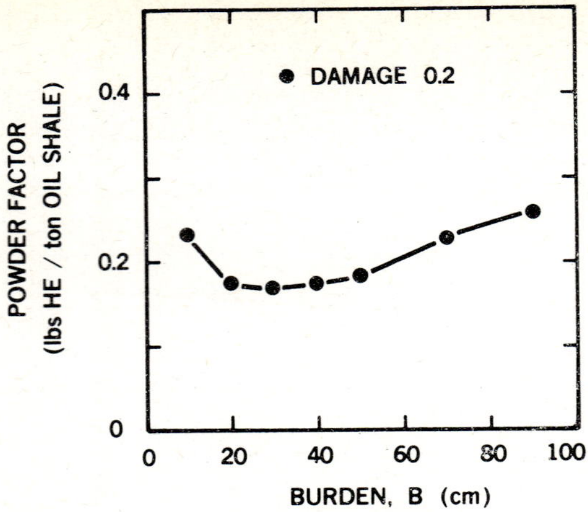

Figure 7. Calculated powder factors for a spherical charge.

(c) COLUMN CHARGES. In blasting operations, spherical charges are rarely of interest, so the charge in the previous example was changed to a column with a length-to-diameter ratio of 8. The center of the 80 gm charge was located 40 cm from the free surface. When the charge is detonated at the end closest to the free surface, the detonation proceeds away from the free surface. The region that exceeds a calculated damage level of 0.2 is outlined in Figure 8. The largest

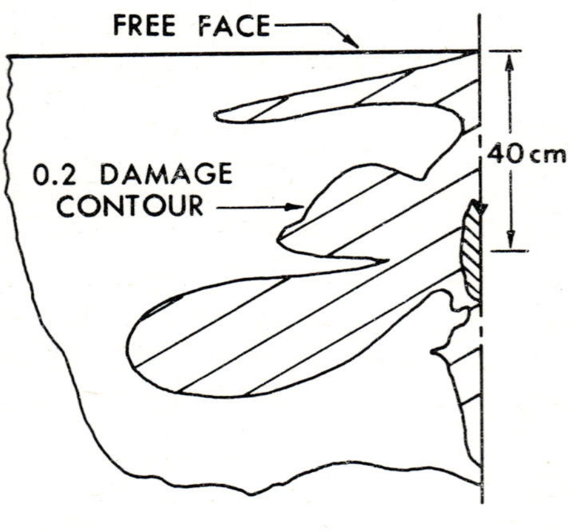

Figure 8. Column: Top detonation.

portion of fractured material lies below the top of the explosive charge. A relatively small fraction of the damage occurs between the charge and the free surface. The total volume of damage is nearly the same as that for the spherical charge at the same location (Figure 6).

When the same charge is bottom-detonated, the detonation proceeds towards the free surface. The calculated damage, shown in Figure 9, extends to nearly twice the volume of the top-detonated case. The fractured oil shale is most heavily concentrated in a region between the top of the column and the free surface.

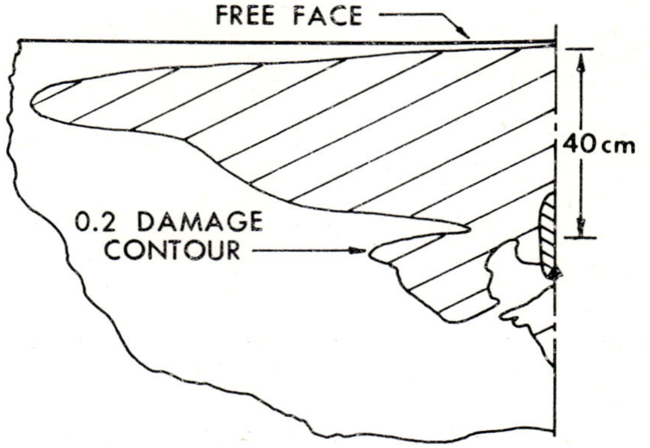

Figure 9. Column: Bottom detonation.

(d) MINE EXPERIMENT. An experiment was conducted in oil shale by Los Alamos Scientific Laboratory [18] at the Colony Mine, Colorado, in which a 0.1 m borehole was drilled into the floor of the mine. A 0.75 m high column charge of 5.2 kg of ANFO (Ammonium Nitrate/Fuel Oil) explosive was centered at a depth of 1.6 m and bottom detonated. After the event, the loosely fractured rock was excavated, and profiles from the resulting crater were determined (Figures 10 and 11). The

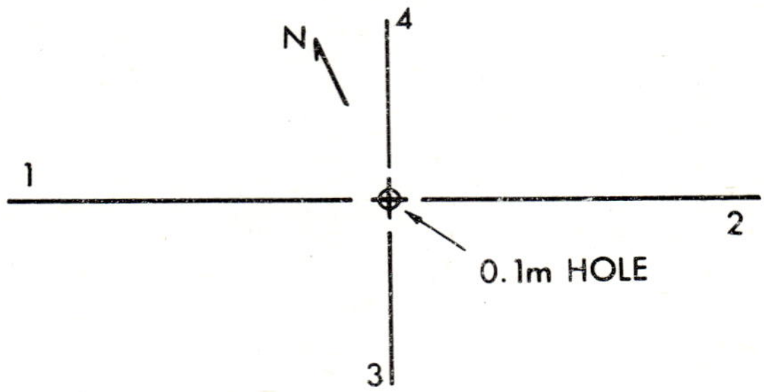

Figure 10. Crater profile planes.

detonation characteristics of ANFO vary substantially with borehole diameter. In a 10 cm diameter hole, the Chapman-Jouguet pressure is 2.4 GPa, and the detonation velocity is 3.5 km/s, for an initial density of nominally 0.9 Mg/m^3 [19].

A calculation of this experiment was made using the above explosive information, and the fracture and fragmentation model

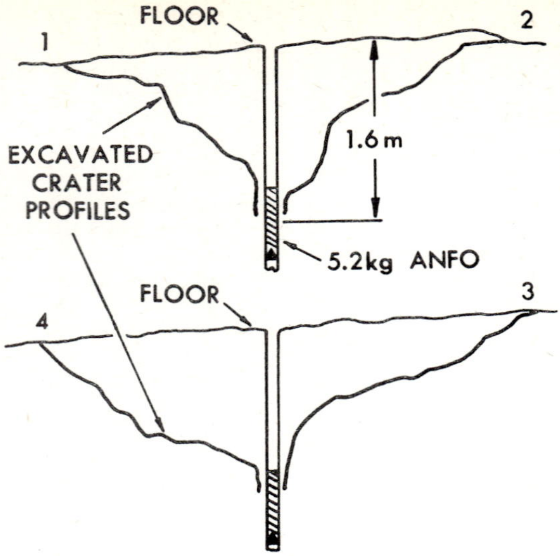

Figure 11. Crater profiles.

with the oil shale parameters determined from the laboratory-scale experiments (Equation 17). In Figure 12, the region calculated to exceed a damage level of 0.2 is plotted. The

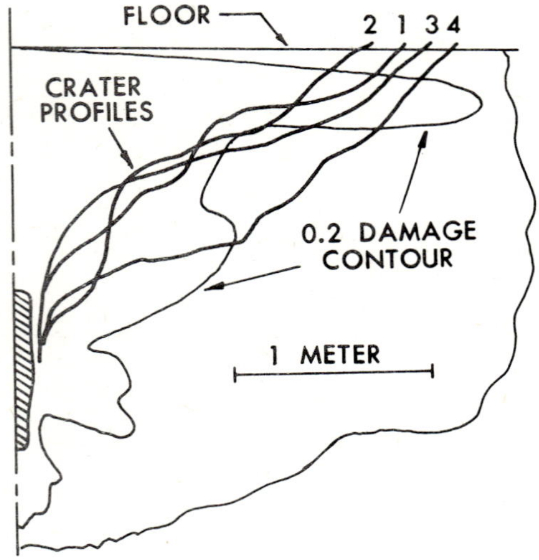

Figure 12. Calculated damage compared with crater profiles.

largest part of the fractured rock lies between the top of the explosive and the free surface. The numbered lines correspond to the crater contours from Figure 11. The calculated damage region appears to be a good representation of the experimental crater.

The average fragment sizes calculated to form in this explosive geometry are shown in Figure 13, along with the four crater profiles. The smallest fragment sizes are in regions A, near the explosive and the free surface. The fragment sizes gradually increase with distance from the explosive. The excavated crater profiles are observed to have spatial variations exceeding 0.5 m, and pass through regions where the predominant average fragment size is calculated to be larger than 0.1 m.

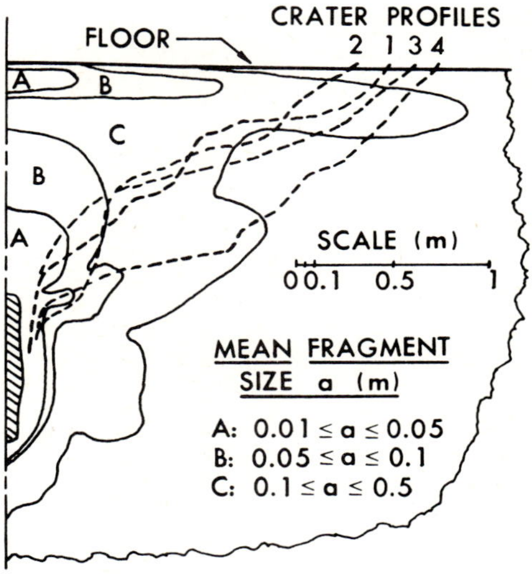

Figure 13. Calculated average fragment sizes compared with crater profiles.

There are some regions, especially near the lower end of the explosive, that appear to be accommodating fragment sizes that are larger than the dimensions of the region. This suggests that a few large cracks have formed, but the rock has remained relatively competent. The damage plot in Figure 12 tends to support this interpretation. Overall, the calculation agrees with these experimental results quite well.

CONCLUSIONS

Essential physical processes of the dynamic fracture and fragmentation of rock have been included in the model outlined in this paper. The overall agreement of the calculations with experiments on different scales indicates that the model correctly accounts for the scale of the event, even though the parameters are based upon laboratory-scale experiments. One of the most useful aspects of the model, when coupled into a wave-propagation code, is the ease with which changes in significant parameters, such as explosive type or detonation position, can be evaluated. Hence, a thorough understanding can be obtained of the wave propagation in a blasting geometry, and an approximate survey of the expected fragment sizes. Although complex geometries involving multiple boreholes cannot yet be calculated, individual borehole studies can reduce the number of uncertainties involved in such designs. This has the potential of greatly enhancing the efficient use of explosives to fragment rock.

REFERENCES

1. WINZER, S. R., MONTENYOHL, V. I., and RITTER, A. - High-Speed Cinematography of Production Blasting Operations. Mining Congress Journal, pp. 46-61, October 1979.

2. GRADY, D. E. and LIPKIN, J. - Criteria for Impulsive Rock Fracture. Geophys. Res. Let. (in press).

3. SHOCKEY, D. A., et al. - Fragmentation of Rock Under Dynamic Loads. Int. J. Rock Mech. Min. Sci., Vol. 11, pp. 303-317, 1974.

4. GRADY, D. E. and KIPP, M. E. - The Micromechanics of Impact Fracture of Rock. Int. J. Rock Mech. Min. Sci., Vol. 16, pp. 293-302, 1979.

5. GRADY, D. E. and KIPP, M. E. - Continuum Modelling of Explosive Fracture in Oil Shale. Int. J. Rock Mech. Min. Sci. (in press).

6. KIPP, M. E., GRADY, D. E. and CHEN, E. P. - Strain-Rate Dependent Fracture Initiation. Int. J. Fracture (in press).

7. CHEN, E. P. and SIH, G. C. - Transient Response of Cracks to Impact Loads. Elasto-Dynamic Crack Problems (Ed., G. C. Sih) Vol. 4 of Mechanics of Fracture, Noordhoff, Netherlands, 1977.

8. TULER, F. R. and BUTCHER, B. M. - A Criterion for the Time Dependence of Dynamic Fracture. Int. J. Frac. Mech., Vol. 4, No. 4, pp. 431-437, 1968.

9. DAVISON, L. and STEVENS, A. L. - Thermomechanical Constitution of Spalling Elastic Bodies. J. Appl. Phys., Vol. 44, pp. 668-674, 1973.

10. SEAMAN, L., CURRAN, D. R. and SHOCKEY, D. A. - Computational Models for Ductile and Brittle Fracture. J. Appl. Phys., Vol. 47, No. 11, pp. 4814-4826, 1976.

11. COCHRAN, S. and BANNER, D. - Spall Studies in Uranium. J. Appl. Phys., Vol. 48, No. 7, pp. 2729-2737, 1977.

12. DAVISON, L., STEVENS, A. L. and KIPP, M. E. - Theory of Spall Damage Accumulation in Ductile Metals. J. Mech. Phys. Solids, Vol. 25, pp. 11-28, 1977.

13. DIENES, J. K. - A Statistical Theory of Fragmentation. Proc. 19th U. S. Symp. Rock Mech., State Line, NV, pp. 51-55, 1978.

14. GRADY, D. E. and KIPP, M. E. - Oil Shale Fracture and Fragmentation at Higher Rates of Loading. Proc. 20th U. S. Symp. Rock Mech., pp. 403-406, Austin, TX, 1979.

15. WALSH, J. B. and BRACE, W. F. - Elasticity of Rock in Uniaxial Strain. Int. J. Rock Mech. Min. Sci., Vol. 9, pp. 7-15, 1972.

16. SWEGLE, J. W. - TOODY IV - A Computer Program for Two-Dimensional Wave Propagation. Sandia Laboratories Report, SAND-78-0552, Sandia Laboratories, Albuquerque, NM, 1978.

17. GRADY, D. E. and KIPP, M. E. - Explosive Fracture Studies on Oil Shale. Proc. 54th Soc. Pet. Eng. Fall Tech. Conf., Las Vegas, NV, 1979.

18. SCHMIDT, S. - Private Communication, Los Alamos Scientific Laboratory.

19. JOHNSON, J. N. - Calculation of Explosive Rock Breakage: Oil Shale. Proc. 20th U. S. Symp. Rock Mech., Austin, TX, 1979.

THE FRACTURE TOUGHNESS OF CONCRETE

B. Barr [1] W.T. Evans [2] R.C. Dowers [3]
B.B. Sabir (4)

SUMMARY

The paper describes a simple test which has been developed to determine the fracture toughness of concrete. The test is carried out using ordinary concrete cubes, modified by the introduction of two slits, and then subjected to a eccentric load.

A finite element analysis has been carried out for the test specimen geometry and stress intensity factors are given in the papers for the geometries used.

The paper discusses the effect of varying notch depth, water/cement ratio and testing system. The test results appear to be independent of notch deptch effects. A simple hydraulic bench-mounted test rig which has been developed to carry out the split-cube test is described.

(1) Principal Lecturer in Civil Engineering The Polytechnic of Wales.

(2) Principal Lecturer in Mechanical Engineering The Polytechnic of Wales

(3) Research Student, The Polytechnic of Wales

(4) Computer Applications Engineer, Mott, Hay and Anderson Croydon, Surrey.

INTRODUCTION

Kaplan [1] was one of the first to investigate the fracture toughness of concrete. In his experiments, Kaplan used rectangular beams containing sharp, uniform V-notches which were subjected to bending. Other workers [2], [3] and [4] have used similar notched beams to determine the fracture toughness of concrete. Brown [5] has performed a series of tests to measure the fracture toughness of cement paste and mortar for varying crack lengths.

Bear and Barr [6] have described two tests which may be used to evaluate the fracture toughness of both rock and fine-grained concrete. The tests used circumferentially notched round bar specimens which were subjected to two loading systems. In the first case the notched round specimens were loaded in four-point bending as shown in Fig 1, and in the second case the specimens were subjected to an eccentric longitudinal load as shown in Fig 2. The loading system shown in Fig 1 has recently been used by Javan and Dury [7] to determine the fracture toughness of fibre concrete.

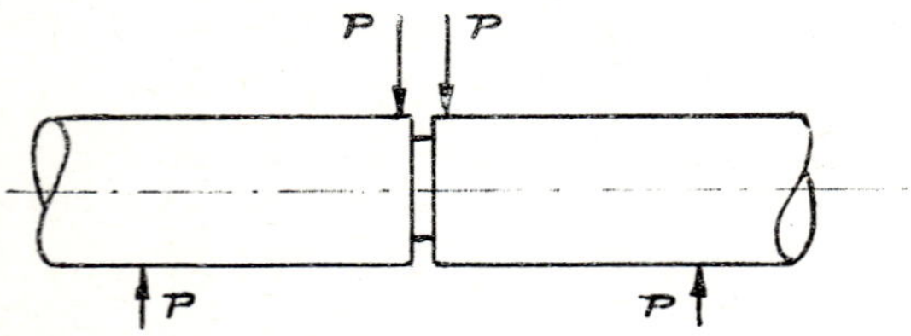

Fig 1. Circumferentially notched round bar subjected to four-point bending.

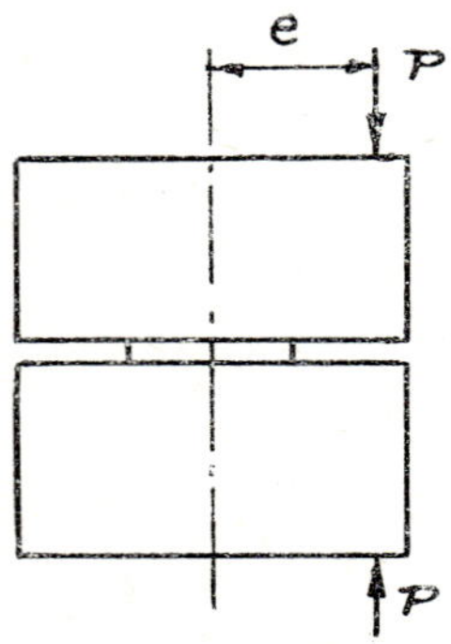

Fig 2. Circumferentially notched round bar subjected to eccentric longitudinal loading.

Fig 1 is similar to the bend specimen used to evaluate the fracture toughness of high-strength metals, and Fig 2 has some similarity with the compact tensile specimen which is the alternative test specimen geometry used to determine the fracture toughness of metals. The test geometry described in this paper is a natural development of that shown in Fig 2. Whereas the tests described in [6] required the use of a fine-grained concrete due to the relatively small dimensions of the test specimen, the testing system described here has been used for ordinary concrete mixes and is currently being used to determine the fracture toughness of polymer concrete.

TEST SPECIMEN GEOMETRY

The fracture toughness of concrete has been determined using split-cubes as shown in Fig 3. The tests were carried out using modified 150 mm and 100 mm cubes. The cubes were cut by means of a 'Clipper' masonry saw along two opposite faces as shown. The modified cubes were then subjected to eccentric loads as shown in Fig 4. The load was applied via two lengths of 6 mm square steel bars and the points of application of the load taken at the inside edge of the steel bars - since the load was concentrated at the inside edges as deformation took place.

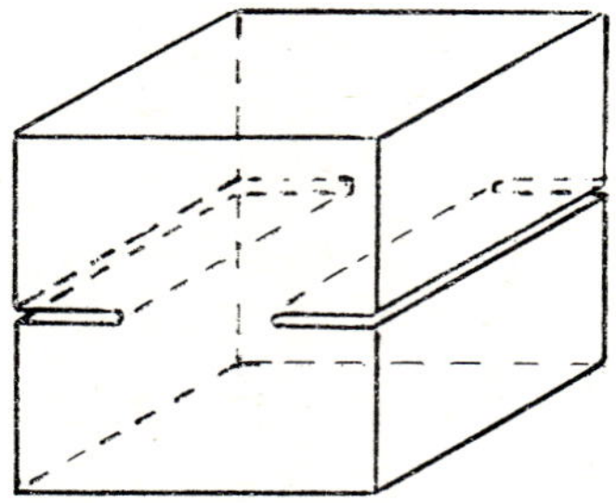

Fig 3. Concrete cube slit along two opposite faces.

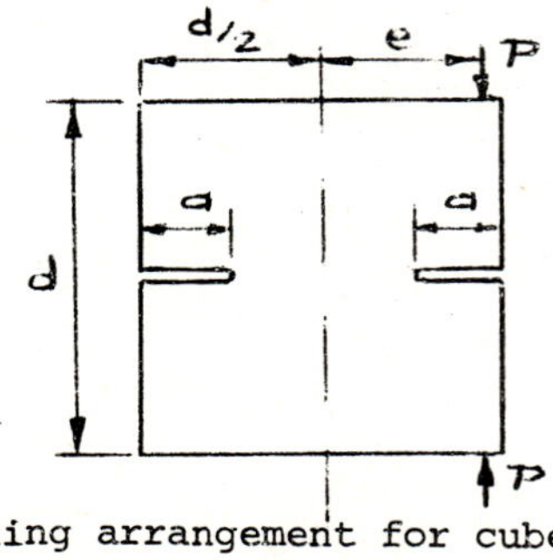

Fig 4. Loading arrangement for cube.

DETERMINATION OF STRESS INTENSITY FACTOR

Finite element solutions for the split-cube shown in Fig 4 have been obtained for both 100 mm and 150 mm cubes. Plane strain conditions were assumed and several notch depths were considered giving a range of a/d ratios. The finite element representations were similar to those described in an earlier paper at the Conference [8].Fig 5 shows the details for the case of a 100 mm cube with a notch depth of 25 mm.

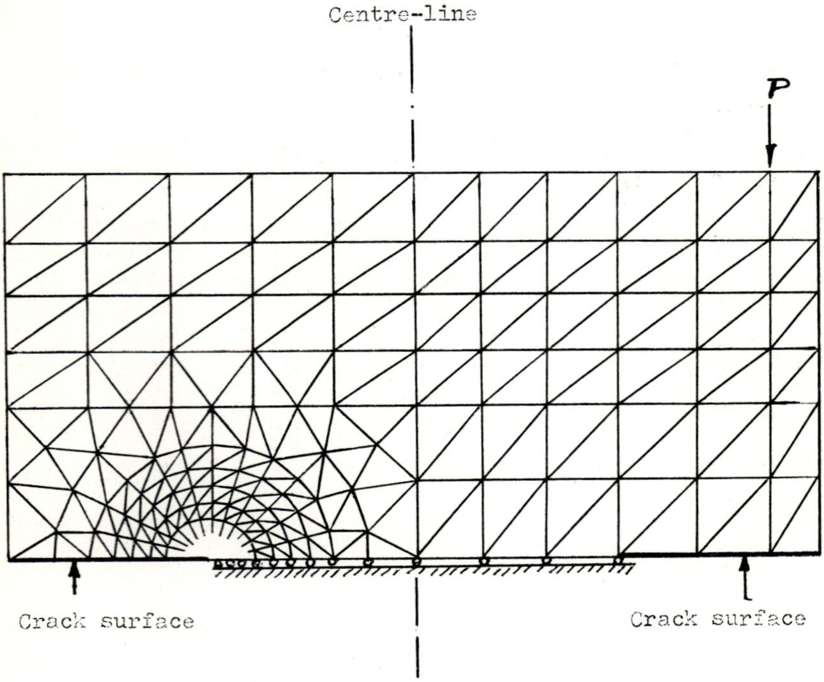

Fig 5. Finite element representation

The region within r = a/70 surrounding the crack tip in the tension zone was covered by elements I and II (for details of these elements see [8] and the remainder of the body was represented by CST elements. Details of the results for the stress intensity factors for the 100 mm cubes obtained from the V-displacements around the tip of the notch are given in [9] .

Stress intensity factors are often expressed in terms of polynomial functions of the ratio of crack size/specimen width. The five a/d ratios considered in the finite element analyses enabled the results to be expressed in terms of five powers of a/d, similar to those used for the compact tensions specimen [10]. For the 100 mm cubes, the following expression was obtained:

$$k_1 = \frac{P}{Bd^{\frac{1}{2}}} \cdot \left[18.3\left(\frac{a}{d}\right)^{\frac{1}{2}} - 430.0\left(\frac{a}{d}\right)^{\frac{3}{2}} + 3445.2\left(\frac{a}{d}\right)^{\frac{5}{2}} - 11075.8\left(\frac{a}{d}\right)^{\frac{7}{2}} + 12966.8\left(\frac{a}{d}\right)^{\frac{9}{2}}\right] \quad ..(1)$$

The corresponding polynomial expression for the 150 mm cubes is given by

$$k_1 = \frac{P}{Bd^{\frac{1}{2}}} \cdot \left[30.5\left(\frac{a}{d}\right)^{\frac{1}{2}} - 612.0\left(\frac{a}{d}\right)^{\frac{3}{2}} + 4441.5\left(\frac{a}{d}\right)^{\frac{5}{2}} - 13404.8\left(\frac{a}{d}\right)^{\frac{7}{2}} + 14930.0\left(\frac{a}{d}\right)^{\frac{9}{2}}\right] \quad ..(2)$$

TEST DETAILS

Some preliminary tests on 100 mm cubes were carried out using a standard mix but with varying water/cement ratio. The cubes were cured in water for 28 days, the slits being introduced, using the 'Clipper' on the 21st day. All the initial cubes were tested in an Instron 1251 Model testing machine at nominal room temperature and at a strain rate of 10×10^{-3}/sec. The load-displacement curves were plotted autographically and the maximum load recorded during a test was taken as the failure load and used to determine fracture toughness.

Three notch depths were investigated for four water/cement ratios. The results for these initial tests are shown in Table 1. The notch depths are nominal values as there was considerable variation in the actual notch depth produced in these first tests, a variation of up to ± 1.5 mm being recorded. However, in the determination of the fracture toughness, values are based on the actual notch deptch.

w/c Ratio	Cube Strength (N/mm^2)	a/d ratio (i.e. crack/width)		
		0.25	0.30	0.35
0.40	62.5	2.32	2.26	2.24
0.47	55.5	2.19	2.08	2.28
0.53	52.5	1.97	1.94	1.87
0.60	40.0	1.71	1.58	1.74

Table 1. Fracture toughness results ($kN/mm^{3/2}$) for 100 mm cubes.

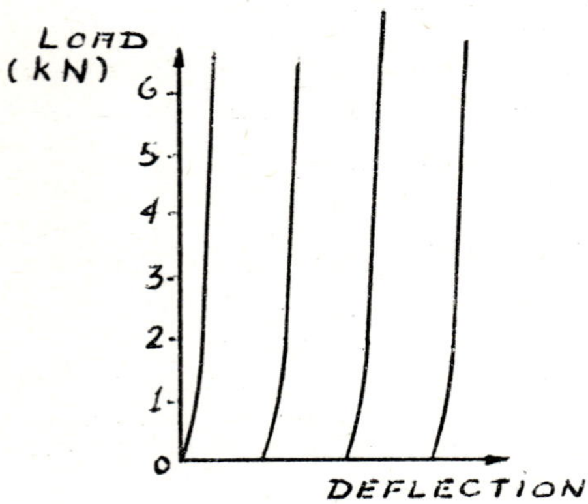

Fig 6. Typical load-deflection graphs

Typical load-deflection curves obtained during the tests are shown in Fig 6. After the initial period when the loading points are bedding in to the specimen, the load deflection curve is linear up to fracture. The load at fracture is noted and used in the expression for k_1, to give the fracture toughness.

A number of cubes, in these initial tests, with a nominal notch depth of 25 mm did not fail by a crack propagating from the tensile zone. Almost half of these cubes failed in shear. The shear occurred in the zone adjacent to the point of application of the load. From these initial tests we could determine the minimum depth of notch to be used for later work. As a result of the variation in notch depth obtained in the first series of tests, a special rig was devised to ensure a more uniform notch depth. A portal frame was used to locate the 'Clipper' blade at the required level to produce a given cut in the concrete cubes. Following these initial development tests, series of tests were carried out to evaluate the effect of variation of :-

(a) notch depth
(b) water/cement ratio
(c) testing system

Three notch depths (25 mm, 30 mm and 35 mm) were investigated, corresponding to a/d ratios of 0.25, 0.3 and 0.35. In order to minimise further the inaccuracies in specimen preparation, a lathe using a tool-post grinder was modified to produce the notches. This system resulted in notch tolerances of ± 0.5 mm for a given batch of specimens and ± 0.25 mm longitudinally along individual notches. The figures compare with 1.5 mm and 0.75 mm respectively for notches cut by the 'Clipper'. Moreover the notch width achieved with a diamond impregnated blade used on the lathe was 1.5 mm compared with 3.25 mm using the carborundum blade of the 'Clipper'.

The second variable considered was water/cement ratio. Water/cement ratios of 0.40, 0.45, and 0.50 were chosen for their low, medium and high workabilities. The water content and hence ratio was based on the free water available for reaction and did not include the extra water which must be added to allow for aggregate absorption. This extra water requirement was taken as 0.35% of the total weight of fine and coarse aggregate.

The initial development work for the test was carried out using an Instron 1251 model testing machine. The tests were conducted in displacement control i.e. the machine was acting as a stiff testing machine. A much simpler testing system was investigated and developed as a part of this work. A hand operated hydraulic jack connected to a small ram and calibrated gauge were mounted in a simple load frame. Cubes were tested using this simple bench test rig and the load at failure was recorded by means of a maximum load indicator on the gauge. The limitations of this simple, adaptable, portable testing system are the uncertain rate of application of load and the contribution to failure of the strain energy stored in the "soft" testing rig.

TEST RESULTS

A typical set of results for various water/cement ratios, notch depths and testing system is shown in Table 2. Of particular interest is the low coefficient of variation (often around 5%) for the failure load. More detailed results for various notch depths and water/cement ratios are given in Tables 3 and 4. Table 3 gives the fracture toughness values as determined by means of the simple hydraulic testing system, while Table 4 gives the corresponding results obtained in tests using the Instron in displacement control. The overall pattern of results are the same in Table 3 and Table 4, although the fracture toughness values obtained using the hydraulic test rig is a little higher than the values obtained in the Instron.

The variation of fracture toughness with notch depth is shown in Fig 7 and the variation of fracture toughness with water/cement ratio is shown in Fig 8. Both Fig 7 and Fig 8 show that tests carried out in the Instron reduce the spread of the results.

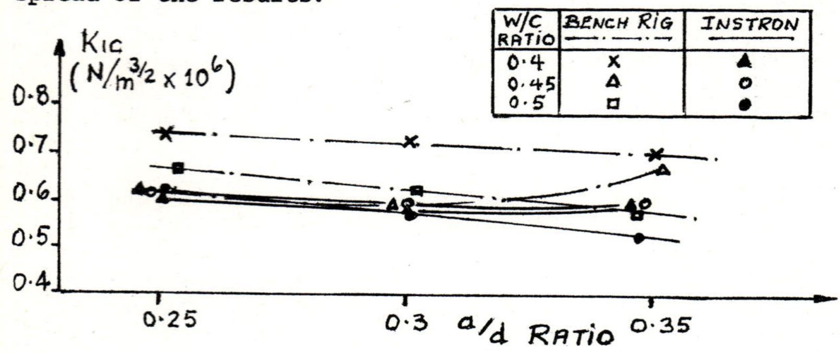

Fig 7. Effect of notch depth on fracture toughness.

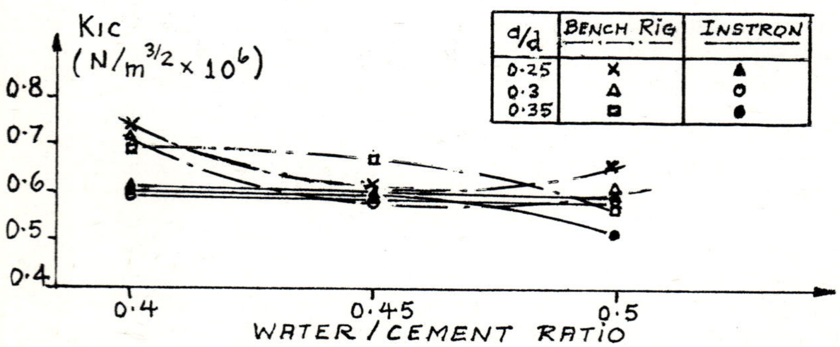

Fig 8. Effect of water/cement ratio on fracture toughness.

MIX NO.	NO. IN SAMPLE	FAILURE LOAD (KN)	AVERAGE FAILURE LOAD (KN)	COEF OF VARIATION (%)	WATER/ CEMENT RATIO	AVERAGE NOTCH DEPTH (mm)	TESTING MACHINE	NOTCH TYPE
25D	9	7.3 7.1 7.5 7.1 6.8 7.9 7.25 7.1 7.45	7.27	4.3	0.40	30.6	BENCH TESTING RIG.	LATHE
26D	10	3.55 3.5 3.7 4.1 3.6 4.05 3.45 3.6 3.5 3.5	3.66	6.4	0.40	35.7	BENCH TESTING RIG.	LATHE
27D	7	13.6 13.5 13.0 12.95 12.9 12.0 12.7	12.95	4.1	0.40	25.2	BENCH TESTING RIG.	LATHE
28D	2	11.6 12.0	11.80	2.4	0.45	25.0	BENCH TESTING RIG.	LATHE
29D	6	10.55 11.0 11.45 10.8 9.85 12.8	11.08	9.0	0.50	26.0	BENCH TESTING RIG.	LATHE
30D	7	12.05 12.1 11.1 12.0 11.0 11.9 11.7	11.69	3.9	0.40	25.9	BENCH TESTING RIG.	LATHE
6D	13	5.0 5.75 5.6 5.8 5.5 5.25 6.05 5.4 6.4 6.0L 5.6 5.45L 6.35	5.70	8.6	0.40	29.9	INSTRON	LATHE
9D	15	9.0 8.5 10.2 8.5 9.1L 9.9 9.5* 8.9 8.9 9.3 9.0 8.9 9.5 9.4 8.9*	9.17	5.6	0.40	25.7	INSTRON	LATHE
10D	15	10.0 9.2 9.8 10.6 10.4 10.1 8.7* 10.1 8.9 9.3 10.2 9.5 10.0 9.8 9.0*	9.84	5.0	0.40	25.6	INSTRON	LATHE
11D	15	8.7 8.9 9.6* 9.0* 9.1* 9.5 9.1 8.4* 9.8 8.9 8.9 8.4* 8.4 8.4 9.4	9.00	5.1	0.40	25.7	INSTRON	LATHE

Table 2. Typical set of raw data.

Nominal Notch Depth (mm)	WATER/CEMENT RATIO								
	0.04			0.45			0.50		
	$K_{1c} \times 10^6$ $(N/m^{3/2})$	Coef. of Var.	No.	$K_{1c} \times 10^6$ $(N/m^{3/2})$	Coef. of Var.	No.	$K_{1c} \times 10^6$ $(N/m^{3/2})$	Coef. of Var.	No.
	7.48	10.5	11						
	7.27	2.4	4						
	7.03	7.1	12						
							6.35	5.0	9
25				6.02	5.8	10			
	7.78	5.6	7						
				6.94	3.0	2			
							7.10	6.4	6
TOTALS	7.35	8.0	41	6.18	7.7	12	6.65	7.9	15
				5.51	4.7	14			
				6.32	10.5	15			
				5.26	7.3	15			
				6.37	3.7	14			
30							6.39	7.9	10
							6.01	5.6	11
				6.08	3.3	11			
				6.25	4.8	11			
	7.27	4.1	10						
	7.21	5.5	9						
TOTALS	7.24	4.7	19	5.95	9.7	80	6.19	7.5	21
				7.12	10.1	15			
							6.08	6.9	11
				7.08	10.1	11			
35				6.14	6.1	11			
							5.39	5.0	10
	6.82	11.2	10						
	7.24	4.1	10						
TOTALS	7.03	8.6	20	6.82	11.1	37	5.75	8.6	21

TABLE 3. Fraction toughness results using bench test rig.

Nominal Notch Depth (mm)	WATER/CEMENT RATIO								
	0 .04			0.45			0.50		
	K_{1c} ×10^5 ($N/m^{3/2}$)	Coef. of Var.	No.	K_{1c} ×10^5 ($N/m^{3/2}$)	Coef. of Var.	No.	K_{1c} ×10^5 ($N/m^{3/2}$)	Coef. of Var.	No.
	5.72	5.0	12						
	6.12	5.4	13						
	5.63	3.1	10						
							5.89	6.0	6
25				5.98	7.1	7			
	6.46	4.8	8						
				6.33	5.7	6			
	6.44	7.7	66						
							6.32	3.7	3
TOTALS	6.02	7.4	49	6.14	6.8	13	6.03	6.2	9
	5.27	7.6	11						
							6.14	6.4	9
							5.63	8.1	10
30				5.80	4.5	8			
				6.04	7.1	10			
	6.18	6.0	10						
	6.26	8.9	9						
TOTALS	5.87	10.6	30	5.93	6.3	18	5.87	8.4	19
							5.79	9.7	10
				6.54	10.2	10			
				5.47	8.9	9			
35							4.63	5.6	9
	5.70	5.6	10						
	6.21	9.3	9						
TOTALS	5.94	8.7	19	6.03	13.2	19	5.24	14.0	19

TABLE 4. Fraction toughness results using Instron.

DISCUSSION AND CONCLUSIONS

The most important feature of the work is the simplicity of the test described. The test requires an ordinary concrete cube, modified by the introduction of two slits, to be subjected to an eccentric load. This load is used (together with the geometrical details of the cube and its notches) to give directly the fracture toughness values.

The load-displacement curves obtained in the tests are linear-elastic up to the point of failure. Furthermore, there appears to be no noticeable notch depth effect. These two points would suggest a valid test for fracture toughness determination. For the results shown, there is very little variation in fracture toughness with water/cement ratio. This may be due to the high cement content of the mixes used, even the 0.5 water/cement ratio mix had an average compressive strength of 45 N/mm^2 at 28 days.

The better results were obtained in the tests carried out in the Instron machine. However, the results obtained in the simple hydraulic test rig were encouraging. The obvious advantage of the hydraulic test rig is that it could be used in most laboratories and even on site.

The test geometry described is currently being used by the authors to determine the fracture toughness of polymer concrete and glass fibre concrete. The test could also be used to evaluate the fracture toughness of rocks and soil-cement mixes.

REFERENCES

1. KAPLAN, M.F. Crack Propagation and the Fracture of Concrete, ACI Journal, Vol.58, Noll, pp 591-610 (1961)

2. NAUS, D.J. and LOTT, J.L. Fracture Toughness of Portland Cement Concretes. ACI Journal, Vol.66 ,pp 481-489 (1969)

3. LOTT, J.L. and KESLER, C.E. Symposium on Structure of Portland Cement Paste and Concrete, Washington, D.C., Highway Research Board, Special Report 90, pp 204-218 (1966)

4. MOAVENZADEH, F and KUGUEL, R. Fracture of Concrete Journal of Materials, Vol4, No. 3 pp 497-519 (1969)

5. BROWN, J.H. Measuring the Fracture Toughness of Cement Paste and Mortar, Magazine of Concrete Research Vol. 24 No. 81, pp 185 - 196 (1972)

6. BEAR, T.J. and BARR B. Fracture Toughness Tests for Concrete, Int. Journal of Fracture, Vol 13 pp.92 - 96 (1977)

7. JAVAN, L. and DURY, B.L. Fracture Toughness of Fibre Reinforced Concrete Concrete, Vol. 13, No. 12 pp 31 - 33 (1979)

8. SABIR B.B. and BARR B. New Finite Elements for Fracture Analysis 2nd Int. Conference on Num. Methods in Fracture Mechanics, Swansea (1980)

9. SABIR.BB. The Application of the Finite Element Method to Fracture Mechanics Problems. Ph.D. Thesis, C.N.A.A. (1980)

10. BROWN W.F. and SRAWLEY, J.E. Plain Strain Crack Toughness Testing of High Strength Metallic Materials, ASTM Special Technical Publication No. 410 (1966)

NUMERICAL SOLUTION OF ECCENTRICALLY LOADED STRUTS IN NON-TENSION MATERIAL

by F. Sawko, D.Sc.[1] and M.A. Rahman[2]

[1]Professor, [2]Post Doctoral Fellow, University of Liverpool.

ABSTRACT

The paper develops a numerical procedure for predicting the load carrying capacity of rectangular eccentrically loaded brick struts taking into account the brittle and non-linear properties of the material.

It is assumed that brickwork is incapable of developing tension. A parabolic stress-strain relationship in compression is assumed. This is based on published experimental evidence. The parabola has a falling branch with the maximum strain being 1.5 times the peak strain.

The procedure is one of successive corrections to the equilibrium and compatibility equations as the strut undergoes deflections. For a particular load and applied eccentricity, a finite difference formulation gives the deflections at a number of sections along the strut. These are then used to modify the eccentricities and the procedure is repeated until two sets of deflections converge. A comparison of the maximum strain along the strut with the limiting strain of the material allows the load to be increased to obtain a new equilibrium state. The process is continued until the maximum strain reaches the limiting strain or the strut loses stability.

The procedure was programmed for a desk-top Commodore PET micro computer possessing a display facility to enable visual monitoring of convergence process. The rate of convergence was found to be very rapid. The results of a number of analyses for various slenderness ratios and material constitutive relationships are presented. A comparison of the results with BS5628, the new Code of Practice on Load Bearing Brickwork, is also made.

INTRODUCTION

The load carrying capacity of brick walls and columns when axial and eccentric loading is of considerable importance in design of load bearing brick structures. The new Code of Practice "Structural Use of Masonry" BS5628 : Part 1 : 1978 (Ref. 1) defines the property in Table 7 using the Capacity Reduction Factor to define the reduction of load which the strut can carry with increased slenderness and eccentricity. These factors have been derived empirically at the Building Research Establishment several years ago, and the authors felt that a theoretical solution to this problem was needed. The paper describes the assumptions made and method adopted for the solution, and draws the general conclusion, that Table 7 cannot fully define the properties of walls built with different brick and mortar properties.

ASSUMPTIONS

It is assumed that the strength of brickwork in tension is small to that in compression, and can be neglected. It is further assumed that in compression the stress-strain relationship is parabolic, as shown in Figure 1. This has a falling branch with $\varepsilon_{max} = 1.5\ \varepsilon_m$. Experimental evidence (Ref. 2) confirms this assumption for a wide range of brick types and strengths.

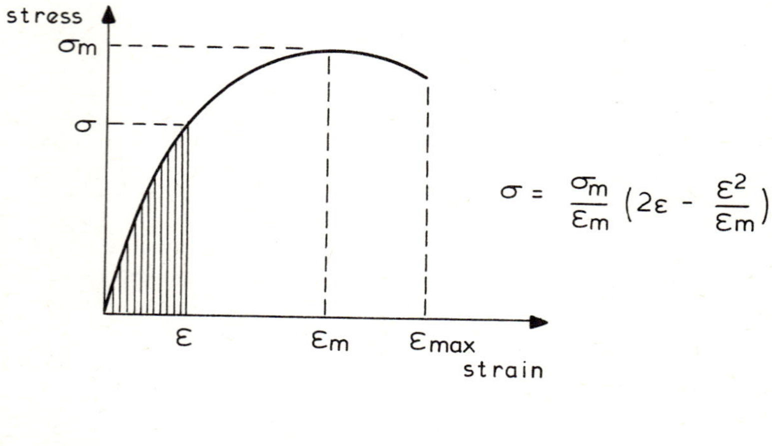

Fig. 1. Assumed stress/strain curve for brickwork in compression.

ANALYSIS OF SECTION

At a typical section the load P will act at an eccentricity of e from geometric centroid of section (Fig. 2).

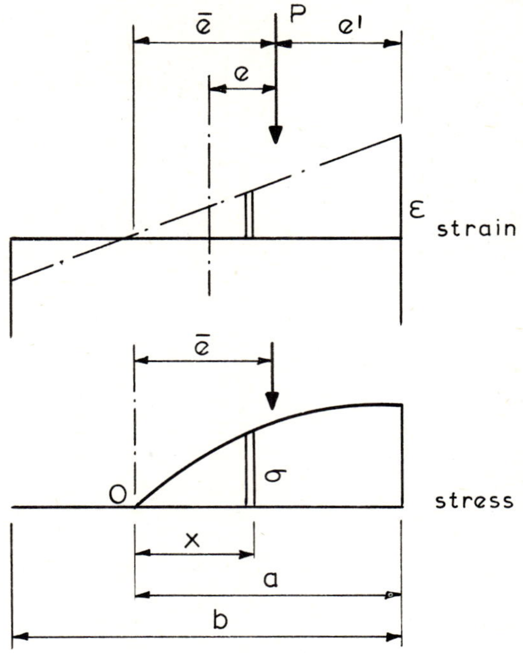

Fig. 2. Section Under Eccentric Load

From Figure 1. stress at distance x from 'zero stress' axis $\sigma = \frac{\sigma_m}{a}\frac{\varepsilon}{\varepsilon_m}(2x - \frac{x^2}{a}\frac{\varepsilon}{\varepsilon_m})$

Total force on section per unit thickness

$$P = \int_0^a \sigma dx = \frac{\sigma_m}{a}\frac{\varepsilon}{\varepsilon_m}(a^2 - \frac{a^2}{3}\frac{\varepsilon}{\varepsilon_m}) \quad \therefore P = \frac{a\sigma_m\varepsilon}{\varepsilon_m}(1 - \frac{1}{3}\frac{\varepsilon}{\varepsilon_m}) \qquad (1)$$

$$\text{Moment of force @ o} = \int_0^a \sigma x \, dx = \frac{\sigma_m\varepsilon}{a\,\varepsilon_m}\int_0^a (2x^2 - \frac{x^2}{a}\frac{\varepsilon}{\varepsilon_m})dx$$

$$= \frac{\sigma_m\varepsilon}{a\,\varepsilon_m}(\frac{2}{3}a^3 - \frac{1}{4}a^3\frac{\varepsilon}{\varepsilon_m}) \quad \therefore M = \frac{a^2\sigma_m\varepsilon}{\varepsilon_m}(\frac{2}{3} - \frac{1}{4}\frac{\varepsilon}{\varepsilon_m}) \qquad (2)$$

Position of centroid of compression

$$\bar{e} = \frac{M}{P} \qquad \bar{e} = a\,\frac{8 - 3\frac{\varepsilon}{\varepsilon_m}}{12 - 4\frac{\varepsilon}{\varepsilon_m}} \tag{3}$$

From geometry $a - \bar{e} = \frac{b}{2} - e \quad \therefore \quad \bar{e} = a + e - \frac{b}{2}$ (4)

Subst. to (3) $\frac{b}{2} - e = a\,\frac{4 - \frac{\varepsilon}{\varepsilon_m}}{12 - 4\frac{\varepsilon}{\varepsilon_m}}$ put $\frac{b}{2} - e = e'$

(i.e. e' is the distance of P from free edge)

$$a = e'\,\frac{12 - 4\frac{\varepsilon}{\varepsilon_m}}{4 - \frac{\varepsilon}{\varepsilon_m}} \tag{5}$$

Subst. to (1) to eliminate a

$$P = \frac{\sigma_m\,\varepsilon}{3\,\varepsilon_m}\left(3 - \frac{\varepsilon}{\varepsilon_m}\right)e'\left(\frac{12 - 4\frac{\varepsilon}{\varepsilon_m}}{4 - \frac{\varepsilon}{\varepsilon_m}}\right) \quad \text{put } \frac{\varepsilon}{\varepsilon_m} = x$$

$$3P = \sigma_m\,e'x(3-x)\,\frac{(12 - 4x)}{(4 - x)}$$

$$4\sigma_m e'\,x^3 - 24\sigma_m e'\,x^2 + (36\sigma_m e' + 3P)x - 12P = 0 \tag{6}$$

This is a cubic in x, all other factors are known constants-

Solution gives $x = \frac{\varepsilon}{\varepsilon_m}$ (See Appendix 1)

Subst. to (5) to obtain a.

Curvature $C = \frac{\varepsilon}{a}$ (7)

ANALYSIS OF STRUT

It has been demonstrated in the proceeding section that for any eccentricity of load it is possible to calculate the curvature C at any section. From curvatures, the deflected profile can be obtained using numerical methods.

Divide the strut into 10 segments. From symmetry (Fig. 3) there will be five unknown deflections $y_1 - y_5$.

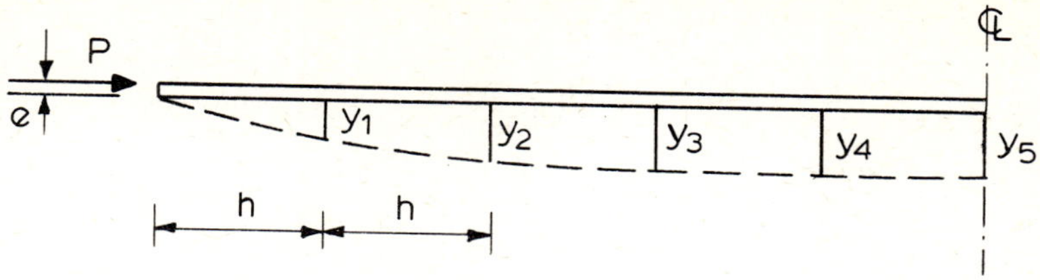

Fig. 3. Division of strut into 10 segments

Curvature $C = \frac{d^2y}{dx^2} = \frac{y_1 - 2y_2 + y_3}{h^2}$ or $y_1 - 2y_2 + y_3 = h^2C$

Since curvatures will be different at stations 1-5, the set of equations will be:

$$
\begin{array}{cccccccc}
-2y_1 & +y_2 & & & & = & C_1h^2 \\
y_1 & -2y_2 & +y_3 & & & = & C_2h^2 \\
 & y_2 & -2y_3 & +y_4 & & = & C_3h^2 \\
 & & y_3 & -2y_4 & +y_5 & = & C_4h^2 \\
 & & & 2y_4 & -2y_5 & = & C_5h^2
\end{array} \tag{8}
$$

There equations are solved explicitly (Appendix 2) to give the following result:

$$
\begin{aligned}
y_1 &= h^2(C_1 + C_2 + C_3 + C_4 + 0.5C_5) \\
y_2 &= h^2(C_1 + 2C_2 + 2C_3 + 2C_4 + C_5) \\
y_3 &= h^2(C_1 + 2C_2 + 3C_3 + 3C_4 + 1.5C_5) \\
y_4 &= h^2(C_1 + 2C_2 + 3C_3 + 4C_4 + 2C_5) \\
y_5 &= h^2(C_1 + 2C_2 + 3C_3 + 4C_4 + 2.5C_5)
\end{aligned} \tag{9}
$$

Thus it is possible to obtain the deflection profile of strut for any eccentric loading. It will be observed that on deflection the eccentricity of loading changes at every station. This introduces nonlinearity, and the overall solution is only possible by a process of successive corrections.

OUTLINE OF SOLUTION PROCEDURE

As a starting point it is assumed that curvatures at all 5 positions are the same, as determined from equ. (7). Solution of equ. (9) gives deflection at the 5 stations. The eccentricity is then modified as shown in Fig. 4.

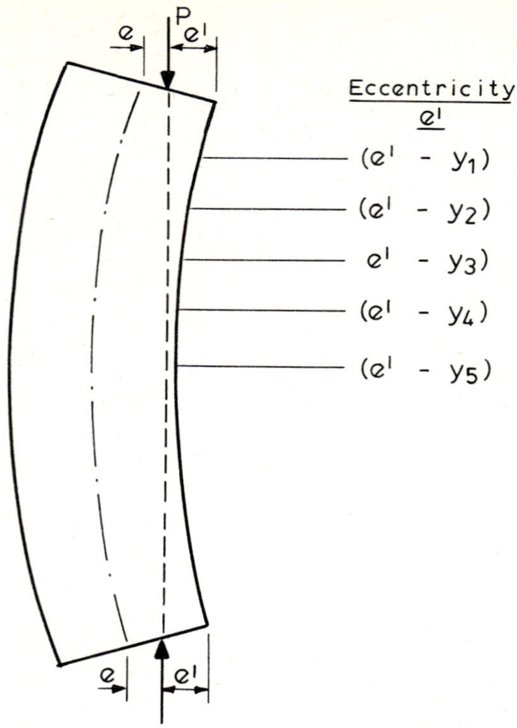

Fig. 4. Deflection Shape of Eccentrically Loaded Strut.

At each section the eccentricity e' in equation (6) is modified by new eccentricity (e' - y_n), leading to new curvatures from equ. (7), new deflections from equ. (9). This is repeated until convergence is obtained.

EVALUATION OF LIMITING VALUES

It is most convenient to select a particular eccentricity e, assume a value for P which is known to be conservative and follow the solution procedure outlined above. Once convergence has been achieved the maximum strain at middle (critical) section is compared with ε_{max} and P increased to obtain a new equilibrium state, until $\varepsilon = \varepsilon_{max}$ which is the limiting value of P for the eccentricity examined.

Using this procedure a complete set of results for a given strut geometry is obtained.

COMPUTER PROGRAM

The method described has been programmed for a desk top Commodore PET micro-computer possessing a visual display facility to facilitate visual monitory of the convergence process. No programming difficulties have been encountered. In a practical run, the rate of convergence was found to be very rapid.

NUMERICAL RESULTS

With a developed computer program, there is no difficulty in carrying out any number of runs on different eccentricities and strut geometries. Some numerical results are presented in figures 5-7. A wall of unit width with a load eccentricity of 90mm was analysed for a whole range of height to width ratios from 1 to 26. Three different types of brickwork of different strengths and elastic properties were considered.

Figure 5 gives directly the ultimate load for all struts. It will be observed, that the maximum stress σ_m is not the only criterion determining the strength of strut (compare B and C). Fig. 6 which reproduces the same results as a ratio of ultimate load of actual strut to ultimate load of stocky strut is even more informative, showing clearly the influence of modulus of elasticity on strut behaviour. Fig. 7 demonstrates that a strut with low ε fails at low strain ratios, and that most of the struts fail by buckling rather than crushing of brickwork.

CONCLUSIONS

It is possible to conclude that the capacity reduction factors contained in the new Code of Practice on "Structural use of Masonry" cannot fully explain the behaviour of struts under eccentric loading, since it ignores elastic properties of brickwork.

ACKNOWLEDGEMENTS

The Authors are grateful to Science Research Council for the research fellowship for the second author and to Brick Development Association for financially supporting this investigation.

REFERENCES

1. STRUCTURAL USE OF MASONRY, British Standard Code of Practice BS 5628: Part 1 : 1978.

2. POWELL AND HODGKINSON - "Determination of Stress-Strain Relationship of Brickwork", The British Ceramic Society Research Association, Technical Note No. 249, Jan. 1976.

3. CHEMICAL RUBBER CO., Standard Mathematical Tables, Sixteenth Edition 1978.

APPENDIX 1

EXPLICIT SOLUTION OF A CUBIC EQUATION (REF. 3)

A cubic equation, $y^3 + py^2 + qy + r = 0$ may be reduced to the form:-

$$x^3 + ax + b = 0$$

by substituting for y the value, $x - \frac{p}{3}$. Here

$$a = \tfrac{1}{3}(3q - p^2) \text{ and } b = \tfrac{1}{27}(2p^3 - 9pq + 27r).$$

For solution let:-

$$A = \sqrt[3]{-\frac{b}{2} + \sqrt{\frac{b^2}{4} + \frac{a^3}{27}}}, \qquad B = \sqrt[3]{-\frac{b}{2} - \sqrt{\frac{b^2}{4} + \frac{a^3}{27}}},$$

then the values of x will be given by,

$$x = A + B, \; -\frac{A+B}{2} + \frac{A-B}{2}\sqrt{-3}, \; -\frac{A+B}{2} - \frac{A-B}{2}\sqrt{-3}.$$

If p,q,r are real, then:-

If $\frac{b^2}{4} + \frac{a^3}{27} > 0$, there will be one real root and two conjugate imaginary roots;

If $\frac{b^2}{4} + \frac{a^3}{27} = 0$, there will be three real roots of which at least two are equal;

If $\frac{b^2}{4} + \frac{a^3}{27} < 0$, there will be three real and unequal roots.

APPENDIX 2 - EXPLICIT SOLUTION OF EQU. (8)

	y_1	y_2	y_3	y_4	y_5	
(1)	2	-1				
(2)	-1	2	-1			
(3)		-1	2	-1		
(4)			-1	2	-1	
(5)				-2	2	
(6)	1.0	-0.5				(1) ÷ 2
(7)		1.5	-1			(2) + (6)
(8)		1.0	-0.667			(7) ÷ 1.5
(9)			1.333	-1		(3) + (8)
(10)			1.0	-0.75		(9) ÷ 1.333
(11)				1.25	-1	(4) + (10)
(12)				1.0	-0.8	(11) ÷ 1.25
(13)					0.4	(5) + 2(12)
(14)					1.0	(13) ÷ 0.4
(15)				1.0		(12) + 0.8(14)
(16)			1.0			(10) + 0.75(15)
(17)		1.0				(8) + 0.667(16)
(18)	1.0					(6) + 0.5(17)

$$y_1 = (1) + (3) + (3) + (4) + 0.5(5)$$
$$y_2 = (1) + 2(2) + 2(3) + 2(4) + (5)$$
$$y_3 = (1) + 2(2) + 3(3) + 3(4) + 1.5(5)$$
$$y_4 = (1) + 2(2) + 3(3) + 4(4) + 2(5)$$
$$y_5 = (1) + 2(2) + 3(3) + 4(4) + 2.5(5)$$

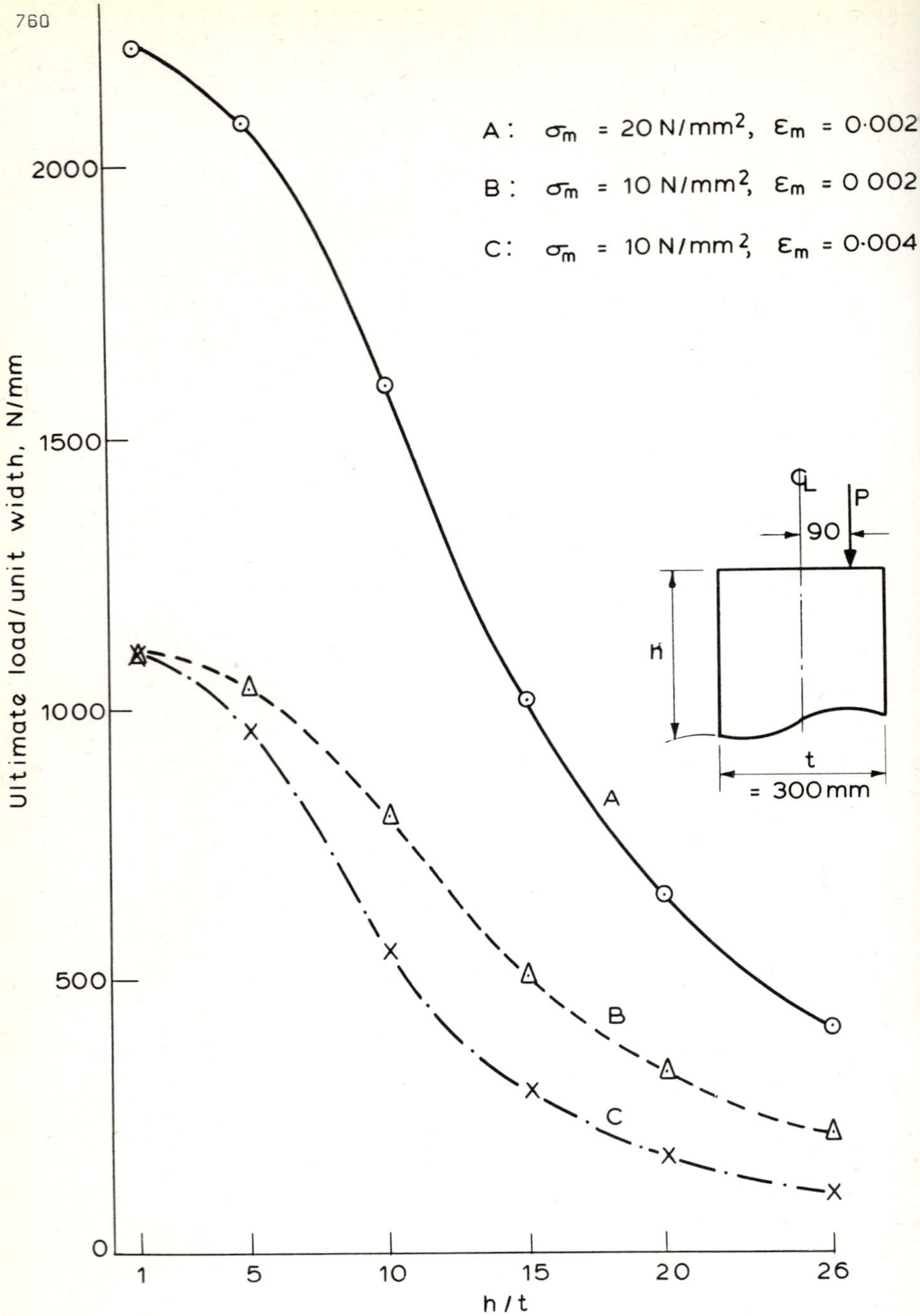

FIG. 5. ULTIMATE LOADS FOR DIFFERENT SLENDERNESS RATIOS & BRICKWORK PROPERTIES.

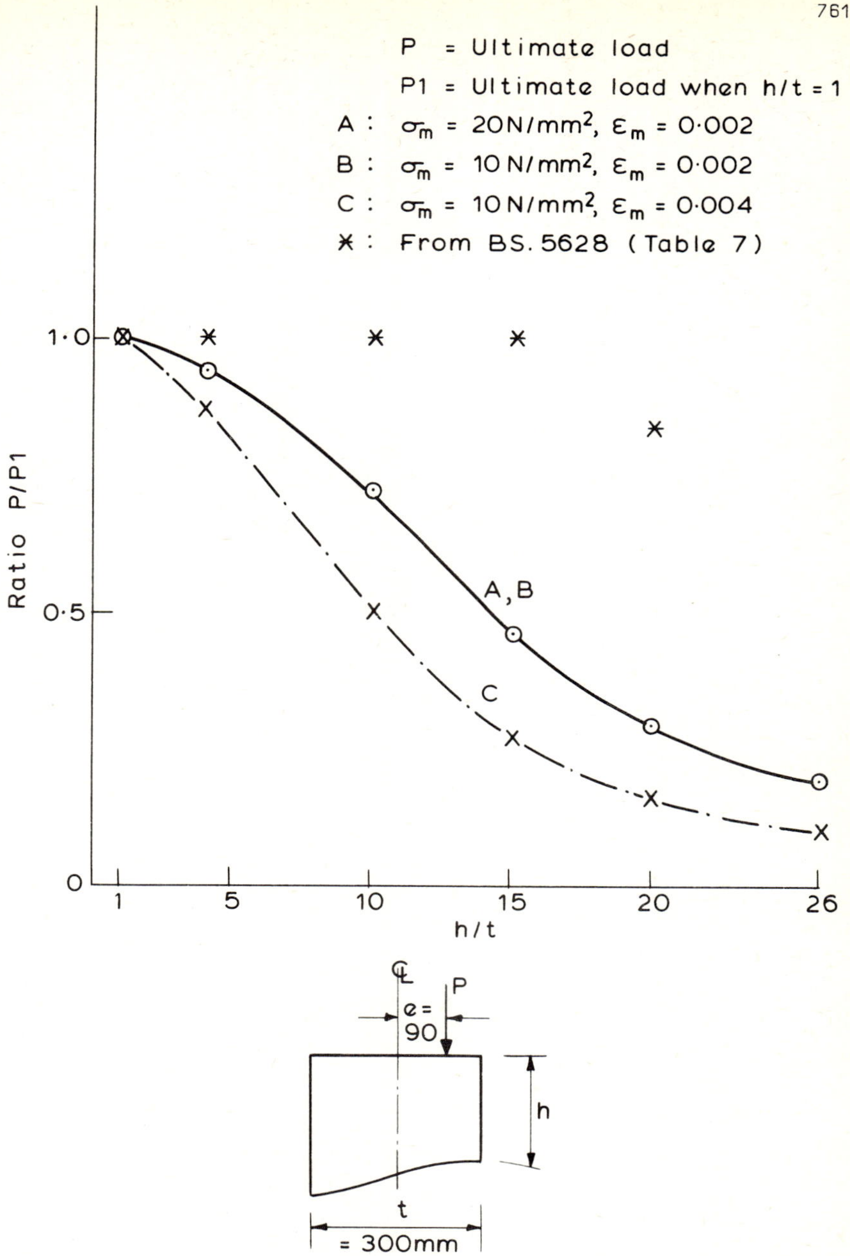

FIG. 6. ULTIMATE LOAD RATIOS FOR DIFFERENT SLENDERNESS RATIOS & BRICKWORK PROPERTIES.

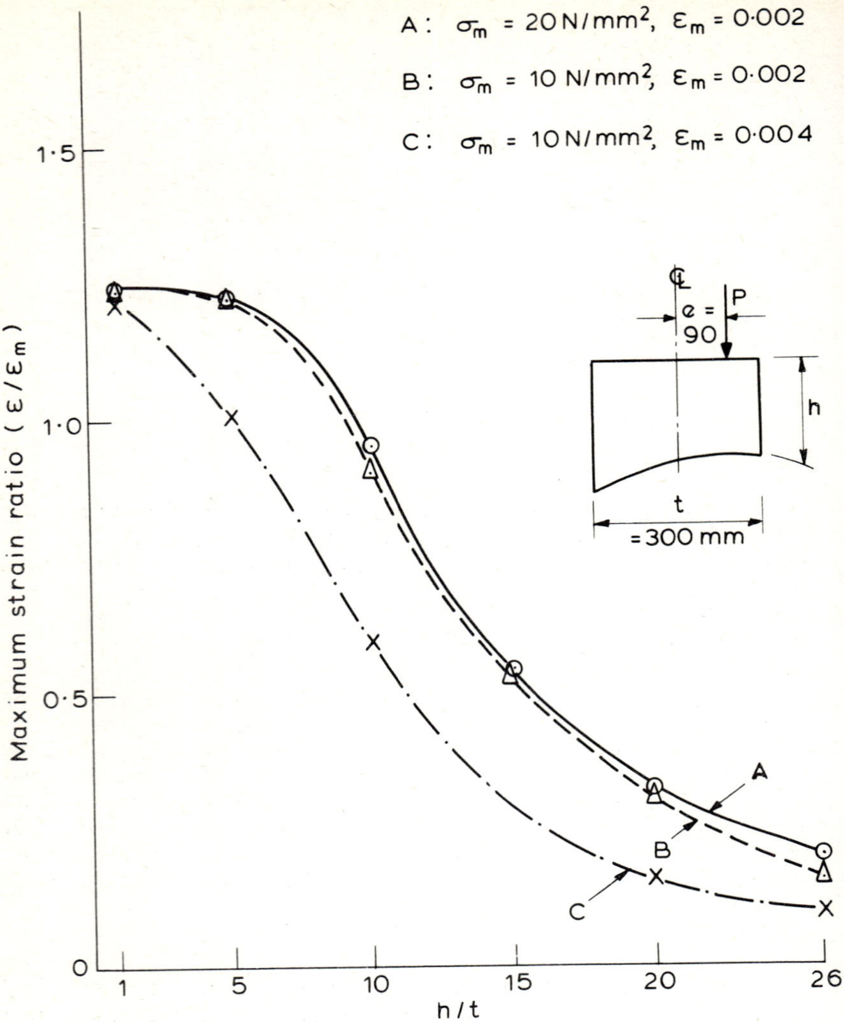

FIG. 7. MAXIMUM STRAIN RATIOS FOR DIFFERENT SLENDERNESS RATIOS & BRICKWORK PROPERTIES.

THE EFFECTS OF REPAIR PATCHES ON THE STRESS DISTRIBUTION IN A CRACKED SHEET

G. Dowrick[1], D.J. Cartwright[1] and D.P. Rooke[2]

SUMMARY

The effect of patch shape, attachment flexibility and patch stiffness on the stress intensity factor of a crack in a thin sheet is investigated analytically. The patch is considered to be made up of a series of strips perpendicular to the crack. Stress intensity factors and the force distribution along the crackline in the patch and along the middle and outer strip patches are determined for a rigidly attached patch of varying size and fixed stiffness, a square patch of fixed stiffness and variable attachment flexibility and finally a rigidly attached square patch of variable stiffness. The distributions of force are determined for two cracklengths, where the crack is completely under the patch and where the crack tip is ahead of the patch edge. It is shown that the effect of patch size is not very important but attachment flexibility and patch stiffness are significant in the reduction of the stress intensity factor.

INTRODUCTION

At present interest is being shown in the use of patches as a method of arresting the growth of cracks in damaged structures. For strong, light structures, made from aluminium sheet, such as used for the construction of aircraft, an efficient repair can be made by bonding a patch of composite material over the region of damage. Subsequent fatigue loading may cause cracks to grow from the region of damage in the aluminium sheet. Under these conditions, if the damage has been repaired by patching, there is a transfer of load from the sheet to the

1. Department of Mechanical Engineering, The University, Southampton SO9 5NH, England.

2. Materials Department Royal Aircraft Establishment, Farnborough, Hampshire, England.

patch which reduces the stress intensity factor of the crack below that for the unpatched panel. This reduction in the stress intensity factor must be known so that estimates of residual life of the repaired structure can be made and the failure stresses determined [1].

The effects of shape, size and orientation of the patch need to be studied so that the design repair patches may be optimised. Some preliminary work on the effect of circular patches on the stress intensity factor of a crack in a uniformly stressed sheet has been undertaken [2]. In the present paper this work is extended to the study of rectangular patches. Three rectangular patches are considered each having a fixed width 2W measured in a direction parallel to the crack. The length 2H of each patch, measured perpendicular to the crack, is taken to be 1.2W, 2W and 4W. The stress intensity factors, load concentrations in the patch at the crackline and the adhesive forces are presented for each of the patches. For a square patch results are presented for the effect of the shear deflection of the adhesive on the stress intensity factor and load transfer between the sheet and the patch: similar results are presented on the effect of patch stiffness.

MODEL OF PATCHED SHEET

The configuration analysed is shown in Figure 1. The infinite sheet of modulus E, Poisson's ratio ν and thickness t contains a crack of length 2a at the origin of coordinates z=0 ($z=x/a + iy/a$). The sheet is considered to be in a state of plane stress and is subjected to a uniform tensile stress perpendicular to the crack. The patch, of width 2W, height 2H and modulus E_p, is simulated with a series of contiguous strips of width W/10 and thickness t_p each connected to the sheet at N attachment points. The attachment points are spaced a distance W/10 apart with the distance from the crackline to the first one equal to W/20. In this relatively simple model there is no allowance for the transfer of longitudinal shear between each of the strip patches, except through the sheet. Also the model is restricted to a symmetrical array of attachments, i.e. the position and number of attachment points is symmetrical with respect to both the crackline and a line which bisects the crack at right angles. An important feature of the present formulation is that the effects of both cohesive flexibility and debonding may be simulated. Adhesive flexibility is simulated by allowing for shear deflection of the attachment points and debonding may be modelled by allowing for failure of the attachment points at a critical stress (or strain).

METHOD OF ANALYSIS

The condition that the displacement in the sheet and the patch be equal at each attachment point leads to the following equation of compatibility for the elongation of the interval

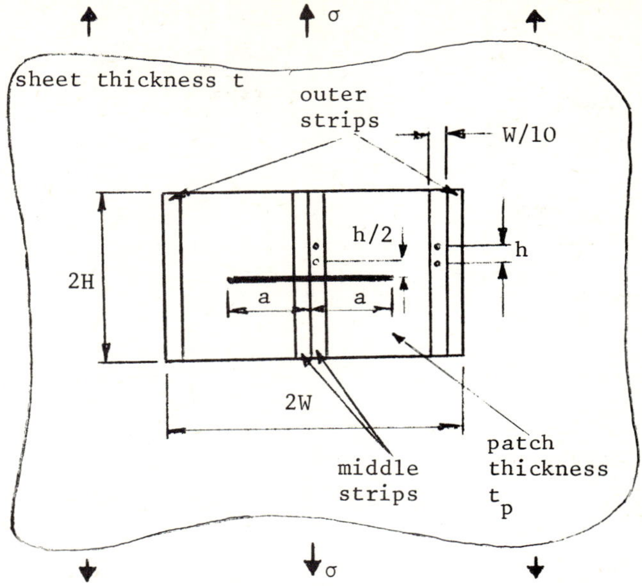

Figure 1: Model of Patch

between the nth and the (n+1)th attachments in the mth strip

$$\Delta v^a_{m,n} + \Delta v^b_{m,n} - \Delta v^c_{m,n} = \Delta v^d_{m,n} \qquad (1)$$

where $\Delta v^a_{m,n}$, $\Delta v^b_{m,n}$, $\Delta v^c_{m,n}$ and $\Delta v^d_{m,n}$ are the elongations of the interval resulting from (a) the attachment forces in the sheet, (b) the uniform stress on the sheet, (c) the attachment forces in the strip and (d) the deflection of the attachments.

The term $\Delta v^a_{m,n}$ is given by

$$\Delta v^a_{m,n} = \frac{(1+\nu)^2}{8\pi Et} \sum_{m=1}^{20} \sum_{\ell=1}^{N} P_{m,\ell} \ \mathrm{Im}\left[\Phi^a(z_{m,n+1}, z_{m,\ell}) - \Phi^a(z_{m,n}, z_{m,\ell})\right] \qquad (2)$$

where Im[ ] denotes 'imaginary part' and where $P_{m,\ell}$ is the force at the ℓth attachment point in the mth strip. The displacement function Φ^a, obtained by Erdogan [3], and used by Arin [4] and Cartwright and Rich [5] is given by

$$\Phi^a(z,z_o) = i\left[-\kappa \log|z-z_o|^2 + 1 - \frac{z-z_o}{\bar{z}-\bar{z}_o} + \frac{R^2}{(\bar{z}-\bar{z}_o)^2}\right] + \kappa U(z) - U(\bar{z}) - (z-\bar{z})\,\overline{\Phi(z)} \qquad (2a)$$

where $\kappa = \frac{3-\nu}{1+\nu}$, R is the rivet radius and U(z) and $\Phi(z)$ are defined in [5].

For the special case of the elongation of the interval between the crackline and the first attachment this becomes

$$\Delta v^a_{m,o} = \frac{(1+\nu)^2}{8\pi E t} \sum_{m=1}^{20} \sum_{\ell=1}^{N} P_{m,\ell} \, \mathrm{Im}\left[\Phi(z_{m,1}, z_{m,\ell})\right] \qquad (3)$$

The summation limit, N, is the total number of attachment points in a strip, above and below the crackline.

The term $\Delta v^b_{m,n}$ is given by

$$\Delta v^b_{m,n} = \frac{\sigma a(1+\nu)}{2E} \mathrm{Im}\left[\Phi^b(z_{m,n+1}) - \Phi^b(z_{m,n})\right] \qquad (4)$$

and

$$\Delta v^b_{m,o} = \frac{\sigma a(1+\nu)}{2E} \mathrm{Im}\ \Phi^b(z_{m,1}) \qquad (5)$$

where the displacement function Φ^b can also be obtained from [4] and [5], and is given by

$$\Phi^b(z) = \kappa\left[-\frac{z}{2} + (z^2-1)^{\frac{1}{2}}\right] - \left[\frac{\bar{z}}{2} + (\bar{z}^2-1)^{\frac{1}{2}}\right] - (z-\bar{z}).$$

$$\left[\frac{\bar{z}^2}{(\bar{z}^2-1)^{\frac{1}{2}}} - \frac{1}{2}\right] \qquad (5a)$$

The term $\Delta v^c_{m,n}$ is given by

$$\Delta v^c_{m,n} = \frac{h}{A_p E_p} \sum_{\ell=n+1}^{N} P^s_{m,\ell} \qquad (6)$$

and

$$\Delta v^c_{m,o} = \frac{h}{2A_p E_p} \sum_{\ell=1}^{N} P^s_{m,\ell} \qquad (7)$$

where $P^s_{m,\ell}$ is the force in the mth strip at the ℓth attachment and h is the attachment spacing.

The term $\Delta v^d_{m,n}$ is given by

$$\Delta v^d_{m,n} = q\ (P^s_{m,n+1} - P^s_{m,n}) \qquad (8)$$

and

$$\Delta v^d_{m,o} = q\ P^s_{m,1} \qquad (9)$$

where q is the compliance of the attachment point. It can be expressed in terms of the adhesive properties as

$$q = \frac{t_a}{\mu_a w_p h} \qquad (10)$$

where t_a and μ_a are respectively the thickness and shear modulus of the adhesive, w_p is the width of a strip patch and h is the attachment pitch.

Equilibrium at each attachment point is given by

$$P_{m,n} + P^s_{m,n} = 0 \quad (m = 1, 20 \quad , \quad n = 1, N) \tag{11}$$

Substituting equations (2) to (11) into (1) gives the following series of equations which can be solved for the unknown non-dimensional forces $\bar{P}_{m,\ell}$ at each attachment point.

$$\frac{(1+\nu)^2}{8\pi}\sum_{m=1}^{20}\sum_{\ell=1}^{N}\bar{P}_{m,\ell}\ \mathrm{Im}\left[\Phi^a(z_{m,n+1},z_{m,\ell}) - \Phi^a(z_{m,n},z_{m,\ell})\right] + \frac{\lambda\alpha\bar{h}}{2}\cdot$$

$$\sum_{\ell=n+1}^{N}\bar{P}_{m,\ell} + Q(\bar{P}_{m,n+1}-\bar{P}_{m,n}) = -\frac{\lambda(1+\nu)}{4}\cdot$$

$$\mathrm{Im}\left[\Phi^b(z_{m,n+1}) - \Phi^b(z_{m,n})\right] \tag{12}$$

where $\alpha = \frac{1}{2}$ when n=0, and $\alpha = 1$ when $n \geq 1$.

$$\bar{P}_{m,\ell} = \frac{P_{m,\ell}}{A_p E_p (\sigma/E)} \quad , \quad \lambda = \frac{2atE}{A_p E_p} \quad , \quad \bar{h} = \frac{h}{a}$$

$$Q = qEt.$$

The stress intensity factor is determined from the attachment forces

$$K_I = \sigma\sqrt{\pi a} + \frac{1}{t\sqrt{\pi a}}\sum_{m=1}^{20}\sum_{n=1}^{N} P_{m,n}\, g(z_{m,n}) \tag{13}$$

where $g(z_{m,n})$ is the non-dimensional Green's function for the stress intensity factor due to a pair of collinear and opposing unit forces in the sheet at $z_{i,j}$ and $\bar{z}_{i,j}$ acting away from and in a direction perpendicular to the line of the crack.

The total force at the crackline in the patch is given for the mth strip by

$$L_m = \sum_{\ell=1}^{N} P_{m,\ell} \tag{14}$$

NUMERICAL RESULTS

The results are divided into three sections, the effect on the stress intensity factor of a change in patch shape, and the effects of both rivet flexibility and stiffener ratio when the patch is square.

1. Effect of patch shape

Figure 2a shows the variation of the stress intensity factor as a function of the ratio of crack length (2a) to patch width (2W), for three sizes of patch; the aspect ratios (i.e. height to width) are 0.6, 1.0 and 2.0. The patch has a relative stiffness S ($=tE/t_pE_p$) equal to unity and the rivet attachments are rigid. It can be seen that the stress intensity factor reduces initially with increasing crack length until the crack tips reach the edge of the patch. While the crack is under the patch edge the stress intensity factor decreases with decreasing patch size. As the crack spreads beyond the edge of the patch the stress intensity factor increases rapidly, but now the stress intensity factor decreases with increasing patch size. It is thus important that the patch be large enough to always cover the crack. Since, if the crack projects beyond the edge of the patch, then the rapid increase in the stress intensity factor will mean a rapid increase in the growth of the crack under fatigue loads: crack growth rates in fatigue are approximately proportional to K^n where n lies between 2 and 4 for aluminium alloys.

Figures 2b and 2c show the attachment forces P_i at the ith attachment point (coordinates x_i, y_i from the patch centre) along each of two strip patches. Figure 2b shows the variation of load for the strip nearest the line perpendicular to the centre of the crack for a crack length a/W=1.3. Since the forces in the middle strip were found to be almost unchanged as the relative crack length is increased from 0.7 to 1.3, the graph for a/W=0.7 has been omitted. In both cases the large loads occur at the centre and ends of the strip, but for intermediate attachments the forces are close to zero. The large forces at the ends and at the crackline imply that there will be large shear stresses in the adhesive at these points. This means that debonding of the adhesive is more likely at these end points.

Figure 2c shows the variation of load in the outermost strip patch, for a relative crack size a/W=1.3. The corresponding graph for a/W=0.7 has been omitted since the load distribution is virtually unchanged. The only difference is that at the first attachment ($y_i/2W=0.05$), the force is zero, because, in this case, the crack has not yet reached the outer strip. Whereas in Figure 2c the forces are similar to those in the middle strip, large at the centre and edge of the strip for all patches and almost zero elsewhere.

The sum of all the attachment forces $\sum P_i$, in the strip patch (i=1 to N/2, where N is the number of attachment points), is the total load at the crackline. Figure 2d shows the normalised total load, $\sum P_i/\sigma Wt$, in the three patches. The relative stiffness S is equal to unity, and the patches are attached to the sheet with a rigid adhesive (Q=0). Figure 2d shows results

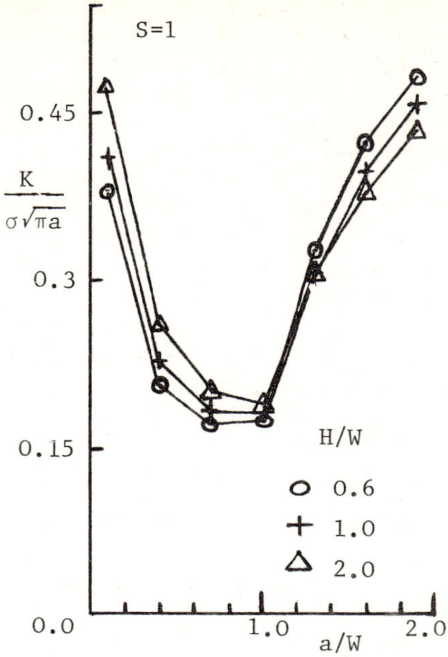

Figure 2a: Stress Intensity Factor

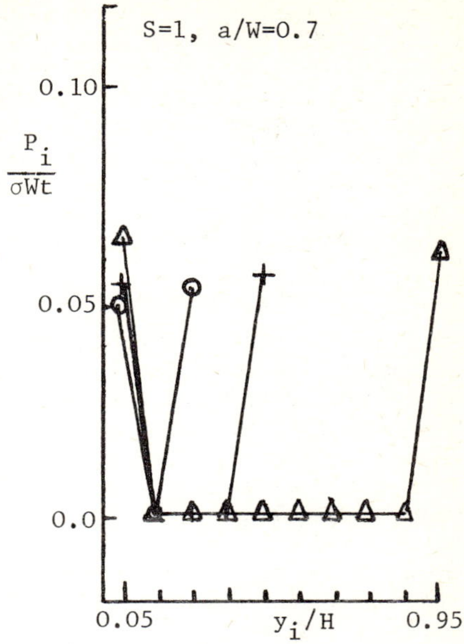

Figure 2b: Attachment loads along Middle Strip

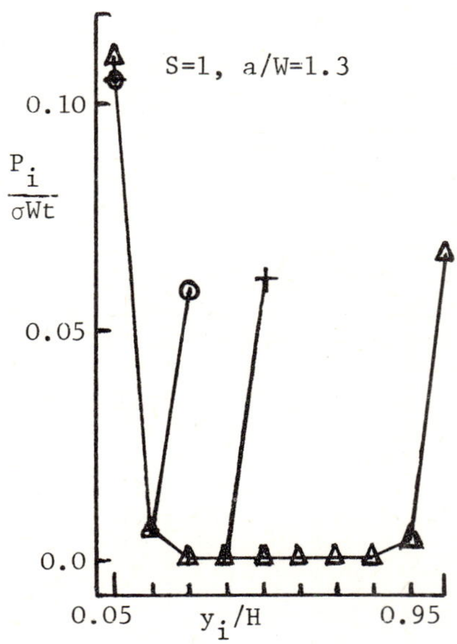

Figure 2c: Attachment Loads along Outer Strip

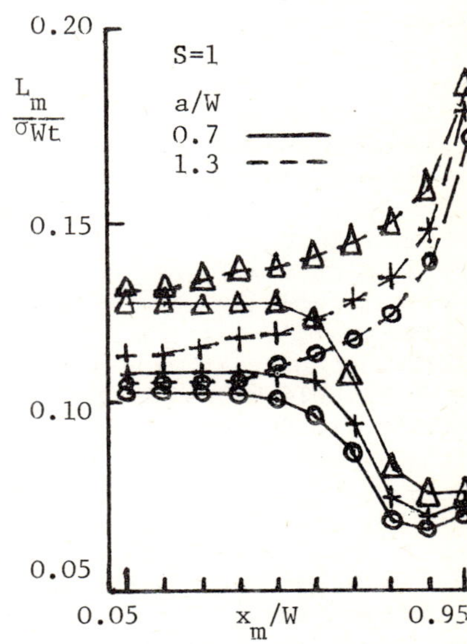

Figure 2d: Total Load at Crackline

Figure 2: Effect of Aspect Ratio for a Rectangular Patch(Q=0) for the different sizes of patch for the two crack lengths $a/W=0.7$ and 1.3. It can be seen that the total loads increase as the patch size increases. For $a/W=0.7$ the loads fall off sharply near the crack tip, whereas for $a/W=1.3$ they increase towards the edge of the patch, since the crack tip is ahead of the patch. For $a/W=0.7$, the distribution of load which occurs across the patch is approximately uniform for $x/W\leq 0.5$, i.e. when the patch covers the crack. This is not so for the longer crack $a/W=1.3$; the load in the patch at the crackline increases towards the edge of the patch. Thus the longer the crack, the larger the loads in the patch.

2. Effect of rivet flexibility

Figure 3a shows the stress intensity factor as a function of crack length to patch width for varying attachment flexibilities (Q). The results shown in Figure 3a and subsequent figures are for a square patch with ten attachment points, either side of the crack, in each strip. The patch is made up of twenty strips as before. The relative stiffness, S, of the patch is equal to unity. The stress intensity factor for Q equal to zero represents the case of a rigid adhesive, and Q equal to 5, 10 and 15 corresponds to increasing flexibility of the adhesive. The rapid increase in the stress intensity factor when the crack tip is beyond the patch is similar to that shown in Figure 2a. It can also be seen that flexibility in the adhesive, reduces the effect of the patch on the stress intensity factor. For typical values of the rivet flexibility, the stress intensity factor for $a/W\leq 1$ is approximately twice that for a rigid adhesive ($Q=0$).

Figures 3b,3c and 3d are of the same form as Figures 2b,2c and 2d, but show the effect of attachment flexibility on the attachment loads. Figure 3b again shows large loads at the centre and ends of the middle strip for a rigid rivet ($Q=0$), but as the attachments become more flexible the loads at the centre and edge of the strip reduce, producing a concomitant increase in the loads at the intermediate attachments; so that the loads are now more uniform. In Figure 3c the load at the first rivet is very small, because the crack has not yet passed under the outer strip since $a/W=0.7$. The distribution of load in the middle and outer strip for $a/W=1.3$ is very similar to that of Fig.3b except that the loads at the crackline are slightly higher.

The total load at the crackline, obtained by summing all the attachment loads P_i, has been obtained for various values of the attachment flexibility and is shown in Figure 3d. It is seen from Figure 3d that as the attachments become more flexible the loads decrease; this accompanied by an increase in the stress intensity factor (see Figure 3a). Figure 3d shows that the total loads are approximately uniform for $x/W\leq 0.35$ and $a/W=0.7$. For the longer crack ($a/W=1.3$), Figure 3d shows that the loads continually increase towards the edge of the patch.

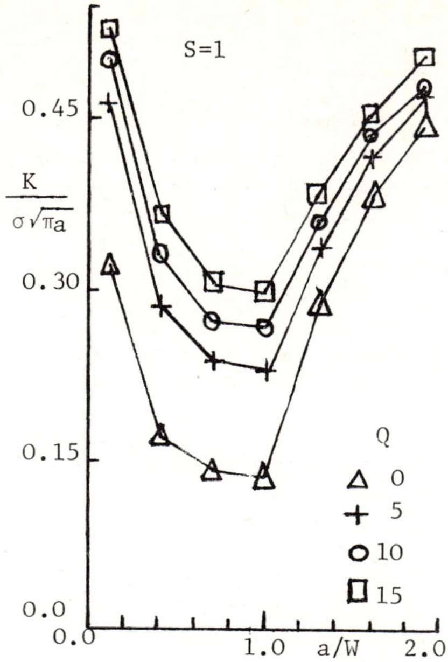

Figure 3a: Stress Intensity Factor

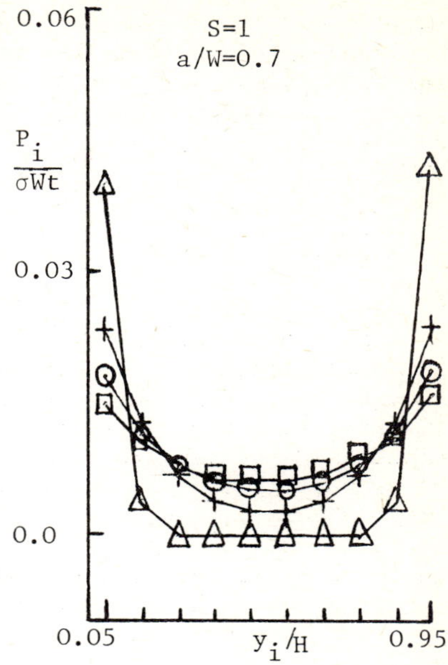

Figure 3b: Attachment Loads in Middle Strip

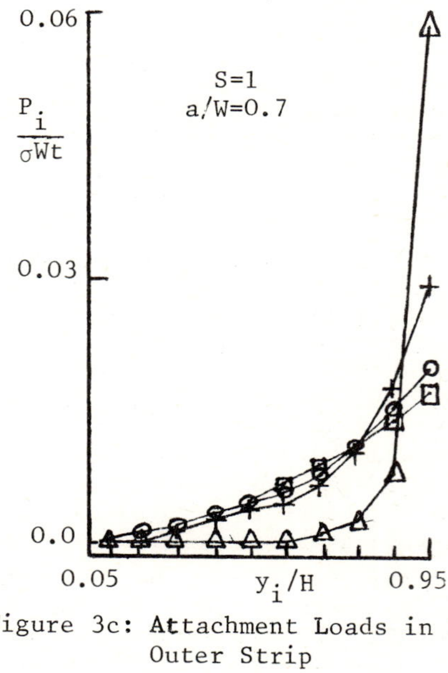

Figure 3c: Attachment Loads in Outer Strip

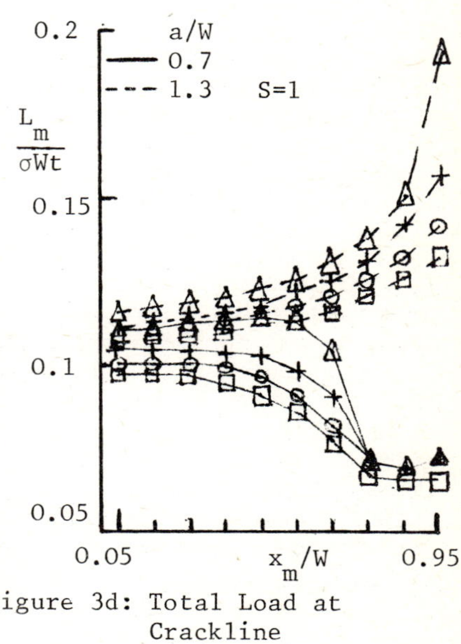

Figure 3d: Total Load at Crackline

Figure 3: Effect of Adhesive Shear Deflection for a Square Patch

3. Effect of stiffness ratio

Stress intensity factors and attachment loads have been calculated for two values of the relative stiffness (S=0.5 and 2.0). The calculations are for the same square patch as in the previous section, but with rigid attachments (Q=0) only. The stress intensity factor is shown in Figure 4a as a function of a/W for the two stiffnesses. It is seen that the stiffer patch (smaller S) produces a greater reduction in the stress intensity factor.

Figures 4b and 4c show the distribution of attachment loads in the middle and outer strips for S=0.5 and 2.0 for two crack lengths a/W=0.4 and 1.3. It is again seen that the large loads occur at the crackline and at the ends of the strips. For the middle strip (Figure 4b) the results for a/W=0.4 were virtually identical with those for a/W=1.3 and are not shown. The stiffer patch (S=0.5) results in larger loads at the end of the strip, but smaller loads at the crackline. This behaviour is also seen in the outer strip (see Figure 4c) for a/W=1.3. For a/W=0.4, the stiffer patch again results in larger loads at the end of the strip but the loads are zero at the crackline, as the crack tip has not reached that strip.

Figure 4d shows the distribution of total load along the crackline for relative crack lengths a/W=0.4 and 1.3, and for stiffness ratios S=0.5 and 2.0. For both a/W=0.4 and 1.3, the stiffer patch has the larger total loads. For a/W=0.4 the loads are approximately constant for $x/W \leq 0.25$. Also, for S=2 the loads are uniform for $x/W \geq 0.55$. For a/W=1.3 the loads are generally higher than for the smaller crack, increasing continually towards the edge of the patch. The loads fall ahead of the crack tip, for a/W=0.4, before rising again. For both crack lengths the total load is larger for the stiffer patch.

DISCUSSION

Effect of patch shape

While the crack is under the patch, the stress intensity factor is reduced for all three patches. In this case (rigid attachments) the smallest patch shows the greatest reduction; this will not necessarily be so for flexible attachments. In general, there is no great difference in the stress intensity factors for the three patches considered. If the crack crosses a strip the forces are large at the ends and middle of the strip, otherwise the forces are virtually zero at the crackline. If the crack is under the patch, strain at the crackline remains approximately constant for those strips which bridge the crack, and is approximately 50% higher than those in the strips ahead of the crack. However, if the crack tip is outside of the patch, large loads occur in the outer strip which could lead to the failure of the patch due to excessive static loads or by fatigue.

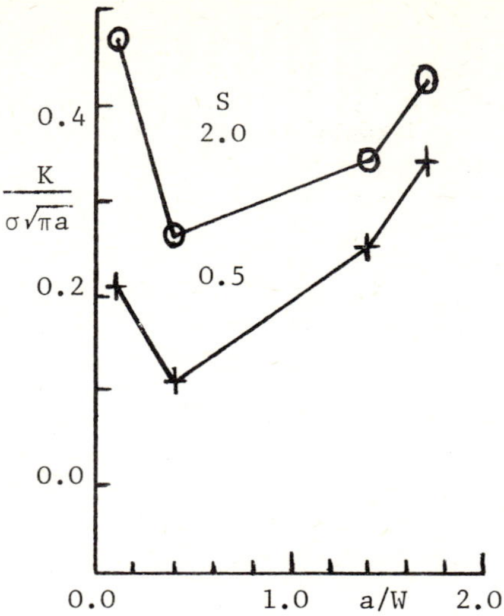

Figure 4a: Stress Intensity Factor

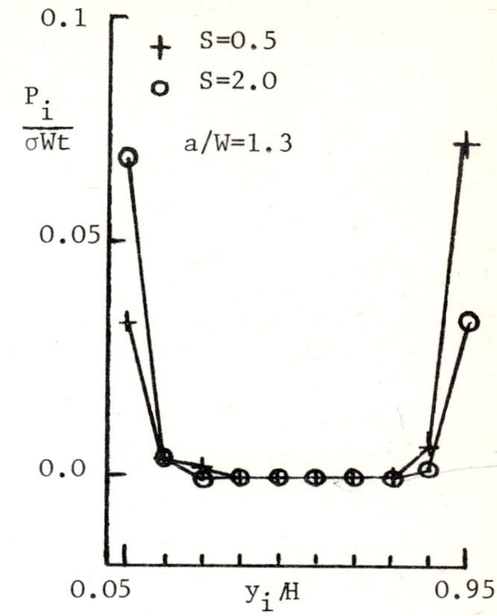

Figure 4b: Attachment Loads in Middle Strip

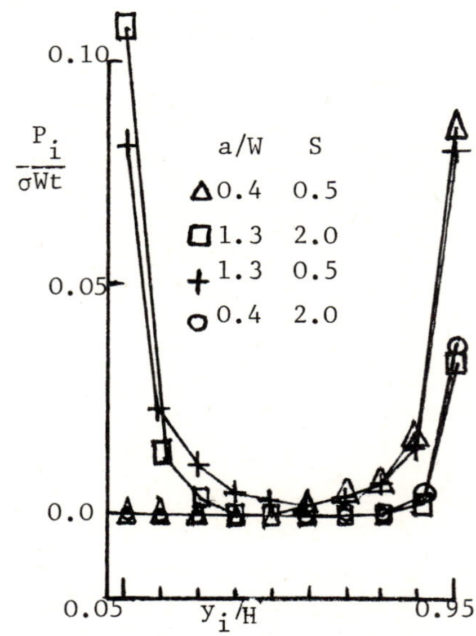

Figure 4c: Attachment Loads in Outer Strip

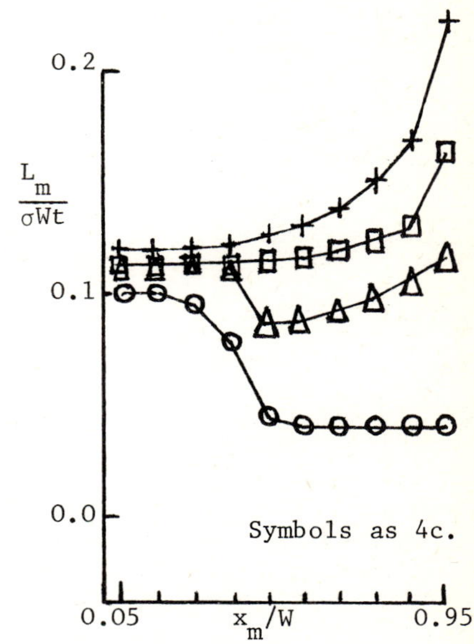

Figure 4d: Total Load at Crackline

Figure 4: Effect of Patch Stiffness for a Square Patch (Q=0)

Effect of attachment flexibility

The more flexible the adhesive the larger the stress intensity factor; the greatest change with crack length occurs for a rigid attachment. The large loads that occur when the crack crosses a strip are significantly reduced with increasing flexibility and the loads at the intermediate attachments raised. This has the effect of making the distribution of load more uniform.

Similar behaviour is seen with the total loads at the crackline, in that they are significantly reduced with greater flexibility of attachment, which is consistent with an increase in the stress intensity factor.

Effect of patch stiffness

A stiffer patch means a lowering of the stress intensity factor at the expense of higher total loads. However, if the strip bridges the crack the force nearest the crackline is smaller for the stiffer patch but the force at the end is larger for the stiffer patch.

CONCLUSIONS

For the patches considered, size does not appear to be very important. But the flexibility of the adhesive and the variation in patch stiffness appear to have a significant effect on the reduction of the stress intensity factor.

ACKNOWLEDGMENT

This work was supported by a grant from the Procurement Executive, Ministry of Defence, under agreement 2040/0203/MAT/RAE.

REFERENCES

1. RATWANI, M.M. - A parametric study of fatigue crack growth behaviour in adhesively bonded metallic structures. Trans. ASME, Journal of Engineering Materials and Technology, Vol.100, No.1, pp1-6, 1978.

2. DOWRICK, G. and CARTWRIGHT, D.J. - The effect of a circular reinforcing patch on a crack in a uniaxially stressed sheet. Report No. ME/80/2, Southampton University, January 1980

3. ERDOGAN, F. - On the stress distribution in plates with collinear cuts under arbitrary loads. Proc.4th U.S.Nat.Cong. on Applied Mechanics, pp.547-553, 1962

4. ARIN, K. - A plate with a crack stiffened by a partially debonded stringer. Engineering Fracture Mechanics Vol.6, pp.133-140, 1974.

5. CARTWRIGHT, D.J. and RICH, T.P. - Plastic strip yielding for a crack in a stiffened sheet. First International Conference on Numerical Methods in Fracture Mechanics, Eds.D.R.J. Owen and A.R. Luxmore, University College, Swansea, 1978.

6. ERDOGAN, F. and CIVELEK, M.B. - Contact problem for an elastic reinforcement bonded to an elastic plate. Trans. ASME J.Appl.Mech. Vol.41, pp.1014-1018, 1974

7. CARTWRIGHT, D.J. and DOWRICK, G. - Effect of attachment deflection on fatigue crack growth of a crack in a stiffened sheet. Engineering Fracture Mechanics, Ed. P. Stanley, Applied Science Publishers, 1976.

8. BLOOM, J.M. and SANDERS, J.L. - The effect of a riveted stringer on the stress in a cracked sheet. J.Appl.Mech. Vol.33, pp.561-570.

9. MUSKHELISHVILI, N.I. - Some Basic Problems of the Mathematical Theory of Elasticity, Noordhoff, 1953

10. DOWRICK, G. - The effect of repair patches on cracks in sheet materials. Report No.ME79/13, Southampton University, July 1979.

11. ROOKE, D.P. and CARTWRIGHT, D.J. - Compendium of Stress in Intensity Factors. Her Majesty's Stationery Office, London, 1976.

12. SWIFT, T. - The effects of fastener flexibility and stiffener geometry on the stress intensity in stiffened cracked sheets. Conference on Prospects of Fracture Mechanics, Delft, 1974.

13. POE, C.C. - Stress intensity factors for a cracked sheet with riveted and uniformly spaced stringers. NASA Tech. Report R-358, 1971.

FINITE ELEMENT ANALYSIS OF DEBONDING IN AN AXISYMMETRIC BIMATERIAL BODY

Per Gradin* and Jan Bäcklund**

Department of Mechanical Engineering, Linköping Institute of Technology, S-581 83 Linköping, Sweden

SUMMARY

In this paper the potential energy release rate in connection with debonding in an axisymmetric bimaterial body is considered.

The energy release rate is determined in three ways. In the first way it is obtained from the work performed by the external loads during a virtual crack extension and in the second and third ways from two path independent integrals derived from general expressions due to Broberg, Knowles and Sternberg. All three methods give results that closely agree. One of the integrals becomes very easy to implement in a FEM program having routines for calculation of the conventional J-integral for plane problems. An attempt was also made to use the result for a plane bimaterial body, to determine the energy release rate from the displacement of the crack faces near the crack front. The results showed, however, to be in very poor agreement with the other results. The reason is probably that the region considered near the crack tip is to large to reflect the plane deformation behaviour that will prevail close to the crack front.

ABSTRACT

The increased use of fibrous composites in various applications has drawn the interest to how and when such materials collapse.

Fibrous composites can collapse in various ways when subjected to some kind of loading. The basic mechanisms are matrix breakage, debonding and fibre breakage, and failure of a structure will probably be due to some combination of these mechanisms.

The topic of this paper is to study debonding and in

---

* Research assistant
** Associate professor

particular the release rate of potential energy, which (cf Salganik and Malyshew [1]) is assumed to govern the onset of debonding.

Restricting ourselves to an axisymmetric case, the potential energy release rate is determined in three different ways. The particular model considered in this paper consists of a cylinder with a bar, representing the fibre, in its centre. A concentrated force is applied to the upper end of the fibre while a part of the boundary of the cylinder is held fixed.

Firstly the energy release rate is determined from the work performed by the external load during a virtual crack extension, and secondly and thirdly by means of two path independent integrals.

Using Smelser's [2] result for a plane bimaterial body an attempt was made to determine the energy release rate from the displacements of the crack faces near the crack front (tip), but the result showed up to be in very poor agreement with the other results.

INTRODUCTION

In order to derive the path independent integrals by which the potential energy release rate can be determined in the case of axial symmetry, the two surface independent integrals R_k and J_k defined by eqs (1) and (2) respectively are considered

$$R_k = \int_S \frac{1}{x_k}(wn_k - n_j\sigma_{ji}u_{i,k})ds - \int_V \left(\frac{\sigma_{ki}u_{i,k}}{x_k^2} - \frac{w}{x_k^2}\right)dV \quad (1)$$

(k is not summed over)

$$J_k = \int_S (wn_k - n_i\sigma_{ji}u_{i,k})ds \quad (2)$$

In eqs. (1) and (2) S is the boundary over which the integration is performed and V is the volume bounded by S, σ_{ij} is the stress tensor, w is the strain energy density, n_i is the outward normal to S, u_i is the displacement vector and x_k finally is the radius vector to a point on or inside S.

R_k and J_k was derived by Broberg [3] and Knowles and Sternberg [4], respectively. Now, it is easy to show that if the stress tensor is given by $\sigma_{ij} = \frac{\partial w}{\partial \varepsilon_{ij}}$ where ε_{ij} is the stress tensor and if there is no singularity inside S, then

R_k and J_k equals zero for every k=1,2,3. Furthermore, Budiansky and Rice [5] show that J_k integrated over a surface enclosing a traction free crack, is equal to the potential energy release rate when the crack is translated in the k-direction.

R_k AND J_k SPECIALIZED TO AXIAL SYMMETRY

Now consider an axisymmetric case with $x_3(=z)$ as rotation axis, and an integration volume V according to Fig 1 (a similar derivation can be found in Bergkvist [6]).

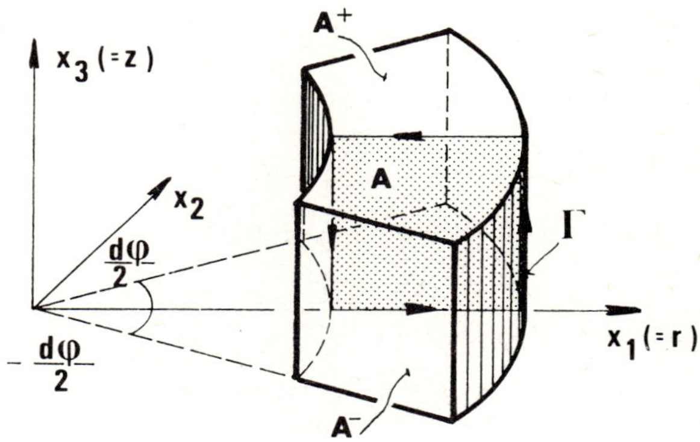

Fig 1 Integration volume.

Using the well known transformation formulas and specializing to small angles $\phi \ll 1$ one obtains

$$\begin{aligned}
\sigma_{11} &\simeq \sigma_r \\
\sigma_{22} &\simeq \sigma_\phi \\
\sigma_{12} &\simeq (\sigma_r - \sigma_\phi)\phi \\
\sigma_{33} &= \sigma_z \\
\sigma_{31} &= \sigma_{rz} \\
\sigma_{32} &\simeq \sigma_{rz} \cdot \phi
\end{aligned} \tag{3}$$

and also $$\frac{\partial u_1}{\partial x_3} \simeq \frac{\partial u_r}{\partial z}$$

$$\frac{\partial u_2}{\partial x_3} \simeq \frac{\partial u_r}{\partial z} \cdot \phi \qquad (4)$$

$$\frac{\partial u_3}{\partial x_3} \simeq \frac{\partial u_z}{\partial z}$$

Putting k equal to three in eq (1) yields

$$R_3 = \int_\Gamma \frac{1}{x_3}(wn_3 - n_j\sigma_{ji}u_{i,3})x_1 \cdot d\phi ds + \int_{A^+ + A^-} \frac{1}{x_3}(wn_3 - n_j\sigma_{ji}u_{i,3})dA - \int_A (\frac{\sigma_{3i}u_{i,3}}{x_3^2} - \frac{w}{x_3^2})x_1 \cdot d\phi dA$$

Now the following must hold

$$\text{On } A^+:\ n_1 \simeq -\frac{d\phi}{2};\ n_2 \simeq -1;\ n_3 = 0$$

$$\text{On } A^-:\ n_1 \simeq -\frac{d\phi}{2};\ n_2 \simeq 1;\ n_3 = 0$$

The above together with eqs (3) and (4) give

$$\frac{R_3}{d\phi} = J_r = \int_\Gamma \frac{r}{z}(wn_z - n_j\sigma_{ij}u_{i,z})_{i,j=r,z}\, ds - \int_A \frac{r}{z^2}(\sigma_{rz}\frac{\partial u_r}{\partial z} + \sigma_z\frac{\partial u_z}{\partial z} - w)dA \qquad (z \neq 0) \qquad (5)$$

In a similar way one obtains from eq (2)

$$\frac{J_3}{d\phi} = J_z = \int_\Gamma r(wn_z - n_j\sigma_{ij}u_{i,z})_{i,j=r,z}\, ds \qquad (6)$$

It should be noted that J_z doesn't contain any surface integral.

The problem to be delt with involves the case of a crack in the interface between two different materials, see Fig 2, and it must be shown that J_r and J_z continues to be path independent also in this situation. (cf Smelser and Gurtin [7]). Consider for this purpose Fig 3.

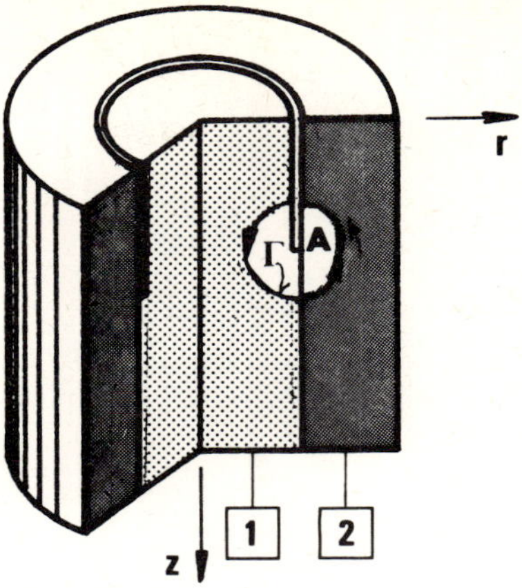

Fig 2 An axisymmetric crack in the interface between two different materials (1) and (2)

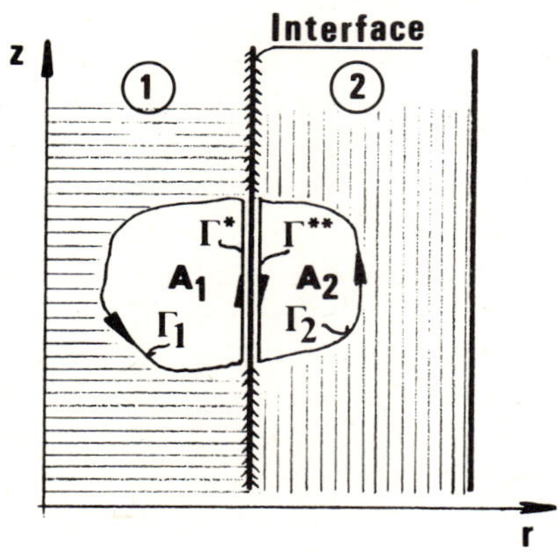

Fig 3 Integration paths used to show the path independency of J_r and J_z

Using square brackets to denote the jump in a quantity across the interface the following conditions must hold, (j=r,z)

$$\begin{aligned} [\sigma_{jr}] &= 0 \\ [u_j] &= 0 \\ [u_{j,z}] &= 0 \end{aligned} \tag{7}$$

The last relation is a consequence of $[u_j] = 0$.
J_R for region (1) becomes

$$J_{R1} = \int_{\Gamma_1} (\quad)ds + \int_{\Gamma^*} (\quad)ds - \int_{A_1} (\quad)dA = 0$$

and for region (2)

$$J_{R2} = \int_{\Gamma_2} (\quad)ds + \int_{\Gamma^{**}} (\quad)ds - \int_{A_2} (\quad)dA = 0$$

Adding these equations and using (7) one obtains

$$\int_{\Gamma_1+\Gamma_2} (\quad)ds - \int_{A_1+A_2} (\quad)dA = 0 \tag{8}$$

Eq (8) shows that J_R is independent of the integration path and considering a traction free crack in the interface between two different materials it is clear that J_R has the same value irrespective of the integration path taken around the crack tip.

In the same way path independency of J_z is shown. Now it will be shown that under the assumption that a state of plane deformation prevails near to the crack tip, J_R and J_z are related to each other in a simple way. Stresses and strains near the crack tip under such an assumption will be of the form (cf Williams [8])

$$\sigma_{ij}, \varepsilon_{ij} \sim \frac{1}{\sqrt{\delta}} f(\theta) \cos(\beta \ln \delta) \tag{9}$$

δ and θ are defined according to Fig 4, β is a constant dependent of the elastic properties and f is some function of the angle θ.

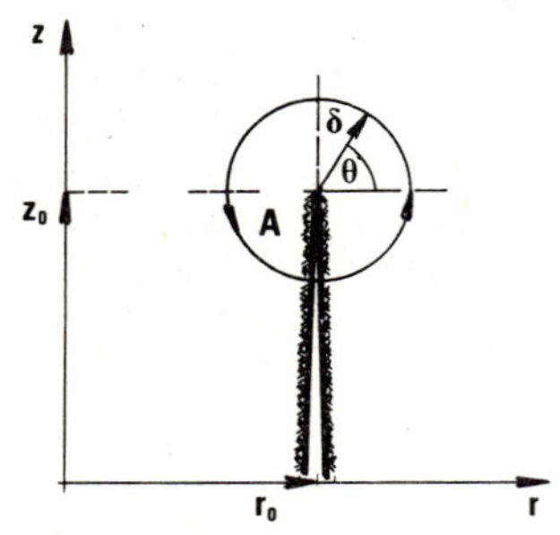

Fig 4 A circular integration path around the crack tip

The integrand in the surface integral of eq (5) is of the type stress times strain. Now, using eq (9), noting that $dA \sim \delta \cdot d\delta$ and letting δ tend to zero, that is, letting r and z tend to r_0 and z_0 respectively, the surface integral contribution will vanish so that

$$J_R = \frac{r_0}{z_0} \int_\Gamma (wn_z - n_j \sigma_{ij} u_{i,z}) \, ds \qquad i,j=r,z$$

J_z for the same integration path will become

$$J_z = r_0 \int_\Gamma (wn_z - n_j \sigma_{ij} u_{i,z}) \, ds \qquad i,j=r,z$$

Comparison of the above expressions yields

$$J_z = z_0 J_R \qquad (z_0 \neq 0) \tag{10}$$

But since J_z and J_R are path independent, eq (10) will hold for every path Γ surrounding the crack tip. As was mentioned in the introduction, J_k was the potential energy release rate when the crack was translated in the k-direction. Following [5] this statement can be shown to hold also for the bimaterial case so that J_R and J_z are equivalent measures of the energy release rate.

NUMERICAL RESULTS

The integrals defined by eqs (5) and (6) were implemented into the FEM program PIFEM [9]. The model treated was the one shown in Fig 5 and the FEM breakdown and integration paths for the crack length a = 40 mm is shown in Fig 6. The side nodes on sides adjacent to the crack tip have been moved to the quarter points in order to obtain the square root singularity at the crack tip.

The potential energy release rate from the work obtained by the external load was calculated in the following way. Holding the external load constant, the displacement of the loaded boundary is calculated for the crack lenghts a and $a + \Delta a$, the work being W_1 and W_2 respectively. The energy release rate $-\partial u/\partial a$ is approximately given by

$$-\frac{\partial u}{\partial a} = \frac{1}{2\Delta a} (W_2 - W_1) \tag{11}$$

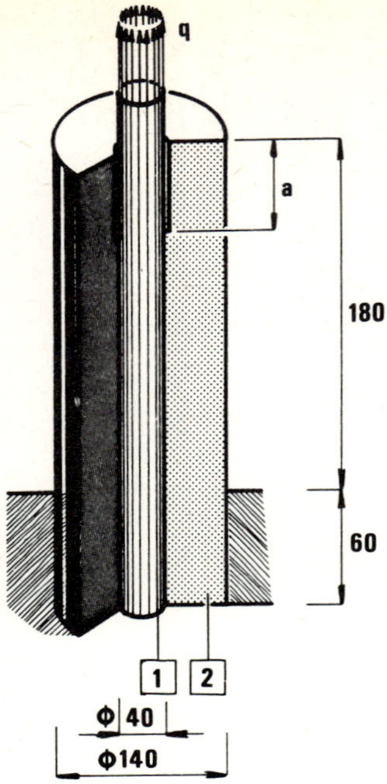

Fig 5 Model used for the calculations

q = 1.o MPa, E_1 = 70000 MPa, ν_1 = .3
E_2= 10000 Mpa, ν_2 = .35

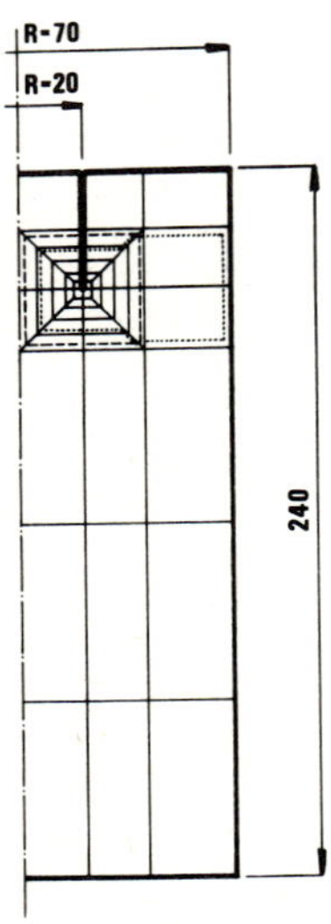

Fig 6 FEM breakdown and integration paths

In Fig. 7, finally, the potential energy release rate versus crack length obtained by use of eqs (5) together with (10), (6) and (11) is given. The results according to eqs. (5) and (6) are taken as the average of the two integration paths, the maximum deviation from the mean value was for J_R about 1 per cent and for J_z 0.5 per cent.

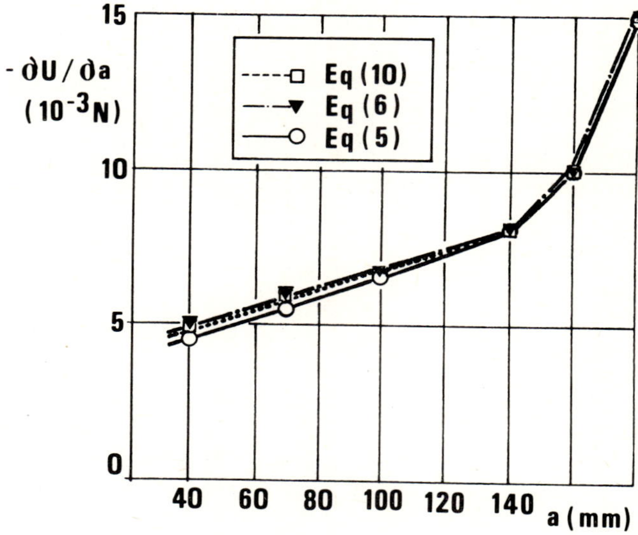

Fig 7 Potential energy release rate

DISCUSSION

It has been shown that the general integrals, due to Broberg and Knowles, Sternberg can be specialized to and used in an axisymmetric bimaterial case to calculate the potential energy release rate, a parameter of primary interest when investigating for example debonding. It was also pointed out that one of the integrals, namely J_z, did not contain any surface integral and is therefore very easily incorporated in a FEM program already having routines for calculating the conventional J-integral. This is done, simply by multiplying the integral with the distance from the rotation axis to the point in question and also adding an extra term due to the circumferential stress and strain, to the strain energy density.

REFERENCES

[1] B M Malyshev and R L Salganik, Int Journal of Fract Mech Vol 1, pp 114-128, 1965

[2] R E Smelser, Int Journal of Fract, Vol 15, pp 135-144, 1979

[3] B Broberg, GALCIT - report No SM 76-9, Pasadena 1976

[4] J Knowles and E Sternberg, Arch Rational Mech and Anal, Vol 44, pp 187-211, 1972

[5] B Budiansky and J Rice, Journal of Appl Mech, Vol 40, Ser E, pp 201-203, 1973

[6] H Bergkvist and Guex-Le Lan Huong, Paper presented at the Int Conf on Fract Mech and Technology, Hong Kong, March 21-25, 1977

[7] R E Smelser and M E Gurtin, Int Journal of Fract, Vol 13, pp382-384, 1977

[8] M L Williams, Bulletin of the Seismological Soc of America, Vol 19, pp 199-204, 1959

[9] J Bäcklund, H Wennerström, K Axelsson, PIFEM (in Swedish), LiTH-IKP-S-067, Linköping Inst of Technology, 1976

Some Computational Studies of the use of the J Contour Integral in Elastic-Plastic Fracture

C.E.TURNER

Professor of Materials in Mechanical Engineering, Imperial College

(Additional paper for Section 4)

SUMMARY

The plausibility of a single parameter description of fracture in the elastic-plastic regime is examined in terms of J. Three particular aspects are discussed: the characterising role of J to describe onset of separation, the energetic use to describe unstable growth and a simple design curve of J versus strain. It is concluded that with incremental behaviour, a one-parameter model is inherently inadequate because of possible changes in triaxiality even in plane strain and that the energy release rate in the presence of plasticity, termed I, is not necessarily either G or J. The distinction between deep and shallow notch cases, between extensive and contained yield and the use of a toughness value beyond initiation, is outlined in relation to the development of a single design curve type approach.

The role of computing in assisting the development of practical solutions for use in fracture mechanics has been of the greatest importance. In lefm such assistance has ranged from numerical solution of analytically formulated problems for awkward configurations or boundary conditions, to the complete re-formulation of problems in finite element, or recently boundary integral terms. Nevertheless, it is probably fair to say that this contribution has resulted in the extension of case studies and accuracy of approximate solutions rather than in a major step of basic understanding, since the latter already existed from the introduction of the concept of the energy release rate, G, and stress intensity factor, K, in conventional studies. Extension of fracture mechanics into plasticity has required and indeed, still requires not only data from case studies, but also improved insight into the fundamentals. As one who has never mastered the intricacies of programming (or indeed even the ever-changing codes and passwords for connecting a terminal to a stored computer program) I lay down the challenge to the numerical analysts, of whether they can offer insight in terms understandable to an engineer more readily and convincingly than the conventional theoretician.

In plasticity most of the modelling has been conducted on the macroscale and my remarks will be confined to that approach. It is self-evident that studies of fracture per se cannot be conducted on the computer. The studies must clearly be of the models proposed for fracture or of the mechanics of the situation. However convincing or otherwise such results may seem, experiment must be the final arbiter of whether some theory or other is useful in practice. This is equally true of many branches of science where physical or mechanical properties dominate the behaviour in question. Nevertheless, it has proben useful if not essential to establish the behaviour of "ideal" cases - perfect gases, inviscid fluids, Newtonian viscosity, linear elasticity, isotropic behaviour and so on, as a datum for further advance. Too simple a model may preclude the very behaviour in question - such as no aerodynamic drag in an inviscid gas, or no dissipation of energy in a perfectly elastic body and one of the key points is to perceive when the model is inherently too simple for the purpose in hand.

In the realms of yielding fracture mechanics two concepts, COD and J, appear to dominate all others. The writer has argued elsewhere {1} that in terms of understanding, or expressed as theoretical concepts, the several proposals such as equivalent energy {2}, non-linear energy {3}, stress concentration zone {4}, tangent modulus {5}, gross yielding {6}, modified COD {7}, two parameters {8}, two -criteria {9}, and of course, lefm with plastic zone correction, can all be expressed in terms of J or related so closely to J that there is no point in discussing each separately. These several theories were summarised in Ref.10. In using J as an umbrella for all these methods two reservations must be noted. The procedure advocated by one or other of the above methods may have great virtue and the development of a concept into a practically useful design methodology may be more advanced in one case than another. Examples are the use or otherwise of full thickness test pieces advocated by the proponents of COD and the design methodology developed for both COD or two criteria procedures. In short, the point is that in so far as all these concepts use or imply a single parameter description of fracture, they are all equally limited. In so far as test procedures or additional analyses bring in extra parameters, then the practical usefulness of the method may be enhanced. However, in the writer's opinion the dominant question is to what extent is a single parameter model of fracture adequate, and in that sense the writer sees no proposal more embracing than J, despite the limitations of J that have emerged.

WHAT MATERIAL BEHAVIOUR?

A rigorous interpretation of J requires non-linear elastic (nle) material. With real plasticity materials at least four distinctions from nle arise. The first is the change in behaviour from Hookean elastic to plastic, implying the introduction of a yield criterion and different laws for the volumetric components of strain. The second is the nature of the flow rule, incremental plasticity being commonly accepted as a more realistic description of real materials than the total or deformation theory descriptive of nle. The third and most obvious different is the dissipative nature of the deformation and the fourth is the linear elastic nature of unloading. Consider first of all only monotonic loading which leads to the initiation of separation from a sharp defect. It may be recalled that even in strict lefm onset of separation is not necessarily unstable, for example, when loaded by a splitting force such that K reduces with crack length. Such a stable event need not be caused by a release of internal energy but can be driven by external work which in fact also increases the internal (strain) energy. Conversely, a simple picture of unstable growth requires a surplus of energy in a static analysis, which surplus provides for the kinetic energy of separation. It therefore seems clear that a general statement for a criterion of onset of separation must relate to a characteristic severity of the crack tip deformation, and even in strict lefm it is only in particular, albeit rather common configurations that the associated rate of change of energy given by $G = K^2/E'$, may be a release that in the absence of other dissipations provides a "driving force" for the crack.

A combination of stress and crack length that reaches a particular value of G equal to the material toughness R may be thought of as causing an incipient instability, but it will not cause actual unstable crack growth, unless also dG/da > dR/da, which for material of constant toughness implies dG/da + ive. The point that is being belaboured, perhaps unnecessarily, is that although the dG/da tangency concept {11} is well accepted for predicting unstable growth in plane stress situations, in which case there is indeed an increasing apparent toughness that is reflected in a rising R curve as the crack tip plastic zone grows in size, the requirement for a + ive gradient in G to cause unstable growth, also holds in any situation, even plane strain lefm. However, for many configurations, particularly when loaded by a constant force, dG/da is inherently + ive and for such cases (with R constant) unstable growth will occur when G = R (i.e. G_{Ic}) because the conditions for providing an increasing energy release rate (i.e. dG/da + ive) are satisfied at any stress level, so that reaching $G = G_{Ic}$ becomes the controlling feature. As these two requirements are extended into plasticity there are two complications: the characteristic severity is not identifiable with the energy release rate (i.e. there is no direct analogue of $K^2 = E'G$) and R ceases to be readily identifiable as a surface dissipation, it now being dominated by the related dissipations in the ever-growing plastic region that for a sufficiently tough material would ultimately extend through the bulk of the component. For nle material McClintock {12} expressed the crack tip stress and strain patterns in terms of the HRR solutions in the form of $J^{1/n+1}$ etc., and the energy release rate is, of course, J. Thus, in the nle case the unique relationship between the characteristic and energetic term remains and since there is no dissipation other than true surface energy, an exact parallel exists with lefm but with G replaced by J.

Although G is usually thought of as G = dU/Bda where U is strain energy (or more generally in terms of P where P is potential energy of both the internal and external systems) it can also be written in terms of the work per unit ligament area, Bb, as

$$G = \eta_{el}\, w/Bb \qquad \text{Eqn.1}$$

where w = Qq/2, the (elastic) work done and η_{el} is a factor that depends upon the geometry in question {13}. As discussed {14} there are several expressions for η such as

$$\begin{aligned} \eta &= (b/w)(dw/da) && \text{Eqn.2a} \\ &= (b/q)\,(\partial q/\partial a|_Q && \text{b} \\ &= -(b/Q)(\partial q/\partial a|_q && \text{c} \\ &= (b/\phi)\,(d\phi/da) && \text{d} \end{aligned}$$

where ϕ is the compliance. For any configuration for which K is known as

$$K = Y\sigma\sqrt{a} \qquad \text{Eqn.3}$$

then (as in Ref.13) $\eta = bY^2a/\int Y^2 ada$ Eqn.4

and the unnotched compliance is used to evaluate the constant of integration in the denominator. A similar expression

$$J = \eta_{nl}\ w/Bb \qquad \text{Eqn.5}$$

can be written for a power law load deflection diagram $Q\alpha q^n$, $w = Qq/(n+1)$. It has been shown {14},{15}, that η_{nl} exists as a constant (i.e. dependent only on a/W and independent of degree of deformation) if the variables of geometry (i.e. a/W) and deformation are separable. Data from incremental plasticity oomputations was given {10},{14}, to show the variation of η_o with configuration and degree of deformation where η_o is defined in relation to the overall values of J and w

$$J = \eta_o\ w/Bb \qquad \text{Eqn.6}$$

as distinct from elastic or plastic components or nle estimates. A little more data is shown Fig.1, to relate plane stress and plane strain values of η_o for the same configuration and to show the near constancy or otherwise of η_o in the contained yield region.

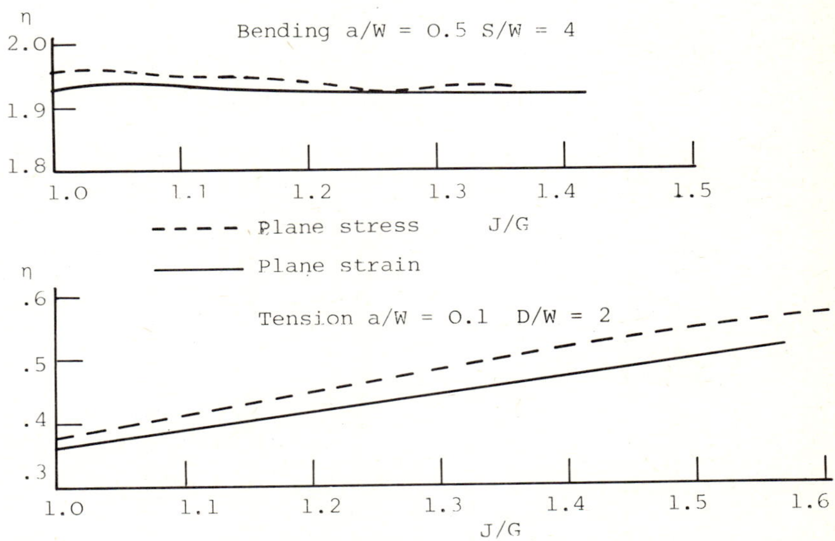

Fig.1 η values in plane stress and plane strain

It can be argued, though not perhaps rigorously, that a significant departure of the value η_o from the original value η_{el} implies a loss of a one-parameter crack tip stress state {16} and Appendix) although there is no direct evidence other than experimental fracture data on how large a change causes a change in the fracture behaviour. Variations of η_o with deformation can also be found experimentally by using the techniques of Ernst et al {17} who, at that time, evaluated Eqn.2c without denoting it by the symbol η. It is suggested that further study of the constancy of otherwise of η_o with degree of deformation and the reality of the suggestion that $\eta_o \simeq \eta_{el}$ is a measure of the existence of a one parameter crack tip system could well be fruitful.

UNSTABLE CRACK GROWTH

If it is argued that energy release rate is relevant to unstable behaviour then, with plasticity, the most obvious difference from elasticity is its irreversible nature, followed by linear elastic unloading. The latter feature is relative easily incorporated and the term I, Fig.2, defined as the "elastic energy release rate in the presence of plasticity" has been introduced {10},{18}.

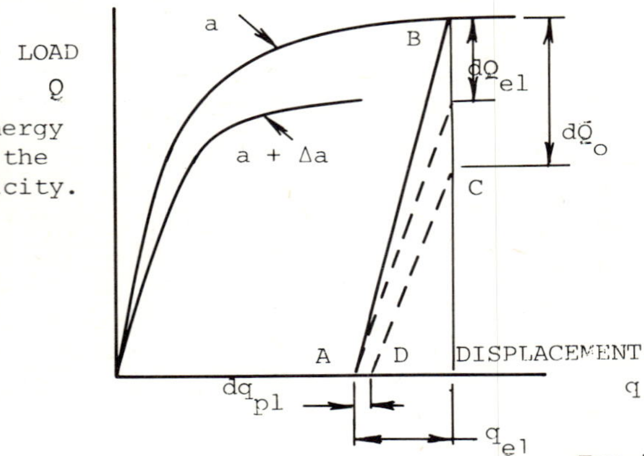

Fig.2. Elastic energy release, ABCD, in the presence of plasticity.

$$I = J_{el} - q_{el}\left(\frac{\partial Q_o}{\partial a} - \frac{\partial Q_{el}}{\partial a}\right)\Big|_q \delta a \qquad \text{Eqn.7}$$

whence, since $q_{el} = 2w_{el}/Q_o$

$$I/J_{el}\ (\equiv I/G) = (2\eta_o/\eta_{el}) - 1 \qquad \text{Eqn.8}$$

In lefm $\eta_{el} = \eta_o$ and $I/G = 1$, of course. In general, the problem is that the change in load $\partial Q_o/\partial a$ or the component of change $(\partial Q_o/\partial a - \partial Q_{el}/\partial a)$ due to non-linear behaviour depends upon the stress-strain behaviour assumed and the configuration of component or nature of loading. Using Eqn.2c, for a given value of Q the change $\partial Q/\partial a$ would be independent of the stress-strain law if η were a function of configuration only, (i.e.

$\eta_o = \eta_{el}$). It has been argued above that constancy of η_o relates to the relevance of a one-parameter system and it is now seen from Eqn.8 that where (or if) this is so, then (for linear unloading), I ≡ G. For nle, of course, I = J, since Fig.2 is inappropriate. If it is also accepted that G corrected for size of plastic zone, here called G_p, is now viewed as an estimate of J, then for those cases where $\eta_o \simeq \eta_{el}$, the energy release rate I is not G_p (unless nle material) but nearer G. In short, in the use of lefm R curves for small-scale yielding, it is not self-evidently desirable to correct for size of plastic zone if instability is seen as a balance of energy rates, and would appear to be wrong to do so for a circumstance where in contained yield $\eta_o \simeq \eta_{el}$. For circumstances where $\eta \simeq \eta_{el}$, then I > G, though there is still no strict argument whether or not I = G_p. As a general statement (with some reservation on the value attached to J according to the flow rule used) G < I < J. The difference between J and either G or J can clearly be greatest in gross yield where J >> G. To the extent that in contained plasticity J ≃ G, then the differences discussed are irrelevant, but if use of G_p is supported as a better estimate of J than is G, then the relevance of the difference between I and G or J becomes significant. Where a J analysis is not used, correction of G for plastic zone size is still legitimate for the work absorption rate, i.e. the material R curve, though logically R is still a function of real crack length a, not apparent crack length (a + r_p). If the whole instability process were viewed as a characteristic rather than energetic argument, then use of G_p would again be rational for the applied severity leading to the use of a tangency between dJ/da and the R curve in the case of plasticity. This viewpoint has been adopted by some {19},{20}. The writer prefers the energetic argument {18} that for instability I > dw/Bda, where w is total work absorbed, be it at the fracture surface or elsewhere. The R curve expressed in terms of J is called J_r and defined in terms of total work dissipation rate {21}. The work absorption rate for any configuration is normalised to this reference J_r-R curve by use of the size factor Bb and shape factor η. That argument is not pursued here beyond the statement that apart from the important effects of anisotropy on the R curve {22} the normalisation through b and η would be adequate in so far as η is independent of degree of deformation, i.e. $\eta_o = \eta_{el}$. In so far as with real material that is not satisfied, then the η factor is inadequate but, it is argued, so is J or any other single parameter characterisation. Such points of adequacy must ultimately be resolved by experiment. The only point pursued here is the evaluation of I. An estimate of η and hence I/G was made for contained yield {23} based on G_p evaluated from the conventional Irwin plastic zone correction

$$E'G_p = Y^{-2}\sigma^2\bar{a} \qquad \text{Eqn.9}$$

where $\bar{a} = a + r_p$; $\bar{Y} = f(\bar{a}/W)$ with $r_p = K^2/2\pi\sigma_Y^2$ (plane stress) or $r_p = K^2/6\pi\sigma_Y^2$ (plane strain). Some sample data is given, Table 1. The trend is the same as for the computed data but the results do not agree closely, apparently because in contained yield (such as where r_p/a = (say) 0.1) G_p overestimates J and hence η. Obviously, in gross yield G_p underestimates J as is well-known but it is not clear whether in contained yield the discrepancy (of some 50% in the correction, though only perhaps 10% of the value of G or J) is due to the simple approximate nature of the lefm plastic zone correction procedure, or due to inaccuracy in the computed data. It is clearly necessary to confirm whether the differences are genuine, particularly in the light of the uncertainties that have recently been thrown on a variety of elastic -plastic computational programs, {24}, and the theoretical implications of the loss of a single parameter variable separable system in some cases.

TABLE 1

Energy release rate in the presence of plasticity

Configuration	State	r_p/a	Estimated G/G_p	Estimated I/G	Computed J/G	Computed I/G
3 point bend	Plane strain					
	non w.h.	.0075	1.019		1.007	1.00
a/W = 0.5		.0505	1.174		1.071	1.00
S/W = 4		.1124	1.453		1.421	.99
3 point bend	Plane stress					
	Mild w.h.	.0097	1.022		1.011	1.00
a/W = 0.5		.0532	1.180	.85	1.076	.99
S/W = 4		.1219	1.496		1.297	.97
Tension	Plane					
	strain	.0089	1.000		1.000	
a/W = 0.1	Mild w.h.	.1025	1.117		1.038	1.07
D/W = 4		.2777	1.354		1.545	1.89
Tension	Plane					
	stress	.015	1.000		1.000	1.00
a/W = 0.1	Mild w.h.	.1363	1.1524		1.077	1.14
D/W = 2		.2619	1.322	1.59	1.187	1.35

A J-BASED DESIGN CURVE

The development of a single J design curve, or at least the examination of whether such a curve is feasible, has also been greatly facilitated by the use of computed elastic-plastic case studies, {25}, {26}, {27}, {10}. The data for relatively shallow notches (a/W small) and stress concentration cases does not follow the same trend of J v displacement as estimated by Bucci et al {28} for deep notch testpiece data. The deep notch cases allow the deformation to be concentrated at the net section, whereas with a quite small degree of work hardening in a shallow notch case the deformation is spread to the gross section. The gradient dJ/dq is a function of the η factor {10}. Such cases with a/W small were the basis of the COD design curve from the early Wells wide plate test {29}, although gross yield is not normally met in practice, except under fault or accident conditions. In the stress concentration cases three effects interact : the quite high local strain of several times the yield strain that can exist even without a crack, whilst the net section stress is well below yield; the decrease of strain away from the immediate region of concentration, and the containment by the outer elastic field, so that J does not increase too markedly above G. By neglecting the first and second effects and taking a nominal strain of $k\sigma/E$ (where k is the elastic stress concentraction factor) a surprisingly good estimate of J can be made that falls close to the non-strain concentration cases for the few configurations so far computed. A curve fitted to the upper edge of the computed data can be represented approximately by

$$J/G_Y = (e/e_Y)^2 \text{ for } e/e_Y \leq 0.85 \qquad \text{Eqn.10a}$$

$$J/G_Y < 5\{(e/e_Y) - 0.7\} \text{ for } 0.85 \leq e/e_Y \leq 1.2 \qquad \text{b}$$

$$J/G_Y < 2.5\ \{(e/e_Y) - 0.2\} \text{ for } e/e_Y > 1.2 \qquad \text{c}$$

where $G_Y = Y^2\sigma^2_Y a/E$

These computations were all in plane strain. One of the difficulties in use is that the whole body and hence the estimated loads and strains in the absence of a crack might well be in plane stress for many plate-like components. The fitting together of a remote field in plane stress with a plastic zone and crack tip in plane strain has been examined by Hu and Liu {30} and until elastic-plastic three-dimensional computations are more readily available it might be worth further examination of their method.

The important case of residual stress has been allowed for by approximate methods {27}. This case cannot of course be calculated by rigorous methods unless the J contour integral is modified by an area term as has been suggested by several workers {31}, {32} and summarised by Hellen {32}.

The suggestion made in {27}, based on estimating J_{el} from the total elastic stored energy using Eqn.1, was that the ordinate J/G_Y be increased by 1 to allow for yield level residual stress.

A real uncertainty remains whether for such cases and for three-dimensional problems the term required is a maximum local value of J or a value averaged along the crack front. Presumably the maximum value will control the local onset of separation but that would not lead to unstable fracture until an energetic balance was reached, and that is likely to be related to an averaged J (or indeed I) value except perhaps in the case of failure by cleavage that might follow initiation immediately because of a reduction in toughness with crack speed.

In the absence of cleavage, the converse effect that, even in two dimensions, unstable growth in the presence of plasticity occurs subsequent to initiation has already been discussed. The question of whether some effective toughness can be derived from a J-R curve of the form

$$J_{eff} = \frac{b}{\eta} \frac{dJ}{da} r \qquad \text{Eqn.11}$$

was discussed {34}. Such a term would allow the simple design curve to be used without a full instability analysis. A similar proposal that a toughness value well beyond initiation has long been marked in connection with the COD method and Garwood and Towers {35} have recently offered a justification of such a proposal.

In the writer's opinion care must be taken that (b/η) for the structure must be larger than (b/η) for the test piece on which the R curve is measured if the same value of dJ_r/da is to be reached, and the data must relate to full thickness of use if there is to be assurance that the component is not in a more severe state of plane strain than the testpiece, since if it were the R curve would be much less well-developed, or in the limit of lefm behaviour, non-existent. Even when these features are accommodated, an R curve method that implies significant plasticity provides assurance on a "one off" basis, since the apparent increase of toughness in a rising R curve is caused by plastic deformation that can only be used once. Crack growth into an already deformed region would "see" a much reduced toughness that would seem to provide little reserve against another overload that caused further deformation. It is possible that use of a toughness beyond initiation can be permitted on a "regular use" basis, i.e. against many load applications whilst still not entering the fatigue regime, if neither unstable crack nor yield occurs. This condition is essentially contained yield for which G-R rather than J-R curves are appropriate and the writer does not see how a J-R curve with a contribution from both shear lip and plastic work "remote from" the crack face can degenerate to a G-R curve that is controlled by shear lip alone (since the remainder of the component is still elastic). Computational studies of stable crack growth have now been made by several workers {36}. It is not yet clear how sensitive the results are to the criterion adopted for crack advance. No doubt that topic will be discussed in the present conference. The final arbiter of the use of a simple design curve and of the use of a toughness value beyond that for onset of crack growth must again be experience, but the development of a rational procedure owes much to the study of the case data that has been generated by elastic plastic finite element computations.

CONCLUSIONS

Derivation of the conce s outlined here followed largely from computational studies. Value of the η factor that relates J to overall work appears closely similar, in contained yield, for plane stress and plane strain and it is argued that a near invariance of η with deformation reflects the acceptability of a one parameter description of the crack tip and stress system. The energy release rate in the presence of plasticity, I, evaluated in terms of η is near G for bending cases but near J for tension cases. This is attributed to the change in constraint with deformation that may occur in some configurations when the incremental flow rule is used. The main characteristics of a simple design curve, rather similar to the COD curve have been derived elsewhere, and also depend on the η factor. For realistic estimates of the strain in the uncracked body with plasticity, some compromise between the extremes of plane stress and plane strain seem desirable until three-dimensional elastic plastic data are more readily available.

SUMMARY OF ARGUMENTS ON CIRCUMSTANCES IN WHICH $\eta_{nl} \simeq \eta_{el}$ (Ref.16)

If the non-linear (nl) stress strain relation is expressed as a power law $\sigma \propto e^n$ and if loading is proportional and boundary conditions fixed, then for either plane stress or plane strain the stress ratios are constant but a function of n i.e. $\sigma_x : \sigma_y : \sigma_z$ $(= r_{1,2}) = g(n)$. By Ilyushin's theorem* the load displacement curve is also power law with the same value of n; i.e. $Q = q^n/\psi$ where ψ is a compliance function, ψ (a/W, n). The variables of geometry, a/W and material, n, are separable and for that case η exists as a constant independent of degree of deformation (14) (15) but $\eta = \eta$ (n) so that in general $\eta_{nl} \neq \eta_{el}$. From the HRR solutions in plane strain for n=1 the stress ratios are 1:1:1 (constant volume ; 1:1:2ν for conventional elasticity) and for $n \to 0$ the stress ratios are π: $2+\pi$: $1+\pi$ ($\simeq$ 0.6:1:0.8, the Prandtl field). Despite the variation of n over the whole range 0 to 1 the change in stress ratios is not large and for cases where a similar (high) constraint is maintained (e.g. deep notch bending) computations (10) (14) and here, show that $\eta_o \simeq \eta_{el}$ so that the variation of n is not per se important even for incremental behaviour. It is inferred, though not perhaps proven vigorously, that it is the loss of constraint in some configurations with incremental behaviour that violates the conditions for which η is independent of n. The computational evidence suggests that departure of η_o from the value of η_{el} is a good guide to the loss of constraint and hence to the loss of a one parameter description of the crack tip.

*Ilyushin, A.A., "The Theory of Small Elastic-Plastic Deformations", Prikadnaia Matematika i Mekhanika, P.M.M.10, 1946, p.347.

1. TURNER, C.E. A review of several theories of elasto-plastic fracture and their relationship to the J integral method. Colloquium on practical application of fracture mechanics. Int . Inst. Weld., 1979.

2. WITT, F.G. The application of the equivalent energy procedure for predicting fracture in thick pressure vessels. Practical application of fracture mechanics to pressure vessel technology. Inst.Mech.Engrs., London, pp.200-8, 1971.

3. EFTIS, J., LIEBOWITZ, M., & JONES, H. On fracture toughness in the non-linear range. Engng.Fract.Mech., 7, pp.491-503, 1975.

4. IRVINE, W.H. An elasto-plastic theory of fracture and its application to the analysis of experimental data. Tolerance of flaws in pressurised components. Inst.Mech. Engrs., London, pp.13-26, 1978.

5. RANDALL, P.N. & MERKLE, J.G. Gross strain crack tolerance of steels. Nucl. Engng. & Design, 17, pp.46-63. 1971.

6. SOETE, W. An experimental approach to fracture initiation in structure of steels. Fracture 1977, ICF-4, 1, Ed. Taplin, D.M.R. Univ. of Waterloo, pp.775-804, 1977.

7. HEALD, P.T., SPINK, W & WORTHINGTON, G. Post-yield fracture mechanics. Mats.Sci. & Engng, 10, pp.129-138, 1972.

8. NEWMAN, J.C. Fracture analysis of various cracked configurations in sheet and plate materials. Properties related to fracture toughness , Am.Soc.Test.Mat., Philadelphia, STP 605, pp.104-123, 1976.

9. DOWLING, A.R. & TOWNLEY, C.H.A. Effect of defects on structural failure: a two-criteria approach. Int.J. Press. Vess., Piping, 3, pp.77 - 107., 1975.

10. TURNER, C.E. Methods for post-yield fracture safety assessment. Post-yield fracture mechanics, Ed. Latzko, D.G.H., Applied Science, London, Chap. 2.

11. KRAFFT, J.M. , SULLIVAN, A.M. & BOYLE, R.W. Effects of dimensions on fast fracture instability in notched sheets. Symp. on Crack Propagation, College of Aeronautics, Cranfield, Paper A1., 1961.

12. McCLINTOCK, F. Plasticity aspects of fracture. Fracture: an advanced treatise. Ed.: H.Liebowitz, Academic Press, New York, 3, Chap. 2., 1968.

13. TURNER, C.E. Fracture toughness and specific fracture energy: a re-analysis. Mat.Sci.Engng., 11, pp.275-82, 1973.

14. TURNER, C.E. The ubiquitous η factor. 12th Nat.Sym. Fract. Mech. ,1979. To be published ASTM.

15. PARIS, P.C., ERNST, H., TURNER, C.E. A J-integral approach to the development of η factors. 12th Nat. Sym. Fract,Mechs., 1979. To be published ASTM.

16. TURNER, C.E., ERNST. H., & PARIS, P.C. Correspondence on η factor. (See Appendix). To be published.

17. ERNST, H., PARIS, P.C., ROSSOW, M., & HUTCHINSON, J. The analysis of load-displacement relationships to determine J-R curves and tearing instability material properties. 11th Nat.Symp. on Fract.Mechs., ASTM, STP 677. pp.581-599, 1979.

18. TURNER, C.E. Description of stable and unstable crack growth in the elastic-plastic regime in terms of J_r resistance curves. ASTM STP 677, pp.614-628. , 1979.

19. SHIH, C.F., DE LORENZI, H.G. & ANDREWS, W.R. Studies on crack initiation and stable crack growth. Elastic-plastic fracture, ASTM, Philadelphia, STP 668, pp.65-120, 1979.

20. GARWOOD, S.J. Effect of specimen geometry on crack growth resistance. ASTM STP 677 pp.511-532., 1979.

21. GARWOOD, S.J., PRATT, P.L. & TURNER, C.E. Measurement of slow stable crack growth in structural steel and prediction of unstable fracture. Elasto-plastic fracture mechanics, CSNI Report 32. UKAEA-OECD (NEA) 1, Paper 11. 1978.

22. WILLOUGHBY, A.A., PRATT, P.L. & TURNER, C.E. The effect of specimen orientation on the R curve. Int.J.Fract., 14, (5), pp. R249-51.,

23. TURNER, C.E. Use of the R curve for design with contained yield. 2nd European Congress on Fracture, London, 1980.

24. WILSON, W.K. & OSIAS, J.R. A comparison of finite element solutions for an elastic-plastic crack problem. Int.J.Fr. 14 (3) 1978, pp. R95-R 108.

25. SUMPTER, J.D.G. & TURNER, C.E. Design using elastic-plastic fracture mechanics, Int.J.Fr. 12 (6), pp.861-871, 1976.

26. TURNER, C.E. A J design curve based on estimates of some two-dimensional shallow notch configurations. Elasto-plastic fracture mechanics, CSNI Report 32 I UKAEA/OECD (NEA) Paper 18, 1978.

27. TURNER, C.E. A J based design curve, Advances in Elasto-Plastic Fracture Mechanics, Ed. Larsson, L.H. Applied Science, pp.301-318, 1979.

28. BUCCI, R.J., PARIS, P.C., LANDES, J.D. & RICE, J.R. J-integral estimation procedures. Fracture Toughness ASTM, Philadelphia, STP 514, pp.40-69, 1972.

29. BURDEKIN, F.M. & DAWES, M.G. Practical use of linear elastic and yielding fracture mechanics with particular reference to pressure vessels. Practical Application of Fracture Mechanics to Pressure Vessel Technology. Inst.Mech.Engrs., London, pp.28-37., 1971.

30. HU, W. & LIU, H.W. Crack tip strains - a comparison of finite element method calculations and Moiré measurements ASTM STP 601, pp. 522-34, 1976.

31. BLACKBURN, W.S. ET AL. An integral associated with the state of crack tip in a non-elastic material. Int.J. Fr.Mech. 13, pp. 183-200, 1977.

32. AINSWORTH, R.A., NEALE, B.K. & PRICE, R.H. Fracture behaviour in the presence of thermal strains. Tolerance of Flaws in Pressurised Components. Inst.Mech.Eng. London, pp.197-204, 1978.

33. HELLEN, T.K. Methods for computing contour integrals. Post-Yield Fracture Mechanics. Ed.: Latzko, D.G.H. Applied Science, 1979, Chap. 4.

34. TURNER, C.E. Remarks on unstable ductile crack growth. Plastic Tearing Instability.Nureg CP0010/CSNI Rep.39,1980.

35. TOWERS, O. & GARWOOD, S.J. The significance of maximum load toughness. Int.J. Fr.16 pp.R85-90, 1980.

36. Numerical Methods in Fracture Mechanics. Ed. Luxmoore,A.R. & Owen, D.R.J. Proc. 1st Int.Conf., Univ.Swansea, 1979, Section 3.